D0862427

LIGHT

R. W. Ditchburn, M.A., B.Sc., Ph.D.
University of Reading

DOVER PUBLICATIONS, INC.
New York

This Dover edition, first published in 1991, is an unabridged republication of the 1961 edition of the work published by Interscience Publishers, New York. The work was first published by Blackie & Son, Ltd., Bishopbriggs, Glasgow, Scotland, in 1953.

Dover Publications, Inc.
31 East 2nd Street
Mineola, New York, 11501

Library of Congress Cataloging-in-Publication Data

Ditchburn, R. W.
Light / R. W. Ditchburn.
p. cm.
Reprint. Originally published: New York : Interscience Publishers, 1961.
Includes bibliographical references and index.
ISBN 0-486-66667-0 (pbk.)
1. Light. I. Title.
QC355.2.D57 1991
535—dc20 91-9783
CIP

PREFACE

Sixty years ago, a single theory of light—the electromagnetic theory—seemed capable of describing all the experimental results. The position was very different thirty years ago when wave theory and quantum theory were each successful in its own field but apparently mutually irreconcilable. In this book I have endeavoured to show the student that we have again a single theory of light, a logical whole. I have tried to describe the wave theory in such a way that the quantum theory may appear as a natural development rather than as an alternative theory. For this reason I have, at an early stage, stressed the concept of a wave-group.

It is expected that the student has studied physics to intermediate standard before commencing to read this book, and that his knowledge of electricity and magnetism, and of mathematics, will advance in parallel with his study of light. The electromagnetic theory is not introduced until Chapter XIII, and the more difficult mathematics is placed in appendices (or in separated paragraphs in small type) which are intended to be omitted on first reading. The whole of Chapter XVI forms a separate section whose reading need not precede that of Chapters XVII–XIX.

I wish to acknowledge the help of many friends with whom I have discussed sections of this book. I particularly wish to thank Dr. O. S. Heavens who has been associated with me in the final stages of preparing the diagrams, proof reading, etc. I am grateful to Prof. K. G. Emeléus with whom the general plan of the book was discussed, Dr. E. H. Linfoot who made suggestions in connexion with Chapter VIII, Mr. T. L. Tippell who read and commented on the first draft of Chapter IX, and to Dr. P. White in regard to Appendix IVB. The responsibility for the final text is, of course, my own. I wish to thank the following for providing material for plates and figures:

Dr. W. M. Gray and Mr. P. J. Jutsum (Plate II *b*, *c*, *d*, *e*, *f*),
Mr. M. E. Haine, A.E.I. Research Laboratories (Plate V *f*, *g*),
Dr. H. G. Kuhn (fig. 8.7 and information in text),
Dr. A. C. Menzies (Plate V *a*, *b*, *c*, *d*),
Prof. R. W. Pearse (Plate II*a*),
Prof. E. T. S. Walton (fig. 3.11).

I also acknowledge with thanks permission to make reproductions of copyright material as follows:

Messrs. Hilger & Watts for figs. 1–7 (inclusive), 9, 11 and 13 of Chapter IX; the Council of the Royal Irish Academy for fig. 8.5; and the Director, N.P.L., the Council of the Royal Society and the authors for fig. 9.16 from a paper by Sears and Barrell.

R. W. DITCHBURN

UNIVERSITY OF READING,
February, 1952

CONTENTS

CHAPTER I

HISTORICAL INTRODUCTION

CHAPTER II

WAVE THEORY—INTRODUCTION

CHAPTER III

WAVE THEORY—COMBINATION OF WAVE MOTIONS

CHAPTER IV

REPRESENTATION OF LIGHT BY WAVE TRAINS OF FINITE LENGTH

CHAPTER V

INTERFERENCE

CHAPTER VI

DIFFRACTION

CHAPTER VII

HUYGENS' PRINCIPLE AND FERMAT'S PRINCIPLE

CHAPTER VIII

THE ACCURACY OF OPTICAL MEASUREMENTS

CHAPTER IX

MEASUREMENTS WITH INTERFEROMETERS

CHAPTER X

THE VELOCITY OF LIGHT

CHAPTER XI

RELATIVISTIC OPTICS

CHAPTER XII

POLARIZED LIGHT

CHAPTER XIII

THE ELECTROMAGNETIC THEORY

CHAPTER XIV

THE ELECTROMAGNETIC THEORY OF REFLECTION AND REFRACTION

CHAPTER XV

THE ELECTROMAGNETIC THEORY OF ABSORPTION AND DISPERSION

CHAPTER XVI

ANISOTROPIC MEDIA

CHAPTER XVII

THE INTERACTION OF RADIATION AND MATTER

CHAPTER XVIII

QUANTUM THEORY OF RADIATION

CHAPTER XIX

INTERACTION PROCESSES IN RELATION TO QUANTUM MECHANICS

CHAPTER I

Introduction

1.1. The Scientific Picture.

New experimental data nearly always cause some alteration in scientific theories, but in certain periods of history the changes are very gradual. The new material is assimilated by extending and modifying the theories while leaving unchanged certain fundamental ideas on which all the theories are based. Progress of this kind went on during most of the nineteenth century, but near the end of the century it became impossible to modify the current theories so as to accept the new experimental results. Certain fundamental difficulties affecting the whole basis of physical science were revealed and, in order to overcome these difficulties, it has been necessary to clarify our views concerning the nature and purpose of scientific inquiry. It would not be appropriate to discuss this matter at length in a book on one branch of science. On the other hand, it would be very difficult to give an adequate account of the modern theory of light without some reference to these general considerations. It therefore appears desirable to state, at the outset, the objective which the author has in mind during the development of the theory of light. Later, the reader may be able to judge for himself whether the objective has been attained. He may also decide whether he feels that the objective is satisfactory both from the practical and from the intellectual point of view.

1.2.—The practical scientific worker makes observations with the senses of sight and hearing, and also with scientific instruments which increase the range, delicacy and number of his observations. The theoretical worker accepts these observations as given data which he has to co-ordinate. In order to be able to reason about them, he first collects them into groups. Each group is then organized in a system which exhibits relations between the members of the group. A system of this kind is called a scientific theory. The whole body of scientific theories and the connections between them constitute a scientific picture of the world. In the process of making scientific theories, words like " electron ", " energy ", " organism " are introduced.

These words are symbols invented in order to create a language capable of describing the results of observation in a logical and elegant way.

1.3.—The construction of a scientific theory may be compared to the preparation of a weather map at a central meteorological station. A large number of observations of pressure, temperature, etc., are received and recorded, at the appropriate places, on a large chart. When all the data are entered, the meteorologist inserts isobars, etc., and proceeds to make predictions. In discussing the map, he uses terms like " depression " or " cold front ". These terms form a convenient way of summarizing certain aspects of the observations. They help him to think quickly and clearly about the meteorological situation. The weather map is, however, primarily a representation of the observations. The isobars are useful only in so far as they represent the observations. In a similar way, in the theory of light, we use terms like " waves " and " particles " for the description and discussion of the results of experiments. We need to remember that the meaning of these words is derived from the experiments which they describe. We must not attempt to deduce the special properties of light waves or light particles from any preconceived ideas about waves or particles in general. All that we can say about light must be deduced from experimental observation.

1.4.—New scientific theories usually begin by relating new observations to familiar concepts, based upon older observations. For a long time the theory of light was discussed in terms of waves or particles, because it is easy to form mental pictures of waves and particles. Recent advances have forced us to accept the fact that a complete theory of light cannot be expressed in terms of simple analogies of this type. We are, however, able to construct a summary of our observations in mathematical terms. This mathematical theory is precisely defined and enables us to make certain kinds of predictions concerning the probable results of future observations. It is logically consistent within itself. We often find it convenient to " translate " part of this theory into words, but the translation is never quite perfect, though it may frequently be very useful. A wave picture of light furnishes an adequate description of a wide range of observations just as a set of isobars expresses the results of certain meteorological observations. The wave theory is unsuited to describe certain other types of observations and these may be discussed in terms of light particles or " photons ". In a similar way certain types of meteorological observations cannot be described simply by drawing isobars, but can be included in the weather map in other ways. Any attempt to make a complete theory of light in terms of waves *or* particles must lead to confusion and error. We must admit that the results of our experiments on light are, in some ways, so different from the results of observations on things like waves on water, or moving particles, that analogies break down. They cease to be useful and become a burden. At this point it is necessary to leave the analogies and revert to the mathematical equations. When all this has been said, it still remains true that most people think more readily in terms of words than in terms of equations. We therefore use the analogies as far as possible —like a man who travels as far as possible by train even though he knows that none of the places he wishes to visit lies exactly on the railway line.

1.5.—In the historical development of a subject ideas are gradually introduced in order to include fresh observations within the theoretical description. In the treatment given in a textbook it is often convenient to disregard the historical order and to introduce many of the current ideas as hypotheses to be tested by experiment. The author knows, in advance, that most of the hypotheses which he introduces are going to be " approved " by the experimental results which he subsequently describes. In this way he avoids the necessity of burdening the reader with details of theories which have been found to be unsatisfactory and are now only of historical interest. The formal treatment of the subject, in this book, follows this plan. It begins in Chapter II. In the remainder of the present chapter our object is to consider the theory of light more from the historical point of view and to show how each of the more important types of experimental observation has been incorporated in, and has led to alterations in, the theory of light. A summary of this process is given in fig. 1.6 which will be considered in detail at the end of the chapter. In the course of this review of the progress of the theory of light we also seek to indicate the general relations of the theory of light to other departments of science.

1.6. **Light in Relation to Biological Science.**

The scientific picture of the world would be seriously incomplete if it did not include an account of the physical and physiological processes by which man makes his observations. The scientific picture must include an account of the link or links between the human brain and the things—atoms, molecules, etc.—whose existence is postulated in order to describe the observations. Historically, it has been recognized from the earliest times that a very important set of our observations involves light and vision. Early theories of light were therefore theories of vision. One school postulated that the eye sends out invisible antennae or sensitive probes and is thus able to feel objects which are too distant to be touched by hands or feet. This theory may be called the " tactile " theory. Another view was that something is emitted by bright objects and that when this thing enters the eye it is able to affect some sensitive part of the eye and so give rise to the sensation of sight. This theory was called the " emission " theory. Both these theories were current among Greek thinkers about 500 B.C.*

1.7.—The tactile theory is inherently simple because it describes the unknown in terms of the known. The more mysterious sense of vision is directly related to the simpler and more obvious sense of touch. The tactile theory has some difficulty in explaining why things can be felt, but not seen, in the dark, and why bodies can be made visible in the dark by heating them. The fact that certain bright bodies are able to make neighbouring bodies visible also receives no obvious explanation.

The tactile theory can include this type of observation by postulating that the visual probes are able to feel only certain kinds of surfaces and then making a series of assumptions that surfaces can be modified under various conditions. When this is done, the simple relation to the sense of touch has been lost. The theory becomes intolerably complicated. These observations are described in a simple and satisfactory way by the emission theory if it be assumed that some bodies emit a radiation to which the eyes are sensitive, and that others are able to reflect or scatter this radiation so that it enters the eye. For these and similar reasons, the emission theory gradually displaced the tactile theory. The process was very slow and it was not until about 1000 A.D. that, under the influence of the Arabian astronomer Alhazen, the tactile theory was finally abandoned.

1.8.—The emission theory being accepted, light may be defined as " visible radiation ", and we may give the following general account of the visual process. Light, being emitted, reflected, or scattered, enters the eye and is focused by the lens of the eye on a surface situated at the back of the eye. This surface is called the *retina*. It

* Many variations and combinations of these theories were also suggested. We need not consider them since they are more complicated than the two theories we have described and have no important compensating advantages.

contains a large number of nerve endings. When light falls on one of them, a chemical and physical action takes place. As a result, a series of electrical impulses is sent along an appropriate nerve fibre to the brain.

A complete theory of vision thus involves many sciences. The description of the emission, reflection and scattering of light and of its transmission to the eye is a part of physical science. The description of the structures of the eye, of the optic nerve and of the associated parts of the brain, belongs to the anatomist. The description of the processes by which the eye lives and transmits its messages to the brain is within the science of physiology. The description of the way in which the mind interprets a pattern of visual sensation and relates it to other visual and non-visual experience falls within the domain of psychology. Most of the theory of vision is clearly outside the scope of a book which is primarily concerned with the physical properties of light, though knowledge of some of the non-physical aspects of vision is necessary in order to understand the subject of photometry and the physical specification and measurement of colour.

1.9. Light in Relation to Physical Science.

Although the theory of light started as the study of vision, this is not, to the physicist, the most important part of the subject. He can detect light through its heating effect on a thermopile or through its electrical effect on a photocell. He can also detect it as an agent capable of causing chemical action or through its effect on a photographic plate. To him, light is a form of energy which travels from one place to another. It can interact with matter and can be transformed into thermal, electrical or chemical energy. The physical equations would be incomplete and the energy conservation law would fail if the transfers and transformations of energy due to the action of light were not taken into account. To the physicist the effect of light on the retina is only one example of photo-chemical action.

1.10. Waves or Corpuscles.

If light is a form of energy which can be transferred from one place to another, it is reasonable to seek to describe it by analogy with other methods of transport of energy. Moving bodies possess kinetic energy. This energy accompanies the body in its movement and thus passes from one place to another. Another mechanical mode of transfer of energy is by means of the propagation of waves. This mode is not, in general, accompanied by any bodily movement of the medium.

Many physicists of the seventeenth and eighteenth centuries sought to describe light either in terms of moving particles or of waves. To them, these forms of moving energy were sharply differentiated, in that particle energy is highly localized. The kinetic energy of a rifle-bullet travels from one well-defined, small region of space to another, and does not spread during transit. If, however, a wave is started by dropping a stone into a pond, the energy quickly spreads over the whole surface, and usually no small region receives a very high proportion of it.

1.11. **Rays of Light.**

In the seventeenth century it was known that the propagation of light could be represented by means of rays. If the light from a very small source was interrupted by an opaque obstacle, a very sharp

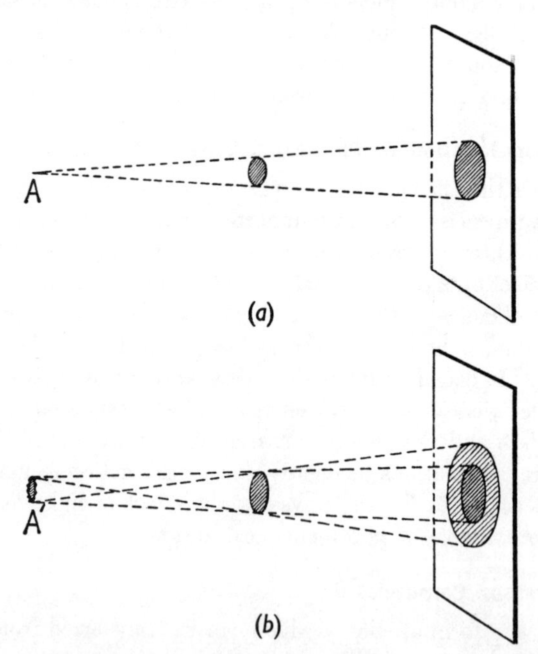

Fig. 1.1.—Formation of shadows by (*a*) a point source A, and (*b*) an extended source A′, in relation to rectilinear propagation of light

shadow was formed (see fig. 1.1*a*). If the source was not very small the edge of the shadow was not so sharp. There was a dark shadow known as the *umbra* and a diffuse edge known as the *penumbra* (see fig. 1.1*b*). These observations are simple examples of a large number which are generally described by the statement that light travels out from the source along rays which are straight lines. On this view, the change of illumination which occurs in the penumbra is due to the fact that a point in this region receives light from only a portion of the source. There is no evidence here that light energy spreads outside the region defined by the rays. A ray may therefore be defined as a path along which light energy travels from a source to a receptor. Transit along this path is prevented if the ray is cut by an opaque obstacle at any point. If all the rays from a given source to a receptor are interrupted by opaque obstacles, no light energy can reach the receptor from the source. Note that two ideas are involved—first, that light is propagated in rays, and, second, that these rays are straight lines. The motion of a small particle inevitably sweeps out a line, but not necessarily a straight line. According to Newton's Laws of Motion, particles travel in straight lines in the absence of any " impressed " force. It therefore appeared to Newton that light should be described as a system of particles following paths described by his Laws of Motion. It is true that, owing to the action of the earth's gravitational field, the trajectories of material particles are not straight lines. The rectilinear propagation of light must therefore be explained, either by assuming that the light-particle has no weight, or that its speed is always so high that the curvature due to the action of gravity is too small to be detected.

1.12. **Interference and Diffraction.**

Although Newton knew of no evidence for the spreading of light energy, he first studied what were later called *interference* phenomena. These are described in an elegant way by the wave theory of light. He placed a convex lens of large radius of curvature (about 50 feet) in contact with a plane piece of glass and viewed the reflected light (see fig. 1.2). A series of coloured rings—alternately bright and dark—was observed (see Plate I*e*, p. 72). These are known as Newton's Rings.

Newton recognized that these rings indicated the presence of some kind of periodicity and that this suggested a wave theory of light. He believed that the rectilinear propagation of light was an insuperable objection to a simple wave theory. He therefore suggested that light consists of corpuscles which either possess an internal vibration of their own or are in some degree controlled by waves or vibrations of the medium through which they travel. The objection to a simple wave theory was removed when it was discovered that the propagation of light is not strictly linear. Light *does* spread, though to a very small extent, from the edges of beams defined by rays (see Plate III, p. 214). For example, the shadow of a straight edge formed by a small source is not *perfectly* sharp when seen under high magnification. Some light penetrates into the region which ought to be completely dark if light were propagated entirely in straight lines, and there is a

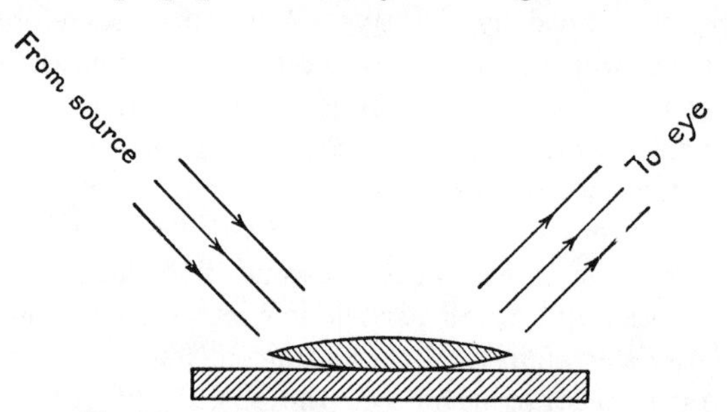

Fig. 1.2.—Apparatus for viewing Newton's rings

series of fine, light and dark bands at the edge of the region outside the shadow. Some observations of this type were made by Grimaldi in Newton's lifetime, but it was not until 150 years later that this phenomenon (which is known as *diffraction*) was clearly understood. The discovery of diffraction showed that the propagation of light is not exactly rectilinear. The concept of the ray of light does not *exactly* correspond to the results of observations. It is only an approximation.

1.13. **Development of the Wave Theory.**

We shall see later (Chapters VI and VII) that whilst the wave theory cannot give a satisfactory account of *exactly* linear propagation, it is well suited to describe *approximately* linear propagation, provided that it be assumed that the wavelength is small in relation to the relevant dimensions of the apparatus. The nineteenth century saw very important advances in the technique of experimental physics, and the number and accuracy of experiments on light increased greatly. The observations obtained were well described by the wave theory which became more exactly defined. We may note three important types of observation.

(i) *The Wavelength of Light.*

Many detailed experiments on interference and diffraction were made. These led to a set of determinations of the wavelength of light.

It was shown that, in a spectrum, the wavelength is related to the colour. The wavelength is about $6{\cdot}5 \times 10^{-5}$ centimetre for red, $5{\cdot}6 \times 10^{-5}$ for green and $4{\cdot}5 \times 10^{-5}$ for blue light. Different methods of measurement gave consistent results.

(ii) *The Velocity of Light.*

In 1676, a Danish astronomer, Römer, made the fundamental discovery that the velocity of light was finite, and estimated it from astronomical observations. About 200 years later the velocity was measured by terrestrial experiments and was shown to be very near to 3×10^{10} centimetres per second (or 186,000 miles per second).

(iii) *The Polarization of Light.*

In 1670, Bartholinus discovered that when a beam of ordinary light passes through certain crystals (such as calcite) each ray splits into two. On passing the two rays into a second crystal, the effect depends on the orientation of the crystal with respect to the beam. For certain orientations the two rays proceed unchanged. For other orientations the two rays each split into two (see fig. 1.3). This phenomenon is known as double refraction. It indicates that a beam of

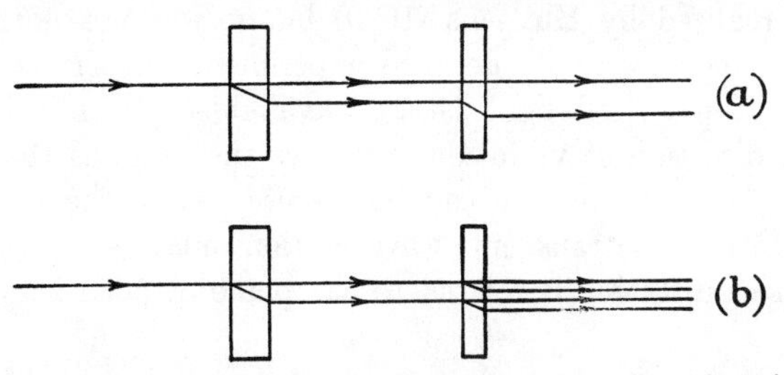

Fig. 1.3.—Double refraction: (*a*) two crystals with similar orientations, (*b*) two crystals with different orientations of crystal axes

light which has passed through a crystal is differentiated in respect of planes including the direction of propagation. The simplest experiment on this property was made much later (1808) and is due to Malus. He reflected a beam of light at the surfaces of two unsilvered pieces of glass (see fig. 1.4). He showed that when the two reflections are in the same plane (as shown in the figure) a high proportion of the light incident on the second mirror (M_2) is reflected. If the mirror M_2 is turned so that the second reflection is directed out of the plane of the paper, the reflected beam becomes weaker. It is of nearly zero brightness when the two reflections are in planes at right angles. This shows that after the first reflection the beam of light has a special property in relation to the plane of the paper. It can be strongly reflected, at a glass surface, in this plane but not in a plane at right angles. A beam of light which possesses this property is said to be plane-polarized. This type of property finds no place in a theory of longitudinal waves. For this reason it was regarded by Newton, who considered only such waves, as an additional important objection to the wave theory. It

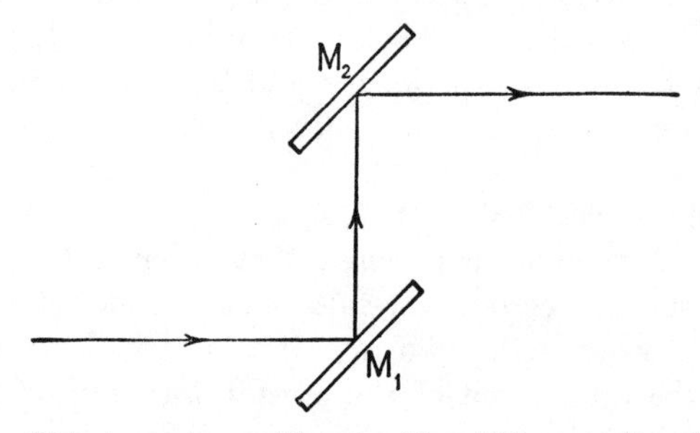

Fig. 1.4.—Malus' experiment. Note that M_1 and M_2 are *unsilvered* mirrors

is, however, adequately represented in a theory of *transverse* waves. This was realized by Huygens (1690) but it was not until the nineteenth century that really detailed experiments on the reflection and refraction of polarized light became available. With a longitudinal wave, the direction of vibration is always the same as that of propagation, so the wave motion can be represented as the variation of a scalar quantity. A transverse wave motion must be represented by a vector whose direction is related to the plane of polarization.

1.14. **Electromagnetic Theory.**

The wave theory of light was formulated before the development of the fundamental laws of electromagnetism. It was assumed that there existed some kind of medium which had properties like that of an elastic solid. This medium pervaded all space but was modified by the presence of matter. A theory of transverse waves in such a medium formed a qualitative description of the fundamental phenomena of interference, diffraction and polarization. In order to fill in the details of the wave theory, it was necessary to make special assumptions concerning the density and elasticity of this medium, and also concerning the conditions obtaining at the surface separating two media such as glass and air. Discussion of these details revealed certain difficulties and appeared to indicate that there were some inconsistencies in the theory. All these difficulties were resolved by Maxwell's electromagnetic theory of light.

Maxwell formulated the equations of electromagnetism in a general form, and he showed that they suggest the possibility of the propagation of transverse electromagnetic waves. The velocity of propagation can be derived from constants measured in laboratory experiments on electricity and magnetism. The value calculated is in close agreement with the directly measured velocity of light. Maxwell's theory included an account of the propagation of electromagnetic waves in media such as glass. He was able to show that it gave a general account of the phenomena of reflection and refraction, including the formation of a spectrum by the dispersion of light. It is important to realize that all this was achieved without introducing any arbitrary assumptions. The theory of light became, in the hands of Maxwell, a part of the theory of electricity and magnetism.

1.15. **The Electromagnetic Spectrum.**

The elastic-solid theory of light could not explain why all the observed waves had wavelengths between about 7×10^{-5} and 4×10^{-5} centimetre. The theory of electromagnetic waves suggested the possibility of producing waves of other wavelengths by electrical means. Success was first obtained by Hertz, who in 1887 succeeded in propagating electromagnetic waves of about 10 metres wavelength. A great deal of the progress of experimental physics since that time has consisted in the discovery of methods of producing electromagnetic waves of different wavelengths. Some of the properties of these waves depend upon their wavelength, but they are all propagated with the same velocity (in free space) and they are all described by the equations of Maxwell. Fig. 1.5 shows them arranged in order of wavelength. Modern technical advances have provided methods of producing or of detecting waves of nearly every wavelength from above 3000 metres to below 10^{-11} centimetre.

There are certain regions in the spectrum where it is still very difficult to excite the waves. These regions are not completely explored but they are not really gaps in the spectrum. The limits of the spectrum at the two ends are not perfectly definite. At the long-wavelength end, methods of producing and detecting radiation gradually become less efficient as the wavelength increases. At the short-wavelength end, an enormous concentration of energy is needed to produce the vibrations of extremely high frequency. Also these radiations are difficult to detect since they are very little absorbed by matter. The range of wavelengths to which the eye is sensitive and to which

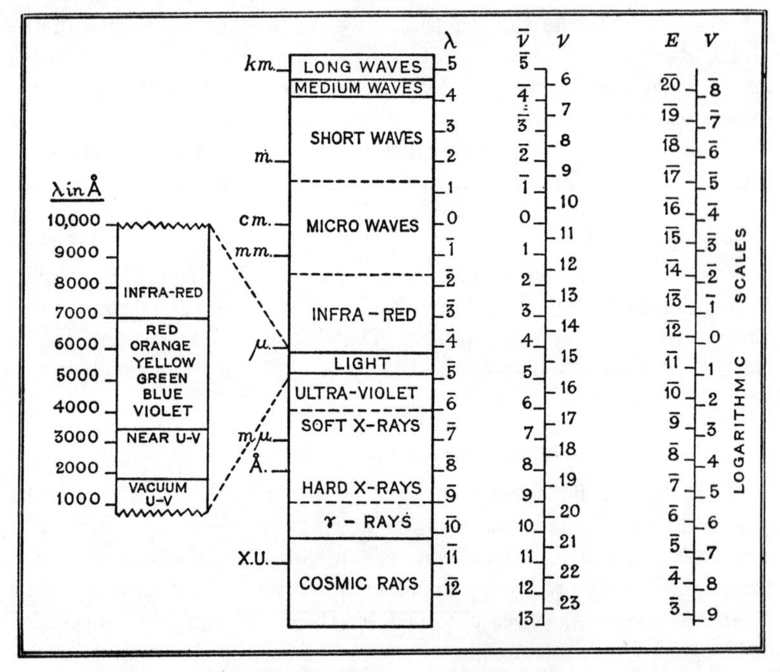

Fig. 1.5.—The Electromagnetic Spectrum

The main diagram shows the names given to electromagnetic waves of different wavelengths. The logarithm (to base 10) of the wavelength in centimetres is shown on the right. The relation between different units of length is shown on the left, e.g. 1 Ångström unit (Å.) = 10^{-8} cm. The data given enable wavelengths to be converted from one unit of length to another, e.g. for red-orange light, $\lambda = 6 \times 10^{-5}$ cm. = 6000 Å. = 0·6 μ.

The inset figure on the left is an enlargement of the region between 10,000 Å. and 1000 Å. The scale for this enlargement shows the wavelength directly. The scales to the right (which are intended for use in connection with later chapters of the book) show

(i) $\log_{10} \bar{\nu}$, where $\bar{\nu}$ is the wave number, i.e. the reciprocal of λ in cm.
(ii) $\log_{10} \nu$, where ν is the frequency in sec.^{-1}
(iii) $\log_{10} E$, where $E = h\nu$ is the energy in ergs of a quantum of frequency ν.
(iv) $\log_{10} V$, where V is the energy in electron-volts (see Example 1(iii)).

we give the name " light " is thus seen as part of a very much wider spectrum. The electromagnetic theory brings light into relation with the other types of electromagnetic radiation and also with the fundamental theories of electricity and magnetism.

1.16. **Photons.**

Let us now return to the fundamental conflict between wave and particle theories of light.

The wave theory, in explaining the approximately rectilinear propagation, seemed to have defeated the particle theory, and all the experimental results of the nineteenth century appeared to be adequately described in terms of wave concepts. Early in the twentieth century, a series of observations on photo-electricity created a really serious difficulty for the wave theory. It was found that light could cause atoms to emit electrons and that, when light released an electron from an atom, the energy possessed by the electron very greatly exceeded that which the atom could, according to the electromagnetic-wave theory, have received. Einstein suggested that, in order to give an adequate description of these observations, it was necessary to assume that the energy of a light beam is not evenly spread over the whole beam, but is concentrated in certain regions. These localized concentrations of energy he called *photons*. They are propagated like particles. It is assumed that there are usually a very large number of them, the energy in any one photon being very small. Thus to most ordinary experiments, the energy of a light beam is evenly distributed, just as a gas exerts a very nearly uniform pressure on the surface of an ordinary vessel, because each molecule is very small and the number of molecules is very large. When very small areas are involved (e.g. when the movement of an ultra-microscopic particle is observed), the irregularities of the Brownian movement show the discontinuous " structure " of the gas. In a similar way, the atom presents to the light beam an area so small that it indicates the presence of " molecules of light " or photons. In order to describe the observation in detail, it is necessary to assume that the photons corresponding to light of one wavelength all have the same energy.

Shortly before Einstein suggested the concept of photons, Planck had found it necessary to use a somewhat similar hypothesis for entirely different reasons. He was concerned with the light emitted by hot bodies. He found that the observations indicated that light energy is emitted by atoms in multiples of a certain energy unit. It is not possible to emit a fraction of a unit. The size of the unit, which is called a *quantum*, depends on the wavelength (λ) of the radiation. Its value is

$$E = hc/\lambda, \quad \quad 1(1)$$

where h is a universal constant, known as *Planck's constant*, and c is the velocity of light. The value of Planck's constant (h) is $6{\cdot}6 \times 10^{-27}$ erg-second. If ν is the frequency of the radiation, we have $c = \nu\lambda$, and hence

$$E = h\nu. \qquad \text{1(2)}$$

Planck's hypothesis did not require that the energy should be emitted in *localized* bundles and it might, though with some difficulty, have been reconciled with the electromagnetic-wave theory of radiation. When Einstein showed that it seemed necessary to assume the existence of *concentrations* of energy travelling through free space, a solution of this kind was excluded. The concept of a particle appeared to be necessary.

1.17. **Relativity Theory.**

Mainly as a result of experiments on the propagation of light in moving media, Einstein investigated the foundations of dynamics. In 1905 he published what is known as the Restricted Theory of Relativity. This theory is a new system of dynamics, modifying and in a certain sense superseding the Newtonian theory. The difference between the relativistic dynamics and the Newtonian dynamics is very important when the particles under consideration are moving with speeds near to that of light. Any satisfactory theory of light must therefore agree with the concepts of the relativistic dynamics. When the theory of light is brought into relation to the theory of relativity, it is possible to give an elegant account of observations on light which is emitted by a source moving with respect to the observer, or on light which passes through a medium moving with respect to either source or observer. In 1915, the theory of relativity was extended to include the dynamics of bodies moving in fields of force. The theory made specific predictions concerning the properties of light emitted in, or passing through, strong gravitational fields. The verification of these predictions by astronomical observations gave support to the general theory of relativity and showed that it, too, is relevant to the theory of light.

1.18. **Modern Theory.**

The modern theoretical physicist is required to invent a unified description of two very different types of experiment. On the one hand stand all the phenomena of interference, diffraction and polarization, which are so well described by the wave theory. On the other hand, modern experiment has greatly increased the number and range of the experiments which are readily described in terms of photons. The electromagnetic picture has no place for the photons, and the particle theory has no place for the waves, yet both are required to give a complete description of the phenomena. In a similar situation, Newton considered the possibility of particles which possessed periodic properties or were guided by waves. Many suggestions of this type were considered during the first quarter of this century but none of these was entirely successful.

1.19.—The solution which is now accepted is more radical. The modern quantum mechanics constitutes one theory including the properties of light and of matter. The description is very closely knit together so that it is not possible to separate one part and call it the "theory of light". The theory is not easy to understand but it is not unnecessarily complicated in view of the wide range of phenomena covered. As indicated at the beginning of this chapter, it can be stated completely only in mathematical form. It must not be thought that the theory is complete and will never require modification. With this qualification, it may fairly be said that the main difficulties of the particle-wave conflict have been resolved, and that a really unified theory has been produced. In this unified theory the particle and wave ideas appear as complementary rather than as rival conceptions. The theory shows, in a systematic and logical way, that wave and particle concepts are each to be used in appropriate contexts, and it shows the relation between them.

1.20.—This introduction to the theory of light is summarized diagrammatically in fig. 1.6, which indicates how the theory has gradually been extended and modified to include, within one description, the increasing range of experimental material. Starting at the top left-hand side, we see that the rectilinear propagation of light was included in both the tactile and the emission theory. The former, which could not explain many phenomena which indicate that light is a form of energy, was abandoned about 1000 A.D. The emission theory is divided into (*a*) the corpuscular emission theory, and (*b*) the wave emission theory. The phenomena of interference and diffraction appeared to give a clear decision in favour of a wave theory. Slightly later it became necessary to use a tranverse-wave theory in order to include the phenomena of double refraction, etc. The transverse (or vector) wave theory was initially stated in terms of an elastic solid theory, but the theory of Maxwell was preferred because it gave an elegant explanation of the relation between the velocity of light and the ratio of the electromagnetic units. Early in this century, experiments on the interaction of radiation and matter led to formulation of the quantum theory which had some features in common with the

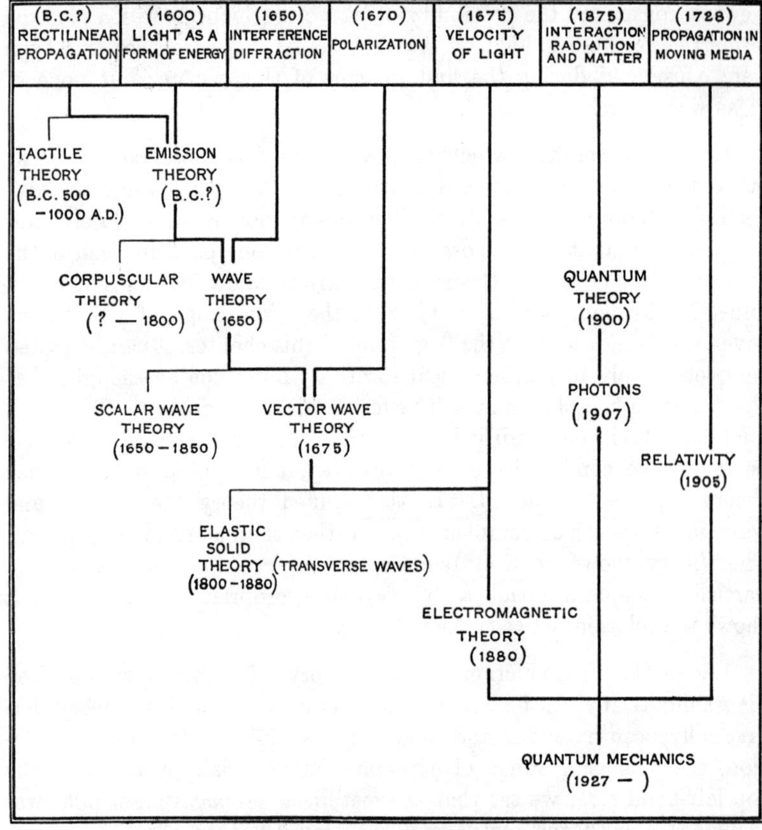

Fig. 1.6.—Development of the theory of light

(The dates given are intended to give a general indication of the times at which the different types of experimental result became available, and the times at which the different theories became current.)

earlier corpuscular theory. The theory of relativity was formulated about the same time. The modern quantum mechanics (developed since 1927) incorporates the appropriate parts of the electromagnetic wave theory, the quantum theory and the relativity theory.

EXAMPLES [1(i)–1(v)]

1(i). What is the energy of a quantum of red light of wavelength $6{\cdot}6 \times 10^{-5}$ cm.? What is the associated frequency?

[$E = 3 \times 10^{-12}$ erg; $\nu = 4{\cdot}5 \times 10^{14}$ sec.$^{-1}$]

1(ii). What are (*a*) the frequency and (*b*) the wavelength of the radiation for which the quantum of energy is equal to the kinetic energy of an electron which has been accelerated by a potential difference of 1·24 volt? The charge of an electron is $4{\cdot}8 \times 10^{-10}$ e.s.u. [$\nu = 3{\cdot}0 \times 10^{14}$ sec.$^{-1}$; $\lambda = 10^{-4}$ cm. $= 1\mu$.]

1(iii). One electron-volt is the energy of an electron which has fallen through a potential difference of one volt. How many electron-volts are equivalent to a quantum of X-rays whose frequency is 3×10^{19} sec.$^{-1}$? [123,800.]

1(iv). What are the dimensions of Planck's constant? [ML^2T^{-1}.]

1(v). If N atoms each emit a quantum of frequency ν, how much energy (in joules) is available? 1 joule $= 10^7$ ergs. [$6{\cdot}6 \times 10^{-34}$ $N\nu$ joules.]

CHAPTER II

Wave Theory—Introduction

2.1. Fundamental Ideas.

The theory of wave motion forms a descriptive system appropriate, with minor modifications, to a wide range of observations in sound and light, as well as to waves propagated along the surface of a liquid. The same system of equations may be used because the different groups of phenomena have many properties in common and it is these common properties which are described by the equations of wave motion. In the general account of wave theory, and in all the applications discussed in Chapters III–IX, it is not necessary to specify the detailed physical properties of the disturbance which represents light. It is even a matter of indifference whether the disturbance considered is a scalar quantity, like the pressure of a gas, or a vector quantity, like the electric or magnetic field vectors.

2.2.—The theory of wave motion involves three distinct concepts:

(*a*) There is some physical property which, at any given instant, has a defined and measurable value at every point.

(*b*) The value of this property at any given point can undergo a periodic fluctuation or disturbance.

(*c*) A disturbance at one point at a given time produces a similar disturbance at a neighbouring point at a slightly later time so that the pattern of the disturbance is continuously transferred from one place to another.

In studying wave motion it is convenient to start with the simple harmonic oscillator. This is the simplest and most fundamental of the many types of vibrating source which may give rise to waves. Moreover, its motion is essentially the same as the motion of any point in a medium through which waves are passing. The picture of a progressive wave is obtained by combining the equations of simple harmonic motion with certain general equations of propagation.

2.3. The Simple Harmonic Oscillator.

The undamped simple harmonic oscillator is a mathematical abstraction, just as a frictionless pulley is a mathematical abstraction.

Although no completely undamped oscillator is found in nature, the motion of an undamped oscillator forms a good first approximation to the motion of many physical systems. By its use certain important results are obtained in a simple way, and it is not difficult to insert the effects of damping as a second approximation.

In studying the simple harmonic oscillator, it is convenient to picture a particular physical system, and we therefore start by considering a torsional pendulum (fig. 2.1). The pendulum consists of two equal masses A and A′ connected by a weightless rod. The rod is suspended from its mid-point M by a thin wire whose upper end E is fixed. Suppose that the system is initially at rest, that the rod is rotated through a small angle about the vertical line EM and is then released. The subsequent motion may be studied by considering the variation of the quantity q which is defined to be the linear displacement of A (or A′) from its mean position. It is to be understood that q is measured along the circular arc through which the masses move. Then

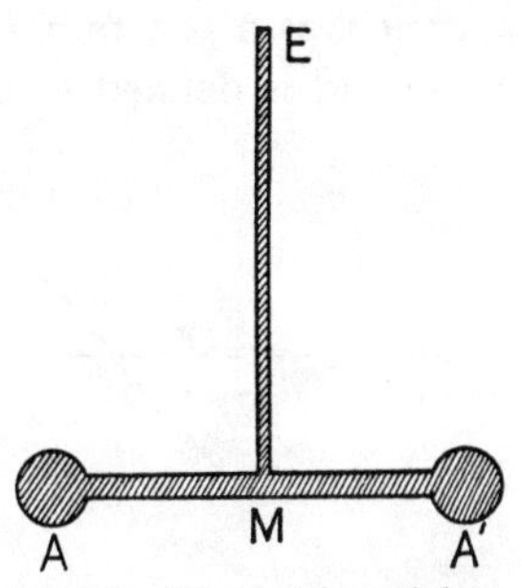

Fig. 2.1.—Torsional pendulum

$$q = f(t), \qquad \ldots \ldots \ldots \quad 2(1)$$

where t is the time measured from any convenient original moment.

2.4. *Experimental Observations.*

The following experimental observations might be made on such a system:

(i) The pendulum oscillates to and fro in a horizontal plane and there is a constant interval between the times when A passes through its equilibrium position, i.e. between the times when $q = 0$. The time between successive transits *in the same direction* is called the *period* (T).

(ii) The value of q varies between two limits $\pm a$. The quantity a is called the *amplitude*. In practice the amplitude slowly decreases, but with a suitable choice of material for the wire (EM) the decrease in one period is small. As stated above, it is neglected in the present discussion.

(iii) The variation of q with t is represented by a graph similar to that shown in fig. 2.2*a*.

This graph can be represented by the equation

$$q = a \sin(\omega t + \delta) = a \sin \phi, \quad . \quad . \quad . \quad . \quad 2(2)$$

where $\omega = 2\pi/T$ and is called the *angular frequency* * or *circular frequency*, δ is a constant called the *epoch* or *epoch angle*, ϕ is called the *phase* and is defined to be equal to $(\omega t + \delta)$. The angular frequency

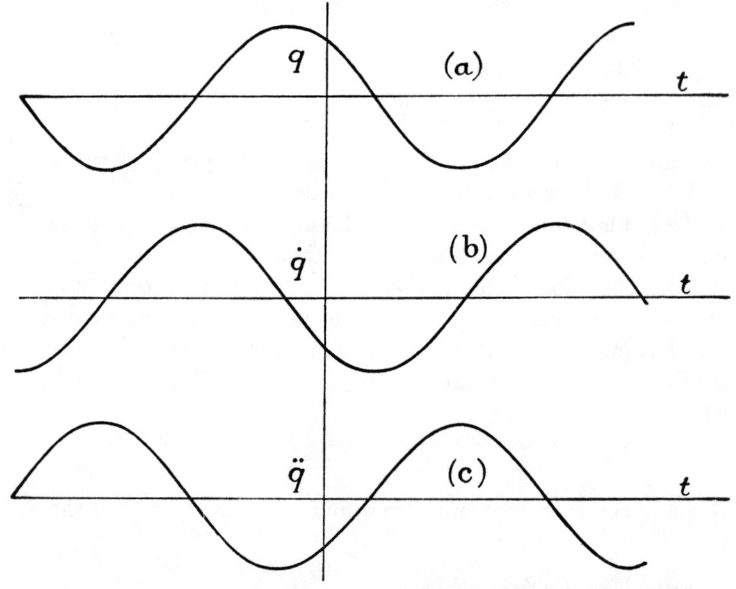

Fig. 2.2.—Variation of displacement, velocity, and acceleration with time, for a simple harmonic oscillator

is closely related to the *frequency* or number of vibrations per second. If ν is the frequency,

$$\nu = \frac{1}{T} = \frac{\omega}{2\pi}. \quad . \quad . \quad . \quad . \quad . \quad . \quad 2(3)$$

The value of δ is determined by observing the value of q when $t = 0$, i.e. $q_0 = a \sin \delta$. If the origin of time is chosen so that $t = 0$ when $q = 0$, then $\delta = 0$ and 2(2) becomes

$$q = a \sin \omega t. \quad . \quad . \quad . \quad . \quad . \quad . \quad . \quad . \quad . \quad 2(4)$$

It is usually convenient to choose the origin of t in this way when dealing with a single system but, when two systems are being considered, they do not in general

* This quantity is sometimes called the *pulsance*.

have the same epoch. While one can be reduced to the form 2(4), the more general relation 2(2) must be used for the other. Although only one system is now under consideration the more general form 2(2) is retained in view of later applications.

(iv) The force required to turn the rod AA′ so as to maintain a deflection q is found to be proportional to q. Write it equal to kq. The work done in deflecting the pendulum (slowly) through a distance q is

$$V = \tfrac{1}{2}kq^2. \qquad \qquad 2(5)$$

(v) The period of oscillation of the system is found to be

$$T = 2\pi\sqrt{\frac{m}{k}}, \qquad \qquad 2(6)$$

where $\frac{1}{2}m$ is the mass of either A or A′.

2.5. *Equations of Motion.*

By differentiating equation 2(2) with respect to time, we obtain

$$\dot{q} = \omega a \cos(\omega t + \delta) = \omega a \cos\phi = \pm\omega(a^2 - q^2)^{\frac{1}{2}}. \qquad 2(7)$$

and

$$\ddot{q} = -\omega^2 a \sin(\omega t + \delta) = -\omega^2 a \sin\phi = -\omega^2 q. \qquad 2(8)$$

These equations are represented in figs. 2.2*b* and 2.2*c*. Both the positive and the negative sign have to be used in 2(7) because each value of q is associated with two equal and opposite values of $\dot{q}$. The system passes through each point twice in one period. On these two occasions the magnitude of $\dot{q}$ is the same but the sign is different. Equation 2(8) may also be obtained directly from the dynamical laws of motion. The torsion of the wire provides a couple which is equivalent to two forces each equal to $\frac{1}{2}kq$ and each acting on a mass $\frac{1}{2}m$.

Thus

$$\ddot{q} = -\frac{k}{m}q. \qquad \qquad 2(9)$$

The minus sign is required because the force always acts in a direction opposite to the displacement (q) and so acts as a *restoring force* tending to return the system to its equilibrium position. Equation 2(9) is equivalent to equation 2(8) provided that

$$\omega = \sqrt{\frac{k}{m}}, \qquad \qquad 2(10a)$$

or

$$T = 2\pi\sqrt{\frac{m}{k}}. \qquad \qquad 2(10b)$$

This equation is verified by direct observation (see § 2.4). The period (or angular frequency) is thus determined by the ratio of the restoring force per unit displacement to the mass.

From 2(7) the kinetic energy ($\frac{1}{2}m\dot{q}^2$) is seen to be $\frac{1}{2}m\omega^2(a^2 - q^2)$, and using 2(5) it may be seen that the total energy (W) is given by

$$W = \tfrac{1}{2}m\omega^2a^2. \qquad \text{2(11)}$$

This energy is the same at all stages of the motion.

EXAMPLES [2(i)–2(vi)]

2(i). Write down the more important steps of the above discussion, using the angle through which the rod AA′ is deflected as the variable. If this angle is θ, show that $\ddot{\theta} = -\omega^2\theta$ and obtain a solution of this equation.

2(ii). What are the dimensions of the constant k? Show that equations 2(5), 2(6) and 2(9) are dimensionally correct.

[The dimensions on each side of 2(5) are $[\mathrm{ML^2T^{-2}}]$ and, on each side of 2(9), $[\mathrm{LT^{-2}}]$.]

2(iii). What are the dimensions of ω, ν, ϕ and δ? Show that equation 2(10) is dimensionally correct.

[The dimensions of ω and ν are $[\mathrm{T^{-1}}]$; ϕ and δ are angles and hence are pure ratios.]

2(iv). Show that the small oscillations of a simple pendulum are simple harmonic and find the angular frequency. [$\omega = \sqrt{(g/l)}$.]

2(v). Find the period of oscillation of a compound pendulum supported at a distance h from the centre of gravity.

[$T = 2\pi\sqrt{\{(h^2 + k^2)/gh\}}$, where k is the radius of gyration about the centre of gravity.]

2(vi). Make a list of some other physical systems whose motion is approximately simple harmonic.

[Any system in which there is a restoring force proportional to the displacement, e.g. a common balance, a magnet suspended in a magnetic field, the coil of a moving-coil galvanometer, etc.]

2.6. *Arbitrary Constants.*

Equation 2(8) is the fundamental differential equation of the motion. Its solution contains two *arbitrary constants* and may be written

$$q = a\sin(\omega t + \delta) = a\sin\phi, \qquad \text{2(2)}$$

or

$$q = -a\cos(\omega t + \delta') = -a\cos\phi', \qquad \text{2(12)}$$

or

$$q = A\sin\omega t + B\cos\omega t. \qquad \text{2(13)}$$

In 2(2) the arbitrary constants are a and δ; in 2(12) they are a and δ', and in 2(13) they are A and B.

These arbitrary constants are determined by the *initial conditions.* They may be derived if the values of q and $\dot{q}$ at some time t_0 are given. For example, if we are given $q = 0$ and $\dot{q} = u$ when $t = 0$, then $\delta = 0$ and $a = u/\omega$, so that $q = (u/\omega) \sin \omega t$ in this case. If 2(2), 2(12) and 2(13) all refer to the same initial conditions, the following relations must hold:

$$\left.\begin{aligned} \delta' - \delta &= \tfrac{1}{2}\pi, \\ \phi' - \phi &= \tfrac{1}{2}\pi, \\ A = a \cos \delta \quad &\text{and} \quad B = a \sin \delta, \\ \text{so that} \quad a^2 &= A^2 + B^2. \end{aligned}\right\} \quad . \quad . \quad . \quad 2(14)$$

The initial conditions may be given in various ways, e.g. the values of q and $\ddot{q}$ (or $\dot{q}$ and $\ddot{q}$) might be stated for a certain value of t, or one of these variables might be given for two values of t. In general, two independent pieces of information must be supplied. Note that ω and the two associated quantities ν and T are not arbitrary constants. They depend on the physical properties of the oscillator and not upon the initial conditions.

EXAMPLES [2(vii) and 2(viii)]

2(vii). Determine the arbitrary constants and write down equations corresponding to 2(2), 2(12) and 2(13) when

(*a*) $q = 0$ and $\dot{q} = u$, when $t = nT$ (n being any integer),

(*b*) $q = q_0$ when $t = 0$, and $\dot{q} = u_0$ when $t = t_1$,

(*c*) $q = q_1$ when $t = t_1$, and $q = q_2$ when $t = t_2$.

[(*a*) $q = \dfrac{u}{\omega} \sin \omega t$,

(*b*) $q = \left\{\dfrac{u_0 + q_0\omega \sin \omega t_1}{\omega \cos \omega t_1}\right\} \sin \omega t + q_0 \cos \omega t$.

(*c*) $q = \left\{\dfrac{q_1 \cos \omega t_2 - q_2 \cos \omega t_1}{\sin \omega(t_1 - t_2)}\right\} \sin \omega t + \left\{\dfrac{q_2 \sin \omega t_1 - q_1 \sin \omega t_2}{\sin \omega(t_1 - t_2)}\right\} \cos \omega t.$]

2(viii). Why do you fail to obtain the arbitrary constants if you are given $q = q_1$, when $t = 0$ and $q = q_1$ when $t = nT$?

[Because the dynamical equations imply that the value of q when $t = nT$ is the same as when $t = 0$. Thus only one piece of information about initial conditions is supplied.]

2.7. *General Equations of Motion.*

The general equation which applies to all simple harmonic oscillators is 2(5). This leads to the equation

$$H = \tfrac{1}{2}\frac{p^2}{m} + \tfrac{1}{2}kq^2, \quad . \quad . \quad . \quad . \quad . \quad . \quad . \quad . \quad 2(15)$$

where p is the generalized momentum and H is the total energy expressed as a function of p and q. Equation 2(9) may then be obtained from one of Hamilton's equations

$$\frac{\partial H}{\partial q} = -\frac{dp}{dt}.$$

2.8. **Vector Representation of Simple Harmonic Motion.**

A straight line joining two points has direction and magnitude. Such a line is called a vector. When we are confined to lines in a given plane, two quantities are needed to define a vector. These may be (i) the length of the line, and (ii) the angle which it makes with some fixed axis. A vector may also be specified by giving its components along two fixed axes, and in various other ways. Always two quantities are involved and, for this reason, a vector (in one plane) is suitable for the simultaneous representation of two quantities. If the vector is given, the two associated quantities can be derived and, conversely, if the two quantities are given, we can draw the vector. The vector representation of simple harmonic motion may be carried out in either of two ways:

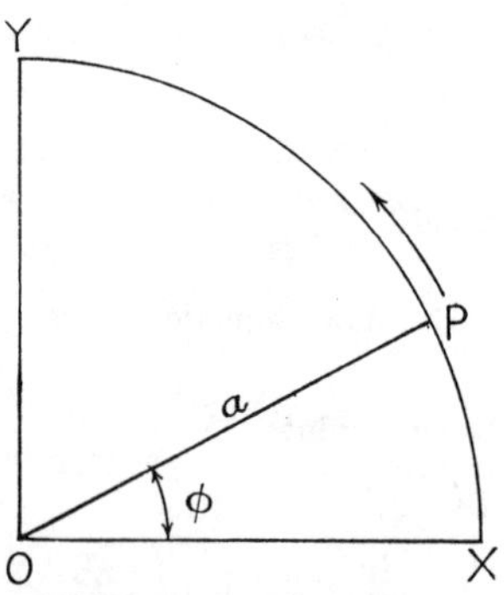

Fig. 2.3.—Representation of simple harmonic motion by a rotating vector.

(i) A *rotating* vector may be used. The line OP (fig. 2.3) is the representative vector. Its length is equal to the amplitude (a) and the angle which it makes with the line OX is equal to ϕ [see equation 2(2)]. The component of the vector in the direction OY is equal to $a \sin \phi$ and is thus equal to q. As ϕ increases with t the vector rotates at a constant rate of ω radians per second. Its rotation represents the progress of the simple harmonic motion. The reader may verify that the components of the velocity and acceleration of the point P, resolved in the direction OY, are equal to $\dot{q}$ and $\ddot{q}$ as given by equations 2(7) and 2(8).

(ii) A *stationary* vector may be used to represent the two arbitrary constants. This vector may be regarded as an instantaneous "snapshot" of the vector OP at the instant for which the initial conditions are given.

Suppose that the length of OQ (fig. 2.4) is equal to a and the angle QOX is equal to δ. Then the vector OQ, together with the axis OX, states the initial conditions. It does not rotate with the motion, but if we are given OQ, and equation 2(2), we can determine the subsequent motion of the system.

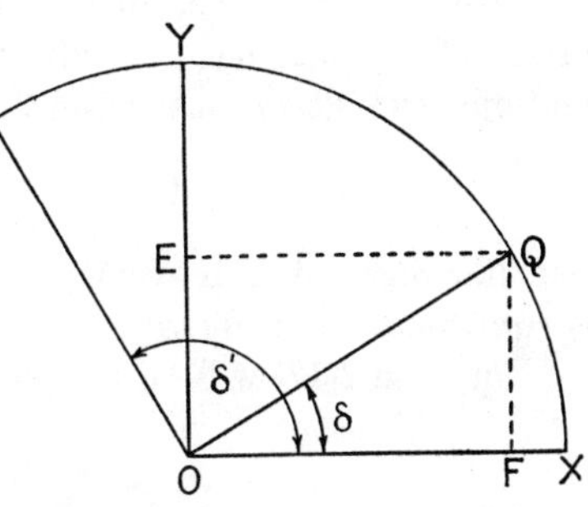

Fig. 2.4.—Representation of simple harmonic motion by a stationary vector

It is important to recognize that the vector representation of simple harmonic motion is not related to any possible vector property of the disturbance which is represented. What has been said about vector representation applies equally whether the magnitude which fluctuates is a scalar quantity or a vector quantity.

The relation between equations 2(2), 2(12) and 2(13) may be seen from fig. 2.4. In 2(2) the vector OQ is specified by its magnitude (a) and its angle (δ) to the axis OX. In 2(12) the angle δ' is given instead of δ. In 2(13) the components of OQ along the two axes are given, for $OE = a \sin \delta = B$ and $OF = a \cos \delta = A$.

2.9. Equation of Propagation—One Dimension.

We shall now consider the propagation of a disturbance ξ which, at any given time t, has a defined value at any point on a given straight line which is taken as the axis of x. We then have

$$\xi = f(x, t). \qquad \qquad 2(16)$$

If the disturbance considered is a ripple running along a stretched string, then ξ will be the displacement of the string (at the point x and the time t) from its equilibrium position. ξ might equally well represent the kind of disturbance which forms a set of ripples on the surface of a liquid, provided that the ripples form a series of lines parallel to each other, and perpendicular to the axis of x. ξ is then a function of x and not of y. In a similar way, ξ might represent a very wide parallel beam of light travelling in the direction of the axis of x. For each of these disturbances 2(16) is satisfied.

The values of ξ at a particular time t_1 form a function of x only.

The curve connecting ξ and x at a given time is called the *profile* of the disturbance. If the disturbance is propagated unchanged in the direction OX, then in a given interval of time (t') all the values of ξ move a certain distance (x') along the axis of x. Algebraically, an increase in t has the same effect as an alteration in the origin of x. If the disturbance is propagated with speed b, in the positive direction of x, then $x' = bt'$, i.e. an increase of t' in t and a movement in the origin of x in the negative direction by an amount bt' produce equal and opposite alterations in the values of ξ. Therefore

$$f(x, t) = f(x + bt', t + t'), \quad . \quad . \quad . \quad . \quad 2(17)$$

remembering that a movement of the origin in the negative direction is equivalent to an *increase* in all the values of x.

Equation 2(17) will be true for all values of x and t if, and only if,

$$\xi = f(bt - x), \quad . \quad . \quad . \quad . \quad . \quad . \quad . \quad 2(18)$$

since $$b(t + t') - (x + bt') = (bt - x).$$

Similarly a disturbance propagated in the negative direction of x is represented by

$$\xi = g(bt + x), \quad . \quad . \quad . \quad . \quad . \quad . \quad . \quad 2(19)$$

where f and g represent any two continuous functions. *If the speed of propagation* (b) *is the same for all values of x and t*, we obtain by differentiation of 2(18),

$$\left.\begin{aligned} \frac{\partial^2 \xi}{\partial x^2} &= f'', \\ \frac{\partial^2 \xi}{\partial t^2} &= b^2 f'', \end{aligned}\right\} \quad . \quad . \quad . \quad . \quad . \quad . \quad . \quad 2(20)$$

where dashes represent differentiation with respect to $(bt - x)$.

Equations 2(20) lead to

$$\frac{\partial^2 \xi}{\partial x^2} = \frac{1}{b^2} \frac{\partial^2 \xi}{\partial t^2}. \quad . \quad . \quad . \quad . \quad . \quad . \quad . \quad 2(21)$$

This is the fundamental differential equation for the propagation of a disturbance with constant velocity and without change of profile.

It may easily be verified by differentiation that 2(19) is a solution of this equation and that any linear combination of 2(18) and 2(19), such as

$$\xi = H_1 f(bt - x) + H_2 g(bt + x) \quad . \quad . \quad . \quad 2(22)$$

(where H_1 and H_2 are constants), is also a solution. Equation 2(22) represents one disturbance propagated in the positive direction, and a second disturbance (not necessarily of the same profile) propagated in the negative direction.

2.10.—Consider a disturbance defined by

$$\xi = a \sin \frac{\omega}{b} (bt + x), \quad . \quad . \quad . \quad . \quad . \quad 2(23)$$

where a, b and ω are constants for all values of x and t. This disturbance has the following three properties:

(i) It is of the same form as 2(19) and is therefore propagated unchanged with velocity b in the negative direction of x.

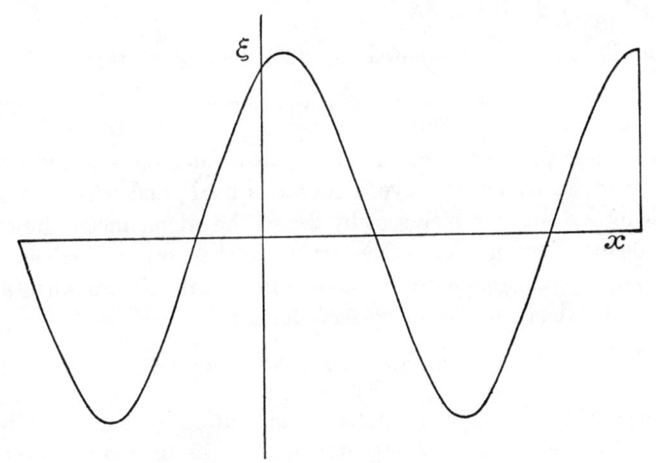

Fig. 2.5.—Profile of a simple sine wave

(ii) At any given point the disturbance is simple harmonic; for, if we put $x = x_0$ in 2(23), and make $\delta = \omega x_0/b$, the resultant expression is the same as 2(2). ω is again the angular frequency of the simple harmonic motion.

(iii) At any given instant the profile forms a simple sine function since, if t is put equal to t_0,

$$\xi = a \sin \frac{\omega}{b} (x + bt_0). \quad . \quad . \quad . \quad . \quad . \quad 2(24)$$

This profile is shown in fig. 2.5. It is of the same shape as that shown in fig. 2.2a, but the variables are now ξ and x instead of q and t.

2.11. *Wavelength and Wavelength Constant.*

The profile shown in fig. 2.5 is space-periodic. It is repeated at distances of $2\pi b/\omega$. This distance is called the *wavelength* and is denoted by λ. An associated constant κ is called the *wavelength constant* (or *propagation constant*) and is defined to be equal to $2\pi/\lambda$.

Using these definitions together with equation 2(3), the following relations may be derived:

$$\left.\begin{aligned} \lambda\nu &= b. \\ \lambda &= bT. \\ b &= \frac{\omega}{\kappa}. \end{aligned}\right\} \qquad 2(25)$$

2.12. *Phase of the Wave.*

The disturbance represented by 2(23) may be written

$$\xi = a \sin(\omega t + \kappa x) = a \sin\phi, \qquad 2(26)$$

and the phase $\phi = (\omega t + \kappa x)$ is seen to be a function of x and t. The phase increases by 2π whenever t increases by T, and whenever x increases by λ. When ϕ increases by 2π all the trigonometric functions which define ξ and its derivatives return to their original values.

Equation 2(26) represents a wave whose phase is zero when $t = 0$ and $x = 0$. A slightly more general form is

$$\xi = a \sin(\omega t + \kappa x + \delta_1) = a \sin\phi_1. \qquad 2(27)$$

When a single wave is under consideration it may usually be reduced to the form of 2(26), but when two or more waves are considered together it is usually necessary to use 2(27).

EXAMPLES [2(ix)–2(xi)]

2(ix). Show that the discussion of §§ 2.10–2.12 applies to the following disturbances:

$$\xi = a \cos(\omega t - \kappa x + \delta_2). \qquad 2(28)$$

$$\xi = a \sin(\omega t - \kappa x + \delta_3). \qquad 2(29)$$

$$\xi = a \cos(\omega t + \kappa x + \delta_4). \qquad 2(30)$$

What kinds of waves do these expressions represent? What is the phase relation between a wave represented by 2(29) and one represented by 2(28)?

[$\delta_3 - \delta_2 = \pi/2$.]

2(x). Differentiate 2(27) twice with respect to t and twice with respect to x. Hence verify directly that equation 2(21) is satisfied.

2(xi). Show that the sum of a number of expressions similar to 2(27), 2(28), 2(29) and 2(30) is always a solution of the wave equation.

2.13. **Propagation of Waves in Three Dimensions.**

When a disturbance is propagated in a three-dimensional space, the value of ξ at any given point in the space undergoes a periodic variation. In the simplest type of wave, the variation of ξ with t at any point is simple harmonic. The phase ϕ varies from point to point and is a continuous function of x, y, z, and t. The variation of ϕ is such that, at any given time, the phase has the same value over the whole of certain surfaces. These surfaces are called *wave surfaces*. They are defined by the relation

$$\phi_{t_0} = g(x, y, z) = \chi_0, \qquad \text{2(31)}$$

where χ_0 is the value of the phase for one particular wave surface at time t_0. In general, the wave surfaces form a family of surfaces, χ being the variable parameter which selects a particular member of the family. Wave surfaces may constitute a family of concentric spheres, or a family of parallel planes, or they may take other forms. The waves are called spherical waves, plane waves, etc., according to the shape of the wave surfaces.

2.14. **Plane Waves.**

When ϕ has the form

$$\phi = \omega t - \kappa(\alpha x + \beta y + \gamma z) + \delta, \qquad \text{2(32)}$$

where α, β, γ are real constants connected by the relation

$$\alpha^2 + \beta^2 + \gamma^2 = 1, \qquad \text{2(33)}$$

then the wave surfaces constitute a family of planes whose direction-cosines are α, β, γ.

Differentiating 2(32) with respect to t we have

$$\dot{\phi} = \omega - \kappa(\alpha\dot{x} + \beta\dot{y} + \gamma\dot{z}). \qquad \text{2(34)}$$

A point which moves with a velocity such that $\dot{\phi} = 0$ will always have the same phase and be in the same wave surface. From 2(34) such a point will have a velocity b whose magnitude is ω/κ and whose components along the co-ordinate axes are αb, βb, and γb. Such a point moves in a direction perpendicular to the wave surfaces. This applies to any point on any of the original wave surfaces and therefore every wave surface advances perpendicular to itself with speed $b = \omega/\kappa$. The plane wave in three dimensions is similar to the wave discussed in §§ 2.10–2.12 and may indeed be reduced to the form of 2(26) by a suitable change of axes. The more general form is required when more than one set of waves (not all travelling in the same direction) have to be considered, as, for example, in the theory of reflection and refraction.

2.15. **The Wave Equation.**

A wave whose phase is of the form given in 2(32) may be represented by

$$\xi = a \sin \phi = a \sin [\omega t - \kappa(\alpha x + \beta y + \gamma z) + \delta]. \qquad 2(35)$$

Differentiation of 2(35) gives

$$\frac{\partial^2 \xi}{\partial t^2} = -\omega^2 \xi, \quad . \quad . \quad . \quad . \quad . \quad . \quad . \qquad 2(36)$$

$$\frac{\partial^2 \xi}{\partial x^2} = -\kappa^2 \alpha^2 \xi, \quad . \quad . \quad . \quad . \quad . \quad . \quad . \qquad 2(37)$$

and two similar equations.

Combining 2(36) and 2(37), using 2(33), we obtain

$$\frac{\partial^2 \xi}{\partial x^2} + \frac{\partial^2 \xi}{\partial y^2} + \frac{\partial^2 \xi}{\partial z^2} = \frac{\kappa^2}{\omega^2} \frac{\partial^2 \xi}{\partial t^2} = \frac{1}{b^2} \frac{\partial^2 \xi}{\partial t^2}. \quad . \quad . \qquad 2(38)$$

This is a general equation of propagation in three dimensions and 2(35) is the particular solution which represents plane waves of angular frequency ω.

A more general solution * is

$$\xi = f(bt - \alpha x - \beta y - \gamma z), \quad . \quad . \quad . \quad . \quad . \qquad 2(39)$$

where α, β, γ are related by 2(33).

This solution represents a plane wave whose profile is, in general, not of the simple sine-wave form.

* Further solutions are discussed in Reference 2.1.

EXAMPLES [2(xii)–2(xv)]

2(xii). Verify by differentiation that both 2(39) and

$$\xi = g(bt + \alpha x + \beta y + \gamma z) \quad . \quad . \quad . \quad . \quad . \quad 2(40)$$

are solutions of 2(38).

Show that any linear combination of these solutions such as

$$\xi = Hf(bt - \alpha x - \beta y - \gamma z) + Kg(bt + \alpha x + \beta y + \gamma z) \quad . \quad 2(41)$$

is a solution. What does this solution represent?

[Show that the wave surfaces, as defined in § 2.14, are planes, and deduce their speed and direction of motion.]

2(xiii). What does the solution

$$\xi = H_1 f_1(bt - \alpha_1 x - \beta_1 y - \gamma_1 z) + H_2 f_2(bt - \alpha_2 x - \beta_2 y - \gamma_2 z) \quad 2(42)$$

represent? Make this solution more general.

[2(42) represents two plane waves travelling in directions represented by $\alpha_1, \beta_1, \gamma_1$ and $\alpha_2, \beta_2, \gamma_2$.]

The more general form—representing waves in various directions—is

$$\xi = \Sigma\, H_n f_n(bt - \alpha_n x - \beta_n y - \gamma_n z). \quad . \quad . \quad . \quad . \quad 2(43)$$

2(xiv). Write down equations similar to 2(32)–2(39) for the propagation of waves in two dimensions. Note that the wave surfaces reduce to lines. Strictly speaking, there are no plane waves in two dimensions, but " straight-line waves " are often called plane waves since they are regarded as sections of three-dimensional plane waves whose wave surfaces are at right angles to the plane in which the line waves are propagated.

2(xv). Show that the expression

$$\xi = f(bt - x\cos\theta - y\sin\theta) \quad . \quad . \quad . \quad . \quad . \quad 2(44)$$

represents a line wave travelling at an angle θ to the x axis. Write down an expression for waves whose lines of constant phase are given by $y = mx + C$ (the variable parameter being C).

[$\xi = f(bt - \alpha m x + \alpha y)$, where $\alpha^2 = (1 + m^2)^{-1}$.]

2.16. **The Velocity of Propagation.**

In § 2.5 it was shown that ω is determined not by the initial conditions but by the physical properties of the oscillator; e.g. for the torsional pendulum ω is equal to $\sqrt{(k/m)}$ and for the simple pendulum it is equal to $\sqrt{(g/l)}$. In a similar way the velocity of wave propagation is determined by the physical properties of the medium. The calculation of the velocity of light in terms of certain fundamental electromagnetic units will be given in Chapter XIII. For the present we may illustrate the problem and show the way in which a velocity of propagation may be determined by considering longitudinal waves transmitted along a rod of elastic material (see fig. 2.6).

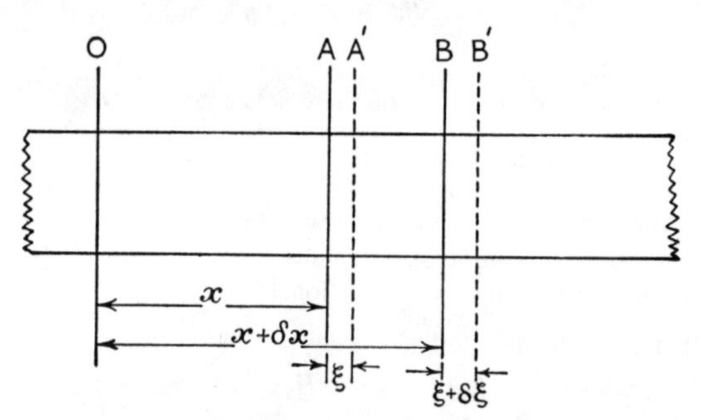

Fig. 2.6.—Transmission of waves along a rod

2.17. *Waves on a Rod.*

Suppose that the area of cross-section of the rod is σ, the density of the material is ρ, and the value of Young's modulus is q. Let x be the distance of the plane A from the plane O when the rod is undisturbed, and let $x + \delta x$ be the corresponding distance for the plane B. Suppose that the rod is subject to a disturbance ξ which is a continuous function of x and t so that at a given instant A is displaced by ξ to A′, and B by $\xi + \delta\xi$ to B′. Then the extension per unit length of the piece of the rod which initially was between A and B is

$$\frac{\mathrm{A'B'} - \mathrm{AB}}{\mathrm{AB}} = \frac{\delta\xi}{\delta x},$$

and the tension Q is given by

$$Q = \sigma q \frac{\delta\xi}{\delta x}, \qquad \text{. . . . . . 2(45)}$$

provided that $\delta\xi/\delta x$ is everywhere so small that Hooke's law is obeyed. The tension Q is also a continuous function of x and t. The net force on the element is the difference between the tension at A′ and the tension at B′. This is equal to

$$\frac{\partial Q}{\partial x}\,\delta x = \sigma q \frac{\partial^2 \xi}{\partial x^2}\,\delta x, \qquad \text{. . . . . . 2(46)}$$

and equating force to mass × acceleration we have

$$\sigma q \frac{\partial^2 \xi}{\partial x^2} \delta x = \sigma \rho \frac{\partial^2 \xi}{\partial t^2} \delta x,$$

which reduces to

$$\frac{\partial^2 \xi}{\partial x^2} = \frac{\rho}{q} \frac{\partial^2 \xi}{\partial t^2}. \qquad 2(47)$$

This is equivalent to 2(21) if

$$b = \sqrt{\frac{q}{\rho}}. \qquad 2(48)$$

Thus the velocity of propagation is determined by the values of density and Young's modulus for the material of which the rod is made.

2.18. **Transport of Energy and Momentum.**

In § 2.5 it was shown that a vibrating system possesses both kinetic and potential energy. In simple harmonic motion, the energy passes to and fro between the kinetic and potential forms but the total remains constant and is proportional to $\omega^2 a^2$. The elastic waves discussed in the last paragraph possess both kinetic energy, because the medium is moving, and potential energy, because it is in a state of strain. It may be shown that the total energy per unit volume is constant, and is proportional to $\omega^2 a^2$ [see Example 2(xvi), p. 41]. Thus the wave motion of the medium implies the existence of an *energy density* D which is proportional to the square of the amplitude.

It is often more convenient to deal with the square of the amplitude rather than with the actual energy density. We shall refer to the square of the amplitude as the *relative energy*, because it is the ratio of the energy density of the given wave to that of a wave of the same frequency with unit amplitude. The importance of the relative energy is that most physical instruments for measuring light of one frequency are devices for comparing the relative energies at different places.

The energy density is constant during the whole time for which a wave of constant amplitude is present in the medium. This constant energy density is maintained by energy being transported into any particular portion of the rod from the direction of the source and an equal amount being transported out in the direction in which the wave is travelling.

2.19.—If a disturbance is exactly represented by 2(26), everywhere and at all times, the wave train must extend from $-\infty$ to $+\infty$. The wave trains which represent light are never infinitely long, and in Chapter IV it will be necessary to consider in detail how equation 2(26) must be modified to take account of the finite length of the wave train. In a general consideration of the transmission of energy it is not necessary to assume any definite limit to the length of the wave train, but it is necessary to assume that there is a source somewhere (even though at an indefinitely great distance in the negative direction of x), and that either the wave is being absorbed (at some far point in the positive direction of x) or that the front of the wave is advancing into hitherto undisturbed portions of the medium.

In either case it is implied that there is a transport of energy. Under the simple conditions which we have so far considered, the energy crossing unit area of a plane parallel to the wave surface per second is equal to bD, because in one second the energy in a tube of length b crosses a surface placed at right angles to the direction of propagation.

A detailed account of the transference of energy and momentum in a beam of light cannot be given until the properties of the waves which represent light have been specified more exactly. For the present it is reasonable to assume as a working hypothesis, from analogy with elastic waves, that the energy density in the medium, and the rate of transfer of energy, are proportional to the square of the amplitude.

It may be shown * that the propagation of elastic waves involves a transfer of momentum, so that a system of waves exerts a pressure on any body which absorbs or reflects it. We shall show later that light also exerts a pressure in similar circumstances and that, for a parallel beam of light, the pressure is numerically equal to the energy density.

2.20. **Spherical Waves—Inverse Square Law.**

It is found by experiment that light is propagated from a small source in such a way that the flow of energy per unit area is proportional to the inverse square of the distance from the source. It is natural to attempt to represent this propagation of radiation from a small source by a system of spherical waves and to see whether this important law is correctly included in the representation. In order to study spherical waves, we return to equation 2(38) and transform it

* Reference 2.2.

into spherical polar co-ordinates. If it be assumed that the solution is spherically symmetrical this may be done as follows:

Let r be the radius vector drawn from the origin O. Then

$$r^2 = x^2 + y^2 + z^2, \qquad \text{2(49)}$$

and hence

$$\frac{\partial \xi}{\partial x} = \frac{\partial \xi}{\partial r} \cdot \frac{\partial r}{\partial x} = \frac{x}{r} \cdot \frac{\partial \xi}{\partial r}. \qquad \text{2(50)}$$

$$\frac{\partial^2 \xi}{\partial x^2} = \frac{\partial}{\partial r}\left(\frac{x}{r} \cdot \frac{\partial \xi}{\partial r}\right)\frac{\partial r}{\partial x}$$

$$= \frac{x}{r}\left\{\frac{x}{r} \cdot \frac{\partial^2 \xi}{\partial r^2} + \frac{\partial \xi}{\partial r} \cdot \frac{\partial}{\partial r}\left(\frac{x}{r}\right)\right\}$$

$$= \frac{x^2}{r^2} \cdot \frac{\partial^2 \xi}{\partial r^2} + \left(\frac{1}{r} - \frac{x^2}{r^3}\right)\frac{\partial \xi}{\partial r}. \qquad \text{2(51)}$$

Two similar expressions may be obtained for derivatives with respect to y and z. Adding these and using 2(49), we find that 2(38) becomes

$$\frac{\partial^2 \xi}{\partial r^2} + \frac{2}{r}\frac{\partial \xi}{\partial r} = \frac{1}{b^2} \cdot \frac{\partial^2 \xi}{\partial t^2}. \qquad \text{2(52)}$$

An alternative form of this equation is

$$\frac{\partial^2}{\partial r^2}(r\xi) = \frac{1}{b^2}\frac{\partial^2}{\partial t^2}(r\xi). \qquad \text{2(53)}$$

A general solution of this equation corresponding to 2(22) is

$$\xi = \frac{A}{r} f(bt - r) + \frac{B}{r} f(bt + r). \qquad \text{2(54)}$$

The expression

$$\xi = \frac{A}{r} \sin(\omega t - \kappa r) = \frac{A}{r} \sin \phi \qquad \text{2(55)}$$

is a special form of 2(54), and is therefore a solution. It represents a spherical wave since at any given time the phase is the same over the whole of any sphere centred on the origin. The phase existing on any sphere of radius r_0 is transferred to a larger sphere of radius $(r_0 + bt_0)$ after a time t_0, and therefore the expression represents a spherical wave diverging from the origin. The amplitude of this wave is not constant but is inversely proportional to r. The rate of transfer of energy across unit area of a wave surface is thus proportional to $1/r^2$

and the inverse square law of propagation is included in the description. The total energy crossing any sphere concentric with the origin is independent of r, since the area of a sphere of radius r is $4\pi r^2$. Once the wave is established, the total amount of energy entering the space between any two spheres centred on the origin is equal to the amount leaving this space in the same time.

The expression

$$\xi = \frac{B}{r} \sin (\omega t + \kappa r) \quad . \quad . \quad . \quad . \quad . \quad 2(56)$$

represents a wave converging towards the origin and has properties analogous to those of the wave * represented by 2(55). Equations 2(55) and 2(56) represent idealized concepts. In practice, light never diverges from, or converges to, a mathematical point and wave surfaces are not exactly spherical.

2.21. **Photometry—Definitions.**

The subject of photometry deals with the measurement of amounts of light chiefly in relation to the uses of light (and especially of artificial light) for visual tasks such as reading. The inverse square law for the rate of transfer of radiant energy across unit area of a spherical surface surrounding a point source is a basic assumption in most photometric calculations. Actual sources are treated by dividing them into small elementary parts, each of which is small enough to be regarded as a point source. The technique of photometric measurements is not considered in this book, but it is convenient to introduce some of the definitions at this stage. Confining ourselves to light of one wavelength, we state these definitions:

(i) *Flux* (across a given surface)—a quantity proportional to the rate at which light energy crosses the surface. In the absence of absorption and similar effects, the flux across any surface surrounding a source is proportional to the rate of emission of energy. Symbol, F.

(ii) *Illumination* (at a given point on a surface) is proportional to the flux per unit area across a small element of area including the point in question. Symbol, E.

(iii) *Intensity* (of a source in a given direction) is equal to the flux per unit solid angle in the given direction from the source. Symbol, I.

(iv) *Brightness* or *luminance* (of a source in a given direction) is equal to the intensity per unit area of the source in the given direction. Symbol, B.

In the discussion of the interference and diffraction of light, the most important quantity is the illumination. When interference

* Other types of spherical wave are of importance in the theory of sound (see References 2.1 and 2.2).

fringes are seen on a screen, we are interested in the variation of illumination over the screen. When they are viewed either directly by eye or through an instrument, we are interested in the variation of illumination over the plane on which the instrument is focused. When a system of fringes is viewed by an eyepiece, we may alternatively regard them as forming a source of light situated in the focal plane. We may then speak of the distribution of brightness in this plane.

It will be seen that, if the above definitions are accepted, it is not correct to speak of the intensity distribution in a system of fringes, or to use the word intensity for the square of the amplitude of a light wave. We have called the latter the " relative energy " (see § 2.18). We may, however, use the term " relative intensity of a spectrum line " to refer to the ratio of the amount of energy emitted by the source in a given spectral region to the total energy emitted by the same source.

In practical photometric measurements, illumination, brightness, etc., for different coloured lights have to be measured in units which take account of the effectiveness of different wavelengths in regard to vision. Since we are here concerned only with physical measurements, we use the above symbols to denote flux, illumination, etc., measured in *energy units*. Our values of these quantities would require to be multiplied by a *visibility factor* (depending on the colour) to convert them to the units used in practical photometry (e.g. candle-power as unit of intensity).

2.22. Doppler-Fizeau Principle.

The sound produced by a source such as a tuning fork may be detected by certain instruments, and its frequency measured. The frequency of the vibrating source may also be measured independently. It is found that, when a source and a receptor have no relative motion, the frequency of the sound received is equal to the frequency of vibration of the source. If, however, the source and receptor are approaching one another, the frequency of the sound received is higher than that of the source. If they are receding from one another, it is lower. The effect is observed as a sudden fall in pitch whenever a rapidly moving source of sound passes an observer. This phenomenon is included in the wave picture in the following way.

Suppose that a source and a receptor are at a distance L apart at time t_0, and that they are approaching one another with velocity v (which is a small fraction of b) so that they pass at time $(t_0 + L/v)$. Then, during the approach, the receptor receives the waves which initially lay between it and the source, as well as those emitted by the source during the interval L/v. The waves initially between the source and receptor were emitted during a time L/b. Therefore the source receives during the interval L/v all the waves emitted during a time $L/v + L/b$. Thus if ν is the frequency of the waves received and ν_0 is the frequency of the waves emitted,

$$\nu = \nu_0\left(1 + \frac{v}{b}\right). \quad \ldots \ldots \quad 2(57)$$

2.23.—This effect was first discovered by Doppler in relation to sound and was later discovered independently by others including Fizeau, who probably made the first correct application to light. It was known that the light emitted from certain gaseous sources could be closely represented by the types of simple wave trains which have been considered in the preceding paragraphs. The wavelengths could be accurately measured by methods to be described later. Since the velocity of light is unaffected by the movement of the source or the observer (see Chapter XI), 2(57) together with 2(25) implies, when v is small compared with b, that

$$\lambda = \lambda_0\left(1 - \frac{v}{b}\right) \quad \ldots \ldots \quad 2(58)$$

and a change of wavelength is to be expected when source and observer are approaching one another or receding from one another. The effect predicted by Fizeau could not immediately be observed in the laboratory because of the technical difficulty of producing a source moving with an appreciable fraction of the velocity of light. In more modern times this difficulty has been overcome in two different ways and the observed change of wavelength agrees well with that predicted.

2.24.—One set of experiments was carried out by Bélopolsky and a later set by Galitzin and Wilip. They used rotating mirrors to produce a virtual source moving at 400 metres per second. The change in wavelength was only one part in a million, but using the delicate methods described in Chapter IX, they were able to measure this change. The change observed agreed with that calculated from 2(58) to within about 5 per cent.

The effect was also found in experiments on the " canal rays ". The apparatus is shown in fig. 2.7. Positively charged atoms or molecules are accelerated by the electric field in a discharge tube at low pressure. They thus acquire a speed which is related to the potential (V) through which they have fallen by the equation

$$\tfrac{1}{2}mv^2 = Ve, \qquad \ldots\ldots \qquad 2(59)$$

where m is the mass and e the charge.

For hydrogen atoms $m = 1{\cdot}67 \times 10^{-24}$ gramme and $e = 4{\cdot}8 \times 10^{-10}$ electrostatic unit. Thus an atom which has fallen through a potential of 30,000 volts (or 100 e.s.u.) has a speed of 2×10^8 centimetres per

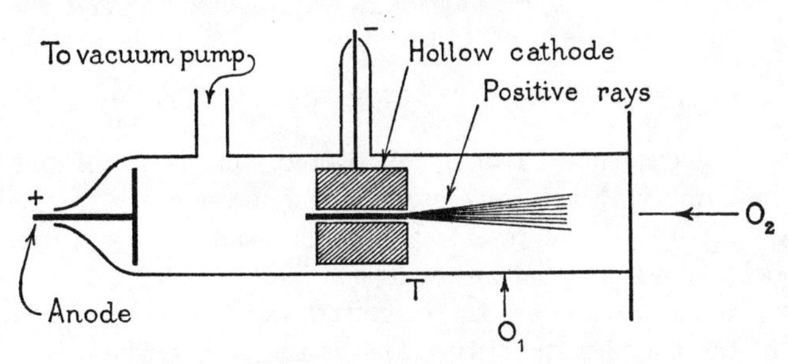

Fig. 2.7.—Apparatus for observing the Doppler effect

second. The fast-moving ions are neutralized during their passage through the tube T without any appreciable loss of speed and emit light in the region to the right of T.

The light is observed first from O_1, and then from O_2. The difference of wavelength (which is now about one part in 150) can be observed with a small spectroscope.

2.25.—The Doppler-Fizeau principle is of interest in two ways:

(i) It is an important experimental result which is satisfactorily included in the wave description of light and has no obvious place in a particle theory.

(ii) Once the principle has been verified by the above experiments it may be used to determine the velocities of sources of light under conditions where any other measurement would be difficult or impossible, e.g. the velocities of stars, or the velocities of atoms in gaseous discharges.

It will be noticed that the above account of the Doppler-Fizeau principle is valid only when the velocity of the source relative to the observer is a small fraction of b. The equation 2(58) is valid only when terms of order v^2/b^2 may be neglected. It will be shown in Chapter XI that when we consider terms of order v^2/b^2, equation 2(58) must be amended.

2.26. Representation of Wave Motion by Complex Quantities.†

The vector representation of the motion of a simple harmonic oscillator shown in fig. 2.3 may be applied to the representation of simple wave motion provided it be understood that ϕ is now a function of both x and t. Another method of representing simple harmonic motion has a special advantage in relation to wave motion because it allows the part of ϕ which varies with x to be separated from the part which varies with t in a convenient way. A solution of the wave equation 2(21) is

$$\xi = a \exp i(\omega t - \kappa x + \delta) = ae^{i\phi}. \quad . \quad . \quad 2(60) \ddagger$$

The displacement at a given place and time is a *real* and not a complex number. Some convention is therefore required to enable us to apply 2(60) to the representation of the results of observations. In this book, we assume that an expression like 2(60) represents a vector whose two components are the magnitudes of the real and imaginary parts of the complex quantity. This is the vector whose magnitude is equal to the amplitude of the wave, and whose angle, with a chosen fixed line, represents the phase. The real part of the quantity ξ, which by itself is a solution of the wave equation, gives the physical displacement at x and t. The sum of the squares of the magnitudes of the real and imaginary parts of the complex number thus represents the square of the amplitude (i.e. the relative energy). Equation 2(60) may be written

$$\xi = ae^{i\delta} \,.\, e^{i\omega t} \,.\, e^{-i\kappa x} \quad . \quad . \quad . \quad . \quad . \quad 2(61)$$

or

$$\xi = Pe^{i\omega t}e^{-i\kappa x}, \quad . \quad . \quad . \quad . \quad . \quad . \quad . \quad 2(62)$$

where

$$P = ae^{i\delta}. \quad . \quad . \quad . \quad . \quad . \quad . \quad . \quad . \quad . \quad 2(63)$$

The complex quantity P represents both the constants a and δ. If P^* is the complex quantity conjugate to P, it follows that PP^* is equal to a^2 and is proportional to the illumination. The quantity P is sometimes called the *complex amplitude.*

2.27.—Let two simple harmonic motions be represented by

$$\xi_1 = P_1 \exp i(\omega t - \kappa x) \quad . \quad . \quad . \quad . \quad 2(64)$$

and

$$\xi_2 = P_2 \exp i(\omega t - \kappa x), \quad . \quad . \quad . \quad . \quad 2(65)$$

† An account of the theory of complex quantities is given in Reference 2.3.

‡ The abbreviation " exp " will be used for " exponential " in all but the simplest expressions.

where P_1 and P_2 are complex. Then the *phase difference* δ is given by

$$\tan \delta = \frac{B}{A}, \qquad \ldots \ldots \qquad 2(66)$$

where

$$\frac{P_1}{P_2} = A + iB = Ce^{i\delta}. \qquad \ldots \ldots \qquad 2(67)$$

This result is obtained directly if we write $P_1 = C_1 e^{i\delta_1}$ and $P_2 = C_2 e^{i\delta_2}$, C_1 and C_2 being real numbers.

EXAMPLES [2(xvi)–2(xviii)]

2(xvi). Obtain a solution of equation 2(47) and write down the expression for the kinetic energy of the rod. Using equation 2(45), obtain the expression for the potential energy of the rod and hence find the total energy per unit volume.

2(xvii). Show that if $P = A + iB$, then 2(62) may be written

$$\xi = (A + iB) \cos (\omega t - \kappa x) + (iA - B) \sin (\omega t - \kappa x),$$

and that A and B have the values given in 2(14).

2(xviii). Show that if $\xi_1 = P_1 \exp i(\omega t + \kappa x)$ and $\xi_2 = P_2 \exp i(\omega t - \kappa x)$ are solutions of 2(21), then $\xi_s = \xi_1 + \xi_2$ is also a solution, and that ξ_s may also be written as

$$(P_1 + P_2)e^{i\omega t} \cos \kappa x + i(P_1 - P_2)e^{i\omega t} \sin \kappa x.$$

REFERENCES

2.1. Coulson: *Waves* (Oliver and Boyd).

2.2. Wood: *Acoustics* (Blackie).

2.3. Green: *The Theory and Use of the Complex Variable* (Pitman).

CHAPTER III

Wave Theory—Combination of Wave Motions

3.1. Principle of Superposition.

The type of wave motion discussed in Chapter II is produced by the action of a single simple harmonic oscillator. We shall now consider the disturbance produced by the simultaneous action of two or more oscillators. The simplest hypothesis which can be made is that if ξ_1, ξ_2, ξ_3, etc., are the disturbances produced by the individual oscillators at a given place and time, and ξ is the resultant disturbance due to them all, then

$$\xi = \xi_1 + \xi_2 + \xi_3 + \dots \qquad 3(1)$$

If the resultant motion is represented by the wave equation, i.e. by equation 2(38), it is necessary that ξ shall be a solution of this equation. The form of the wave equation is such that solutions are additive [see § 2.9, and Examples 2(xi), 2(xii), 2(xviii)], and therefore 3(1) is a solution of the wave equation. It is important to recognize that this mathematical result does not by itself guarantee that 3(1) correctly represents the effect of the simultaneous action of several wave motions at a given point. The *principle of superposition* is a physical hypothesis which states that for light waves the disturbance (at a given place and time) due to the passage of a number of waves is equal to the algebraic sum of the disturbances produced by the individual waves. Equation 3(1) is a mathematical formulation of this principle. This hypothesis is confirmed in so far as calculations based upon it give a satisfactory description of the relevant observations on light.

3.2.—With sound waves it is found that, for waves of large amplitude, the velocity of propagation is not independent of the amplitude. It is also found that when two loud sources of different frequencies operate simultaneously, sum and difference tones are heard.* In order to describe this type of phenomenon it is necessary to assume

* Reference 2.2.

that sound waves of finite amplitude are not exactly represented by the simple waveforms discussed in Chapter II, and that the disturbance ξ due to the simultaneous operation of two sources is given by

$$\xi = \xi_1 + \xi_2 + \alpha_1 \xi_1^2 + \alpha_2 \xi_2^2 + \alpha_{12} \xi_1 \xi_2 + \dots \quad . \quad 3(2)$$

where α_1, α_2, and α_{12} are constants which are small compared with $1/\xi_1$.

Some hypothesis of this type would be necessary if corresponding phenomena were observed with light, but so far all attempts to observe these phenomena have produced negative results. Schrödinger has considered the effect of introducing certain non-linear terms (of a type suggested by Born) into the equations of propagation of electromagnetic waves. The calculation indicates that, for very high intensities, the speed of light should depend on the amplitude but, under practical conditions, the effect is far too small to be observed.

3.3. **Addition of Simple Harmonic Motions.**

In the theory of light, many calculations have to be made of the resultant produced by the superposition of two or more simple harmonic motions of the same frequency. Three different methods are available for carrying out these calculations. The choice of method to be used in solving a particular problem is a matter of mathematical convenience since each of the three methods must yield the same result if used correctly. In order to show how the three methods are applied, two simple problems will now be solved by each method in turn. These problems are (*a*) the addition of two simple harmonic motions whose amplitudes and epochs are different, and (*b*) the addition of several simple harmonic motions whose amplitudes are equal and whose epoch angles are in arithmetical progression.

3.4. *Algebraic Method.*

Consider two simple harmonic motions represented by

$$\xi_1 = a_1 \sin(\omega t - \kappa x + \delta_1) \quad . \; . \; . \; . \quad 3(3)$$

and

$$\xi_2 = a_2 \sin(\omega t - \kappa x + \delta_2). \quad . \; . \; . \; . \quad 3(4)$$

Then

$$\begin{aligned} \xi &= \xi_1 + \xi_2 \\ &= a_1 \sin(\omega t - \kappa x + \delta_1) + a_2 \sin(\omega t - \kappa x + \delta_2) \quad . \quad 3(5) \\ &= (a_1 \cos \delta_1 + a_2 \cos \delta_2) \sin(\omega t - \kappa x) \\ &\quad + (a_1 \sin \delta_1 + a_2 \sin \delta_2) \cos(\omega t - \kappa x). \end{aligned}$$

This is identical with

$$\xi = a \sin(\omega t - \kappa x + \delta), \quad \ldots \quad 3(6)$$

provided that

$$\left.\begin{aligned} a^2 &= (a_1 \cos \delta_1 + a_2 \cos \delta_2)^2 + (a_1 \sin \delta_1 + a_2 \sin \delta_2)^2 \\ \text{and} \quad \tan \delta &= \frac{a_1 \sin \delta_1 + a_2 \sin \delta_2}{a_1 \cos \delta_1 + a_2 \cos \delta_2}. \end{aligned}\right\} \quad 3(7)$$

This calculation might have been simplified a little by choosing the origins of x and t so as to make $\delta_1 = 0$ and by putting $(\omega t - \kappa x) = \chi$. Then instead of 3(5) we have

$$\xi = a_1 \sin \chi + a_2 \sin(\chi + \delta_2), \quad \ldots \quad 3(8)$$

and the resultant is then

$$\xi = a \sin(\chi + \delta), \quad \ldots \quad 3(9)$$

where

$$a^2 = (a_1 + a_2 \cos \delta_2)^2 + a_2{}^2 \sin^2 \delta_2$$

$$\left.\begin{aligned} &= a_1{}^2 + a_2{}^2 + 2a_1a_2 \cos \delta_2 \\ \text{and} \quad \tan \delta &= \frac{a_2 \sin \delta_2}{a_1 + a_2 \cos \delta_2}. \end{aligned}\right\} \quad \ldots \quad 3(10)$$

The simpler forms of equations 3(9) and 3(10) would be used in a practical example, but equation 3(7) is given here in order to show the symmetrical character of the result. This form will also be useful for comparison with a calculation based on the vector representation. The form of equations 3(6) and 3(9) implies that the resultant of two simple harmonic motions of the same frequency is itself a simple harmonic motion. Also the frequency of the resultant is the same as that of the component motions. By repeated application of this process it may be shown that *the resultant of any number of simple harmonic motions of the same frequency is itself a simple harmonic motion of this frequency*. The general algebraic formula for the addition of m simple harmonic motions, all of the same frequency, but differing in amplitude and epoch, is

$$\left.\begin{aligned} \xi &= \sum_{r=1}^{m} \xi_r \\ &= \sum_{r=1}^{m} a_r \sin(\omega t - \kappa x + \delta_r). \end{aligned}\right\} \quad \ldots \quad 3(11)$$

Since ξ represents a simple harmonic motion of angular frequency ω, this expression is equivalent to

$$\xi = a \sin(\omega t - \kappa x + \delta),$$

$$\left.\begin{aligned} \text{where} \quad a^2 &= \left(\sum_{r=1}^{m} a_r \cos\delta_r\right)^2 + \left(\sum_{r=1}^{m} a_r \sin\delta_r\right)^2 \\ \text{and} \quad \tan\delta &= \frac{\sum_{r=1}^{m} a_r \sin\delta_r}{\sum_{r=1}^{m} a_r \cos\delta_r}. \end{aligned}\right\} \quad . \quad 3(12)$$

When the amplitudes are all equal, and the epoch angles are in arithmetic progression, we have

$$a_1 = a_2 = a_3 = \ldots = a_0 \text{ and } \delta_r = r\delta_0,$$

so that $\sum_{r=1}^{m} a_r \cos\delta_r = a_0(\cos\delta_0 + \cos 2\delta_0 + \ldots + \cos m\delta_0).$

But $2\sin\tfrac{1}{2}\delta_0 \cos\delta_r = \sin\tfrac{1}{2}(2r+1)\delta_0 - \sin\tfrac{1}{2}(2r-1)\delta_0,$

and $2\sin\tfrac{1}{2}\delta_0 \cos\delta_{r+1} = \sin\tfrac{1}{2}(2r+3)\delta_0 - \sin\tfrac{1}{2}(2r+1)\delta_0,$

so that $2\sin\tfrac{1}{2}\delta_0 \sum_{r=1}^{m} \cos\delta_r = \sin\tfrac{1}{2}(2m+1)\delta_0 - \sin\tfrac{1}{2}\delta_0.$

Hence
$$\begin{aligned} \sum_{r=1}^{m} a_r \cos\delta_r &= a_0\{\sin\tfrac{1}{2}(2m+1)\delta_0 - \sin\tfrac{1}{2}\delta_0\}/2\sin\tfrac{1}{2}\delta_0 \\ &= a_0 \cos\tfrac{1}{2}(m+1)\delta_0 \cdot \frac{\sin\tfrac{1}{2}m\delta_0}{\sin\tfrac{1}{2}\delta_0}. \quad . \quad . \quad . \quad 3(13) \end{aligned}$$

Similarly it may be shown that

$$\sum_{r=1}^{m} a_r \sin\delta_r = a_0 \sin\tfrac{1}{2}(m+1)\delta_0 \cdot \frac{\sin\tfrac{1}{2}m\delta_0}{\sin\tfrac{1}{2}\delta_0}; \quad . \quad 3(14)$$

and, substituting in 3(12), we obtain for the relative energy

$$\left.\begin{aligned} a^2 &= a_0^2\left(\frac{\sin\tfrac{1}{2}m\delta_0}{\sin\tfrac{1}{2}\delta_0}\right)^2 \\ \text{and} \quad \tan\delta &= \tan\tfrac{1}{2}(m+1)\delta_0. \end{aligned}\right\} \quad . \quad . \quad . \quad . \quad 3(15)$$

3.5. *Vector Method.*

Since ξ_1 and ξ_2 may both be represented by vectors, in the way described in § 2.8 the sum $\xi = \xi_1 + \xi_2$ should be represented by the vector sum of these two vectors. Using the second type of vector representation in fig. 3.1, it may be seen from the trigonometry of the figure that the law of vector addition does give the same result as equations 3(7). The algebraic calculation is, indeed, equivalent to resolving each vector parallel to the lines OX and OY, adding the components and recombining the resultant components to give the final resultant. In fig. 3.1, $OP_1 = a_1$, $OA_1 = a_1 \cos \delta_1$, and

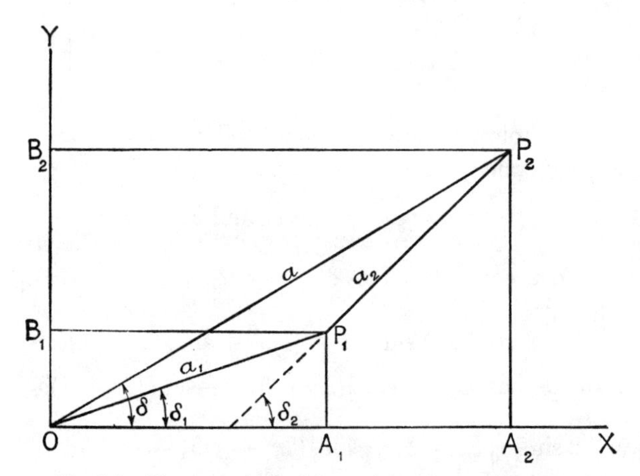

Fig. 3.1.—Vector method of determining the resultant of two simple harmonic motions

$A_1A_2 = a_2 \cos \delta_2$. Hence $OA_2 = a_1 \cos \delta_1 + a_2 \cos \delta_2$ and similarly $OB_2 = a_1 \sin \delta_1 + a_2 \sin \delta_2$. Since $OP_2^2 = OA_2^2 + OB_2^2$, the resultant amplitude is that given in equation 3(7). The epoch of the resultant is obtained in a similar way.

The use of the vector polygon to add a number of simple harmonic motions is equivalent to equations 3(11) and 3(12). The case of several motions all of the same amplitude and with epoch angles in arithmetic progression is shown in fig. 3.2*a*. The representative vectors OA_1, A_1A_2, A_2A_3, etc., are so arranged that O, A_1, A_2, etc., all lie on the same circle (with centre C) and the vectors each subtend an angle δ_0 at the centre of the circle. The resultant subtends an angle $m\delta_0$, and hence if R is the radius of the circle $a_0 = 2R \sin \frac{1}{2}\delta_0$ and $a = 2R \sin \frac{1}{2}m\delta_0$, and this leads at once to equation 3(15) for the relation between a_0 and a.

If the number of sides of a regular polygon increases indefinitely it approximates to a circle (see fig. 3.2*b*). Thus if $m \rightarrow \infty$ while $a_0 \rightarrow 0$

and $\delta_0 \to 0$ in such a way that the products ma_0 and $m\delta_0$ tend to the finite limits A and δ, the resultant becomes

$$a = \frac{A \sin \frac{1}{2}\delta}{\frac{1}{2}\delta}, \quad . \quad . \quad . \quad . \quad . \quad . \quad . \quad 3(16)$$

where δ is the phase difference between the infinitesimal elements at the two ends of the series.

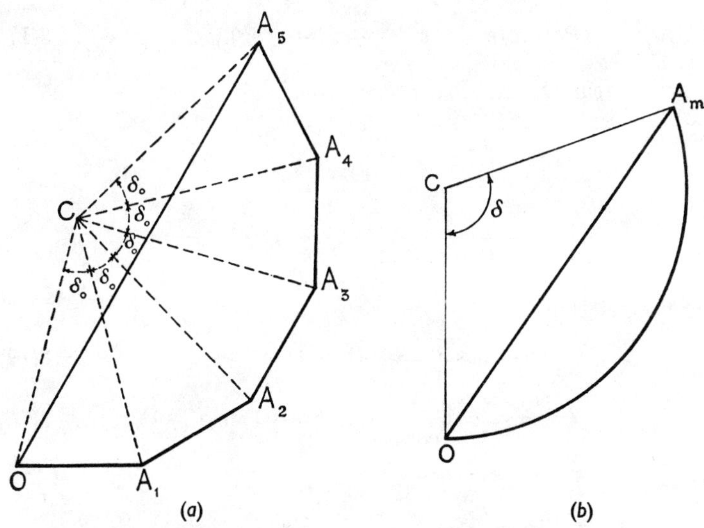

Fig. 3.2.—(*a*) Vector diagram showing the combination of a finite number of simple harmonic motions whose amplitudes are all equal and whose epoch angles are in arithmetic progression. (*b*) Limiting case of continuously varying phase.

3.6. *Calculation using Complex Quantities.*

Using the convention stated in § 2.26, a simple harmonic motion of angular frequency ω may be represented by

$$\begin{aligned} \xi_1 &= a_1 \exp i(\omega t - \kappa x + \delta_1) \\ &= P_1 \exp i(\omega t - \kappa x), \end{aligned}$$

where $\qquad P_1 = a_1 e^{i\delta_1}.$

The sum of a number of simple harmonic motions of the same frequency is then given by the relation

$$\begin{aligned} \xi = \Sigma\, \xi_r &= (P_1 + P_2 + P_3 + \ldots + P_r) \exp i(\omega t - \kappa x) \\ &= P \exp i(\omega t - \kappa x), \end{aligned}$$

where $\qquad P = \Sigma\, P_r.$

Thus we have the simple rule that the complex amplitude of the resultant displacement is equal to the sum of the complex amplitudes of the individual motions. The resultant relative energy (i.e. the square of the real amplitude of the resultant) is equal to the square of the modulus of P. When all the simple harmonic motions have the same amplitude a_0 and the epoch angles are δ_0, $2\delta_0$, $3\delta_0$, etc., we have

$$P_1 = a_0 e^{i\delta_0},\ P_2 = a_0 e^{2i\delta_0},\ P_3 = a_0 e^{3i\delta_0},\ \text{etc.},$$

and

$$P = a_0(e^{i\delta_0} + e^{2i\delta_0} + \ldots + e^{mi\delta_0}). \qquad 3(17)$$

The sum of this geometrical progression is

$$P = a_0 e^{i\delta_0} \cdot \frac{e^{im\delta_0} - 1}{e^{i\delta_0} - 1} \qquad 3(18)$$

$$= \frac{a_0 e^{i\delta_0}}{2(1 - \cos \delta_0)} (e^{im\delta_0} - 1)(e^{-i\delta_0} - 1),$$

using the relation

$$(1 - \cos \theta) = \tfrac{1}{2}(e^{i\theta} - 1)(e^{-i\theta} - 1); \qquad 3(19)$$

similarly,

$$P^* = \frac{a_0 e^{-i\delta_0}}{2(1 - \cos \delta_0)} (e^{-im\delta_0} - 1)(e^{i\delta_0} - 1), \qquad 3(20)$$

so that

$$a^2 = PP^* = 4\left[\frac{a_0}{2(1 - \cos \delta_0)}\right]^2 (1 - \cos m\delta_0)(1 - \cos \delta_0), \qquad 3(21)$$

using 3(19) again.

Hence

$$E = a^2 = a_0^2 \left(\frac{\sin \tfrac{1}{2} m\delta_0}{\sin \tfrac{1}{2}\delta_0}\right)^2 \qquad 3(22)$$

in agreement with 3(15).

3.7.—From the above, it may be seen that the three methods are equivalent. The vector method is probably the most elegant and gives an especially clear insight into the physical conditions. From the vector diagram it is usually possible to see which members of a set of vectors are opposing the resultant and which ones are in phase with the resultant. It is often convenient to carry out the calculation of the resultant by algebraic methods or using the complex amplitudes. When this is done, it is usually worth while to draw a rough vector diagram in order to obtain a general view of the problem as a whole.

EXAMPLES [3(i)–3(vi)]

3(i). Show that the resultant relative energy for an infinite series of simple harmonic motions whose amplitudes are a, $\frac{1}{2}a$, $\frac{1}{4}a$, etc., and whose epochs are 0, $\pi/2$, π, $3\pi/2$, etc., is $4a^2/5$.

[The calculation is very similar to equations 3(15)–3(22). The vector diagram is shown in fig. 3.3*a*.]

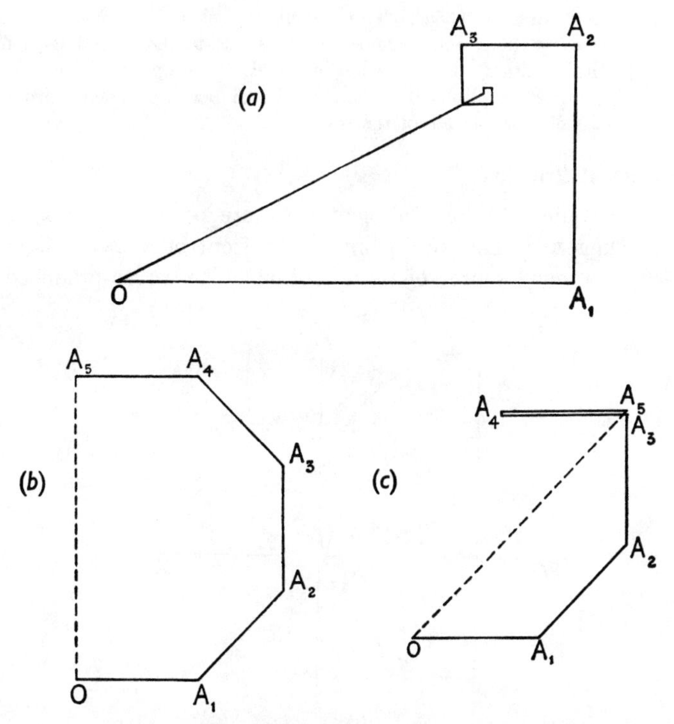

Fig. 3.3.—Vector diagrams showing the resultants of certain combinations of simple harmonic motions [see Examples 3(i) and 3(ii)]

3(ii). Calculate the resultant relative energy for 5 simple harmonic motions of equal amplitude, (a) when the epochs are 0, $\frac{1}{4}\pi$, $\frac{1}{2}\pi$, $\frac{3}{4}\pi$, π, and (b) when the epochs are 0, $\frac{1}{4}\pi$, $\frac{1}{2}\pi$, π, 2π. Draw the appropriate vector diagrams and compare your diagrams with figs. 3.3*b* and 3.3*c*. [(a) $a^2(3 + 2\sqrt{2})$; (b) the same as (a).]

3(iii). Find the resultant amplitude for n simple harmonic motions of equal amplitudes, (a) when the epochs are π/n, $2\pi/n$, $3\pi/n$, etc., and (b) when the epochs are $2\pi/n$, $4\pi/n$, $6\pi/n$, etc. [(a) a cosec $(\pi/2n)$; (b) zero.]

3(iv). Find the resultant relative energy for $(2n + 1)$ simple harmonic motions of equal amplitudes when the epochs are

$$\pi, \frac{n-1}{n}\pi, \frac{n-2}{n}\pi, \ldots, \frac{\pi}{n}, 0, \frac{\pi}{n}, \frac{2\pi}{n}, \ldots, \frac{n-1}{n}\pi, \pi.$$

3(v). Would the resultant relative energy in Example 3(iv) be increased or decreased by removing (*a*) the first or (*b*) the last or (*c*) the centre member of the series? [(*a*), (*b*), decreased, (*c*) increased.]

3(vi). Calculate the resultant relative energy for Example 3(i) when the second term of the series is removed.

[The resultant is $0{\cdot}65a^2$. This calculation should be made by *subtracting* the simple harmonic motion due to the second term from the resultant, due regard being paid to the epoch angles, i.e. the complex amplitude for the second member is subtracted from the complex amplitude of the resultant.]

3.8. **Huygens' Principle.**

In an attempt to form a mental picture of wave propagation, Huygens suggested that each point at the front of a wave might be regarded as a small source of wave motion. The waves produced by

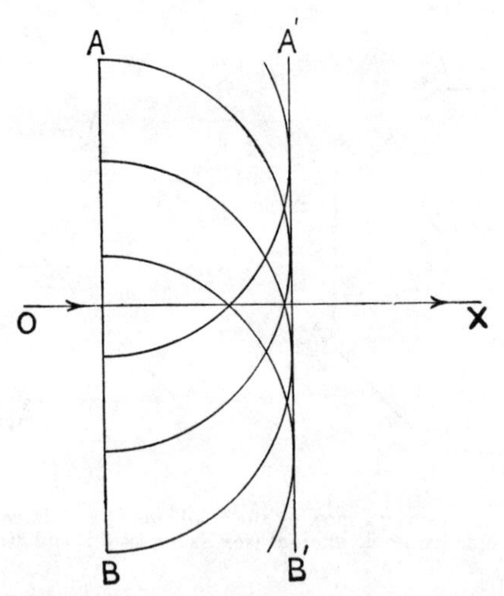

Fig. 3.4.—Huygens' principle applied to a plane wave in uniform medium

these small sources were called *secondary waves*, and it was assumed that the position of the main (or primary) wave at some later time was the envelope of the secondary waves.

We may illustrate this method by applying it to a plane wave (see fig. 3.4). The direction of propagation is OX and the position of a certain wave surface at time t_0 is represented by a plane through AB and perpendicular to the plane of the paper. We draw a series of spheres with a common radius bt_1, and with centres at various points on AB. Clearly the plane through A′B′ situated at a distance bt_1 in front of, and parallel to, the plane through AB is part of the envelope of these spheres and the phase on this plane when $t = t_1 + t_0$ will be the same as the phase at AB when $t = t_0$. Hence, in this simple case, the construction based on Huygens' principle does enable us to deduce the position of one wave surface from that of another, and the result is correct. It is not difficult to verify that the same construction gives the correct result when non-planar wave surfaces are involved.

Although, in a certain sense, the principle gives the correct result, it carries with it implications which require further consideration. Huygens postulated that the action of the secondary wavelets was confined to the points at which they touched their envelope, and he considered only those parts of the envelope which lay in the forward direction of propagation. There is no direct physical or mathematical justification for this arbitrary decision to ignore all the unwanted parts of the secondary waves.

3.9.—Fresnel later attempted to provide a physical justification for a modified form of Huygens' principle. He assumed that a wave surface could be divided into a very large number of small elements of area and that each of these elements was a source of secondary waves. He assumed that these secondary waves were effective everywhere, but that the amplitude of the disturbance at any point Q due to the secondary wave from an element dS of the wave surface (situated at P) was a function both of the distance QP and of the angle θ between the line QP and the normal to the element dS (see fig. 3.5). The variation with angle is called the *obliquity function* or the *inclination factor*. Fresnel and later workers attempted to adjust the obliquity function so that the resultant disturbance due to all the secondary waves was zero everywhere except at the points where they touched the envelope. These attempts were never completely successful.

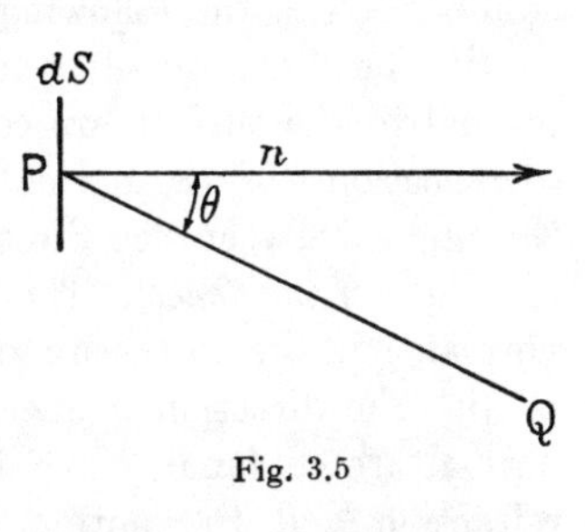

Fig. 3.5

3.10.—The detailed theory of wave propagation, which is given in Chapters VI and VII, shows that it is possible to use a knowledge of the disturbance on one wave surface to derive the amplitude and phase of the disturbance at a later time, and at a place situated a suitable distance ahead. The distribution of light energy thus calculated agrees with that observed. This more fundamental method (mainly due to Kirchhoff) determines the obliquity function from the wave equation instead of introducing it as a new *ad hoc* hypothesis. It also requires a special adjustment in the phases of the secondary wavelets. The exact calculation justifies Fresnel's methods rather than Huygens' principle. It shows that the construction of wave envelopes by Huygens' method gives a satisfactory description of the progress of a wave. We shall therefore use this construction in § 3.11 and in other appropriate places.

3.11. Reflection and Refraction at Plane Surfaces.

It is found that when a parallel beam of light is incident upon a surface separating two transparent media, such as air and glass, part of the light is reflected back into the medium in which it originated, and another part is transmitted into the second medium. The direction of propagation of this second part is not in the same line as the original direction of propagation and the light is said to be refracted. Observations on the angular relations between the directions of propagation of the incident, the reflected and the refracted beams are summarized in the following laws:

(i) The direction of incidence, the direction of reflection and the normal to the surface are coplanar. The angle between the direction of reflection and the normal (called the *angle of reflection*) is equal to the angle between the direction of incidence and the normal (called the *angle of incidence*). The direction of reflection is on the side of the normal opposite to the direction of incidence.

(ii) The direction of incidence, the direction of refraction and the normal are coplanar. The sine of the angle between the direction of refraction and the normal (called the *angle of refraction*) bears a constant ratio to the sine of the angle of incidence.

Thus, if θ_1 = angle of incidence (see fig. 3.6),
θ_1' = angle of reflection,
θ_2 = angle of refraction,

then
$$\theta_1 = \theta_1' \qquad 3(23)$$

and
$$\frac{\sin\theta_1}{\sin\theta_2} = \mu_{12}. \qquad 3(24)$$

The constant μ_{12} is characteristic of the two media. It is found that the constants for different pairs of media are subject to the relation

$$\mu_{13} = \mu_{12} \cdot \mu_{23}. \qquad \text{3(25)}$$

The constant μ_{12} is called the *index of refraction of the two media*, it being understood that the light travels from the medium 1 into the medium 2. The value of the constant obtained when light travels from a vacuum into a medium is called the *index of refraction of that medium*

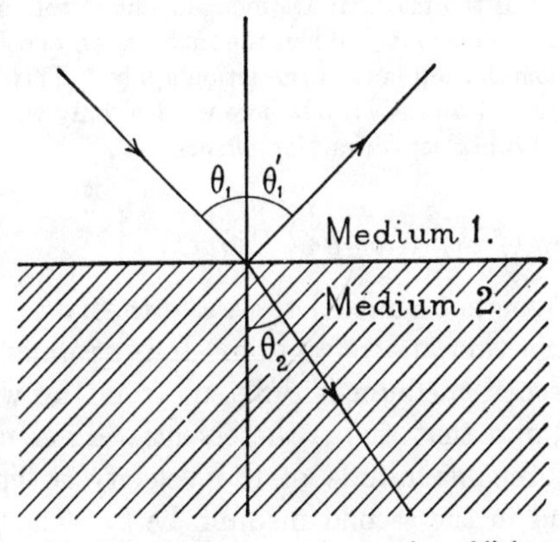

Fig. 3.6.—Reflection and refraction of light

and is denoted in this book by the symbol n (with a suffix when necessary). The index of refraction from air to a medium is denoted by μ (with a single suffix to specify the medium when necessary). Since the refractive index of air is nearly unity, the values of μ_g and n_g for glasses are nearly equal, and it is only in a few applications that it is necessary to distinguish them.

3(25) implies that 3(24) may be written in the symmetrical form

$$n_1 \sin \theta_1 = n_2 \sin \theta_2, \qquad \text{3(26)}$$

and that

$$\frac{n_2}{n_1} = \mu_{12}. \qquad \text{3(27)}$$

Indices of refraction for common gases (at standard temperature and pressure) range from 1·000035 for helium to 1·00030 for nitrogen. Indices for liquids and solids include 1·33 for water, 1·48 for soda glass and 1·7 for heavy flint glass. A few substances have appreciably higher indices. A medium of high refractive index is said to be of greater *optical density* than a medium of low refractive index.

3.12.—The law of reflection was certainly known to the Greek philosophers and was probably discovered independently by many different observers. Ptolemy and others drew up tables connecting the angles of incidence and of refraction. Many attempts were made to enunciate a law of refraction but they all failed, partly because the necessary mathematics was not sufficiently developed. The law given above was first discovered by Snell (1621) though it was not published until after his death. This law of refraction applies only when the media are *isotropic*. (A material is said to be isotropic when the physical properties of a thin slice cut from a mass of the material are independent of the original orientation of the slice before it was cut out.) When the medium on one side of the surface is *anisotropic* more complicated laws of refraction apply.* This type of medium is discussed in Chapters XII and XVI, but here we deal only with isotropic media. In this chapter we also neglect diffraction effects.

3.13. *Wave Theory of Reflection and Refraction.*

Huygens' construction enabled the wave theory to give the following account of reflection and refraction. It is assumed that the wave velocity in the second medium is different from the wave velocity in the first medium, and that this is the essential difference between the two media. Let the magnitude of the velocity of light in the first medium be b_1, and in the second medium be b_2. The plane wave AB which represents the parallel beam of incident light falls on the plane surface OP (see fig. 3.7). The line NON′ is normal to this surface. When the wave reaches a point O on the surface it becomes a source of secondary waves which spread both in the first medium and in the second medium. Suppose that the wave surface AB reached the line OO′ at time $t = 0$ and that it would reach the line PP′ at time t if there were no reflection and refraction. Then

$$\mathrm{O'P} = b_1 t = \mathrm{OP}\sin\theta_1 = \mathrm{OP'}. \quad . \quad . \quad . \quad 3(28)$$

At time t, the secondary waves from O form a hemisphere of radius $b_1 t$ ($= \mathrm{OP}\sin\theta_1 = \mathrm{OP}_1$) in medium 1 and a hemisphere of radius $b_2 t$ [$= (b_2/b_1)\,\mathrm{OP}\sin\theta_1 = \mathrm{OP}_2$] in medium 2. The secondary waves are then just starting from P and it may easily be seen that tangent planes PP_1 and PP_2 from P to the above hemispheres also touch the hemispheres corresponding to the secondary waves from points such as Q and R, intermediate between O and P. The new wavefronts A′B′ and A″B″ are parallel to the planes PP_1 and PP_2. The new

* Such media are sometimes called *aeolotropic*.

directions of propagation of the waves make angles θ_1' and θ_2 with the normal to the surface of OP, where

$$\sin \theta_1' = \frac{\mathrm{OP}_1}{\mathrm{OP}} = \frac{b_1 t}{\mathrm{OP}} = \sin \theta_1. \quad . \quad . \quad . \quad . \quad 3(29)$$

and

$$\sin \theta_2 = \frac{\mathrm{OP}_2}{\mathrm{OP}} = \frac{b_2 t}{\mathrm{OP}} = \frac{b_2}{b_1} \sin \theta_1. \quad . \quad . \quad . \quad 3(30)$$

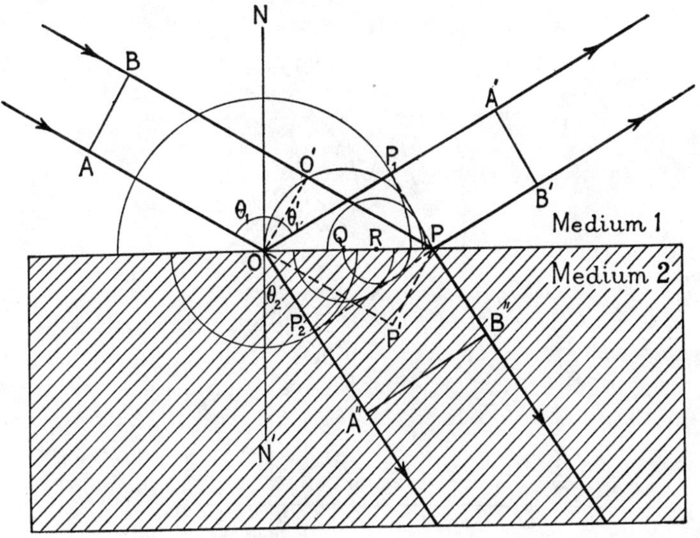

Fig. 3.7.—Huygens' principle applied to the reflection and refraction of light at a plane surface

Thus 3(29) agrees with 3(23), and 3(30) agrees with 3(24) and 3(27) provided that

$$\mu_{12} = \frac{b_1}{b_2} = \frac{n_2}{n_1}. \quad . \quad . \quad . \quad . \quad . \quad . \quad 3(31)$$

Equation 3(31) is in agreement with direct measurements of the velocity of light in different media (see Chapter X).

The wave-theory picture of reflection and refraction gives a simple description of the observations on the relation between the values of μ for different pairs of media; for

$$\mu_{13} = \frac{b_1}{b_3} = \frac{b_1}{b_2} \cdot \frac{b_2}{b_3} = \mu_{12} \cdot \mu_{23},$$

which agrees with 3(25).

3.14. Reflection and Refraction at Spherical Surfaces: Mirrors and Lenses.

When a beam of light diverging from a point O on the axis of a spherical mirror is reflected, it forms a real or virtual image of O at some point I also on the axis, i.e. after reflection the beam converges

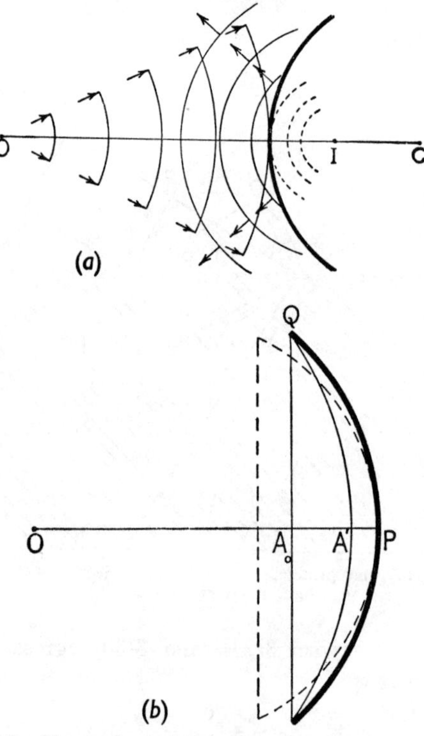

Fig. 3.8.—Huygens' principle applied to the reflection of light at spherical surfaces

towards I or appears to diverge from I (see fig. 3.8*a*). If the distances of O and I from the mirror are u and v respectively, and if r is the radius of curvature of the mirror, then it is found that

$$\frac{1}{v}+\frac{1}{u}=\frac{2}{r}=\frac{1}{f}=F. \quad . \quad . \quad . \quad . \quad . \qquad 3(32)$$

The constant f is called the *focal length* and the constant F is called the *power* of the mirror.

Distances are always measured from the mirror and the direction of the incident light is taken to be positive.

Similarly, for a lens with spherical surfaces of radii r_1 and r_2, it is found that

$$\frac{1}{v} - \frac{1}{u} = (\mu - 1)\left(\frac{1}{r_1} - \frac{1}{r_2}\right) = \frac{1}{f} = F, \quad . \quad . \quad 3(33)$$

where μ is the index for refraction from the surrounding material (usually air) to the material from which the lens is made (usually glass). These empirical formulæ should be regarded as approximations which apply when the diameter of the mirror or lens is small compared with its radius of curvature.

3.15.—The wave theory gives the following account of these observations. The beam of light approaching the mirror is represented by a spherical wave diverging from the point O (fig. 3.8*a*). As each part of the wave reaches the mirror, it gives rise to a system of secondary waves and Huygens' construction may be applied to find the position of the wavefront after reflection. When this is done it is found that, to a first approximation, the reflected wave is spherical and is centred upon the point I, the relation of the distances of O and I from the mirror being given by 3(32).

To show that application of Huygens' construction leads to equation 3(32), we use the well-known geometrical relation between the sagittal distance and the curvature of a *small* circular arc. If in fig. 3.8*b* we put $A_0P = s_r$ and $A_0Q = b$, we have, as a sufficient approximation when $b \ll r$,

$$2rs_r = b^2,$$

or

$$s_r = \tfrac{1}{2}b^2R, \quad . \quad . \quad . \quad . \quad . \quad . \quad 3(34)$$

where $R\ (= 1/r)$ is the curvature. The sagittal distance (s_r) is proportional to b^2 and to the curvature. A wavefront advancing on the mirror from O (see fig. 3.8*b*) has a curvature $U\ (= 1/u)$. One part of the wavefront reaches Q when another part is at A′, where

$$A'P = \tfrac{1}{2}b^2(R - U).$$

Thus when the central part of the wavefront reaches the mirror and gives rise to a wavelet from P, the corresponding wavelet from Q has travelled out a distance A′P from Q. The secondary waves from P and

Q thus both touch a curve whose sagittal distance s_v exceeds s_r by A′P, i.e. we have for this curve

$$s_v = s_r + \mathrm{A'P} = \tfrac{1}{2}b^2(2R - U).$$

Since the sagittal distance is proportional to b^2, the curve is a circle whose curvature is given by

$$V = 2R - U,$$

i.e.
$$\frac{1}{v} + \frac{1}{u} = \frac{2}{r},$$

which agrees with 3(32). It is important to remember that this result is valid only when PQ is small compared with r, v and u.

EXAMPLES [3(vii) and 3(viii)]

3(vii). Show, by a discussion similar to that of § 3.15, that, when a plane wave normal to the axis is incident upon a paraboloidal mirror, the reflected wavefront is *exactly* spherical, and has its centre at the focus.

[Use the polar equation of the parabola.]

3(viii). Show that the wavefront obtained in the preceding Example is the envelope of the wavelets emitted from the paraboloidal mirror.

[Obtain the equation of the family of surfaces constituting the reflected wavelets in terms of the constant of the parabola as parameter. Differentiating this equation partially with respect to the parameter, and eliminating the parameter between the resultant equation and that of the family of surfaces, yields the envelope of the reflected wavelets.]

3.16.—By similar applications of Huygens' construction the wave theory is able to describe the behaviour of lenses and mirrors generally. On this view the effect of refraction at a spherical surface is to change the curvature of the incident wave surfaces. For a given lens the change in curvature is a constant and is equal to the reciprocal of the focal length of the lens. This constant change in curvature is called the *power* of the lens. The effect of reflection is both to change the curvature of the wave surfaces and to reverse the direction of propagation. This second effect is responsible for the difference of sign between equations 3(32) and 3(33). It corresponds to the fact that a lens of zero power (e.g. a very thin sheet of glass) has no effect on a beam of light, while a mirror of zero power (i.e. a plane mirror) alters the direction of propagation without altering the magnitude of the curvature of the wave surfaces. Thus for a lens of zero power 3(33) gives $v = u$ and the image and object coincide. For a mirror of zero power 3(32) gives $v = -u$, and the image and object are on opposite sides of the mirror.

3.17. Dispersion.

Newton showed that, when an approximately parallel beam of white light falls upon a glass prism, the emergent light is spread out into a coloured strip which he called a *spectrum*. This phenomenon is called *dispersion*. It implies that the refractive index from air to glass is not the same for all colours of the spectrum, and hence that the ratio of the speed of light in air to that in glass depends on the colour of the light. This is confirmed by direct measurements of the velocity of light in different media. It is found that in a vacuum the speed of light is the same for all colours but that in a medium such as water blue light travels more slowly than red (see § 10.18). Experiments on interference and diffraction make it necessary to assume that the different colours in the spectrum must be represented by waves which have different values of λ and therefore of κ, ω and ν. The frequency is highest at the blue and lowest at the red end of the spectrum. The dispersion of light then implies that:

$$b = f(\lambda) \qquad 3(35)$$

and

$$n = F(\lambda). \qquad 3(36)$$

In the various differentiations, etc., by which the wave equation was derived it was not assumed that b was independent of ω or κ. Therefore the existence of dispersion does not invalidate the proof that the pure sine wave (represented by 2(2)) is propagated without change of profile. It does, however, imply that a composite wave, made up of two or more sine waves of different frequency, changes profile as it advances, because each component is propagated with its own speed, and the phase differences between the different components alter as the wave advances. These effects will be considered in Chapter IV.

The *speed of light in vacuo* is one of the fundamental physical constants. It is usually represented by the symbol c. This velocity has been found to be very close to $3{\cdot}00 \times 10^{10}$ centimetres per second (see § 10.12). Since the velocity in air (at standard temperature and pressure) is only about one part in 1000 less than that *in vacuo*, it is often sufficient to use the value c for the velocity of propagation in air, though it must be understood that it is used as an approximation.

A medium in which b varies with λ is called a *dispersive medium.* The only truly non-dispersive medium for light waves is a vacuum.

3.18.—The form of the function in equations 3(35) or 3(36) is of considerable practical importance. Cauchy proposed the following empirical formula:

$$n - 1 = A(1 + \frac{B}{\lambda^2} + \frac{C}{\lambda^4} + \ldots) \quad . \quad . \quad . \quad 3(37)$$

where A, B, C, etc., are constants whose magnitudes are such that each term of the series in the bracket is much less than the preceding one. The quantity $(n - 1)$ is called the *refractivity.* In *normal* dispersion the refractive index increases regularly along the spectrum from red to blue, i.e. A and B are both positive. For a few substances the dispersion differs very greatly from the form given by 3(37) and it may happen that over a short part of the spectrum the refractive index increases when the wavelength increases. This phenomenon is called *anomalous* dispersion. For most gases the variation of the refractive index with wavelength is fairly well represented by a formula including the first two terms of 3(37). Formulæ of this type also form a moderately good representation of the properties of optical glasses. The following alternative formula for the refractive indices of glasses is due to Hartmann:

$$n = n_0 + \frac{\alpha}{(\lambda - \lambda_0)^{1\cdot 2}}. \quad . \quad . \quad . \quad . \quad . \quad 3(38)$$

This formula gives a better agreement than a three-constant formula of the type proposed by Cauchy.

3.19.—The effect of dispersion on lenses may be seen from equation 3(33). The focal length of a lens is smaller, and its power greater, for blue light than for red. This means that the image formed by blue light is not quite the same in size or in position as the image formed by red light. Thus the images formed with white light appear coloured at the edges and are not so clear as images formed with monochromatic light. This effect is known as *chromatic aberration.*

3.20. **Stationary Waves.**

When a parallel beam of light represented by

$$\xi_1 = a \sin(\omega t - \kappa x) \quad . \quad . \quad . \quad . \quad . \quad 3(39)$$

is reflected at a perfect reflector, the reflected wave is represented by

$$\xi_2 = a \sin(\omega t + \kappa x + \delta), \quad . \quad . \quad . \quad . \quad 3(40)$$

where the constant δ depends on the position of the reflector. There is also a possibility that the process of reflection itself may be accompanied by a change of phase. Let this change, if any, be included in δ.

If the origins of x and t are moved to the right by $\delta/2\kappa$ and $\delta/2\omega$ respectively, equation 3(39) is unchanged and 3(40) becomes

$$\xi_2 = a \sin(\omega t + \kappa x). \qquad 3(41)$$

In this way the origin is moved so that $x = 0$, $t = 0$ corresponds to the same phase for both waves.

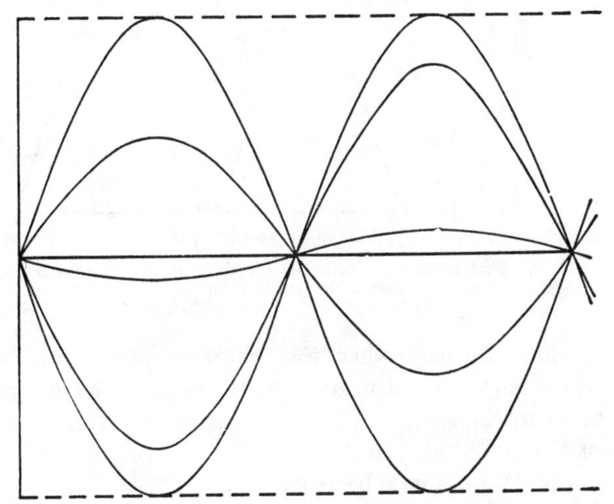

Fig. 3.9.—Profile of a stationary wave at different times

Applying the principle of superposition we have

$$\xi = \xi_1 + \xi_2 = a \sin(\omega t - \kappa x) + a \sin(\omega t + \kappa x), \qquad 3(42)$$

and this may be written

$$\xi = 2a \cos \kappa x \sin \omega t. \qquad 3(43)$$

This type of wave is represented in fig. 3.9. At every point the motion is simple harmonic but the amplitude of the motion varies from point to point. The profile of the wave expands and shrinks as shown in the figure, but does not move forwards or backwards. Waves of this type are known as *stationary waves*.

3.21.—In a stationary wave there is no net transference of energy in either direction, though there is an energy density in the medium with a mean value proportional to a^2. This energy density is not uniformly distributed. It is a maximum at points for which $\cos^2 \kappa x = 1$, i.e. at $x = 0$, $x = \pi/\kappa$, $x = 2\pi/\kappa$, etc., and a minimum at points where $\cos^2 \kappa x = 0$, i.e. at $x = \pi/2\kappa$, $3\pi/2\kappa$, $5\pi/2\kappa$, etc. The points where the energy is a maximum are called *antinodes* or *loops*. Those where it is a minimum are called *nodes*. If a detector is moved along the x axis it gives a maximum reading at the antinodes and zero at the nodes (see fig. 3.10). The distance from a node to the nearest antinode is $\pi/2\kappa = \lambda/4$ and the distance between successive nodes is $\lambda/2$.

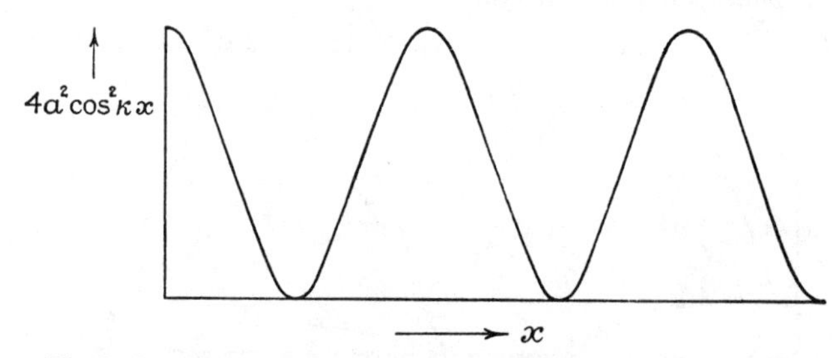

Fig. 3.10.—Distribution of energy in a medium in which there is a simple stationary wave

Now consider the case when the reflection is not perfect. The *coefficient of reflection* (ρ) is defined to be the ratio of the energy of the reflected beam to the energy of the incident beam. This implies that the ratio of the amplitudes is $\rho^{\frac{1}{2}}$.

Equation 3(41) must now be replaced by

$$\xi_2 = a\rho^{\frac{1}{2}} \sin (\omega t + \kappa x), \qquad . \quad . \quad . \quad . \quad 3(44)$$

and 3(43) is replaced by

$$\xi = 2a\rho^{\frac{1}{2}} \cos \kappa x \sin \omega t + a(1 - \rho^{\frac{1}{2}}) \sin (\omega t - \kappa x). \qquad 3(45)$$

The wave is now partly stationary and partly progressive. A detector moved along the x axis will still record maxima and minima. The readings at minima will not be zero since at points where the first term of 3(45) is zero the detector gives a reading proportional to the square of the amplitude of the second term. Waves which are mainly stationary, though containing a progressive component, occur commonly in acoustics in connection with the theory of organ pipes, etc. A series of observations made by Walton on stationary electromagnetic waves is shown in fig. 3.11. These waves are produced by allowing radiation emitted by a small high-frequency oscillator to fall upon a large sheet of metal placed at $x = 0$. The waves are measured by means of a small crystal detector connected to an amplifier. The metal has a high coefficient of reflection for these waves and the waves are mainly

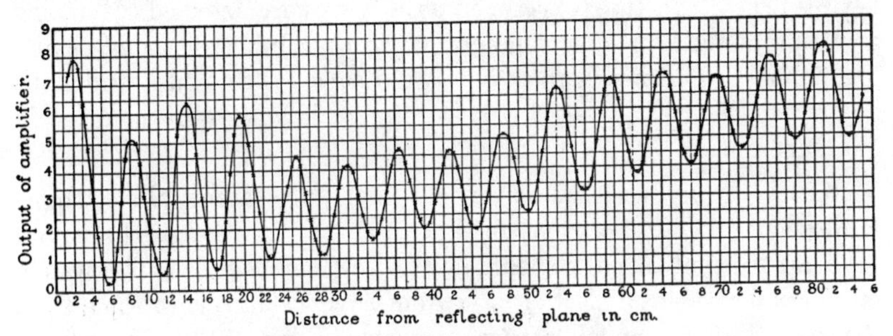

Fig. 3.11.—Experimental observations on stationary electromagnetic waves

stationary waves with a small progressive component. From the distance between the nodes the wavelength is found to be 11.6 centimetres.

3.22. **Wiener's Experiment.**

The detection of stationary light waves is difficult because the nodes and antinodes are very close together, the distance from node to antinode being only of order 10^{-5} centimetre. The experimental difficulties were first overcome by Wiener (1890). Before discussing

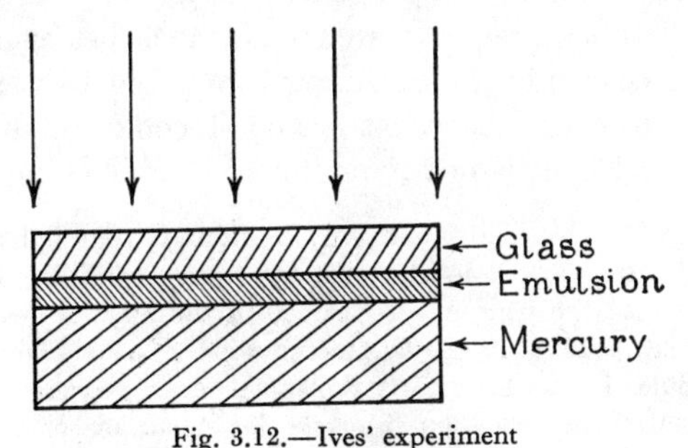

Fig. 3.12.—Ives' experiment

his work we shall describe the experiments of Ives which though more difficult to carry out are theoretically simpler. Ives prepared special photographic plates with a thin fine-grained emulsion. The emulsion side of the plate was placed in contact with a film of mercury and a parallel beam of monochromatic light was directed normally upon the glass side of the plate (see fig. 3.12). The light passed through the glass and the film and was reflected at the mercury surface, standing waves being formed in the film. The nodes form a series of laminæ distant $\frac{1}{2}\lambda$ apart with the antinodes half-way between them. In Ives' experiments the waves are detected by cutting a section of the film in a plane normal to the surface of the glass and examining it under a high-power microscope. The nodes and antinodes can be clearly seen; 250 successive laminæ were counted in one experiment.

3.23.—Wiener used an extremely thin transparent photographic film, about 2×10^{-6} centimetre thick, for the detection of the waves. He arranged that this film should lie at a very small angle to the system of waves formed by reflection at a front-silvered mirror, so that it cut the successive antinodal laminæ as shown in fig. 3.13. The photographic emulsion is blackened along a series of lines where the plane

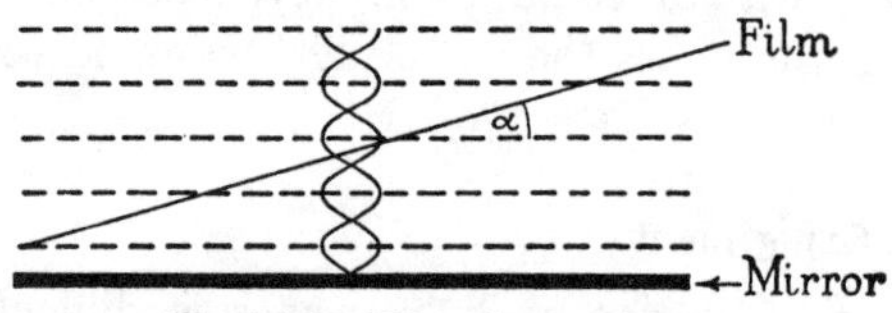

Fig. 3.13.—Wiener's experiment

of the film cuts the antinodal laminæ. The angle α (see fig. 3.13) was made about 10^{-3} radian and the distance between successive fringes was then 1000 times greater than the distance between the laminæ. From these experiments the wavelength of green light could be estimated to be $5{\cdot}5 \times 10^{-5}$ centimetre, and it could be shown that the wavelength of red light is nearly double that of blue light.

Wiener's technique was criticized on the grounds that light reflected from the photographic film, as well as that reflected from the mirror, should be taken into account, so that the fringes were formed in the manner described in § 5.12. Wiener proved that this was not so by putting a film of benzol between the mirror and the film. Since benzol has nearly the same refractive index as the gelatine, it almost eliminates the reflection from the lower surface of the photographic film. Wiener still obtained the fringes under these conditions.

3.24.—An interesting addition to Wiener's experiment is due to Drude and Nernst. They first repeated the main experiment using a fluorescent film as the detector instead of a photographic plate. They then took a glass plate, silvered half of the plate and coated the whole with a very thin film of fluorescent material. A parallel beam of radiation was allowed to fall normally upon the whole plate, and it was found that there was strong fluorescence in the unsilvered region and none in the silvered region. This indicates that the surface of the reflector forms a node of the stationary waves, and implies that reflection of light at a silvered surface produces a change of phase of approximately π. Thus if the *origin of x be taken at the reflector* and if the incident wave is represented by 3(39), the reflected wave must be represented by

$$\xi_2 = -a \sin(\omega t + \kappa x), \qquad . \quad . \quad . \quad . \quad 3(46)$$

and the system of stationary waves by

$$\xi = -2a \sin \kappa x \cos \omega t, \qquad . \quad . \quad . \quad . \quad 3(47)$$

so that the displacement is always zero at $x = 0$.

3.25.—Wiener's experiment was later again repeated by Ives, using a photo-electric method for detecting the nodes and antinodes. Although Wiener's experiment does not form an accurate method of measuring the wavelength of light, it is of considerable theoretical interest. The formation of stationary light waves is probably the simplest application of the principle of superposition and forms a very direct, though not a very exact, test of the principle. Stationary light waves are the foundation of a method of colour photography. They have also important applications in connection with photo-electric cells and in the theory of polarized light.

Stationary waves in three-dimensional enclosures are discussed in connection with the theory of temperature-radiation.

3.26. **Coefficient of Reflection—Normal Incidence.**

The coefficient of reflection at the boundary separating two transparent media can be calculated if the *boundary conditions* are known. Consider a parallel beam of light travelling in the OX direction and a boundary formed by a plane normal to the direction of propagation. For all ordinary kinds of waves the value of ξ is the same at any two points which are an infinitesimal distance apart, but on opposite sides of the boundary, at all times. This forms one boundary condition.

The second boundary condition depends upon the type of wave involved and on the difference between the two media. Fresnel, using an elastic-solid theory of light, deduced the condition that $\partial\xi/\partial x$ should have the same value on the two sides of the boundary. Hence he calculated the coefficient of reflection for normal incidence and obtained the value given in 3(56). In the next paragraph, this boundary condition will be used. From our point of view, it is justified, for the present, by its success in giving a formula for the coefficient of reflection which is in agreement with experimental observations. It will require to be re-examined when we come to consider the reflection of electromagnetic waves.

3.27.—Suppose that a parallel beam of light represented by

$$\xi_1 = a_1 \sin(\omega t - \kappa_1 x) \quad . \quad . \quad . \quad . \quad . \quad 3(48)$$

is incident normally on a plane surface separating two media. Suppose that the light is passing from medium 1 to medium 2 and that the index of refraction is μ_{12}. Then if ξ_1' represents the reflected and ξ_2 the refracted wave, the boundary conditions are

$$\xi_1 + \xi_1' = \xi_2, \text{ for all } t, \quad . \quad . \quad . \quad . \quad 3(49)$$

$$\frac{\partial \xi_1}{\partial x} + \frac{\partial \xi_1'}{\partial x} = \frac{\partial \xi_2}{\partial x}, \text{ for all } t. \quad . \quad . \quad . \quad . \quad 3(50)$$

Equations 3(49) and 3(50) hold at the boundary which we take to be the plane $x = 0$.

Equation 3(49) requires that reflected and refracted waves shall have the same frequency as the incident wave. Since the velocities are different in the two media, the values of λ and κ for the refracted wave must be different from the corresponding values for the incident wave, in order to comply with equations 2(25). We therefore write

$$\xi_1' = a_1' \sin(\omega t + \kappa_1 x) \quad . \quad . \quad . \quad . \quad 3(51)$$

and

$$\xi_2 = a_2 \sin(\omega t - \kappa_2 x), \quad . \quad . \quad . \quad . \quad 3(52)$$

where

$$\kappa_2 = \mu_{12}\,\kappa_1. \quad . \quad . \quad . \quad . \quad . \quad . \quad . \quad . \quad . \quad 3(53)$$

In writing 3(51) and 3(52) we assume that the phase differences between the reflected and refracted waves and the incident wave are zero or π.

The boundary conditions now give

$$a_1 + a_1' = a_2 \quad . \quad . \quad . \quad . \quad . \quad 3(54a)$$

and

$$\kappa_1 a_1 - \kappa_1 a_1' = \kappa_2 a_2, \quad . \quad . \quad . \quad . \quad . \quad 3(54b)$$

and from 3(53) we obtain

$$a_1 - a_1' = \frac{\kappa_2}{\kappa_1} a_2 = \mu_{12} a_2,$$

and hence
$$a_1' = -a_1 \frac{\mu_{12} - 1}{\mu_{12} + 1}, \quad \ldots \ldots \quad 3(55)$$

$$\rho = \left(\frac{\mu_{12} - 1}{\mu_{12} + 1}\right)^2. \quad \ldots \ldots \quad 3(56)$$

When reflection takes place at an air-glass surface, equation 3(56) gives a reflection coefficient of 4 per cent for a glass of index 1·5.

3.28.—Equation 3(55) indicates that there is a reversal of phase (i.e. a change by π) when light is reflected, if the reflection takes place in the medium of lower optical density, but no change if it takes place in the medium of higher optical density. This change of phase is verified by direct experiment (see § 5.10). Equation 3(56) predicts that the coefficient of reflection is the same on whichever side the beam strikes the surface since the value of ρ is unchanged when μ_{21} is substituted for μ_{12}. If these coefficients were not equal, it would be possible to construct a thermodynamic cycle which would infringe the Second Law.

3.29.—In the derivation of 3(56) it was assumed that the boundary between the two media was mathematically sharp. Most real surfaces are not perfectly clean. Glass surfaces commonly contain traces of polishing material and also occluded air. In practice, it appears that ordinary clean surfaces are sufficiently sharp to give results substantially in accord with 3(56). The sizes of irregularities are probably a little greater than the diameters of atoms and molecules (i.e. only 1/100 of a wavelength of light). The presence of thin films of grease, etc., or of any slight roughness in the surface seriously affects the reflection.

3.30. **Optical Path Difference.**

In many optical problems it is necessary to calculate the phase difference between two beams of light which have started from the same source and reached the same point by different paths, having been reflected or refracted by systems of mirrors or prisms (see fig. 4.1). The phase difference between such beams is made up of two parts. The first part, which has just been considered, consists of changes of phase on reflection; the second is due to a possible difference in length of the two paths. A sine wave undergoes a phase change of 2π when the wave advances through a distance equal to a wavelength. Thus the phase change $\delta\phi$ when the wave advances a distance δs in a medium of refractive index n_1 is given by

$$\delta\phi = \frac{2\pi}{\lambda_1} \delta s, \quad \ldots \ldots \quad 3(57)$$

where λ_1 is the wavelength in the medium.

If λ and κ are the wavelength and wavelength constant for a wave of the same frequency in a vacuum we have

$$\delta\phi = \frac{2\pi}{\lambda} n_1 \, \delta s = \kappa n_1 \, \delta s. \quad . \quad . \quad . \quad . \quad 3(58)$$

The phase difference is thus proportional to $n_1 \, \delta s$ and the *optical distance* between two points is defined to be the integral of this function along the path traversed.

The *optical path difference* is the difference between the values of this integral taken along the two alternative paths. The phase difference is κ times the optical path difference. If one beam of light passes through a plate of glass of thickness e and another follows a parallel path in air, the optical path difference is $(n_1 - 1)e$, although there is no difference in the lengths of the geometrical paths.

3.31. **Corpuscular Theory of Reflection and Refraction.**

In a corpuscular theory of light, the law of reflection, equation 3(23), is easily pictured as a form of elastic reflection. The corpuscles impinge on a perfectly smooth and perfectly elastic surface; the component of their velocity perpendicular to the surface is reversed, while that normal to the surface is unchanged.

In order to be able to explain refraction, the corpuscular theory has to assume that some, but not all, of the corpuscles are able to penetrate the surface. This constitutes a fundamental difficulty, because the natural assumption is that the corpuscles representing one kind of light are all the same. If this is so, it would be possible to explain why some particles penetrate the surface and others do not, by assuming that the surface has some form of structure—due perhaps to the atomic structure of matter. On this view a corpuscle would be reflected if it struck one of the surface atoms directly (see fig. 3.14), and would be refracted if it fell on one of the spaces, just as some particles penetrate a thin sheet of metal while others do not.

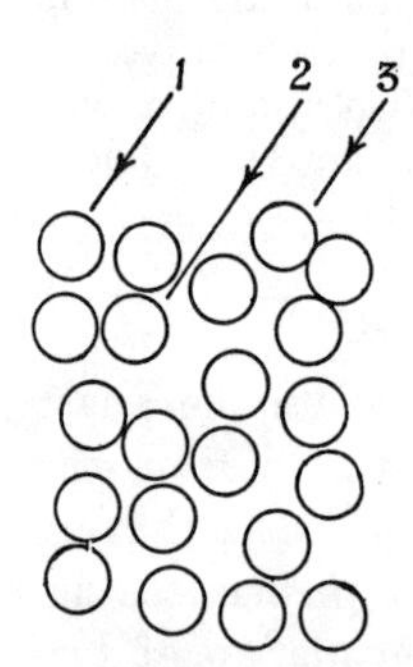

Fig. 3.14.—Corpuscular theory of reflection and transmission of light.

This type of theory involves many difficulties. If it were correct the corpuscles would almost certainly be scattered at the surface and there would be diffuse instead of regular reflection and refraction. Also it would be very difficult to explain rectilinear propagation in a medium whose atoms exerted such powerful forces on the corpuscles. It is therefore necessary to consider the possibility that the corpuscles may not be identical.

3.32.—If a beam of monochromatic light falls normally on a sheet of transparent glass, a certain fraction is transmitted. If the transmitted light falls upon a second sheet of glass similar to the first, the same fraction is transmitted. Thus it appears that, in this case, the light transmitted does not differ essentially from the incident light—both contain the same proportion of corpuscles capable of being transmitted. This implies that there are not some corpuscles permanently capable of transmission and others which must be reflected. On the contrary, all the corpuscles must be capable both of transmission and of reflection. As Newton suggested they must have " fits of reflection " and " fits of transmission ", and must pass periodically from one to the other. There is also no way of predicting when any one corpuscle will be in a " fit of reflection ". Thus the corpuscular theory would appear to involve a certain element of *indeterminacy*—to admit that the behaviour of the corpuscles cannot be followed in detail and described in terms of cause and effect: it also seems to involve a kind of periodicity —a characteristic feature of the wave theory.

3.33.—Even if the corpuscular theory should succeed in giving an account of the fact that some light is reflected and some refracted, it would still have to provide means for calculating (i) the proportion of light transmitted, and (ii) the relation between the angle of incidence and the angle of refraction (Snell's law). It has never been able to attack these problems with any degree of success. It is well known that an attempt was made to explain Snell's law by the special assumption that the speed of the corpuscles increased in a certain ratio in passing from a light to a dense medium, the component of velocity parallel to the surface remaining unchanged. This is a very artificial assumption since normally energy would be needed to increase the speed of the corpuscles. It is directly opposed to the results of measurements of the velocity of light.

3.34.—It thus appears that while the wave theory can give an elegant account of the basic phenomena of reflection and refraction, the corpuscular theory can give only a fragmentary and unsatisfactory description of these phenomena. In order to do even this, it has to import special assumptions which lead towards a wave theory. In considering the great superiority of the wave description, it is desirable to remember that later we shall have to discuss a group of phenomena where a corpuscular theory is entirely successful, and the wave theory has correspondingly great difficulties.

EXAMPLES [3(ix)–3(xv)]

3(ix). Show that if the refractive index were *exactly* represented by the first two terms of 3(37) it would be possible to construct a lens free from chromatic aberration by combining a positive and negative lens made of different glasses.

[Let the refractive indices be given by

$$n' - 1 = A' + A'B'/\lambda^2$$

and

$$n'' - 1 = A'' + A''B''/\lambda^2.$$

Let the powers of the two lenses be F' and F'' respectively and let $\rho' = (1/r_1 + 1/r_2)$ for the first lens and ρ'' the corresponding function for the second.

Then

$$F = F_1 + F_2 = (n' - 1)\rho' + (n'' - 1)\rho''$$
$$= A'\rho' + A''\rho'' + (A'B'\rho' + A''B''\rho'')/\lambda^2,$$

and if $A'B'\rho' + A''B''\rho'' = 0$, then F is independent of λ.]

3(x). Show that the process suggested in the last example fails if $A'/B' = A''/B''$.

3(xi). The refractive indices of a flint glass and a crown glass are as follows:

	Red light	Blue light
Flint	1·644	1·664
Crown	1·514	1·524

Find the focal lengths of two lenses, one made of flint and the other of crown glass, such that, when combined, the resultant focal length is 10 metres for each kind of light. [−5·784 m. and 3·664 m. for blue light.]

3(xii). Calculate the ratio of the amount of light transmitted normally through a piece of glass whose refractive index is 1·5, to that transmitted through a piece of fused quartz whose index is 1·55.

[1·019. Allow for reflections at two surfaces.]

3(xiii). Insert epoch angles δ_1' and δ_2 in 3(51) and 3(52) and apply the conditions 3(49) and 3(50). Hence show that δ_1' and δ_2 are either zero or integral multiples of π.

3(xiv). In an experiment similar to Wiener's, the film was one centimetre long. One end was in contact with the reflecting surface and the other was separated from it by a piece of mica 10^{-3} centimetre thick. The distance between successive dark lines on the plate was 0·025 centimetre. Find the wavelength of the light used. [5000 Å.]

CHAPTER IV

Representation of Light by Wave Trains of Finite Length

4.1. Sources of Light. Types of Spectra.

The classification of sources of light is an important preliminary to the application of wave theory. From this point of view, the most significant classification is made by means of *spectra*. Modern instruments, developed from the simple apparatus by which Newton first discovered the dispersion of light, enable the spectra produced by different sources to be examined in detail. These instruments are called *spectroscopes* when they are designed for visual examination of the spectrum, and *spectrographs* when they are arranged so that it may be photographed. Using these instruments, three main types of spectra have been discovered. These are called *line spectra*, *band spectra*, and *continuous spectra*. Each type may be observed either as an emission spectrum or as an absorption spectrum. Plate I (p. 72) and Plate II (p. 126) include some typical spectra.

4.2. *Line Spectra and Continuous Spectra.*

An emission-line spectrum consists of a number of fairly narrow lines with dark regions between them (see Plate I*a*, *b*, *c*). There may be only a few, or there may be several thousand lines. Narrow (or " sharp ") lines are produced when atoms are able to radiate light without being greatly affected by collisions with other atoms. Sharp lines are usually produced by an electrical discharge in gases at low pressure. Each line is characteristic of the kind of atom by which it is emitted. Sodium emits two lines fairly close together in the yellow region of the spectrum (Plate I*a*); cadmium emits a strong red and a strong green line, as well as many weaker lines (Plate I*b*); mercury emits several strong lines (Plate I*c*). When an electrical discharge is produced in a gas at a pressure of a few atmospheres, the lines become less sharp (Plate I*d*), and when the pressure is made higher still the lines run together to form a continuum. The spectra of light from hot solids are also continuous. This type of spectrum is usually produced under conditions such that each atom is strongly influenced by its neighbours.

4.3. *Band Spectra.*

A band spectrum consists of a very large number of lines which crowd together in certain regions of the spectrum to form characteristic "heads" (Plate II*a*). Spectra of this type are due to molecules and a given system of bands is characteristic of the molecule which emits them. The theoretical description of these spectra is more complicated than the corresponding description of line spectra, but does not involve any essentially different principles.

4.4. *Infra-red and Ultra-violet Radiation.*

Using suitable photographic plates, it is possible to photograph lines, bands, etc., which lie outside the limits of the visible spectrum. This indicates that there are certain types of radiation to which the plate is sensitive but the eye is not. This is confirmed by taking a very sensitive thermopile and moving it through the spectrum. The thermopile gives a reading in the visible region and its reading increases greatly when it passes through a region where a bright line can be seen. The reading does not, however, fall to zero at the ends of the visible spectrum. Beyond the red end it usually increases; beyond the violet end the reading is usually small but still sufficient to indicate the presence of radiant energy. The radiation beyond the red end of the visible spectrum is called *infra-red radiation*; that beyond the violet end is called *ultra-violet radiation** (Plate II*b*, p. 126. See also fig. 1.5).

4.5. *Absorption Spectra.*

When light from a source which normally gives a continuous spectrum is passed through certain vapours or gases, it is found that a number of dark lines appear in the spectrum (Plate II*c*, *d*). Continuous absorption may also be produced by vapours, or by gases, or by liquids. Absorption lines or bands are characteristic of the absorbing gas or vapour, and it is found that their positions in the spectrum coincide with the positions of some of the lines or bands which are emitted by the same gas or vapour under the influence of an electrical discharge (Plate II*c*, *d*, *e*). These absorption lines or bands enable us to detect the

* This radiation is sometimes called *ultra-violet light*. The word "light" should be reserved for "visible radiation".

(*a*) Emission spectrum (sodium).

(*b*) Emission spectrum (cadmium).

(*c*) Emission spectrum (mercury, low-pressure).

(*d*) Emission spectrum (mercury, higher-pressure).

(*e*) Newton's rings.

(*f*) Fabry-Pérot fringes (mercury, blue only).

(*g*) Fabry-Pérot fringes (mercury, blue and green).

PLATE I

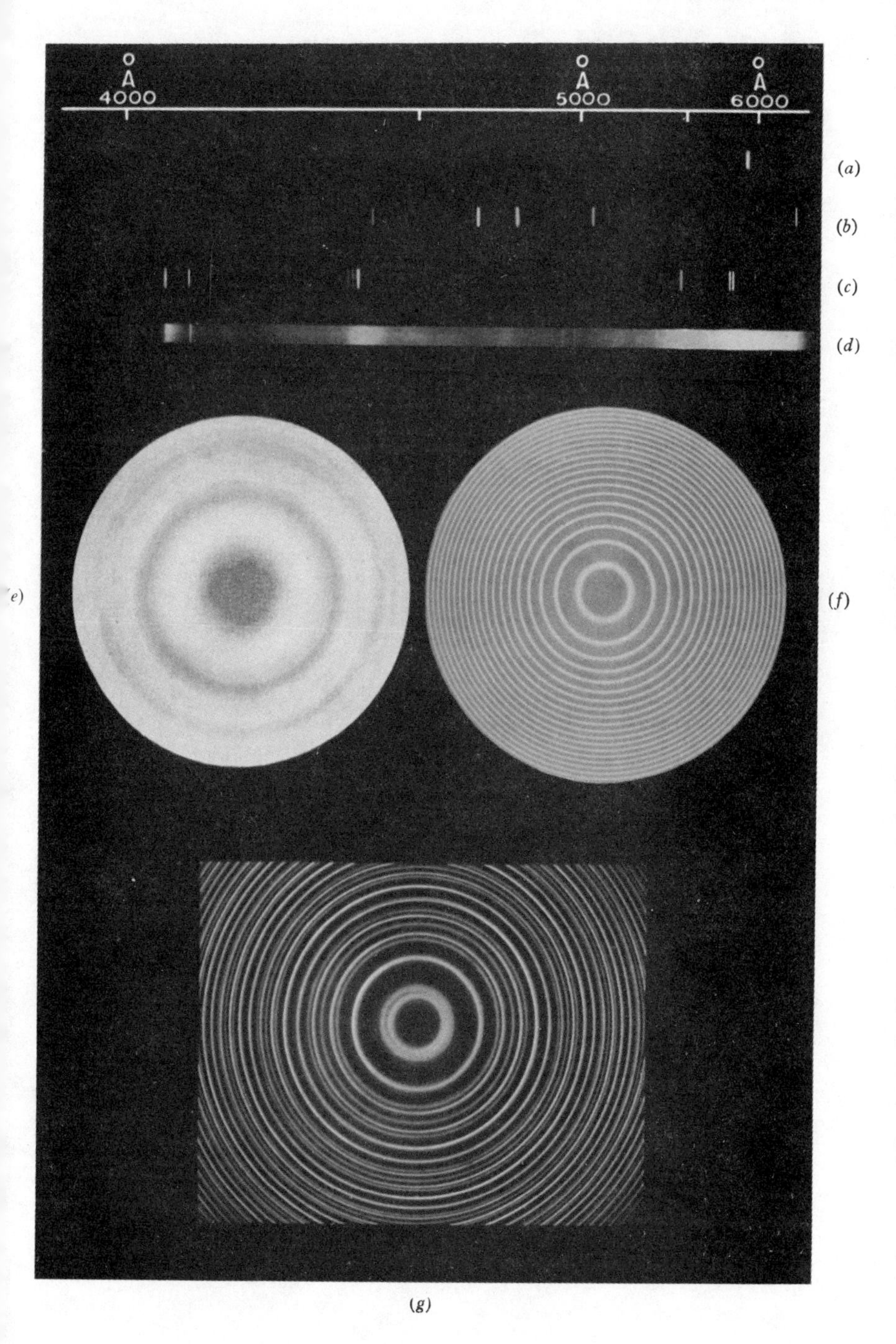

This plate is reproduced in full color on the inside front cover.

presence of the corresponding atom or molecule in the gas or vapour. This type of spectrum was first discovered by J. von Fraunhofer (1787–1826), who showed that the spectrum of light from the sun is crossed with a number of dark lines. The position of these lines in the spectrum coincide with the positions of some of the emission lines obtained in laboratory spectra (Plate II*c*, *e*). These lines originate in the following way. The centre of the sun is a dense mass of very hot gas which emits a continuous spectrum. This light passes through the cooler and less dense outer layers of the sun, where part of the light is absorbed, producing the dark lines. These lines show the presence of certain of the terrestrial elements in the outer layers of the sun. The central region is called the *photosphere*, the main absorbing region is called the *chromosphere*. The *corona* is a much less dense region of the sun which extends far beyond the chromosphere. It is normally seen only at eclipses when the main light of the sun is excluded by the moon. In this region a number of weak but very sharp lines are emitted.

4.6. *Atomic Oscillators.*

Let us now assume, on the basis of certain experiments which will be described later, that each line in a spectrum corresponds to a definite wavelength. It also corresponds to a definite frequency, the relation between wavelength and frequency being given by equation 2(25). The wavelengths and frequencies for different regions of the spectrum are given in fig. 1.5. The emission and absorption of spectra which consist of sharp lines suggest that the atom may be regarded as a system of simple harmonic oscillators. Each oscillator emits light of the wavelength corresponding to its natural frequency and thus produces a line in the spectrum. When white light passes through a gas or vapour, the oscillators in the various atoms resonate and absorb light of the wavelengths which correspond to their own natural frequencies of oscillation. Thus the absorption lines coincide with the emission lines. When an atom is subject to strong interaction with its neighbours (as in a gas at high pressure, or in a solid or liquid) the oscillators are continually being disturbed. They emit irregular pulses instead of simple harmonic waves, and these pulses (which have no well-defined frequency) make up a continuous spectrum. Some of the natural periods of atoms or molecules correspond to wavelengths greater or less than those to which the eye is sensitive, and they give lines in the infra-red and ultra-violet respectively.

4.7.—This general description of the emission and absorption of radiation by atoms and molecules includes many of the observations, but there are certain difficulties. It is not easy to understand why some atoms should emit so many lines if each line corresponds to a distinct mode of oscillation which has its own natural frequency. This difficulty is greater when we consider that even the hydrogen molecule, which consists of only four particles, emits an extremely complicated spectrum containing tens of thousands of lines. It is also found that normally only *some* of the emission lines appear in the absorption spectrum.

In the emission spectra of atoms some lines appear only in the spark spectrum and not in the arc spectrum. Others appear only in the gaseous discharge. These observations indicate that under a given set of conditions some of the oscillators are not available, and the theory does not suggest any simple reason why this should be so. Although there are many difficulties, the simple picture of the atom as a set of harmonic oscillators is still very useful. At a later stage it will be incorporated in a more detailed theory of the emission and absorption of light. At present it may be regarded as a working hypothesis to be amended when more detailed experimental data are available. It suggests that it is desirable to see whether all the properties of the light belonging to one of the sharp lines in the spectrum are characteristic of the long trains of sine waves which would be emitted by a simple harmonic oscillator. For this purpose it is necessary to isolate one of the lines of the spectrum. This may be done by placing a mask over the spectrum with a slit placed so as to allow only a narrow region, containing one line, to pass. Such an arrangement is called a *monochromator*. Sometimes the same result can be more easily achieved by the use of colour filters which transmit only a part of the spectrum. By passing the light through a suitable combination of filters, it is possible to arrange that only one line is transmitted, provided that the source does not have too complicated a spectrum. When the light corresponding to one line of the spectrum has been isolated, its properties may be studied in detail by means of the instrument which will be described in the next paragraph.

4.8. **The Michelson Interferometer.**

The apparatus shown in fig. 4.1 was invented by A. A. Michelson (1852–1931). S_e is an extended source of light such as a gaseous discharge-tube or a sodium flame. Light from S_e passes through the filter F and is made roughly parallel by the lens L. It then falls upon the mirror M_1. This mirror is a plate of glass half-silvered on the side remote from S_e. M_2 and M_3 are fully silvered on the front surfaces. Some of the light from S_e passes through M_1 and is reflected, first by M_3 and then by M_1, so as to enter the telescope T. Another portion of the light is first reflected by M_1, then by M_2, from which it passes through M_1 to the telescope. The light entering the telescope is viewed through the eyepiece E. The compensator C is a plate of unsilvered glass equal in thickness to M_1. Light reflected from M_3 passes twice

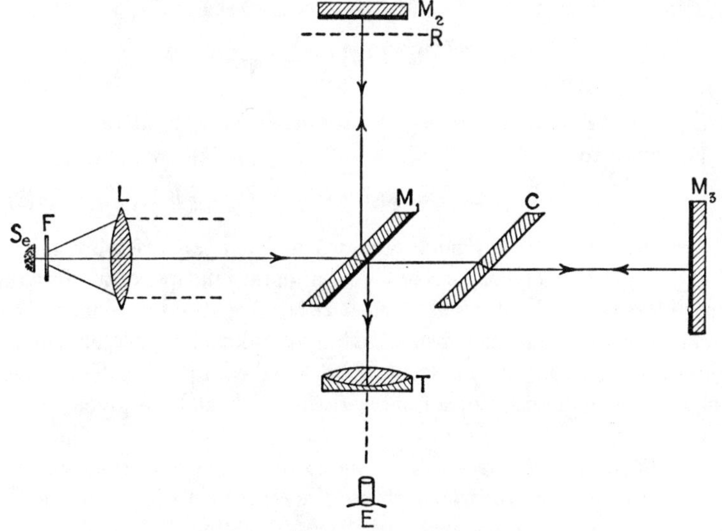

Fig. 4.1.—The Michelson interferometer

through C and thus passes through the same thickness of glass as the beam from M_2, which passes through M_1 three times. The mirror M_2 is mounted on a carriage which can be moved by a screw in a direction parallel to the axis of the telescope. It moves along accurately constructed guides, so that no rotation occurs during its movement. Let R be a plane coincident with the image of the mirror M_3, reflected in the mirror M_1. This plane is called the reference plane. Then the difference in phase between the beam reflected from M_2 and that reflected from M_3 is the same as if the latter beam had been reflected from R.* In the usual adjustment of the instrument M_2 is set parallel

* For studying the interaction between *two* beams of light, the Michelson interferometer is superior to a simple thin film arrangement (such as would be produced by placing a half-silvered mirror at the plane R), because (*a*) there are no beams due to multiple reflections (such as reflection from M_2 to R, thence to M_2 again and from M_2 to the telescope); (*b*) M_2 can be made to coincide with R; and (*c*) it is fairly easy to make the two beams of equal amplitude.

to and about a millimetre from R. Circular fringes are seen in the telescope when it is suitably adjusted.*

4.9.—We account for these fringes in the following way. Since the light from S_e is not accurately parallel some beams reach T having been reflected normally from M_2, others after reflection at a small angle. If e is the distance from M_2 to the reference plane R, the path difference † for beams reflected at an angle θ to the normal is $2e \cos \theta$, and the phase difference is

$$\delta = \frac{2\pi}{\lambda}(2e \cos \theta) = \frac{4\pi e \cos \theta}{\lambda}.$$

If the two beams are of equal amplitude, the relative energy is proportional to

$$2a_1^2 + 2a_1^2 \cos \delta = 4a_1^2 \cos^2 \tfrac{1}{2}\delta. \quad \text{(See § 3.4)}. \quad . \quad . \quad 4(1)$$

The relative energy is a maximum when δ/π is an even integer, i.e. when the path difference $2e \cos \theta$ is an integral multiple of λ. It is zero when δ/π is an odd integer, i.e. when $2e \cos \theta$ is an odd number of half wavelengths. A parallel beam of light falling on a telescope is focused at a point in the focal plane of the objective. The hollow cone of light consisting of all these rays which make a given angle θ with the axis of the telescope is seen as a circular ring. The phase difference for all points on the ring is the same and the ring constitutes a bright fringe if this phase difference is an integral multiple of 2π.

The *order of interference* between two beams for which the phase difference is $2\pi p$ (corresponding to a path difference of $p\lambda$) is p. In this definition of order of interference p is not necessarily integral. The maximum which corresponds to an integral value (p_1) of p is called the *maximum of order* p_1 or the bright fringe of order p_1. In a similar way we may speak of the minimum of order $(p_1 + \frac{1}{2})$ or the dark fringe of order $(p_1 + \frac{1}{2})$.

Let p_0 (not necessarily integral) be the order of interference at the centre, and let θ_{p_1} be the angular *radius* of the bright fringe of order p_1 (where p_1 is an integer). Then

$$2e = p_0\lambda \quad . \quad . \quad . \quad . \quad . \quad . \quad . \quad 4(2a)$$

and

$$2e \cos \theta_{p_1} = p_1\lambda. \quad . \quad . \quad . \quad . \quad . \quad . \quad . \quad 4(2b)$$

* Instructions for adjusting the Michelson interferometer are given in Appendix IV A.

† See § 5.13.

Note that the order at the centre (p_0) is always *greater* than the order of any of the rings (p_1). When e is constant the loci of maximum relative energy depend only on θ, and are therefore circular rings.* If M_2 is moved away from R, fresh rings appear at the centre. Each ring moves outwards but the central rings move more rapidly so that, in any given part of the field, the angular separation of the fringes becomes smaller.

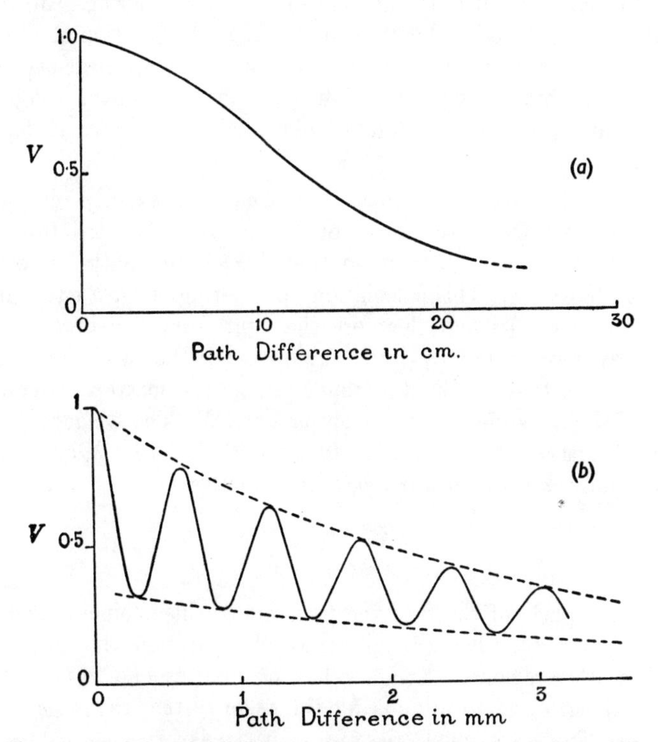

Fig. 4.2.—Variation of visibility with path difference for (*a*) cadmium red line (6438 Å.) and (*b*) sodium yellow lines (5890 Å. and 5896 Å.)

4.10. *Visibility of the Fringes.*

The reduction of the separation of the fringes when the path difference is large makes them more difficult to see if they are viewed with the same telescope, i.e. under constant magnification. Michelson found that apart altogether from this effect the clearness of the fringes

* It is shown in Chapter V that this is still true for sources of finite size.

varies in a way characteristic of the source used. He defined the *visibility* of a fringe to be

$$V = \frac{E_{(max)} - E_{(min)}}{E_{(max)} + E_{(min)}}, \quad . \quad . \quad . \quad . \quad . \quad . \quad 4(3)$$

where $E_{(max)}$ is the relative energy for a bright fringe and $E_{(min)}$ is the relative energy for the neighbouring dark fringes. The visibility so defined is independent of the angular diameters of the rings. Michelson devised an ingenious method * of measuring the variation of V with e. Results of measurements of the value of V for fringes near the centre of the field, obtained with cadmium red light, are shown in fig. 4.2*a*. Corresponding results for sodium yellow light are shown in fig. 4.2*b* (p. 77).

4.11.—If the radiation from the source S_e is exactly represented by a pure sine wave, the value of V may be calculated from 4(1). This equation gives $E_{min} = 0$, so that $V = 1$, no matter how large the path difference. This calculation does not agree with either of the experimental results and therefore the light from the sources cannot be exactly represented by pure sine waves. The next most simple assumption is that the light is represented by a mixture of two sine waves of slightly different wavelengths λ and λ'. The fringes produced by wavelength λ' are then similar to those produced by λ, but if θ'_{p_1} is the angular diameter for order p_1 of λ', we have

$$\frac{\cos \theta'_{p_1}}{\cos \theta_{p_1}} = \frac{\lambda'}{\lambda}. \quad . \quad . \quad . \quad . \quad . \quad . \quad . \quad 4(4)$$

In some parts of the field the two sets of rings coincide and reinforce one another. Owing to difference of separation, they go " out of step " in other regions. Certain values of θ correspond to maxima for one wavelength and to minima for the other so that the rings become confused. There are thus series of alternations of rings " in step " and " out of step " leading to a variation of visibility. Just as the ratio of the periods of two pendulums may be determined by observing the frequency of coincidences, so the ratio of the wavelengths may be determined by observing the fluctuations of the visibility of the fringes.

4.12.—Observation of the variation of the visibility may also be used to determine the ratio of the amplitudes of the two waves and hence the relative energies. Two sine waves of equal amplitude give zero visibility when the waves are exactly out of step. If the ampli-

* See Reference 4.1.

tudes are not equal the visibility is never zero since the weaker fringes do not completely obliterate the stronger ones. If the amplitudes of the sine waves are a_1 and a_2, the maximum visibility is unity and the minimum visibility is

$$V_{\min} = \frac{a_1^2 - a_2^2}{a_1^2 + a_2^2}.$$

Thus by measuring (*a*) the separation of the maxima, and (*b*) the ratio of the maxima to the minima in the visibility curve, it is possible to determine both the ratio of the wavelengths and the ratio of the amplitudes. Let us now apply this method of analysis to fig. 4.2*b*, temporarily ignoring the gradual decrease of visibility with increasing values of *e*, and considering only the alternations. We find that yellow sodium light contains two components whose wavelengths differ by about 1 part in 1000. One contributes about twice as much as the other to the total intensity of the source.

Michelson's method is very powerful in the detection of small amounts of inhomogeneity in the source. He was able to show that the red Balmer line of hydrogen contained two components with a separation of 0·14 Å. or about one part in 40,000. Modern methods have shown the presence of weaker components and have given a value of 0·1358 Å. for the separation of the two strong components. As a method for the analysis of sources of light, Michelson's procedure suffers from two defects. On the theoretical side Rayleigh (J. W. Strutt) showed that the analysis of the visibility curve does not always give a unique result. In complicated cases, different sets of components lead to the same visibility curve. On the practical side, Michelson's procedure is laborious and demands very great experimental skill. The accuracy of the result he obtained for the hydrogen line is an indication of Michelson's own genius as an experimental scientist as well as of a considerable capacity for taking pains! The analysis of sources of light can now be carried out by methods which show directly what he had to infer (see Chapters V and IX). Michelson's interferometer is no longer used for this purpose.

4.13.—The visibility curve for cadmium red light does not show an alternation of intensity but a gradual decrease of approximately exponential form (fig. 4.2*a*). The visibility curves for sodium light also show a gradual decrease superimposed upon the alternations. Two explanations of this gradual decrease will now be considered. In the first place let us suppose that cadmium red light is represented by a very large number of sine waves, differing slightly in wavelength but with energies falling off sharply on either side of a certain wavelength for which the energy is a maximum (fig. 4.3). Zero path difference gives a maximum relative energy for all wavelengths. Since most of the light is nearly of the same wavelength, the fringes are clear for small path differences. They become less clear when the path difference is increased, because the positions of the maxima due to wavelengths on the edge of the line (i.e. to wavelengths which have a comparatively large difference from the central wavelength) deviate from those due to the mean wavelength. As the path difference is further increased, more and more of the light goes to form fringes which are not " in register " either with the fringes due to the central wavelength or with each other. Thus the fringes gradually become less and less clear and there are no alternations.

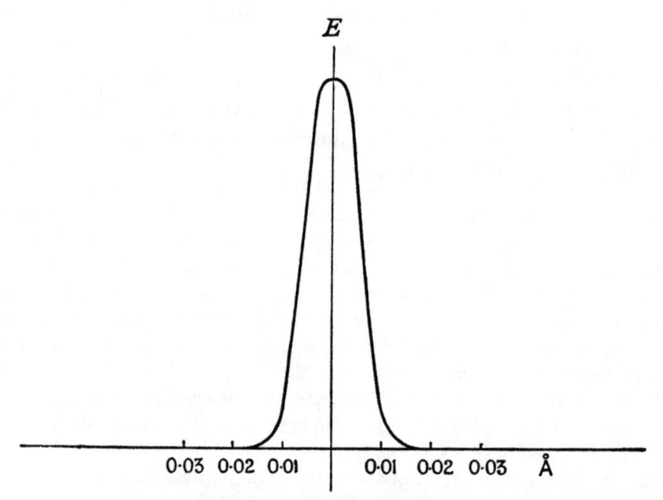

Fig. 4.3.—Distribution of energy with wavelength for the cadmium red spectrum line (low-pressure lamp)

The situation may be illustrated by a simple analogy. If a number of men, with slightly different lengths of stride, start off walking in step, they will gradually fall out of step. They will all be in step again at a distance which is the least common multiple of all the pace lengths, and at 2, 3 . . . times this distance. If the number of men becomes indefinitely large and the differences of pace length become infinitesimal, there are no alternations; the group falls gradually out of step and never comes together again as a whole.

Michelson showed that his results for the cadmium red line were explained if the light was represented by a distribution (see fig. 4.3) in which the *energy* falls to half its maximum value for wavelengths differing from the mean by $\pm 0{\cdot}0065$ Å. This distance is called the *half-width* of the line.* He showed that, in a similar way, sodium light could be represented by two components, neither of which is strictly monochromatic.

4.14.—As an alternative explanation of the decrease in visibility, we now suppose that cadmium red light is represented, not by a single continuous train of waves but by a series of wave trains of finite length. Each train of waves is divided into two trains of equal length by the mirror M_1 (see fig. 4.1). These return from M_2 and M_3 to the telescope. When the path difference is small, the telescope receives the returning trains of waves nearly simultaneously, and they can interact. When M_2 is at a sufficient distance behind R, the train from M_2 enters the telescope after that from M_3 has passed. It is as though light from two independent sources was being received in the telescope. There is no interaction and no fringe system. For intermediate value of the path difference, the wave trains partially overlap and fringes of reduced visibility are formed. Thus Michelson's results may be explained, at least qualitatively, either by postulating a group of waves of slightly varying frequency or by postulating a wave train of finite length.

In order to discuss this matter further, it is necessary to examine the properties of groups of waves and to employ a mathematical method for the analysis of such groups. The complete theory involves rather lengthy mathematical calculations. In the following paragraphs, the results of these calculations are given without proof. The proofs are given in outline in Appendix IV B.

4.15. Waves of Irregular Profile.

The starting point for our discussion is a mathematical theorem due to J. B. J. Fourier (1768–1830), who showed that a wave of irregular profile may always be regarded as the sum of a series of simple harmonic waves. This analysis is very helpful in solving particular problems when the irregular profile has some special form that gives a simple series. It is also useful because we can prove certain general propositions for a single simple harmonic wave and then, using the principle of superposition, show that they apply to the sum of a set of simple harmonic motions. Fourier's method then shows that they apply generally to waves of irregular profile.

In connection with the Fourier analysis it is important to remember that when we write a mathematical equation without qualification,

* Some writers define a *half-value width* which equals twice the *half-width* defined above.

we imply that it is true for all values of the variables. The displacement for a simple harmonic wave is given by 3(3) and this expression is valid for all space and time. The wave whose profile is shown in fig. 4.4 is not a pure simple harmonic wave. Over a considerable region the profile is that of a simple harmonic wave, but to the left and to the right of this region the displacement is zero. Therefore the profile as a whole is not that of a simple harmonic wave. When the

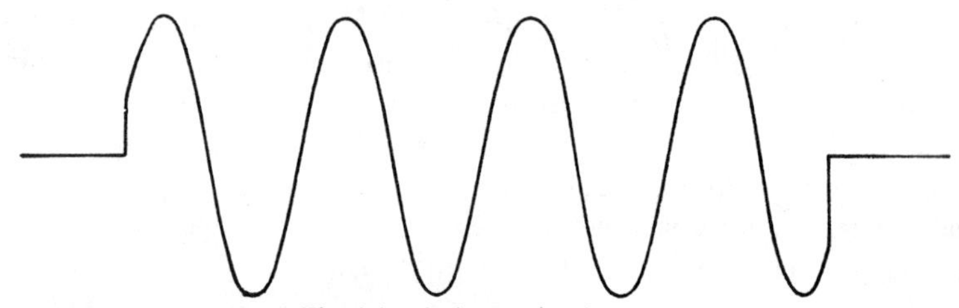

Fig. 4.4.—A short train of waves

profile of a wave follows a simple sine curve (with constant or slowly varying amplitude) over an appreciable region, and the displacement is zero elsewhere, we call it a *wave train*. When the region over which it simulates a simple harmonic wave is large compared with the wave-length we speak of a *long wave train*, even though in our ordinary measure the length may be rather short. A train of light waves one millimetre long would contain 2000 waves and would be a fairly long wave train.

4.16.—In Chapter III we described methods of calculating the resultant of a number of simple harmonic waves of the same frequency. The Fourier mathematical method enables us to calculate the resultant of a large number of simple harmonic waves, not necessarily of the same frequency, provided that the amplitudes and epoch angles are given. One important case to which the Fourier method has been applied is that of the *wave group*. So long as we are confined to one-dimensional problems, we may define a wave group as the resultant of a set of simple harmonic waves closely grouped round a certain mean frequency. If any members of the group have frequencies which differ from the mean frequency by more than a small fraction of this frequency, then their amplitudes are a small fraction of the amplitudes of waves whose frequencies are near the mean frequency. Thus nearly all the energy is concentrated in frequencies near to the central frequency.

We have defined the term " wave train " by reference to the profile, and the term " wave group " by reference to the distribution of energy among different frequencies. We shall show that a long train of waves with constant or slowly varying amplitude is associated with a narrow range of frequency. *A long wave train is a wave group.* When the wave train is *very* long, the frequency range of the corresponding wave group is correspondingly narrow. The reverse relation is also true; the synthesis of a group of waves whose frequencies cover a narrow range is a long wave train. We may deduce this result in a qualitative form from very elementary arguments. In a very short wave train no one frequency is dominant. We should expect any analysis in terms of simple harmonic waves to yield a wide distribution of frequencies. It is well known that irregular sounds give such a distribution. If a wave train covers a large number of wavelengths, one frequency becomes well established. When the train becomes very long it approaches very nearly to a simple harmonic wave, and we should expect the analysis to represent this by a gradual narrowing of the frequency range. In the limit, the frequency range tends to zero as the length of the wave train tends to infinity. Whether we consider the frequency distribution or the profile, the limit is a simple harmonic wave of exactly constant amplitude extending over all space.* If we consider the other extreme, we may imagine a wave train which is so short or so irregular that no one frequency predominates markedly. This is called a *pulse.*

4.17. **Fourier's Series.**

Fourier's method enables any suitable mathematical function of one variable to be expressed either (*a*) as the sum of a series of cosine functions whose wavelengths are all submultiples of a certain chosen wavelength or (*b*) as an integral involving a set of cosine functions whose wavelengths vary continuously from 0 to ∞. The series expansion is valid over only a finite range of the variable but the integral is valid over the whole range. Generally, the theorem applies to any function which can be shown by a graph and not merely to functions which are given by a single algebraic expression. Thus both continuous functions and functions which have discontinuities of slope or magnitude may be expressed in terms of Fourier's series, provided that the number of discontinuities in the relevant range is finite. Fourier's series is especially convenient for the representation of functions which cannot be expressed by one simple algebraic function

* In this chapter we apply the Fourier method to a one-dimensional problem. Later we shall have to generalize the concept of a wave group to deal with problems in two and three dimensions.

but consist of parts each of which can be so expressed. Examples of such functions are shown in figs. 4.5 and 4.6.

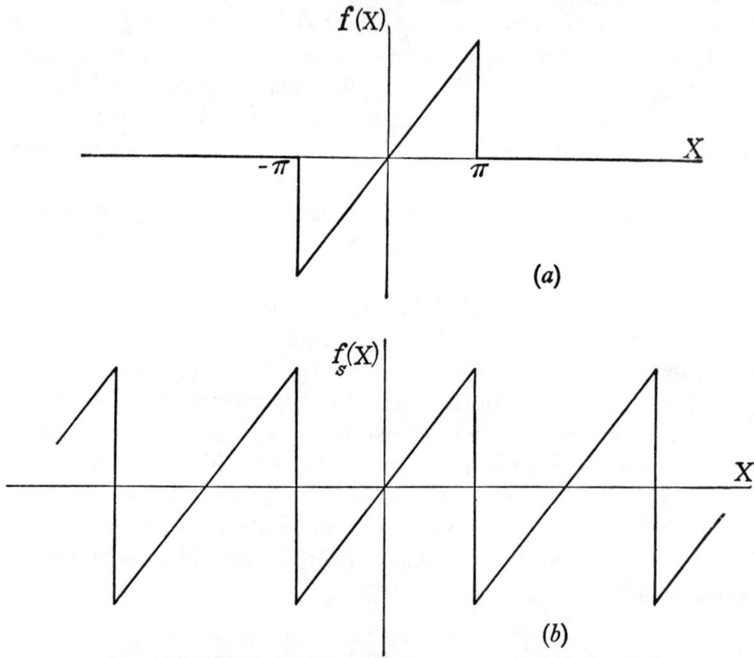

Fig. 4.5.—The relation between (*a*) a non-periodic function and (*b*) the sum of a related Fourier series.

Suppose that it is desired to replace $\xi = f(x)$ by a series of sine functions and that *the expansion is to apply to a range from* $-x_0$ *to* $+x_0$. Then if $X = \pi x/x_0$, Fourier's series is defined by the relation

$$\begin{aligned} f(X) = a_0 + a_1 \cos X + a_2 \cos 2X + \ldots \\ + b_1 \sin X + b_2 \sin 2X + \ldots \end{aligned} \quad . \quad . \quad 4(5)$$

It may be shown that *

$$\left.\begin{aligned} a_0 &= \frac{1}{2\pi}\int_{-\pi}^{\pi} f(X)\, dX, \\ a_m &= \frac{1}{\pi}\int_{-\pi}^{\pi} f(X)\cos mX\, dX, \\ b_m &= \frac{1}{\pi}\int_{-\pi}^{\pi} f(X)\sin mX\, dX. \end{aligned}\right\} \quad . \quad . \quad . \quad . \quad 4(6)$$

* See Appendix IV B and Reference 4.2.

By introducing new constants A_0, A_1, etc., and δ_1, δ_2, etc., defined by

$$A_0 = a_0, \quad A_1 \sin \delta_1 = a_1, \quad \text{and} \quad A_1 \cos \delta_1 = b_1, \quad . \quad 4(7)$$

we may write 4(5) in the form

$$f(X) = A_0 + A_1 \sin (X + \delta_1) + A_2 \sin (2X + \delta_2) + \ldots \quad . \quad 4(8)$$

Each term in 4(8), except the first, represents a pure sine wave. In a similar way 4(5) may be expressed as the sum of a series of complex exponentials (see § 2.26 and Appendix IV B).

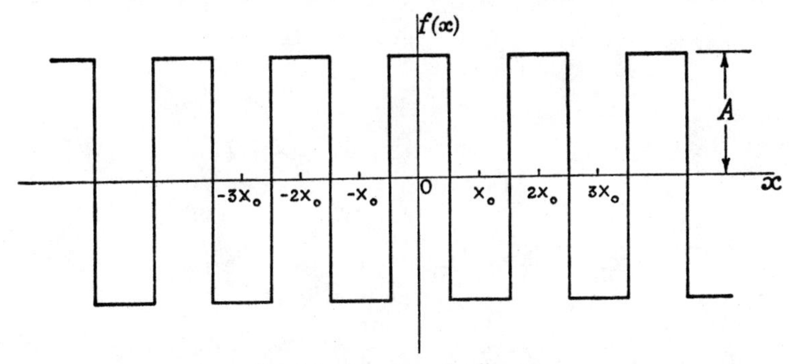

Fig. 4.6.—The " top-hat " curve

4.18.—The values of functions such as $\sin mX$ and $\cos mX$ for $X = X_0 + 2\pi$ are the same as the values of the corresponding functions for $X = X_0$. Let $f_s(X)$ be the sum of the series 4(5) for any value of X. Then, since $f_s(X)$ is the sum of a series of terms such as $\sin mX$ and $\cos mX$, we shall have

$$f_s(X_0 + 2\pi) = f_s(X_0),$$

i.e. the sum of the series must lead to a function which repeats its values when X_0 changes by 2π. The sum of the series $f_s(X)$ always agrees with the original function $f(X)$ in the range $-\pi$ to $+\pi$, but does not agree with it outside this range unless $f(X)$ happens to be a periodic function for which $f(X + 2\pi) = f(X)$.

Fig. 4.5a shows a function which is proportional to X in the range $-\pi$ to π, and zero elsewhere. The Fourier series

$$f_s(X) = 2 \sum_{m=1}^{\infty} (-1)^{m+1} \frac{\sin mX}{m},$$

shown in fig. 4.5b, agrees with the corresponding function in the range

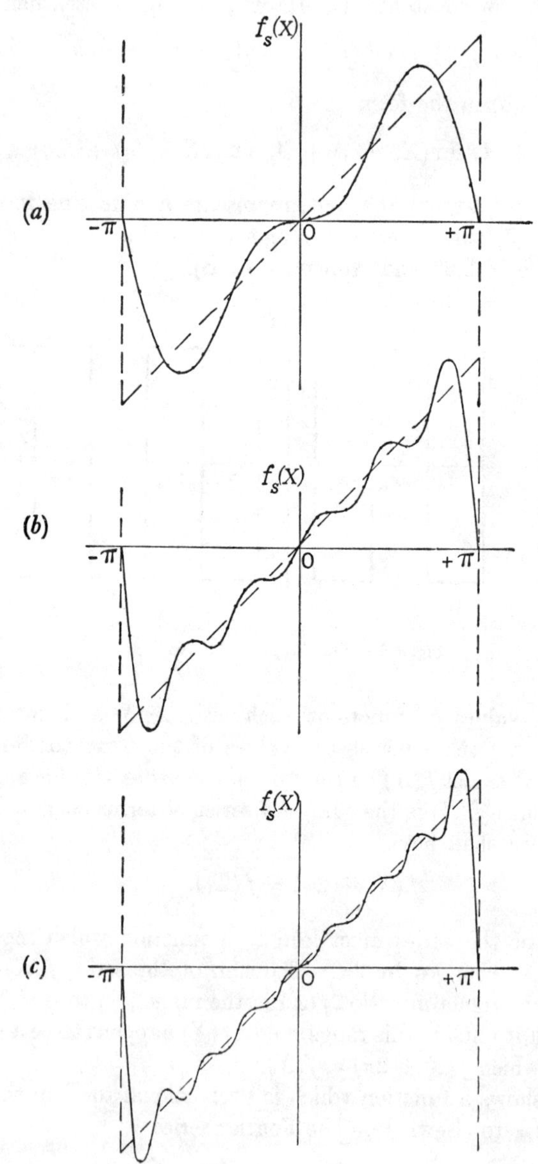

Fig. 4.7.—Representation of a function by a Fourier series: (*a*) sum of 3 terms of the series, (*b*) sum of 6 terms of the series, (*c*) sum of 9 terms of the series.

In (*a*), (*b*), (*c*) the dotted line shows the original function $f(X) = X$.

$-\pi$ to $+\pi$ but not outside this range. On the other hand, the "top-hat curve" shown in fig. 4.6 can be represented by an appropriate Fourier series [see answer to Example 4(iii), p. 101] for all values of x_0 because the curve to be represented is periodic. The relation between a limited number of terms of a Fourier series and the corresponding function is shown in fig. 4.7. [See also Example 4(ii) and fig. 4.8.]

4.19. **Fourier's Integral**

In the preceding paragraphs we have discussed the analysis of a wave whose profile is given, into a number of simple harmonic waves. This process is the inverse of the combination of a number of simple harmonic waves, each of known frequency and amplitude, which we discussed in Chapter III. Using the principle of superposition, we may calculate the resultant of a large number of simple harmonic waves whose amplitudes are $a_1, a_2, \ldots a_n$, and whose circular frequencies are $\omega_1, \omega_2, \ldots \omega_n$. Fourier's integral theorem, which we shall now state, enables us to extend this calculation to the limit in which the number of simple harmonic waves to be combined is indefinitely large, and in which the differences of frequency and of amplitude between successive members of the set are infinitesimal. It also shows the relation between the processes of analysis and of synthesis in an elegant and symmetrical form.

Let us first consider the combination of n simple harmonic motions, of which one is represented by

$$\xi_r = a_r \exp i(\omega_r t - \kappa_r x) + a_{-r} \exp - i(\omega_r t - \kappa_r x).$$

Then, from the principle of superposition, the resultant is *

$$\xi = \sum_{r=-n}^{n} a_r \exp i(\omega_r t - \kappa_r x), \quad . \quad . \quad . \quad . \quad 4(9)$$

where $\omega_{-r} = -\omega_r$. Let us define a function $a(\kappa)$ by the relation

$$a_r = a(\kappa)\,(\kappa_{r+1} - \kappa_r),$$

where κ has some value between κ_r and κ_{r+1}.

We now imagine the number of waves to increase without limit and the intervals to become infinitesimal. Then we put $\kappa_{r+1} - \kappa_r = d\kappa$ and write

$$\xi \equiv f(x, t) = \int_{-\infty}^{\infty} a(\kappa) \exp i(\omega t - \kappa x)\, d\kappa, \quad . \quad . \quad 4(10)$$

* The interpretation of the "negative harmonics" is discussed on p. 105.

i.e. we make the usual transition from the sum of a large, but finite, number of elements to the definite integral.* In general $a(\kappa)$ will be a complex quantity of the form

$$a(\kappa) = a'(\kappa) \exp i\delta_\kappa,$$

where δ_κ represents an epoch angle (§§ 2.4 and 2.26). The profile at $t = 0$ is given by the *real part* of

$$\xi = f_0(x) = \int_{-\infty}^{\infty} a(\kappa) e^{-i\kappa x}\, d\kappa. \quad . \quad . \quad . \quad 4(11)$$

This equation gives the profile when $a(\kappa)$ is known; this is equivalent to knowing the distribution of energy among the different frequencies (see Appendix IV B).

The complementary problem, to find the energy distribution when the profile is given, can be solved by an integral theorem which Fourier derived as an extension of the method of analysis described in §§ 4.17 and 4.18. In the form relevant to the present problem, this theorem states that

$$a(\kappa) = \frac{1}{2\pi} \int_{-\infty}^{+\infty} f_0(x) e^{+i\kappa x}\, dx. \quad . \quad . \quad . \quad 4(12)$$

This equation is valid for all values of x and κ. Note that 4(11) and 4(12) differ, not only in the factor $1/2\pi$ (which is unimportant), but also in that 4(12) contains a positive exponential and 4(11) contains a negative exponential.†

4.20.—Fourier's series enables us to analyse a wave of irregular profile into a discrete set of harmonic waves whose wavelengths are all submultiples of a certain fundamental wavelength. This analysis is valid over a range $-x_0$ to $+x_0$ equal to the fundamental wavelength. If we choose a different range, say $-x_1$ to $+x_1$, the composite wave appears to be made up of a different set of harmonic waves [see Example 4(iv), p. 101]. If we extend the range from $-\infty$ to $+\infty$, the analysis is in terms of a continuous distribution. It is important to realize that there is no inconsistency in the fact that different methods of analysis give different results when applied to the same wave-profile. If x_1 is

* The discussion of the conditions under which this transition can be made is given in standard works on the integral calculus. The physical interpretation of the "negative values of κ" in 4(10) is discussed in Appendix IV B.

† The function $g(\kappa)$ used in the Appendix is equal to $a(\kappa)\sqrt{(2\pi)}$. Using this function the factor outside the integrals is the same [see equations 4(60) and 4(61)]. It is also shown that equation 4(11) may be replaced by one containing $e^{+i\kappa x}$, and that the negative exponential then appears in the equation corresponding to 4(12).

greater than x_0, the analysis in terms of a set of waves based on the fundamental wavelength x_1 applies in the range $-x_0$ to $+x_0$. *Within this range*, the two sets of harmonic waves are equivalent. They give different resultants in the ranges $-x_1$ to $-x_0$ and x_0 to x_1, but here the analysis based on x_0 does not apply. The equivalence of the two results of analysis exactly represents the results of experiment.

4.21. The Gaussian Wave Group.*

We shall now use the method of Fourier to discuss the properties of the particular kind of wave group for which

$$a(\kappa) = A' \exp\{-\alpha(\kappa - \kappa_0)^2\}, \quad . \quad . \quad . \quad 4(13)$$

where A', α, and κ_0 are constants. This type of distribution results when an oscillator is subject to irregular disturbances which cause a large number of small random variations in its period. This is called a Gaussian wave group because many physical problems of random variation were investigated by Gauss. The shape of the wave profile when $t = 0$ [obtained by using equation 4(11)] is represented by

$$f_0(x) = \int_{-\infty}^{+\infty} A' \exp\{-\alpha(\kappa - \kappa_0)^2 - i\kappa x\}\, d\kappa. \quad . \quad 4(14)$$

In Appendix IV B this is shown † to give

$$f_0(x) = A' \sqrt{\frac{\pi}{\alpha}} \cdot e^{-x^2/4\alpha} \cdot e^{-i\kappa_0 x}. \quad . \quad . \quad . \quad 4(15)$$

This profile is shown in fig. 4.8*a* for the case where $\alpha\kappa_0^2 = 2000$; in fig. 4.8*b* for the case where $\alpha\kappa_0^2 = 200$; and in fig. 4.8*c* for $\alpha\kappa_0^2 = 20$. Note that the amplitude of the wave is proportional to $e^{-x^2/4\alpha}$. This curve is shown by the dotted line in fig. 4.8*b*.

When $\alpha\kappa_0^2$ is very large compared with unity, the expression 4(15) represents a close approximation to a pure sine wave whose wavelength is $2\pi/\kappa_0$. The amplitude remains nearly constant over a region which covers many wavelengths and only begins to decrease at distances from the origin such that x becomes comparable with $\sqrt{\alpha}$. Even there the decrease in a region of one wavelength is small. Thus, when $\sqrt{\alpha}/\lambda$ is large compared with unity, i.e. when $\alpha\kappa_0^2$ is large, 4(15) represents a wave group with a well-defined frequency. As $\alpha\kappa_0^2$

* See Appendix IV B, p. 112, and Reference 4.3.

† Use equations 4(99) and 4(103). The constant $A = 2A'$.

becomes smaller the train becomes shorter and shorter until when $\alpha\kappa_0^2$ is of the order of unity, the profile is that of a pulse.

4.22.—Figs. 4.9*a*, *b*, *c* show the function $a(\kappa)$ for different values of $\alpha\kappa_0^2$ corresponding to figs. 4.8*a*, *b*, *c*. From this figure, or by comparison of equations 4(15) and 4(13), it may be seen that when the

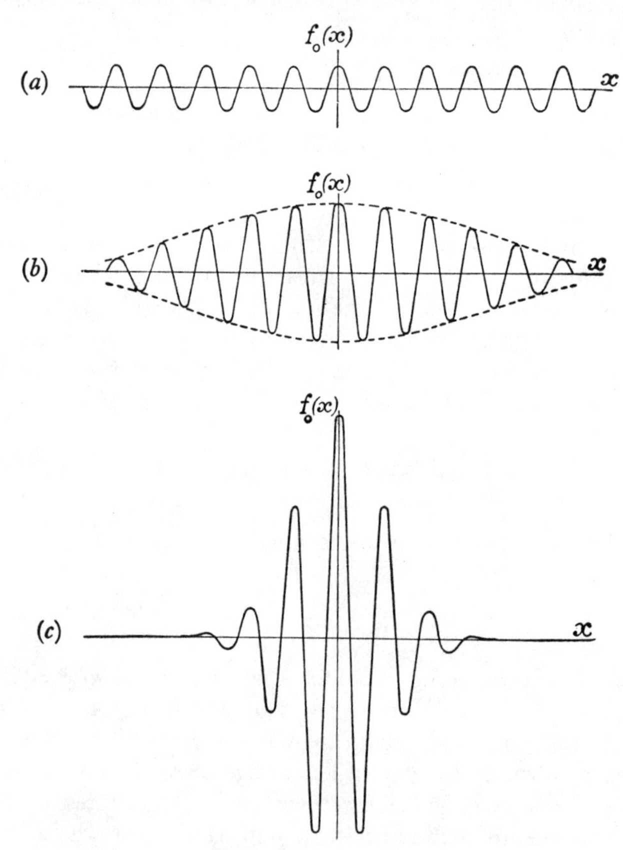

Fig. 4.8.—Gaussian wave groups, (*a*) $\alpha\kappa_0^2 = 2000$; (*b*) $\alpha\kappa_0^2 = 200$; (*c*) $\alpha\kappa_0^2 = 20$. The maxima are adjusted for equal total energy. The real part of $f_0(x)$, which is the physical variable, is shown.

train of waves is short, the energy is spread over a wide range of frequency, but that as the wave train becomes longer the energy becomes concentrated within a narrower and narrower range of frequencies. In the limiting condition the energy all passes into an indefinitely narrow range of frequency as the train of waves becomes indefinitely long. Then, and only then, we have a pure sine wave. If we had considered a wave group of any other suitable mathematical form, defined by some function $a(\kappa)$, the details of the calculation would have been different but, qualitatively, the final result would have been the same. When the parameters of $a(\kappa)$ were adjusted

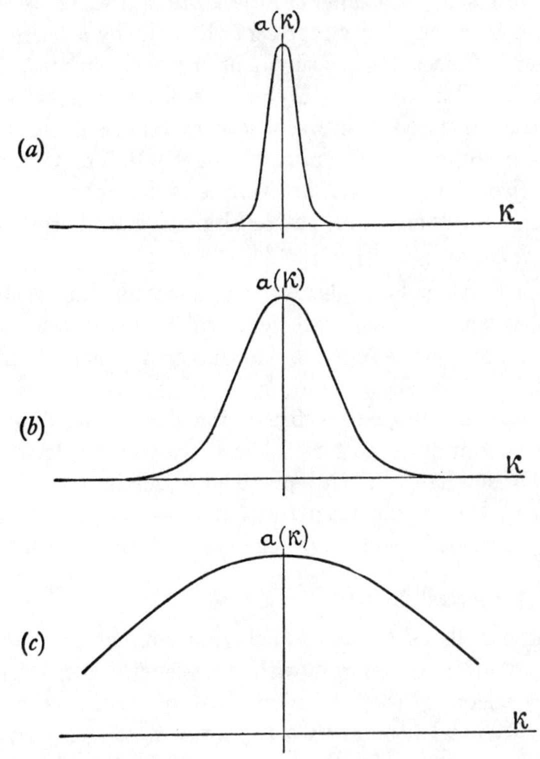

Fig. 4.9.—Gaussian wave groups. Distribution of energy with κ for (a) $\alpha\kappa_0^2 = 2000$, (b) $\alpha\kappa_0^2 = 200$, and (c) $\alpha\kappa_0^2 = 20$. The scales of $a(\kappa)$ have been adjusted to give equal values at the maximum. The actual maxima for (a), (b), (c) are in the ratio 10: 3: 1. The scale of κ is arbitrary but is the same for each curve.

so that the frequency distribution became more and more restricted, the length of the group would increase and the amplitude would become more nearly constant. The pure sine wave would appear as the limiting case. Some forms of the function $a(\kappa)$ are discussed in Appendix IV B.

4.23.—In § 4.14 we considered two possible theories of the gradual decrease of visibility with increasing path. One possibility was to represent the light emitted from the cadmium atom by wave trains of finite length, the other was to represent it by a series of infinitely long wave trains having slightly different frequencies. The foregoing discussion has shown that these are not two alternative theories such that if one is right the other must be wrong; on the contrary, they are two ways of stating the same theory, so that if either is right the other must also be right. If we represent the light by a beam of finite length, a sufficiently accurate measurement of wavelength will show the presence of a distribution of energy over a finite interval of wavelength. On the other hand, if we choose to represent the beam by a series of infinitely long wave trains of slightly different frequencies, we shall find that the resultant disturbance is zero except in a finite region of space. In other words, the resultant is a wave train of finite length.

4.24.—Michelson analysed his observations on the visibility of fringes obtained with cadmium red light and he found that this light could be adequately represented by a Gaussian distribution with a value of α such that the half-width is 0·0065 Å. He obtained values of the same order for the half-width of the lines emitted by sodium and other sources at low pressure. The corresponding length of the wave train is about half a metre—or about a million times the wavelength. Although the light is not perfectly homogeneous, it does possess a remarkably high degree of homogeneity.

4.25. **Width of Spectral Lines.**

Experiments such as those of Michelson can, at best, determine only the distribution of energy with wavelength and cannot lead directly to any theory of the way in which atoms emit spectral lines of finite width. General atomic theory, however, suggests three main causes, operating together, for the broadening of spectral lines. These are (*a*) natural damping, (*b*) Doppler effect, (*c*) pressure broadening.

(*a*) *Natural damping.*

Dirac's theory (see Chapter XIX) of the emission of radiation shows that the atom is to be regarded as a damped oscillator. The emission of energy causes the amplitude of oscillation to fall and therefore the amplitude of the emitted wave is falling while it is being emitted. In the simplest case, the wave train emitted is represented by

$$\xi = \exp\left[-\frac{1}{2c\tau}(x + ct) + i(\omega t + \kappa x)\right], \quad . \quad . \quad . \quad 4(16)$$

where τ is a constant. The energy is proportional to $\xi\xi^*$ and its damping constant is $\gamma = 1/\tau$. Equation 4(16) represents the wave due to an oscillator situated at

$x = x_0$ which began to oscillate with unit amplitude at $t = -x_0/c$. The profile (for $t = 0$) is given by

$$f_0(x) = e^{-x/2c\tau}\, e^{-i\kappa_0 x}. \qquad \qquad 4(17)$$

This profile is shown in fig. 4.10. The wave is travelling to the left. It corresponds to a damped wave. Fourier's method * shows that the energy distribution is proportional to $[a(\kappa)]^2$, where

$$[a(\kappa)]^2 = \frac{A'}{\gamma^2 + 4c^2(\kappa - \kappa_0)^2} \qquad \qquad 4(18)$$

and A' is a constant.

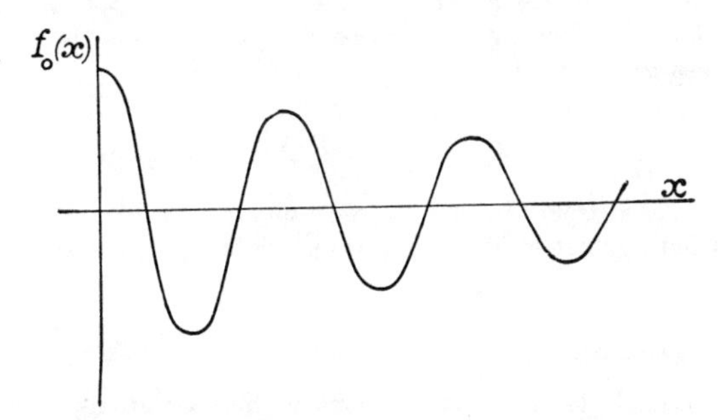

Fig. 4.10.—Profile of a damped wave [the real part of $f_0(x)$]

The half-width of the line (see § 4.13) is given by

$$\Delta\kappa = \frac{\gamma}{2c} = \frac{1}{2c\tau}. \qquad \qquad 4(19)$$

It is inversely proportional to τ. The value of τ may be calculated from Dirac's theory. The values obtained differ a great deal according to the emitting state. For the type of line we have considered τ is of order 10^{-8} second. It is seldom much less than this, though it is sometimes much greater. Since about 10^{15} waves are emitted per second the damping is very small. The wave trains are at least 10^7 waves long. The half-widths of the lines may be calculated when γ is known and are found to be of order 0·0005 Å. or less. Natural damping can therefore account for only a small part of the observed widths.

(*b*) *Doppler effect.*

Owing to thermal motion, the emitting atoms are moving in different directions relative to the measuring apparatus. Even if the light emitted by stationary atoms were perfectly homogeneous, the light observed would not be homogeneous. The observed frequency (ν) for light received from an atom which is approaching the observer with velocity v is given by

$$\nu - \nu_0 = \frac{v}{c}\,\nu_0. \qquad \qquad 4(20)$$

* See Appendix IV B, p. 111.

where ν_0 is the frequency for a stationary atom [see equation 2(57)], or

$$\kappa - \kappa_0 = \frac{v}{c}\,\kappa_0. \qquad \qquad 4(21)$$

It may be shown that the number of atoms whose component of velocity in a given direction lies between v and $(v + dv)$ is proportional to

$$\exp\left(-\frac{mv^2}{2kT}\right) dv, \qquad \qquad 4(22)$$

where m is the mass of the atom, T is the temperature, and k is Boltzmann's constant.* Hence the energy distribution due to this cause is a Gaussian distribution whose parameter is

$$2\alpha = \frac{mc^2}{2kT\kappa_0^{\,2}}. \qquad \qquad 4(23)$$

The value calculated from the thermal constants gives a half-width of 0·0038 Å. for cadmium red light at room temperature. Note that 2α is the parameter for the energy.

(*c*) *Pressure broadening.*

It is not difficult to see that collisions with other atoms may disturb the vibrations of an emitting atom so as to cause it to emit a wave of irregular profile (i.e. an imperfect simple harmonic wave). The complete theory is very complicated, especially when the two atoms are similar, so that a certain type of resonance between them occurs during the interaction. Detailed calculation shows that the distribution of energy is given by an expression somewhat similar to 4(18), but with a value of γ proportional to the pressure. The value varies a good deal from one gas to another. Ten atmospheres' pressure of argon increases the half-width of the mercury line (2537 Å.) by 0·12 Å. Collisions also cause a certain asymmetry in the energy distribution and the centre is moved usually in the direction of shorter wavelengths.†

4.26.—From the above it may be seen that, at low pressures, the most important cause of inhomogeneity is the Doppler effect. Michelson's results give a half-width only a little larger than that calculated from the Doppler effect. Also the variation of visibility is consistent with a Gaussian distribution which would be given by the Doppler effect, but too much importance should not be attached to this last point since the measurement of visibility is not sufficiently accurate to detect small departures from a Gaussian distribution. The width due to natural damping can be detected only when special methods have been used to eliminate the Doppler broadening.‡

* See Chapter III, § 11 of Reference 4.4. † See Reference 4.5.
‡ Reference 4.7.

4.27.—It is interesting to compare the distribution given by 4(18) with the Gaussian distribution given by 4(13). When κ is nearly equal to κ_0 they both reduce to the form

$$a(\kappa) = A - B(\kappa - \kappa_0)^2, \quad . \quad . \quad . \quad . \quad . \quad . \quad . \quad 4(24)$$

where A and B are constants; i.e. for any given value of α, it is possible to choose a value of γ which gives the same distribution of energy for wavelengths near to the wavelength of maximum energy. Thus measurements on the distribution of energy near the centre of the line cannot directly distinguish between a curve represented by 4(18) and one represented by 4(13). In regions remote from the centre of the line the energy given by 4(13) is very much less than that given by 4(18). Thus, if a line has a half-width of 0·005 Å., the energy at a distance of 0·02 Å. from the centre of the line should be $1{\cdot}5 \times 10^{-5}$ of the maximum energy according to 4(13), and 6×10^{-2} of the maximum energy according to 4(18) (see fig. 4.11). Moderately accurate measurements on the edges of a line, combined with measurements near the centre, distinguish between a distribution due to Doppler effect and one due to natural damping or collision broadening. Certain observations on the shape of absorption lines in the solar spectrum show that the shape of the line near the centre is mainly due to Doppler effect, but that in the " wings " of the line the distribution of energy is determined by natural damping.*

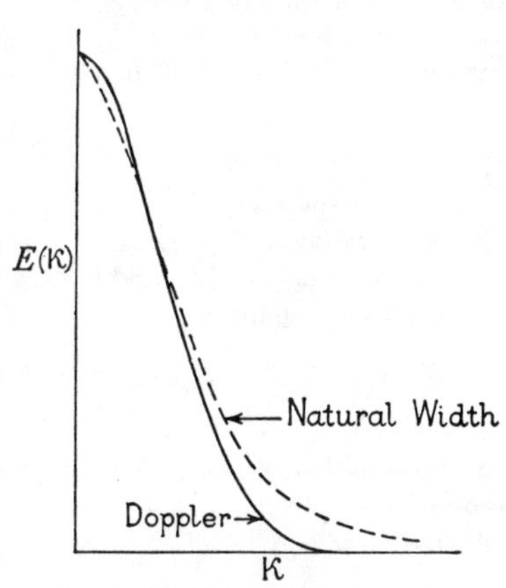

Fig. 4.11.—Energy distribution for a spectrum line

4.28. **Propagation of a Wave Group in a Dispersive Medium.**

In a dispersive medium, the components of a wave group move with different speeds, and the phase relations between the components are altered. We shall now show that, provided the group includes only a fairly narrow range of frequencies, the alteration of the profile owing to dispersion is fairly slow. Over considerable regions the group is propagated as a whole so that the positions of well-marked features of the profile, such as the point of maximum amplitude, remain well defined. Under these conditions the group has a definite velocity which is called the *group velocity*. This is not, in general, equal to the wave velocity.

* Reference 4.7.

Let us first consider the Gaussian wave group. The displacement at time t is obtained by inserting into 4(10) the expression for $a(\kappa)$ which is given in 4(13). The result is

$$\xi = \int_{-\infty}^{\infty} A' \exp\{-\alpha(\kappa-\kappa_0)^2 + i(\omega t - \kappa x)\}d\kappa. \quad . \quad . \quad 4(25)$$

In integrating this equation, we need to remember that ω is in general a function of κ, and not usually a very simple function. We assume that over the range of wavelengths included in the group, the relation between ω and κ is given by taking the first three terms in a Taylor series, i.e. we put

$$\omega = \omega_0 + U(\kappa - \kappa_0) + W(\kappa - \kappa_0)^2, \quad . \quad . \quad . \quad . \quad 4(26)$$

where ω_0 is the value of ω when $\kappa = \kappa_0$, and U and W are the corresponding values of $d\omega/d\kappa$ and $\frac{1}{2}d^2\omega/d\kappa^2$. It is shown in Appendix IV B that, when this expression for ω is substituted in 4(25) and the integration carried out, ξ is given [see 4(104) and 4(105)] by

$$\xi = A'' \exp\left\{-\frac{(x-Ut)^2}{4\alpha'} + i\phi'\right\}. \quad . \quad . \quad . \quad . \quad 4(27)$$

In this equation ϕ' is a rather complicated function of x and t which represents the fact that the phase is now varying in a complicated way with x and t. The value of α' is

$$(\alpha^2 + W^2t^2)/\alpha. \quad . \quad . \quad . \quad . \quad . \quad . \quad . \quad . \quad 4(28)$$

Thus the boundary curves of the profile (shown by the dotted line in fig. 4.8*b*) is still a Gaussian curve, but the width has been increased in the ratio given by 4(28). When the group is confined to a narrow range of frequencies, α is large and if Wt is small compared with α the group spreads only very slowly. Yet it does spread and, given sufficient time, α' increases without limit. From 4(27) it may be seen that the group as a whole travels forward with velocity $U = d\omega/d\kappa$.

4.29. Group Velocity.

The way in which a group spreads as it advances depends on the form of the group. The velocity with which the group moves, so long as it retains its form, will now be shown to be independent of the distribution of energy between the different components, provided that the group is confined to a fairly narrow range of frequencies. Consider first two components whose wavelength constants are $\kappa - \frac{1}{2}d\kappa$ and $\kappa + \frac{1}{2}d\kappa$ and whose amplitudes are equal. Then

$$\begin{aligned}\xi &= a_1 \sin[(\omega - \tfrac{1}{2}d\omega)t - (\kappa - \tfrac{1}{2}d\kappa)x] \\ &\quad + a_1 \sin[(\omega + \tfrac{1}{2}d\omega)t - (\kappa + \tfrac{1}{2}d\kappa)x] \\ &= 2a_1 \cos\tfrac{1}{2}(t\,d\omega - x\,d\kappa)\sin(\omega t - \kappa x). \quad . \quad . \quad 4(29)\end{aligned}$$

This is represented in fig. 4.12. The vibration is enclosed between the "beat wave" curves shown as dotted lines. This beat wave is represented by the cosine factor in 4(29) and travels with a speed

$$U = \frac{d\omega}{d\kappa}. \qquad \qquad 4(30)$$

This is the same as the speed obtained for the movement of the outer curve which encloses the profile of the Gaussian group.

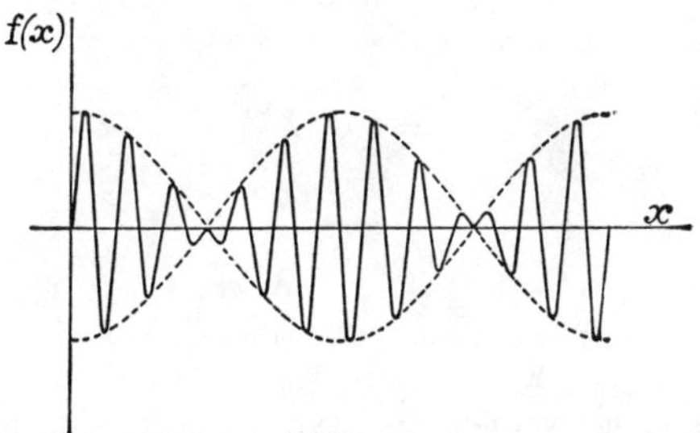

Fig. 4.12.—Simple beat wave

4.30.—Generally the phase of any component of a complex group is given by

$$\phi = \omega t - \kappa x + \delta, \qquad \qquad 4(31)$$

where δ is a constant.

ϕ varies with x and t, as well as with ω and κ. If we consider a variation of ϕ with x and t (κ and ω constant), we obtain

$$d\phi_1 = \omega\, dt - \kappa\, dx; \qquad \qquad 4(32)$$

whereas, if we consider ϕ to vary with ω and κ (x and t constant), we have

$$d\phi_2 = t\, d\omega - x\, d\kappa. \qquad \qquad 4(33)$$

The points for which $d\phi_1 = 0$ move with velocity ω/κ. This is the velocity of a component of one frequency, i.e. it is the phase velocity (b). The points for which $d\phi_2 = 0$ move with velocity U. This is the velocity of the points where the phases of different members of the group are in the same phase relation. In particular, it is the velocity of the point of maximum agreement of phase.

4.31.—The expression 4(30) for U may be put in various forms, most of which will be required in later applications. Since $\omega = 2\pi\nu$ and $\kappa = 2\pi/\lambda$, we have

$$U = -\lambda^2 \frac{d\nu}{d\lambda}; \qquad \text{4(34)}$$

and since $\omega = b\kappa$, we may write

$$U = \frac{d(b\kappa)}{d\kappa} \qquad \text{4(35)}$$

$$= b + \kappa \frac{db}{d\kappa} \qquad \text{4(36)}$$

$$= b - \lambda \frac{db}{d\lambda}. \qquad \text{4(37)}$$

Since $n = c/b$, we have

$$U = \frac{c}{n}\left(1 + \frac{\lambda}{n}\frac{dn}{d\lambda}\right). \qquad \text{4(38)}$$

It is sometimes convenient to have an expression for $1/U$, and we put

$$\frac{1}{U} = \frac{1}{c}\frac{d(n\omega)}{d\omega} \qquad \text{4(39)}$$

$$= \frac{n}{c} + \frac{\omega}{c}\frac{dn}{d\omega},$$

i.e.

$$\frac{1}{U} = \frac{1}{b} + \frac{\omega}{c}\frac{dn}{d\omega}. \qquad \text{4(40)}$$

From 4(37) it may be seen that in a non-dispersive medium $U = b$. In a dispersive medium U may be greater or less than b. Note that in equations 4(29)–4(40) the symbols λ and κ represent the wavelength and the wavelength constant in the medium of index n [see Example 4(x), p. 102].

4.32. **Representation of Light by Wave Groups.**

In experiments on light, we never deal with infinitely long wave trains or with light of one wavelength. It is possible to give a general account of the results of many experiments by means of a picture in which sharp spectral lines are represented by simple harmonic waves of one wavelength and white light is represented by a combination of such waves. The concept of a simple harmonic wave is valuable because it enables us to express many of our data in a convenient way, but it is usually necessary to represent "monochromatic" light by a wave group in order to give a satisfactory account of the details of any experiment. For example, in Michelson's experiments, the concept of pure sine waves is sufficient to account for the presence of the fringes and to explain the relation between the diameters of the different rings. It does not, however, give any account of the variation of the visibility when the path difference is altered.

4.33. **White Light.**

Towards the end of the last century there was a considerable controversy concerning the nature of white light. One side held that white light "really consisted" of superimposed trains of pure sine waves and the other that it "really consisted" of irregular pulses from which it was possible to create trains of waves by suitable experimental arrangements. Thus the one side held that Newton's famous experiment showed that white light "was composed of" various colours; on the other view, the colours were produced by the prism. Various attempts were made to discover a crucial experiment which would show which view was correct. These all failed since, whenever an experiment was "explained" on one view, the protagonists of the other view were always able to produce an equally good "explanation". Sometimes one explanation was more complicated than the other, but the more complicated one was equally logical.

4.34.—Rayleigh and Schuster showed that the two ways of representing light are equivalent. By means of Fourier's theorem, a pulse may be analysed (mathematically) into a series of simple harmonic waves. Therefore, if white light is adequately represented by pulses, it is equally well represented by a series of sine waves. Thus there are not two theories but only two different ways of stating the same theory. The difference is one of mathematical form. Consequently, there is no possibility of a crucial experiment in which one side would be proved to be right and the other wrong.

In considering the representation of white light by a series of simple harmonic waves, it is necessary to remember that our experimental arrangements are *never* able to produce infinitely long wave trains of one frequency. In practice, when we seek to analyse light, we always analyse it into wave groups—not into infinitely long wave trains. With suitable experimental arrangements we can produce wave trains which are over a million waves in length, but they are not *infinitely* long, i.e. the light is never of *exactly* one frequency.

As an example of the sort of difficulty that arose through neglecting the group concept we may take the paradox suggested by Carvallo. He said that if white light " consisted of " a combination of infinitely long wave trains, these wave trains would be separated by a spectroscope, so that it would be possible to see the spectrum both before the source was lighted and after it had gone out. His mistake lay in the assumption that the spectroscope selected light of exactly one wavelength. If we put a fine slit in a spectrum we select a group of waves covering a small but finite range of frequency. Such a group may be regarded as made up of an infinite series of pure harmonic waves. These interact in such a way that the total amount of energy passing through the slit before the source is lighted, and shortly after it has gone out, is zero. Thus the " analytic " view is quite correct provided that we remember that practical analysis produces wave groups, not pure sine waves.

We shall not give a detailed account of the various arguments put forward during this controversy, though in §§ 8.32–8.36 it is shown how the dispersion of white light (*a*) by a prism, and (*b*) by a grating, can be calculated on the two views. It will be understood that these two calculations are, from our point of view, simply different mathematical devices. Although the controversy concerning the " nature of white light " is now chiefly of historical interest, it was important in that it called attention to the necessity for representing actual beams of light by wave groups rather than by infinite wave trains. We shall see later that this concept of wave groups plays an important part in the modern quantum theory of radiation.

EXAMPLES [4(i)–4(ix)]

4(i). Show that when e is large compared with λ, the angular diameters of the rings near the centre (see § 4.9) are given by

$$2\theta_{p_1} = \left[\frac{4(p_0 - p_1)\lambda}{e}\right]^{\frac{1}{2}}.$$

[Put $\cos\theta_{p_1} = 1 - \frac{1}{2}\theta_{p_1}{}^2$.]

4(ii). Show that $f(x) = x^2$ may be represented in the range $-x_0$ to $+x_0$ by a series in which

$$a_0 = \tfrac{1}{3}x_0{}^2; \quad a_m = \frac{(-1)^m 4x_0{}^2}{\pi^2 m^2}; \quad b_m = 0.$$

Fig. 4.13 shows the sum of 2, 4, and 8 terms of the series compared with the function.

4(iii). Show that the " top-hat curve " (fig. 4.6) is represented by

$$f(x) = \frac{4A}{\pi} \sum_{m=1}^{\infty} (-1)^{m-1} \frac{\cos(2m-1)\pi x/x_0}{2m-1}$$

for all values of x.

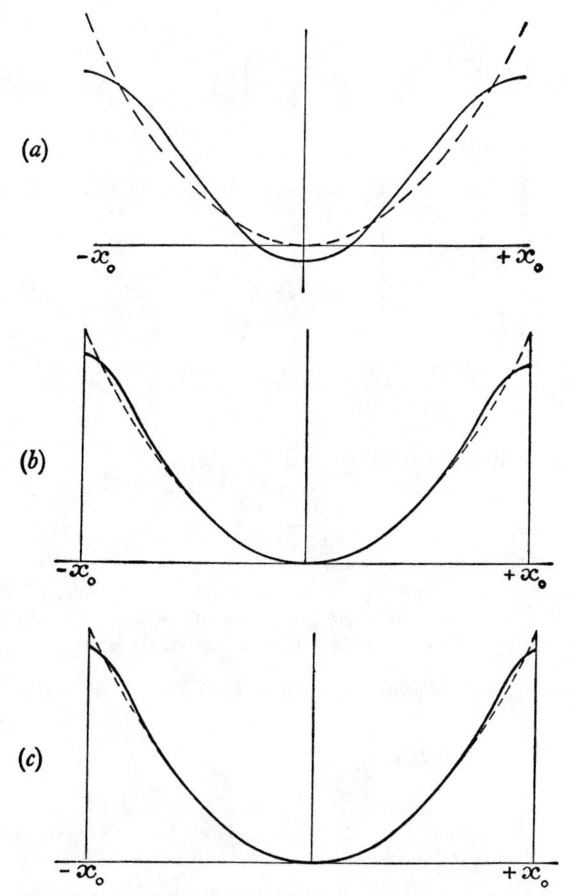

Fig. 4.13.—Representation of a function by a Fourier series: (*a*) sum of 2 terms of the series, (*b*) sum of 4 terms, (*c*) sum of 8 terms. In (*a*), (*b*), and (*c*) the dotted line shows the original function $f(x) = x^2$.

4(iv). Show that if $f(x) = x$ from $-x_1$ (fig. 4.14) to $+x_1$ and $f(x) = 0$ when $x^2 > x_1^2$, the function may be represented *in the range* $-x_0$ *to* x_0 (where $x_0 < x_1$) if

$$f_{0s}(x) = 2\frac{x_0}{\pi} \sum_{m=1}^{\infty} (-1)^{m+1} \frac{\sin(m\pi x/x_0)}{m}$$

Show that the series defined by

$$b_m = -\frac{2}{\pi^2}\sum_{m=1}^{\infty}\left\{\frac{\pi x_1}{m}\cos\frac{m\pi x_1}{x_2} - \frac{x_2}{m^2}\sin\frac{m\pi x_1}{x_2}\right\}$$

would represent the function from $-x_2$ to $+x_2$ if $2x_1 > x_2 > x_1$.

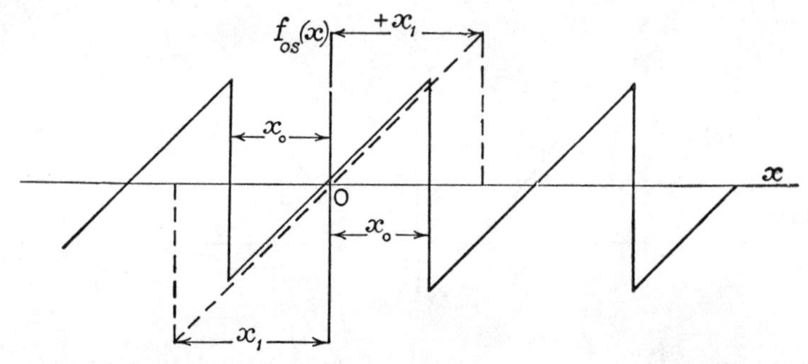

Fig. 4.14.—Representation of a function by a Fourier series: (*a*) dotted line shows the function to be represented, (*b*) full line shows the sum of the series $f_{0s}(x)$. (See Example 4(iv).)

4(v). For waves on the surface of a liquid* of depth h

$$b^2 = \frac{g}{\kappa}\frac{e^{\kappa h} - e^{-\kappa h}}{e^{\kappa h} + e^{-\kappa h}}.$$

Show that when h is small the group velocity is nearly equal to the wave velocity, and that when h is large the group velocity is approximately half the wave velocity corresponding to the central frequency of the group.

4(vi). Show that if the relation between n and κ is given by $n^2 = A' + B'\kappa^2 + C'\kappa^4$, then

$$U = \frac{A' - C'\kappa^4}{A' + B'\kappa^2 + C'\kappa^4}\,b.$$

4(vii). Find the group velocity when (*a*) $db/d\lambda = b/\lambda$, and (*b*) when $b = A + B\lambda$ (with A and B constant). [(*a*) $U = 0$ and (*b*) $U = A$.]

4(viii). Show that in a medium which has normal dispersion (§ 3.18) U is less than b.

4(ix). Show that, if $\lambda_v = 2\pi c/\omega$ is the wavelength in vacuum, then

$$\frac{1}{U} = \frac{1}{b} - \frac{\lambda_v}{c}\frac{dn}{d\lambda_v}. \qquad 4(41)$$

* See Chapter V, equation (32) of Reference 4.3.

REFERENCES

4.1. MICHELSON: *Light Waves and their Uses* (University of Chicago Press).

4.2. CARSLAW: *Introduction to the Theory of Fourier Series and Fourier Integrals* (Macmillan).

4.3. COULSON: *Waves* (Oliver and Boyd).

4.4. ROBERTS: *Heat and Thermodynamics* (Blackie).

4.5. *Reviews of Modern Physics*: Vol. 8, pp. 22 and 398.

4.6. TOLANSKY: *High Resolution Spectroscopy* (Oxford University Press).

4.7. HEITLER: *Quantum Theory of Radiation* (Oxford University Press).

4.8. CAMPBELL and FOSTER: *Fourier Integrals for Practical Application* (Bell System Technical Publication Monograph B–584, 1931).

APPENDIX IV A

ADJUSTMENT OF THE MICHELSON INTERFEROMETER

(1) Set the movable mirror M_2 (fig. 4.1) so that its distance from the half-silvered side of M_1 is about equal to that of M_3. This brings M_2 near to the reference plane R. At this stage, M_2 and R will not necessarily be parallel and the centre of M_2 may be up to two millimetres from the centre of R.

(ii) Taking the interferometer alone (i.e. without S_e, F, L and the telescope shown in fig. 4.1), place a small bright source, such as a pea-lamp, in the position indicated for S_e. Look into the instrument in a direction roughly normal to M_2 with the eye in the position indicated by E. Several images of the bright source may be seen. If the silvering of M_1 is in good condition, two images will be much brighter than the others. These are formed by rays which have followed the paths shown in fig. 4.1, the others being due to internal reflections in M_1. The two strong images should be brought into coincidence by adjusting the screws at the back of M_3 (and if necessary M_2). M_3 and R are now parallel within about a minute of arc.

(iii) Replace the pea-lamp by the source S_e and F (conveniently a low-pressure mercury source with a green filter). The lens L must be placed so as to allow an approximately parallel beam to fall on M_1 in a direction approximately normal to M_3. With the eye placed at E, fringes should be seen crossing the field of view. The screws at the back of M_3 are adjusted so as to broaden the fringes. At a certain stage the fringes are seen to be circular but with the centre of the system outside the field of view. For the final adjustment of the Hilger instrument two special screws, which alter the angular setting of M_3, are provided. The centre of the fringes is now brought into the field of view. M_2 is now traversed a little until the diameter of the fringes does not change when the eye is moved from side to side. In the final adjustment, it may be necessary to traverse M_2 a little, and make small alterations in the tilt of M_3, successively.

(iv) If white-light fringes are desired, a lamp is placed at the position S_e leaving L in position. M_2 is then traversed slowly until it moves into coincidence with the reference plane R. A movement of not more than about one tenth of a millimetre should be sufficient.

Note.—If a sodium lamp is used instead of a mercury source, the fringes have good visibility when the path difference is 0, 1000λ, 2000λ, etc., and very poor visibility at half-way between these positions (see fig. 4.2*b*). If fringes are not seen in the early stages of (iii) above, the mirror M_2 should be traversed a little.

APPENDIX IV B

Fourier's Series and Fourier's Integral Theorem

1.—The complete mathematical investigation of the validity of the methods by which certain limits are derived is difficult and very lengthy. These limit processes are essential to any formal logical proof of the Fourier integral theorem, and their investigation is needed for a complete understanding of the Fourier series. A reader who is interested in the mathematical side should consult one of the standard mathematical texts.* Other readers may be willing to accept the general result of this investigation, namely that the Fourier expansions, in the forms in which we shall derive them, are applicable to a wide range of mathematical functions, including all the functions in which we are interested. The derivation given below shows that the theorem is at least plausible, and should help the reader to understand and to apply the Fourier methods. It does not claim to be a complete proof.

2.—Let us assume that, in the range $-\pi$ to π, a mathematical function $f(X)$ may be represented by a series of sines and cosines so that

$$f(X) = a_0 + \sum_{n=1}^{\infty} a_n \cos nX + \sum_{n=1}^{\infty} b_n \sin nX. \quad . \quad . \quad . \quad 4(42)$$

On integrating each side, we find that

$$a_0 = \frac{1}{2\pi}\int_{-\pi}^{\pi} f(X)\,dX. \quad . \quad . \quad . \quad . \quad . \quad . \quad 4(43)$$

Multiplying each side by $\cos nX$ and integrating, we obtain

$$a_n = \frac{1}{\pi}\int_{-\pi}^{\pi} f(X)\cos nX\,dX, \quad . \quad . \quad . \quad . \quad . \quad 4(44)$$

since $\int_{-\pi}^{\pi} \cos nX \cos mX\,dX = 0$ when $m \neq n$, and $= \pi$ when $m = n$;

and $\int_{-\pi}^{\pi} \cos nX \sin nX\,dX = 0$ for all values of m and n.

* See References 4.2 and 4.8.

Similarly $$b_n = \frac{1}{\pi}\int_{-\pi}^{\pi} f(X) \sin nX \, dX. \quad \quad 4(45)$$

The expansion may also be written

$$f(X) = A_0 + \sum_{n=1}^{\infty} A_n \sin (nX + \delta_n), \quad \quad 4(46)$$

where $$A_0 = a_0; \; A_n \sin \delta_n = a_n; \; A_n \cos \delta_n = b_n. \quad \quad 4(47)$$

In a similar way, the function may be represented by a series consisting of a constant term plus a series of cosines. Essentially the analysis is in terms of a constant plus a number of simple harmonic waves, each of which is represented by two terms in 4(42) and by one term in 4(46). Two constants, a_n and b_n in 4(42), and A_n and δ_n in 4(46), are associated with each harmonic wave. The relative energy associated with the nth harmonic is $a_n^2 + b_n^2$ if we use the representation of 4(42), and A_n^2 if we use that of 4(46).

3.—If we make the substitutions

$$2 \cos nX = (e^{inX} + e^{-inX}) \quad \quad 4(48a)$$

and $$2 \sin nX = i(e^{-inX} - e^{inX}), \quad \quad 4(48b)$$

we may express the series in the form

$$f(X) = \sum_{n=-\infty}^{\infty} c_n e^{inX}, \quad \quad 4(49a)$$

where $$\left.\begin{aligned} 2c_n &= a_n - ib_n, \\ 2c_{-n} &= a_n + ib_n, \end{aligned}\right\} \quad \quad 4(49b)$$

except that $$c_0 = a_0.$$

These relations imply that

$$c_n = c_{-n}^{*} \quad \quad 4(50)$$

and $$c_n = \frac{1}{2\pi}\int_{-\pi}^{\pi} f(X)\, e^{-inX} \, dX. \quad \quad 4(51)$$

Equation 4(51) is valid when n is zero or negative, as well as when it is positive.

4.—So far we have considered $f(X)$ simply as a mathematical function. If we now assume that it is a wave profile, we know that it is always a real function and, while we can attach a meaning to the nth harmonic when n is positive, we give no physical meaning to a negative harmonic. In § 2.26 we stated a convention, according to which a wave motion is represented by a complex quantity, with the understanding that the real part of the complex gives the actual physical displacement. When we apply this convention to the interpretation of 4(49), we see that every negative term in the expansion must be taken with the corresponding positive term. By virtue of 4(50) the two together give a real number which represents one pair of terms in 4(42) or one term in 4(46). This term (or pair of terms) refers to " the nth harmonic "—with n always positive. The relative energy associated with the nth harmonic is

$$A_n{}^2 = a_n{}^2 + b_n{}^2 = 4c_nc_n{}^* = 4c_{-n}c_{-n}{}^*. \quad . \quad . \quad . \quad . \quad 4(52)$$

5.—We may extend the range over which a given type of function is represented by a Fourier series in the following way. Let $x = lX/\pi$,

then
$$f(x) = \sum_{n=-\infty}^{n=\infty} c_n \exp\left(\frac{in\pi x}{l}\right), \quad . \quad . \quad . \quad 4(53)$$

where
$$c_n = \frac{1}{2l}\int_{-l}^{+l} f(x') \exp\left(-\frac{in\pi x'}{l}\right) dx'. \quad . \quad . \quad . \quad 4(54)$$

Hence
$$f(x) = \sum_{n=-\infty}^{\infty} \frac{1}{2l}\int_{-l}^{+l} f(x') \exp\frac{in\pi}{l}(x - x')\, dx'. \quad . \quad . \quad 4(55)$$

We write x' instead of x in 4(54) in order to distinguish the variable to be integrated in 4(55). The expression on the right-hand side of 4(55) represents the function on the left-hand side between the limits $-l$ and $+l$. Since l may be made as large as we please, the representation is available for any finite limits. Subject to the necessary investigation of limit processes we may extend the limits to $\pm\infty$. Consider the nth harmonic of the series, for which the wavelength $\lambda_n = 2l/n$. The corresponding wavelength constant $\kappa_n = 2\pi/\lambda_n = n\pi/l$. Similarly for the $(n + 1)$th harmonic, we have

$$\kappa_{n+1} = \frac{(n + 1)\pi}{l}.$$

Hence
$$\kappa_{n+1} - \kappa_n = \pi/l = \Delta\kappa. \quad . \quad . \quad . \quad . \quad . \quad . \quad 4(56)$$

Equation 4(55) may therefore be written

$$f(x) = \frac{1}{2\pi}\sum_{n=-\infty}^{\infty} \Delta\kappa \int_{-l}^{+l} f(x') \exp i\{n\,\Delta\kappa(x - x')\}\, dx'. \quad . \quad 4(57)$$

It is usual to approach the concept of a definite integral by considering the sum of the areas of narrow strips. The integral is the limiting value of the area when the strips are made indefinitely narrow. Thus we have

$$\int_{-\kappa_1}^{\kappa_1} G(\kappa)\,d\kappa = \lim_{\Delta\kappa\to 0}\sum_{n=-N}^{N} G(n\,\Delta\kappa)\,\Delta\kappa, \quad . \quad . \quad . \quad 4(58)$$

where
$$\lim_{\Delta\kappa\to 0}(n\,\Delta\kappa) = \kappa,$$

as the definition of the integral. Thus we have

$$f(x) = \frac{1}{2\pi}\int_{-\infty}^{+\infty} d\kappa \int_{-\infty}^{+\infty} f(x') \exp i\kappa(x - x')\, dx'. \quad . \quad . \quad 4(59)$$

If
$$g_0(\kappa) = \frac{1}{\sqrt{2\pi}}\int_{-\infty}^{+\infty} f(x)e^{-i\kappa x}\, dx, \quad . \quad . \quad . \quad . \quad 4(60)$$

then
$$f(x) = \frac{1}{\sqrt{2\pi}}\int_{-\infty}^{+\infty} g_0(\kappa)e^{+i\kappa x}\, d\kappa. \quad . \quad . \quad . \quad . \quad 4(61)$$

The accent on x is omitted in 4(60) since it is no longer needed to distinguish one part of the expression. By returning to 4(53) and replacing c_n by c_{-n} and $\exp(in\pi x/l)$ by $\exp(-in\pi x/l)$, we could derive the relations

$$g_1(\kappa) = \frac{1}{\sqrt{2\pi}} \int_{-\infty}^{+\infty} f(x) e^{+i\kappa x}\, dx, \qquad 4(62)$$

and

$$f(x) = \frac{1}{\sqrt{2\pi}} \int_{-\infty}^{+\infty} g_1(\kappa) e^{-i\kappa x}\, d\kappa. \qquad 4(63)$$

When a pair of functions is connected either by 4(60) and 4(61) or by 4(62) and 4(63), each is said to be the "Fourier transform" of the other. Note that in one relation of the pair there is always a *positive* exponent and in the other a *negative* exponent.

6.—Equation 4(63) may be written

$$f(x) = \frac{1}{\sqrt{2\pi}} \int_0^\infty g_1(\kappa) e^{-i\kappa x}\, d\kappa + \frac{1}{\sqrt{2\pi}} \int_0^\infty g_1(-\kappa) e^{+i\kappa x}\, d\kappa \qquad 4(64)$$

$$= \frac{1}{\sqrt{2\pi}} \int_0^\infty \{g_1(\kappa) + g_1(-\kappa)\} \cos \kappa x\, d\kappa - \frac{i}{\sqrt{2\pi}} \int_0^\infty \{g_1(\kappa) - g_1(-\kappa)\} \sin \kappa x\, d\kappa, \qquad 4(65)$$

and similarly we obtain

$$g_1(\kappa) = \frac{1}{\sqrt{2\pi}} \int_0^\infty \{f(x) + f(-x)\} \cos \kappa x\, dx + \frac{i}{\sqrt{2\pi}} \int_0^\infty \{f(x) - f(-x)\} \sin \kappa x\, dx, \qquad 4(66)$$

and

$$g_1(-\kappa) = \frac{1}{\sqrt{2\pi}} \int_0^\infty \{f(x) + f(-x)\} \cos \kappa x\, dx - \frac{i}{\sqrt{2\pi}} \int_0^\infty \{f(x) - f(-x)\} \sin \kappa x\, dx. \qquad 4(67)$$

Hence

(i) If $f(x)$ is real for all values of x,

$$g_1(\kappa) = g_1^*(-\kappa). \qquad 4(68)$$

(ii) If $f(x)$ is real everywhere and is an even function [i.e. $f(x) = +f(-x)$], we have also

$$g_1(\kappa) = g_1(-\kappa) \qquad 4(69)$$

and $g_1(\kappa)$ is real.

Then 4(65) and 4(66) become

$$f(x) = \sqrt{\frac{2}{\pi}} \int_0^\infty g_1(\kappa) \cos \kappa x\, d\kappa \qquad 4(70)$$

and

$$g_1(\kappa) = \sqrt{\frac{2}{\pi}} \int_0^\infty f(x) \cos \kappa x\, dx. \qquad 4(71)$$

(iii) If $f(x)$ is real everywhere and is odd [$f(x) = -f(-x)$],

$$g_1(\kappa) = -g_1(-\kappa) \qquad 4(72)$$

and $g_1(\kappa)$ is a pure imaginary.

Equations 4(65) and 4(66) then become

$$f(x) = -i\sqrt{\frac{2}{\pi}}\int_0^\infty g_1(\kappa)\sin\kappa x\,d\kappa \quad . \quad . \quad . \quad . \quad 4(73)$$

and

$$g_1(\kappa) = i\sqrt{\frac{2}{\pi}}\int_0^\infty f(x)\sin\kappa x\,dx. \quad . \quad . \quad . \quad . \quad . \quad 4(74)$$

7.—In one kind of optical measurement we place a receptor in a system of interference fringes and record its reading at various points. From these readings we deduce a wave profile, i.e. we calculate a wave profile which would lead to the observed distribution of energy in the fringe system. A typical experiment of this kind is the measurement of the visibility of fringes obtained with the Michelson interferometer. In a second kind of experiment, we place a receptor in the focal plane of a spectrograph and observe the variation of reading as it is moved along the spectrum. We usually interpret the result in terms of a distribution of energy with wavelength (or wavelength constant). The purpose of many of the applications of Fourier analysis to optics is to correlate the results of these two kinds of experiment. We may be given a profile, i.e. a space function $f(x)$, and we wish to calculate a relative-energy function $E(\kappa)$ such that the relative energy corresponding to the range $(\kappa - \frac{1}{2}d\kappa)$ to $(\kappa + \frac{1}{2}d\kappa)$ is $E(\kappa)\,d\kappa$. Alternatively, we may be given $E(\kappa)$ and wish to calculate $f(x)$. Directly, the analysis deals not with $E(\kappa)$ but with $g(\kappa)$, and we must now consider the relation between $g(\kappa)$ and $E(\kappa)$. Let us first consider cases when $f(x)$ is real. When equation 4(62) is used we shall obtain a function $g(\kappa)$ which satisfies 4(68), but is not necessarily real, and which has values corresponding to negative values of κ. Physically we must regard $g(\kappa)$ and $g(-\kappa)$ as both belonging to the frequency range $(\kappa - \frac{1}{2}d\kappa)$ to $(\kappa + \frac{1}{2}d\kappa)$ just as, in 4(52), c_n and c_{-n} both belong to the nth harmonic. By analogy with 4(52) [since $g(\kappa)$ is analogous to $\sqrt{2\pi}\,.\,c_n$] we expect $E(\kappa)$ to be given by

$$E(\kappa)\,d\kappa = \frac{2}{\pi}\,g_1(\kappa)\,g_1{}^*(\kappa)\,d\kappa. \quad . \quad . \quad . \quad . \quad . \quad 4(75)$$

4(75) gives the energy associated with the frequency range $d\kappa$ and, in it, κ is an essentially positive quantity. Detailed examination of the process by which an expression for the energy based on 4(52) passes into the integral form justifies 4(75).

8.—Consider a wave profile represented by

$$f(x) = h(x)\cos\kappa_0 x, \quad . \quad . \quad . \quad . \quad . \quad . \quad 4(76)$$

where $h(x)$ is a real function of x.

We may put

$$f(x) = \tfrac{1}{2}h(x)e^{i\kappa_0 x} + \tfrac{1}{2}h(x)e^{-i\kappa_0 x}. \quad . \quad . \quad . \quad . \quad 4(77)$$

Then $\sqrt{2\pi}g_1(\kappa) = \frac{1}{2}\int_{-\infty}^{\infty} h(x)\exp i(\kappa+\kappa_0)x\,dx + \frac{1}{2}\int_{-\infty}^{\infty} h(x)\exp i(\kappa-\kappa_0)x\,dx$

$$g_1(\kappa) = \tfrac{1}{2}g_1{}'(\kappa+\kappa_0) + \tfrac{1}{2}g_1{}'(\kappa-\kappa_0), \quad . \quad . \quad . \quad . \quad . \quad . \quad 4(78)$$

where

$$g_1{}'(\kappa) = \frac{1}{\sqrt{2\pi}}\int_{-\infty}^{\infty} h(x)e^{i\kappa x}\,dx, \quad . \quad . \quad . \quad . \quad . \quad 4(79)$$

i.e. $g_1'(\kappa)$ is the Fourier transform of the function $h(x)$ itself. Many physical problems involve functions of the type on the right-hand side of 4(77). It is often convenient to proceed by first finding $g_1'(\kappa)$, using 4(79), and then $g_1(\kappa)$ from 4(78). When this function is inserted in 4(75) we obtain the relative energy. It often happens that the first term on the right-hand side of 4(78) is numerically small compared with the second term and may be ignored in the final calculation.

*9.—The following procedure is sometimes used but is not correct in principle, though it does lead to the correct result in some cases of practical importance. According to the convention discussed in § 2.26, the right-hand side of 4(76) is replaced by $h(x)e^{-i\kappa_0 x}$, with the understanding that the real part of the complex function represents the physical displacement. Equation 4(62) is then used to obtain

$$g_1''(\kappa) = \frac{1}{\sqrt{2\pi}}\int_{-\infty}^{\infty} h(x) \exp i(\kappa - \kappa_0)x \, dx$$

$$= g_1'(\kappa - \kappa_0), \qquad \text{4(80)}$$

i.e. we obtain twice the second term of 4(78). When, as explained in the preceding paragraph, this is the predominant term, the function $g_1''(\kappa)$ differs from $g_1(\kappa)$ only by the constant factor 2. If this function is inserted in 4(75) we obtain a sufficiently good approximation to the *relative* energy. It will, however, be understood that, if we had taken $f(x)$ to be the real part of $h(x)e^{+i\kappa_0 x}$, we should have obtained the *first* term on the right-hand side of 4(78), i.e. the term which is usually negligibly small. The procedure described in this paragraph involves multiplying the complex function $h(x)e^{-i\kappa_0 x}$ by a complex function $e^{i\kappa x}$, integrating, and then extracting a real function at the end. This procedure does not give the Fourier transform of the real part of $h(x)e^{i\kappa_0 x}$, and, although it can sometimes be manipulated to give correct results in an apparently simple way, the procedure of § 8 should always be used.

10.—Analysis of a sharply limited Wave Train.

We may first consider a particular case which is so simple that it can be solved first by elementary methods and then by the procedure described in § 8. Suppose that the profile, at $t = 0$, of a wave is represented by

$$f(x) = A \cos \kappa_0 x \quad \text{from } -x_1 \text{ to } +x_1, \qquad \text{4(81)}$$

and $f(x) = 0$ outside the above range (see fig. 4.4). Since $f(x)$ is real and even, we use 4(71) to obtain

$$g_1(\kappa) = \sqrt{\frac{2}{\pi}}\, A\int_0^{x_1} \cos \kappa_0 x \cos \kappa x \, dx \qquad \text{4(82)}$$

$$= \frac{A}{\sqrt{2\pi}}\int_0^{x_1} \{\cos(\kappa_0 + \kappa)x + \cos(\kappa_0 - \kappa)x\}\, dx,$$

i.e.

$$g_1(\kappa) = \frac{A}{\sqrt{2\pi}}\left\{\frac{\sin(\kappa_0 + \kappa)x_1}{\kappa_0 + \kappa} + \frac{\sin(\kappa_0 - \kappa)x_1}{\kappa_0 - \kappa}\right\}, \qquad \text{4(83)}$$

* This paragraph may be omitted on first reading.

and the energy distribution is given by substituting from 4(83) into 4(75). Only positive values of κ need be considered. Using the procedure described in § 8, we note that 4(81) is of the same form as 4(76), with $h(x) = A$ from $-x_1$ to x_1, and zero elsewhere. We have

$$g_1'(\kappa) = \frac{A}{\sqrt{2\pi}}\int_{-x_1}^{x_1} e^{i\kappa x}\,dx$$

$$= \sqrt{\frac{2}{\pi}}\,A\,\frac{\sin \kappa x_1}{\kappa}. \qquad \ldots \ldots \qquad 4(84)$$

Hence using 4(78) we obtain $g_1(\kappa)$, and the result is identical with 4(83). It is easy to verify that, if we had used the incorrect procedure described in § 9, we should have obtained one or other term in 4(83), but not both. Also there would have been an incorrect factor of 2.

11.—We are usually interested in cases where there is a fairly large number of waves in the train, i.e. $\kappa_0 x_1$ is large compared with 2π. We note that the second term in the bracket of 4(83) is equal to x_1 when $\kappa = \kappa_0$. Since κ may take only positive values, the first term never reaches a comparable value and may be neglected. We may put

$$g(\kappa) = \frac{A}{2}\sqrt{\frac{2}{\pi}}\,\frac{\sin(\kappa_0 - \kappa)x_1}{\kappa_0 - \kappa} = \frac{A}{2}\sqrt{\frac{2}{\pi}}\,\frac{\sin 2\Delta x_1}{2\Delta}, \qquad \ldots \qquad 4(85)$$

where $$2\Delta = \kappa_0 - \kappa.$$

The energy distribution is shown in fig. 6.9*a* (p. 177). A high proportion of the energy is included in the range $\kappa = \kappa_0 - \pi/x_1$ to $\kappa = \kappa_0 + \pi/x_1$, i.e. between the two minima on either side of the central maximum. The width of this range is inversely proportional to x_1 and we see that, for this wave form, a long wave-train implies that the energy is concentrated in a narrow range of κ.

12.—Profile for sharply limited Wave Band.

Let κ_0 and κ_1 be two fixed values of κ with $\kappa_1 > \kappa_0$.

Suppose $g(\kappa) = 1$ when $\kappa_0 < \kappa < \kappa_1$, and $g(\kappa) = 0$ outside this range. Then, using 4(61), we have

$$f(x) = \frac{1}{\sqrt{2\pi}}\int_{\kappa_0}^{\kappa_1} e^{i\kappa x}\,d\kappa$$

$$= \frac{1}{\sqrt{2\pi}}\cdot\frac{1}{ix}\left(e^{i\kappa_1 x} - e^{i\kappa_0 x}\right), \qquad \ldots \ldots \qquad 4(86)$$

and the real part is

$$f_r(x) = \frac{1}{\sqrt{2\pi}}\,\frac{\sin \kappa_1 x - \sin \kappa_0 x}{x}$$

$$= \sqrt{\frac{2}{\pi}}\,\frac{\sin \frac{1}{2}(\kappa_1 - \kappa_0)x \cos \frac{1}{2}(\kappa_1 + \kappa_0)x}{x} \qquad \ldots \qquad 4(87)$$

$$= \sqrt{\frac{2}{\pi}}\,\frac{\sin \Delta x \cos \kappa_m x}{x}, \qquad \ldots \ldots \ldots \qquad 4(88)$$

where $2\Delta = \kappa_1 - \kappa_0$ and $2\kappa_m = \kappa_1 + \kappa_0$. When Δ is small compared with κ_m, the first factor in the numerator of 4(88) varies only slowly, and we have a train of waves whose profile is bounded by the curve $\sqrt{\frac{2}{\pi}}\frac{\sin \Delta x}{x}$. The effective length of the wave train increases as Δ decreases, i.e. as the effective range of κ decreases.

13. Distribution of Energy for a Damped Harmonic Wave.

It is sometimes convenient to consider the disturbance as a function of t and the energy distribution as a function of ω. Mathematically the relation between $f(t)$ and $g(\omega)$ is similar to that between $f(x)$ and $g(\kappa)$. As an example we may consider the wave whose disturbance at $x = 0$ is given by

$$f(t) = 0 \text{ from } t = -\infty \text{ to } t = 0,$$

$$f(t) = Ae^{-\gamma t/2} \cos \omega_0 t \text{ from } t = 0 \text{ to } t = +\infty. \qquad 4(89)$$

Such a wave would be emitted by an oscillator which commenced to radiate at time $t = 0$ and was subject to logarithmic damping with constant γ. This is the same form as 4(76) with $h(t) = Ae^{-\gamma t/2}$ from 0 to ∞ and zero from $-\infty$ to 0. We have

$$g_1'(\omega) = \frac{A}{\sqrt{2\pi}}\int_0^\infty \exp\{-\gamma/2 + i\omega\}t\,dt \qquad 4(90)$$

$$= -\frac{A}{\sqrt{2\pi}}\frac{2}{(2i\omega - \gamma)}. \qquad 4(91)$$

Using 4(78) we obtain an expression with two terms, one containing $(\omega + \omega_0)$ in the denominator and the other containing $(\omega - \omega_0)$. Under most practical conditions the former is negligible and we write

$$g_1(\omega) = -\frac{A}{\sqrt{2\pi}}\frac{1}{2i(\omega - \omega_0) - \gamma}.$$

The relative energy is equal to

$$\frac{2}{\pi}g_1(\omega)g_1^*(\omega) = \frac{A^2}{\pi^2}\frac{1}{4(\omega - \omega_0)^2 + \gamma^2}. \qquad 4(92)$$

When the damping is small, the energy is concentrated in a narrow range of frequency round ω_0. Thus, again, a long wave train is associated with a narrow range of frequency. In this example the problem has been worked out in terms of t and ω; it could equally well have been worked in terms of x and κ. For some purposes it is desirable to introduce a constant A_0^2 equal to the total energy for all frequencies. We have, in this case,

$$A_0^2 = \frac{2}{\pi}\int_0^\infty g_1(\omega)g_1^*(\omega)\,d\omega = \int_0^\infty \frac{A^2}{\pi^2}\frac{d\omega}{4(\omega - \omega_0)^2 + \gamma^2}$$

$$= \frac{A^2}{2\pi^2\gamma}\left(\frac{\pi}{2} + \tan^{-1}\frac{2\omega_0}{\gamma}\right). \qquad 4(93)$$

If γ is small compared with ω_0, then $\tan^{-1}(2\omega_0/\gamma) = \frac{1}{2}\pi$ and $2\pi\gamma A_0^2 = A^2$. We then have

$$g_1(\omega)g_1(\omega)^* = \frac{\gamma A_0^2}{\gamma^2 + 4(\omega - \omega_0)^2}. \quad \quad 4(94)$$

14. The Gaussian Wave Group.

Consider the wave whose profile at time $t = 0$ is given by

$$f_0(x) = A\sqrt{\frac{\pi}{\alpha}}\, e^{-x^2/4\alpha} \cos \kappa_0 x. \quad \quad 4(95)$$

This again is of the form of 4(76) with

$$h(x) = A\sqrt{\frac{\pi}{\alpha}}\, e^{-x^2/4\alpha}.$$

Then

$$g_1'(\kappa) = \frac{A}{\sqrt{2\alpha}}\int_{-\infty}^{\infty} \exp\left[-\left(\frac{x^2}{4\alpha} - i\kappa x\right)\right] dx. \quad \quad 4(96)$$

Now

$$\int_{-\infty}^{\infty} \exp\,(au - bu^2)\, du = e^{a^2/4b}\int_{-\infty}^{\infty} \exp\left[-b\left(u - \frac{a}{2b}\right)^2\right] du$$

$$= e^{a^2/4b}\int_{-\infty}^{\infty} e^{-bv^2}\, dv = \sqrt{\frac{\pi}{b}}\, e^{a^2/4b}. \quad \quad 4(97)$$

Putting $a = i\kappa$ and $b = 1/4\alpha$, we have

$$g_1'(\kappa) = A\sqrt{(2\pi)}e^{-\alpha\kappa^2}. \quad \quad 4(98a)$$

We neglect the part of $g_1(\kappa)$ which contains $\exp\,[-\alpha(\kappa_0 + \kappa)^2]$ and obtain

$$g_1(\kappa) = A\sqrt{(\tfrac{1}{2}\pi)}\, \exp\,[-\alpha(\kappa_0 - \kappa)^2]. \quad \quad 4(98b)$$

The relative energy for a range $d\kappa$ is proportional to $[g_1(\kappa)]^2\, d\kappa$. Figs. 4.8 and 4.9 show the relation between the length of the wave train and the distribution of energy.

15.—Progress of the Wave Group in a Dispersive Medium.

Consider the system of waves for which the displacement is given by the real part of

$$\xi = \frac{1}{\sqrt{2\pi}}\int_{-\infty}^{\infty} g_1(\kappa) \exp i(\omega t - \kappa x)\, d\kappa, \quad \quad 4(99)$$

where $g_1(\kappa)$ is given by 4(98b). (The physical problem includes only the part of $g(\kappa)$ from 0 to ∞, but since the part from $-\infty$ to 0 is negligibly small, we may take the whole range—and we do so for convenience of manipulation.)

In a dispersive medium ω is a function of κ and, for a small range of κ near to κ_0, we have

$$\omega = \omega_0 + U(\kappa - \kappa_0) + W(\kappa - \kappa_0)^2$$
$$= \omega_0 + 2U\Delta + 4W\Delta^2, \quad \ldots \ldots \quad 4(100)$$

where $$2\Delta = \kappa - \kappa_0; \; U = d\omega/d\kappa \text{ and } W = \tfrac{1}{2}d^2\omega/d\kappa^2, \quad . \quad 4(101)$$

the values of U and W being taken at κ_0.

Then 4(99) becomes

$$\xi = A \exp i(\omega_0 t - \kappa_0 x) \int_{-\infty}^{\infty} \exp\{2i(Ut - x)\Delta - (4\alpha - 4iWt)\Delta^2\}\, d\Delta. \quad 4(102)$$

As a first approximation we take only the first two terms of 4(100), i.e. we assume $W = 0$. In 4(97) we put $a = 2i(Ut - x)$ and $b = 4\alpha$. We obtain

$$\xi = \frac{A}{2}\sqrt{\frac{\pi}{\alpha}} \exp\left[-\frac{(x - Ut)^2}{4\alpha}\right] \exp i(\omega_0 t - \kappa_0 x), \quad . \; . \; . \quad 4(103)$$

i.e. to this approximation, the form of the wave is unchanged, but the curve which bounds the profile moves forward with velocity $U = d\omega/d\kappa$. In the second approximation we integrate, putting $a = 2i(Ut - x)$ and $b = 4(\alpha - iWt)$ in 4(97). The difference is that α in 4(103) is replaced by $(\alpha - iWt)$, and we write

$$\xi = \frac{A}{2}\sqrt{\left[\frac{\pi}{(\alpha - iWt)}\right]} \exp\left[\frac{-(x - Ut)^2(\alpha + iWt)}{4(\alpha^2 + W^2t^2)}\right] \exp i(\omega_0 t - \kappa_0 x). \quad 4(104)$$

The imaginary parts can be separated to form a complicated phase term which varies slowly with the time. The important effect in which we are interested is that the curve which bounds the profile is still a Gaussian curve, but the parameter α is replaced by

$$\frac{\alpha^2 + W^2t^2}{\alpha}, \quad \ldots \ldots \ldots \quad 4(105)$$

i.e. the half-width is increasing with time, and the group is spreading slowly as it advances.

CHAPTER V

Interference

5.1. **Law of Photometric Summation.**

The Michelson interferometer, which we have described in the preceding chapter, is one of a large number of experimental arrangements which produce interference fringes. In the most general sense, interference fringes are variations of resultant amplitude (and hence of the relative energy and the illumination) from point to point in the field of view. The variations are due to varying path differences between the interfering beams of light. These produce varying phase differences, so that the waves reinforce one another at some points and oppose each other elsewhere. The variations of relative energy may be calculated by applying the principle of superposition. We shall later have to make many calculations on fringe systems, but first we consider an important related problem.

It is a matter of common observation that we do *not* obtain interference fringes due to interaction between light from two different sources (such as a ceiling light and a reading lamp). Photometric measurements show that, for such sources, the resultant relative energy at any point is the sum of the relative energies produced by the individual sources, each acting alone. This empirical law, which we shall call the *law of photometric summation*, applies also to energy received from different parts of a large extended source of light. Any satisfactory wave theory must include an account of this law, which is obviously of great practical importance. The theory must be able to show what conditions must be fulfilled in order to obtain fringes, and must be able to describe what happens under the more common conditions of everyday experience, when we do not observe fringes.

5.2. **Interaction of Independent Sources of Light.**

Suppose that light is being received at a point Q (fig. 5.1) from an extended source of light S_e. Some light will come from an atom at A_1, some from another at A_2, and so on. At any one moment the disturbance at Q is the sum of the disturbances which would be produced by each atom acting independently. If each atom were emitting an infinitely long train of pure sine waves, the resultant relative energy would be given directly by equation 3(12), which is based upon the principle of superposition. The experiments described in Chapter IV show that the light from a given atom must be represented by trains of waves which are seldom more than 10^6 waves long, and which take about 10^{-9} second to pass a given point. The shortest practical period

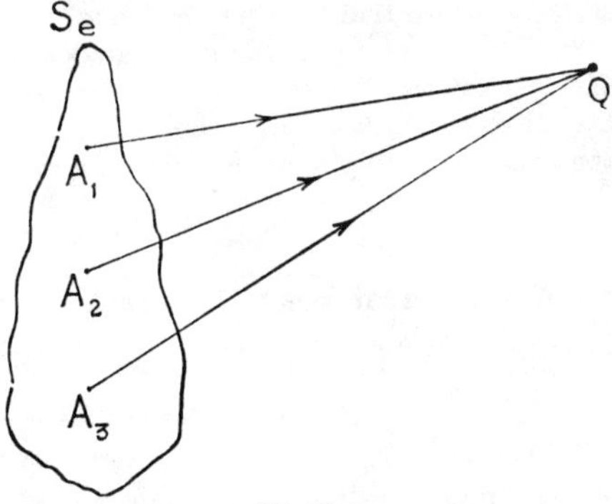

Fig. 5.1

of observation is of the order of a few microseconds. During this time the train of waves from a given atom is interrupted frequently, and each interruption constitutes an arbitrary variation of the relation between the phase of the wave from the given atom and the phases of the waves from other atoms. The illumination at Q due to a number of sources may be calculated if it can be assumed that the phase differences vary in a purely random way.

From 3(12) we have

$$E_R = \left(\sum_{r=1}^{m} a_r \cos \epsilon_r\right)^2 + \left(\sum_{r=1}^{m} a_r \sin \epsilon_r\right)^2$$

$$= \sum_{r=1}^{m} a_r^2 + \sum_{r=1}^{m} \sum_{\substack{s=1 \\ (s \neq r)}}^{m} a_r a_s (\cos \epsilon_r \cos \epsilon_s + \sin \epsilon_r \sin \epsilon_s)$$

$$= \sum_{r=1}^{m} a_r^2 + \sum_{r=1}^{m} \sum_{\substack{s=1 \\ (s \neq r)}}^{m} a_r a_s \cos (\epsilon_r - \epsilon_s). \quad \quad 5(1)$$

If the phase differences are varying in a purely random way, the average value of the second summation is zero, since to every possible positive value of any term there corresponds an equally probable negative value. Therefore the mean value is

$$E_R = \sum_{r=1}^{m} a_r^2 = \sum_{r=1}^{m} E_r. \quad . \quad . \quad . \quad . \quad . \quad 5(2)$$

The resultant illumination at Q is the sum of the illuminations due to the individual sources.

It is important to recognize that equation 5(2) applies only to an average taken over a time sufficiently long to include many random variations of phase. If a series of extremely short-period observations of the illumination at Q could be made, widely varying fluctuations would be obtained. Rayleigh showed that, under such conditions, the probability of observing values very different from the average is not small.

5.3. **Coherent and Non-coherent Beams of Light.**

Two beams of light are said to be *coherent* when the phase difference between the waves by which they are represented is constant during the period normally covered by observations. Two beams are said to be *non-coherent* when the phase difference changes many times and in an irregular way during the shortest period of observation.

The terms " coherent " and " non-coherent " describe idealized concepts never exactly realized in any practical situation. It follows from the discussion of Chapter IV that perfectly monochromatic waves of exactly the same frequency, i.e. infinitely long wave trains, are always completely coherent. No other waves are coherent in the strictest sense of the term. It is reasonable to apply the term to wave groups which have geometrically similar profiles since such wave groups may be regarded as the sums of coherent monochromatic waves. In practice we go a stage further and apply the term to a set of wave groups whose profiles are not quite the same—provided the differences are so small that they do not produce appreciable effects under ordinary conditions of observation. Non-coherence, i.e. completely arbitrary variation in the sense understood by statisticians, is also rare, but in practice any two beams from two different lamps are effectively independent. It is usual to apply the terms " coherent " and " non-coherent " to sources of light according to the properties of the radiation which they emit. Coherent sources of light are always images of one original source.

5.4.—We have shown (§ 5.2) that the resultant relative energy at a given point, due to a number of non-coherent sources, is equal to the sum of the relative energies which would be produced at this point by the individual sources acting separately. A similar rule applies to the illumination because this quantity is proportional to the relative energy. With n equal, non-coherent sources each capable of producing a relative energy E_r, the resultant relative energy is nE_r.

When a certain area is irradiated simultaneously by a number of coherent sources, the relative energy usually varies from point to point, giving rise to fringes. In Chapter III it was shown that the relative energy due to a number of coherent sources is $(\Sigma a_n)^2$ when all the waves have the same phase. It was also shown that, when the representative vectors form a closed polygon, the resultant relative energy is zero (§ 3.5). Thus n coherent sources each capable of producing a relative energy $E_r = a^2$ acting jointly will produce a resultant relative energy of

$$(na)^2 = n^2 E_r$$

at any point where their waves are all in phase. At another point their resultant may be zero. The interaction between coherent sources alters the distribution of energy but does not alter the total amount. Whether the sources are coherent or not, the resultant relative energy integrated over the whole region covered by the waves is equal to the sum of the integrals of the relative energies due to the individual sources. It will be understood that the total energy received at a given point from two non-coherent sources is equal to the energy which would be received at the same point from two coherent sources acting successively, so that the wave trains do not overlap. For this reason non-coherent sources are said to be independent. The discussion of § 5.2 shows that it is misleading to say that they do not interact.

5.5. **Formation of Interference Fringes.**

It will now be shown that both interference and photometric summation are involved in the theoretical description of fringes such as those of the Michelson interferometer.

Consider the light entering the telescope of the Michelson interferometer within a small solid angle. This light has come from many different atoms (A_1, A_2, etc.) in the extended source S_e (fig. 5.1). It includes many *pairs* of wave trains. When the mirror M_2 (fig. 4.1) is near to the reference plane R, the phase difference between the two members of any pair is constant. The members of the pair are mutually coherent. The relative energy due to the pair is given by 4(1). Also the phase difference between the pair of wave trains from A_1 is the same as that between the pair of wave trains from A_2.

Let θ be the angle between a given direction for light entering the telescope and the axis of the instrument. Then if light from A_1 is a maximum in a direction defined by θ, that from A_2 is also a maximum in that direction. Similarly, if the relative energy due to light from A_1 is zero in a direction defined by θ', that from A_2 is also zero. The resultant of the two wave trains from A_1 has a phase which depends on that of the light emitted from A_1. Similarly the phase of the resultant of the two trains from A_2 depends on that of the light emitted from A_2. Since emissions of light from A_1 and A_2 occur independently, the resultant of a pair of trains from A_1 has no permanent phase relation to the resultant of a pair from A_2, i.e. the resultants are non-coherent. The relative energy propagated in a given direction is, therefore, obtained by first calculating the resultant for a pair of wave trains from one atom, using 4(1), and then summing the relative energies for all the resultants, using 5(2). Fringes are seen only because all the resultants have their maxima in the same directions (such as θ) and their minima in another set of congruent directions (such as θ'). When the path difference is increased so that it is much greater than the average length of the wave trains emitted by A_1, etc., the pairs of wave trains no longer enter the telescope simultaneously. *All* the light entering the telescope at any given moment is then *non-coherent* and no fringes are seen. Obviously similar conditions would hold in an optical system designed to bring light from a source by three or more paths to the same point instead of by only two paths.

5.6.—The general conditions for the observation of interference fringes include the following:

(*a*) The light reaching the observing apparatus must contain sets of wave trains, such that all members of a given set have come from the same atom by different paths.

(*b*) The path difference must be short enough to make these sets at least partially coherent.

(*c*) The location of the fringes produced by light from different parts of the source must be the same or nearly the same.

This last condition is automatically fulfilled if the two sources are both images of one very small source. Conditions (*b*) and (*c*) are not rigidly defined conditions. Obviously the maximum visibility will be obtained when the path difference is very short, so that the sets are exactly coherent, and when the fringes formed by different parts of the same source exactly coincide. The observation of fringes when these conditions are not exactly fulfilled depends greatly on the system of observation. Visual observation is greatly affected by the brightness of the field. Fringes will not usually be observed if the maxima due to light from the edges of an extended source are separated by more than a quarter of the width of a fringe from the maxima due to the central part of the source.

5.7. Interference between Two Sources Side by Side.

One of the first attempts to observe interference was made by Grimaldi, who allowed the light from a source S_e to pass through a pair of slits (a_1 and a_2) in the screen B (fig. 5.2*a*). He observed fringes

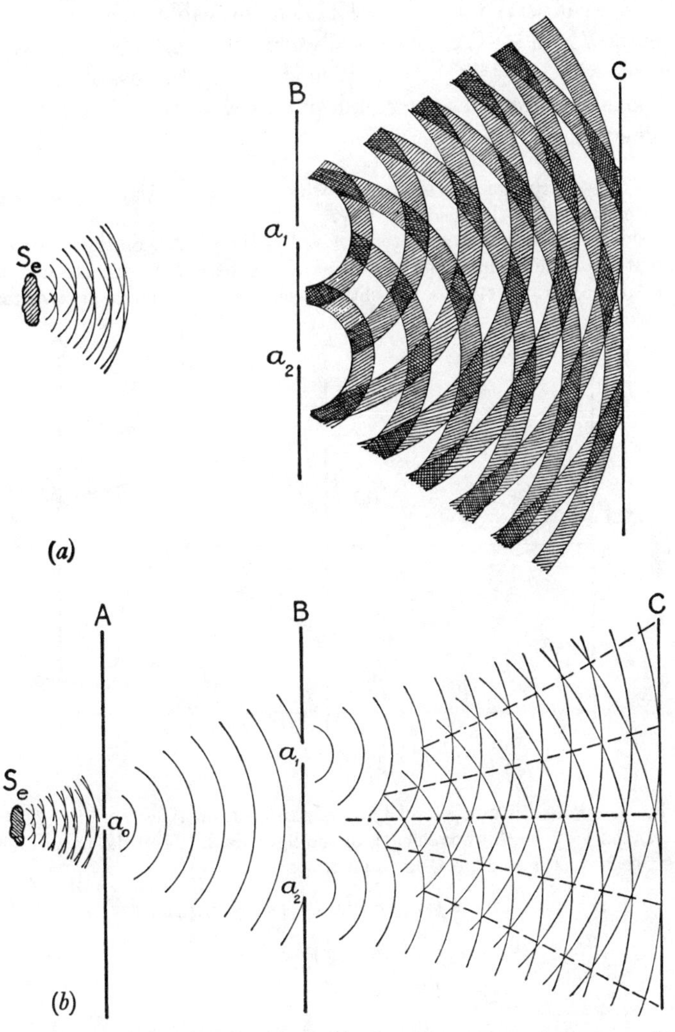

Fig. 5.2.—(*a*) Grimaldi's experiment, and (*b*) Young's experiment. In the lower diagram is shown Young's experiment with fringes due to interference of nearly spherical waves from a_1 and a_2. In the upper diagram the wavefronts are blurred because an irregular system of waves from S_e falls on screen B. In practice the blurring would be greater than that shown and the fringes would be obliterated.

on the screen C. It was later shown by Young that interference fringes should be seen only if the source S_e is fairly small. He used an extra screen A with a single slit a_0 to reduce the effective size of the source, and observed interference fringes on the screen C (fig. 5.2*b*). This is generally regarded as the first true observation of interference fringes. The fringes observed by Grimaldi were probably due to diffraction (Chapters VI and VII). Essentially similar fringes could have been obtained with one slit. In any modern repetitions of Young's or Grimaldi's experiments, a medium-power eyepiece would be used to view the fringes.

5.8.—The maximum size of the slit which will allow fringes to be observed may be calculated in the following way.

Consider the interference of light received at the point Q on the screen C from the point P on the source (fig. 5.3.) Let X′OX be a line perpendicular to the screens and halfway between the slits, which are each placed at a distance h

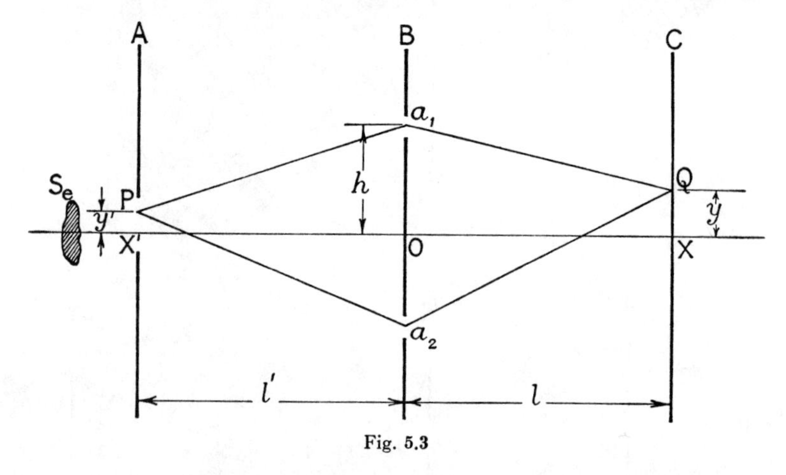

Fig. 5.3

from it. Let P be distant y' and Q be distant y from this line. Let $X'O = l'$, $XO = l$, and let λ be the wavelength of the light used. Then the path difference of the two beams from P to Q is equal to

$$\Delta s = (Pa_1 + a_1Q) - (Pa_2 + a_2Q). \quad . \quad . \quad . \quad . \quad . \quad 5(3)$$

Also
$$(Pa_1)^2 = l'^2 + (h - y')^2$$

and
$$(Pa_2)^2 = l'^2 + (h + y')^2.$$

Hence
$$(Pa_1)^2 - (Pa_2)^2 = 4hy'$$

or
$$(Pa_1 - Pa_2) = \frac{4hy'}{2l'} = \frac{2hy'}{l'} \text{ approximately.}$$

The value of $(a_1Q - a_2Q)$ is found in a similar way, so that 5(3) becomes

$$\Delta s = \frac{2hy'}{l'} + \frac{2hy}{l}. \qquad 5(4)$$

The positions of the maxima are obtained by putting $\Delta s = n\lambda$. From 5(4), it may be seen that the fringes due to light from P are equidistant, the separation being $l\lambda/2h$. If y' varies, the fringes move as a whole without altering their separation. The change δy in y due to a small variation $\delta y'$ in y' is given by

$$\frac{\delta y'}{l'} = -\frac{\delta y}{l}. \qquad 5(5)$$

If δy is not to exceed $l\lambda/8h$, to fulfil the condition stated at the end of § 5.6, it is necessary that

$$\frac{\delta y'}{l'} \not> \frac{\lambda}{8h}.$$

Thus if the slits a_1 and a_2 are one millimetre apart (so that $h = 0{\cdot}5$ millimetre) and the wavelength of the light used is 6×10^{-4} millimetre, the angle which the source subtends at the point O should not exceed $1{\cdot}5 \times 10^{-4}$ radian. If the screen A is two metres from the screen B, the hole in A should not be more than about three-tenths of a millimetre wide.

5.9.—Other arrangements for observing interference with two sources side by side are:

(*a*) *Fresnel's Mirrors.*

Two images (S′ and S″) of a slit source S are formed by two plane mirrors M′ and M″ set at a small angle (fig. 5.4*a*, p. 122).

(*b*) *Fresnel's Biprism.*

Two images (S′ and S″) of a slit source S are formed by refraction in two prisms (P′ and P″) of small angle (fig. 5.4*b*, and Plate III*b*, p. 214).

(*c*) *Lloyd's Mirror.*

One image (S′) is formed by reflection in a plane mirror M, and light from this interferes with light coming directly from S (fig. 5.4*c*).

(*d*) *The Billet Split Lens.*

The two images are formed by the parts of a lens which has been bisected and the parts separated a little (fig. 5.4*d*). This is equivalent to the use of one lens and two small prisms. In each case the fringes are usually observed with a low-power eyepiece.

With the arrangement used in fig. 5.2*b*, no light reaches the centre of screen C (where the interference fringes are seen) by uninterrupted rectilinear paths. The methods illustrated in fig. 5.4 do bring light by rectilinear paths (of the type to which geometrical optics applies) to the region of interference. It has been suggested, therefore, that Young's fringes are due to a mixture of diffraction and interference, while the Fresnel and Lloyd experiments depend on interference alone. The difference is not very important. The central fringes in Young's experiment disappear if a_1 or a_2 is covered and are clearly due to interference, but the distribution of light in any of the fringe systems cannot be accurately calculated without applying the theory of diffraction. In the arrangements of Fresnel, Lloyd, etc., some diffraction occurs at the edges of the prisms or mirrors.

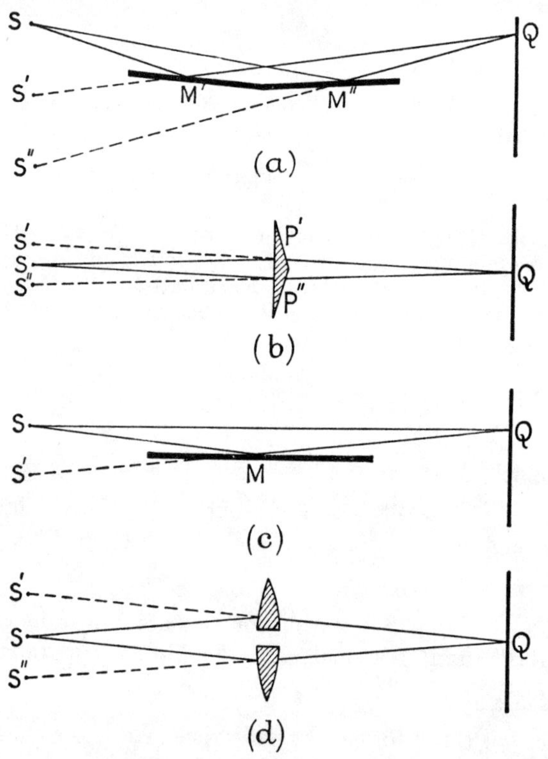

Fig. 5.4.—(*a*) Fresnel's mirrors. (*b*) Fresnel's biprism. (*c*) Lloyd's mirror. (*d*) Billet's split lens.

5.10.—The most interesting of these arrangements is that due to H. Lloyd (1800–81). With the arrangement shown in fig. 5.4*c*, only half the system of fringes is seen. By interposing a thin transparent plate in the path of the direct beam the bands can be shifted so that the central part of the system can be seen. With roughly monochromatic light, such as that produced by a sodium flame, the fringe of zero path difference is not distinguishable from any other fringe. Using white light, the central fringe is seen clearly, a few fringes on either side of the centre are coloured, and farther out the fringes overlap to such an extent that they cannot be distinguished. The fringe corresponding to zero path difference is clear because, with zero path difference, the phase difference has the same value (π) for all wavelengths. The distance between successive bright fringes is, in general, a function of the wavelength. For example, in the simple arrangement of Young, it is proportional to the wavelength. As we go away from the centre, the fringes of different colours become more and more displaced relative to one another.

For path differences which are small, but not zero, the fringes appear coloured, but for larger path differences they become completely confused. If there is no change of phase on reflection, the centre of the system should be bright, since zero path difference would then mean zero phase difference for all wavelengths. Lloyd found that the centre is dark, thus indicating a change of phase of π due to a reflection in a less dense medium. This result (which is confirmed by experiments of Fresnel and others) is in agreement with the theory given in § 3.27, and justifies the choice of the boundary condition.

5.11.—The distribution of light in the above fringe system cannot be calculated without making allowance for the diffraction of light by the slit (Chapter VI). One effect of the slit width may be noted here. In the experiments of Fresnel and Billet the virtual sources are images of the same source and are formed in such a way that the upper part of one source (fig. 5.4*a*, *b*, and *d*) corresponds to the upper part of the other. Thus, with a source S of finite width the distance between different corresponding points on S′ and S″ is constant. The fringe width is the same for all parts of the source, but the fringe systems due to different parts of the source are displaced relative to one another. In Lloyd's experiment, the centres of the fringe systems due to different parts of the source coincide, but the outer bands are confused, because the width of the bands is not the same for different parts of the source. Lloyd's arrangement is thus more suitable for studying the central achromatic fringe. The other arrangement gives clearer fringes in the outer parts of the field with monochromatic light.

5.12. **Interference produced by Thin Films.**

Coloured fringes are often seen when a thin film of transparent material is viewed by reflected light. The film may be a layer of oil on water or on the surface of a road. Similar colours are seen in the wall of a soap bubble or of a very thin glass bulb. The films may be of lower refractive index than the surrounding medium, e.g. a film of air or liquid between two glass plates. Fringes are often seen between the components of an old cemented lens when the cement has failed locally, leaving a small inclusion of air. These colours are due to interference between beams which, having been reflected at least once in the film, reach the observer by paths of different lengths (fig. 5.5). When the surfaces of the film have a low coefficient of reflection, the fringes are best seen by observing the light which emerges on the

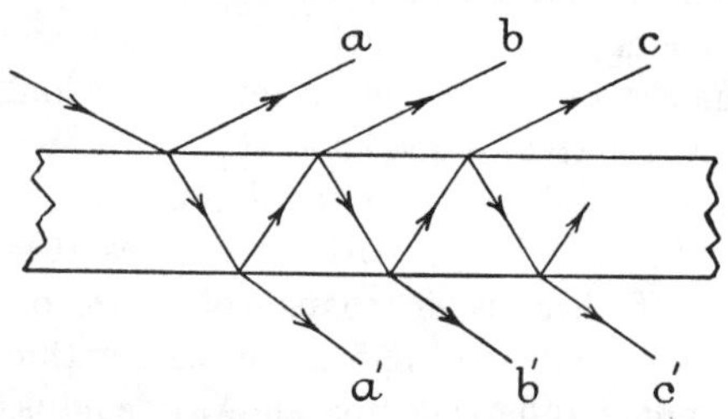

Fig. 5.5.—Beams of light emerging from a thin film. (Each beam is represented by a single ray.)

side of incidence. Such fringes (due to beams *a*, *b*, *c* in fig. 5.5) are said to be seen by reflected light. The fringes due to interference between beams which emerge on the opposite side of the film (beams *a*′, *b*′, *c*′ in fig. 5.5) are said to be viewed by transmitted light. They are very weak and not easy to observe when the coefficients of reflection are small, but they become very sharp and clear when the film has a high reflection coefficient.

Fringes may be obtained by taking two pieces of good plate glass, cleaning them carefully, and then sliding two surfaces into contact. A film of air of varying thickness is formed between the two plates. If the glass is moderately good, the film will be nowhere more than a few wavelengths thick. If the plates are placed on a dark cloth and viewed by reflected light from a window, brilliant coloured fringes are obtained. It is inadvisable for an inexperienced person to carry out this procedure with good optical glass owing to the risk of scratching the surfaces if the plates are not perfectly clean. Also if the plates are of very high quality, the air film may be too thin to show good fringes.

5.13.—The path difference, for a film of constant thickness, may be calculated as follows. Let e be the thickness of the film and μ the index of refraction from air to the material of which the film is composed. Then, in fig. 5.6, the difference in optical path is $\mu(\mathrm{AB}+\mathrm{BC})-\mathrm{AD}$, and if θ_1 is the angle of incidence and θ the angle of refraction, $\mathrm{AE}=\mathrm{AB}\sin\theta$ and

$$\mathrm{AD}=2\mathrm{AB}\sin\theta\sin\theta_1=2\mu\mathrm{AB}\sin^2\theta.$$

The path difference (l) is then $2\mu AB(1 - \sin^2\theta)$ and this gives

$$l = 2\mu e \cos\theta. \quad . \quad . \quad . \quad . \quad . \quad . \quad . \quad . \quad 5(6)$$

It may easily be shown that the path differences between b and c, between a' and b' and between b' and c' (fig. 5.5) are all equal to l. It may also be shown that when the film is a wedge of small angle α, the path difference is approximately $2\mu e(\cos\theta - \sin\theta\tan\alpha)$, where e is the thickness at a point midway between rays a and b. This may be taken to be equal to $2\mu e\cos\theta$ except when the light is near grazing incidence.

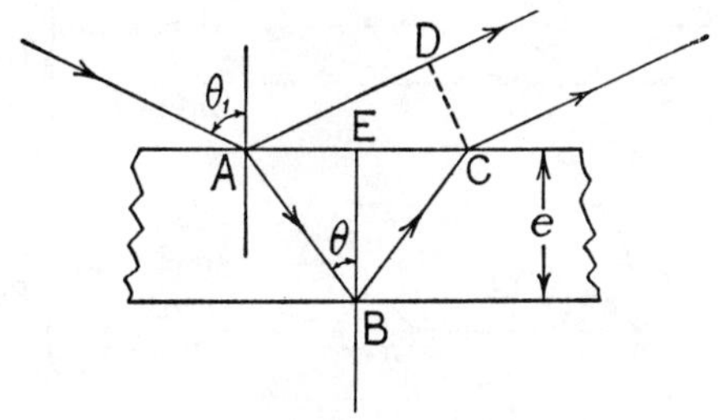

Fig. 5.6.—Path difference between successive beams

There is a change of phase of π at the upper surface if the film is optically denser than the surrounding medium, and at the lower surface if it is optically less dense. Thus the phase difference between the first two beams is

$$\delta = \frac{2\pi}{\lambda}(2\mu e\cos\theta) \pm \pi = \frac{2\pi}{\lambda} l \pm \pi. \quad . \quad . \quad . \quad 5(7)$$

Since a phase difference of 2π is immaterial, it does not matter which sign is used in 5(7), and we shall therefore use always the plus sign. The second term in 5(7) is usually omitted in calculations of the separation of the fringes. If, however, this term is not included when calculating the amount of light reflected from an infinitesimally thin film, we should conclude that such a film would give strong reflection. When the phase change at one, and only one, surface is included, the calculated reflected energy tends to zero when the thickness of the film tends to zero.

5.14. Visibility of the Fringes.

When the two surfaces are unsilvered, the reflection coefficient is low except for very large angles of incidence. We may assume a reflection coefficient of 0·05 at each surface and calculate the fraction of the incident energy in each of the beams shown in fig. 5.5. We find that for the beams a, b, c, which emerge on the upper side of the film, the relative energies are 0·05, 0·045, 0·0001 respectively. The amplitudes of the waves which represent beams a and b are nearly equal and are very much greater than the amplitudes of all the other beams which emerge from the upper surface. Thus we need only consider these two beams in an approximate calculation of the distribution of energy in the fringes seen by reflected light. Moreover, we

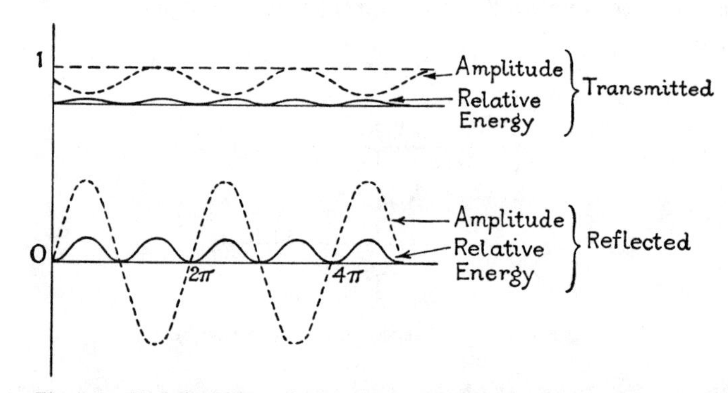

Fig. 5.7.—Variation of amplitude and of relative energy for transmitted and reflected light. (Unsilvered plates, $r^2 = 0{\cdot}05$.) The negative amplitude corresponds to a resultant whose phase is opposite to that of the incident beam.

shall expect these two nearly equal beams to give fringes of high visibility, provided the path difference is small. The approximate relative energies of beams a', b', and c' are 0·90, 0·002 and 0·000 006 respectively. Again, we need take account of only two beams, but since these beams are represented by waves of very unequal amplitude, the fringes will have a very low visibility. Fig. 5.7 shows the variation of the *amplitude* and the *relative energy* of the resultant with phase difference for the fringes seen by reflected light and the corresponding functions for the fringes seen by transmitted light. (See also Plate II*g* and II*h*, p. 126.)

5.15.—Interference fringes are observed only if at least two of the beams enter the eye. For moderate angles of incidence the displacement of beam b relative to beam a is of the same order of magnitude as the thickness of the film. When this thickness is a good deal smaller than the diameter of the pupil (i.e. when it is 0·5 millimetre or less) it is easy to place the eye so as to receive both these beams. For films of greater thickness it is necessary to use an optical instrument to

(*a*) and (*b*) are negatives, (*c*), (*d*), (*e*), and (*f*) are positives.

(*a*) Nitrogen emission bands in the region 4000–3000 Å.

(*b*) Hydrogen emission lines with sodium absorption lines in the ultra-violet (3000–2000 Å).

(*c*) Solar spectrum showing Fraunhofer absorption lines.

(*d*) Sodium absorption spectrum (visible region).

(*e*) Sodium emission spectrum (visible region).

(*f*) Channelled spectrum.

The wavelength scale applies only to (*c*), (*d*), and (*e*).

(*g*) Contour fringes (wedge-shaped air film between unsilvered plates viewed by reflection).

(*h*) As (*g*) but wedge of smaller angle viewed by transmitted light.

(*i*) Contour fringes (wedge-shaped air film between silvered plates viewed by reflection).

(*j*) As (*i*) but viewed by transmission.

(*k*) Contour fringes (air film between a piece of mica and an optical flat viewed by reflection).

(*l*) As (*k*) but viewed by transmission.

PLATE II

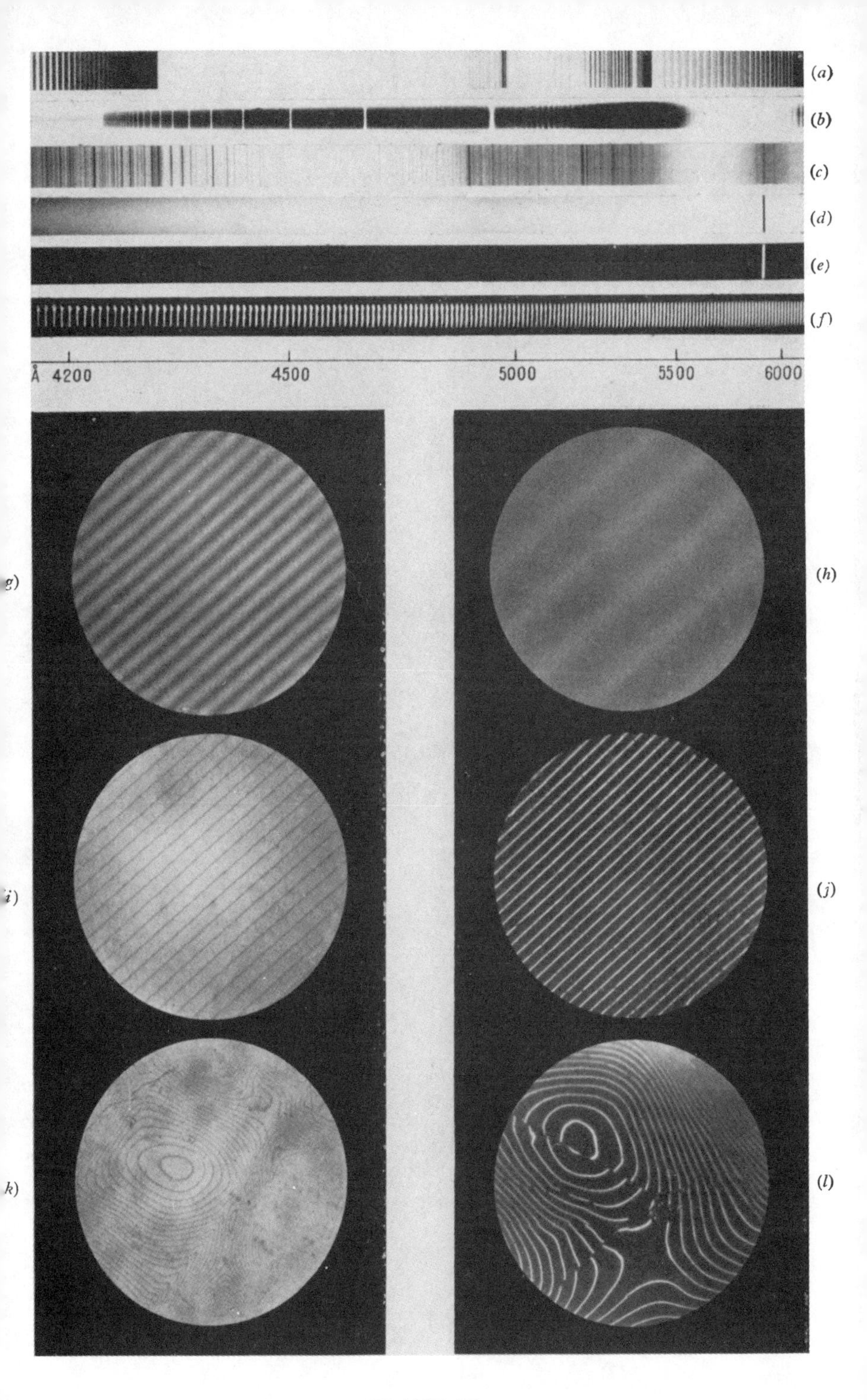

PLATE II

collect the beams or else to arrange to view the film by rays which are nearly normal to the surface of the film (fig. 5.8). The above condition for observing the fringes is necessary, but it is not the only condition which must be fulfilled. When white light is used, coloured fringes are obtained with films which are only a few wavelengths

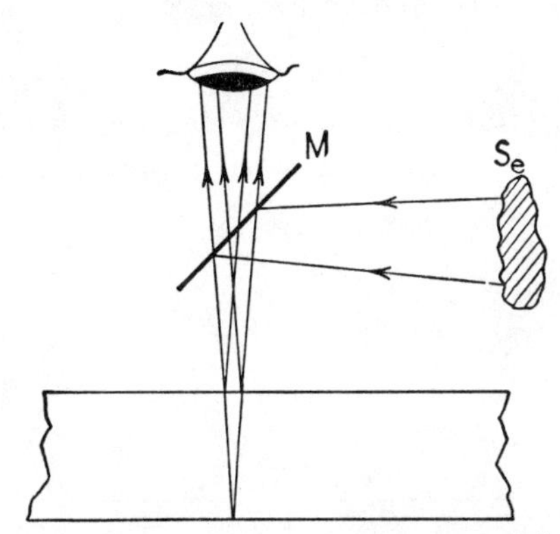

Fig. 5.8.—Thick plate viewed by nearly normal rays. The mirror M may be an unsilvered piece of glass

thick, but no fringes are seen under ordinary conditions of observation with thicker films.* With monochromatic light, fringes may be obtained with plates of thickness up to a few centimetres, provided that the eye, or the optical instrument, is focused on the correct plane (§ 5.20).

5.16. **Fringes as Loci of Constant Path Difference.**

A fringe is a locus of points, in the plane on which the eye is focused, for which the phase difference $\kappa\mu e \cos\theta$ has some constant value, say ϕ_1 for one fringe, ϕ_2 for the next, and so on. The phase difference may vary owing to a change † in any one of the three quantities μ, e, and θ. Consider the arrangement shown in fig. 5.9, where a thin film is viewed with the naked eye, using an extended source of light such as the sky or a sodium flame. The film may be of constant optical thickness, or there may be a slow variation of μe from point to point, due to variation of either μ or e. If the film is a few wavelengths thick, fringes will be seen when the eye is focused on the film. Fringes are

* We shall later discuss special methods of producing achromatic fringes (see § 5.34).

† We shall see later that it is also possible for variations in one of the quantities to compensate variations in one of the others (§ 5.20).

seen under these conditions because the eye subtends only a very small angle at a given point on the film.

For example, suppose that the film is being viewed at 30° from the normal with a pupil of 3 millimetres diameter situated 50 centimetres from the film. The angle of incidence for light from S_e entering one side of the pupil is 30·17° and for light entering the other side is 29·83°.

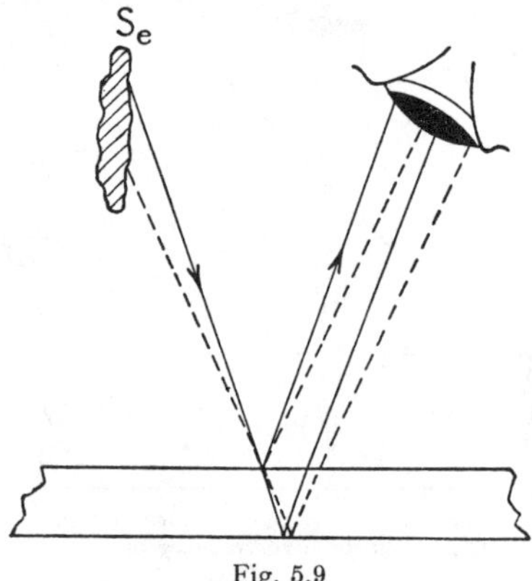

Fig. 5.9

The difference between the cosines is only 0·003. Thus cos θ is effectively constant. With sodium light, fringes may be seen when the thickness is more than a few wavelengths, provided that the thickness does not vary too rapidly. If it does vary by more than about 5λ per millimetre, measured along the surface of the film, then the fringes become too close to be observed with the naked eye.

5.17. **Fringes of Constant Inclination.**

It is not possible to make precise deductions from observations obtained with the above arrangement, because the variation from one fringe to another is partly due to variation of θ, and partly to variation of optical thickness. In laboratory experiments, we usually keep *either* the optical thickness *or* the angle θ constant. The fringes observed with a film of constant optical thickness are loci of constant θ and are known as *fringes of constant inclination*. Such fringes are obtained with the Michelson interferometer when M_2 is set accurately parallel to the reference plane. They may also be seen with a plate of glass several millimetres thick, which is of moderately good quality, using the arrangement shown in fig. 5.8 to obtain nearly normal incidence. With thicker plates or larger angles of incidence, a telescope must be used to collect the beams of light.

The fringes of constant inclination obtained with thick plates are called Haidinger's fringes, though they were first systematically studied by Mascart and by Lummer. Since the light in a given fringe refers to a constant value of θ, the telescope should be focused for infinity and, when the axis of the telescope is normal to the surface, the fringes are circular rings in the focal plane of the objective. When fringes of constant inclination are viewed with the naked eye, it is desirable to focus the eye for infinity but, owing to the small range of angles included by the eye pupil, there is great depth of focus and the fringes may still be seen when the eye is focused near the film, provided it is very thin.

5.18. **Fringes of Constant Optical Thickness.**

When the sides of a thin film are not quite parallel, we may use an arrangement which makes the angle θ constant over the whole field of view, and may then observe fringes which are essentially contours of constant μe. These are known as *fringes of constant optical thickness.*

For example, consider a wedge of angle 1 in 1000 viewed at nearly normal incidence by an arrangement such as that shown in fig. 5.8. Dark fringes will then occur when $2e = n\lambda$ and the distance between successive dark fringes will be $1000\lambda/2$ or about a quarter of a millimetre. If the eye is placed even a few centimetres from such a film, the variation of θ from one fringe to the next (or even over ten fringes) is very small. The fringes are conveniently viewed with a magnifying glass or a low-power microscope, unless the wedge angle is very small. Fringes of this type may be produced with the Michelson interferometer by setting the mirror M_2 (fig. 4.1) at a small angle to the reference plane. When the wedge is more than a few wavelengths thick, approximately monochromatic light must be used. Using sodium light and a long wedge of angle about 1 in 100, variations in visibility may be observed. The fringes are clear when e is small, disappear when $2e$ is about 500λ, and are very clear again when $2e$ is about 1000λ. The explanation of these variations, in terms of the superposition of the fringe systems due to the two sodium yellow lines, is similar to that given in Chapter IV for the corresponding effects observed with the Michelson interferometer.

5.19. **Newton's Rings.**

Fringes of constant optical thickness are formed when two spherical surfaces of unequal curvature are placed in contact (see Plate I*e*, p. 72). The loci of equal thickness are then circles. Such fringes were first observed by Hooke (1635–1703) but were first studied by Newton and are known as Newton's rings. They may be observed by placing a

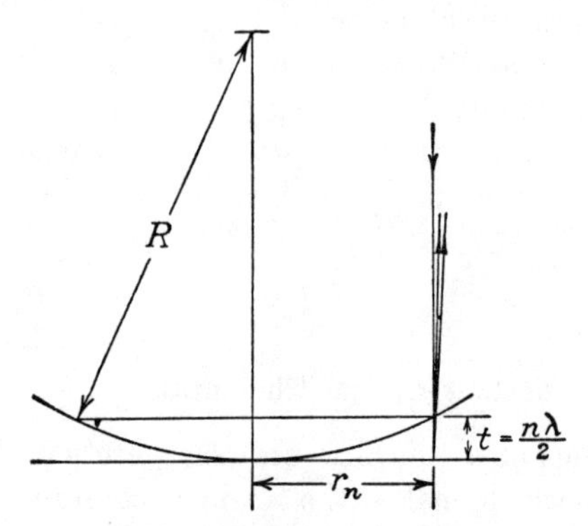

Fig. 5.10.—Newton's rings

convex lens on a flat plate and using the reflex arrangement shown in fig. 5.8. The rings are of convenient size for viewing with a low-power microscope when the radius of curvature of the lower surface of the lens is about one metre. From the geometry of fig. 5.10 it may be seen that, if d_n is the diameter of the nth dark ring and R is the radius of curvature of the lens,

$$d_n^2 = 2n\lambda(2R - n\lambda/2)$$

and approximately $$d_n = 2\sqrt{(Rn\lambda)}. \qquad 5(8)$$

Owing to the change of phase of π at the lower surface, the centre is dark. The radii of the rings are thus proportional to the square roots of the natural numbers and the distances between successive rings decrease as n increases. With white light, a few fringes near the point of contact may be seen.

EXAMPLES [5(i)–5(ix)]

5(i). Let r be the ratio of the amplitude of the reflected beam to the amplitude of the incident beam which is taken to be unity. Write down the amplitudes of the beams shown in fig. 5.5; also of the nth beam which emerges on the upper side and the nth beam which emerges on the lower side, assuming that the film is perfectly transparent. Note that the reflection coefficient $= r^2$.

[Amplitude for beam a is r, for beam b is $r(1 - r^2)$, for beam c is $r^3(1 - r^2)$, and for the nth beam on the upper side $r^{2n-3}(1 - r^2)$. The amplitude for beam a' is $(1 - r^2)$, for beam b' is $r^2(1 - r^2)$, for beam c' is $r^4(1 - r^2)$, and for the nth beam on the lower side $r^{2n-2}(1 - r^2)$. The first formula does not apply when $n = 1$. Note that the relative energy of the beam transmitted through one surface is $1 - r^2$, and its amplitude is $(1 - r^2)^{1/2}$.]

5(ii). Find an expression for the resultant amplitude due to all beams $b, c, \ldots$ etc. up to the nth beam when the phase difference is a multiple of 2π. To what limit does this expression tend when n becomes indefinitely large? What will be the phase relation between this last resultant and the wave which represents beam a?

[The resultant amplitude is $r\{1 - r^{(2n-1)}\}$ (except when $n = 1$) which tends to r as $n \to \infty$, since $r < 1$. Resultant will differ in phase from beam a by π.]

5(iii). If for a given thickness and angle of incidence the film absorbs a fraction f of the energy which passes through it once, write down expressions corresponding to those required in Example 5(i). Assuming that only two beams need be considered, calculate the visibility for fringes seen by reflected light when $r^2 = 0{\cdot}05$, (a) when $f = 0$, and (b) when $f = 0{\cdot}8$.

[The amplitude for beam a is r, for beam b is $r(1 - r^2)(1 - f)$, for beam c is $r^3(1 - r^2)(1 - f)^2$, and for the nth beam from the upper side $r^{2n-3}(1 - r^2)(1 - f)^{n-1}$. The amplitude for beam a' is $(1 - r^2)(1 - f)^{1/2}$, for beam b' is $r^2(1 - r^2)(1 - f)^{3/2}$, for beam c' is $r^4(1 - r^2)(1 - f)^{5/2}$, and for the nth beam on the lower side $r^{2n-2}(1 - r^2)(1 - f)^{(2n-1)/2}$. When there is no absorption, the visibility of the fringes (as defined in § 4.10) is practically unity. When $f = 0{\cdot}8$, it is approximately $0{\cdot}36$.]

5(iv). Show that the diameters of Newton's rings, when two surfaces of radii r_1 and r_2 are placed in contact, are related by the equation

$$\frac{1}{r_1} \pm \frac{1}{r_2} = \frac{4n\lambda}{d_n{}^2}.$$

Consider in what circumstances the sign on the left-hand side is positive.

5(v). Three surfaces A, B, C are placed in contact in pairs successively, and, using light of wavelength 5000 Å., the diameter of the 25th bright ring is found to be:

20 mm. when A and B are in contact,
26 mm. when B and C are in contact,
16 mm. when C and A are in contact.

Find the three radii of curvature. Give the answer to two significant figures.
[8·3 m., 560 m., 14 m.]

5(vi). A film of oil of refractive index 1·7 is placed between an equiconvex lens and a flat plate. The refractive index of the glass is 1·5 and the focal length of the lens is one metre. Find the radius of the 10th dark ring when light of wavelength 6000 Å. is used. (An equiconvex lens is one in which both surfaces are convex and the radii of curvature are numerically equal.)
[Radius of 10th ring = 1·88 mm. Refer to equation 3(32).]

5(vii). Why would the rings be more difficult to see with the arrangement of Example 5(vi) than with an air film?

[The fringes would be less bright in the ratio 10 : 1 because the index of refraction oil/glass is much smaller than the index glass/air. Some light reflected from the top surface of the lens "dilutes" the fringes and this is the same in both cases.]

5(viii). Show that elliptical rings are obtained when a convex cylindrical surface of radius r_c is placed in contact with a convex spherical lens of radius r_s.

[If the axis of the cylinder is the x axis, the thickness of the air film at point (x, y) is $\dfrac{y^2}{2r_c} + \dfrac{x^2 + y^2}{2r_s}$.]

5(ix). Show that the loci of points at which the path difference of beams from two point sources is constant, are hyperboloids of revolution about the line passing through the sources.

5.20. **Localization of Interference Fringes.**

Fringes due to the interference of light from two small sources (as in Young's experiment) may be received upon a screen placed at any convenient distance from the sources. If the screen is too close to the sources the fringes are inconveniently small, and if it is too far away the illumination is too weak. Over a considerable range, however, the fringes may be obtained either by photography or by viewing the light through an eyepiece. There are, in fact, a set of surfaces of constant phase difference. It may be shown that when point sources are used these surfaces are hyperboloids. The fringes observed form the intersection of these hyperboloids with the plane of observation. The beams of light from the sources to any point on the plane of observation subtend very small angles and no focusing process is involved in the formation of the fringes. The interference of light from extended sources must involve a certain type of focusing. Clear fringes will be observed only when the phase differences of pairs of beams received at a given point from different parts of the source are the same or nearly the same. This condition will, in general, be satisfied only when the point of observation lies on a certain surface, i.e. they will be seen only if the observer's eye (or his instrument of observation) is focused upon this surface. Such fringes are said to be *localized.*

It is found that fringes of equal inclination are best seen with an instrument focused for infinity. Fringes of equal thickness are found to be localized near the film and, for nearly normal incidence, they are most clearly seen by means of a microscope focused on the film. When fringes are formed by narrow pencils of light there may be considerable depth of focus.

5.21.—In a system of fringes of equal inclination a given phase difference corresponds to a definite direction of observation no matter from what part of the source the light originates. The light at a given point in the focal plane of a telescope objective will thus corre-

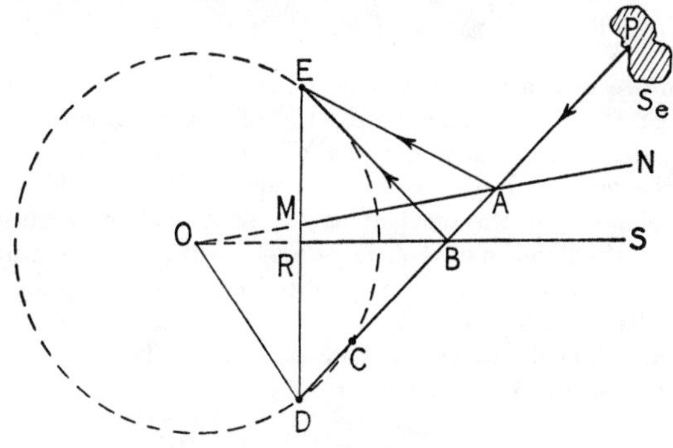

Fig. 5.11.—Localization of fringes

spond to a definite phase difference, and the fringes will be clearly seen through an eyepiece focused upon this plane. The localization of fringes of equal thickness formed by reflections at the surfaces of an air-film between two glass plates may be investigated by the following method which is due to G. F. C. Searle.*

Suppose that the two surfaces of reflection are planes normal to the plane of the diagram (fig. 5.11) which they cut in the lines MN and

* Reference 5.1.

RS, and let these lines meet at the point O. Consider a ray from a particular point P on the extended source (S_e) which is reflected by the upper surface at the point A, and by the lower surface at the point B. Suppose that these two rays meet at the point E. Then E is the image (formed by reflection in the surface MN) of some point C and also the image (formed by reflection in the surface RS) of some point D, both C and D lying on the original direction of the ray PAB. Since MN is the perpendicular bisector of CE and RS of DE, it follows that E, C and D lie on a circle whose centre is O. The path difference of the two beams of light received at E is equal to AB + BE − AE and this is equal to AD − AC, i.e. to CD. If θ is the angle between the plates and r is the radius of the circle, CD is equal to $2r \sin \theta$. Thus the phase difference is the same for all pairs of paths leading from any points on the source to a point on the circle.

Suppose that a microscope is focused on E with its axis directed along the tangent to the circle. The light focused at the centre of the field of view comes from E and, owing to the finite depth of focus, from points on the circle near to E. All this light has the same phase difference and, if $2r \sin \theta = n\lambda$, a maximum is seen at E. The system of fringes is traversed by moving the microscope perpendicular to its axis (i.e. parallel to the radius OE).

One advantage of Searle's treatment is that, since the fundamental relation is derived from the image-forming property of a reflecting surface, it applies even when the ray considered is not in the plane of the diagram. It is also of interest that the separation of the fringes is equal to $\lambda/(2r \sin \theta)$ and is independent of the inclination of the incident light. This result may easily be verified experimentally. The location of the fringes may be studied by viewing the system with the naked eye and moving a pin until coincidence is obtained, using the method of parallax. In this way the circle appropriate to a given fringe may be plotted. For a further discussion of the localization of fringes of constant thickness, see References 5.1, 5.2, and 5.3.

5.22. **Non-reflecting Films.**

The reflection of light at glass/air surfaces is very undesirable in camera lenses and in many other optical components. In a compound lens with four glass/air surfaces, about 20 per cent of the light is lost by reflection. This reduces the light-gathering power of the lens. Also some of the light reaches the image plane after multiple reflections. This reduces the contrast in the picture. The reflection of light may be reduced by coating the lens surfaces with a film of crystalline material whose refractive index is less than that of the glass (fig. 5.12). The light reflected from the air/crystal surface interferes with the light reflected from the glass/crystal surface. If the index of refraction and the thickness of the film are correctly chosen, the two reflected waves exactly annul one another for one particular wavelength and one angle of incidence.

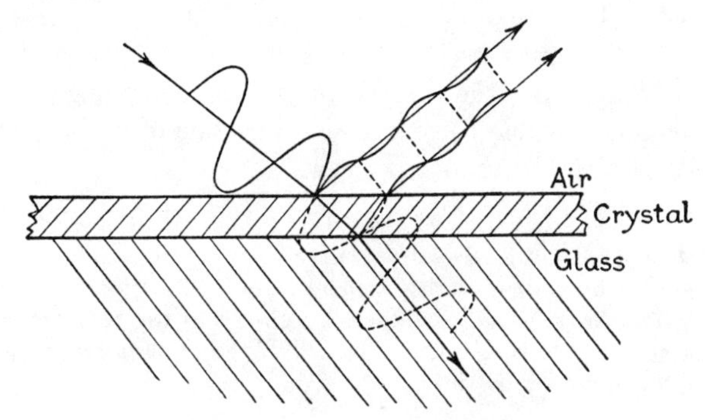

Fig. 5.12.—Non-reflecting film

In order that the two beams shall annul one another two conditions must be fulfilled: (*a*) the amplitudes must be equal, and (*b*) the phase difference must be π. Consider light at normal incidence. If μ_g is the refractive index for air/glass, and μ_c the index for air/crystal, then $\mu_{cg} = \mu_g/\mu_c$, and from 3(55) the condition that the amplitudes shall be equal in magnitude is

$$\frac{\mu_c - 1}{\mu_c + 1} = \frac{\mu_{cg} - 1}{\mu_{cg} + 1}.$$

This is satisfied if

$$\mu_c = \sqrt{\mu_g}. \qquad \text{5(9)}$$

Both reflections take place in the less dense medium so the change of phase on reflection is the same for both. Thus in order to produce a phase difference of π it is necessary that

$$2\mu_c e = (2n + 1)\tfrac{1}{2}\lambda. \qquad \text{5(10)}$$

In addition to satisfying 5(9), the ideal crystalline material should be very hard, should not be affected by moisture, etc., and should adhere well to the glass surface. No crystalline material meets all these requirements. Magnesium fluoride or cryolite (sodium aluminium fluoride) is generally used. The refractive indices are 1·38 and 1·36 respectively. When used with heavy flint glasses ($\mu = 1{\cdot}7$) they satisfy 5(9) approximately, but when used with crown glasses ($\mu = 1{\cdot}51$) their refractive indices are too high to give the best effect.

It is usual to make the optical thickness equal to a quarter wavelength of green light ($\lambda = 5500$ Å.). The reflection for green light is then very low, while the reflection at the ends of the spectrum is larger. Thus, if white light is incident on the composite surface, the reflected light is mainly red and blue. It has a purple colour rather like the " bloom " on some kinds of ripe plum. For this reason the process of applying the film is known as " blooming ". By this process, it is now possible to reduce the reflection of visible light from about 5 per cent to less than 1 per cent.

5.23.—The films are now usually applied by evaporating the crystals *in vacuo* on to the glass surfaces. It is necessary to have a fairly good vacuum and to clean the glass surfaces by special methods in order to secure good adhesion. The thickness of the film is usually adjusted by observing the reflected light and watching for the characteristic purple colour. If the amplitudes are equal the relative energy is given by putting

$$\delta_1 = \frac{2\pi}{\lambda}(2\mu_c e \cos\theta)$$

in 4(1). This gives

$$E = 4a_1^2 \cos^2\left(\frac{2\pi}{\lambda}\mu_c e \cos\theta\right). \quad . \quad . \quad . \quad . \quad . \quad 5(11)$$

By differentiating 5(11) it may be seen that when 5(10) is satisfied, $\partial E/\partial\theta$, $\partial E/\partial\lambda$, and $\partial E/\partial e$ are all zero at $\theta = 0$. For normal incidence and one wavelength the effectiveness of the film is not greatly reduced if the thickness is not exactly right. Also if the thickness is right for normal incidence and one wavelength, it is nearly right for a considerable range of angle about normal incidence and for a considerable range of wavelength.

5.24. High-efficiency Reflecting Films.

In optical instruments it is sometimes desirable to divide a beam of light into two beams of nearly equal energy. If partly reflecting mirrors are made by coating glass with a thin film of aluminium, a considerable part of the light is absorbed by the metal (a typical film gives 35 per cent transmission, 35 per cent reflection and 30 per cent absorption).* Composite reflecting films of high efficiency may be produced by coating a glass surface with a thin film of low refractive index and then laying down a thin film of high refractive index. If the thicknesses are correctly chosen, the waves from the three reflecting surfaces are all in phase. An overall reflection coefficient of nearly 50 per cent can be obtained and nearly all the light which is not reflected is transmitted. Higher reflection coefficients may be obtained using many layers. With many layers, however, the films become strongly coloured because the condition for strong reflection is exactly satisfied for only one wavelength.

* For a given reflecting power, silver films show much lower absorption when new. They are liable to tarnish rapidly unless protected.

5.25.—Attempts have been made to produce non-reflecting films by treating glass surfaces with chemical reagents which dissolve part of the material. This leaves a surface layer whose refractive index is lower than that of the bulk material. The process is very difficult to control. Films are often seen on old glass or metal surfaces. Sometimes these give low reflection and sometimes high reflection. They are due to the chance production of thin surface layers of abnormal refractive index either during manufacture or by the subsequent action of the air or of cleaning materials on the surfaces.

EXAMPLES [5(x)–5(xii)]

5(x). Calculate the thickness which gives (*a*) maximum and (*b*) minimum reflection for light of wavelength 5000 Å. when a crystal of index 2·0 is deposited upon a glass whose index is 1·50. Consider normal incidence.

[(*a*) $6{\cdot}25 \times 10^{-6}$ cm., (*b*) $12{\cdot}5 \times 10^{-6}$ cm. Note that reflection at the glass/crystal surface takes place in the denser medium].

5(xi). A film of index 1·4 is deposited on a glass of index 1·6, and the thickness is adjusted to give minimum reflection for $\lambda = 5000$ Å. and for normal incidence. Calculate the effective reflection coefficient for (*a*) 5000 Å., $\theta = 0°$; (*b*) 6000 Å., $\theta = 0°$; (*c*) 5000 Å., $\theta = 30°$; (*d*) 6000 Å., $\theta = 30°$; (*e*) 4000 Å., $\theta = 30°$. Give two significant figures in the answers. Assume that, to the accuracy required, the reflection coefficient for one surface at 30° is the same as that at normal incidence.

[(*a*) 1·0%, (*b*) 1·3%, (*c*) 1·2%, (*d*) 1·8%, (*e*) 1·1%.]

5(xii). Show that the reflecting power for white light of a glass surface coated with a film many wavelengths thick is approximately half that of an uncoated surface.

[Passing from one end of the spectrum to the other, the cosine in 5(11) goes through many periods. The mean value of the square is 0·5.]

5.26. **Interference with Multiple Beams.**

If the surfaces bounding a film have a fairly high reflection coefficient, fringes of transmission with very sharp bright maxima on a dark ground may be produced (Plate I*f*, p. 72). These fringes are produced by the mutual interference of several beams which have suffered different numbers of reflections. Let us first consider an air-film bounded by two plane, parallel, half-silvered surfaces. Several parallel beams of light are produced by an initial beam incident at an angle θ (fig. 5.13). The transmitted beams are collected by the lens L and brought to a focus at a point Q in its focal plane. The lens brings all the beams together at Q with the phase differences which they had when they crossed the plane AB which is normal to OQ. Suppose that the two films are similar, and that a fraction σ of the light is transmitted and a fraction ρ reflected at each incidence. Let the change of phase on reflection be ϵ. With metallic films, owing to absorption, $\rho + \sigma$ is not equal to unity. Also ϵ is not exactly 0 or π. The ratios of the amplitudes of the transmitted and reflected waves to the amplitude of the incident wave are $\rho^{1/2}$ and $\sigma^{1/2}$ respectively.

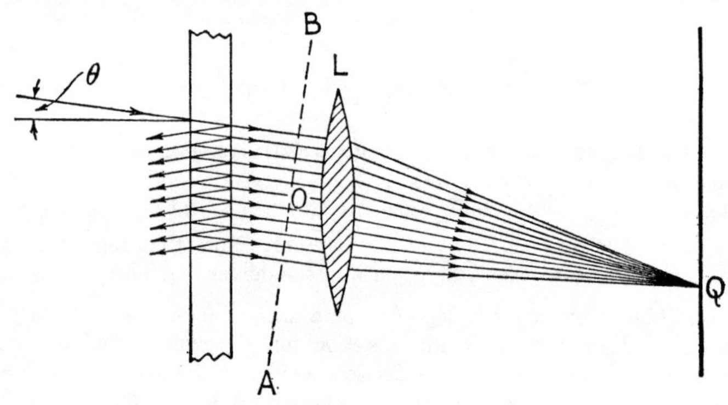

Fig. 5.13.—Multiple-beam interference

The phase difference between two successive beams is

$$\delta = \frac{2\pi}{\lambda} 2e \cos \theta + 2\epsilon. \quad . \quad . \quad . \quad . \quad . \quad 5(12)$$

If the incident beam is represented by

$$a_1 \exp i(\omega t - \kappa x) = a_1 e^{i\varphi},$$

the resultant transmitted wave is represented by

$$\sigma a_1 e^{i\varphi} + \rho\sigma a_1 e^{i(\varphi-\delta)} + \rho^2 \sigma a_1 e^{i(\varphi-2\delta)} + \ldots.$$

If P is the complex amplitude of the resultant (§ 2.26), we have

$$P = a_1\{\sigma + \rho\sigma e^{-i\delta} + \rho^2\sigma e^{-2i\delta} + \ldots\}$$

$$= \frac{a_1\sigma}{1 - \rho e^{-i\delta}}. \quad . \quad . \quad . \quad . \quad . \quad . \quad . \quad . \quad . \quad . \quad 5(13)$$

The complex conjugate P^* is given by

$$P^* = \frac{a_1\sigma}{1 - \rho e^{i\delta}},$$

and by a calculation similar to that given in § 3.6 we have

$$E = PP^* = \frac{a_1{}^2\sigma^2}{1 + \rho^2 - 2\rho\cos\delta},$$

which may be written

$$E = \frac{a_1{}^2\sigma^2}{(1-\rho)^2} \cdot \frac{1}{1 + \dfrac{4\rho}{(1-\rho)^2}\sin^2\tfrac{1}{2}\delta}. \quad . \quad . \quad 5(14)$$

Maxima occur when $\cos\delta = 1$ (i.e. when $\delta = 2n\pi$) and the minima are half-way between the maxima. The positions of the maxima and minima are thus independent of the values of ρ and σ. The maximum relative energy (E_{max}) is $a_1{}^2\sigma^2/(1-\rho)^2$ and the minimum (E_{min}) is $a_1{}^2\sigma^2/(1+\rho)^2$. Using equation 5(12), it may be seen that the maxima occur when

$$2e\cos\theta = \lambda\left(n - \frac{\epsilon}{\pi}\right). \quad . \quad . \quad . \quad . \quad 5(15a)$$

This reduces to

$$2e\cos\theta = n\lambda, \quad . \quad . \quad . \quad . \quad . \quad 5(15b)$$

if the change of phase on reflection may be neglected.

Circular fringes are obtained with the arrangement shown in fig. 5.14. The order of interference decreases from the centre outwards, because the path difference is smaller when the rays impinge at an

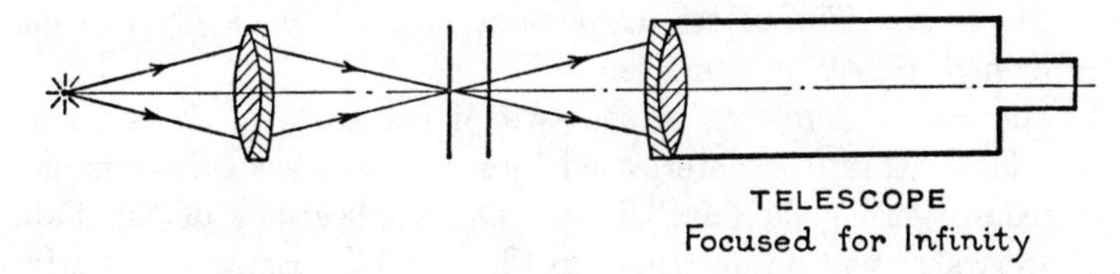

Fig. 5.14.—Plate of constant thickness viewed by convergent light

angle than when they are normal to the plates. The fringes are localized at infinity and are usually viewed with a telescope. The light reaching a given point in the focal plane of the objective corresponds to a certain direction of incidence. It comes from all parts of the film.

5.27.—From equation 5(14), it may be seen that a high reflection coefficient produces sharp maxima. If δ differs a little from $n\pi$, so that $\sin^2 \frac{1}{2}\delta$ has some value η, then

$$E_\eta = \frac{E_{max}}{1 + L\eta}, \quad . \quad . \quad . \quad . \quad . \quad . \quad . \quad 5(16)$$

where

$$L = \frac{4\rho}{(1 - \rho)^2}.$$

When ρ is of the order of 0·05 corresponding to unsilvered plates $L =$ 0·2, but when ρ is 0·8, corresponding to fairly heavily silvered plates, $L = 80$. Thus with strongly silvered plates the illumination in directions which differ only a little from the directions of the maxima is very low and the fringes appear as sharp bright lines on a dark ground. The value of ρ chosen is determined by the necessity of compromising between the increase of sharpness due to high ρ and the reduction of overall brightness due to the consequent reduction of σ. The effect of multiple beams in producing sharp maxima is discussed in §§ 9.15 and 9.50.

5.28. **Fabry-Pérot Interferometer.**

Several different methods of using the fringes formed by multiple reflections in plates of constant thickness have been designed. The Fabry-Pérot *interferometer* consists of two glass plates, one fixed and the other mounted on a carriage similar to that used in the Michelson interferometer. This plate can be slowly moved in a direction perpendicular to itself. The guides are so accurately constructed that the moving plate remains parallel to the fixed plate to within less than a second of arc. The plates are coated with a film of silver or other metal of high reflection coefficient.

In the Fabry-Pérot *etalon* the two plates are at a fixed distance apart. This distance is determined by separators accurately made so as to secure very good parallelism. The applications of the Fabry-Pérot apparatus will be described in Chapter IX. Examples of fringes obtained with a Fabry-Pérot etalon are shown in Plate I*f* and I*g* (p. 72).

5.29. **Lummer-Gehrcke Plate.**

The interference apparatus shown in fig. 5.15 was invented by Lummer and developed by him in collaboration with Gehrcke. Light from an extended source is totally reflected at the hypotenuse of the right-angled prism and is incident on the lower surface of the plate at an angle slightly less than the critical angle. A small fraction of the light is refracted to form beam 1. The rest is reflected and is incident on the upper surface where again part is refracted (forming beam 2) and part is reflected. Thus the light travels along the plate and at each incidence on the surface a portion escapes. The refracted light emerging from the lower surface forms a series of parallel beams with a constant phase-difference. The amplitudes decrease in geometrical progression. Fringes due to multiple reflections are thus formed. A similar set of fringes is produced by the beams 2, 4, 6, etc., which emerge from the upper surface.

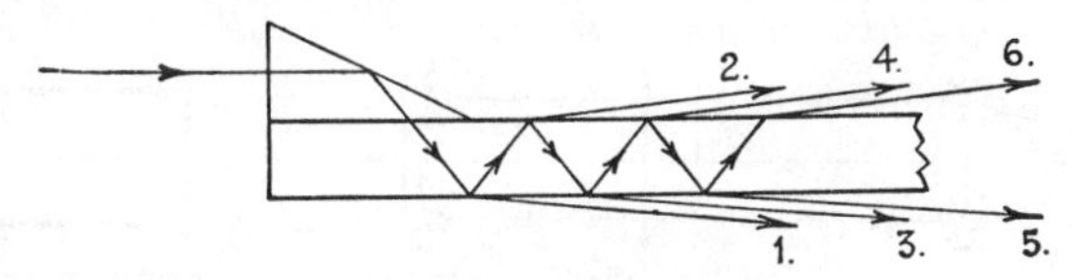

Fig. 5.15.—Lummer-Gehrcke plate

With the Fabry-Pérot arrangement, using nearly normal incidence, all the beams which have any appreciable amplitude are collected by the lens (fig. 5.13). When the Lummer plate is used, the number of beams available is normally limited by the length of the plate. The resultant amplitude is calculated from the sum of a series similar to that given in equation 5(13), but restricted to a finite number of terms. Partly for this reason the formulæ giving the positions of the bright fringes and the variation of brightness in the interference pattern are rather more complicated with the Lummer-Gehrcke plate than with the silvered plate etalon.*

5.30. **Channelled Spectrum.**

When a parallel beam of white light is allowed to fall upon a Fabry-Pérot etalon and the transmitted light is viewed in a spectroscope (fig. 5.16), the spectrum is crossed by a number of dark bands (Plate II*f*, p. 126). It is therefore called a *channelled* or *banded* spectrum though such spectra have, of course, no connection with *band* spectra due to molecules (§ 4.3). From equation 5(14), it is clear that maximum brightness occurs at wavelengths for which $\sin \frac{1}{2}\delta = 0$, i.e. when δ is an integral multiple of 2π. If the phase change on reflection [equation

* See p. 94 of Reference 9.1 for details of theory.

5(12)] may be neglected, maxima occur at wavelengths λ_0, λ_1, λ_2, etc., which satisfy the relation

$$2e \cos \theta = p\lambda_0 = (p + 1)\lambda_1 = (p + 2)\lambda_2, \quad . \quad 5(17)$$

where p is an integer.

By a simple extension of the discussion given in § 5.27, it may be shown that the maxima are sharp when the reflection coefficient is high, but a very high reflection coefficient implies a low transmission and makes the bands difficult to observe. In practice very sharp bands are obtained only if the beam of light is well collimated and if the plates of the etalon are accurately plane and parallel.

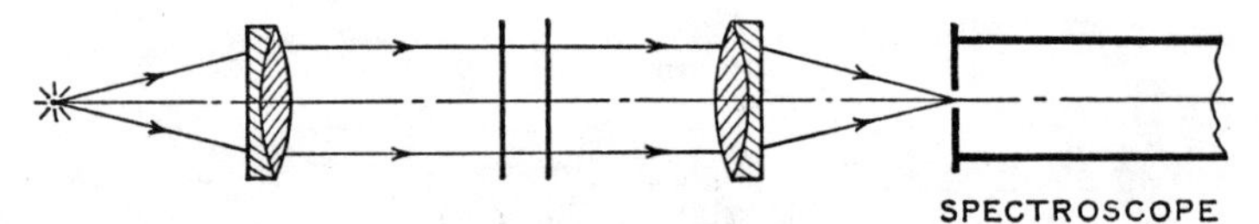

Fig. 5.16.—Edser-Butler method

Channelled spectra may also be observed by spectroscopic examination of light in the outer parts of a fringe system produced by any of the arrangements described in §§ 5.9–5.10. The bands so obtained are not so sharp because only two interfering beams are involved.

5.31. **Edser-Butler Method of calibrating the Spectrometer.**

If the thickness of the etalon has been measured (by any of the methods described in Chapter IX), the bands may be used to determine a curve, or formula, giving the relation between wavelength and position in the spectrum. The method is a convenient one because the bands provide a series of calibration marks suitably spaced throughout the spectrum. By choosing the right thickness of etalon, the spacing may be made as close as is desirable for the particular spectrograph or spectroscope which is being calibrated. Fringes of this type were first discovered by Fizeau and Foucault (1850) and were first used in the above way by Esselbach (1856). The method of calibration did not come into common use until it was rediscovered in 1896 by Edser and Butler and it is generally known in England by their name.

The banded spectrum may be used, in conjunction with two known wavelengths, to determine the thickness of the etalon. The light from the etalon is allowed to form a banded spectrum and, by means of a small mirror, the line spectrum including the known lines λ_1 and λ_2 is made to appear above or below the channelled spectrum. The number of maxima between the two lines is counted. If this number is m we have approximately

$$2e = p_1\lambda_1 = (p_1 + m)\lambda_2, \qquad 5(18)$$

and hence

$$2e = m\left(\frac{\lambda_1\lambda_2}{\lambda_1 - \lambda_2}\right). \qquad 5(19)$$

Equation 5(19) is exactly correct only if two bright bands happen to coincide with the lines. When this is not so, the positions of the maxima on either side of λ_1 and λ_2 may be measured. Using these readings and interpolating by proportional parts, two fractional parts to be added to m may be determined, and 5(19) may then be applied. This method of determining the thickness of the etalon forms an instructive class experiment, but is generally less accurate and less convenient than the method described in § 9.32.

EXAMPLES [5(xiii)–5(xvi)]

5(xiii). Compare and contrast the fringes formed by the Fabry-Pérot etalon using convergent monochromatic light with Newton's rings.

5(xiv). Show that the Edser-Butler method provides a set of marks spaced at equal intervals of wavelength constant (κ).

5(xv). Show that for a given direction of incidence the phase change on reflection is equivalent to a small correction to the separation of the plates of a Fabry-Pérot etalon.

5(xvi). The circular fringes formed by a Fabry-Pérot etalon are numbered 1, 2, 3, ... p from the centre outwards. Show that if the separation of the plates is a large integral multiple of the wavelength, the angular diameter of the pth ring is approximately proportional to $\sqrt{(p - 1)}$, when p is fairly small.

5.32. **Fringes of Superposition.**

Consider a beam of light incident successively upon two etalons (fig. 5.17). There will be a considerable number of emergent beams corresponding to different internal reflections. Under suitable conditions certain sets of emergent beams may have small path differences and give rise to interference fringes with white light. Consider, for example, two etalons of thicknesses e_1 and e_2, inclined at a small angle α to one another. Suppose that the emergent light is observed in a telescope and consider light which makes an angle θ with the normal to the first etalon. The path difference between (a) light which has made two reflections in the first etalon and four in the second, and (*b*) light which has made six reflections in the first etalon and two in the second is

$$\begin{aligned} 6e_1 \cos\theta + 2e_2 \cos(\theta+\alpha) - 2e_1\cos\theta - 4e_2\cos(\theta+\alpha) \\ = 4e_1\cos\theta - 2e_2\cos(\theta+\alpha) \\ = (4e_1 - 2e_2\cos\alpha)\cos\theta + 2e_2\sin\theta\sin\alpha. \quad . \quad 5(20) \end{aligned}$$

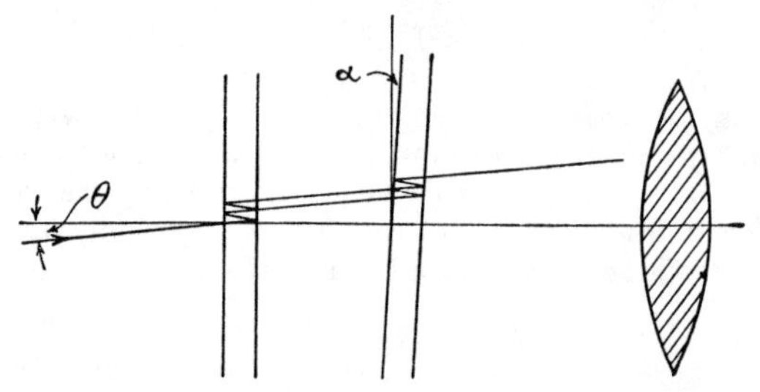

Fig. 5.17.—Fringes of superposition

If the second etalon is slightly more than twice as thick as the first, this path difference will be zero for some fairly small value of θ. Suppose that the inclination (α) has been adjusted so that the path difference is zero for $\theta = 0$. We then have

$$2e_1 = e_2 \cos\alpha \quad . \quad . \quad . \quad . \quad . \quad . \quad 5(21)$$

or, approximately,

$$e_2 - 2e_1 = \tfrac{1}{2}e_2\alpha^2 \quad . \quad . \quad . \quad . \quad . \quad . \quad 5(22)$$

and bright fringes will be obtained when

$$n\lambda = 2e_2 \sin\theta \sin\alpha. \quad . \quad . \quad . \quad . \quad . \quad . \quad 5(23)$$

Since the maxima correspond to a definite value of θ, the fringes are localized at infinity and may be seen in the telescope. If the telescope is normal to the first etalon, the central achromatic fringe will be in the centre of its field and the angular separation θ' of the fringes will be given approximately by

$$\theta' = \frac{\lambda}{2e_2\alpha}. \quad . \quad . \quad . \quad . \quad . \quad . \quad . \quad . \quad 5(24)$$

A measurement of the angular separation of the fringes determines α and equation 5(22) gives $(e_2 - 2e_1)$, provided an approximate value of e_2 is available. Fringes of this type are called fringes of superposition. They were first observed by Brewster in 1815. He used two glass plates of nearly equal thickness. The above method of using the fringes to compare the thicknesses of two etalons was suggested by Fabry and Buisson and was used by Sears and Barrell (§ 9.45).

5.33.—An alternative method of using the fringes to compare the thicknesses of two etalons is to set the two etalons so that $\alpha = 0$, and to allow the light to pass also through a thin wedge. In these circumstances fringes localized in the wedge are obtained. The centre of the fringe system is located at the point where the path in the wedge exactly compensates for the difference in path due to the etalons. This method of comparing two etalons was used by Benoît, Fabry and Pérot (§ 9.41).

Either of the methods described may be used to compare etalons the ratio of whose thicknesses is approximately $p : q$, where p and q are two small whole numbers. Fabry and Buisson state that etalons of ratio 10 : 1 may be compared. If the numbers p and q are not small, large numbers of internal reflections are involved with a corresponding loss of light. The fringes then become indistinct because they are seen against a brighter background due to beams which have not made the correct series of reflections to give interference fringes. Although, in the above treatment, we have considered only two beams, fringes of superposition are formed by multiple-beam interference when silvered plates are used. The reflection coefficient affects their sharpness just as it does with the circular fringes.

5.34. **Achromatic Fringes.**

If the geometrical path difference between two routes from a source of light to a point of observation is the same for all wavelengths, the phase difference will depend on the wavelength. Fringes are then obtained with white light only when the path difference is small. If,

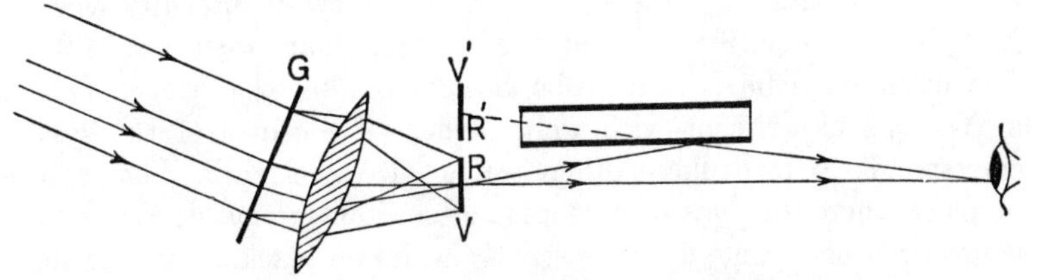

Fig. 5.18.—Lloyd's arrangement for achromatic fringes

however, the geometrical path is different for different wavelengths, then there is a possibility that the geometrical path difference may be made proportional to the wavelength, so that the phase difference is the same for all wavelengths. If this condition is fulfilled, achromatic fringes may be obtained with white light even when the path difference is fairly large. A simple arrangement for producing fringes of this type is shown in fig. 5.18. The slit source used with Lloyd's mirror is replaced by a short spectrum RV formed by the slits and a prism or grating placed at G. The grating is to be preferred since it produces a spectrum in which distances are directly proportional to wavelength (§ 6.31). The separation of the "sources" in Lloyd's mirror is then proportional to the wavelength and the separation of the fringes is the same for all wavelengths [equation 5(4)].

5.35.—The example we have just considered shows the possibility of the formation of a system of achromatic fringes, but is too restricted to show the general conditions required for their observation. In discussing the wider problem let us start by considering two extreme cases. With the Michelson interferometer, using cadmium red light, fringes corresponding to orders of interference up to about a million have been obtained. There is a very gradual decrease in visibility at high orders due to the imperfect homogeneity of the light, but apart from this effect all fringes are identical. This, as

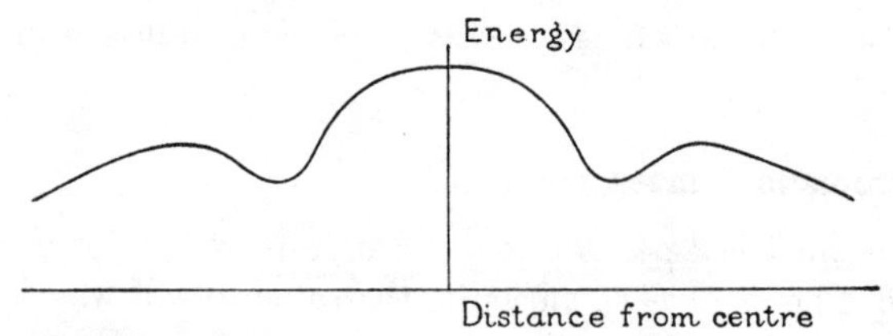

Fig. 5.19.—Energy distribution as measured by bolometer

we shall see later, sometimes occasions a practical difficulty when it is desired to identify the fringe of a particular order (see § 9.32). The other extreme case may be represented by the fringes formed in Young's experiment with white light. The centre of the system corresponds to zero phase difference for all wavelengths, but there is no place where the phase difference is π for all wavelengths. Thus there is not even one fringe which is strictly white on black. By eye, about half a dozen fringes can be seen.

If a bolometer is used to measure the energy arriving at different places on the screen, a curve of the type shown in fig. 5.19 is obtained. The bolometer is a non-selective detector of radiation; it measures the energy and is equally sensitive to all wavelengths. The two minima on either side of the maximum are present because ordinary white light contains more energy of some wavelengths than of others (Chapter

XVII). A source of light for which (in the notation of § 4.19) $a(\kappa)$ is independent of κ shows no trace of interference. If, however, a filter transmitting a small region of the spectrum is placed in front of the bolometer, fringes are recorded just as they would be if the same filter were placed in front of the source. If the light from a certain narrow region is analysed in a spectroscope, bands are seen in the spectrum (§ 5.30). Thus interference is observed when either

(*a*) the source emits radiation in which some wavelengths predominate, or

(*b*) the receptor is able to analyse the energy it receives into a spectrum.

To observe high orders of interference, it is necessary either to have a source of very homogeneous light or to use a receptor which can make a very fine analysis. A selective receptor (such as the bolometer covered with a filter) from this point of view makes a very simple analysis of the light, distinguishing between (*a*) light to which it is sensitive and (*b*) light to which it is not sensitive. The eye makes an analysis of light in terms of three primary colours and is therefore able to see a few fringes where the non-selective bolometer could detect none.

5.36. Achromatic Systems of Fringes.

In Young's experiment, there is one place in the field of view where the phase difference is the same for all wavelengths. There is thus one achromatic fringe. In the experiment described in § 5.34, in which Lloyd's mirror is combined with a grating, the phase difference is independent of the wavelength at all parts of the field. We then have a *system of achromatic fringes*. It is important to distinguish these two cases. In the one we have a single place where the fringes of all wavelengths coincide, but the separation of the fringes is a function of the wavelength, and therefore the system is achromatized for only one fringe. In the second case the separation is independent of the wavelength so that, if the fringes are made to coincide at one point, they do so everywhere.

5.37.—Analytically, we may suppose the order of interference (p) to be a function of λ and of a co-ordinate x which locates the position of the fringe in the field of view. p is not necessarily integral (§ 4.9). Then for an achromatic fringe we require

$$\frac{\partial p}{\partial \lambda} = 0 \quad . \quad . \quad . \quad . \quad . \quad . \quad . \quad . \quad . \quad . \quad 5(25)$$

at some point in the field.

For an achromatic system we require 5(25) and also

$$\frac{\partial}{\partial x}\left(\frac{\partial p}{\partial \lambda}\right)=\frac{\partial^2 p}{\partial x\,\partial \lambda}=0. \qquad 5(26)$$

For *perfect achromatism*, 5(25) should be true for all values of λ, and 5(26) for all values of λ and x. In practice we usually call a lens " achromatic " when the variation of focal length with wavelength has a maximum or minimum for some wavelength near the middle of the visible spectrum (fig. 5.20). In a similar way, we call a *fringe* achromatic if 5(25) is satisfied for some value

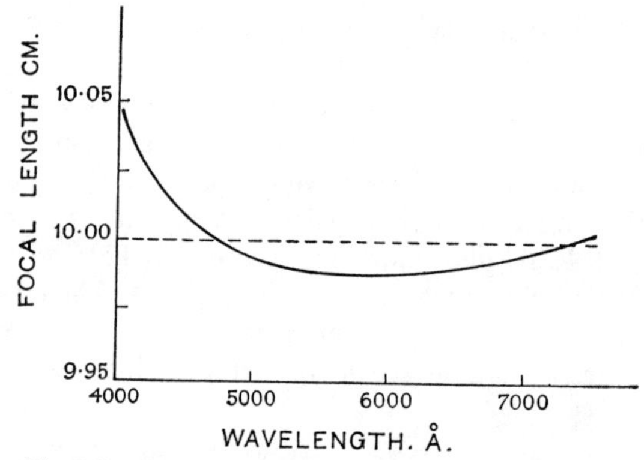

Fig. 5.20.—Variation of focal length with wavelength for an achromatic lens combination

of λ near the middle of a range in which we are interested, and a *system* achromatic if 5(25) and 5(26) are satisfied for one wavelength and at one point. The central fringe in Young's experiment is *perfectly achromatic*, to the extent to which we can neglect secondary effects due to finite slit width, etc. Subject to a similar limitation, the system described in § 5.34 is *perfectly achromatic* when a grating is used. It is achromatic, in the less restricted sense, when a prism is used.

5.38.—It is possible to shift the centre of a system of fringes, such as those obtained by Young, Fresnel or Lloyd, by inserting a plate of optically dense material in front of one of the real or virtual sources of light, e.g. to the right of slit a_1 in fig. 5.2*b*. If this material is non-dispersive, the central fringe will move to the position where the difference of optical path is zero, i.e. where the paths from the two sources to the point are equal when measured in wavelengths. In general the material has some dispersion. One of the optical paths is a function of the wavelength and the other is not. Therefore, the path difference is a function of the wavelength, and there is no point for which this difference is zero for all wavelengths. In these circumstances the central fringe is located at the point where 5(25) is satisfied, i.e. at the point of maximum agreement of phase difference. This does not, in general, coincide with the place where the path difference is zero for the central, or indeed for any one, wavelength.

5.39.—It was shown in § 4.29 that when a non-homogeneous beam of light travels through a dispersive medium, the point at which the phases of the disturbance, of different wavelengths, are in best agreement, travels with the group velocity. Thus, in the type of experiment which we are considering, the maximum agreement of *difference* of phase will occur at a point P such that the time taken for the light to travel from S to P by the non-dispersive path is equal to the time taken for a wave group to travel from S to P by the dispersive path. We shall now show that when this condition is satisfied, 5(25) is also satisfied.

Suppose that the non-dispersive path from S to P is of length l and that the other path from S to P consists of a distance $L - t_g$ in the non-dispersive medium, which for simplicity we take to be a vacuum, and t_g in the dispersive medium. Then the phase difference between light arriving at P from S by the two routes is

$$p = \frac{2\pi}{\lambda}\{L - l + (n-1)t_g\}, \qquad 5(27)$$

where λ is the wavelength *in vacuum.*

Differentiating with respect to λ, condition 5(25) gives

$$(L - l) + \left(n - \lambda\frac{dn}{d\lambda} - 1\right)t_g = 0. \qquad 5(28)$$

This equation determines the position of the new achromatic fringe.

The condition that the wave groups shall arrive together is

$$\frac{l}{c} = \frac{L - t_g}{c} + \frac{t_g}{U}, \qquad 5(29)$$

where U is the group velocity. Inserting the value of U from 4(41) we have

$$(L - l) + \left(n - \lambda\frac{dn}{d\lambda} - 1\right)t_g = 0, \qquad 5(30)$$

in agreement with 5(28).

It was at one time thought that the achromatic fringe was located at the point where the path difference is zero (presumably for the mean wavelength). The condition so calculated differs from 5(28) in that it does not include the term $\lambda\, dn/d\lambda$. The essentials of the present theory were originally put forward by Sir G. B. Airy (1801–92) in reply to certain critics who, basing their argument on the incorrect formula, thought that the wave theory did not give the right value for the shift of the fringes. The modern development of the theory is due to Rayleigh (Reference 5.4).

5.40. **Interference Filters.**

Coloured glasses or dyed gelatine filters may be used to isolate a region of the spectrum covering a range of 500 Å. but, for many experiments in photo-chemistry and for other purposes, it is desirable to isolate a band 50 Å. wide centred on a chosen wavelength; and occasionally much narrower bands are desired. *Interference filters* may conveniently be used for these purposes. We have seen in § 5.30 that the spectrum of a parallel beam of light which has passed through a Fabry-Pérot etalon consists of a series of sharp bands separated by wider dark regions (Plate II*f*, p. 126). If the distance between the reflecting surfaces is reduced, then the wavelength difference between the transmission maxima is increased. Equation 5(17) shows that for an optical thickness of $\lambda_0/2$, and for normal incidence, the transmission maxima are λ_0, $\lambda_0/2$, $\lambda_0/3$, etc. If λ_0 is placed in the visible spectrum then all transmission bands, except the first, lie in the ultra-violet and are absorbed by a suitable glass. If the silver surfaces have a reflection coefficient of 94 per cent and $\lambda_0 = 5000$ Å., the transmission band-width (measured between wavelengths for which the transmission is half the maximum transmission) is 50 Å. (fig. 5.21*a*).

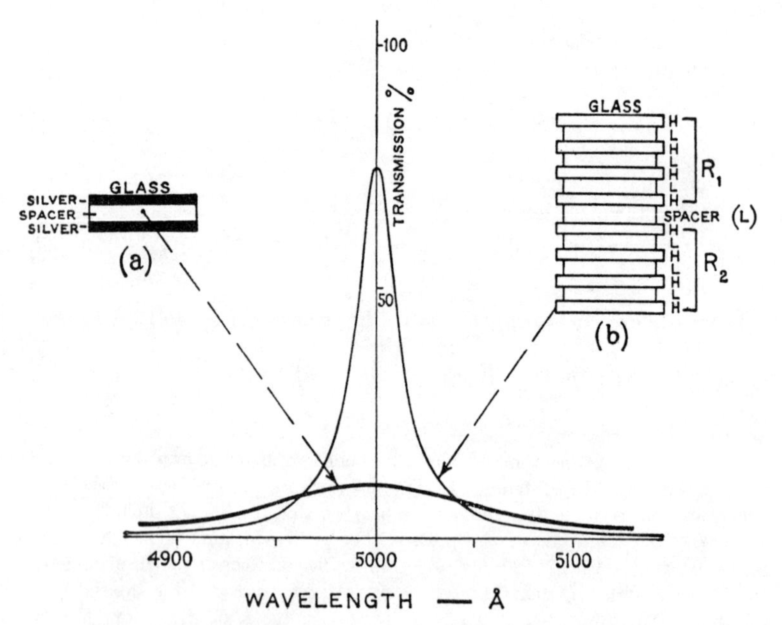

Fig. 5.21.—Interference Filters. Typical transmission curve for (*a*) filter consisting of two layers of silver (94% reflectivity) separated by a dielectric "spacer" whose optical thickness is 2500 Å., and (*b*) a filter consisting of two reflectors R_1 and R_2 separated by a similar spacer. The reflectors consist of layers of high (H) and low (L) index material and the optical thickness of each layer is 1250 Å.

When silver films are used the high reflectivity is obtained only by increasing the thickness of the film and so reducing the maximum transmission (§ 5.30). This limitation may be overcome by using, in place of the silver layers, high-efficiency reflectors constructed in the way described in § 5.24. These high-reflecting " stacks " consist of layers of dielectric materials. The layers are of equal optical thickness and alternate layers are of high and of low refractive index. Since the highest reflectivity is needed only in a narrow range of wavelength, it is possible to use stacks of many layers, and reflectivities of 98 per cent may be obtained. The loss of light by absorption is very small. The layers are deposited by vacuum evaporation and there is a small loss of light by scattering since perfectly homogeneous films are not obtained by this process. It is possible to obtain a band-width of 25 Å. with a maximum transmission of 75 per cent (fig. 5.21*b*). In addition to band-pass filters for desired regions of the visible spectrum, many other devices for giving selective transmission or reflection may be made by suitable combinations of high and low index layers. For further details the reader should consult Reference 5.5.

EXAMPLES [5(xvii)–5(xix)]

5(xvii). Two etalons have separations of 19·9990 mm and 40·000 mm. Find the angular separation of fringes formed in the way described in § 5.32 when light of wavelength 5000 Å. is used. [$6{\cdot}25 \times 10^{-4}$ radian.]

5(xviii). Discuss the achromatization of Lloyd's mirror fringes by means of a prism of material which obeys Cauchy's law (§ 3.18).

5(xix). What effects are observed when the fringes formed in a thin film enclosed between two plane surfaces are viewed through a prism?

[The fringes are shifted but not made achromatic. The position of the new central fringe is obtained by applying the discussion of § 5.39.]

REFERENCES

5.1. SEARLE: *Phil. Mag.*, 1946, Vol. XXXVII, p. 361.

5.2. ARNOT: *Proc. Camb. Phil. Soc.*, 1938, Vol. XXIV, p. 150.

5.3. GUILD: *Proc. Phys. Soc.*, Vol. XXXIII, p. 40.

5.4. RAYLEIGH: *Scientific Papers*, Vol. III, p. 288.

5.5. HEAVENS: *Optical Properties of Thin Solid Films* (Butterworth).

CHAPTER VI

Diffraction

6.1. General Character of the Observations.

When a beam of light passes close to the edge of an opaque obstacle, propagation is not truly rectilinear. If certain conditions (which will be discussed later) are fulfilled, fringes are observed near the edge of the geometrical shadow, and it is found that some light penetrates into the shadow. Typical effects are shown in Plate III (p. 214). Phenomena of this type were first reported by Grimaldi in 1665 and were studied by Newton. An attempt to relate these observations to wave theory was made by Young in 1802. He thought that the fringes were due to interference between light reflected at the edge of the obstacle and the direct beam. Fresnel showed that the fringes were nearly the same whether a razor edge or a rounded edge was used. Since the rounded edge would give a much larger reflection at grazing incidence than the razor edge, Fresnel's observation showed that Young's theory was unsatisfactory. Fresnel calculated the positions of the fringes from the mutual interference of the Huygens wavelets, taking into account the fact that part of the wavefront is obstructed by the obstacle, and he obtained results in general agreement with observation. In detail his theory is not entirely satisfactory, but his main ideas have now been incorporated by Kirchhoff in a logical development of the wave theory.

6.2.—The name *diffraction* is given to all departures from rectilinear propagation of light. The most obvious diffraction effects are produced by opaque obstacles, although diffraction is also produced by obstacles which are not opaque. For example, diffraction fringes may be produced by air bubbles imprisoned in a lens. Diffraction is produced by any arrangement which causes a change of amplitude or phase which is not the same over the whole area of the wavefront. Diffraction thus occurs when there is any limitation of the width of a beam of light. In all optical experiments the width of the beam is limited by the dimensions of the apparatus, so that some diffraction is always present. Diffraction effects are often masked by imperfections in optical images owing to lens defects and in other similar ways. It is only when these other effects have been reduced by suitable design of apparatus that diffraction becomes of great importance. It then sets an inescapable limit to the sharpness of optical images and to the accuracy of certain types of measurement.

6.3. **Fresnel and Fraunhofer Diffraction.**

Fig. 6.1 shows an arrangement in which a parallel beam of light passes through a slit in a screen S_1 (called the *diffraction screen*) and the light is received on a second screen S_2. Fig. 6.2 shows, in a general way, the variation of illumination across the screen S_2 for different distances between S_1 and S_2. When the screens are very close (fig. 6.2*a*) the illumination is uniform within the geometrical image, and zero outside. Within the accuracy of observation we have rectilinear propagation. As the screen S_2 is moved away from S_1, there is

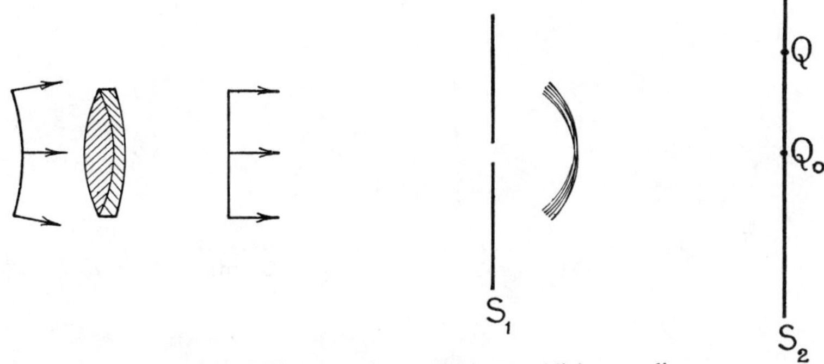

Fig. 6.1.—Diffraction of a parallel beam of light at a slit

a region in which the geometrical image is still easily recognizable although fringes appear on its edges (fig. 6.2*b*). This is known as the region of *Fresnel diffraction*. When the screen S_2 is removed to a very large distance the fringes spread so that they obliterate the geometrical image. A system of fringes is produced on S_2 and the distribution of illumination is determined by the shape and position of the openings in the screen S_1, but it does not indicate the shape of these openings in any simple way. This type of diffraction is called *Fraunhofer diffraction.*

To illustrate this last point, fig. 6.2*c* may be compared with fig. 6.3, which shows the variation of illumination across the screen S_2 when there are two openings of equal size in S_1. The pattern differs from that shown in fig. 6.2, but not in such a way as to show immediately that there is more than one slit. The kind of diffraction produced when the screen S_2 is placed at a very great distance from S_1 is also found when a positive lens of wide aperture is placed between S_1 and S_2, at such a position that S_2 is in its focal plane. The image of S_2 seen from S_1 is then infinitely distant from S_1.

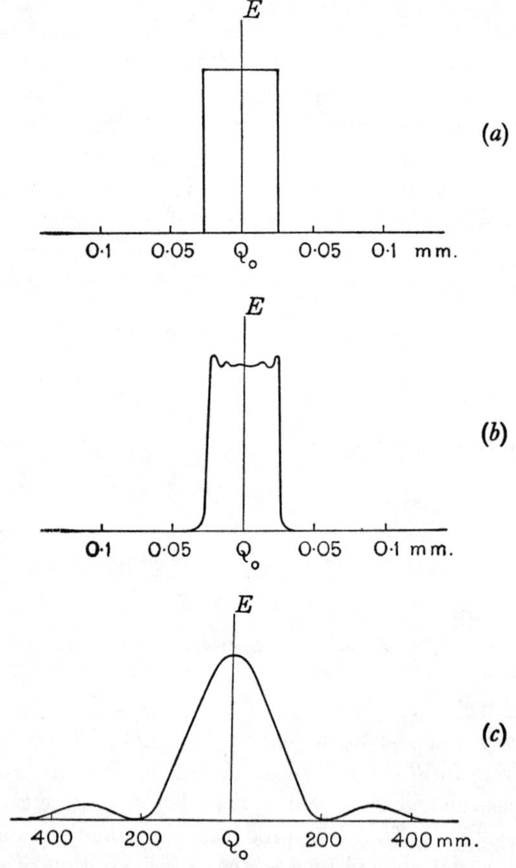

Fig. 6.2.—Variation of illumination across the screen S_2 in fig. 6.1 for slit 0·05 mm. wide: (*a*) when S_2 is in contact with S_1, (*b*) when S_2 is a few centimetres from S_1, (*c*) when S_2 is at 20 metres from S_1.

In this type of diffraction the source of light is in focus on the screen where the fringes are obtained. Provided this condition is fulfilled, Fraunhofer patterns are formed when convergent waves are obstructed, or when a plane or divergent wave which passes an obstruction is rendered convergent, i.e. it does not matter whether the obstacle is in front of, behind, or within the optical system which forms the image of the source (fig. 6.4). Fraunhofer patterns are located in

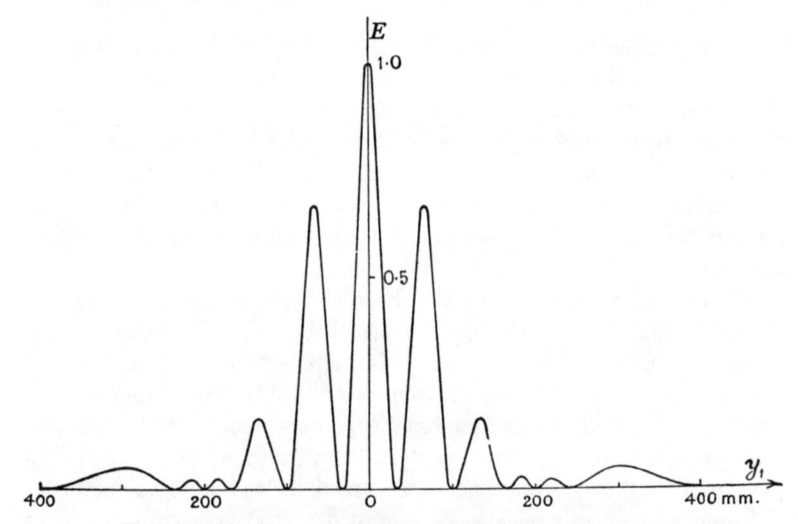

Fig. 6.3.—Distribution of illumination in pattern due to two equal slits

the plane which contains the centre of curvature of the wave surfaces. When a plane wave is received upon a screen at a great distance from the source and from the obstacles, an approximation to this condition is obtained.

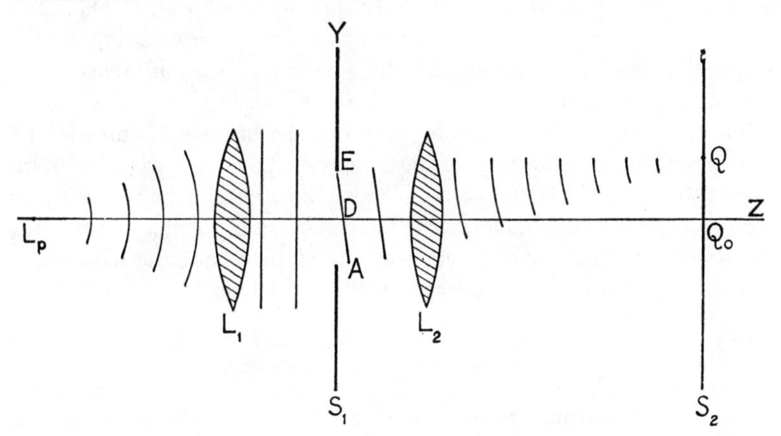

Fig. 6.4.—Arrangement for the production of Fraunhofer diffraction patterns

6.4.—Three regions are thus distinguished:

(1) Sharp image—effectively rectilinear propagation.
(2) Fringed image of the obstacle—Fresnel diffraction.
(3) Fringed image of the source—Fraunhofer diffraction.

It is convenient to distinguish these regions because they are observed with rather different experimental arrangements, and because the mathematical methods used in the calculations are also different. It is, however, important to recognize that there is no sharp distinction between them. On the experimental side, as S_2 is moved away from S_1, we pass gradually from one stage to another and, while three regions are easily recognizable, there is no well-defined boundary between them.

Diffraction phenomena of the Fresnel type were discovered and studied in detail before the Fraunhofer phenomena. They are of very great importance in the historical development of the wave theory and still have some interesting applications. The Fraunhofer diffraction phenomena are of greater practical importance, not only in providing a theory of the action of diffraction gratings, but also in connection with the general theory of optical instruments. The theory of Fraunhofer diffraction may be regarded as an extension of the idea of wave groups discussed in Chapter IV. It is of great importance in relating the wave theory and quantum theory.

6.5. **Theory of Diffraction. The General Problem.**

The distribution of illumination in diffraction patterns of either type may be obtained in principle by finding a solution of the wave equation, subject to a set of boundary values corresponding to the physical conditions obtaining at the surfaces of the obstacles. Unfortunately the mathematical difficulties are very great, and an exact solution has been worked out for only one special case. Sommerfeld * calculated the distribution of light energy produced when a beam passes the edge of a thin screen which is perfectly reflecting. Assuming that $\xi = 0$ over the whole area of the screen, it is possible to solve the wave equation and to pick out portions of the solution which correspond to the forward and backward diffracted light.

6.6.—A summary of Sommerfeld's calculation is given in Reference 6.1. Exact treatment of other diffraction problems involves detailed assumptions concerning the behaviour of the waves at, and near, the boundaries. For this purpose, it is necessary to make detailed assumptions about the type of wave. A general

* Reference 6.1.

theory of the diffraction of light must be based on a theory of transverse waves. Modern calculations are referred to the electromagnetic theory of light, and have been made by Stratton and others.*

6.7.—Let us now return to the diffraction problem illustrated in fig. 6.1 in order to see how approximate solutions may be obtained. Under the conditions of Fresnel diffraction, the illumination at a point Q on the screen S_2 may be computed by calculating the resultant amplitude due to all the Huygens wavelets emanating from different parts of the aperture. In this treatment the plane wave on the left of S_1 is replaced, on the right-hand side of S_1, by a system of *spherical* wavelets. Special assumptions have to be made concerning the phases and amplitudes of these wavelets but, if these assumptions are accepted, the rest of the calculation is merely an application of the principle of superposition. For purposes of calculation Fresnel divided the area of the aperture into a series of zones known as Fresnel zones (§ 7.2).

6.8.—Now consider an approximate treatment of the corresponding Fraunhofer problem. Let us suppose that two lenses are inserted as shown in fig. 6.4. Light from the point source L_p is focused into a parallel beam by the lens L_1, and the screen S_2 is situated in the focal plane of the lens L_2. It is assumed that the properties of the wave to the right of S_1 may be calculated by imagining the incident wave to be replaced by a system of *plane* waves. These plane waves originate in the aperture in the screen S_1 and their phase relations are determined by the phase of the incident wave in the plane of the aperture. In the case chosen, the phase of the incident wave is the same (at a given time) at all points in the aperture. It is therefore assumed that the distribution of phases on a plane such as ADE is determined by the distances from points on ADE to corresponding points of the apertures. In the case shown this implies that if ADE makes an angle θ with the screen S_1, the phase at E is in advance of that at A by an amount $(2\pi/\lambda)2d \sin \theta$, where $2d$ is the width of the slit. For points intermediate between A and E, the phase varies uniformly. The action of the lens † is such that a series of wavelets reaches a certain point Q on the screen S_2 with the same phase relations as they had when they left the plane ADE.

The relative energy may be obtained by the use of the vector diagram (§ 3.5). The diagram for a large number of infinitesimal

* See References 6.1, 6.2, and 6.3.

† We neglect aberrations of the lens.

elements whose phases vary uniformly is an arc of a circle, the resultant amplitude being the chord (fig. 3.2*b*).

If $2u_1$ is the difference of phase between the elements at the two ends, the resultant amplitude is proportional to $\sin u_1$. If $E(y)$ represents the ratio of the energy at the point Q to the energy at the centre of the diffraction pattern, which is at Q_0, we have

$$E(y) = \left(\frac{\sin u_1}{u_1}\right)^2 = \left\{\frac{\sin [(2\pi d \sin \theta)/\lambda]}{2\pi d \sin \theta/\lambda}\right\}^2. \quad . \quad . \quad 6(1)$$

This function is shown in fig. 6.9*a* (p. 177).

6.9.—Comparing the approximate methods of calculation described in §§ 6.7 and 6.8, we see that in one case the incident wave is replaced by a system of *spherical* waves and in the other by a system of *plane* waves. The difference between these substitutions is purely one of mathematical convenience. In Fresnel diffraction problems we wish to calculate the amplitude at a point fairly near the diffraction aperture. It is natural to draw radii from this point to elementary areas on the aperture and consider the interference of spherical wavelets propagated along these lines. In Fraunhofer problems we are concerned with light diffracted in a given *direction*. It is then convenient to consider plane waves. There is only one basic theory of light behind both the Fresnel and the Fraunhofer methods of calculation. Both of these methods need to be justified by reference to the fundamental hypotheses of the wave theory.

Kirchhoff has made an important analysis which justifies the formulæ we shall use both in the case of Fresnel and in the case of Fraunhofer diffraction *subject to certain approximations being accepted.* The conditions of many diffraction problems are such that these approximations are valid. Owing to mathematical difficulties no general derivation of exact diffraction formulæ has yet been found. Kirchhoff's analysis and its limitations are discussed in Appendix VI A (p. 191). In the text of this chapter we discuss the problems of Fraunhofer diffraction, using Fourier methods, and obtain, in a somewhat different way, some of the formulæ which will be derived in the Appendix.

6.10. Extension of the Concept of a Wave Group.

In Chapter IV it was shown that a train of waves of finite length does not constitute a pure simple harmonic wave, but can be analysed by Fourier's theorem into a set of simple harmonic waves of different wavelengths. When the wave train is fairly long, the effective range of wavelength is small and the system is called a *wave group*. In a similar way, we shall now show that a beam of light which is of finite width cannot be represented by a single plane wave, but may be regarded as the resultant of a system of plane waves travelling in different directions. When the width of the beam is very large in comparison with the wavelength, the effective angular spread is small, and the waves form a group of plane waves. All beams of light which occur in real problems are limited both in the direction of propagation and in the two transverse directions. There are certain problems for which the effects due to the limitation in the direction of propagation are unimportant. It is then sufficient to assume for the time being that the wave train is infinitely long in the z direction, which we take to be the centre of the small solid angle which contains the beam. This implies that we have a simple harmonic wave of one sharply defined frequency. In order further to simplify the initial stages of our calculation we consider, for the present, a beam which is limited in the y direction and unlimited in the x direction.

6.11.—An unlimited plane wave propagated in a direction perpendicular to the x axis and making an angle θ with the z axis may be represented by

$$a \exp i(\omega t - \kappa y \sin\theta - \kappa z \cos\theta), \qquad 6(2a)$$

which may be written

$$a \exp i(\omega t - \kappa_y y - \kappa_z z), \qquad 6(2b)$$

where $$\kappa_y = \kappa \sin\theta \quad \text{and} \quad \kappa_z = \kappa\cos\theta. \qquad 6(3)$$

Consider a group of waves whose directions lie within the range of angles which correspond to values of κ_y between $(\kappa_y - \frac{1}{2}d\kappa_y)$ and $(\kappa_y + \frac{1}{2}d\kappa_y)$. The amplitude due to all the waves is assumed to be proportional to $d\kappa_y$ and to a function of κ_y which we denote by $a(\kappa_y)$. Hence, if $d\xi$ is the resultant disturbance for the infinitesimal group, we have

$$d\xi = a(\kappa_y) \exp i(\omega t - \kappa_y y - \kappa_z z)\, d\kappa_y. \qquad 6(4)$$

Consider a system of waves whose mean direction is along the z axis, and which lies within the range of θ corresponding to $\kappa_y = -\Delta$ to $\kappa_y = +\Delta$ where Δ is small but finite.

The resultant disturbance for $t = 0$ and $z = 0$ is

$$\xi_0(y) = \int_{-\Delta}^{\Delta} a(\kappa_y) \exp(-i\kappa_y y)\, d\kappa_y. \qquad 6(5)$$

If it is understood that $a(\kappa_y) = 0$ for all values of κ_y outside the range $\pm\Delta$, which corresponds to the angular spread of the beam, no alteration is caused by extending the limits of integration to infinity; thus we may write

$$\xi_0(y) = \int_{-\infty}^{\infty} a(\kappa_y) \exp(-i\kappa_y y) \, d\kappa_y. \quad . \quad . \quad . \quad 6(6)$$

By Fourier's theorem this relation implies that

$$a(\kappa_y) = \frac{1}{2\pi} \int_{-\infty}^{\infty} \xi_0(y) \exp(i\kappa_y y) \, dy. \quad . \quad . \quad . \quad 6(7)$$

For a strictly parallel beam travelling in the direction of the z axis, $a(\kappa_y)$ would be zero except when $\kappa_y = 0$; 6(6) shows that $\xi_0(y)$ would then be independent of y. The displacement would then be the same at all points on the y axis and the beam would be infinitely wide in the y direction. If the beam is not strictly parallel, but $a(\kappa_y)$ is given, then the disturbance at different points on the y axis for $t = 0$ can be calculated from 6(6).

If, for example, $a(\kappa_y) = A$ when κ_y is between $\pm\Delta$, and zero outside this range,* then

$$\xi_0(y) = \frac{2A \sin y\Delta}{y}. \quad . \quad . \quad . \quad . \quad . \quad 6(8)$$

From this equation we see that $\xi_0(y)$ has appreciable values only when y is in the range π/Δ to $-\pi/\Delta$; if Δ is very small the beam *must* be very wide. If Δ is fairly large, then the main energy of the beam may be confined within a narrow range of y.

6.12. Beam of Finite Width—One Dimension.

Let us now assume that we are given that $\xi_0(y) = A$ when y is between $-d$ and $+d$, and that $\xi_0(y) = 0$ outside this range. The beam has constant amplitude on the line $y = 0$ over a range which is physically small, though still large compared with the wavelength. Equation 6(7) then gives

$$a(\kappa_y) = \frac{A}{\pi} \cdot \frac{\sin \kappa_y d}{\kappa_y}. \quad . \quad . \quad . \quad . \quad . \quad 6(9)$$

* See Appendix IV B, equation 4(88). Remember that the mean value of κ_y corresponding to κ_m is zero.

If E is the ratio of the energy in the direction defined by κ_y to the energy in the direction of the z axis [for which $\kappa_y = 0$ and $(\sin \kappa_y d)/\kappa_y = d$], we have

$$E = \left\{\frac{\sin \kappa_y d}{\kappa_y d}\right\}^2 = \left\{\frac{\sin [(2\pi d \sin \theta)/\lambda]}{2\pi d \sin \theta/\lambda}\right\}^2. \quad . \quad . \quad 6(10)$$

From 6(10) we see that if d is very large compared with λ, the value of E is small except when θ is small. The beam is then nearly a parallel beam. As d decreases, the energy is effectively spread over a wider and wider range of angle. The width of the beam and the angular spread are connected by a definite mathematical relation.*

6.13. **St. Venant's Hypothesis.**

In the preceding discussion we have assumed that if $a(\kappa_y)$ is to be calculated, $\xi_0(y)$ is given. If we are required to calculate the distribution with angle, we are given the disturbance for $t = 0$ and $z = 0$ at all points on the line $y = 0$. In practical problems we are given that a screen with an aperture (or a series of apertures) is placed so as to intersect † the line $y = 0$. It is usual to assume that the disturbance at all points corresponding to clear apertures in the screen is the same as if the screen were not present, and that the disturbance at all points covered by the screen is zero. This assumption is known as St. Venant's hypothesis. St. Venant's hypothesis cannot be exactly correct because, if it were, the value of $\partial\xi/\partial x$, etc., at the edge of an opaque obstacle would be infinite and this would be inconsistent with the wave equation.‡ Nevertheless calculations based on this hypothesis do lead to results which are very closely in agreement with experimental results, except when the sizes of the apertures are of the same order as the wavelength. This is usually taken as sufficient reason for accepting the hypothesis as approximately correct. Once it is accepted we can apply the foregoing discussion to actual diffraction problems.

* The relation we have derived applies when a plane wave is incident upon the slit. This gives the *minimum* spread of waves emerging from the slit. A greater angular spread will be obtained if converging or diverging waves are incident.

† The general problem involves restriction by a screen which is in the xy plane. At present we consider only one dimension.

‡ See Reference 6.1 for a general discussion of the mathematical difficulties involved.

6.14. Beam restricted in Two Dimensions.

Before proceeding to particular problems it is convenient to obtain equations which correspond to 6(6) and 6(7), but which apply to two dimensions. We consider a beam restricted both in the x direction and in the y direction.

A direction of propagation is now defined by the two variables κ_x and κ_y, where

$$\kappa_x = \alpha\kappa \text{ and } \kappa_y = \beta\kappa, \qquad 6(11)$$

α and β are the cosines of the angles between the chosen direction and the axes of x and y. A group of waves now covers a small solid angle $d\kappa_x \, d\kappa_y$, and equations 6(6) and 6(7) are replaced by

$$\xi_0(x, y) = \int_{-\infty}^{\infty}\int_{-\infty}^{\infty} a(\kappa_x, \kappa_y) \exp\{-i(\kappa_x x + \kappa_y y)\} \, d\kappa_x \, d\kappa_y, \qquad 6(12)$$

and

$$a(\kappa_x, \kappa_y) = \frac{1}{4\pi^2}\int_{-\infty}^{\infty}\int_{-\infty}^{\infty} \xi_0(x, y) \exp i(\kappa_x x + \kappa_y y) \, dx \, dy. \qquad 6(13)$$

The last equation may also be written

$$a(\kappa_x, \kappa_y) = \frac{1}{4\pi^2}\int_{-\infty}^{\infty}\int_{-\infty}^{\infty} \xi_0(x, y) \exp \frac{2\pi i}{\lambda}(\alpha x + \beta y) \, dx \, dy. \qquad 6(14)$$

This equation is of the form $a(\kappa_x, \kappa_y) = C + iS, \qquad 6(15)$

where C and S are real functions of κ_x and κ_y.

The ratio of the flux of radiation in a direction defined by κ_x and κ_y to that in the centre of the pattern is

$$E(\kappa_x, \kappa_y) = \frac{C^2 + S^2}{C_0^2 + S_0^2}, \qquad 6(16)$$

where C_0 and S_0 correspond to $\kappa_x = \kappa_y = 0$.

When the clear parts of the diffraction screen are symmetrical with respect to the axes of co-ordinates, $S = 0$. Equation 6(13) then reduces to

$$a(\kappa_x, \kappa_y) = C = \frac{A}{4\pi^2}\iint \cos(\kappa_x x) \cos(\kappa_y y) \, dx \, dy \qquad 6(17)$$

(where A is the constant value of ξ_0 over the domain of integration), and, in this case,

$$E = \frac{C^2}{C_0^2}. \qquad 6(18)$$

The calculation of the angular distribution for a beam of rectangular cross-section (width equal to $2d_x$ in the x direction and to $2d_y$ in the y direction) is carried out by inserting the appropriate limits in 6(17). The result is

$$E(\kappa_x, \kappa_y) = \left(\frac{\sin(\kappa_x d_x)}{\kappa_x d_x}\right)^2 \left(\frac{\sin(\kappa_y d_y)}{\kappa_y d_y}\right)^2. \qquad 6(19)$$

From equation 6(19) we see that a restriction in the width of the beam in the x direction involves an angular spread in the xz plane, and a restriction on the width in the y direction involves an angular spread in the yz plane. In the type of problem with which we are now concerned the finite length of the wave train does not produce any important effects. We still use the assumption stated in § 6.10 that there is no restriction in the z direction.

6.15. **Diffraction at a Rectangular Aperture.**

The distribution of illumination on the screen S_2 (fig. 6.4), when a point source of light is used with a rectangular slit aperture in S_1, is given by equation 6(19). If the screen S_2 is distant z_1 from the lens L_1 and if x_1 and y_1 are given by

$$\frac{2\pi}{\lambda}\frac{x_1}{z_1} = \kappa_x \quad \text{and} \quad \frac{2\pi}{\lambda}\frac{y_1}{z_1} = \kappa_y, \qquad \text{6(20)}$$

then all portions of the wave whose direction of propagation is defined by κ_x and κ_y are focused at a point Q whose co-ordinates are x_1 and y_1. The distribution of illumination on the screen is given by

$$E(x_1, y_1) = \left\{\frac{\sin\left(\frac{2\pi}{\lambda}\frac{d_x}{z_1}x_1\right)}{\frac{2\pi}{\lambda}\frac{d_x}{z_1}x_1}\right\}^2 \left\{\frac{\sin\left(\frac{2\pi}{\lambda}\frac{d_y}{z_1}y_1\right)}{\frac{2\pi}{\lambda}\frac{d_y}{z_1}y_1}\right\}^2. \qquad \text{6(21)}$$

The illumination is zero whenever $d_x x_1/(2\lambda z_1)$ or $d_y y_1/(2\lambda z_1)$ is a whole number (other than zero). Thus the diffraction pattern contains two sets of mutually perpendicular lines on which the illumination is zero. Regions of maximum illumination lie between these lines.

The pattern is shown in Plate III*m* (p. 214). It will be noticed that the long side of the rectangles in the pattern corresponds with the short side of the aperture. If the aperture is a very long vertical slit, the horizontal lines in the pattern are so close that they become confused. The resulting pattern is similar to that shown in Plate III*e*.

6.16. **Diffraction at a Circular Aperture.**

In view of the symmetry of the aperture, equation 6(17) may be used. The integral is to be evaluated over the area of a circle whose radius we shall take to be R. It is sufficient to investigate the variation of illumination along any radius from Q_0; we therefore choose the radius through Q_0 parallel to the x axis, for which $\kappa_y = 0$. Equation 6(17) is integrated over the region covered by the aperture, the limits of integration for y being $\pm(R^2 - x^2)^{1/2}$:

$$C = \frac{A}{4\pi^2}\int_{-R}^{R}\int_{-(R^2-x^2)^{1/2}}^{(R^2-x^2)^{1/2}} dy \,.\, \cos(\kappa_x x)\, dx \qquad \text{6(22)}$$

$$= \frac{A}{2\pi^2}\int_{-R}^{R}(R^2 - x^2)^{1/2}\cos(\kappa_x x)\, dx. \qquad \text{6(23)}$$

If we put $x = R\cos\chi$ and $\rho = \kappa_x R$, then

$$C = \frac{AR^2}{2\pi^2}\int_0^{\pi}\sin^2\chi\cos(\rho\cos\chi)\,d\chi. \quad . \quad . \quad . \quad . \quad 6(24)$$

Bessel's integral of order unity is defined by the equation

$$J_1(\rho) = \frac{\rho}{\pi}\int_0^{\pi}\sin^2\chi\cos(\rho\cos\chi)\,d\chi. \quad . \quad . \quad . \quad . \quad 6(25)$$

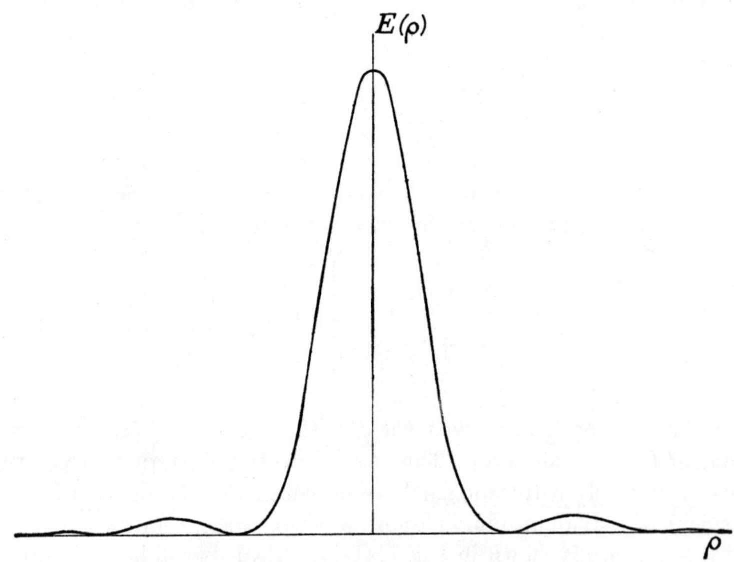

Fig. 6.5.—Distribution of illumination in the Airy diffraction pattern for a circular aperture

This expression cannot be integrated in the form of a finite algebraic expression, but values have been calculated by series and tables are available. From 6(24) and 6(25) we have

$$C = \frac{AR^2}{2\pi}\cdot\frac{J_1(\rho)}{\rho}, \quad . \quad . \quad . \quad . \quad . \quad . \quad . \quad . \quad 6(26)$$

and, using the fact that $J_1(\rho)/\rho$ tends to the value $\frac{1}{2}$ when ρ tends to 0, we have

$$E(\rho) = 4\left(\frac{J_1(\rho)}{\rho}\right)^2, \quad . \quad . \quad . \quad . \quad . \quad . \quad . \quad . \quad . \quad 6(27)$$

$$E(x_1) = 4\left\{\frac{J_1\left(\frac{2\pi}{\lambda}\cdot\frac{Rx_1}{z_1}\right)}{\frac{2\pi}{\lambda}\cdot\frac{Rx_1}{z_1}}\right\}^2. \quad . \quad . \quad . \quad . \quad . \quad 6(28)$$

$E(x_1)$ is the illumination at a distance x_1 from P_0. It is computed by calculating the value of ρ in terms of x_1, z_1, and R, and then referring to the tables. The result is shown graphically in fig. 6.5 and a picture of the diffraction pattern is shown in Plate III*k* (p. 214). The pattern consists of a bright central disc surrounded by a series of alternate bright and dark rings. The radii corresponding to maxima and minima are given in Table 6.1, which shows also the values of the maximum illumination. It may be seen that all except the innermost rings are very weak. Computation shows that 84 per cent of the whole light is concentrated in the central disc. This is known as Airy's disc, after Sir G. B. Airy (1801–92), who first made a general investigation of this problem.

TABLE 6.1

Ring number	Radius (in units of $z_1\lambda/2R$)		Maximum illumination (relative to centre)
	Dark ring	Bright ring	
1	1·22	1·64	0·0174
2	2·23	2·69	0·0041
3	3·24	3·72	0·0016
4	4·24	4·72	0·0008
5	5·24	5·72	0·0004

6.17. **Diffraction with a Slit Source.**

In most experiments on diffraction by slits, wires, etc., a slit source is used. Each point on the slit source acts as an independent source of spherical waves. These sources are effectively non-coherent, and the resultant beam is *not* represented by a cylindrical wave. The illumination at any point in the diffraction pattern is the sum of the illuminations (not the square of the sum of the amplitudes) due to the different parts of the slit. Suppose that the rectangular aperture, whose diffraction pattern with a point source is shown in Plate III*m*, is illuminated by a vertical slit source. The resulting pattern is the same as that which would be obtained by superposing a series of patterns individually similar to that shown in Plate III*m* but separated in a vertical direction. The result is a series of vertical bands (Plate III*e*). The horizontal lines have nearly disappeared. The variation of illumination in a horizontal direction is given by equation 6(21), i.e. it is equal to

$$E(x_1) = \left\{ \frac{\sin\left(\dfrac{2\pi}{\lambda}\dfrac{d_x}{z_1}x_1\right)}{\dfrac{2\pi}{\lambda}\dfrac{d_x}{z_1}x_1} \right\}^2. \qquad \text{6(29)}$$

Suppose that the width of the slit source is initially small and is increased until the width of its image on the screen S_2 is comparable with the distances between the vertical bands. The effect will be the same as superposing a series of patterns separated in the *horizontal* direction. The fringes will gradually become less clear and finally disappear. Thus, in order to obtain clear fringes, it is necessary that the source of light should be a *narrow* slit. It is also necessary that the axis of the slit source be fairly accurately parallel to the slit or wire which forms the diffraction object. These considerations apply with minor modifications to Fresnel, as well as to Fraunhofer, diffraction by slits, wires, etc. For similar reasons a clear pattern is obtained with a circular aperture only when the size of the source is so small that its image is smaller than the Airy disc.

6.18. **Diffraction by a Number of Similar Apertures.**

We shall now discuss the diffraction patterns produced by a distant point source and a large number of geometrically similar apertures. We shall show that the illumination at any point in the diffraction pattern is the product of (*a*) the illumination which would be given by one of the apertures, and (*b*) a factor depending on the arrangement. It will be shown that when the apertures are arranged in a completely irregular way, the second factor is simply equal to the number of apertures. The relative illumination at a given point from all the apertures is the same as that found at the same point in the pattern given by a single aperture. When the apertures are arranged in a completely regular way, they are said to constitute a diffraction grating, and the second factor (*b*) depends on the size of the unit which is repeated. Obviously, there are various types of arrangements intermediate between complete regularity and random spacing, but only the two extreme cases will be considered.

6.19.—Let O_1, O_2, O_3, etc. (fig. 6.6), be a series of points, one in each of the apertures, and let each be similarly placed with respect to its aperture. Regard these as a series of local origins and let the co-ordinates of a point in a given aperture (referred to its own local origin) be (x', y'). Let the co-ordinates of O_1, O_2, etc., referred to a given co-ordinate system lying in the diffraction screen be (X_1, Y_1), (X_2, Y_2), etc. Let (x, y) be the co-ordinates of any point, in any aperture, referred to the main co-ordinate system. Then we have, for the fth aperture,

$$x = X_f + x' \text{ and } y = Y_f + y'. \quad . \quad . \quad . \qquad 6(30)$$

For the moment, let us restrict the treatment to one dimension. The apertures then become slits. Let the width be $2d$. We may then insert the above value of y in 6(7) and obtain

$$a(\kappa_y) = \frac{1}{2\pi} \sum_{f=1}^{N} \int_{-d}^{+d} \xi_0(y) \exp\,[i\kappa_y(Y_f + y')]\, dy' \quad . \quad 6(31)$$

for the pattern produced by N slits. The integrals are to be taken over the areas of the slits.

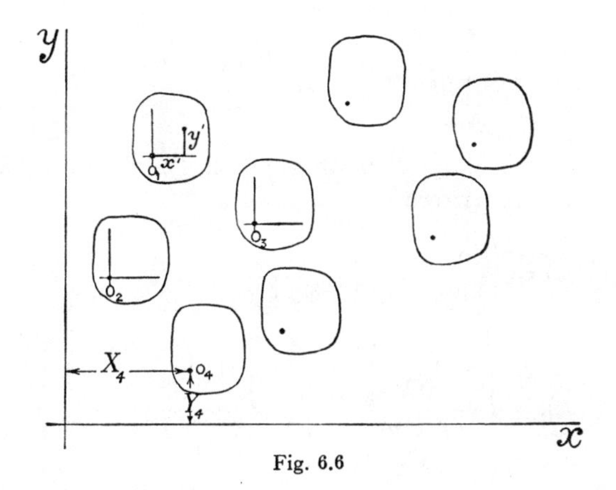

Fig. 6.6

If we assume that O_1, etc., are situated in the centres of the slits, and that the amplitude has the constant value A over the clear parts of the screen, then 6(31) becomes

$$a(\kappa_y) = \frac{A}{2\pi} \sum_{f=1}^{N} \exp\,(i\kappa_y Y_f) \int_{-d}^{+d} \exp\,(i\kappa_y y')\, dy'. \quad . \quad 6(32)$$

Carrying out the integration,

$$a(\kappa_y) = \frac{A}{\pi} \frac{\sin \kappa_y d}{\kappa_y} \sum_{f=1}^{N} \exp\,(i\kappa_y Y_f). \quad . \quad . \quad 6(33)$$

The first factor, obtained from the integral, is the amplitude due to one slit alone [equation 6(9)]. The second one shows that the phase of the disturbance arriving at a given point on the screen S_2 from the fth aperture depends on Y_f. When the slits are irregularly arranged the phases given by the exponential factor vary in an irregular way, and the vibrations are non-coherent. Consequently the total illumination is equal to the sum of the illuminations for the individual apertures

(§ 5.2). Since these are all equal it is equal to N times the illumination due to one aperture. The ratio of the illumination at any point to that at the centre of the pattern is the same as that for one aperture; it is given by 6(10).

6.20.—When the apertures are separated by equal distances, the disturbances become a series of equal amplitude with phase angles increasing in arithmetic progression. The summation was carried out in § 3.4. If the distance between the centres of successive slits is $2e$, the result is

$$a(\kappa_y) = \frac{A}{\pi}\frac{\sin \kappa_y d}{\kappa_y}\frac{\sin N\kappa_y e}{\sin \kappa_y e}, \qquad 6(34)$$

and the ratio of the illumination in the direction defined by κ_y to that in the centre of the pattern is

$$E(\kappa_y) = \left\{\frac{\sin \kappa_y d}{\kappa_y d}\right\}^2 \left\{\frac{\sin N\kappa_y e}{N \sin \kappa_y e}\right\}^2, \qquad 6(35)$$

and

$$E(y_1) = \left\{\frac{\sin\left(\frac{2\pi}{\lambda}\frac{d}{z_1}y_1\right)}{\frac{2\pi}{\lambda}\frac{d}{z_1}y_1}\right\}^2 \left\{\frac{\sin\left(\frac{2\pi}{\lambda}\frac{Ne}{z_1}y_1\right)}{N\sin\left(\frac{2\pi}{\lambda}\frac{e}{z_1}y_1\right)}\right\}^2. \qquad 6(36)$$

The general case of apertures of any shape and spacing may be solved by the following method, which is due to Drude. Insert the values of x and y from 6(30) into 6(13), and regard x' and y' as the independent variables. Then

$$a(\kappa_x, \kappa_y) = \frac{1}{4\pi^2}\sum_{f=1}^{N}\iint \xi_0(x', y') \exp i[\kappa_x(x' + X_f) + \kappa_y(y' + Y_f)]\, dx'\, dy'. \qquad 6(37)$$

Put $\xi_0(x', y') = 4\pi^2 A$ within the area of an aperture. Separate the real and imaginary parts. Let c and s refer to the real and imaginary parts of the amplitude for one aperture, and C and S to those for the whole set. Then

$$\left.\begin{aligned} C &= A\sum_{f=1}^{N}\iint \cos\,[\kappa_x(x' + X_f) + \kappa_y(y' + Y_f)]\, dx'\, dy',\\ S &= A\sum_{f=1}^{N}\iint \sin\,[\kappa_x(x' + X_f) + \kappa_y(y' + Y_f)]\, dx'\, dy', \end{aligned}\right\} \qquad 6(38)$$

$$\left.\begin{aligned} c &= A\iint \cos\,(\kappa_x x' + \kappa_y y')\, dx'\, dy',\\ s &= A\iint \sin\,(\kappa_x x' + \kappa_y y')\, dx'\, dy', \end{aligned}\right\} \qquad 6(39)$$

all the integrals being taken over the area of one aperture. Write

$$\left.\begin{aligned} c' &= \sum_{f=1}^{N} \cos(\kappa_x X_f + \kappa_y Y_f), \\ s' &= \sum_{f=1}^{N} \sin(\kappa_x X_f + \kappa_y Y_f). \end{aligned}\right\} \quad . \quad . \quad . \quad . \quad . \quad 6(40)$$

Then from 6(38), 6(39) and 6(40)

$$\left.\begin{aligned} C &= c'c - s's \\ \text{and} \qquad S &= s'c + c's, \end{aligned}\right\} \quad . \quad . \quad . \quad . \quad . \quad . \quad . \quad 6(41)$$

so that $$C^2 + S^2 = (c^2 + s^2)(c'^2 + s'^2). \quad . \quad . \quad . \quad . \quad . \quad . \quad 6(42)$$

Equation 6(42) is a general equation for summing the effect of a large number of similar apertures. $(c'^2 + s'^2)$ is the factor determined by the arrangement. We may write it

$$\begin{aligned} c'^2 + s'^2 = \sum_{f=1}^{N} \cos^2(\kappa_x X_f + \kappa_y Y_f) + \sum_{f=1}^{N} \sin^2(\kappa_x X_f + \kappa_y Y_f) \\ + \sum_{f=1}^{N} \sum_{g=1}^{N} \cos(\kappa_x X_f + \kappa_y Y_f) \cos(\kappa_x X_g + \kappa_y Y_g) \\ + \sum_{f=1}^{N} \sum_{g=1}^{N} \sin(\kappa_x X_f + \kappa_y Y_f) \sin(\kappa_x X_g + \kappa_y Y_g). \quad 6(43) \end{aligned}$$

In the double summations, terms for which $f = g$ are to be omitted. When X_f and Y_f are varying arbitrarily, the terms involving a double summation vary irregularly between plus and minus one and the sum is nearly zero (§ 5.2). Hence

$$c'^2 + s'^2 = N. \quad . \quad . \quad . \quad . \quad . \quad . \quad . \quad . \quad . \quad 6(44)$$

A new treatment of diffraction problems especially intended for irregular screens has recently been given by Booker, Ratcliffe and Shinn (Reference 6.9). They consider the auto-correlation function

$$G(\alpha) = \frac{\int_{-\infty}^{\infty} \xi^*(x)\,\xi(x+\alpha)\,dx}{\int_{-\infty}^{\infty} \xi^*(x)\,\xi(x)\,dx}.$$

They show that while the *amplitude* of the light diffracted in a given direction is proportional to the Fourier transform of $\xi(x)$ [see equation 6(7)], the *energy* is proportional to the Fourier transform of the auto-correlation function. If we know the auto-correlation function but do not know ξ, it is possible to calculate the angular distribution of energy but not the phase distribution in the diffracted light. Conversely, if the energy distribution, but not the phase, is given, we cannot calculate ξ but we can calculate the auto-correlation function $G(\alpha)$. This type of analysis leads very readily to the result given in 6(44).

6.21. Babinet's Theorem.*

Suppose that the diffraction screen is removed from the apparatus shown in fig. 6.4. Then the pattern formed on the screen S_2 will consist of the central image of the source together with very faint fringes due to diffraction at the edges of the aperture (i.e. at the edges of the lenses). Now suppose that two diffraction screens S_1' and S_1'' have the property that the clear regions in S_1' exactly correspond with the opaque regions in S_1'' and *vice versa.* These screens are said to be *complementary* screens.

A theorem due to J. Babinet (1794–1872) states that, except in the region of the central image, the distributions of illumination in the diffraction patterns due to two complementary screens are identical. This theorem is proved by applying the principle of superposition. The disturbance at any point, when the screens are removed, is the sum of the disturbances produced by the two complementary screens. Therefore at points where the amplitude without the screens is nearly zero, the amplitudes due to the complementary screens must be nearly equal, and the phases must differ by π. Thus the values of the illumination given by two screens must be equal. It is important to note that the theorem applies only at points where the illumination with the screens removed is nearly zero. There is no *simple* relation between the illuminations or amplitudes produced at other points by complementary screens. The theorem applies to Fresnel diffraction patterns, but since in these patterns there are not usually large regions where the illumination without the screens is nearly zero, it is of very restricted application.

6.22. Diffraction by a Number of Circular Apertures or Obstacles.

The discussion of §§ 6.18–6.20 shows that the diffraction pattern due to a large number of *irregularly arranged* circular apertures is similar to that calculated for a single circular aperture and consists of a central bright disc surrounded by a series of alternate light and dark rings. The application of Babinet's principle shows that the diffraction pattern due to a set of irregularly arranged dark obstacles is similar except for a very small region in the centre. Diffraction rings of this type may sometimes be seen surrounding street lights, when small water-droplets are present in the atmosphere. They may also be formed by small particles on the surface of the eye. Under suitable

* The theorem is often called " Babinet's *Principle* ".

meteorological conditions, the moon is seen to be surrounded by halos owing to diffraction by large numbers of small ice crystals in the upper atmosphere (though not all lunar and solar " halos " are formed in this way).

6.23. **Young's Eriometer.**

It is possible to apply this result to measure the diameters of small particles such as blood corpuscles. A simple apparatus for carrying out the experiment is formed by taking a sheet of metal and drilling about 12 holes, each of 1 millimetre diameter, at equal intervals round a circle of diameter 15 centimetres. A hole of 3 millimetres diameter is drilled in the centre of the circle (fig. 6.7). A source of roughly monochromatic light (such as a sodium lamp or a filament covered with a green filter) is placed behind the screen. A microscope slide containing the corpuscles, or a piece of glass dusted with lycopodium powder, is held close to the observer's eye. Looking towards the screen he sees the central hole surrounded by a series of light and dark rings. By adjusting his distance from the screen he may bring one of the rings into coincidence with the circle defined by the small holes. If the distance from the screen is then measured, the diameter of the particles on the slide is obtained by reference to Table 6.1 (p. 165). This device is known as Young's Eriometer.

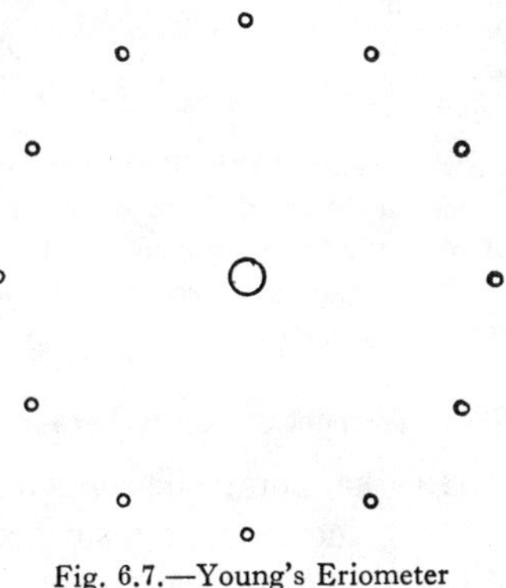

Fig. 6.7.—Young's Eriometer

The diameters of blood corpuscles from one individual are not all exactly equal, but are distributed about a mean. This reduces the sharpness of the diffraction rings. Even with this limitation, readings of the diameter of a given ring on one specimen are reproducible to within 3 per cent, and it is easy to detect any variation from the norm which is sufficiently large to be of clinical importance.

6.24. **Diffraction by Reflecting Screens.**

It is found that if the apertures in a diffraction screen are covered with reflecting material, and the opaque spaces with non-reflecting material, the reflected light includes a diffraction pattern similar to those discussed above.

This result may be included in the theory developed in § 6.10, provided we specify more closely the conditions to be fulfilled at a reflecting surface. In § 6.10 we assumed that the incident wave might be replaced by a system of plane waves with different values of κ_y, but all with a component of velocity in the positive direction of z. The same value of ξ in the plane of the screen would have been given by a similar set of waves having the corresponding velocity components in the negative direction of z. To justify the use of the first set rather than the second, when the spaces are clear, we now assume that, in the plane of the diffraction screen, the incident wave must be replaced by a system which gives the same value of $\partial\xi/\partial z$ as well as the same value of ξ. In a similar way, referring to § 3.27, we see that when a reflecting screen is used we must have

$$\left.\begin{aligned} \xi + \xi' &= 0, \\ \text{and} \qquad \frac{\partial \xi}{\partial z} + \frac{\partial \xi'}{\partial z} &= 0, \end{aligned}\right\} \qquad \qquad 6(45)$$

where ξ refers to the disturbance due to the incident wave, and ξ' is the disturbance due to the diffracted waves. These conditions are satisfied only by a system of waves with components in the negative direction of z.

It should be noted that we always assume $\xi = 0$ on the opaque (or non-reflecting) spaces.

6.25. **Diffraction by a Screen not Coincident with a Wave Surface.**

In the above discussion we have assumed that the diffraction screen coincides with a surface of constant phase for the incident wave. When this condition is not fulfilled, the fundamental equations such

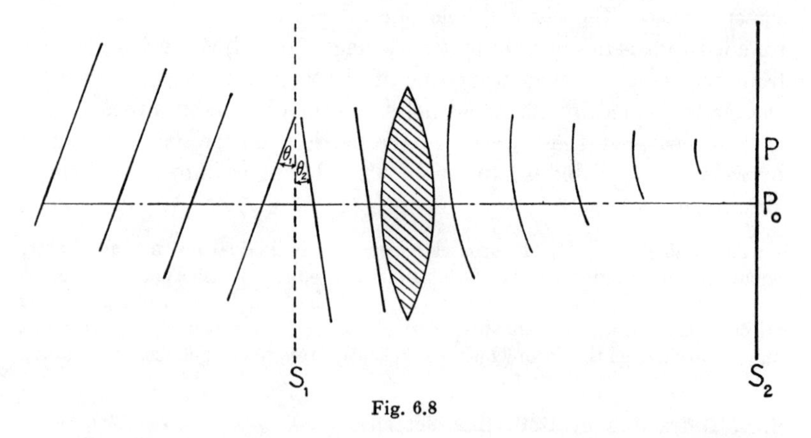

Fig. 6.8

as 6(7) and 6(12) are still valid but ξ_0, instead of being a real function of x and y, includes an imaginary part representing the variation of phase. In order to see how this occurs, we may consider the case illustrated in fig. 6.8.

Suppose that the incident light falls on the diffraction screen at an angle θ_1 to the normal, and we wish to calculate the amount of light diffracted in the direction θ_2. We write

$$\kappa_y' = \kappa \sin \theta_1 \quad \text{and} \quad \kappa_y = \kappa \sin \theta_2, \qquad 6(46)$$

and the incident wave is represented by

$$\xi = A \exp i(\omega t - \kappa_y' y - \kappa_z' z). \qquad 6(47)$$

The disturbance on the line $z = 0$ is given by

$$\xi = Ae^{-i\kappa_y' y} e^{i\omega t} = \xi_0 e^{i\omega t}, \qquad 6(48)$$

where

$$\xi_0 = Ae^{-i\kappa_y' y}. \qquad 6(49)$$

ξ_0 is thus a complex quantity which is a function of y. Its value gives both the real amplitude and the phase. Inserting this value of ξ_0 in 6(7), we have

$$a(\kappa_y) = \frac{1}{2\pi} \int Ae^{-i\kappa_y' y} \,.\, e^{+i\kappa_y y}\, dy,$$

$$= \frac{A}{2\pi} \int \exp \{i(\kappa_y - \kappa_y')y\}\, dy. \qquad 6(50)$$

The integral has to be taken over the region covered by the apertures. For a single aperture 6(10) is replaced by

$$E = \left\{ \frac{\sin (\kappa_y - \kappa_y')\, d}{(\kappa_y - \kappa_y')\, d} \right\}^2, \qquad 6(51)$$

or

$$E = \left\{ \frac{\sin \dfrac{2\pi}{\lambda} d(\sin \theta_2 - \sin \theta_1)}{\dfrac{2\pi}{\lambda} d(\sin \theta_2 - \sin \theta_1)} \right\}^2, \qquad 6(52)$$

where E is the ratio of the energy diffracted in direction κ_y to that diffracted in direction κ_y'.

Corresponding substitutions may be made in equations 6(19) and 6(37). Thus we see that the distribution of illumination in these patterns is affected by the change in the direction of incidence only in that the centre has been transferred from the direction defined by $\kappa_y = 0$ to the direction defined by $\kappa_y = \kappa_y'$.

Other types of diffraction screen may

(*a*) alter the amplitude over part of the wavefront (local absorption without complete opacity);

(*b*) insert a partially reflecting surface in part of the wavefront;

(*c*) cause a local phase-change over part of the wavefront (e.g. a bubble in a lens or a local indentation or " hump " on the surface of a lens).

The distribution of illumination may, in all these cases, be obtained by first calculating the value of ξ_0 as a complex function of the co-ordinates, and then inserting it in equation 6(13). A further extension of diffraction theory to cases where the obstacles are not all in one surface may also be made. This " three-dimensional diffraction " is of great importance in connection with the diffraction of X-rays by crystals and has been studied by Laue and others.

In the theory of X-ray diffraction the energy diffracted in a given direction is shown to be proportional to the product of two factors. One factor (called the " structure factor ") depends on the distribution of the atoms in the unit cell; the other factor depends on the arrangement of the cells (i.e. upon the space-group). The problem is analogous to the optical problem discussed in §§ 6.18–6.20.

6.26. **Laws of Rectilinear Propagation, Reflection and Refraction.**

In § 6.11 we saw that when the width of the beam was very large compared with the wavelength of the radiation, the energy was very small except in the direction defined by $\kappa_y = 0$. Thus, after passing a very wide aperture, nearly all the energy continues to travel in the direction of the incident wave. Equation 6(52) shows that this is still true when the wave strikes the aperture obliquely. Applying the argument of § 6.24 to a beam falling obliquely upon a reflecting aperture, it may be shown that the favoured direction is now that for which $\kappa_y = -\kappa_y'$, i.e. a direction which makes the same angle with the normal as the incident direction, but on the opposite side of the normal. Finally, referring to equation 6(51), we see that when the diffraction screen separates two media of different refractive indices (n_1 and n_2), the centre of the pattern is located in a direction defined by

$$\kappa_y = \kappa_y',$$

or

$$n_1 \sin \theta_1 = n_2 \sin \theta_2. \qquad 6(53)$$

Thus in each case the centre of the diffraction pattern lies in the direction predicted by the ordinary laws of rectilinear propagation, of reflection and of refraction. The theory of diffraction thus includes these laws as limiting relations, valid for indefinitely wide beams.

6.27. **Diffraction Gratings.**

Any regular arrangement of similar apertures constitutes a diffraction grating, but one particular type of diffraction grating is of special importance in optics. This grating consists of a series of parallel lines ruled on glass or metal and spaced at equal intervals. A simple grating of this type may be made by covering a glass plate with an opaque layer of silver and then ruling parallel lines in the silver. The silver is removed by the needle, and clear spaces are left. This method produces a grating in which the clear spaces have a very high transmission, while the opaque spaces have nearly zero transmission, provided that the lines are not too close. Following the procedure explained in § 6.25 and substituting $(\kappa_1 - \kappa_2)$ for κ_y in equation 6(35), we have

$$E(\kappa_2) = \left\{\frac{\sin(\kappa_1 - \kappa_2)\,d}{(\kappa_1 - \kappa_2)\,d}\right\}^2 \left\{\frac{\sin N(\kappa_1 - \kappa_2)e}{N\sin(\kappa_1 - \kappa_2)e}\right\}^2, \qquad 6(54)$$

$$E(\theta_2) = \left\{\frac{\sin\dfrac{2\pi}{\lambda}(\sin\theta_1 - \sin\theta_2)\,d}{\dfrac{2\pi}{\lambda}(\sin\theta_1 - \sin\theta_2)\,d}\right\}^2 \left\{\frac{\sin\dfrac{2\pi N}{\lambda}(\sin\theta_1 - \sin\theta_2)e}{N\sin\dfrac{2\pi}{\lambda}(\sin\theta_1 - \sin\theta_2)e}\right\}^2, \qquad 6(55)$$

where $2d$ is the width of a clear space and $2e$ is the distance between the centres of two consecutive clear spaces.

We may write 6(55) in the form

$$E(\theta_2) = E(U, W) = f(U)\,F(NW), \qquad 6(56)$$

where

$$\left.\begin{aligned} f(U) &= \frac{\sin^2 U}{U^2}, \\ F(NW) &= \frac{\sin^2 NW}{N^2 \sin^2 W}, \end{aligned}\right\} \qquad 6(57)$$

$$\left.\begin{aligned} U &= \frac{2\pi}{\lambda}\,d(\sin\theta_1 - \sin\theta_2), \\ W &= \frac{2\pi}{\lambda}\,e(\sin\theta_1 - \sin\theta_2). \end{aligned}\right\} \qquad 6(58)$$

$f(U)$ is the factor depending on the shape of the individual aperture (now defined by a single parameter $2d$) and $F(NW)$ represents the factor depending on the arrangement of the apertures (i.e. the unit of separation, $2e$). From the point of view adopted in § 6.8, $2U$ is the difference of phase between wavelets from two sides of the same line, and $2W$ is the difference of phase between wavelets from the centres of two successive lines.

6.28. **The Functions $f(U)$ and $F(NW)$.**

These two functions are of fundamental importance in connection with the theory of the diffraction grating and also in other problems. The first appears whenever we sum a series of elements of constant amplitude with continuously and uniformly increasing phase. The second appears when we sum a number of elements of equal amplitude whose phases vary discontinuously in uniform steps. $f(U)$ represents a limiting form to which $F(NW)$ tends when N becomes indefinitely large. It is convenient to consider the two functions together. They are shown graphically in fig. 6.9. Both functions are essentially positive, and both have a central maximum equal to unity when U or $W = 0$, together with a series of side maxima and minima. It is in relation to these side maxima and minima that they differ. $f(U)$ is zero when $U = m\pi$ (if m is integral and not equal to zero), and it has a series of maxima *approximately* half-way between any two successive zero values. These maxima rapidly decrease in magnitude as we go away from the centre of the pattern [Example 6(vi), p. 190]. About 93 per cent of the area below the curve lies within the region between the two central zeroes. $F(NW)$ has two types of maxima. The first occur when $W = m\pi$ (m having any integral value including zero). At these points $F(NW) = 1$. These are known as *principal maxima.* $F(NW)$ has also a series of *subsidiary maxima* flanking each of the principal maxima. The magnitudes of these maxima are small except for those near to the principal maxima. $f(U)$ differs from $F(NW)$ chiefly in that the former has only one principal maximum. The positions of the maxima and minima of these functions are summarized in Table 6.2 (p. 179).

The maximum occurring when $W = m\pi$ is called the principal maximum of order m. The maximum occurring when U and $W = 0$ is called the " central image " or the maximum of " zero order ".

6.29.—The relation between the two functions may be seen by considering the vector diagrams. Fig. 6.9*b* shows the vector diagrams corresponding to the values of $f(U)$ marked on fig. 6.9*a*. It may be seen that as U increases, the vector system is rolled up into a circular curve of continually decreasing radius of curvature. When the curve forms a complete circle (after one or more revolutions), the resultant is zero, since the curve is closed. Subsidiary maxima occur when the vector curve is very nearly an integral number of semicircles. As U increases, each subsidiary maximum is smaller than the preceding one, because the resultant

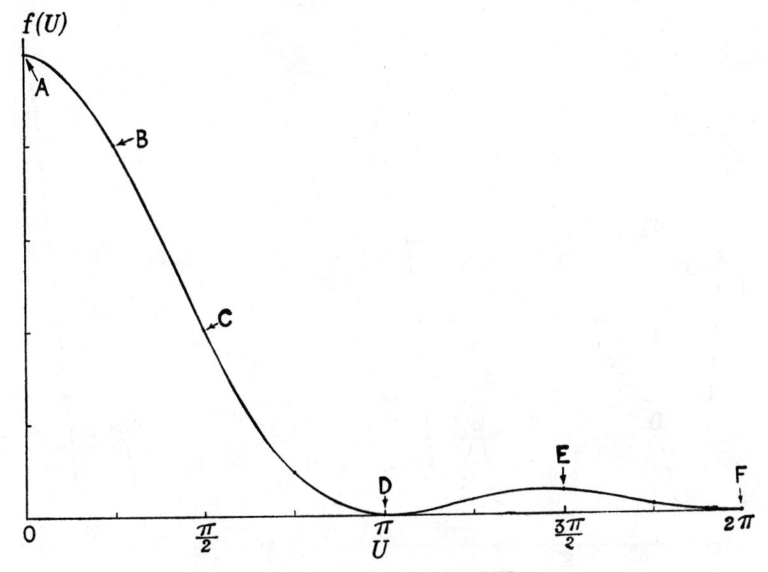

Fig. 6.9*a*.—The function $f(U)$

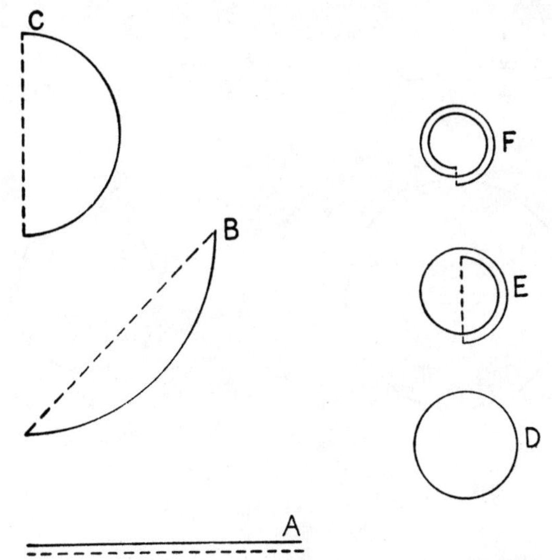

Fig. 6.9*b*.—Vector diagrams corresponding to the points indicated on fig. 6.9*a*

Fig. 6.9*c* and 6.9*d* appear on next page.

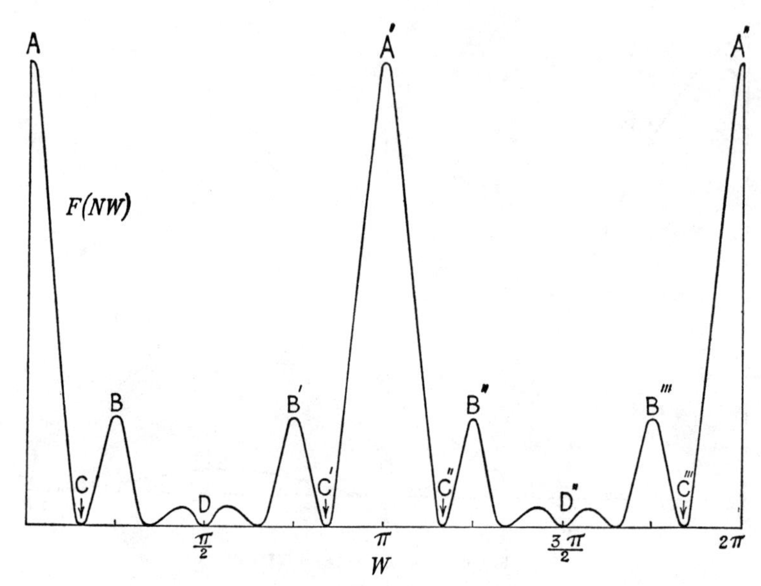

Fig. 6.9*c*.—The function $F(NW)$

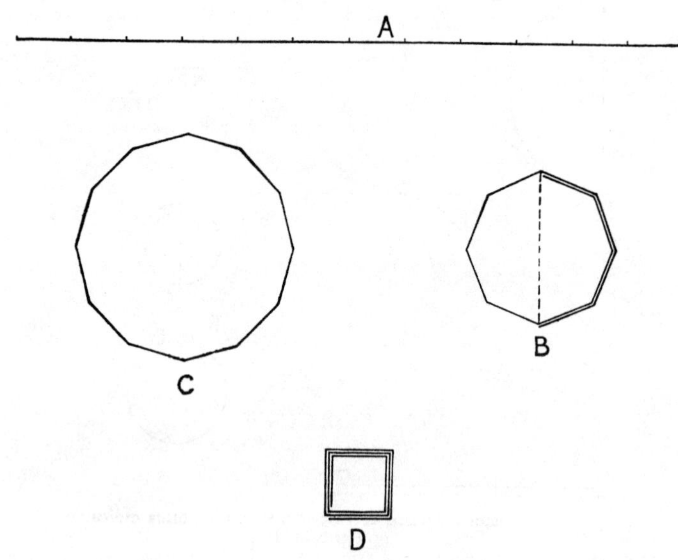

Fig. 6.9*d*.—Vector polygons corresponding to the points indicated on fig. 6.9*c*. (Note that the vector polygons for A′, and A″, etc., are the same as those shown for A, B, etc.)

is proportional to the square of the diameter of a smaller circle. The principal maximum occurs when all the elements are in the same phase, i.e. when the vector system forms a straight line. This happens only when $U = 0$. The vector diagram corresponding to $F(NW)$ forms part of a regular polygon—or of a series of overlapping polygons (fig. 6.9*d*). When the polygon is closed the resultant is zero. Subsidiary maxima occur when there is a whole number of overlapping polygons plus approximately half a polygon. Principal maxima occur whenever the elements are all in phase and the vector system forms a straight line. This happens not only when the variable is zero, as it did with $f(U)$, but also when the first of the N elements has rotated through 2π. When this occurs, all the other elements will have rotated through multiples of 2π, and therefore they will all be in the same phase again. This condition is never obtained when the elements form a continuous curve. When U tends to infinity, the vector elements corresponding to $f(U)$ roll up into a circle of indefinitely small radius.

TABLE 6.2.—$f(U)$

Principal maximum	When $U = 0$,	$f(U) = 1$.
Secondary maxima	When $U \doteqdot \left(\dfrac{2m+1}{2}\right)\pi$,	$f(U) = \dfrac{1}{\pi^2}\left(\dfrac{2}{2m+1}\right)^2$.
Minima	When $U = m\pi$,	$f(U) = 0$.

$F(NW)$

Principal maxima	When $W = 0$ or $m\pi$,	$F(NW) = 1$.
Secondary maxima	When $W \doteqdot \left(\dfrac{2m+1}{2N}\right)\pi$,	$F(NW) = \dfrac{1}{N^2\pi^2}\left(\dfrac{2}{2m+1}\right)^2$.
Minima	When $W = \dfrac{m}{N}\pi$, $\left(\text{except } \dfrac{m}{N} \text{ integral}\right)$	$F(NW) = 0$.

m stands for an integer (other than zero) and the sign $\doteqdot$ means "is *approximately* equal to".

6.30. Distribution of Light among the Principal Maxima.

When monochromatic light from a slit source falls upon a grating, the distribution of energy in the diffracted light is given by the function E of equation 6(56). When the width of the clear spaces is fairly small in comparison with the distance between them, the distribution is that shown in fig. 6.10. When $F(NW)$ has a principal maximum, the distribution function becomes equal to $f(U)$. At other points it is less than $f(U)$. Thus the curve of $f(U)$ forms a limiting curve for the function.

The angles at which principal maxima of $E(U, W)$ occur are determined by e (i.e. by the spacing of the grating), but the relative energies of the different principal maxima are determined by d (i.e. by the width of each clear space), and by the relation between d and e.

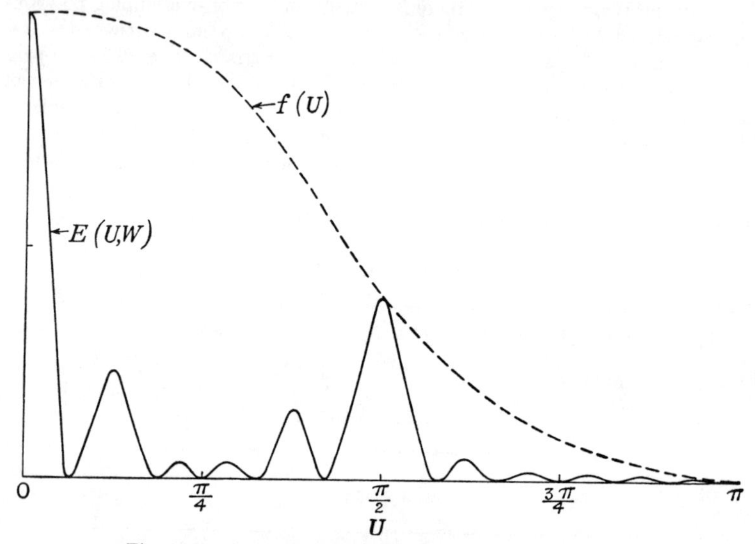

Fig. 6.10.—The distribution of light diffracted by a grating

From eqn. 6(55) and Table 6.2 we see that the mth principal maximum occurs when

$$\sin \theta_1 - \sin \theta_2 = \frac{m\lambda}{2e}, \qquad 6(59)$$

and that the ratio of the energy of the mth maximum to that of the central maximum is

$$F_m = \left\{\frac{\sin (\pi md/e)}{\pi md/e}\right\}^2. \qquad 6(60)$$

At the centre of the diffraction pattern there is complete agreement of phase, as there would be if the grating were not present, but the amplitude is reduced in the ratio d/e by the interposition of the opaque spaces of the grating. Hence, if we write F_T for the illumination in the image formed when the grating is removed, and F_0 for the illumination at the centre of the diffraction pattern, we have

$$F_0 = \frac{d^2}{e^2} F_T, \qquad 6(61)$$

and

$$\frac{F_m}{F_T} = \frac{1}{m^2\pi^2} \sin^2 \left(\frac{\pi md}{e}\right). \qquad 6(62)$$

Since the sine factor can never exceed unity, a grating composed of alternate transparent and opaque parts can never give more than $1/(m^2\pi^2)$ of the incident light in the mth order. Thus at most about one-tenth of the incident light goes into the first order and about one-fortieth into the second order. It should be remembered that these formulæ for the distribution of light between different orders are obtained from calculations which depend on the use of St. Venant's hypothesis. They can be used with a reasonable degree of confidence only when the widths of the spaces are large compared with the wavelength of the radiation.

6.31. **Diffraction Grating Spectra.**

If light of two different wavelengths λ' and λ'' is incident normally upon a grating, the angles corresponding to the principal maxima are given by

$$\left.\begin{aligned} \sin\theta_m' &= \frac{m\lambda'}{2e} \\ \text{and} \qquad \sin\theta_m'' &= \frac{m\lambda''}{2e}, \end{aligned}\right\} \qquad 6(63)$$

where θ_m' corresponds to the mth principal maximum for λ' and θ_m'' to the corresponding maximum for λ''.

If white light is incident, each principal maximum except the central maximum is drawn out into a spectrum. The spectrum corresponding to the first principal maximum is called the first-order spectrum, and the other spectra are classified in a similar way. The spectra may be received upon a screen placed at a great distance from the grating if no lens or mirror is used. If an optical system of lenses or mirrors is used to focus an image of the slit source (as shown in fig. 6.4) at a convenient distance from the grating, the spectra will be in focus at approximately the same distance. If the angles of diffraction are fairly small (so that $\sin\theta$ is nearly equal to $\tan\theta$), the ratio of the distance of the principal maximum corresponding to λ' from the centre of the pattern, to the corresponding distance for λ'', is equal to λ'/λ''. A spectrum for which this relation holds is called a *normal spectrum*, and the grating is said to give *normal dispersion.*

The simple relation between distance or angle on the one hand and wavelength on the other makes it convenient to use grating spectra for comparing the wavelengths of light corresponding to different spectrum lines. In grating spectroscopes the angle is measured directly, and in spectrographs the comparison is made by means of distances measured on a photographic plate. The accuracy of the measurements is discussed in detail in Chapter VIII. From 6(63) we see that for a given grating the dispersion is proportional to the order.

6.32. **Overlapping of Orders.**

From equation 6(63) it may be seen that to some extent the spectra overlap. If $m'\lambda' = m''\lambda''$ (where m' and m'' are integers), the principal maximum of order m' for light of wavelength λ' will coincide with the principal maximum of order m'' for λ''. When measurements are confined to the visible spectrum, there is only very slight overlapping with the second order. Under favourable conditions the eye can detect radiation of wavelengths greater than 7000 Å. and less than 3800 Å., but for most practical purposes the visible spectrum may be taken to lie within these limits (Chapter I, fig. 1.5) The second order of 3800 Å. would coincide with the first order of 7600 Å., and thus the violet end of the second-order visible spectrum just fails to overlap the red end of the first-order spectrum. When photographic plates sensitive to wider ranges of wavelength are used, even the first-order spectrum is not free from overlap by other orders. It is then possible to restrict the range of wavelengths received by the instrument by the use of filters. In this way the lines of different orders can be identified.

6.33. **Gratings Ruled on Glass or Metal.**

For the measurement of wavelengths of spectrum lines, it is desirable to produce spectra of high dispersion. Since some of the lines are very faint, it is also desirable that a high proportion of the light received by the spectroscope or spectrograph should go into one spectrum. When light is dispersed by a prism, all the light goes into one spectrum, but it is not possible to obtain very high dispersion in this way. Gratings with a very close spacing (of the order of a few wavelengths) give a very high dispersion, but, if the grating is formed of alternate opaque and transparent strips, only a small fraction of the incident light is found in any one spectrum. It is therefore of interest to inquire whether any other type of grating might be expected to give a better concentration of light in one spectrum.

Returning for this purpose to fig. 6.9, we see that the maximum of $f(U)$ coincides with the central maximum of $F(NW)$, i.e. with the central image, and not with one of the spectra which can be used for the measurement of wavelengths. If the width of a clear space ($2d$)

is small compared with the distance between successive spaces ($2e$), then an appreciable proportion of the light *transmitted* by the grating goes into the first order, but the transmitted light is then a small fraction of the incident light. If the width of a clear space is large compared with the width of an opaque space, then the fraction of the incident light transmitted is high, but nearly all the light goes into the central image. Thus, as shown in § 6.30, at most about 10 per cent of the *incident* light goes into one spectrum. The only practical way to obtain improvement is to depart from the simple grating of alternate clear and opaque strips, and to alter the *unit* of the grating in such a way as to make the maximum of $f(U)$ coincide with one of the lateral maxima of $F(NW)$. In the case of reflecting gratings this may be done by altering the shape of the groove.

6.34.—The following quotation from Lord Rayleigh's article on "The Wave Theory of Light" (*Encyclopædia Britannica*, 1888; *Scientific Papers*, Vol. III, p. 108) is very interesting in view of later developments:

> "If it were possible to introduce at every part of the aperture of the grating an arbitrary retardation, all the light might be concentrated in any desired spectrum. By supposing the retardation to vary uniformly and continuously we fall upon the case of an ordinary prism; but there is then no diffraction spectrum in the usual sense. To obtain such it would be necessary that the retardation should gradually alter by a wavelength in passing over any element of the grating, and then fall back to its previous value, thus springing suddenly over a wavelength. It is not likely that such a result will ever be fully attained in practice; but the case is worth stating, in order to show that there is no theoretical limit to the concentration of light of assigned wavelength in one spectrum."

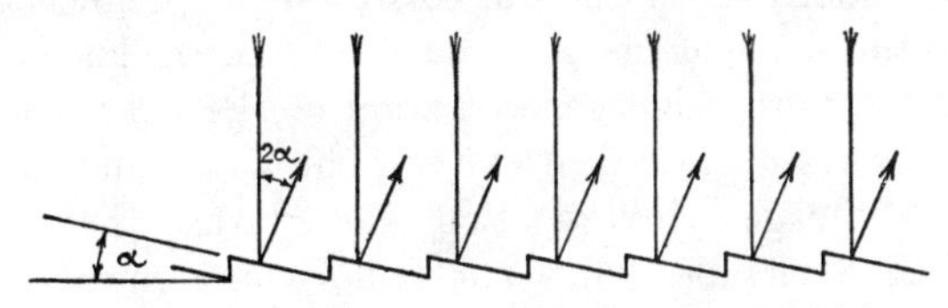

Fig. 6.11.—Reflecting grating suggested by Rayleigh

Fig. 6.11 shows a section of a reflecting grating of the type suggested by Rayleigh. Suppose each elementary facet makes an angle α with the plane which defines the macroscopic surface of the grating. When the light is incident in a direction normal to this plane, the maximum illumination for one element of the grating considered by itself is found in a direction making an angle 2α with the normal. If, for a given wavelength λ, a principal maximum of $F(NW)$ occurs in that direction, then most of the light will be thrown into this maximum. The relation between $E(U, W)$ and $f(U)$ has been changed from that shown in fig. 6.10 to that shown in fig. 6.12.

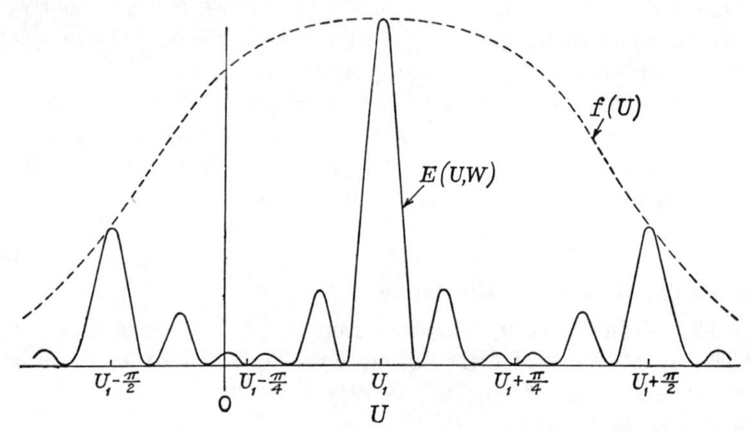

Fig. 6.12.—Distribution of light for the grating of fig. 6.11

6.35.—At the time when Rayleigh wrote what has been quoted above, Rowland had succeeded in ruling gratings on metal and glass, using a diamond point and an accurately constructed ruling engine. When the gratings were ruled on speculum metal, up to 100,000 lines could be ruled before the ruling point broke down, and the spacing could be made as close as 15,000 lines to the inch. Each unit in the pattern was then about 3 wavelengths wide. The form of the groove cut was not under control and varied from one specimen to another, according to the shape of the diamond point. It was known that some gratings gave a fairly high concentration of the light in one or two orders, but always a great deal of the light was found in the undispersed central image. Investigation showed that the ruling point, when working at its best, did not remove any metal but created depressions and elevations which approached, though not very closely, the pattern described by Rayleigh. The only effective control was in regard to the depth of the groove, since the shape of the point was regarded as unalterable.

6.36. Echelette Gratings.

In 1910, R. W. Wood * succeeded in producing gratings with grooves of controlled shape. They were ruled on gold-plated copper

* Reference 6.4.

plates using the natural edge of a selected carborundum crystal as the ruling point. The angle between the planes forming the edge is 120°. In the following year, Wood and Trowbridge made similar gratings with a diamond ground to the desired angle by Brackett. These gratings were of a coarser structure than those made by Rowland. They were suitable for the production of infra-red spectra and for a practical study of the effect of different orientations of the groove angle. For a wavelength of about 30,000 Å. they concentrated nearly all the energy into one or two orders on one side of the central image. They were called *echelette* gratings * because they were regarded as intermediate between ordinary gratings and the echelon gratings devised by Michelson which will be described in § 6.39. Copies of the original gratings could be made by flowing collodion on to the copper plates, allowing it to harden, and then stripping it. The material removed was mounted on glass plates and sputtered with gold. The copies so made were better than the originals because the copper plates were not perfectly flat, whereas the copies became flat when the collodion was pressed down on to optically flat glass plates.

6.37.—For optical gratings the corresponding problem was not solved until 1935. By then, it had been found possible to deposit highly reflecting films of aluminium on to glass plates by evaporation in vacuum. When the vacuum is sufficiently good and other technical points are controlled, the films are durable and adhere to the glass so strongly that the material is " moulded " rather than removed by the ruling point. Using diamond points of controlled form and orientation, R. W. Wood † has succeeded in ruling gratings with 15,000 lines to the inch, in which 80 per cent of the light of the green mercury line (5461 Å.) is concentrated in the first order. The process is sufficiently under control so that the light can be concentrated in first, second or higher order as desired. These gratings have already proved very useful in obtaining spectra of faint stars and nebulæ.

6.38.—With normal incidence the maximum concentration of light is obtained for only one wavelength. By altering the angle of incidence, the wavelength which gives maximum concentration can be altered. The reason for this may be seen from the following relations:

If α is the angle between the facets and the macroscopic surface, and θ is the angle between the incident beam and the normal to the macroscopic surface, then the maximum concentration occurs in a direction β given by

$$\beta = \theta + 2\alpha, \qquad \qquad 6(64)$$

* Reference 6.5. † Reference 6.6. See also Reference 6.7.

while the mth principal maximum for wavelength λ occurs in a direction θ_m given by

$$2e(\sin\theta_m - \sin\theta) = m\lambda. \qquad \qquad 6(65)$$

If $\theta_m = \beta$ when $\theta = 0$ for wavelength λ, concentration will occur in the mth order of λ when incidence is normal. By altering θ a little, it is obviously possible to make $\theta_m' = \beta$ for some neighbouring wavelength λ'.

6.39. **The Michelson Echelon Grating.**

In 1898, Michelson designed an apparatus which he called an *echelon diffraction grating*. This apparatus consisted of a pile of plates of equal thickness superimposed as shown in fig. 6.13. When a parallel beam of light is transmitted by the echelon, the light from successive " steps " of the echelon forms a series whose retardations increase in arithmetic progression. If t is the thickness of a step, the difference of retardation between two successive beams is $(\mu - 1)t$ for normal incidence.

In one echelon constructed for Michelson there were 20 plates, each 18 millimetres thick. The width of each step was about one millimetre. The retardation is equivalent to about 20,000 wavelengths. Since the width of each element is large compared with the wave-

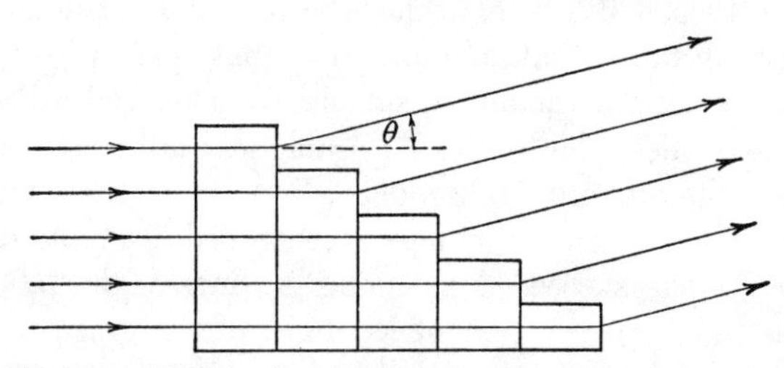

Fig. 6.13.—Michelson's transmission echelon diffraction grating

length, all the light will be concentrated in angles very close to the direction of incidence, which we have taken to be the normal. This implies that it will all be concentrated in (at most) a few spectra of very high order.

The dispersion obtained is higher than that given by any ruled grating and, with suitable adjustment, practically all the light is concentrated into one spectrum. Owing to the high order of interference used, overlapping of orders constitutes a serious problem (§ 9.50). The large difference of path between successive beams suggests that the instrument should be regarded as an interferometer, and indeed the problems involved in its construction and use are very similar to those which apply to the Fabry and Pérot interferometer. The instrument is, however, a true grating since the elements which are brought to interference are elements of wavefront placed side by side, and not behind one another as in the Fabry and Pérot instrument. It is the diffraction effects, due to the finite *width* of each step, that give the instrument its special property of concentrating all the light into one or two fringes.

6.40. The Michelson-Williams Reflecting Echelon.

Michelson realized that there would be many advantages in using the instrument by reflected instead of by transmitted light (fig. 6.14). With a reflecting echelon, the optical paths, whose differences constitute the retardations, are entirely in air, or, if desired, in vacuum. All faults due to inhomogeneity of the optical material are thereby eliminated. The instrument is also available for use in parts of the spectrum to which the material is not transparent. For a given size

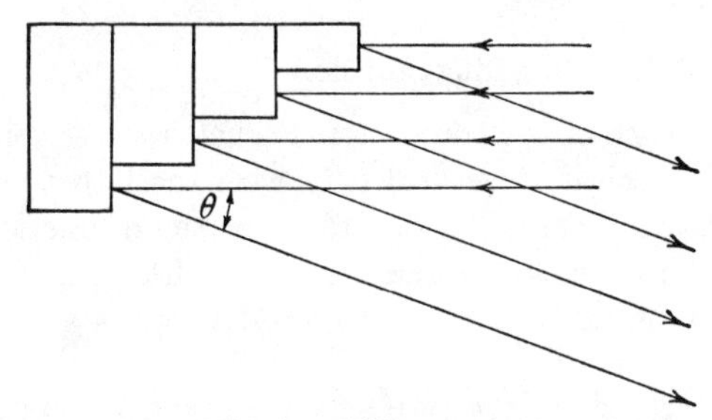

Fig. 6.14.—Michelson-Williams reflecting echelon

of echelon the order of interference is approximately four times higher with the reflection than with the transmission type. In order to attain satisfactory performance with a reflecting echelon, it is necessary that all the plates shall be in perfect contact (or be separated by films of air of identical thickness). This condition must be fulfilled because the retardations depend on the *positions* of the reflecting steps. The thicknesses of the plates determine these positions only when the contact is perfect. With the transmission echelon, on the other hand, the retardations depend directly on the optical thickness of the plates and are not affected by small irregularities in the assembly.

The problem of constructing a satisfactory reflecting echelon was solved by W. E. Williams,* who used optically contacted plates of

* Reference 6.8.

fused quartz. The echelon is constructed by first making the requisite number of plates of identical geometrical thickness (to within a tolerance of less than $0{\cdot}1\lambda$). They are tested for equality by an interferometric method (§ 9.11). The manufacture of a large number of plates of equal thickness, within such a very narrow tolerance, makes the construction very expensive. If one plate is polished so that it falls below the desired thickness, it cannot be used unless all the others are correspondingly thinned. When the plates have been brought to the right thickness, the surfaces are carefully cleaned and placed in contact. They are then heated to a suitable temperature (which is well below the softening temperature of the material). Surfaces which fit exactly adhere, and after cooling can be separated only with considerable force. The final effect is similar to that which would be produced if the whole echelon could be carved out of a solid block of fused quartz. The faces of the steps are coated with a uniform film of aluminium by evaporation in vacuum.

6.41. Theory of the Reflecting Echelon.

Since the theory of the transmission echelon is very similar to that of the reflecting echelon, we deal only with the latter. Suppose that the light is incident normally, and consider the diffraction in a direction θ. If t is the thickness and s is the width of a step, the phase difference between the beams from successive steps is (fig. 6.14)

$$\delta = \frac{2\pi}{\lambda} n_a(t + t\cos\theta - s\sin\theta), \quad . \quad . \quad . \quad 6(66)$$

where n_a is the refractive index of air.

When θ is very small this may be written

$$\delta = \frac{2\pi}{\lambda} n_a(2t - s\theta). \quad . \quad . \quad . \quad . \quad . \quad 6(67)$$

The variation of illumination with θ is given by an expression which is the product of two factors corresponding to $F(NW)$ and $f(U)$ of § 6.28.

Principal maxima of the function corresponding to $F(NW)$ occur when

$$m\lambda = n_a(2t - s\theta), \quad . \quad . \quad . \quad . \quad . \quad 6(68)$$

and the angle between the maxima of order m and $(m+1)$ is λ/s approximately (since n_a is nearly unity).

The function corresponding to $f(U)$ is

$$\left(\sin\frac{\pi s\theta}{\lambda}\right)^2 \Big/ \left(\frac{\pi s\theta}{\lambda}\right)^2.$$

It is zero when $\theta = \pm\lambda/s$ and nearly all the light is concentrated within these limits. If $2n_a t = m\lambda$, principal maxima of the function corresponding to $F(NW)$ occur at $\theta = 0, \pm\lambda/s, \pm 2\lambda/s, \ldots$, etc. The illumination is high for the central maximum, and zero or nearly zero for all the other maxima. If $2n_a t$ is not an integral number of wavelengths, two maxima, separated by an angle λ/s and situated between the limits $\theta = +\lambda/s$ and $\theta = -\lambda/s$, are seen. These are of equal magnitude when they occur at $\theta = \pm\lambda/2s$.

The condition in which only a single maximum appears is called the *single-order position.* The condition giving two equal maxima may be called the *symmetrical position.* With the transmission echelon, it is possible to change from the single order to the symmetrical position by slightly turning the echelon in order to change the angle of incidence. It is not convenient to use this method with the reflection echelon because the reflected beam might pass out of the field of view. The necessary control can be obtained by placing the echelon in an air-tight box with a quartz or fluorite window, and adjusting the pressure (and hence the value of n_a) until the desired condition is obtained. This method has the advantage that once the pressure has been adjusted, and the tap connecting the box to the outside closed, variations in the pressure or temperature of the atmosphere during an exposure do not matter. The value of n_a depends essentially on the mass of air in the box, and this does not change. Also the thermal expansion of fused quartz is so low that the changes in t due to ordinary laboratory temperature variations are too small to affect the positions of the fringes.

EXAMPLES [6(i)–6(x)]

6(i). Calculate the radius of the Airy disc when $\lambda = 5000$ Å. and (i) $R = 1$ cm., $z_1 = 1$ metre; (ii) $R = 3$ mm., $z_1 = 50$ cm.; (iii) $R = 1\mu$, $z_1 = 1$ metre (one micron $= 1\mu = 10^{-6}$ metre).

[(i) $3{\cdot}05 \times 10^{-3}$ cm., (ii) $5{\cdot}08 \times 10^{-3}$ cm., (iii) $30{\cdot}5$ cm.]

6(ii). Derive an equation giving the exact positions of the lateral maxima of $f(U)$. [$\tan U = U$.]

6(iii). Solve the equation mentioned in example (ii) graphically and determine the value of U at which the first lateral maximum of $f(U)$ occurs. Compare the value of $f(U)$ at this maximum with its value when $U = 3\pi/2$.

[Maximum of $f(U)$ when $U = 2{\cdot}86\pi/2$, at which $f(U) = 0{\cdot}0472$. When $U = 3\pi/2$, $f(U) = 0{\cdot}0451$.]

6(iv). Show that as $N \to \infty$ the form of $F(NW)$ becomes identical with that of $f(U)$. Examine what happens to the different maxima of $F(NW)$.

$$\left[\text{Put } F(NW) = \left[\frac{(\sin NW)/NW}{(\sin W)/W}\right]^2, \text{ and let } N \to \infty \text{ while } NW \text{ remains finite.}\right]$$

6(v). How many subsidiary maxima of $F(NW)$ lie between the mth and $(m+1)$th principal maxima? [$N - 2$.]

6(vi). Draw graphs for $f_a(U) = \sin U/U$ and for $F_a(NW) = \sin NW/(N \sin W)$ (when $N = 4$), and compare them with corresponding graphs for $f(U)$ and $F(NW)$. The functions $f_a(U)$ and $F_a(NW)$ represent amplitudes.

6(vii). Show that when the relation between d and e is such as to give as much light as possible in the first order, the brightnesses of all even orders are zero.

6(viii). What ratio of e to d makes the mth order (a) as bright as possible, and (b) of zero brightness? [(a) $2m$, (b) m.]

6(ix). Show that if d is very small compared with e, all the maxima of low orders tend to the same brightness.

$$\left[\frac{F_m}{F_T} = \frac{d^2}{e^2}\,\frac{\sin^2(\pi md/e)}{(\pi md/e)^2} \to \frac{d^2}{e^2} \text{ if } \frac{md}{e} \ll 1.\right]$$

6(x). If the refractive index of air is 1·00029 for 5893 Å. at S.T.P., calculate the change in pressure required to pass from the single-order to the symmetrical position with an echelon of step 2 millimetres. [193 mm.]

REFERENCES

6.1. Baker and Copson: *The Mathematical Theory of Huygens' Principle* (Oxford University Press.)

6.2. Stratton: *Electromagnetic Waves* (McGraw-Hill).

6.3. Pidduck: *Phil. Mag.*, 1946, Vol. 37, p. 280.

6.4. Wood: *Phil. Mag.*, 1910, Vol. XX, 6th Series, p. 770.

6.5. Trowbridge & Wood: *ibid.*, p. 886.

6.6. Wood: *J.O.S.A.*, 1944, Vol. 34, p. 509.

6.7. Babcock: *J.O.S.A.*, 1944, Vol. 34, 1.

6.8. Williams: *Applications of Interferometry* (Methuen). See also Brit. Pat. 315234.

6.9. Booker, Ratcliffe and Shinn: *Phil. Trans. Roy. Soc.*, 1950, Vol. 242, p. 579.

6.10. Sawyer: *Experimental Spectroscopy* (Prentice-Hall).

APPENDIX VI A

KIRCHHOFF'S DIFFRACTION FORMULA

1.—Consider a simple harmonic wave represented by

$$\xi = \psi(x, y, z)e^{i\omega t}, \quad . \quad . \quad . \quad . \quad . \quad . \quad . \quad . \quad 6(69)$$

where ψ is a function of position (not of time) which gives the form of the wave surfaces, e.g. for plane waves ψ has the form

$$\psi = \exp i\kappa(lx + my + nz), \quad . \quad . \quad . \quad . \quad . \quad 6(70)$$

and there is a corresponding form for spherical waves [2(56)]. We do not specify ψ at present, because we do not wish the analysis to be restricted to any

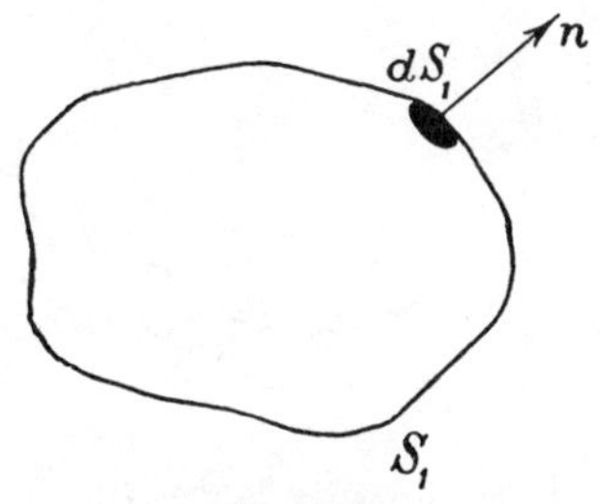

Fig. 6.15

particular type of wave surface. Inserting the value given by 6(69) in the wave equation [2(38)] and remembering that $\kappa b = \omega$, we see that ψ must satisfy the equation

$$\Delta\psi + \kappa^2\psi = 0. \quad . \quad . \quad . \quad . \quad . \quad . \quad . \quad . \quad . \quad 6(71)$$

Green's theorem states that if ψ_1 and ψ_2 are any two functions of position, then

$$\int(\psi_2\,\Delta\psi_1 - \psi_1\,\Delta\psi_2)\,d\tau = \int\left(\psi_2\frac{\partial\psi_1}{\partial n} - \psi_1\frac{\partial\psi_2}{\partial n}\right)dS_1, \quad . \quad . \quad 6(72)$$

where the left-hand integral is taken over the volume enclosed by the surface S_1 (fig. 6.15) and the right-hand integral is taken over the surface. n is the *outward* normal from the surface. The theorem is valid provided that ψ_1 and ψ_2 do not have singularities within the volume.

2.—If ψ_1 and ψ_2 are two solutions of 6(71), then

$$(\psi_2\,\Delta\psi_1 - \psi_1\,\Delta\psi_2) = -\psi_2\kappa^2\psi_1 + \psi_1\kappa^2\psi_2 = 0, \quad . \quad . \quad 6(73)$$

so that the left-hand side of 6(72) is zero. We now choose ψ_1 to be equal to ψ [i.e. to any unspecified solution of 6(71)] and ψ_2 to be equal to $e^{-i\kappa r}/r$. It may easily be verified that this function, which represents a spherical wave diverging from the origin, is a solution of 6(71). Inserting these functions for ψ_1 and ψ_2, we obtain

$$\int\left\{\frac{e^{-i\kappa r}}{r}\frac{\partial\psi}{\partial n} - \psi\frac{\partial}{\partial n}\left(\frac{1}{r}e^{-i\kappa r}\right)\right\}dS_1 = 0, \quad . \quad . \quad . \quad . \quad 6(74)$$

provided that neither ψ nor $e^{-i\kappa r}/r$ have singularities within the volume enclosed by S_1. This condition prevents us from applying 6(74), directly, to a surface which includes the origin since $e^{-i\kappa r}/r$ has a singularity at the origin. Since we wish to deal with surfaces surrounding the origin, we now choose to let S_1 consist of two parts:

(i) The surface S in which we are chiefly interested, and

(ii) a small sphere S_0 surrounding the origin, which we take as the point P (see fig. 6.16).

The volume to which Green's theorem is now applied is that which lies between S and S_0, i.e. the origin is excluded from the domain of integration, and on S_0

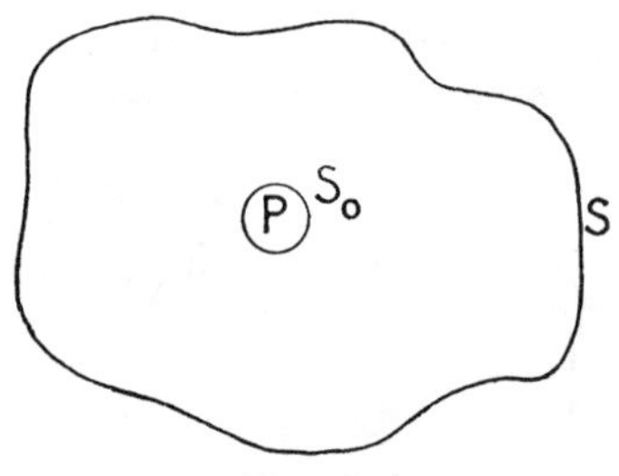

Fig. 6.16

the outward normal is directed towards P, i.e. $\partial/\partial n = -\partial/\partial r$ for this surface. That part of the integral in 6(74) which refers to S_0 may be written

$$-\int\left\{\frac{e^{-i\kappa r}}{r}\frac{\partial\psi}{\partial r}-\psi\frac{\partial}{\partial r}\left(\frac{1}{r}e^{-i\kappa r}\right)\right\}r^2\,d\Omega=-\int e^{-i\kappa r}\left\{\psi+i\kappa r\psi+r\frac{\partial\psi}{\partial r}\right\}d\Omega, \quad 6(75)$$

where $d\Omega$ is an element of solid angle round the origin P. When $r\rightarrow 0$ the right-hand side of 6(75) tends to the value $-4\pi\psi_P$ and 6(74) becomes

$$\psi_P=\frac{1}{4\pi}\int\left\{\frac{e^{-i\kappa r}}{r}\frac{\partial\psi}{\partial n}-\psi\frac{\partial}{\partial n}\left(\frac{e^{-i\kappa r}}{r}\right)\right\}dS. \quad . \quad . \quad . \quad 6(76)$$

Thus, if we are given ψ and $\partial\psi/\partial n$ over the surface S, which encloses P, the value of ψ_P can be calculated. In the preceding discussion we have considered a surface which encloses the point of interest. It may also be shown that, if the values of ψ and $\partial\psi/\partial n$ are given for a surface which excludes the point P but includes all the sources of light, then ψ_P can still be calculated. Equation 6(76) is still valid but n is in this case the inward normal.

3.—Huygens assumed that, if we knew the disturbance on a certain surface, we could calculate the disturbance at a point in advance of the surface. The above analysis only partly supports this view. In the first place it appears that we need to know the disturbance on a given surface, and also to know $\partial\psi/\partial n$ (effectively to know the disturbance on a neighbouring surface). Also, since we assume that the frequency is given, the value of $\partial\psi/\partial t$ is implicitly involved. It should, however, be stated that $\partial\psi/\partial n$ and $\partial\psi/\partial t$ are not completely independent functions, and, for this reason, we shall be able to develop later an approximate form [6(80)] which depends only on ψ and comes near to Huygens' original idea.

4.—Another way of considering the relation of Kirchhoff's formula [6(76)] to Huygens' principle is to introduce the idea of double sources. The amplitude at P due to a point source at O (distant r from P) is $Ae^{-i\kappa r}/r$. Now consider two sources of equal intensity and opposite phase separated by a short distance. Let the line of separation be n and the distance between the sources be dn. The line n is not, in general, coincident with OP. Then the amplitude at P due to the two sources acting together is

$$A \frac{\partial}{\partial n} \left\{ \frac{1}{r} e^{-i\kappa r} \right\} dn.$$

If the separation dn is allowed to diminish indefinitely while the product $A\,dn$ retains the finite constant value B, we arrive at the concept of a double source—closely analogous to an electric or magnetic doublet. This concept is of practical importance in sound. Huygens (and later Fresnel) replaced the wave at S by a series of elementary single sources distributed over S. Kirchhoff's formula [6(76)] allows us to express the effect of an element dS as equivalent to that of a single source of strength $-(\partial\psi/\partial n)\,dS$ plus a double source of strength $\psi\,dS$.

5.—It will be remembered that Huygens' concept applied to an advancing wave. We have, so far, assumed one frequency which implies a steady state (infinite wave train). To apply Kirchhoff's equation to an advancing group of waves in a non-dispersive medium we need only note that the form of 6(76) is such that we could use it to sum the effects at P of a series of waves of different frequencies. Since any pulse, or non-permanent wave, may be analysed by Fourier methods into a series of simple harmonic waves, the Kirchhoff method must apply to the pulse.

6.—*Approximate Form of Kirchhoff's Equation.*

Let us now consider the situation shown in fig. 6.17. O is a point source of light emitting spherical waves and S is a closed surface surrounding P. Consider

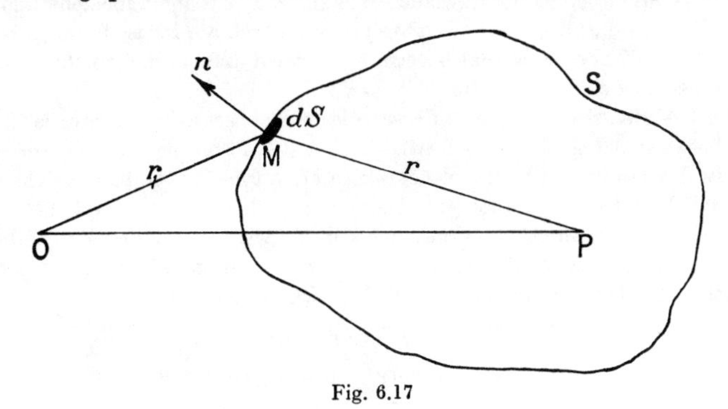

Fig. 6.17

an element dS situated at M, and let OM $= r_1$ and PM $= r$. The outward-drawn normal from S is n. The approximation now under discussion is valid when r

and r_1 are both large compared with the wavelength, i.e. κr and κr_1 are large compared with unity. The disturbance at M may be represented by

$$\xi_M = \frac{a}{r_1} \exp i(\omega t - \kappa r_1). \quad \ldots \ldots \quad 6(77)$$

so that

$$\psi_M = \frac{a}{r_1} e^{-i\kappa r_1}. \quad \ldots \ldots \ldots \quad 6(78)$$

Inserting this value of ψ_M for ψ in 6(76), we obtain

$$\psi_P = \frac{1}{4\pi}\int\left\{\left(\frac{e^{-i\kappa r}}{r}\cos(n, r_1)\frac{\partial}{\partial r_1}\left[\frac{a}{r_1}e^{-i\kappa r_1}\right]\right) - \left(\frac{a}{r_1}e^{-i\kappa r_1}\cos(n, r)\frac{\partial}{\partial r}\left[\frac{e^{-i\kappa r}}{r}\right]\right)\right\}dS. \quad 6(79)$$

Differentiating term by term, and neglecting all terms in which the amplitude is of order a/r_1r^2 or $a/r_1{}^2r$, we have

$$\psi_P = \frac{1}{4\pi} i\kappa \int\left\{\frac{a}{rr_1} e^{-i\kappa(r+r_1)}\left[\cos(n, r) - \cos(n, r_1)\right]\right\}dS. \quad . \quad 6(80)$$

The terms neglected are less than the terms retained by factors of κr and κr_1, i.e. by factors of order r/λ. The expression 6(80) may therefore be used when both O and P are separated from the surface S by distances which are large compared with the wavelength.

APPENDIX VI B

The Concave Grating

1.—The properties of a grating ruled upon the surface of a concave spherical mirror were investigated by Rowland. He showed that light incident upon the grating from a suitably placed slit gives spectra which are focused upon a certain curve. We shall first show that spectra are formed and focused in this way and then discuss how concave gratings are used.

Let AOA′ (fig. 6.18) be part of a section of a sphere whose centre is C. The region between A and A′ is ruled with lines which form the intersections of the sphere with a family of parallel planes which are normal to the plane of the paper and equidistant from one another. The common distance is $2e$ and AA′ $= 4Ne$ (where N is a large integer). Construct a circle with CO as *diameter* and let Q and Q′ be two points on this circle. Let us now find the difference of path for two paths from Q to Q′, i.e. (QA + Q′A) − (QO + Q′O) $= \Delta s$.

2.—Let

$$\text{QO} = u; \quad \text{Q'O} = v; \quad \text{CO} = 2a;$$

$$\angle\text{COQ} = \theta; \quad \angle\text{COQ'} = \theta_m; \quad \angle\text{OCA} = 2\alpha.$$

Then, from triangle QOA,

$$\text{QA}^2 = \text{QO}^2 + \text{OA}^2 - 2\text{OA} \,.\, \text{OQ}\cos\text{QOA}. \quad \ldots \quad 6(81)$$

In fig. 6.19, M bisects the straight line OA and from this figure we see that angle QOA $= (\frac{1}{2}\pi - \alpha) + \theta$ and that OA $= 4a \sin \alpha$. Hence 6(81) gives

$$QA^2 = u^2 + 16a^2 \sin^2 \alpha - 8au \sin \alpha \sin (\alpha - \theta). \quad 6(82)$$

With rearrangement of terms 6(82) becomes

$$QA^2 = (u + 2a \sin 2\alpha \sin \theta)^2 - 4a^2 \sin^2 2\alpha \sin^2 \theta + 8a\,(2a - u \cos \theta) \sin^2 \alpha. \quad 6(83)$$

The equivalence of 6(82) and 6(83) may be verified by expanding both expressions.

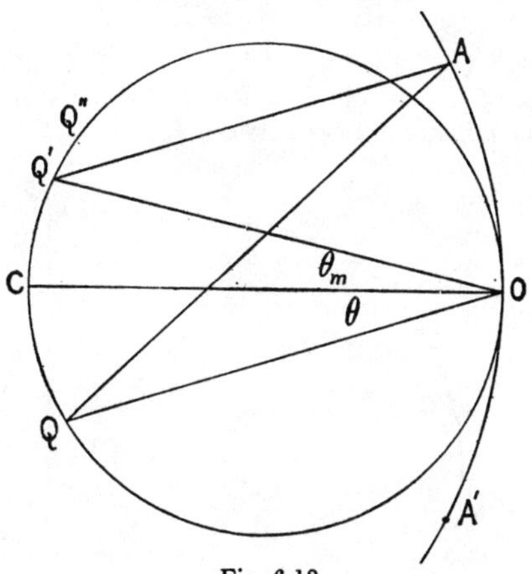

Fig. 6.18

Let us assume that α is a fairly small angle so that we may neglect terms of order α^4. We may then put $\sin^2 2\alpha = 4 \sin^2 \alpha$ and, to the same approximation, 6(83) gives

$$QA^2 = (u + 2a \sin 2\alpha \sin \theta)^2 - 2a\,(u - 2a \cos \theta) \sin^2 2\alpha \cos \theta. \quad 6(84)$$

Q is a point on the circle whose diameter is $2a$, so that $u = 2a \cos \theta$ and 6(84) gives

$$QA - u = 2a \sin 2\alpha \sin \theta. \quad 6(85)$$

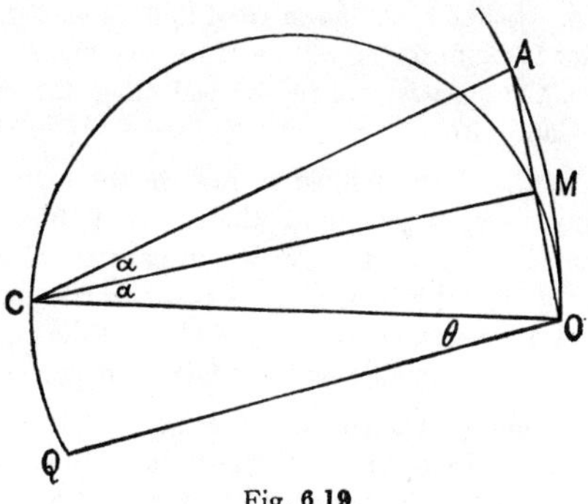

Fig. 6.19

We may obtain a similar expression for Q'A $- v$ and hence

$$\Delta s = (\text{QA} - u) + (\text{Q'A} - v) = 2a \sin 2\alpha \,(\sin\theta - \sin\theta_m). \qquad 6(86)$$

But, from the triangle A'CA,

$$4Ne = \text{AA'} = 4a \sin 2\alpha.$$

Hence the path difference $\Delta s = Nm\lambda$, if

$$2e\,(\sin\theta - \sin\theta_m) = m\lambda. \quad . \quad . \quad . \quad . \quad . \quad 6(87)$$

This is equivalent to 6(65) since m may be either a positive or a negative integer.

3.—We have shown that the path difference for the two rays from Q to Q' is proportional to the number of lines between A and O. When 6(87) is satisfied,

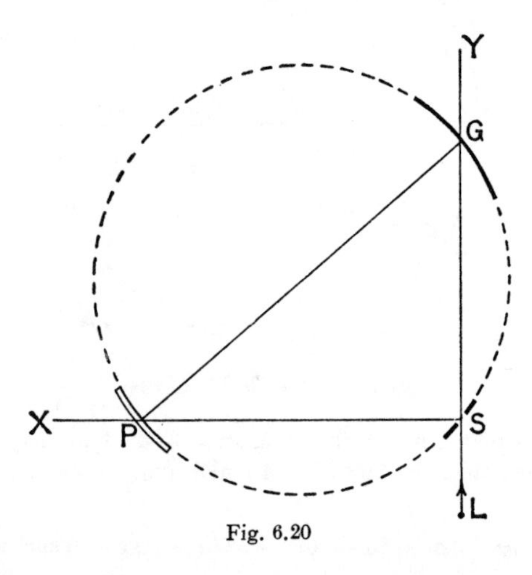

Fig. 6.20

the phase difference for light of wavelength λ increases by $2m\pi$ as we go from one line to the next. Thus the mth order for wavelength λ is focused at Q'. The same order for wavelength λ'' will be focused at some point Q'' on the same circle and near to Q'. If then a point source of light is placed at a point Q on the circle whose diameter is equal to the radius of the grating, and which touches the grating at its centre, the spectra will be focused along the circle. This circle is known as the " Rowland circle ".

4.—The above discussion is confined to rays in the plane of the paper. The focusing for these rays is very good since the terms of third order in α are zero. Further investigation shows that the rays which pass above and below the plane of the paper give an astigmatic image, i.e. a point source placed at Q gives rise to an image which is a line perpendicular to the plane of the paper, passing through Q'. The length of this line increases as the angle of incidence increases.

5.—A method of mounting a concave grating due to Rowland is shown in fig. 6.20. The mounting is designed to allow spectra of different orders or different parts of one spectrum to be photographed in turn, without moving the slit Q or the source L. The grating G and the photographic plate P are placed on carriages which move along mutually perpendicular rails SX and SY. The two carriages are connected by a rod so that they always face one another, and so that the distance from the centre of the grating to the centre of the plate is equal to the diameter of the Rowland circle (i.e. to the radius of curvature of the grating). Thus θ_m is always zero and θ is varied by moving the carriages. Different spectra are then brought on to the plate and, if the grating and mounting are perfect, the instrument is always in focus.

6.—A large grating has a resolving power of order 250,000. To obtain sufficient dispersion to make this resolving power effective it is necessary to use a grating of about 6 metres radius of curvature. Smaller gratings are made with radii of curvature down to 1 metre. This is the smallest value normally used, partly

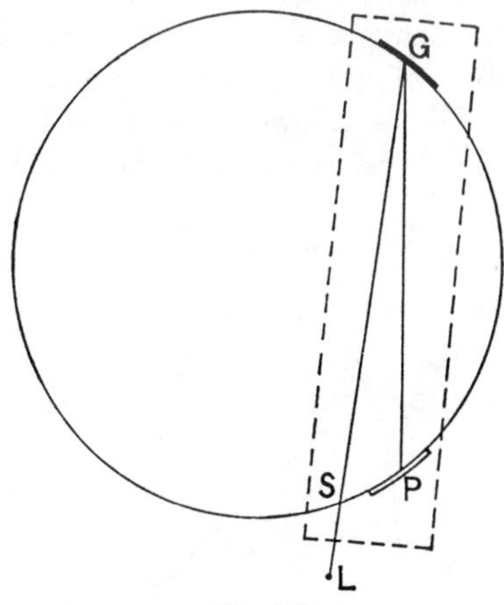

Fig. 6.21

because the photographic plate has to be bent to fit the Rowland circle. The space occupied by a large grating using a Rowland mounting is so large that it is difficult to maintain a constant temperature during the long exposures which are often required. The Rowland mounting is also rather expensive. For these and other reasons, alternative methods of mounting the grating have been devised. In one, due to Eagle, Q′ is brought as close to Q as possible, i.e. the slit is placed beside the plate (see fig. 6.21). Sometimes it is placed below the plane of the paper, and the plate is a little above the plane of the paper. This means that the region between the dotted lines (fig. 6.21) can be enclosed in a box of reasonable size. Temperature control is relatively easy. This mounting has the advantage that it reduces θ_m and θ (which are of opposite sign) and so reduces the astigmatism.

A detailed account of methods of mounting gratings is given in Reference 6.10. It is possible to obtain automatic refocusing with mountings of this type, but the mechanical linkages (levers or cams) are more complicated than the simple arrangement used by Rowland.

CHAPTER VII

Huygens' Principle and Fermat's Principle

7.1. Development of Huygens' Principle.

It was shown in Chapter III that Huygens' principle, in its original form, was able to give a satisfactory account of the laws of reflection and of refraction. It enabled a series of wave surfaces to be constructed when one was given. Huygens' principle by itself was insufficient to enable the distribution of illumination in diffraction patterns to be calculated. Fresnel and his followers completed the Huygens' theory by assuming detailed properties for the wavelets. Fresnel was guided

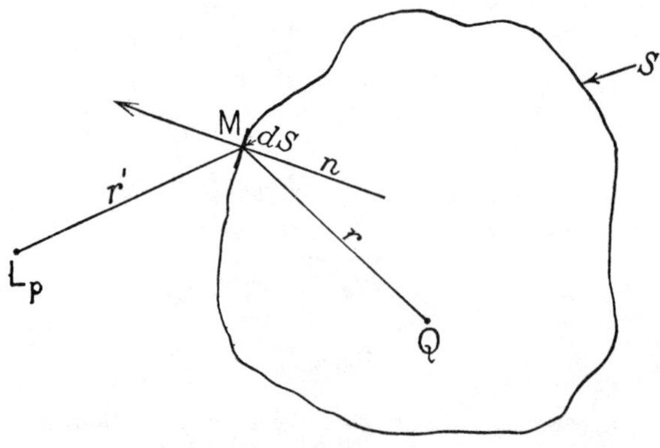

Fig. 7.1

in his choice of assumptions by the requirement that, in the absence of obstacles, the wavelets must interfere in such a way that they reconstruct the forward wave, not only in regard to position but also as regards amplitude. Later Kirchhoff showed that it is unnecessary to use specific assumptions concerning the wavelets since the whole calculation can be made directly from the wave equation. The three stages are thus:

(i) *Huygens* expressed an intuitive rather than a logical conviction that a knowledge of the disturbance produced by a wave motion for all points on a suitably placed surface S at a time t_0 is sufficient to determine the disturbance at a point Q at a later time t (fig. 7.1).

(ii) *Fresnel* made detailed assumptions concerning the amplitude of a wavelet arriving at Q from an elementary area dS. He was then able to calculate the distribution of illumination in diffraction patterns. His results agreed with observations.

(iii) *Kirchhoff* showed that the effect of an elementary area can be derived from the wave equation without making special assumptions. He showed that the assumptions which had been made by Fresnel are satisfactory provided that neither the source nor the point Q is very near to the surface S.

7.2. Fresnel's Method.

To illustrate the method used by Fresnel, it will be sufficient to consider the diffraction of a uniform plane wave by thin laminar obstacles which lie in a wavefront (fig. 7.2). Suppose that the illumination at a point Q is to be calculated and that O is the nearest point

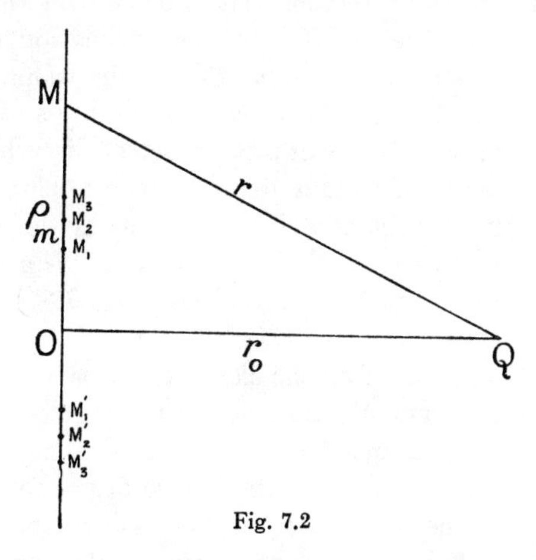

Fig. 7.2

on the wavefront to Q. Let $OQ = r_0$. Then the wavefront is imagined to be divided into a series of zones bounded by circles. M_1 and M_1' in the figure are points on the smallest of these circles, M_2 and M_2' on the second, and so on. The radii of the circles are chosen so that $M_1Q = r_0 + \lambda/2$, $M_2Q = r_0 + 2\lambda/2$, $M_3Q = r_0 + 3\lambda/2$, and so on. Then the radius (ρ_m) of the mth circle is given by

$$\rho_m{}^2 + r_0{}^2 = (r_0 + \tfrac{1}{2}m\lambda)^2,$$

or

$$\rho_m{}^2 = \lambda m r_0, \qquad \qquad 7(1)$$

provided that $m\lambda$ is small compared with r_0.

The area of the mth zone is $\pi\rho_m^2 - \pi\rho_{m-1}^2$ and, so long as 7(1) is valid, this area is equal to $\pi\lambda r_0$, i.e. it is independent of m, so that all zones are of equal area. Fresnel assumed that the amplitude at Q of the wavelets emanating from any infinitesimal area dS of the wave-front is equal to $kA\,dS/r$, where r is the distance from dS to Q. A is the amplitude of the incident wave and k is a constant. He also assumed that the amplitude depends on the angle between r and the normal (n) to dS (fig. 7.1). The factor which expressed this last effect was called the inclination factor. Fresnel found that, for his purposes, it was sufficient to assume that the inclination factor decreases as the inclination increases, and he did not specify the factor exactly.

7.3.—With these assumptions the amplitude at Q may be calculated in the following way. First imagine any one zone to be divided into a series of equal infinitesimal areas by circles centred on O, and differing infinitesimally in radius. Then, since over the small region covered by one zone the variation of the inclination factor may be ignored, the amplitudes at Q due to these infinitesimal areas are all equal. The phases vary uniformly and cover a range of π. The resultant of a set of disturbances of equal amplitude and uniformly varying phase has previously been calculated by vector summation (Chapter III). In this case, the vector diagram is a semicircle with the diameter as resultant (fig. 7.3). The resultant amplitude is $2/\pi$ times the value which would be obtained if all the elements had the same phase. The resultant phase is that of the middle of the zone, i.e. it is a quarter-period behind that of the wavelet arriving from the inner edge of the zone. The phases of the resultants from successive zones thus differ by a half-period, and they are known as *half-period zones*.* The effect at P, when there is no obstruction, is found by summing the resultants from all the zones. The vectors representing the resultants of successive zones are in the same straight line, but in opposite directions. The resultant of all the zones is thus the sum of an infinite series whose terms alternate in sign, but gradually diminish in magnitude. The sum of such a series is equal in magnitude to half the first member. This is shown in fig. 7.4, where S_0S_1 represents the resultant of the first zone, S_1S_2 that of the second, and so on. Thus the effect of all the zones is equal

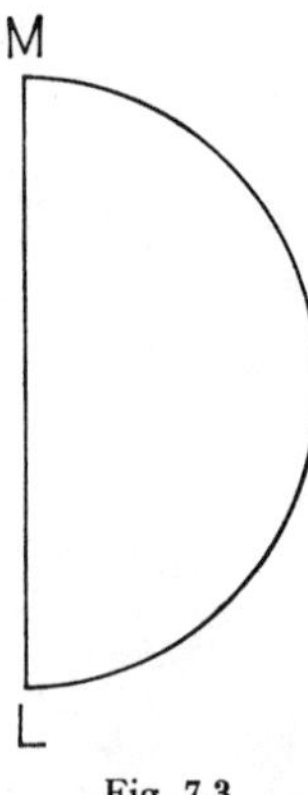

Fig. 7.3

* They are also called " Fresnel zones " or, less commonly, " Huygens zones ".

to half of that which would be produced by the first zone acting alone. It is equal to $\frac{1}{2} \times 2/\pi = 1/\pi$ times that which would be produced by all elements of the first zone acting together in the same phase.

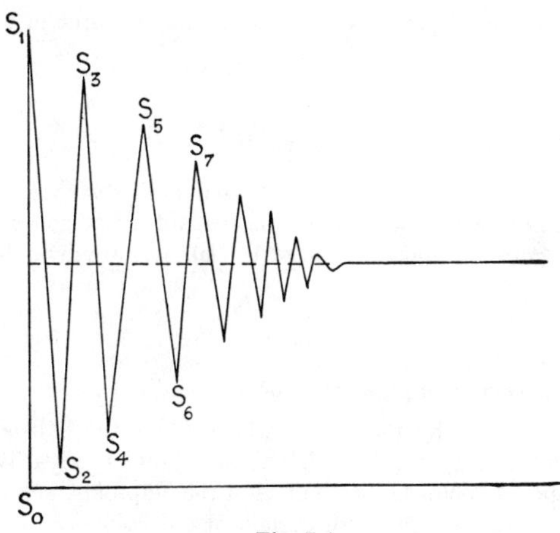

Fig. 7.4

The disturbance at Q is thus equal to

$$\psi_Q = \frac{1}{\pi}\int \frac{kA\,dS}{r}, \quad . \quad . \quad . \quad . \quad . \quad . \quad 7(2)$$

the integral being taken over the first zone. r may be taken as equal to r_0 over the small area involved, and hence the disturbance is

$$\frac{kA}{\pi} \cdot \frac{\pi\lambda r_0}{r_0} = kA\lambda. \quad . \quad . \quad . \quad . \quad . \quad . \quad 7(3)$$

But we know that for a uniform unobstructed plane wave the amplitude at Q is the same as that at O. Hence k must be equal to $1/\lambda$.

7.4.—The resultant of all the zones has the same phase as the resultant of the first zone, i.e. it is a quarter-period behind the phase of wavelets arriving at Q directly from O. From the ordinary wave equation we know that, for the unobstructed wave, the phase of the wave at Q is the same as that of a wave arriving directly from O. Thus if the Huygens wavelets are to give the correct phase at Q, it is necessary to assume that they start a quarter-period in advance of the phase of the wave which they replace.

This assumption appears at first sight rather strange. It is, however, supported by both theoretical and experimental observations. Gouy* showed that if a spherical sound wave is emitted by a small pulsating sphere, the wave which is effective at distances great compared with the wavelength is a quarter-period ahead of the phase of the pulsations. It can also be shown that there is an effective change of phase of half a period when a wave passes through a focus (§ 7.26).

EXAMPLES [7(i)–7(iv)]

7(i). Show that when the source is situated at a distance r' from a plane surface S, the radii of the circles which define the half-period zones, for an observer who is distant r_0 from the opposite side of the surface, are given by

$$\rho_m{}^2 \simeq m\lambda \frac{r_0 r'}{r_0 + r'}. \quad \ldots \ldots \quad 7(4)$$

(Assume that ρ_m is small compared with both r_0 and r'.)

7(ii). Show that generally the curves which define the half-period zones are the intersections of an ellipsoid with the surface S (not necessarily plane). Show that the source and the point Q are the foci of the ellipsoids, and derive an equation for the mth ellipsoid referred to an origin placed half-way between the source L_P and Q.

[Equation of ellipsoid is $\dfrac{x^2}{A^2} + \dfrac{y^2}{A^2 - l^2} + \dfrac{z^2}{A^2 - l^2} = 1$,

where $2A = L_PQ + m\lambda/2$, and $2l = L_PQ$.]

7(iii). If ρ_1 is the radius of the first half-period zone, find the amplitude and phase of the resultant due to the portion of the zone within a circle of radius $f\rho_1$, where f is less than 1.

$$\left[\text{Amplitude} = \frac{\sin f^2\pi/2}{f^2\pi/2}. \quad \text{Phase} = f^2\pi/2.\right]$$

7(iv). Derive a general formula for the resultant at Q due to an annulus between two radii ρ and ρ' which are within the same half-period zone.

$$\left[\text{Amplitude} = \frac{\sin\left(\dfrac{\rho'^2 - \rho^2}{2\lambda r_0}\right)\pi}{\left(\dfrac{\rho'^2 - \rho^2}{2\lambda r_0}\right)\pi}.\right]$$

7.5. Kirchhoff's Analysis.

Fresnel and his followers thus made three special assumptions concerning (i) the constant, (ii) the inclination factor, (iii) the relation between the phase of the wavelets and that of the wave.

These assumptions were all introduced *ad hoc*, in order to give the correct result for the unobstructed wave, but if they are accepted we

* Reference 7.1.

also obtain correct results for a wide range of diffraction problems. This indicates that Fresnel's method constituted an important advance in the wave theory of light. A further advance was made by Kirchhoff, who showed that the main results obtained by Fresnel could be derived from the wave equation without the introduction of special assumptions. If the wave emitted by the source L_P (fig. 7.1) is represented by

$$\xi = \frac{A'}{r'} \exp i(\omega t - \kappa r'),$$

Kirchhoff's analysis leads to the following expression for the amplitude ($d\psi_Q$) at Q due to an element dS at M whose normal is n:

$$d\psi_Q = \frac{iA'\,dS}{2\lambda r r'} \left\{ \cos(n, r) - \cos(n, r') \right\} \exp[-i\kappa(r + r')], \qquad 7(5)$$

where r' is the distance of dS from the source, and r is its distance from Q. This expression * is derived as an approximation which is valid when r and r' are both large compared with λ. The exponential term takes account of phase differences due to differences in the total length of path from L_P to M, and from M to Q. The term in brackets gives the inclination factor to a sufficiently good approximation for all ordinary problems. The factor i represents the phase difference of a quarter of a period between the wave at Q and the "wavelets" which originate at dS. When angles are small, and the direction from the surface S to Q is the forward direction of propagation, the inclination factor is approximately 2. If the surface S is a wave surface (i.e. a sphere with centre at L_P), then we may put $A'e^{-i\kappa r'}/r' = A$ and 7(5) reduces to

$$d\psi_Q = \frac{iA\,dS}{r} e^{-i\kappa r}. \qquad 7(6)$$

This approximate expression is in agreement with the assumptions stated in §§ 7.2 and 7.4.

7.6. Elimination of the Reverse Wave.

An early objection to Huygens' principle was that his wavelets should give rise to a disturbance travelling backwards from any surface on which they originate, as well as to one travelling forwards. In reply to this criticism it was suggested that perhaps the wavelets

* See Appendix VI A, equation 6(80).

interfered in such a way that the reverse wave was of zero amplitude, but no satisfactory proof could be given. A simple proof is possible if the inclination factor is assumed to have the value given in equation 7(5). The argument used in § 7.3 may be applied to the reverse wave. It shows that the amplitude of the reverse wave at any given point is half of the amplitude which would be produced by the first zone acting alone. The inclination factor for the wave travelling in the backward direction from the first zone is zero. Hence the amplitude of the reverse wave is zero.

7.7. Diffraction at a Circular Aperture.

Certain results concerning the distribution of illumination in diffraction patterns may be obtained in a very simple and elegant way by use of the properties of the half-period zones. The calculation of the change of illumination along the axis when a plane wave is diffracted by a small circular aperture may be taken as an example.

When the point under consideration is at a great distance from the aperture, the zones are very large, and only the wavelets from a small portion of the central zone are transmitted. These wavelets are in phase agreement, but their amplitude is proportional to $1/r$, so that the total effect is small. As the point of interest moves in towards the aperture, the zones shrink and the illumination gradually increases until the stage is reached where the first zone exactly fills the aperture. The amplitude is then twice that which would be produced at the same point if the screen were removed and the whole wave transmitted. The illumination is four times that given by the whole wave. If the point moves nearer to the aperture the illumination soon begins to decrease, and is zero when two zones fill the aperture. If the point is moved still nearer, the illumination goes through a series of maxima and minima. The minima occur when an even number of zones fills the aperture. The maxima are very near to the positions where an odd number of zones fills the aperture. The illumination reaches a constant value at points so near to the aperture that small irregularities in the shape of the aperture are comparable with the width of the outermost zone transmitted. The positions of the maxima and minima are calculated by applying equation 7(1).

In order to calculate the amplitude at an axial point intermediate between a maximum and a minimum, it is necessary to estimate the effect of the portion of a zone which is not balanced. In doing this, it is not sufficient to assume that the effect is simply proportional to the area of the fraction. Allowance must be made for phase relations [Examples 7(iii) and 7(iv), p. 202.]

7.8. Diffraction by a Circular Obstacle.

It is found that if a circular obstacle is placed at a moderate distance from a point source, the illumination at the centre of the geometrical shadow is the same as if the obstacle were removed. The following simple argument shows that this should be so.

Let SAO represent the path from S to O which just grazes the edge of the disc, and let SA_1O represent a path which is $\frac{1}{2}\lambda$, SA_2O a path which is λ longer, etc. A half-period zone will lie between the edge of the disc and a circle through A_1 (fig. 7.5), a second between this circle and one through A_2, and so on. In this way we construct a series of

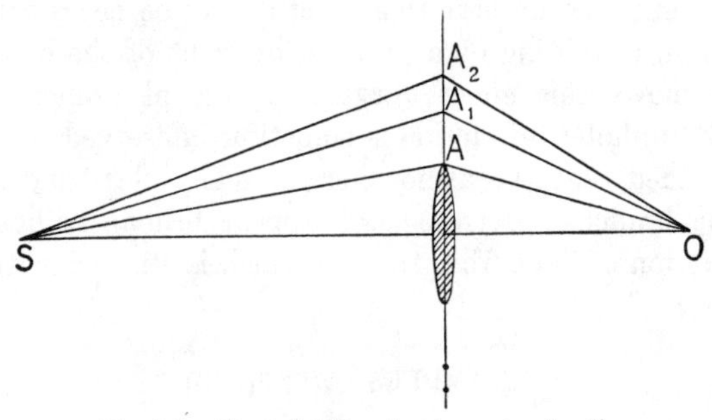

Fig. 7.5.—Fresnel diffraction by a circular disc

half-period zones beginning at the circumference of the disc. The resultant amplitude is equal to half that which would be produced by the first of this series of zones. If the point of interest is not too near the disc, the inclination factor may be ignored, and the amplitude is the same as that which would be produced by the whole wave. When Fresnel's paper on diffraction was presented to the French Academy, Poisson suggested that the above result was implied. He believed that he had disproved the whole theory by a *reductio ad absurdum.* The central bright spot in the shadow had indeed been observed by Delisle about half a century earlier, but his observation attracted little notice since it was not connected with any theory. His experiment was repeated by Fresnel and Arago in order to meet Poisson's criticism.

7.9.—The bright central spot is observed only when the edge of the circular object is free from irregularities of a size comparable with the width of the first zone in the unobstructed portion of the wave. If the obstacle subtends a moderately large angle, either at the point of observation or at the source, the width of this zone becomes very small and the size of the irregularities which can be tolerated becomes very small. The phenomenon can be conveniently observed using a pinhole in front of the image of a Pointolite as source, and a small ball-bearing (about $\frac{1}{8}$ or $\frac{1}{4}$ inch diameter) as the obstacle. The ball-bearing may be attached to a piece of plate glass with a little wax. The source may conveniently be placed beside the obstacle and reflected in a mirror about four metres away. Very near to the obstacle the shadow is dark, owing to the effect of irregularities, but two or three metres away the central spot is clearly seen. Even if no irregularities were present, the illumination very near to the obstacle would be zero owing to the effect of the inclination factor.

7.10.—In the above discussion of Fresnel diffraction it has been assumed that the fringes are received upon a screen or focused by an eyepiece. If the eye is placed in the plane of the screen and is focused upon the obstacle, the edges of the obstacle appear to be ringed with light. To observe this effect it may be necessary to use an artificial pupil, consisting of a pinhole, in front of the eye. Delisle's experiment shows this effect very well. A phenomenon similarly related to Fraunhofer diffraction is sometimes observed slightly before dawn. Provided that the atmosphere is really clear and steady, the edges of the branches of trees, etc., appear brilliantly lighted owing to the diffraction of light from the sun which is still below the horizon.

EXAMPLES [7(v)–7(viii)]

7(v). A plane wave is incident upon a screen containing a circular aperture 1 mm. in diameter. Derive an expression for the distances of the maxima of the axial illumination (§ 7.7) and derive numerical results for the three maxima which are farthest from the screen. Assume a wavelength of 5000 Å.

[Axial maxima are at distances $50/(2n + 1)$ cm., where n is an integer.]

7(vi). Using the data of the previous example, find the illumination at a point which is twice as far away (from the aperture) as the farthest maximum.

[$(8/\pi^2) \times$ illumination produced by unobstructed plane wave.]

7(vii). A point source of light is placed 50 cm. from a screen containing a circular aperture 0·5 mm. in diameter. Find the position of that maximum in the illumination (on the axis) which lies farthest from the aperture. Assume that the wavelength of the radiation is 5000 Å. [17 cm.]

7(viii). A disc, which is 0·5 cm. in diameter and has irregularities of order 10 μ, is placed 1 metre from a point source. Assume that the central spot is visible when the irregularities do not cut into any zone to a depth equal to more than a quarter of its width, and calculate the shortest distance at which the central spot can be seen. [67 cm.]

7.11. **The Zone Plate.**

The resultant amplitude at a given point due to the whole wave-front from a small source has been shown to be approximately half of the amplitude produced by the first zone, because the effects of the odd-numbered zones are in opposition to those of the even-numbered zones. The resultant amplitude due to either the odd-numbered zones or to the even-numbered zones would clearly be much greater than that of the whole wave. This may be tested by making a diffraction screen in which one set of zones is covered by opaque material while the other is clear (fig. 7.6). Such an obstacle, called a *zone plate*, may be made by photographing (on a reduced scale) a diagram prepared

Fig. 7.6.—Zone plate

by drawing circles of radii proportional to the square roots of the natural numbers, and blackening alternate spaces. There is a practical limit to the number of rings which can be drawn, because the outer rings are very close to one another. Plates containing as many as 250 rings have been constructed. A still greater increase of illumination is produced if the phases of either the odd- or the even-numbered zones are retarded by half a period, so that the effects of all zones co-operate. These zone plates are called *phase-reversal zone plates.* R. W. Wood has succeeded in making zone plates of this type.*

7.12.—A zone plate is capable of producing images like a lens or mirror. The exact theory of these images is complicated and not of great practical importance. A general discussion is, however, of some interest and may help in the understanding of the action of lenses and gratings. A perfect lens transforms a plane wave into a spherical wave whose centre of curvature is the focus. The lens

* Reference 7.2. R. W. Wood also devised an interesting way of making ordinary zone plates by direct engraving. The position of the tool for cutting the rings was set by means of a Newton's ring pattern in which the radii of the rings are proportional to the square roots of the natural numbers.

introduces just those differences of phase which are required to make the wavelets, from all parts of the wavefront which it receives, arrive at the focal point in the same phase. The phase-reversal zone plate does not produce quite such a good effect because all the parts of one zone do not co-operate perfectly. The resultant amplitude due to a whole zone is only $2/\pi$ times that which would be produced if all parts of the zone co-operated exactly.

7.13.—Whereas a lens has only one focus (on each side), the zone plate has a series of foci and associated focal lengths. If the radius of the inner edge of the mth zone is ρ_m, and if the source is placed at a very great distance from the plate, the primary focus is found at a distance

$$f_1 = \frac{\rho_m^2}{m\lambda} \quad . \quad . \quad . \quad . \quad . \quad . \quad . \quad . \quad . \quad . \quad 7(7)$$

[equation 7(1)], and a bright point of light will be found at this distance. With an ordinary zone plate having alternate clear and opaque spaces, the path difference between disturbances arriving from successive clear spaces at a point distant f_1 from the plate is λ. With the same plate the path difference between resultant disturbances arriving at a point $\frac{1}{2}f_1$ from successive clear spaces is 2λ. These resultants have agreement of phase, but each resultant is very small because the spaces now cover two half-period zones each. The resultant disturbances arriving at a point $\frac{1}{3}f_1$ have agreement of phase, the path difference being 3λ. Each space now covers three zones. Two of these oppose one another, but the unbalanced third zone gives an appreciable effect and the energy at this point is about one-ninth that of the energy at the primary focus. In a similar way, it may be shown that the zone plate has a series of foci at $f_1/5$, $f_1/7$, etc. There is also a series of foci on the opposite side of the plate, so that it is capable of acting as a concave lens. When the source is not placed at a very great distance from the plate, the relation between the distances of the source and image from the zone plate is similar to that obtained for a lens [equation 3(33)].

7.14.—In the above discussion, we have considered the relation between the zone plate and a lens. There is also a relation between the zone plate and the diffraction grating. Any small region of the plate is effectively a diffraction grating. The spacing of the lines in the small gratings which make up the zone plate varies systematically from point to point. The foci of the plate may be regarded as points where the directions corresponding to Fraunhofer diffraction spectra from different parts of the plate intersect. The primary foci correspond to the coincidence of first-order spectra, the secondary foci to the coincidence of second-order spectra, and so on. With a grating consisting of equal light and dark spaces, the spectra of even order are of zero brightness, and in a similar way the images corresponding to foci $f_1/2$, $f_1/4$, etc., are of zero brightness. The fact that the zone plate forms a *focused* image suggests that a grating in which the spacing increases at a suitable rate from the centre towards the ends, or from one end to the other, should have a similar focusing property. These effects have been observed by Mascart (1837–1908) and are discussed by Cornu* and by Rayleigh.† The focal property of gratings of non-uniform interval ruled on plane surfaces is not of practical importance, but the focusing of spectra due to gratings ruled on curved surfaces is of importance in the design of spectrographs (see Appendix VI B).

* Reference 7.3. † Reference 7.4.

7.15. **Fresnel's Integrals.**

When the diffraction apertures do not have circular symmetry about an axis through the point at which the illumination is to be calculated, the method of half-period zones is still applicable, but it loses a good deal of its simplicity. Usually there are a good many fractions of zones as well as completed zones. Sometimes it is possible to show that the total amplitude due to the fractions is zero, or is very small compared with the amplitude due to the completed zones. When the effect of the fractions of zones is not zero, it is necessary to allow for the fact that the phase of the resultant of a fraction of a zone is not the same as that of the resultant of the whole zone [Examples 7(iii) and 7(iv), p. 202]. The method which will now be developed is of more general application. It is particularly convenient when the obstacles or apertures are bounded by two sets of lines which are mutually perpendicular (e.g. slits, wires, etc.). The treatment given is restricted to the case where the incident wave is a plane wave. The extension to a spherical wave from a point source which is fairly close to the obstacle presents no difficulty. The extension to a line source at a finite distance is covered by an argument similar to that stated in § 6.17.

7.16.—Returning to equation 7(6) and putting $dS = dx_0\,dy_0$,

$$d\psi_Q = \frac{iA}{\lambda}\frac{\exp(-i\kappa r)}{r}\,dx_0\,dy_0, \qquad . \quad . \quad . \quad . \qquad 7(8)$$

provided the angles involved are small so that the inclination factor may be neglected. If the apertures are in the screen S_1 and the point Q lies in screen S_2 of fig. 7.7, then the co-ordinates of Q will be (x_1, y_1, z_1), and the co-ordinates of a point on the aperture will be $(x_0, y_0, 0)$. The distance of Q from a point on the aperture is given by

$$r^2 = z_1^2 + (x_1 - x_0)^2 + (y_1 - y_0)^2, \qquad . \quad . \quad . \qquad 7(9)$$

and this leads to the approximation

$$r = z_1 + \tfrac{1}{2}\frac{(x_1 - x_0)^2}{z_1} + \tfrac{1}{2}\frac{(y_1 - y_0)^2}{z_1}. \qquad . \quad . \qquad 7(10)$$

It is now assumed that the constant value z_1 may be used for r in the denominator of 7(8), but that the approximation 7(10) must be used in the exponential term. This is equivalent to assuming that differences of distances from different parts of the aperture to the point Q may be ignored when applying the inverse square law to determine the amplitude of the wavelets, but that they must be taken into account to the approximation implied in 7(10) when computing the effects of phase differences. In order to see that this procedure is reasonable, we take account of the fact that z_1 is always at least a

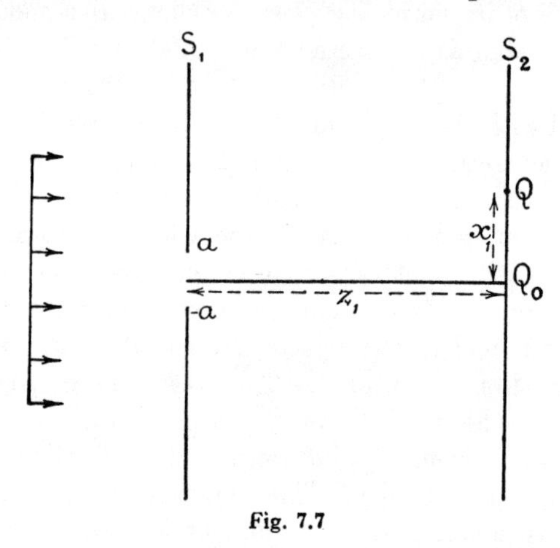

Fig. 7.7

few millimetres (i.e. several thousand wavelengths at least). Variations of a few wavelengths make nearly no difference to the amplitude, but a variation of $\frac{1}{2}\lambda$ is sufficient to reverse the sign of the factor which depends on phase difference. If we are concerned with relative illuminations, we may omit constant factors and write

$$\psi_Q = \iint \exp\left[-\frac{i\kappa}{2z_1}\{(x-x_0)^2+(y-y_0)^2\}\right] dx_0\, dy_0. \quad 7(11)$$

The integral is to be taken over the areas of the apertures in the screen. St. Venant's hypothesis (§ 6.13) is assumed to apply. In the case of diffraction at a slit aperture, equation 7(11) contains the product of two integrals one of which is

$$\psi_{QX} = \int_{a_1}^{a_2} \exp\left[-\frac{i\kappa}{2z_1}(x-x_0)^2\right] dx_0, \quad . \quad . \quad 7(12)$$

where the lines $x_0 = a_1$ and $x_0 = a_2$ are the edges of the slit. Let us introduce a new variable v, defined by

$$\sqrt{\frac{\kappa}{2z_1}}(x-x_0) = \sqrt{\frac{\pi}{\lambda z_1}}(x-x_0) = \sqrt{\frac{\pi}{2}}\,v. \quad . \quad . \quad 7(13)$$

We then have $$\psi_{QX} = \int_{v_1}^{v_2} \exp(-\tfrac{1}{2}i\pi v^2)\, dv, \quad \text{. . . . . 7(14)}$$

and the variation of illumination in the x direction is given by

$$E(x) = \psi_{QX}\psi_{QX}^* = C_F^2 + S_F^2, \quad \text{. . . 7(15)}$$

where $$C_F = \int_{v_1}^{v_2} \cos(\tfrac{1}{2}\pi v^2)\, dv,$$

and $$S_F = \int_{v_1}^{v_2} \sin(\tfrac{1}{2}\pi v^2)\, dv. \quad \text{. . . . . . 7(16)}$$

The functions C_F and S_F are known as Fresnel's integrals.

It is necessary to resort to special series to obtain the values of these integrals between finite limits.* When the limits of integration are $\pm\infty$, the value of either integral is $\pm\frac{1}{2}$. On account of the rapid variation of sign, the part of the range for which v exceeds a fairly small integer (e.g. 10) contributes little to the integral.

7.17. Cornu's Spiral.

The following geometrical method of using the integrals was devised by Cornu. Suppose that a curve is plotted with $u = C_F$ and $w = S_F$ (the limits of integration being 0 and v) as ordinates and abscissæ (fig. 7.8). The square of the length of a line joining two points defined by v_1 and v_2 is equal to the sum of the squares of the integrals, and is therefore proportional to the energy of the light diffracted in the corresponding direction. The direction of this line also gives the phase of the resultant vibration. Let s be the length, measured along the curve, from O to a given point A and let $\mathscr{I}$ be the angle which the tangent to the curve at A makes with the axis of u. Then

$$(ds)^2 = (du)^2 + (dw)^2 = (\cos^2 \tfrac{1}{2}\pi v^2 + \sin^2 \tfrac{1}{2}\pi v^2)(dv)^2 = (dv)^2.$$

Hence $$s = v \quad \text{. . . . . 7(17)}$$

(the constant of integration being zero since $s = 0$ when $v = 0$). Let

$$\mathscr{I} = \tan^{-1}\frac{dw}{du} = \tfrac{1}{2}\pi v^2,$$

i.e. $$\mathscr{I} = \tfrac{1}{2}\pi s^2. \quad \text{. . . . 7(18)}$$

* Methods of evaluating the integrals have been studied by Fresnel, Knockenhauer, Cauchy, Gilbert and Lommel. A summary is given in Reference 7.5.

Then the curvature

$$\frac{\partial \mathscr{I}}{\partial s} = \pi s. \qquad . \quad . \quad . \quad . \quad . \quad . \quad . \quad 7(19)$$

The last two equations show that, when $s = 0$, both $\mathscr{I}$ and $\partial \mathscr{I}/\partial s$ are zero, so that the curve touches the u axis at the origin. From this point it proceeds in the positive direction of s as a curve of continually

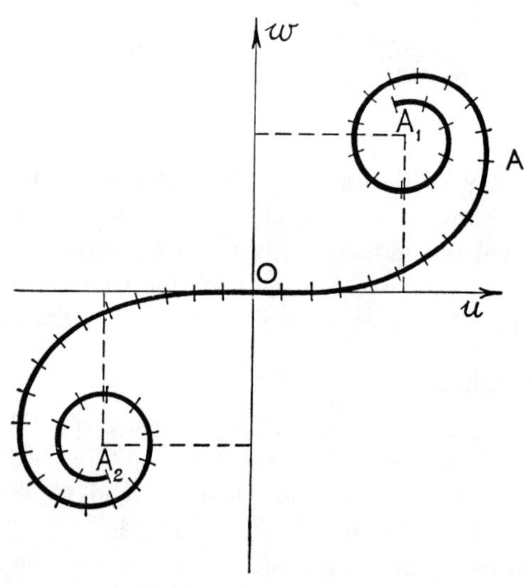

Fig. 7.8.—Cornu's spiral

increasing curvature, i.e. it forms a spiral whose successive turns enclose one another. The curve continually approaches the asymptotic point A_1 ($u = \frac{1}{2}$, $w = \frac{1}{2}$) which corresponds to $v = \infty$. A similar branch in the negative direction revolves about the point A_2 $(-\frac{1}{2}, -\frac{1}{2})$ as shown in fig. 7.8.

7.18.—We may now apply Cornu's spiral to investigate the Fresnel diffraction of a plane wave by a long narrow slit bounded by the lines $x_0 = a$, $x_0 = -a$, $y_0 = b$, and $y_0 = -b$. Suppose that the illumination at Q [whose co-ordinates are (x_1, y_1, z_1)] is to be calculated (fig. 7.7). If the long dimension of the slit is in the y direction, b will be large compared with a. We assume that this is so, and start by considering the variation of illumination with x_1 when $y_1 = 0$, i.e. if the slit is vertical we consider the variation of illumination along a horizontal line which is on the same level as the centre of the slit. The two integrals in 7(16) have to be taken between the limits

$$\left.\begin{aligned} v_1 &= \sqrt{\frac{2}{\lambda z_1}}\,(x_1 + a) \\ \text{and} \qquad v_2 &= \sqrt{\frac{2}{\lambda z_1}}\,(x_1 - a). \end{aligned}\right\} \quad . \quad . \quad . \quad . \quad 7(20)$$

The difference $(v_1 - v_2)$ is equal to

$$\sqrt{\frac{8a^2}{\lambda z_1}}.$$

It is independent of x_1, and hence of the position of Q. The variation of illumination which occurs when Q moves horizontally across the screen S_2 may be calculated by placing an inextensible thread of length $(v_1 - v_2)$ over the spiral. As the thread is moved along the spiral its ends are always at the points corresponding to the limits of integration. This is so because, as Q moves, the limits of integration change, but always in such a way as to keep the distance between the corresponding points (measured along the spiral) constant. The square of the length of the straight line joining the ends of the thread is proportional to the illumination at Q.

Suppose Q is first located at a point far on one side of Q_0, and that it moves in past Q_0 and out on the other side. Then initially the thread is wound up on the close turns of the spiral about one of the asymptotic points (say A_1), and the illumination is nearly zero. As Q approaches Q_0, the thread moves on to larger circles and the illumination oscillates. It has a small value when the ends of the thread are close to one another (but on different turns of the spiral), and a larger one when they are on opposite sides of different turns of the spiral. Each successive maximum is larger than the previous one and the illumination increases rapidly as one end of the thread approaches O. When one end of the thread coincides with O, the point Q is directly opposite one edge of the slit. If the width of the slit is small compared with $\sqrt{(\lambda z_1)}$, the other end of the thread will now be at some point such as A on the first half turn of the spiral. In this case there is only a small change of illumination from the point where Q is opposite one edge, to the point where it is opposite the middle of the slit (and the thread is symmetrically placed on the spiral). From the symmetrical position the thread moves on to the other half of the curve, and the second half of the pattern is similar to the first. This type of pattern is shown in fig. 6.2*c* and Plate III*e*. It really represents an approach towards the conditions of Fraunhofer diffraction (cf. fig. 6.9*a*).

7.19.—When the width of the slit is large compared with $\sqrt{(\lambda z_1)}$, the variations take the same general form when Q is far from Q_0, though the oscillations are smaller. When one end reaches O, a considerable portion of the thread is still wound up on several turns of the spiral. As Q moves towards the centre of the slit there are now considerable oscillations of illumination, though the minima are not nearly zero. If the two ends are both on the sides of the spirals nearest the origin, the illumination will be low. If they are both on the opposite sides, it will be high. There will be a general tendency for the illumination to increase as Q approaches Q_0, but the central point is not of necessity a maximum in the pattern. The pattern will be symmetrical about the centre. A diffraction pattern of this type is shown in Plate III*c*.

7.20.—In the above discussion we have ignored the variation of illumination in the y_1 direction. If the slit is long and narrow and the y_1 co-ordinate of Q is small compared with b (i.e. Q is not too far in a vertical direction from the middle of the slit), then the integrals for the y direction corresponding to C and S have to be taken between limits corresponding to points very near to the asymptotic points, and the values of the integrals are nearly constant, and equal to unity. Thus under these conditions there is no variation in the y_1 direction. The y_1 integrals do, however, introduce one significant effect. From fig. 7.8 it appears that if we took limits for $v = \pm\infty$ (corresponding to the complete wave) we would obtain a phase difference of $\frac{1}{4}\pi$ between the phase of the resultant and the phase of the central element. (The phase of the latter is represented by the direction of the u axis since, as we have seen in § 7.17, the curve touches the u axis at the origin.) When we consider only the variation with x, we are effectively starting with a series of strips, infinitely extended in the y direction. The phase of the resultant from one of these strips is $\frac{1}{4}\pi$ behind the phase of the central portion. In this way, the integration with respect to y introduces another phase difference of $\frac{1}{4}\pi$, giving a total phase difference of $\frac{1}{2}\pi$. This phase difference is cancelled by the factor i [in equation 7(6)], which we omitted when the discussion was confined to relative illuminations.

7.21. **Diffraction at a Straight Edge.**

The Cornu spiral may be applied in a similar way to determine the variation of illumination near the boundary of the shadow of a straight edge. We may imagine the point Q to be initially in the lighted region and remote from the edge of the geometrical shadows. The limits of integration are then so wide that the line whose square represents the illumination has to be drawn between two points which are extremely near to the asymptotic points of the spirals. As Q moves

(*a*) Fresnel diffraction at a straight edge.

(*b*) Interference fringes obtained with Fresnel biprism.

(*c*) Fresnel diffraction pattern (wide slit).

(*d*) Fresnel diffraction pattern (narrower slit).

(*e*) Fresnel diffraction pattern (very narrow slit).

(*f*) Diffraction pattern (needle).

(*g*) Diffraction pattern (wire).

(*h*) Diffraction pattern (fine wire).

(*i*) Fraunhofer diffraction pattern for small circular aperture showing the Airy disc.

(*j*) As (*i*) but with increased exposure to show outer rings.

(*k*) Enlargement of (*j*) for comparison with (*l*).

(*l*) Pattern due to two point sources which are just resolved (circular aperture).

(*m*) Fresnel diffraction obtained with point source and rectangular aperture whose shape is shown in bottom right-hand corner.

PLATE III

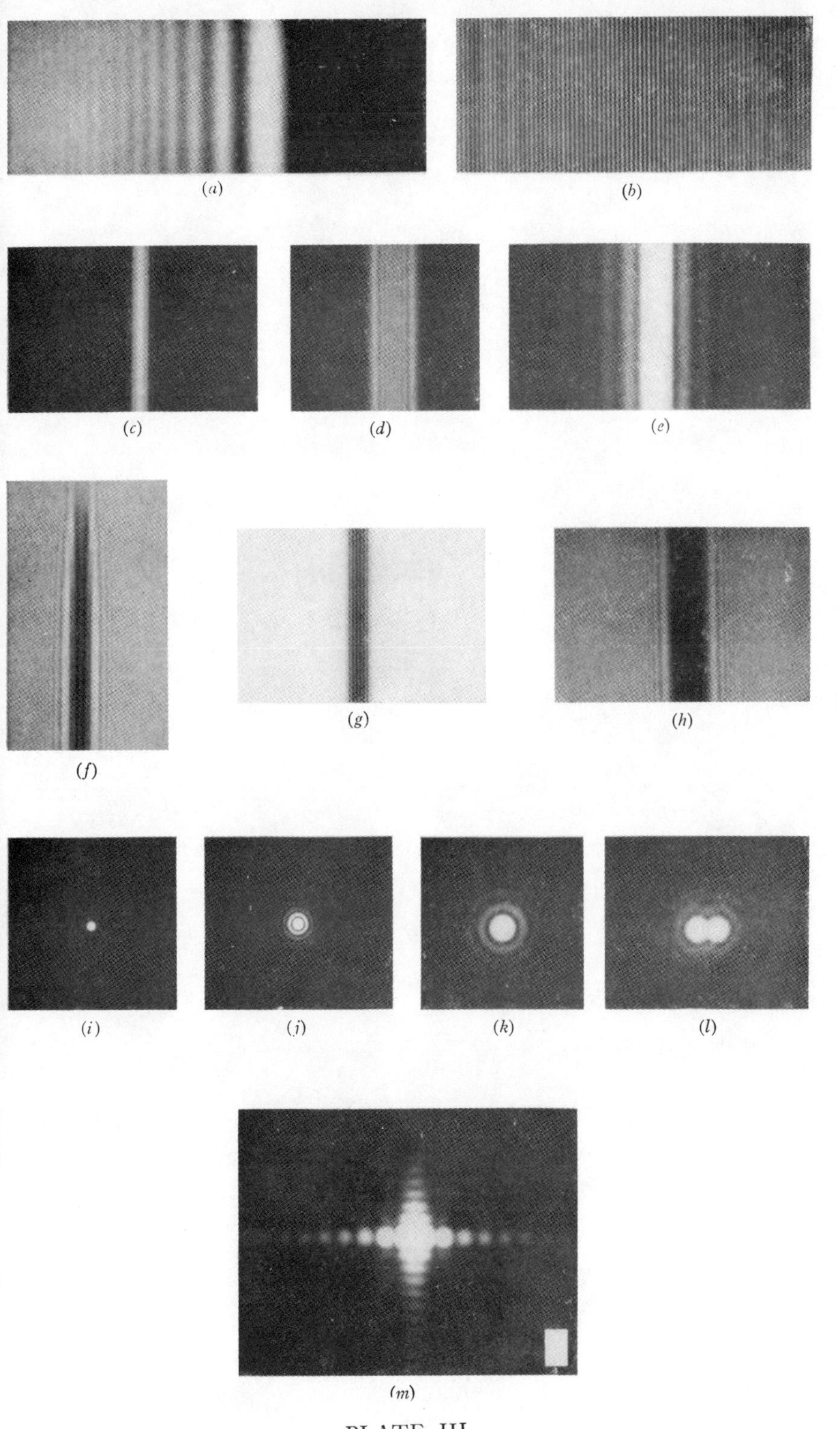

PLATE III

towards the geometrical shadows, one end of this line remains fixed very near to the asymptotic point, since on one side the wavefront is unbounded. The other end moves along the spiral from the asymptotic point towards O. At first this produces very little effect on the resultant illumination because the convolutions of the spiral lie within a narrow radius. As Q comes near to the edge of the geometric shadow, the point on the spiral moves on curves of large radius; the illumination oscillates between values larger and smaller than that which would be obtained if the whole wave were unobstructed. When Q is very near the edge of the geometrical shadow, the illumination decreases sharply as the representative point moves along the last half-turn of the spiral before reaching O. From this stage onward there are no further fluctuations in the illumination; it decreases smoothly and fairly rapidly to an insignificant value. Thus the last fringe is found slightly outside the geometrical shadow and there are no fringes within the shadow. The general system of fringes is shown in Plate III*a*.

The above discussion suffers from the defect that we have included a portion of wavefront corresponding to large angles of diffraction, whereas the equations from which the spiral was constructed are exactly valid only for small angles of diffraction. The effect of zones remote from the centre is very small. The factors we have omitted would, if included, merely deform the spiral a little in the region where it is near the asymptotic point. If it is desired to remove the formal inconsistency, we may suppose that we consider the diffraction at one edge of a very wide slit.

7.22. **Rectilinear Propagation.**

The preceding discussion shows that, for an object such as a slit or straight edge, diffraction effects are confined to a small region near the edge of the geometrical shadow. We shall now show that this restriction applies when the object is of irregular shape. Fig. 7.9 shows a system of half-period zones, partially obscured by an obstacle. In the absence of the obstacle, the amplitude at a point Q, for which the zones are drawn, is the sum of a series of terms alternately positive and negative, and gradually decreasing in magnitude. We have seen that the sum of such a series is equal to half that of the first term, and is independent of the exact rate at which the higher terms decrease. When the obstacle is in the position shown in fig. 7.9*a*, Q is outside the geometrical shadow by a distance equal to about three times the radius of the first zone. The only effect of the obstacle is to make the higher terms in the series decrease more rapidly than before. The resultant is nearly the same as that obtained in the absence of the obstacle. If the obstacle is moved into the position shown in fig. 7.9*b*, Q is inside the shadow, and the same distance as before from the edge. The resultant is the sum of a series of terms which are alter-

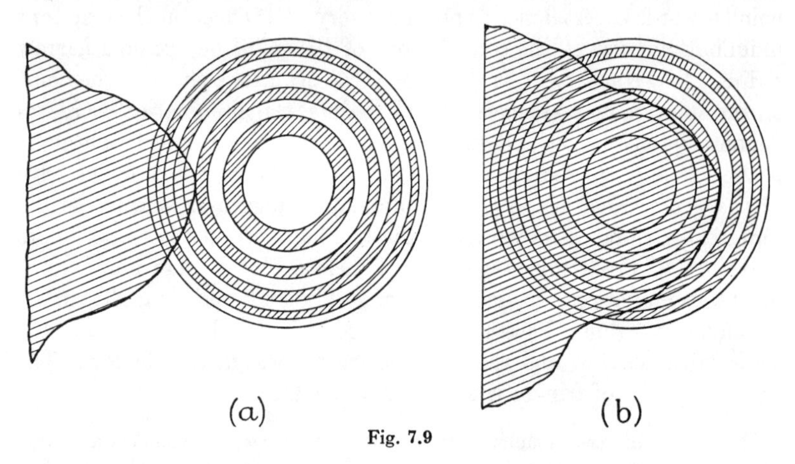

Fig. 7.9

nately positive and negative, and decrease gradually to zero at both ends. The sum of such a series is nearly zero. Thus for both the positions of the obstacle, the illumination is that given by the laws of geometrical optics.

Diffraction effects occur only when either:

(*a*) the edge of the obstacle is in, or near, the central zone; or

(*b*) the obstacle has a special shape so that its boundary follows the edges of one zone.

In either of these cases, the series is interrupted sharply at a certain term, and the resultant amplitude depends greatly on whether it is interrupted at an odd- or even-numbered term. The circular obstacle and aperture are the only simple arrangements in which the boundary may follow the edge of a zone.

7.23. **Fermat's Principle.**

Fermat (1601–65) suggested that the time taken by light to travel from one point on a ray to another is less than the time which would be required for transit between the same points by any neighbouring path. He regarded this " principle " as an example of the essential economy of nature! He showed that his statement is true for rays which have been reflected or refracted at plane surfaces.* Later investigation has shown that the principle as stated by Fermat is not always true when rays are reflected or refracted by curved surfaces.

* Essentially the same idea, in relation to *reflection* at plane surfaces, had been suggested by Hero of Alexandria (150 B.C.).

The principle has been modified and is now stated in the following way: " The difference between the time required for light to travel along the ray (i.e. the actual path) differs only by the second order of small quantities from the time required for light to travel along any neighbouring path." An alternative statement is: " For the true path, the first variation of the path length (measured in wavelengths) is zero." In the notation of the calculus of variations, these statements become

$$\delta \int_A^B \frac{ds}{b} = 0, \qquad \text{7(21)}$$

and

$$\delta \int_A^B \frac{ds}{\lambda} = 0. \qquad \text{7(22)}$$

A third form is

$$\delta \int_A^B n\, ds = 0, \qquad \text{7(23)}$$

where ds is an element of length along the path, b is the speed of light, and n the refractive index appropriate to the place where the element is situated. These forms may be seen to be equivalent by applying equations 2(25) and 3(31). The principle, stated in any of these ways, applies to rays which are reflected or refracted at plane or curved surfaces, and to rays which pass through a medium in which the refractive index varies continuously from point to point.

7.24.—We shall first show that the principle is obeyed in two simple cases: (1) rectilinear propagation in one uniform medium, (2) refraction at a plane surface.

(i) *Propagation in a Uniform Medium.*

We know that the ray from A to C is a straight line ABC joining those points (fig. 7.10). A neighbouring path is AB′C, and this path differs from the actual path by $AB'(1 - \cos\theta_1) + B'C(1 - \cos\theta_2)$. If θ_1 and θ_2 are small quantities, the difference is approximately $\frac{1}{2}(AB\,.\,\theta_1^2 + BC\,.\,\theta_2^2)$. If θ_1 and θ_2 are infinitesimal quantities, the path difference is of the second order of small quantities. Note that a similar proposition does not apply to a path such as AB_1C. If $AB_1'C$ is a path near to AB_1C, the difference of path is approximately

$$\Delta s = \left\{\frac{AD}{\cos\phi_1} + \frac{DC}{\cos\phi_2}\right\} - \left\{\frac{AD}{\cos(\phi_1 + d\phi_1)} + \frac{DC}{\cos(\phi_2 + d\phi_2)}\right\}$$

$$= \frac{AD \tan\phi_1}{\cos\phi_1}\, d\phi_1 + \frac{DC \tan\phi_2}{\cos\phi_2}\, d\phi_2. \qquad \text{7(24)}$$

ϕ_1 and ϕ_2 have the same sign and Δs is of the *first* order of small quantities unless $\tan \phi_1/\cos \phi_1$ and $\tan \phi_2/\cos \phi_2$ are themselves infinitesimal. They vanish only for the ray ABC. Thus this path has a special property [stated in equation 7(21)] which does not apply to any other path.

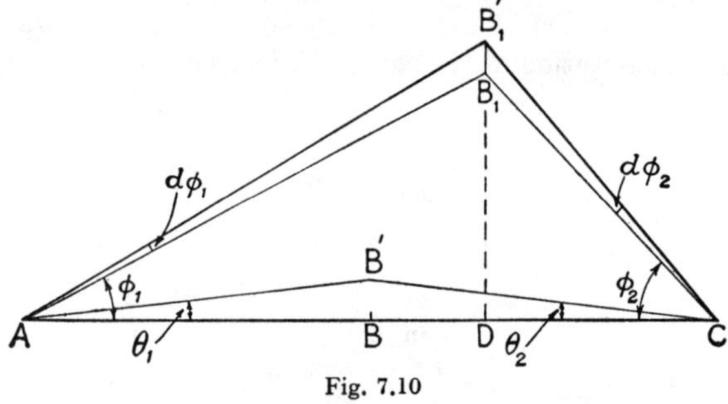

Fig. 7.10

(ii) *Refraction at a Plane Surface.*

Consider the path difference between PQR and PQ′R (fig. 7.11) which are neighbouring rectilinear paths from P to Q, giving slightly

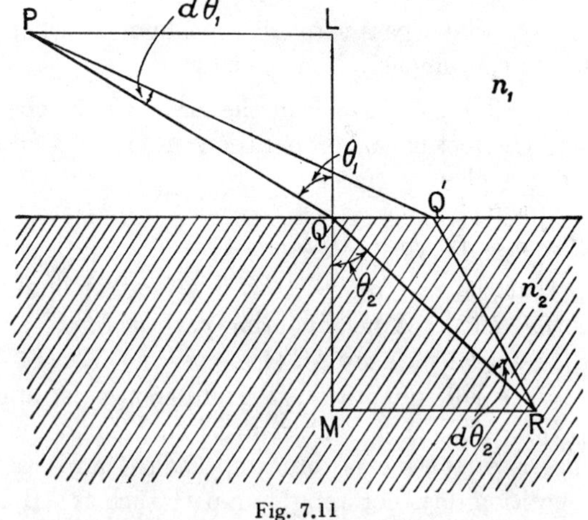

Fig. 7.11

different angles of refraction at the surface between two media of indices n_1 and n_2. For the path PQR the value of the integral of equation 7(23) is

$$n_1(\mathrm{PQ}) + n_2(\mathrm{QR}) = \frac{n_1(\mathrm{QL})}{\cos \theta_1} + \frac{n_2(\mathrm{QM})}{\cos \theta_2},$$

and for the path PQ′R it is

$$\frac{n_1(\text{QL})}{\cos(\theta_1 + d\theta_1)} + \frac{n_2(\text{QM})}{\cos(\theta_2 + d\theta_2)}.$$

The difference is

$$\Delta s = \frac{n_1(\text{QL}) \sin \theta_1}{\cos^2 \theta_1} d\theta_1 + \frac{n_2(\text{QM}) \sin \theta_2}{\cos^2 \theta_2} d\theta_2. \quad . \quad 7(25)$$

But $$(\text{QQ}') \cos \theta_1 = (\text{PQ})\, d\theta_1 = \frac{\text{QL}\, d\theta_1}{\cos \theta_1}.$$

Hence $$\text{QQ}' = \frac{(\text{QL})\, d\theta_1}{\cos^2 \theta_1},$$

and similarly $$\text{QQ}' = -\frac{(\text{QM})\, d\theta_2}{\cos^2 \theta_2}.$$

Thus the path difference is equal to

$$\Delta s = \text{QQ}'(n_1 \sin \theta_1 - n_2 \sin \theta_2),$$

and this difference vanishes, if and only if

$$n_1 \sin \theta_1 = n_2 \sin \theta_2,$$

i.e. if Snell's law is obeyed.

7.25.—When the rays of light emitted by a point source are reflected or refracted at a suitably curved surface, they converge towards a real or virtual image. In general, lenses or mirrors with spherical surfaces do not produce perfect images. The refracted or reflected rays from a point source pass near to a certain point, and the deviations are called *aberrations*. It is, however, possible to calculate certain surfaces for which all the rays from a certain object point P pass through an image point Q. A surface of this type has the property that the sum of the optical paths from P to R, and from R to Q, is constant if R is any point on the surface. When the points P and Q are situated in a medium of constant refractive index, an ellipsoid (formed by revolving an ellipse whose foci are P and Q about the line PQ) brings all the rays from one point to a focus at the other (fig. 7.12). The sum (PR + RQ) is constant. If a surface S′ of larger radius of curvature touches the correct aspherical surface S at R (fig. 7.13), a ray from P which strikes S′ at R is reflected to Q, because the direction of the ray reflected at this point depends only on the angle between the incident ray PR and RN, which is the normal to the plane tangential to all three surfaces. If the radius of curvature of S′ is greater than that of S, the optical path P to R, plus the optical path R to Q, is less

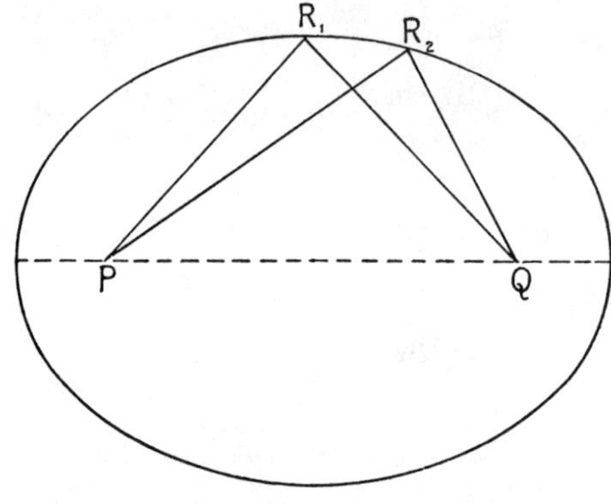

Fig. 7.12

than a path which goes from P to Q via a point R′ which is on S′ and near to R. If the radius of curvature of S″ is less than that of S, then the path from P to Q via R is a maximum. In either case equations 7(21) to 7(23) are satisfied.

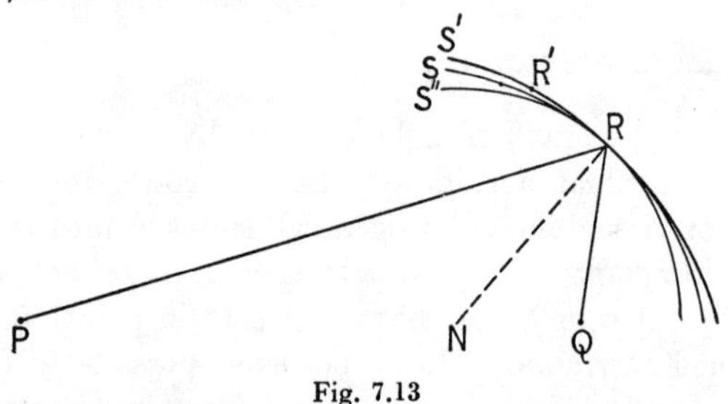

Fig. 7.13

7.26. Gouy's Experiment.

Suppose that a point source of light is placed at P and that systems of half-period zones are constructed on the surfaces S′ and S″ by an observer at Q. For the surfaces S′ and S″ the centre of the zone system will be at R, but for S′ the central zone will be the zone of minimum path, and for S″ the central zone will be the zone of maximum path. No zone can be constructed on S since all paths from P to Q via S are equal.* The phase of the resultant wave reaching Q from P via S′ is the same as the mean phase over the central zone drawn on S′, i.e. it is

* In a certain sense every point on S may be regarded as the centre of a zone system for which the central zone is of infinite area. The whole surface S then covers only an infinitesimal fraction of the central zone.

a quarter-period behind the phase of the wave from R. Similarly the phase of the wave received at Q from P via S″ is a quarter-period ahead of the phase of the wave from R. These phase differences were demonstrated experimentally by Gouy* and others. The experimental arrangement is shown in fig. 7.14. A parallel beam of *white* light is reflected from a plane mirror M_1 and a curved mirror M_2. The curved mirror focuses the light at Q. It has a greater radius of curvature than the correct aspherical surface corresponding to Q′ (since a mirror of smaller radius would be needed to focus the light at Q′). It is of smaller radius of curvature than the correct surface corresponding to Q″. The plane mirror is of greater radius of curvature than the correct surface for all three points.

Now suppose that the mirror M_1 is tilted slightly so as to allow interference fringes to be observed on a small screen placed near Q. When the screen is placed at Q′, both beams have been reflected from a surface whose radius of curvature is higher than that of the correct surface. The fringe system should therefore have a white centre at the point where the paths are geometrically equal. When it is placed at Q″, the beam from M_1 has been reflected at a surface of higher radius of curvature than the correct surface, and that from M_2 has been reflected at a surface of lower radius of curvature. There should therefore be a half-period difference of phase, so that the fringe system should show a dark centre. These effects were observed. More elaborate experiments by Fabry † and by Sagnac ‡ have verified Gouy's results and shown that the phase difference oscillates a number of times as the screen is moved through the focus. From our present point of view the reason for this last effect is that, as the screen approaches the focus, the half-period zones (drawn on the mirror M_2 in respect of a point in the centre of the screen) expand rapidly. So long as the surface of M_2 includes a large number of zones, the phase is the mean phase of the central zone. When, however, the number of zones is small, the phase depends on the exact number of zones included. When an even number of zones plus a small fraction of a zone is included, the phase of the resultant is the same as the phase at the centre of the central zone. If the number of zones increases, the phase gradually alters until, when the next larger even number of zones (minus a small fraction) is included, the phase of the resultant is opposite to that of the central zone. Thus the phase slowly alters by nearly π and then suddenly springs back as the screen passes through a position in which the number of zones on the mirror M_2 is a small even number. When the screen is moved well away from the focus, the size of the oscillations of phase decreases rapidly, because the amplitude due to the outer zones becomes very small.

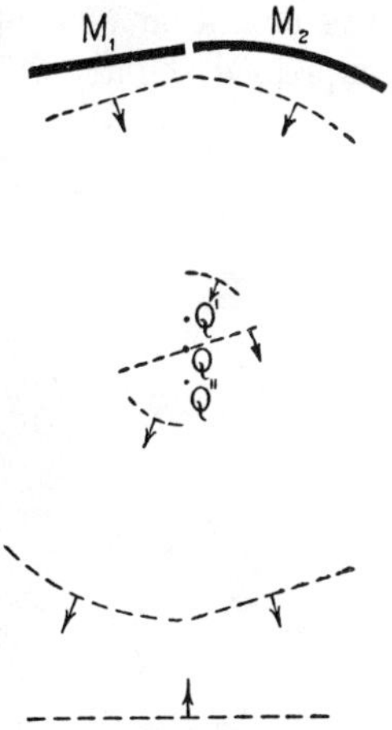

Fig. 7.14.—Gouy's experiment

* Reference 7.1. † Reference 7.6. ‡ Reference 7.7.

7.27. Relation between Wave and Ray Optics.

Fermat's principle is connected both to wave and to ray optics, and clarifies the relation between them. Suppose that equation 7(21) is satisfied for a given path between two points A and B. Then the phase difference between waves arriving at B from A by this path and by neighbouring paths is vanishingly small. Such waves have agreement of phase at B, and energy will travel from A to B if no obstacle is placed on or near a path defined by 7(21). This is just the definition of a ray adopted in Chapter I. This concept may be expressed slightly differently by considering the half-period zones.

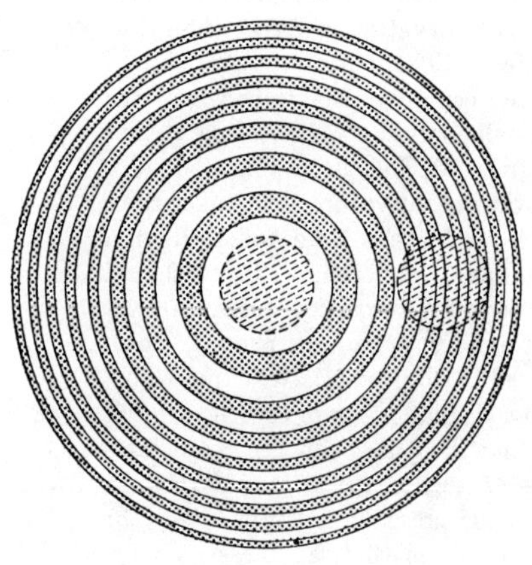

Fig. 7.15

Suppose that the half-period zones have been drawn on any surface intersecting any chosen line drawn from A to B, and suppose an imaginary cylindrical tube, whose cross-section is smaller than the area of the central zone, surrounds the line. This tube cuts the surface in a small circle. If the chosen line intersects the zone system at the centre, the phases of the unobstructed wavelets are all very nearly the same. If, however, it intersects at some point not very far out in the zone system (fig. 7.15), wavelets from a similar area surrounding the point of intersection have widely different phases. If this is the only clear path, very little energy travels from A to B. From this point of view the ray is defined as the locus of the centre of the half-period zone systems on all surfaces which lie between A and B.

7.28. **Rays and Wave Normals.**

In §§ 3.13 and 6.26, it was shown that the laws of reflection and refraction are obeyed by the wave normals of plane waves. The wave normals and the rays then coincide. This relation may be extended to wave surfaces of other shapes. The most general extension is to show that equation 7(21) is obeyed by paths taken along wave normals. This may be shown to be so, provided that the media are isotropic, though they need not necessarily be homogeneous, i.e. the speed of light may vary from point to point, but at any one point it must be independent of the direction of propagation. A wave surface may be defined by putting the phase (at a given time) $\phi(x, y, z)$ equal to some parameter χ (§ 2.13). Suppose light travels from a source placed at a point A, situated on the wave surface defined by $\phi = \chi_1$, to a receptor placed at B, situated on the wave surface defined by $\phi = \chi_2$. Imagine an infinite series of intermediate wave surfaces to be constructed. Provided that the wave surfaces are continuous, successive surfaces are nearly parallel curves, and a path which follows the wave normal is clearly defined. For such a path, the integral of equation 7(23) becomes

$$\int n\,ds = \chi_2 - \chi_1. \qquad \qquad 7(26)$$

In general, a path which is near to this path will be parallel to it part of the way, but will include some portions which lie near to the original path but inclined to it at small angles. These portions will give contributions to the integral equal to $n\,ds/\cos\alpha$, where α is the small angle between the path defined by 7(23) and the varied path. These contributions will differ from the corresponding terms in 7(26) by quantities of the second order. Thus Fermat's principle applies to a path defined by the wave normal. When waves are reflected or refracted at curved surfaces, the wave surfaces intersect, and the above simple treatments do not apply. In order to show that Fermat's principle applies to wave normals under these conditions, it is necessary to show that the wave normals change their direction in the same way as the rays. This has been done in § 6.26. Thus, *in isotropic media*, the wave normals and the rays coincide.

7.29. **Rays in Relation to Wave Groups.**

It was shown in Chapter IV that even "monochromatic" light must be represented by a wave group. In order to show that Fermat's principle applies under practical conditions, it is therefore necessary to show that the above results apply to wave groups. This problem has been investigated * and it has been shown that:

(i) In a dispersive medium the energy in a wave group travels along the ray defined by applying equation 7(23), inserting the values of n appropriate to the mean wavelength.

(ii) The group travels along this ray with the group velocity given by inserting the mean wavelength in equation 4(34).

7.30. Fermat's Principle as a General Statement of the Laws of Ray Optics.

Fermat's principle includes and summarizes the following laws of ray optics:

(i) *Rectilinear propagation* in a medium of constant refractive index.

(ii) *The laws of reflection and refraction* of rays at surfaces where the index changes discontinuously.

(iii) *Curvilinear propagation* in a medium where the index varies continuously along a path for which 7(23) is satisfied.

(iv) *The law of reversibility of path* according to which any line which is a possible path for light energy travelling in one direction is also a possible path in the reverse direction.

The last of these relations is implied in equations 7(21) to 7(23) because, if the variation of the integral is zero when the limits are A to B, it is also zero when they are interchanged to become B to A. It is important to recognize that *all* the above relations apply only under the limiting conditions of ray optics.

EXAMPLES [7(ix)–7(xvi)]

7(ix). Using equation 7(4), verify the statement made in the last sentence of § 7.13.

7(x). A zone plate is made by arranging that the radii of the circles which define the zones are the same as the radii of Newton's rings formed between a plane surface and a surface whose radius of curvature is 3 metres. Find the primary focal length of the zone plate. [300 cm.]

7(xi). Discuss the diffraction of light by a single slit by the method of half-period zones. Show that the main qualitative results given in § 7.19 may be obtained.

7(xii). Discuss the diffraction of light at a straight edge by the method of half-period zones. Show that there are alternations of illumination just outside

* See Reference 7.8.

the shadow. Show that a treatment which ignored phase relations would predict alternations of illumination within the shadow, but that when allowance is made for the phase of a fraction of a zone the correct result is obtained.

7(xiii). Apply the Cornu spiral in a qualitative discussion of the diffraction pattern produced by a fine wire. [See Plate III *g* and *h*, p. 214.]

7(xiv). Use the Cornu spiral to calculate the illumination at the edge of the geometrical shadow of a straight edge when the source and the screen are each at 2 metres from the obstacle.

[One quarter of the illumination at a point well outside the shadow.]

7(xv). Show that Fermat's principle is obeyed when light is reflected at a plane mirror.

7(xvi). Show that equation 7(21) is obeyed when light is reflected from a source placed at the centre of curvature of a concave mirror, but that the path is neither a maximum nor a minimum.

REFERENCES

7.1. Gouy: *Ann. de Chimie et de Physique*, 1891, Ser. VI, Vol. 24, p. 145.

7.2. Wood: *Phil. Mag.*, 1898, Vol. 45, p. 511.

7.3. Cornu: *Comptes Rendues*, 1875, Vol. LXXX, p. 645.

7.4. Rayleigh: *Scientific Papers*, Vol. III, p. 112.

7.5. Rayleigh: *Scientific Papers*, Vol. III, p. 128.

7.6. Fabry: *Journal de Physique*, 1892, Vol. 2, p. 22.

7.7. Sagnac: *Journal de Physique*, 1903, Vol. 2, p. 721.

7.8. Kemble: *Foundations of Quantum Mechanics*, Chap. II, Sec. 12. (McGraw-Hill.)

CHAPTER VIII

The Accuracy of Optical Measurements

8.1. **Imperfections in Images due to Diffraction.**

As explained in § 7.25, an ideal optical system provides a set of optically equal paths from an object point to an image point. In accordance with Fermat's principle, a set of equal optical paths may be represented by a set of rays from an object point which pass through the optical system and meet at the image point. The image is then perfect from the point of view of geometrical optics. It does not, however, form an exact representation of the object. The image of a point object is not a point but a diffraction pattern. The ideal optical system arranges that all portions of the wave which pass through the system arrive at the image point in phase agreement. To form a perfect image, it would also be necessary to arrange that, at every other point in the image plane, the waves interfered to produce zero illumination. The theory of Fraunhofer diffraction is an expression of the fact that no optical system of finite aperture can achieve this result. No optical system transmits the whole wave emitted by the object and there is therefore always some diffraction. When the aperture is circular, the image of a point source consists of the Airy disc and the surrounding ring pattern (§ 6.16). When the angular separation of two point sources (e.g. two stars seen through a telescope) is sufficiently great, the diffraction patterns which form the images do not effectively overlap. Such images are said to be " resolved ". On the other hand, if the angular separation of the two objects is much less than the angular radius of the Airy disc, the two images will be superimposed to such an extent that they cannot be clearly distinguished from an image due to one object alone. Such images are " unresolved ".

8.2. **The Rayleigh Criterion.**

From the above discussion, it is clear that the condition under which the images are seen separately cannot be defined with precision. If two object points, initially very close together, could be moved apart slowly, there would be (i) a condition under which a person observing the image thought there was only one point, and

(ii) a condition under which he could see two points clearly separated. These two conditions would be separated, not by a definite boundary, but by an intermediate region in which the observer suspected the existence of two points, but could not be sure that there were two, and could not determine their relative positions. In the first stage of the study of this phenomenon, it is desirable to adopt some arbitrary criterion which gives a mathematically defined boundary between

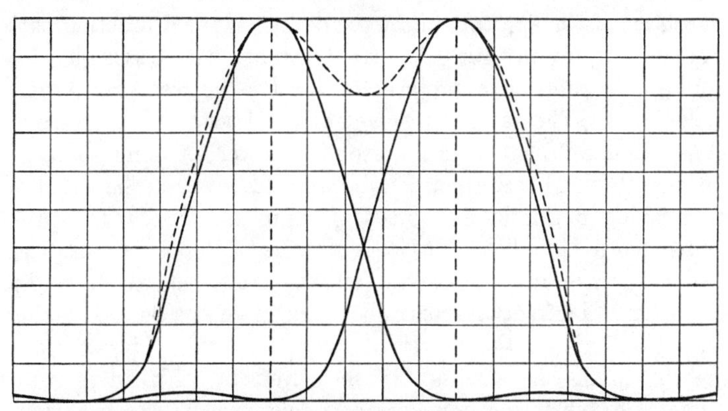

Fig. 8.1.—Energy distribution at the Rayleigh limit for two similar images. (Full lines show the distributions due to the separate sources; the dotted line shows the distribution when both act together.)

resolution and non-resolution. Rayleigh * suggested the following criterion. Two images are regarded as just resolved when the central maximum in the diffraction pattern due to one is situated at a point corresponding to the first minimum in the diffraction pattern due to the other (Plate III*k* (p. 214) and fig. 8.1). Images which have a smaller separation are unresolved. This criterion enables the *resolving power* of an optical instrument to be exactly defined. It has the special merit of being applicable to a wide range of instruments, including telescopes, microscopes, and spectroscopic instruments.

8.3.—Up to a certain point in the history of optical instrument design, diffraction effects were of no practical importance. Imperfections of image due to other causes, such as lens aberrations or defects in manufacture, were so large that any additional unsharpness due to diffraction was negligible. During the nineteenth century, technique advanced so that diffraction became an important factor, and often the limiting factor, in determining the performance of an instrument. In many modern spectrographs, telescopes, and microscopes, imperfections of image due to causes other than diffraction have been made very small so that, *when the instrument is used under precisely the conditions for which it was designed,*

* Reference 8.1.

diffraction forms the practical, as well as the theoretical, limit. The resolving power of such an instrument forms one of the more important constants of the instrument. In §§ 8.5 to 8.13 formulæ enabling the resolving power to be computed from the optical dimensions of an instrument are derived. The significance of the results, in relation to the general theory of physical measurements, will appear later (see Chapter XVIII).

8.4—In the above discussion, we referred to the Airy disc which is obtained with a circular aperture and a point source. Alterations in the shape of the aperture affect the details of the diffraction pattern, but not usually to any very important extent. The Rayleigh criterion can be applied and an appropriate value for the resolving power can be calculated. The resolving power of a telescope or microscope for line objects is different from its resolving power for point sources, but again the Rayleigh criterion can be applied. When the question of distinguishing detail in an object of finite size arises, the problem becomes more complicated. It is not usually possible to use the Rayleigh criterion in a direct way, but the same general physical principles apply. Diffraction blurs the edge of every detail, setting an inescapable limit to the power of discrimination. This, directly or indirectly, limits the accuracy of measurement with an instrument. For example, when the crosswire of a telescope is set on a star, the observer does not see a point image and a mathematical line. He sees diffraction images of the star and of the line, and he has to attempt to superimpose the centre of one on the centre of the other. The accuracy with which this can be done is limited, when all other sources of error have been removed, by the " spread " of the diffraction patterns. It is important to remember that when the limit is set by diffraction, no advantage is gained by increasing the magnification of the instrument. When the images are magnified, the width of the diffraction pattern increases in proportion because all Fraunhofer diffraction effects have a constant *angular* width. There is, in general, an optimum magnification (§§ 8.8 and 8.31).

8.5. Limit of Resolution for a Telescope.

The limit of resolution for a telescope may be determined by applying Airy's results for diffraction at a circular aperture. It is shown in § 6.16 that the angular radius (θ) of the first dark ring in the diffraction pattern is given by

$$\sin \theta = 1{\cdot}22 \frac{\lambda}{d}, \qquad \qquad 8(1)$$

where d is the diameter of the lens.

The angle involved is usually small enough for θ to be used in place of $\sin \theta$, so that $1{\cdot}22\lambda/d$ is then the angular separation of two stars which are just resolved. Taking $\lambda = 5500$Å. (corresponding to the middle of the visible spectrum), putting d in centimetres and θ in minutes of arc, we have

$$\theta = \frac{0{\cdot}231}{d}. \qquad 8(2)$$

Thus a telescope with an object glass 10 centimetres in diameter will separate two stars whose angular separation is $0{\cdot}023' = 1{\cdot}4''$. The angular diameters of stars range up to about 0·05 seconds of arc. The diameter of the image is thus always a small fraction of that of the Airy disc and the star is effectively a point source.

8.6.—An approximate value for the limit of resolution for a lens covered with a rectangular aperture may be obtained by direct application of the Rayleigh criterion. In fig. 8.2 let AB and AB′ represent two wave surfaces received from two very distant objects O and O′. The angle $\alpha = \angle$BAB′ is the angular separation of the objects. All parts of the wavefront AB reach a certain point P in the

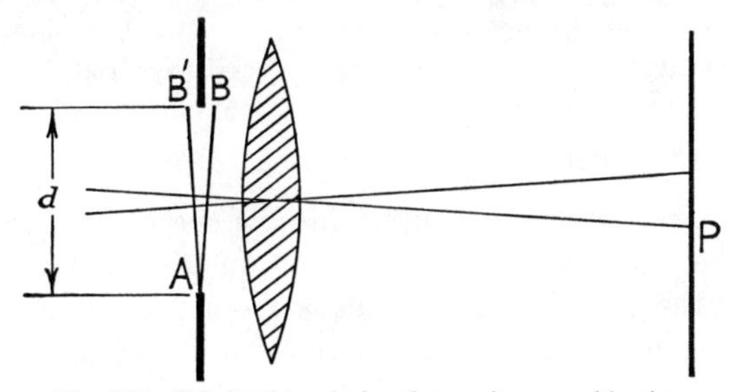

Fig. 8.2.—Limit of resolution for a telescope objective

focal plane of the lens with phase agreement. This point is the centre of the diffraction pattern formed by the point source O. If BB′ = λ, the phases of the wavelets arriving at P from O′ will range uniformly over a period. Their resultant is zero (see § 3.5). Thus the Rayleigh criterion indicates that the minimum angle of resolution is λ/d. In making this calculation we have assumed that the line of separation is parallel to one side of the aperture and have neglected certain small effects (due to the finite length of the other side). It will be noted that the result obtained differs from that given by equation 8(1) only by the factor 1.22.

EXAMPLES [8(i)–8(iii)]

8(i). Find the minimum angle of resolution (*a*) for a telescope with a mirror of 200 in. diameter, (*b*) the lens of the eye when the pupil is 3 mm. diameter. Assume $\lambda = 5500$ Å. [(*a*) 0·00045 minute, (*b*) 0·77 minute.]

8(ii). Find the distance between the images of two stars which are just resolved by a lens of focal length 3 m. and diameter 10 cm. Take $\lambda = 5500$ Å. [0·02 mm.]

8(iii). Assuming that the length of the exposure does not matter, obtain an expression for the optimum value of the size of the hole in a pinhole camera.

[The diameter of the image of a point source estimated by geometrical optics increases as the diameter of the hole increases. If, however, the hole is made very small, the size of the image is determined by the diameter of the Airy disc. This latter is *inversely* proportional to the diameter of the hole. The optimum condition is attained when both sources of " unsharpness " are equal. If d is the diameter of the pinhole, this happens when $\frac{1}{2}d = 1{\cdot}22(\lambda/d)f$, where f is the distance of the plate from the pinhole and the object is assumed to be at an infinite distance. Show that when the object is at a distance u, the optimum diameter of the pinhole is $\{2{\cdot}44\lambda uf/(u+f)\}^{\frac{1}{2}}$. This problem has been investigated in detail by Rayleigh. (See Reference 8.2.)]

8.7. Limit of Resolution for the Eye.

A theoretical limit of resolution for the eye may be calculated by inserting the diameter of the pupil in equation 8(1). The average diameter of the pupil in full daylight is about 2·5 millimetres, and the corresponding minimum angle of resolution (for light of wavelength 5500 Å.) is $2{\cdot}7 \times 10^{-4}$ radian or 56 seconds of arc. Practical tests show that people with good sight can, under the most favourable conditions of observation, distinguish two point sources of light when the separation is a little less than one minute of arc. Thus under these conditions the ability of the eye to distinguish detail in an object is almost entirely determined by diffraction. If the pupil of the eye is made larger, either by the use of drugs or by lower illumination, the practical limit of resolution is no longer equal to the theoretical value calculated from 8(1). This is because lens aberrations and the structure of the retina then have important effects. Average laboratory conditions of observation are not ideal, and it is reasonable to assume a limit of resolution for point objects of about $3{\cdot}4 \times 10^{-4}$ radian or 1·25 minutes of arc for laboratory work. If the point objects are situated at the nearest point of distinct vision (25 centimetres), the minimum linear separation required is a little less than a tenth of a millimetre. Objects which are separated by 0·2 millimetre are resolved comfortably.

8.8. **Useful and Empty Magnification.**

If the image formed by a telescope or microscope is not magnified enough, some detail of the object which is resolved by the instrument may not be seen by the eye. The detail may be correctly represented in the image formed by the instrument, but its size in that image may be so small that it cannot be resolved by the eye. For this reason it is desirable that the image shall be of sufficient size to make the smallest detail resolved by the instrument occupy a space of about 0·2 millimetre in the image seen by the eye. Magnification which gives an image up to this size is called *useful magnification.* Any magnification in excess is called *empty magnification* since it does not reveal any fresh detail in the object. Empty magnification is undesirable because it is usually accompanied by an increase in lens aberration and a reduction in the illumination of the field. The maximum useful magnification is therefore usually the optimum magnification. When an image is formed by a single lens, the resolving power depends on the diameter, and the size of the image depends upon the focal length. We have seen in § 8.5 that a lens whose diameter is 10 centimetres will resolve two stars whose angular separation is 1·4 seconds of arc. The distance between the images of these stars is equal to $6{\cdot}7 \times 10^{-6}$ times the focal length of the lens. Thus, if no eyepiece were used, it would be necessary to have a lens of 30 metres focal length in order to obtain optimum viewing conditions. This length is so great that it is convenient to use an objective of much smaller focal length (say 3 metres) and then to use a magnifying eyepiece to view the image.

When an object is to be photographed, full detail will be represented in the photograph only if the distance between the images of points which are just resolved exceeds the spacing of the grains in the photographic plate. For rapid plates, with a fairly coarse grain, the desirable size of image is about the same as that given above. When fine-grain emulsions are used it is possible to use much lower magnification. The resulting photograph must, however, be viewed under magnification in order that the eye may perceive all the detail which is represented therein.

8.9. **Resolving Power of a Prism Spectroscope.**

When a line spectrum is formed by a spectroscope or spectrograph, the dispersing system (i.e. the prism or grating) forms an image of the slit corresponding to each line. Neighbouring lines give rise to images whose angular separation is small. Lines will be separated only if the angular separation of their images exceeds the limit of resolution of the telescope. The resolving power of a spectroscopic instrument is defined to be equal to $\lambda/\Delta\lambda$ when lines whose wavelengths are λ and $(\lambda + \Delta\lambda)$ are just resolved. Fig. 8.3 shows a simple prism spectroscope. Light from the slit S is collimated by the lens L_1 and plane waves (with wave surfaces parallel to AB) fall upon the prism LMN. The rays are refracted by different amounts in passing through the

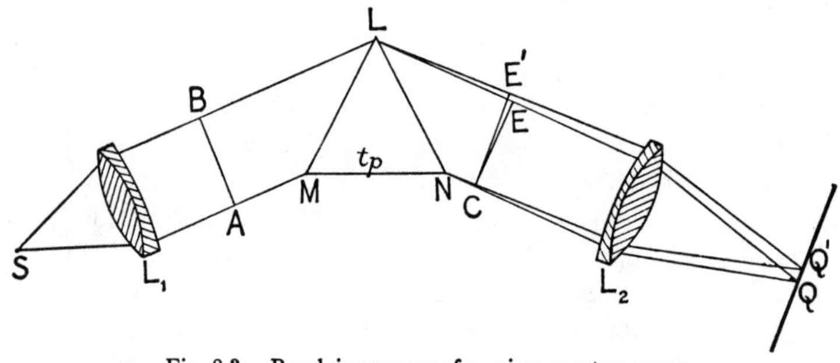

Fig. 8.3.—Resolving power of a prism spectroscope

prism and the emergent wave surfaces are parallel to CE for wavelength λ, and to CE′ for wavelength λ'. The lens L_2 focuses the wavefront CE at Q and CE′ at Q′. The thickness of the base of the prism is t_p and the refractive index of the material is μ for λ and μ' for λ'. The light is supposed to be restricted by a rectangular aperture whose width (CE) is equal to d. By Fermat's principle the total path from S to Q is the same for all rays of one wavelength, and the focal properties of the lenses imply that the paths S to A and S to B are equal. Similarly the paths C to Q and E to Q are equal. Hence we have for wavelength λ

$$\mathrm{BL} + \mathrm{LE} = \mathrm{AM} + \mu(\mathrm{MN}) + \mathrm{NC}, \quad . \quad . \quad . \quad 8(3)$$

and similarly for λ'

$$\mathrm{BL} + \mathrm{LE}' = \mathrm{AM} + \mu'(\mathrm{MN}) + \mathrm{NC}. \quad . \quad . \quad . \quad 8(4)$$

Hence $$\mathrm{LE} - \mathrm{LE}' = (\mu - \mu')\mathrm{MN} = t_p(\lambda - \lambda')\frac{d\mu}{d\lambda}. \quad . \quad . \quad 8(5)$$

The angle between the emergent wave surfaces is $(\mathrm{LE} - \mathrm{LE}')/d$

and the minimum angle which the telescope can resolve is λ/d. Hence the two wavelengths are just resolved if

$$\frac{\lambda}{d} = (\lambda - \lambda')\frac{t_p}{d} \cdot \frac{d\mu}{d\lambda}, \qquad \text{8(6)}$$

and the minimum difference of wavelength for resolution is given by the relation

$$R_p = \frac{\lambda}{\Delta\lambda} = t_p \frac{d\mu}{d\lambda}. \qquad \text{8(7)}$$

It will be seen that although the resolving power of the prism is determined by reference to the aperture of the telescope, the thickness of the base of the prism is the only geometrical factor in the final result. This is so because the wavelength resolving power is determined partly by the aperture and partly by the angular dispersion. When the prism is not completely filled with light, t_p must be put equal to the difference of thickness at the places where the extreme rays traverse the prism. The resolving power is slightly different when the rays are restricted by a circular stop. This case has been treated by Struve, who has also considered the effect of finite length of the slit. In practice the stop is often approximately elliptical in shape.

8.10. **Resolving Power of a Grating Spectroscope.**

Suppose light of wavelengths λ and $(\lambda + \Delta\lambda)$ is diffracted in a certain direction θ by a grating containing N lines and of total width D. Then the path difference between portions of the wavefront which reach the focal point from opposite ends of the grating is $D \sin \theta$. If this direction corresponds to the principal maximum of order m for wavelength λ, and to one of the two neighbouring minima for $(\lambda + \Delta\lambda)$, then

$$mN\lambda = D \sin \theta = (mN - 1)(\lambda + \Delta\lambda),$$

and hence

$$R_g = \frac{\lambda}{\Delta\lambda} = mN. \qquad \text{8(8)}$$

The above method of calculating the resolving power shows the similarity between the action of the grating and that of the prism. The result might also have been obtained directly from Table 6.2 (p. 179).

8.11.—Gratings up to 10 inches wide with 15,000 lines per inch have been constructed. A large grating thus has a resolving power of 150,000 in the first order. It is often practicable to use the third-order spectrum in which the resolving power is nearly half a million. With a reflecting echelon grating of 40 steps, each 15 millimetres high, spectra of order 30,000 are obtained. The resolving power is then nearly $1\frac{1}{4}$

million. The resolving power of prisms of convenient size is very much lower. It varies a good deal with wavelength and with the type of glass as well as with the thickness of the base. A flint glass prism of 5 centimetres base gives a resolving power of about 5000 at 6000 Å. and of about 12,000 at 4500 Å. A quartz prism of 5 centimetres base has a resolving power of 2000 at 6000 Å. and of 75,000 at 2000 Å. Owing to the very high dispersion of quartz in the ultra-violet, large quartz prisms have a resolving power which is not greatly inferior to that of a grating in the region 2500 Å. to 1850 Å. The prism instrument has the advantage that there is only one spectrum and no overlapping of orders. For certain types of work this is an important advantage.

8.12. **The Rayleigh Limit of Aberration.***

In calculating the limits of resolution, no account was taken of imperfections of image due to imperfect *definition* in the instrument. It was assumed that all rays from the object points passed through the corresponding image points and hence that all optical paths from an object point to a corresponding image point were equal. In practice, this condition is not fulfilled exactly. It is therefore desirable to estimate how great the departures from this condition may become before they begin to reduce the sharpness of the image by a detectable amount. Rayleigh suggested that defects of image caused by inequalities of path up to a quarter of a wavelength should be small in comparison with the inescapable unsharpness due to diffraction, and that if the phase of the wavelets arriving from any part of the wavefront differed from that of the resultant by more than a quarter-period, then the image would be noticeably improved by correcting the corresponding part of the optical system. This estimate of the permissible tolerance may be applied to calculate limits for spherical and for chromatic aberration. It may also be used to estimate tolerances in the working of optical surfaces and in the homogeneity of optical materials (see §§ 9.11 and 9.12). It is now known that the permissible tolerance is not the same for all parts of a lens, and may be greater or less than the Rayleigh value. This value still forms a good initial working estimate.

8.13. **Accuracy of Measurements with Mirror and Scale.**

The limiting accuracy of reading with a mirror-and-scale arrangement (see fig. 8.4) may be calculated in a similar way. Suppose that a point Q on the scale

* Reference 8.3.

is at the centre of the image of a slit when the mirror is in a certain position. Then if the mirror rotates so that one edge of it advances through $\frac{1}{4}\lambda$ and the other retreats by $\frac{1}{4}\lambda$, the phases of wavelets arriving at Q from different parts of the mirror will cover a range of one wavelength (because the movement of the edge of

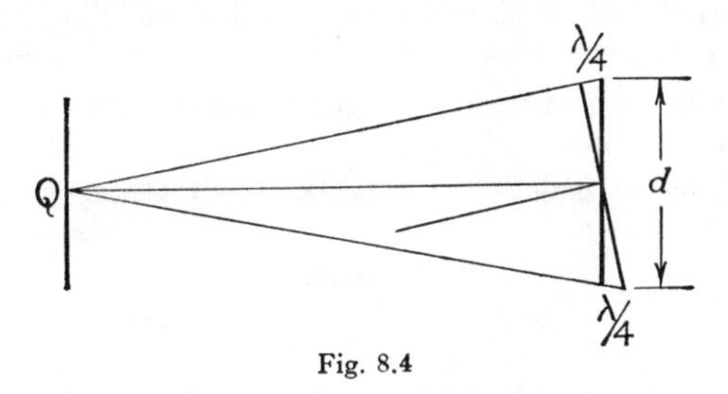

Fig. 8.4

the mirror through $\frac{1}{4}\lambda$ introduces a path difference of $\frac{1}{2}\lambda$). After the rotation Q will be at the first minimum in the diffraction pattern. The minimum angular rotation which can be detected with a mirror of width d is therefore $\lambda/2d$. It is not generally realized that this limit is occasionally approached in practice [see Example 8(x)].

EXAMPLES [8(iv)–8(xi)]

8(iv). Calculate the resolving power of a prism of rocksalt of 4 cm. base at 4000 Å., 5000 Å., and 6000 Å., given the following data:

λ (Å.)	μ
6708	1·5400
6438	1·5412
5461	1·5477
4861	1·5537
4047	1·5665
3034	1·5988
2144	1·6737

[8600; 4400; 2500].

8(v). Derive an expression for the resolving power of a material which obeys Cauchy's law of dispersion (see § 3.18). When $A = 27 \times 10^{-5}$ and $B = 5{\cdot}0 \times 10^{-11}$, plot a curve for the resolving power against wavelength of a prism whose base thickness is 6 cm.

$$\left[R_p = \frac{2t_p AB}{\lambda^3}.\right]$$

8(vi). Using the data given in Example 8(iv), derive values of A and B applicable to the region 4000 Å. to 6000 Å. Hence, using the formula derived in Example 8(v), check the results calculated in Example 8(iv).

[$A = 0{\cdot}525$; $B = 1{\cdot}30 \times 10^{-10}$.]

8(vii). The ruled space on a grating is as wide as the base of a 60° rocksalt prism. It has the same resolving power in the first order as the prism has at 5000 Å. Calculate the size of the grating interval. [$9{\cdot}1 \times 10^{-4}$ cm.]

8(viii). Show that if the diameter of the objective of a telescope is equal to that of the central half-period zone, little advantage is gained by using a lens because the telescope would function like a pinhole camera. Show that the useful length of the " pinhole telescope " tube increases in proportion to the square of the diameter of the " objective ". Show that for a hole of 0·1 in. diameter the useful length is about 10 ft. and for one of 4 in. diameter the appropriate focal length is 3 miles!

[The useful length is that which gives the maximum useful magnification (see § 8.8)].

8(ix). Applying the limit stated in § 8.12, show that the depth of focus of a lens of diameter d (when the object is very far from the lens) is

$$\delta f = \frac{4f^2\lambda}{d^2}. \qquad 8(9)$$

[Assume that all the paths to the true focus are equal. Calculate the difference of path between a ray passing through the centre and one coming from the edge of the lens to a point distant δf from the true focus and situated on the axis of the lens. The calculation is given by Rayleigh.*]

8(x). A short-period galvanometer has a mirror 2 mm. wide. So long as the edge of the image is sharp, it is possible to read a movement of 0·1 mm. on the scale. What is the maximum scale distance which may usefully be employed when light of mean wavelength 5600 Å. is used? [70 cm.]

8(xi). Show that the result given in § 8.13 may also be derived by regarding the mirror as a diffraction aperture restricting a telescope which observes two distant objects whose angular separation is equal to twice the angular movement of the mirror.

8.14. Development of the Theory of Resolving Power.†

Some modern instruments, such as the Fabry and Pérot interferometer, give interference fringes in which the light distribution is very different from the distribution in the diffraction patterns considered by Rayleigh. It is not surprising that the Rayleigh criterion does not apply to these instruments. In order to give a logical definition of resolving power that may apply to them, it is desirable to consider in more detail what is meant by the resolution of two objects. This investigation is also of practical interest in other connections. In the following discussion we shall consider the resolution of two spectrum lines, although the basic ideas are more generally applicable. We shall start with the discussion of an ordinary prism or grating spectroscope, and the difference of wavelength between two lines which are just resolved (according to the Rayleigh criterion) will be called $\Delta\lambda_R$.

* Reference 8.4. † See References 8.1 and 8.5.

8.15.—The diffraction pattern due to the combined effect of two lines of equal intensity whose difference of wavelength is $\Delta\lambda_R$ is shown in fig. 8.5 C. The pattern due to a single line is shown on a larger scale in fig. 6.9*a*, and is indicated by the dotted lines in fig. 8.5. The variation in the combined pattern when the wavelength separation varies

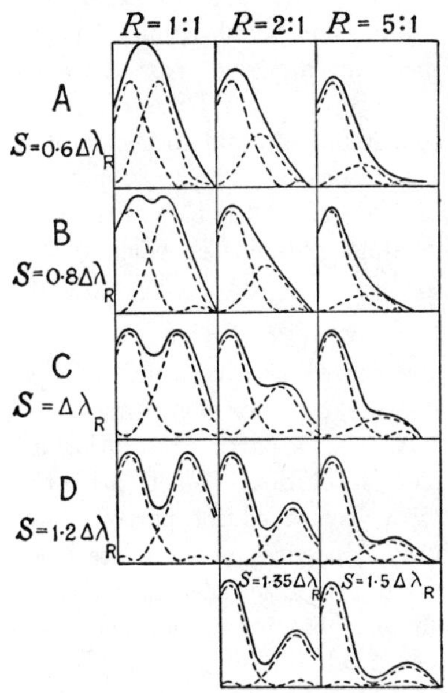

Fig. 8.5.—The dependence of resolution on the relative intensities of two spectrum lines

from 0·6 to $1{\cdot}25\Delta\lambda_R$ is shown in figs. 8.5 A, B, C, D. The remainder of fig. 8.5 shows the combined diffraction patterns for different separations and for lines whose intensity ratios are 2 : 1 and 5 : 1. As the separation increases we may distinguish the following stages:

(A) The lines are so close together that the combined diffraction pattern appears to be the same as that due to a single line of normal width: the lines are then " completely unresolved ".

(B) The lines are so close that only one image can be distinguished, but it is blurred so that it indicates that the radiation is not homogeneous while giving no indication of the number or the separations of the components.

(C) The lines are separated sufficiently to be seen as a double line but not sufficiently to enable the separation and the relative intensity to be measured: we call this stage " partial resolution ".

(D) The lines are completely separated. Each can be measured as though the other were not present. The lines are then " completely resolved ".

8.16.—When two lines are of equal intensity and their separation is $\Delta\lambda_R$, the intensity at a point half-way between the two maxima is 80 per cent of that at the maxima. This condition is often described as a " 20 per cent dip ", and we shall use this phrase with the understanding that when the two lines are of unequal intensity, the minimum is to be 80 per cent of the *smaller* maximum. From the diagrams (and from more detailed discussion which cannot be included here) it may be seen that a 20 per cent dip gives partial resolution and a 60 per cent dip gives complete resolution. These criteria are generally applicable to all types of instruments and to the resolution of components of equal or of unequal intensity. The most important stage of resolution is the one which we have called partial resolution, and it must be understood that this stage is indicated when two lines or objects are said to be " resolved " or " just resolved " without any further qualification.

8.17.—For instruments which give diffraction patterns of the type shown in fig. 8.5, we see that for components of equal intensity a separation equal to $\Delta\lambda_R$ gives partial resolution and a separation of $1{\cdot}25\Delta\lambda_R$ gives complete resolution. When the intensity ratio is 5 : 1, a separation of $1{\cdot}25\Delta\lambda_R$ is needed for partial resolution and a separation of about $1{\cdot}5\Delta\lambda_R$ is needed for complete resolution. In certain types of work (e.g. studies of the Zeeman effect) it is necessary to look for weak satellites whose intensities may be less than 0·001 of that of the main line. Such satellites cannot be detected unless their wavelength separation is several times greater than that required for the resolution of lines of equal intensity.

8.18. **Resolving Power of the Fabry-Pérot Etalon.**

The energy distribution in the pattern produced by this instrument is given in equation 5(14) and is shown graphically in fig. 8.6.

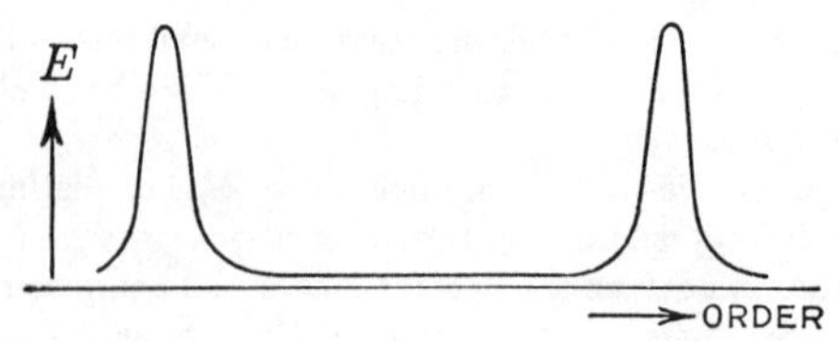

Fig. 8.6.—Energy distribution in Fabry-Pérot pattern

The distribution differs from that given by a diffraction grating in that the subsidiary maxima of the diffraction pattern have been smoothed out. Reference to the theory of the etalon (§ 5.26) shows that this occurs because the interfering beams from the etalon are of gradually decreasing amplitude, whereas those from the grating are of equal amplitude. From equation 5(14) it is possible to calculate the difference in wavelength between two lines when there is a "20 per cent dip" in the pattern which they jointly produce. Reference to equation 5(12) shows that the whole scale of the pattern (including this critical distance) is inversely proportional to $2e/\lambda$ (where e is

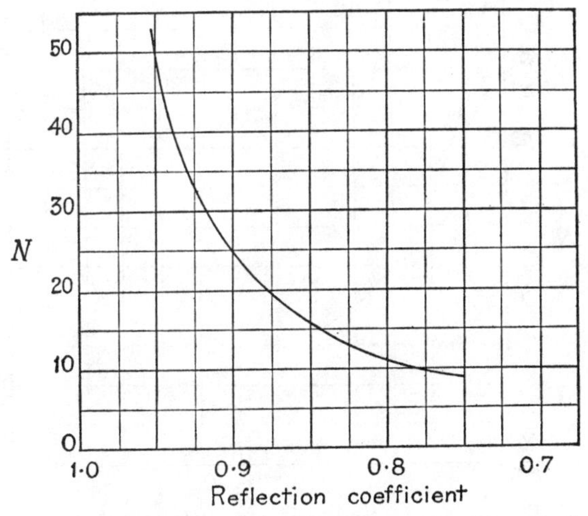

Fig. 8.7*a*.—Resolving power of Fabry-Pérot etalon (theoretical)

the separation of the plates). The resolving power $(\lambda/\Delta\lambda)$ is thus proportional to the separation, i.e. to the order of interference. The resolving power also depends, in a rather complicated way, on the reflection coefficient of the plates. It is possible to regard the series of beams produced by the etalon as equivalent to a certain number of beams of *equal* amplitude. The equivalent number N is defined by the relation $N(2e/\lambda) = R$, where R is the resolving power of the etalon. N is equal to the number of steps in a reflecting echelon which would give the same resolving power and the same order of interference. The equivalent number N is a function of the reflection coefficient. It can be calculated from 5(12), but the calculation is not very simple and graphical methods of computation are useful. Some results calculated by Hansen * are shown in fig. 8.7*a*. The theoretical resolving power

* Reference 8.7.

will be obtained in practice only if the plates are plane to within a deviation of order λ/N, i.e. for high-reflection films $\lambda/50$.

Fig. 8.7*b* shows the results of some measurements for films of different transmissions.* These results cannot be directly compared with the calculated values because the absorption is not known exactly. They do, however, show that very sharp fringes, i.e. very high effective reflection coefficients, can be obtained and they agree with

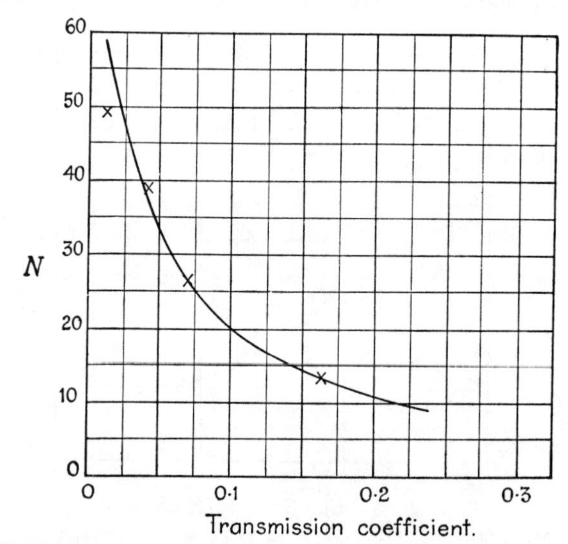

Fig. 8.7*b*.—Resolving power of Fabry-Pérot etalon (crosses represent experimental results; the line represents theoretical values, calculated for an absorption of 4 per cent).

the calculated values of N if we assume an absorption of about 4 per cent. It may be seen that the resolving power increases rapidly as the reflection coefficient increases. If, however, the thickness of the film is increased beyond a certain point in order to increase the reflection coefficient, the amount of light transmitted becomes inconveniently small. There is thus an optimum thickness of film for practical purposes. In practice films of reflection coefficient about 0·85 are used for measurement in the visible region of the spectrum.

* Reference 8.8.

8.19. **Resolving Power of a Microscope.**

In the preceding discussion of resolving power it has been assumed that the light from the two objects is non-coherent, so that the illumination at any point due to the two sources acting jointly is the sum of the illuminations due to the separate sources (see § 5.2). This assumption is clearly justified when the two objects are stars, and when they are images of a slit formed by light of slightly different wavelength. When a microscope is focused on a self-luminous object (such as the incandescent filament of an electric light), the radiation from different parts of the object is non-coherent. An object which is not self-luminous must be illuminated by a source of light and no real source is confined

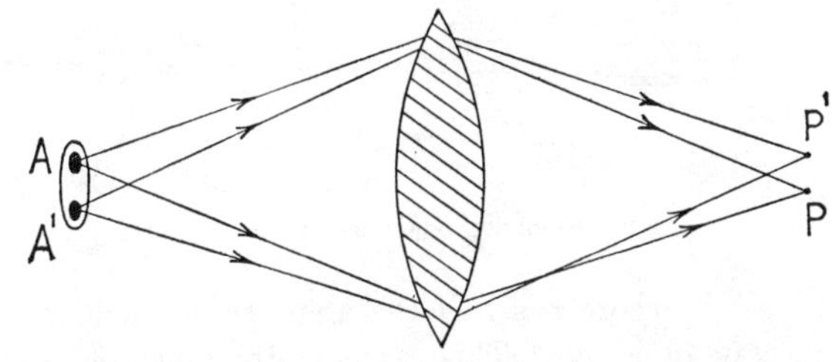

Fig. 8.8.—Illumination of an object by a condenser

to a mathematical point. A condenser is used to form a more or less sharply focused image of the source in the object plane. Owing to diffraction, a point P of the object receives light from a finite area (A) of the source (see fig. 8.8). The light vector at P is the resultant of wavelets from all parts of A. Similarly, the light vector at P′ is the resultant of wavelets from an area A′. If the source is sharply focused, and if P and P′ are well separated, the areas A and A′ do not overlap, and the light at P′ is not coherent with that at P. In considering the performance of microscopes, we are usually concerned with points which are close together, and the areas A and A′ then overlap to a considerable extent so that the light is at least partially coherent. When P and P′ are so close that they are just resolved, the light is effectively coherent. We shall consider the resolution of two objects (*a*) when the light is completely non-coherent, and (*b*) when it is completely coherent. Condition (*a*) applies to self-luminous objects. The illumination of ordinary microscopic objects is not completely coherent, but under practical conditions it may be assumed that the illumination at two points which are just resolved is coherent.

8.20. **Resolution with Non-coherent Illumination.**

When the illumination is non-coherent, the calculation of the minimum distance between two points whose images are just resolved differs very little from the corresponding calculation for the telescope. For a rectangular aperture the result may be obtained in the following way. Let A and B (fig. 8.9) be two points in the field of a microscope which is represented in the diagram by the lens CD. Let A′ and B′ be the image points corresponding to A and B which are at equal distances from the axis of the lens CD. Then all paths from A to A′ are equal and the extreme difference of paths from A to B′ is equal to 2(AC–AD). From the geometry of the figure, this difference is

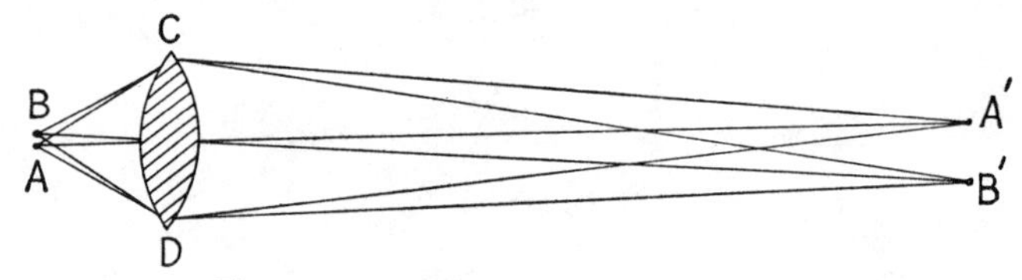

Fig. 8.9.—Resolving power of the microscope

equal to 2AB sin α, where α is half the angle subtended at the object by the microscope objective. This extreme difference of path is equal to the wavelength (λ') when the distance between the objects is equal to $\lambda'/(2 \sin \alpha)$. The entire difference of path is situated in the medium between the objective and the object. Let μ be the refractive index of this medium (with respect to air) and λ the wavelength of the light (in air). Then the distance between points which are just resolved (according to the Rayleigh criterion) is given by

$$y = \frac{\lambda'}{2 \sin \alpha} = \frac{\lambda}{2\mu \sin \alpha}. \quad . \quad . \quad . \quad . \quad 8(10)$$

When the aperture is circular, the size of the Airy disc has to be calculated, allowing for the fact that the object is near to the lens. When this is done,* the distance between two points which are just resolved is found to be about 20 per cent greater than that given by 8(10).

8.21. **Abbe Theory of Resolution with Coherent Illumination.**

The theory of resolution with coherent illumination was developed by Ernst Abbe (1840–1905). His original statements of the theory

* See Rayleigh: *Journal of Royal Microscopical Society*, 1903, Vol. 23, p. 460. Rayleigh does not quite complete the computation, but the result follows from his equations 17–21.

were not very clear and there was a certain amount of controversy chiefly due to misunderstanding of his ideas. This was particularly

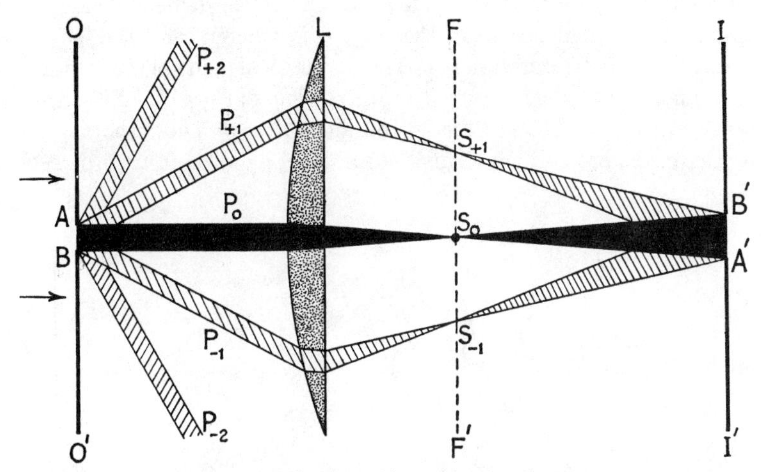

Fig. 8.10*a*.—Formation of images by the microscope

unfortunate because through this theory, and in other ways, Abbe contributed more than any other worker to the development of the modern high-power microscope.

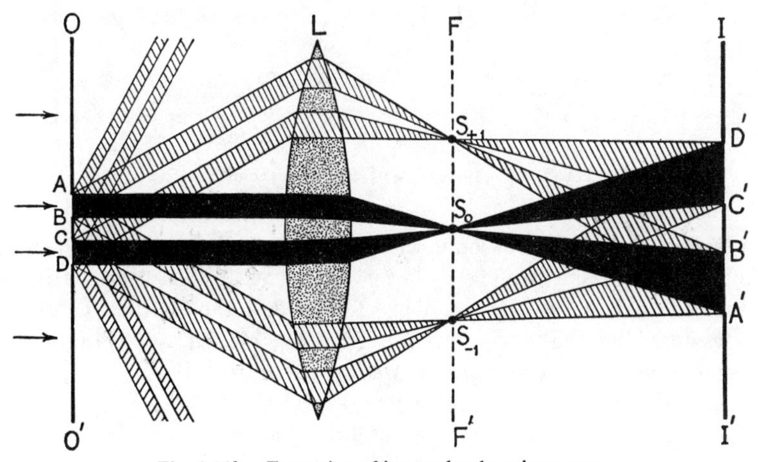

Fig. 8.10*b*.—Formation of images by the microscope

As a preliminary to the discussion of resolving power, let us consider the formation of an image by a lens. Let the object be a plane grating, consisting of alternate transparent and opaque spaces. In fig. 8.10*a* the grating is in the plane OO′ and the image in the plane II′. A parallel beam of light is incident upon the grating from the left-hand side. The light from different spaces of the grating is coherent and the lens produces a Fraunhofer diffraction pattern in the focal plane of the lens (FF′). The same light which forms the diffraction pattern on FF′ gives a focused image on II′. In figs. 8.10*a* and 8.10*b* we show the rays corresponding to light which forms the principal maxima in the diffraction

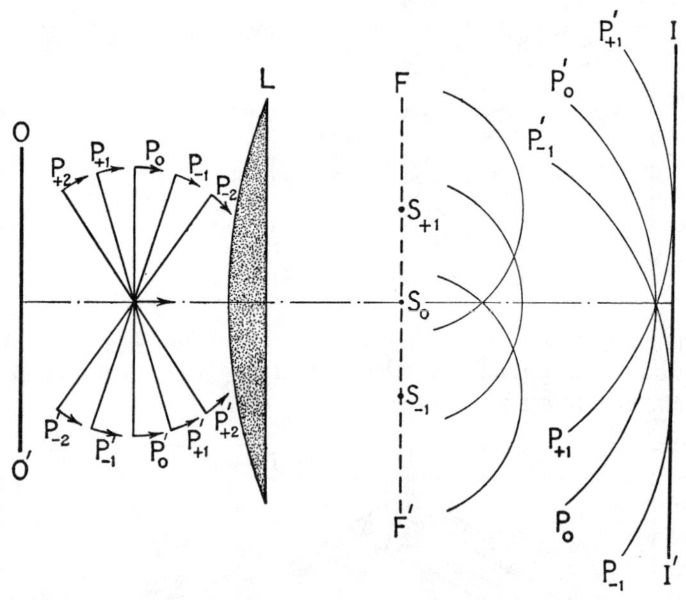

Fig. 8.10*c*.—Formation of images by the microscope

pattern. For simplicity only the rays from one space are shown in fig. 8.10*a* and from two spaces in fig. 8.10*b*. In fig. 8.10*c* the grating is assumed to have a large number of lines and the diffracted light is represented by the wavefronts P_0, P_{-1}, P_{+1}, etc. These plane waves become approximately spherical waves, centred on S_0, S_{-1}, S_{+1}, after passing the lens L. They interfere to form the focused image on the plane II′. These diagrams are useful in helping to form a mental picture of the way in which the lens forms both a Fraunhofer diffraction pattern in the plane FF′ and an image in the plane II′, but it must be understood that they show only the most important rays and wavefronts. The diffracted light is not confined to the rays shown and does not all pass through the points S_0, S_{-1}, S_{+1}. Since, however, a large part of the energy does pass near to these points, we may regard the light as divided into a set of spectra represented by the plane wavefronts, provided that we understand that the spectrum of each order includes all the energy passing near to S_0, S_{-1}, S_{+1}, etc.

8.22.—Abbe's theory is based on the fact that an observer viewing the plane II' from the right can obtain information about the object on the plane OO' only through light which has passed the plane FF', i.e. through the light included in the spectra. He says that, if some of the spectra corresponding to waves P_0, P_{-1}, P_{+1}, etc., are not transmitted by the instrument, then the image will correspond to a grating for which the absent spectra have zero energy. To take an extreme case, suppose that only the spectrum of zero order is transmitted. This maximum would be given by a wide aperture, uniformly illuminated, in the plane OO'. Accordingly the observer sees uniform illumination on the image plane. The central maximum alone gives no resolution of detail.

Pursuing this idea Abbe suggested that, if the aperture of a microscope is not sufficiently wide to include all the diffraction spectra from an object, details in the object may be absent and, under certain conditions, false detail may appear. This last conclusion was strongly resisted by microscopists. Abbe proved his case by a long series of experiments in which he inserted stops into the microscope so as to exclude certain spectra, and showed that false detail appeared. For example, suppose that the object is a grating with a spacing of a few wavelengths and that the clear spaces are fairly narrow. Several spectra contain appreciable energy and stops may be inserted to cut out the spectra of odd orders. The remaining spectra correspond to an object of half the spacing and the image plane contains an image in which each true line appears to be divided into two. When a microscope is working at a high numerical aperture, very small errors in focus will cause some of the spectra to fall on to stops. This may not only lead to loss of detail but may also cause the appearance of false detail. This point is of some practical importance because with certain types of object it is not possible to know when the object is truly in focus. The assumption that the best focus is the one which shows the most detail is not always justified.

8.23.—Abbe's experiments were criticized on the ground that he used stops of special shape and position, but A. B. Porter showed that similar effects could be produced with circular stops such as are commonly used by microscopists. He also gave a striking practical demonstration using a piece of wire gauze of about 0·3 millimetre spacing. The gauze G (fig. 8.11) forms a series of spectra in a rectangular pattern. These spectra are formed on a screen S_1 very close together, and by cutting small holes in the screen different sets of spectra may be allowed to pass. An image is formed by the lens L_2 on the screen S_2. If a slit is used to exclude all the spectra except those on a horizontal line through the central image, then only vertical

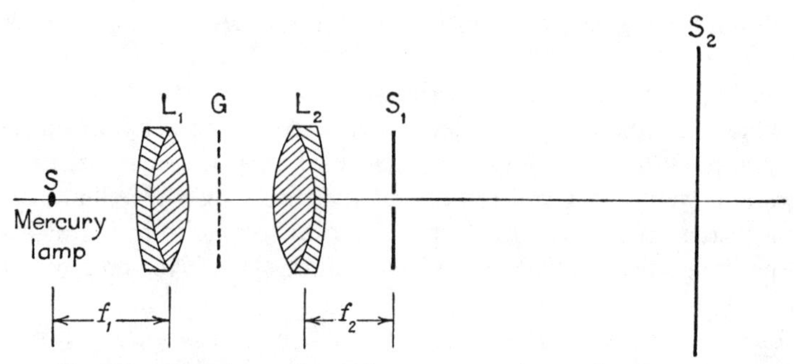

Fig. 8.11.—Porter's experiment (L_1 acts as a condenser. L_2 gives spectra on S_1 and image on S_2)

wires are seen. If the slit is turned through a right angle, then only the horizontal wires are seen. Perhaps the most striking effect is produced by using two slits in the form of a cross. If this is set with its lines at 45° to the lines of the gauze, it allows the diagonal spectra to pass. The eye then " sees " the gauze but with its lines turned through 45°.

8.24.—We have considered Abbe's theory in its application to objects which have a periodic structure. We obtained a limiting distance of resolution by considering the angular separation of the " spectra " produced by a grating of alternate transparent and opaque spaces. The principal maxima occur in directions given by $\sin\theta = n\lambda'/y$ when the source of illumination is a parallel beam of light travelling in the direction of the axis of the microscope, where y is the distance between successive lines of the grating, λ' is the wavelength between the grating and objective, and λ the corresponding wavelength in vacuum. We assume that the points are not resolved unless at least two spectra enter the microscope. This will occur only if

$$y > \frac{\lambda'}{\sin\alpha}, \quad \text{i.e.} \quad y > \frac{\lambda}{\mu\sin\alpha}. \qquad . \quad . \quad . \quad 8(11)$$

When the illumination is at an angle θ', the spectra occur in directions given by

$$\sin\theta - \sin\theta' = \frac{n\lambda'}{y} \quad \text{(see §§ 6.27 and 6.31).}$$

If the illumination is such that the direct light (spectrum of zero order) and the first-order spectrum on one side enter the microscope, then $\alpha = \theta = -\theta'$ and

$$y > \frac{\lambda}{2\mu \sin\alpha}. \qquad \qquad 8(12)$$

Thus, according to the Abbe theory in its simplest form, the minimum distance between resolved points is reduced by a factor of two when oblique illumination is substituted for axial illumination.* This conclusion is verified approximately by experiment. Following Abbe, the number $\mu \sin\alpha$ is called the *numerical aperture* (N.A.).

8.25.—The following account of the delineation of detail in an image formed by a microscope was given by Johnstone Stoney † in the course of an exposition of the Abbe theory. We may imagine the object plane to be crossed by a series of fine gratings set at various angles to one another. The amount of light in the different gratings varies in such a way that together they make up an approximation to the pattern of light in the object plane. Each grating is approximately reproduced (on an enlarged scale) in the image plane if the corresponding spectra are collected. The finest grating for which this can occur is one whose spacing is given by 8(11) or 8(12), according to the type of illumination. Accordingly the image includes only such details as can be formed by the superposition of gratings of this (or greater) spacing. Obviously, the order of magnitude of the detail which can be seen clearly is given by 8(11) or 8(12). If the object is of low contrast, resolution will be more difficult than if it is "black on white". It will be understood that there is no limit to the smallness of an object which can be seen when sufficient light is available. Objects whose sizes are much smaller than the limit of resolution can be seen by the light which they scatter, but they are seen as discs of light. No detail is observed, and in so far as one of these discs has a defined boundary its limits are defined by 8(11) or 8(12) and not by the size of the object.

* For a detailed investigation of the effect on the resolving power of the direction of illumination, see Reference 8.15.

† Reference 8.6.

8.26. **Representation of Detail in an Object seen through a Microscope.**

We shall now consider the action of a microscope in a more general way. Light which is incident upon a material object may be (*a*) absorbed, (*b*) reflected, (*c*) scattered, and (*d*) transmitted with a difference of phase. When we see a non-self-luminous object either with the eye or by means of a physical instrument, our immediate observation refers to the alteration which the object produces in a beam of light which is incident upon it. We infer the properties of the object from these alterations. The most general type of object has extension in three dimensions and all its optical properties vary, in an arbitrary way, from point to point. The general problem of microscopy is to produce an image of a very small object so as to provide a faithful copy on an enlarged scale of small-scale variations of any of the optical properties. This ideal solution is unattainable and the technique of microscopy consists in the development of instruments which give the best possible representation of certain features of practical importance. Obviously a wide range of instruments is needed to deal with different kinds of objects. The design and the study of the best ways of using these instruments is a subject in itself rather than a branch of optics. We are concerned only with certain basic ideas, and in order to discuss these we consider idealized objects which are much more simple than any real object.

In § 8.20 we discussed the resolution of two self-luminous points, and in §§ 8.21–8.24 we have been considering a certain type of grating. We now proceed to a more general discussion. Suppose that a plane wave of unit amplitude is incident upon a plane OO′ (fig. 8.10) and that a microscope is focused on this object plane. The phase and amplitude of the light which has passed through the object (or has been reflected by the object) vary from point to point in the plane OO′. We may represent this light by a disturbance $\xi_0(x, y)$ where x and y are two co-ordinates in the plane. For the present let us consider ξ_0 as a function of y only. Then, according to the discussion of § 6.11, the angular distribution of this light is given by

$$a(\kappa_y) = \frac{1}{2\pi}\int_{-\infty}^{\infty} \xi_0(y) \exp{(i\kappa_y y)}\, dy, \quad . \quad . \quad . \quad 8(13)$$

where

$$\kappa_y = \kappa \sin\theta = \frac{2\pi}{\lambda}\sin\theta. \quad . \quad . \quad . \quad . \quad . \quad 8(14)$$

Now each *point* on the plane FF′ corresponds to a certain *direction* of propagation, i.e. to a certain value of κ_y so that the distribution of light on this plane constitutes a Fourier analysis of the function $\xi_0(y)$. Each *point* on the plane II′ corresponds to a certain *point* on the object, i.e. to a certain value of y. If the lens is free from aberrations, it brings all the wavelets which it receives from a certain point O to a point I without introducing phase differences. If the magnification is $-M$, let us introduce a co-ordinate Y which is equal to $1/M$ times the measured co-ordinate of a point on the image. Then the image may be represented by a Fourier synthesis:

$$\xi_0'(Y) = \int_{-\kappa_1}^{\kappa_1} a(\kappa_Y) \exp(-i\kappa_Y Y) \, d\kappa_Y \quad . \quad . \quad . \quad 8(15)$$

[see equation 6(6)].

The synthesis does not exactly reproduce the object because we must insert limits $\pm\kappa_1$ in 8(15) for the range of κ_Y which corresponds to the numerical aperture of the microscope. Note that κ_Y cannot exceed $2\pi/\lambda'$ with any geometrical arrangement, so that a " perfect " reconstruction is not obtained even if $\alpha = \frac{1}{2}\pi$. Any detail of the object which produces a sharp variation of amplitude or phase in a distance $\frac{1}{2}\lambda'$ will be represented in 8(13) by that part of the integral for which κ_Y is numerically greater than $2\pi/\lambda'$. This is just the part of the integral which is not included in 8(15) and therefore this type of detail is not represented in the image.

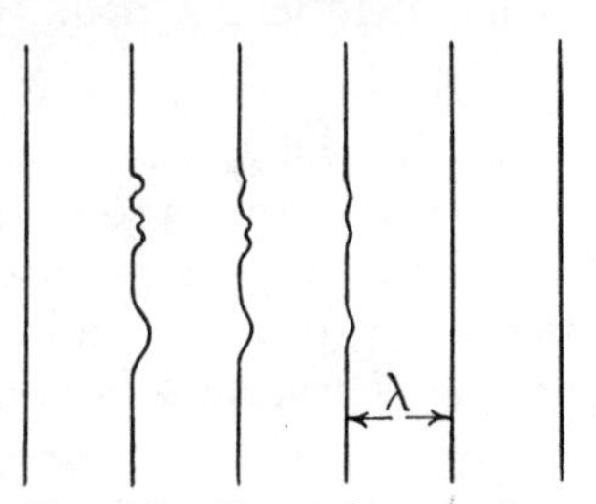

Fig. 8.12.—Decay of corrugations in a plane wave

This result has been derived by Rayleigh in a rather different way. He discusses the propagation of a plane wave with small local variations of amplitude or phase, and he applies the term " corrugation " to a variation which extends over much less than a wavelength. He shows that as a corrugated plane wave advances, the corrugations die out exponentially. After the wave has advanced through a distance of a few wavelengths, the corrugations have effectively been smoothed out and the wave is plane again (fig. 8.12). Thus a plane wave which has passed through an object with spacing much less than λ cannot retain and transmit any record of the structure.*

8.27. **Use of the Fourier Series.**

In the preceding paragraph we have considered the analysis of the light from an object in terms of a Fourier integral. When we are concerned with a variable which extends over a finite range, it is often

* Reference 8.9.

convenient to use a Fourier series rather than a Fourier integral (see §§ 4.17 and 4.19). Consider a region in the field of view of the microscope extending from $y = -d$ to $y = +d$, and replace y by $y' = \pi y/d$. Then the integral of 8(15) may be replaced by the series

$$\xi_0(y') = \sum_{n=-\infty}^{\infty} c_n \exp(iny'), \qquad \qquad 8(16)$$

where

$$c_n = \frac{1}{2\pi}\int_{-\pi}^{\pi} \xi_0(y') \exp(-iny')\, dy', \qquad 8(17)$$

and n may be any integer—positive, negative or zero [see equation 4(51)]. This analysis replaces the effects of the real object by a series of periodic effects each corresponding to a pair of terms (c_n and c_{-n}) in 8(16). We call objects producing effects of this special type " sinusoidal gratings ". It would, in practice, be rather difficult to construct a perfect sinusoidal grating, but that is no reason why we should not, for mathematical convenience, represent the effect of any real object as a superposition of the effects of a number of sinusoidal gratings. The term c_0 in 8(16) corresponds to a uniform distribution of light on the plane OO′ and may be regarded as belonging to a sinusoidal grating of infinitely long period.

8.28.—As before, we expect that the distribution of light on the plane FF′ will represent an analysis of the light from the object. We may see that this is so by considering the nth sinusoidal grating. The space width for this grating is $2d/n$ in y, which corresponds to a variation of $2\pi/n$ in y'. We know that the first principal maximum for a grating of spacing $2d/n$ lies in a direction corresponding to

$$2d \sin\theta = n\lambda,$$

or

$$\kappa_y = \pi n/d. \qquad \qquad 8(18)$$

Substituting the value of κ_y given by 8(18) in 8(13), and making the necessary change from the variable y to the variable y', we can see that c_n represents the light diffracted in the direction corresponding to one first-order principal maximum for a grating of spacing $2d/n$. c_{-n} represents the light in the other first-order principal maximum. Thus each of the sinusoidal gratings into which we have analysed the object is represented on FF′ by the two points corresponding to its first-order principal maxima. A perfect lens will bring together at an image point all the light which it receives from a given object point without change of amplitude or phase. From 8(16) we see that the light received at some image point Y = A (corresponding to $y' = a$) is represented by

$$\xi_0(Y) = \sum_{n=-n_1}^{n_1} c_n \exp(ina). \quad . \quad . \quad . \quad . \quad 8(19)$$

The limits n_1 are determined, as before, by the angle which the lens can accept. The light from the object is represented by a complete Fourier series; that in the image corresponds only to the superposition of gratings whose first-order principal maxima are transmitted through the instrument, so that the representation is imperfect.

8.29. **Phase-contrast Microscope.**

The eye, and all physical receptors, cannot detect differences of phase but only differences of energy. In order to see a transparent object, it is not sufficient to magnify it. We must produce an image in which differences of phase in the object have been converted into differences of energy. For ease of interpretation it is desirable that the differences of energy in the image shall bear a simple relation to the differences of phase which they represent. We shall now describe a device, due to Zernike, which enables *small* differences of phase in an object to be represented by *proportionate* differences of real amplitude in the image plane. Consider two objects:

(i) An "*amplitude* object" for which $\xi_0(y)$ is real for all values of y.

(ii) A *phase* object for which $\xi_0(y)$ is complex but the modulus of $\xi_0(y)$ is unity for all values of y.

For a phase object we may put

$$\xi_0(y) = e^{i\delta}, \quad . \quad . \quad . \quad . \quad . \quad . \quad . \quad 8(20)$$

where δ is some function of y. *When δ is small*, this may be written

$$\xi_0(y) = 1 + i\delta. \quad . \quad . \quad . \quad . \quad . \quad . \quad . \quad 8(21)$$

Now consider three objects for which the *transmitted* light is represented by:

$$(a) \quad \xi_p = 1 + if(y), \quad . \quad . \quad . \quad . \quad . \quad . \quad 8(22a)$$

$$(b) \quad \xi_+ = 1 + f(y), \quad . \quad . \quad . \quad . \quad . \quad . \quad 8(22b)$$

$$(c) \quad \xi_- = 1 - f(y), \quad . \quad . \quad . \quad . \quad . \quad . \quad 8(22c)$$

where $f(y)$ is real for all values of y and is $\ll 1$. Of these ξ_p represents a transparent object (phase object), ξ_+ and ξ_- represent objects which absorb the light in some areas more than in others but which introduce no phase differences (amplitude objects). (*b*) and (*c*) are complementary in that the lighter parts of (*b*) correspond with the darker parts of (*c*) and vice versa. According to 8(22) the total amount of energy transmitted is approximately the same for each object. This is possible if we assume that the energy incident upon the absorbing objects is greater than that incident upon the transparent object. From 8(17) we see that, when $f(y)$ is small compared with π (for all values of y), c_0 is nearly unity for all three objects, the imaginary component being negligibly small. The values of c_n for the three objects satisfy the relation

$$i(c_n)_p = (c_n)_+ = -(c_n)_-. \quad . \quad . \quad . \quad . \quad 8(23)$$

The energy at a typical point on FF′ is given by $c_n c_n^*$ and is the same for all three gratings. The relation of phases between the central light and the non-central light is different for the three gratings. Owing to the different relation of *phases* on FF′, the distribution of *energy* on II′ is different for the three objects, and for (*a*) we obtain uniform illumination on II′. Now suppose that a thin plate is introduced into the plane FF′ so as to retard the phase in the centre by $\frac{1}{2}\pi$. Then the distribution of phases on FF′ given by (*a*) with the plate is the same as that given by (*b*) without the plate. Thus the object (*a*) is visible when the plate is used (*positive phase-contrast*). If the phase in the centre is advanced by $\frac{1}{2}\pi$ the object (*a*) is again visible, but the regions which were dark before are now light (*negative phase-contrast*). Photographs showing positive and negative phase-contrast with transparent objects are given in Reference 8.10. In practice, to obtain good resolution, the object is illuminated with a hollow cone of light. The central image is then replaced by an annular ring on the plane FF′. It is usual to insert in this place a plate which has been coated with an annular ring of transparent material of optical thickness slightly less than $\frac{1}{4}\lambda$. The ring is applied by vacuum evaporation. It is thus necessary for the condenser and the phase plate to match one another.

8.30.—The phase-contrast method is not the only way of making transparent objects visible. If the central image is excluded by a stop (dark-ground illumination) or if all the light on one side of the centre is excluded (Schlieren method), then transparent objects become visible. Several devices of this type were known before the advent of phase-contrast methods. Zernike gave the first really satisfactory theory of these methods, and his own phase-contrast device is the only one that makes contrasts in the image proportional to phase differences in the object plane. Zernike's method does this only when the phase differences are small. We can see that it will not work so well when phase differences are large by considering an object of many strips of equal area with a phase difference π between alternate strips. The energy in the central image is then zero and insertion of a central phase-plate is useless. When phase differences are fairly large (but not π), it is an advantage to be able to alter the energy of the central image as well as its phase. Devices which enable this to be done have recently been developed.*

8.31. Optimum Magnification.

With a dry lens, the maximum value of the numerical aperture is 1·0, and values up to 0·95 have been attained. Oil-immersion objectives with numerical apertures up to 1·65 have been constructed. The minimum distance between points which are just resolved is thus $2{\cdot}7 \times 10^{-5}$ centimetre for a dry lens and $1{\cdot}6 \times 10^{-5}$ centimetre for an oil-immersion lens using oblique illumination and a wavelength of 5500 Å. The *maximum useful magnification* (see § 8.8) is thus about 800 for a dry lens, and 1200 for an oil-immersion lens.

It is possible to reduce the limit of microscopic resolution by photographing the object with ultra-violet radiation of wavelength about 2200 Å. The limit of resolution is then about 1000 Å. or 0·1 μ. The microscope and slides must be constructed of materials transparent to radiation of this wavelength and the optical system must be specially corrected for the wavelength used. The photograph may be enlarged to give an overall magnification of about 3000. Very much higher magnifications have been obtained using electron microscopes. In these instruments the effective wavelength may be of the order of 0·1 Å., but the numerical aperture of models so far available is very low, being only of the order of 0·01. The theoretical limit of resolution is thus of the order of 10 Å. The practical limit is not quite so low and the useful magnification is in the region of 100,000. There are many difficulties in connection with the use of these instruments, and many technical problems concerning the preparation of specimens and the interpretation of results remain to be solved. Further discussion of these matters is outside the scope of this book.†

8.32. Purity of a Spectrum obtained with White Light.

Suppose that a vertical slit is placed in the spectrum of a white-light source formed by means of a prism or grating spectroscope, and that the light which passes through the slit is examined in a second spectroscope of higher dispersion and resolving power. It is found that the light transmitted by the first instrument covers a certain range of wavelength. When the slit is fairly wide the range of wavelength is approximately proportional to the width of the slit. When this width is gradually decreased, the range of wavelength transmitted reaches a constant value not dependent on the width of the slit, which then affects only the total amount of light transmitted. This constant

* References 8.10 and 8.11. † Reference 8.12.

minimum range of wavelength is an inverse measure of the *purity* of the spectrum. If the instrument is free from aberrations, it is determined by the resolving power. Each wavelength in the light incident upon the first spectroscope forms a Fraunhofer diffraction pattern of the slit. The light of any one wavelength is spread over a finite region in the spectrum. Consequently, the light reaching any one point in the spectrum is not entirely of one wavelength, but includes a range of wavelengths. The range of wavelength reaching a given point (with appreciable energy) is of the order of $2\Delta\lambda_R$, where $\Delta\lambda_R$ is the minimum wavelength difference which can be resolved. The purity of the spectrum is inversely proportional to $\Delta\lambda_R$ and is proportional to the resolving power. Thus a *pure spectrum* in which every point corresponds to one and only one wavelength, is an ideal concept which cannot be realized in practice. An instrument of infinite resolving power would be required to produce such a spectrum. In any actual spectrum the light passing any given point is represented by a wave group, not by an infinitely long simple-harmonic wave train. The effective number of wavelengths in the wave train produced by putting a very narrow slit into the spectrum is approximately $\lambda/\Delta\lambda_R$, i.e. it is equal to the resolving power of the instrument (see § 4.21 and also Appendix IV B, §§ 10–14).

8.33.—In the two preceding paragraphs we have assumed that white light may be " analysed " by Fourier's methods into components of different wavelengths. As explained in § 4.33, it was at one time contended that white light " consisted " of a series of irregular pulses, and the validity of analysis was questioned. It is of historical interest to see how the formation of a spectrum with white light can be discussed without analysis. The formation of monochromatic light from an irregular pulse by a grating is explained on the assumption that each clear space of the grating transmits a single small pulse. At a given place in the spectrum these pulses arrive at regular intervals. These intervals correspond to the differences of path from the different elements to the given point, and in a first-order spectrum the path differences between paths from successive spaces are equal to the wavelength whose principal maximum is situated at the given point. Thus a disturbance of the frequency corresponding to that wavelength is produced. Also the number of waves in the wave train is equal to the number of elements in the grating, i.e. to the resolving power of the grating in the first order. The formation of an approximately simple-harmonic wave train in this way has been compared to the reflection of an irregular noise, such as a handclap, from a flight of steps. An observer situated in a suitable position hears, as an echo, not the original sound but a definite note produced by the successive arrivals of reflections from the different steps.

8.34.—In the above discussion, the periodic structure of the grating acting upon an irregular pulse gives rise to regular trains of waves of defined frequency. It might appear, at first sight, that the action of a prism, which possesses no periodic structure of its own, on an irregular pulse could not give rise to regular trains of waves. Rayleigh * has treated this problem in a very elegant way by means of an analogy with the production of a train of waves when a jet of air impinges on the surface of a liquid. We may briefly consider the application to refraction by a prism. Suppose that a pulse travels through distance t_p in a medium where the group velocity is U but waves of frequency ν travel with a higher velocity b. During the passage of the group through the dispersive medium, waves of frequency ν are continually emerging from the front of the group. The first wave of this frequency has passed right through the dispersive medium in a time t_p/b. By the time the group has passed through the medium there are

$$N = \left(\frac{t_p}{U} - \frac{t_p}{b}\right)\nu \qquad . \quad . \quad . \quad . \quad . \quad . \quad . \quad 8(24)$$

waves of frequency ν ahead of it. Inserting the values $\lambda = c/\nu$ and $\mu = c/b$, and the value of U from 4(41), we obtain

$$N = t_p \frac{d\mu}{d\lambda}. \qquad . \quad . \quad . \quad . \quad . \quad . \quad . \quad . \quad . \quad 8(25)$$

Thus a train of waves of wavelength λ containing N waves emerges from the thicker part of the prism. Shorter trains emerge from the thinner parts, but the phase relations are such that all wavelets of this wavelength combine at a certain point (P) in the focal plane of the telescope objective. At other points the trains of this wavelength interfere to destroy one another. At the point P in the spectrum, light of wavelength λ is seen, and the number of waves in the wave train is equal to the resolving power of the instrument.

8.35.—In Chapter V it was shown that in general interference fringes are obtained with white light only when the path difference is small (or when some special achromatizing arrangement is used). It was stated that, if a spectroscope is placed to receive light from a point corresponding to a high order of interference, channelled spectra may be obtained even though no interference effects can be observed with a non-selective receptor. Obviously the " bands " will be observed only if the wavelength difference between successive maxima is greater than the minimum difference resolved by the spectroscope. From equations 5(17) and 5(18) we see that the wavelength difference is λ/n, where n is the order of interference at the point from which light is taken. It follows that the bands are seen only when the resolving power is greater than the order of interference.

8.36. **Talbot's Bands.**

It was shown by Fox Talbot that when a thin plate of glass is inserted in a prism or grating spectroscope so as to cover one-half of the aperture of the lens of the telescope (see fig. 8.13), a series of bands appears in the spectrum. The bands are obtained only if the plate be inserted on the side of the lens towards the blue end of the spectrum.† Thus with a grating the bands appear in spectra on only

* Reference 8.13.

† The plate may be inserted in various positions, either before or after the prism or grating. In the case of a liquid prism it may be inserted within the prism. It may also be inserted in front of the observer's eye so as to cover half the pupil. All these positions are equivalent, for our present purpose, to the one shown in fig. 8.13.

one side of the central image. This is the same side as that on which the plate has been inserted. There is an optimum thickness of the plate which gives the clearest bands and, if the thickness exceeds twice this value, bands are not seen no matter how high the resolving power. It was originally suggested that the bands were due to a simple type of interference between light from the two halves of the aperture. This explanation is not correct because the effect of the plate is not to produce two virtual sources but to displace part of the wave surface. Thus the problem is one of diffraction. The original explanation also fails to show why the fringes are found only when the plate is on one side of the aperture, and why there is an optimum thickness.

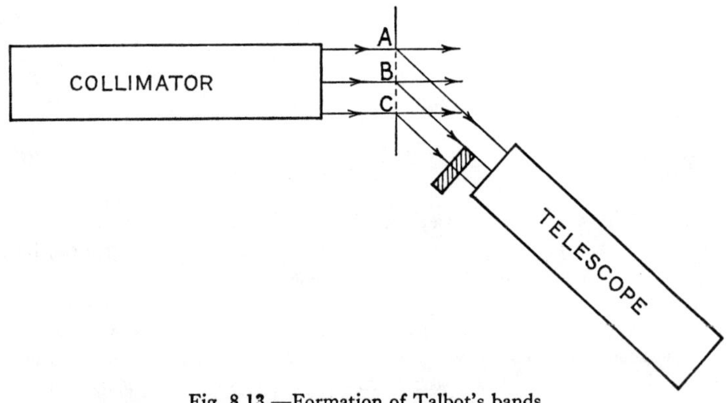

Fig. 8.13.—Formation of Talbot's bands

8.37.—A simple explanation of Talbot's bands may be based on the " irregular pulse " view of white light. In § 8.33 a grating of N lines was assumed to send a series of N wavelets to arrive successively and give rise to a vibration of wave-length λ at a certain point (P) in the spectrum. When the plate is inserted, two wave trains each $N\lambda/2$ in length will arrive at this point. One has arrived directly from the upper half of the grating (see fig. 8.13). The other has passed through the lower half of the grating and through the plate whose thickness we may put equal to t_g. Maximum interference will occur when the two wave trains completely overlap at the point P, i.e. when they start to arrive at that point simultaneously. If the plate were not present the one train would begin to arrive $N\lambda/2c$ seconds before the other. Owing to the presence of the plate it has to travel a distance t_g in a dispersive medium with group velocity U, while the other beam travels a distance t_g in air. We neglect the dispersion of air and put its refractive index equal to unity. The difference of time caused by the insertion of the plate is therefore (from 4(41))

$$\frac{t_g}{U}-\frac{t_g}{c}=\frac{t_g}{c}\left(\mu-\lambda\frac{d\mu}{d\lambda}-1\right). \qquad 8(26)$$

Thus for fringes of maximum visibility we must have

$$\tfrac{1}{2}N\lambda-t_g\left(\mu-\lambda\frac{d\mu}{d\lambda}-1\right)=0. \qquad 8(27)$$

When the above relation holds exactly the two wave trains interfere to produce a maximum at P. If the value of the left-hand side of 8(27), when λ' is substituted for λ, is $\frac{1}{2}\lambda'$, then for this wavelength the amplitude is zero. Thus the relation between the wavelengths corresponding to maxima and minima is given by

$$\tfrac{1}{2}N\lambda - t_g\left(\mu - \lambda\frac{d\mu}{d\lambda} - 1\right) = \tfrac{1}{2}(N-1)\lambda' - t_g\left\{\mu' - \lambda'\left(\frac{d\mu}{d\lambda}\right)' - 1\right\}, \qquad 8(28)$$

where μ' and $(d\mu/d\lambda)'$ are the values corresponding to λ'. Over short ranges of the spectrum we may neglect the difference between μ and μ', and we then have

$$\frac{\lambda - \lambda'}{\lambda} = \frac{2}{N} \quad . \; . \; . \; . \; . \; . \; . \; . \; . \quad 8(29)$$

for the wavelength difference between a maximum and a minimum. If the thickness of the plate has some value between zero and twice the value given by 8(27), the wave trains partly overlap and some fringes are present, though the visibility is very low when t_g is near one of the extreme values. When, however, t_g is greater than twice the value given by 8(27), the wave trains from the part AB of the grating will have passed P before that from BC has commenced to arrive. No fringes will then be produced. Also no fringes will be produced if the plate is placed so as to retard the part AB of the wave train coming from AC. This part is already retarded with respect to the part BC before the plate is inserted and further retardation obviously does not produce interference.

8.38.—If we consider the problem from the point of view of the Fourier analysis, equation 6(7) may be integrated between limits corresponding to the two halves of the grating, the appropriate retardation being introduced. The integration is carried out for one wavelength, and then the resulting expression integrated over all wavelengths. This process is rather lengthy but it does give a complete solution, including the energy distribution both for the simple case considered in § 8.37 and for cases where the plate covers more or less than half the aperture, etc. Detailed treatments have been given by Airy, Stokes and Struve.*

The following is a brief treatment of the case of a plate covering half of a rectangular aperture, from the analytic point of view. Then the path difference, from the two ends of the grating to the point P, is $N\lambda$ and the path difference between the resultant wave arriving at P from AB and that from BC is $N\lambda/2$ when the plate is absent. It is

$$\tfrac{1}{2}N\lambda - (\mu - 1)\,t_p$$

when the plate is inserted.

Maximum illumination will occur at P if the phase difference is equal to an integral multiple of 2π not only for wavelength λ but also for all other wavelengths which are sufficiently near to contribute appreciably to the illumination at P. We have seen above (§ 8.32) that this includes a small but finite range of wavelength approximately equal to twice the smallest difference of wavelength

* A summary of Airy's analysis and full references are given by Rayleigh (Reference 8.14).

which can be resolved. The phase difference for a wavelength $(\lambda + \Delta\lambda)$ arriving at the point P is

$$\frac{2\pi}{\lambda + \Delta\lambda}\{\tfrac{1}{2}N\lambda - (\mu + \Delta\mu - 1)t_g\}$$

$$= \frac{2\pi}{\lambda}\left(1 - \frac{\Delta\lambda}{\lambda}\right)\left\{\tfrac{1}{2}N\lambda - \left(\mu + \Delta\lambda\frac{d\mu}{d\lambda} - 1\right)t_g\right\}. \quad 8(30)$$

This phase difference must be the same for all wavelengths in the neighbourhood of d, i.e. for which $\Delta\lambda/\lambda \ll 1$. Neglecting the second-order term in 8(30), this condition will be met if the coefficient of $\Delta\lambda$ is zero. Thus

$$\tfrac{1}{2}N\lambda - \left(\mu - \lambda\frac{d\mu}{d\lambda} - 1\right)t_g = 0 \quad . \quad . \quad . \quad . \quad . \quad 8(31)$$

in agreement with 8(27).

8.39.—The condition for maximum sharpness of the bands is thus the same whether we adopt the " irregular pulse " or the " analytic " view of white light. In either case, the essential condition is that two wave-groups, one from each half of the grating, shall pass the point P simultaneously. This problem is in many ways similar to the problem of achromatic fringes discussed in Chapter V. Its solution shows once again that the two " theories " of white light are really only one theory. The wave-group concept clarifies both and shows their essential agreement.

EXAMPLES [8(xii)–8(xv)]

8(xii). What is the effect of using a phase-contrast plate with an object which introduces absorption but no phase differences?

> [If the variation of absorption is small, then the object becomes invisible. There are differences of phase in the plane II′ but no differences of energy. This no longer holds if there are large variations of transparency between different parts of the object.]

8(xiii). Write down equations for the passage of a wave group first through a prism (of base thickness t_p and refractive index μ_p) and then through a plate (of thickness t_G and refractive index μ_G). The plate covers half the prism. Consider in order the cases where (1) neither material is dispersive, (2) the prism material is non-dispersive and the plate is not, (3) both materials are dispersive and have different indices and dispersions. Show that the condition for maximum clearness of Talbot's bands in the third case is

$$\left(\mu_p - \lambda\frac{d\mu_p}{d\lambda} - 1\right)\tfrac{1}{2}t_p - \left(\mu_G - \lambda\frac{d\mu_G}{d\lambda} - 1\right)t_G = 0, \quad . \quad 8(32)$$

and that the results for the other two cases are the same as those obtained by putting one or both of the quantities $d\mu_G/d\lambda$ and $d\mu_p/d\lambda$ equal to zero in 8(32).

8(xiv). Baden-Powell obtained bands by inserting a thin plate of dispersive material in a liquid prism, the length of the plate being half the height of the prism. Derive an expression for the maximum clearness of the bands. Under what conditions must the plate be placed (*a*) to touch the apex of the prism and (*b*) to touch the base of the prism?

8(xv). The apparatus for the production of Talbot's bands may be considered as a transmission echelon grating (of two " steps ") and a spectroscope acting together. Show that the echelon is in the single-order position for a wavelength corresponding to a maximum. Calculate the wavelength corresponding to double-order positions (when the plate is of optimum thickness for the production of Talbot's bands). Hence construct a " theory " of the bands. How does this theory explain the fact that bands are not obtained when the plate is inserted in the wrong position?

REFERENCES

8.1. Rayleigh: *Scientific Papers*, Vol. I, p. 415.

8.2. *Ibid.*, p. 513.

8.3. *Ibid.*, p. 436.

8.4. *Ibid.*: Vol. III, p. 103.

8.5. Ditchburn: *Proc. Roy. Irish Ac.*, 1930, Vol. XXXIX, A, p. 58.

8.6. Johnstone Stoney: *Phil. Mag.*, 1896, Vol. 43, p. 332.

8.7. Williams, W. E.: *Applications of Interferometry* (Methuen).

8.8. Bright, R. J., Jackson, D. A., and Kuhn, H.: *Proc. Phys. Soc.* A. 1949, Vol. 62, p. 225.

8.9. Rayleigh: *Scientific Papers*, Vol. III, p. 117.

8.10. Taylor, E. W., *Proc. Roy. Soc.*, 1947, Vol. 190, p. 422.

8.11. Osterberg, H.: *J.O.S.A.*, 1947, Vol. 37, p. 726.

8.12. Cosslett: *The Electron Microscope* (Oxford University Press).

8.13. Rayleigh: *Scientific Papers*, Vol. V, p. 272.

8.14. *Ibid*: Vol. III, p. 123.

8.15. Hopkins, H. H., Barham, P. M.: *Proc. Phys. Soc.* B, 1950, Vol. 63, p. 737.

CHAPTER IX

Measurements with Interferometers

9.1.—Interference methods of measurement provide many examples of experiments in which an accurate result is obtained conveniently and without making excessive demands on the skill of the observer. The accurate result is made possible because the experiment is well designed and the apparatus accurately made. The initial observation is usually a measurement of a displacement in a system of interference fringes. This shows a change in the phase difference between two interfering beams, and indicates a change in optical path difference due either to a mechanical displacement or to a difference of refractive index.

9.2. Classification by Type of Interference.

Interferometers may be divided into the following types:

(*a*) Instruments in which the wavefront is divided into two or more parts. The division may be made by an opaque screen with a number of apertures (as in Young's experiment, § 5.7) or by means of prisms or mirrors as in Fresnel's experiments (§ 5.9).

(*b*) Instruments in which the light is divided by means of semi-reflecting surfaces. Part of the light is reflected, and part transmitted. These parts are brought together again by suitable arrangements of mirrors and prisms, and, being coherent, they interfere. The Michelson interferometer is a typical instrument of this type.

Mirrors may be used in either type of instrument as reflectors. The essential feature of the instruments of Type (*b*) is that at least one mirror is partially reflecting, and that the reflected and the transmitted beams are reunited.

9.3.—An entirely different method of classification is also possible. Instruments may be divided into those which operate by the interference of two beams, and those which employ more than two beams. The Fresnel biprism and the Rayleigh refractometer are examples of instruments of Type (*a*) which use two beams. The echelon diffraction grating is an instrument of Type (*a*) employing more than two beams.

The Michelson interferometer is an instrument of Type (*b*) which uses two beams. The Fabry-Pérot etalon and the Lummer-Gehrcke plate are multiple-beam devices of Type (*b*).

The fringes produced by instruments of Type (*a*) are essentially diffraction fringes, and there would be some justification for calling them " diffractometers ". The ruled diffraction grating is an instrument of this type. The usual practice is to apply the name " interferometer " to many instruments of Type (*a*) as well as to instruments of Type (*b*). This is done because the methods of manufacture and technique of use of many instruments of Type (*a*) are similar to those of instruments of Type (*b*). They are therefore called interferometers, but this name is not applied to the ruled diffraction grating which, from the purely technical point of view, may be regarded as a dispersive system alternative to a prism.

9.4. **Classification of Uses of Interferometers.**

The observations and measurements made with interferometers may be divided into five main classes:

(i) Geometrical measurements.

(ii) Measurements of refractive index.

(iii) Measurement of the ratio between the wavelength emitted by a standard source and a mechanical length.

(iv) Comparison of two wavelengths.

(v) Investigations of theoretical importance.

The first class includes the testing of optical components such as lenses or prisms, the comparison of mechanical gauges used in machine-shop work, and the measurement of small mechanical displacements. The fourth class includes the observation of the very fine structure in a spectrum line, as well as the comparison of lines of different intensities. The fifth class includes experiments like that of Gouy (§ 7.26), as well as a number of experiments which will be described in later chapters (especially in Chapter XI).

9.5. **The Testing of Optical Components.**

From the point of view of wave theory, the action of an optical component is to alter the direction or shape of wave surfaces. When a parallel beam of light falls on a plane mirror the wave surfaces change in direction but remain plane. When a similar beam falls upon an ideal lens, the plane wave surfaces are changed into spherical wave surfaces whose centres are located at a point in the focal plane. In practice, most optical components do not fulfil their function perfectly. They cause some deformation of the wave surfaces. If a beam of light is reflected by a mirror which is plane except for small local deviations,

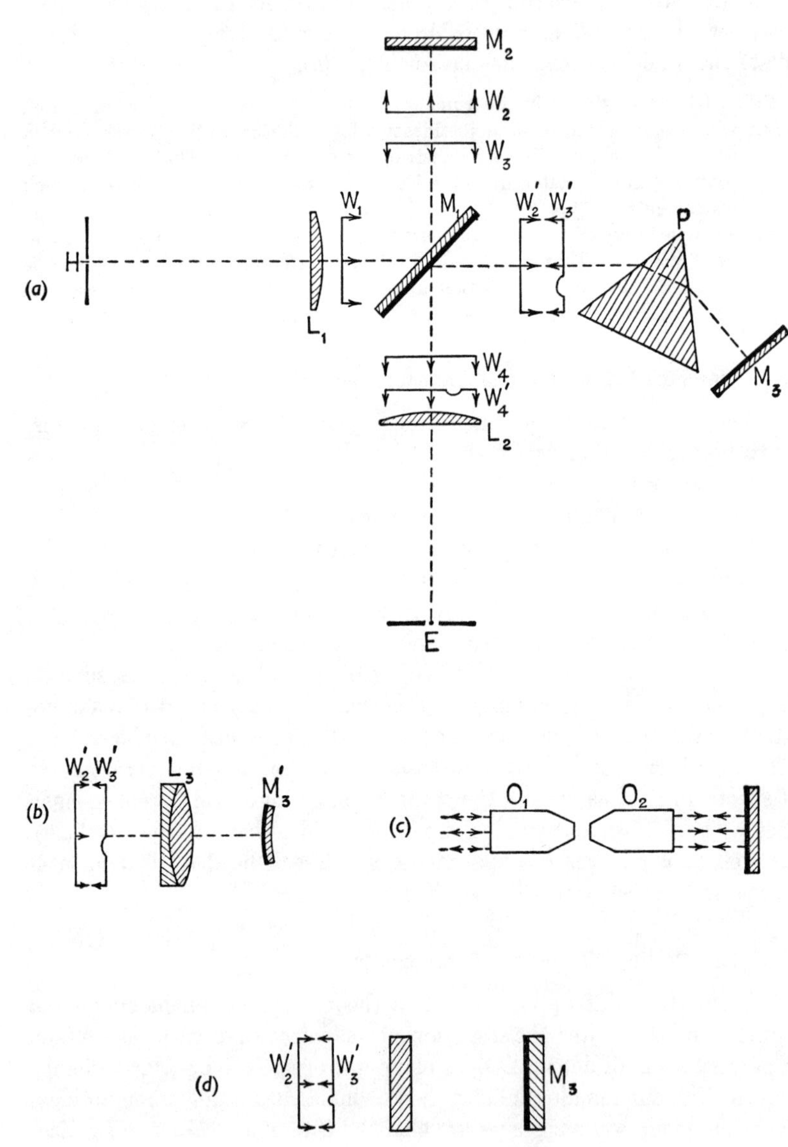

Fig. 9.1.—Twyman-Green interferometer: (*a*) arrangement for testing a prism, (*b*) testing a lens, (*c*) testing a microscope objective, (*d*) testing a block of glass

these irregularities cause corresponding irregularities in the wave surfaces which represent the reflected waves. A lens may suffer from local irregularities and may also show a somewhat different type of error. The emergent wave surface may be smooth, but may not be of the desired shape. Both these errors may be included in the statement that all paths from the object point to the image point through the lens are not optically equal. It is thus desirable to have an instrument which will test directly the total effect which an optical component, or assembly of components, has upon a wave surface. Such a test is also a test of equality of optical path from object point to image point.

9.6. The Twyman-Green Interferometer.

The instrument shown in fig. 9.1 was designed by Twyman and Green for testing prisms. Light from a monochromatic source is focused by an auxiliary lens upon a small hole in a diaphragm placed at H. The point source so formed is at the focus of the lens L_1. Plane waves emerging from this lens are partly reflected and partly transmitted by the half-silvered mirror M_1. The reflected light is again reflected by the mirror M_2, and is focused by the lens L_2 upon a hole in a diaphragm placed at E. The other part of the light passes through the prism, is reflected by M_3 and, after another passage through the prism, reaches the diaphragm at E. The observer's eye is placed immediately behind the diaphragm. If the prism is perfect, he sees a uniformly illuminated field. The illumination is a maximum if the position of the mirror M_2 is adjusted so as to make the difference between the optical paths of the two beams equal to an integral multiple of the wavelength. When the prism is imperfect, the wave surfaces which reach L_2 after reflection at M_3 are not plane. They interfere with the plane wave surfaces from M_2 so as to produce a set of fringes which may be regarded as a contour map of the imperfections in the prism. A typical map is shown in fig. 9.2. Q represents the highest point of a " hill ". The contour lines are usually drawn in wet rouge on the surface of the prism with a paint brush. The excess material may be removed by

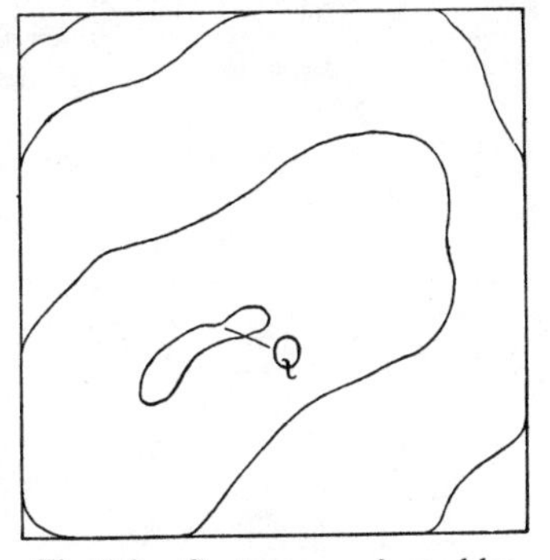

Fig. 9.2.—Contour map formed by fringes of equal thickness

polishing first at the point Q, and then gradually extending the area of polishing to the other contour lines.

9.7—If it is uncertain whether Q represents the top of a hill or the bottom of a depression, the doubt may be resolved by pressing on the frame of the instrument so as to displace the mirror M_3, and thereby to increase the corresponding path. If the contours expand so as to enclose a larger area, a hill is indicated and *vice versa.* The fringes give the total effect on the wave surface produced by double passage through the prism, and show (in wavelengths) the departure of the wave surface from flatness. These imperfections in the wave surfaces may arise from want of flatness of either or both surfaces of the prism, or from local variations in the refractive index of the material. The instrument does not distinguish between these possibilities. In an optically perfect prism the quantity $(\mu - 1)t$ increases linearly from the apex to the base. Optically perfect prisms are usually obtained by polishing the surfaces so that small differences of thickness compensate for inhomogeneity of the optical material. This process is called *figuring.*

9.8.—It should be noted that variations in the contour lines may be obtained by a tilt of the plane of reference. A small adjustment of either M_2 or M_3 may change the contour map from that shown in fig. 9.3 to that shown in fig. 9.4. The

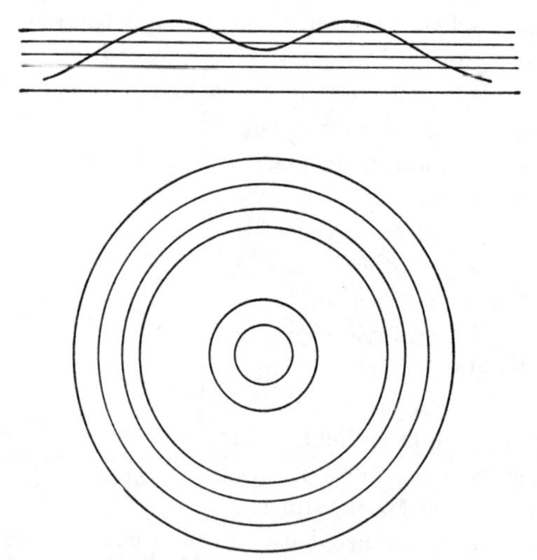

Fig. 9.3.—Contours of local depression (section shown above)

form of the surface is the same in both cases (see sectional diagrams at the tops of the figures) but the plane of section is different. If the operator bases his polishing on either diagram he will obtain a perfect prism, but the angle of the prism obtained by using one diagram will differ slightly from that obtained when the other is used. The skilled operator will adjust the instrument before marking the contours, so as to choose an aspect which requires the minimum polishing to achieve a corrected prism.

9.9.—Although the Twyman-Green interferometer is superficially similar to the Michelson interferometer, its invention marked a radically new departure in interferometry and in optical testing. It differs from the Michelson instrument in the use of the point source which, together with the lens L_1, produces a plane wave surface; it differs

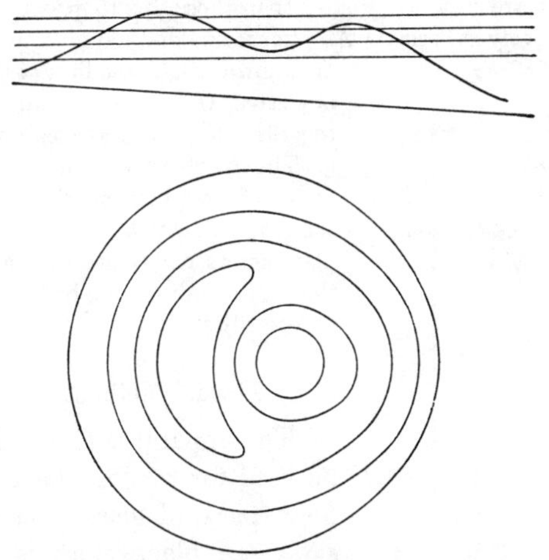

Fig. 9.4.—Change in contours when reference plane is tilted (as shown above)

also in the use of the lens L_2 which arranges that all parts of the emergent wave surfaces enter the observer's eye. It should be noted that the lens L_2 does not act as a telescope and form an image of fringes localized at infinity. The observer's eye is placed *at* its focus, not so as to *view* its focus.

The function of the lens L_2 is to make light from all parts of the field enter the observer's eye. The beams which interfere are nearly parallel, i.e. we are concerned with only narrow pencils of rays. This gives great depth of focus and the fringes are not sharply localized. If the observer puts the marking brush near either surface of a prism and focuses his eye upon it, he will see the fringes in focus, and can mark the part of the surface which must be worked to give correction.

9.10.—The Twyman-Green interferometer may be adapted to test lenses or lens systems (fig. 9.1*b*). If the lens L_3 which is under test changes the incident plane wave into an exactly spherical wave, then a convex mirror M_3' suitably placed will return all the rays along their own paths and the field will be uniformly bright in one adjustment.

The procedure for locating and removing errors is similar to that described for the prism.

There is some difficulty in making spherical mirrors of radius of curvature small enough to test high-power microscope objectives. This difficulty has been overcome by using the surface of a minute drop of mercury as the reflecting surface. When the drop is of the order of a millimetre in diameter, the forces due to surface tension are so much greater than those due to gravity that the form of the drop is effectively spherical. An alternative method is to use an objective O_3 of known high quality, together with the objective O_2 which is under test, so that together they form a telescopic combination.

The interferometer may also be used to test pieces of glass with nearly parallel sides, using the arrangement shown in fig. 9.1*d*. When a plate shows no fringes the value of $(\mu - 1)t$ is the same at all points, but the surfaces are not necessarily flat.

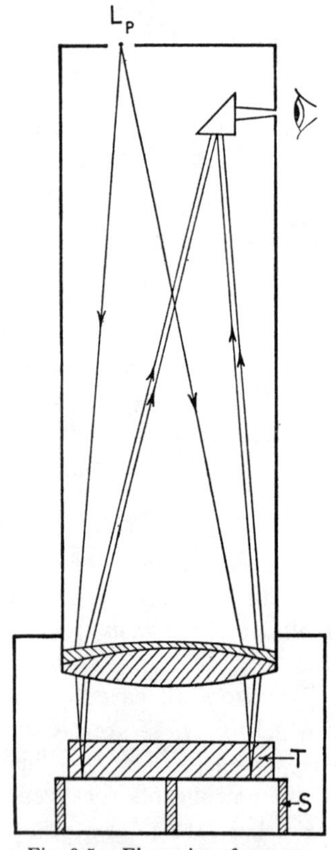

Fig. 9.5.—Fizeau interferometer

9.11. **Fizeau Method.**

The apparatus shown in fig. 9.5 may be used for testing the constancy of μt for a piece of glass with nearly parallel sides. A black cloth is placed at the base of the instrument and interference is obtained between beams of light reflected from the top and bottom faces of the test piece T. These beams are of approximately equal amplitude. The function of the lens L is to arrange that a plane wavefront falls nearly normally on the plate. Thus light reaching the observer's eye from any part of the plate has passed through it nearly normally. The fringes seen are fringes of equal optical thickness. They are localized in the plate (§ 5.20), though owing to the smallness of the source they are not very sharply localized. In many ways this arrangement is like a simplified form of the Twyman-Green interferometer. The lens in fig. 9.5 fulfils the functions both of L_1 and of L_2 in fig. 9.1. It does not act as a telescope to focus the fringes. The essentials of this arrangement were originally due to Fizeau, but the form shown is a more recent development.

The homogeneity of an optical material may be tested by cutting a sample and polishing the sides to be plane and parallel. The Twyman-Green interferometer may then be used to determine the variation of $(\mu - 1)t$, and the Fizeau apparatus to determine the variation of μt. When both of these are known the variations of μ and of t may be calculated.

9.12.—An *optical flat* is a piece of material having one surface plane to within one-tenth of a wavelength or better.* A surface may be tested against a glass optical flat by placing the surfaces in contact, and viewing the fringes of equal thickness described in § 5.18. An extended source of light—such as the light of the sky or a diffusing

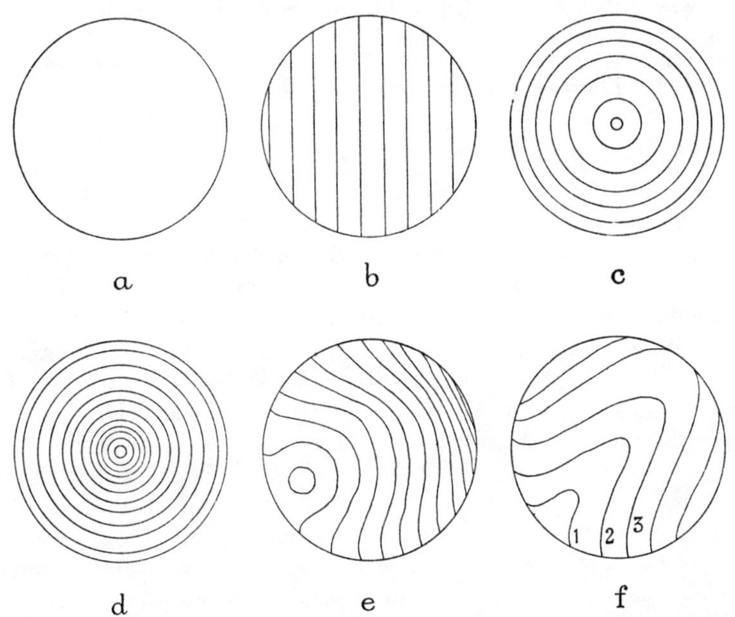

Fig. 9.6.—Fringe systems obtained with Fizeau-type interferometer: (*a*) plane parallel surfaces, (*b*) plane surfaces inclined at a small angle, (*c*) central " hill ", (*d*) central " depression " at top of " hill ", (*e*) small " hill " or " depression " at side, (*f*) " hill " or " valley " running diagonally across plate.

screen illuminated by a mercury arc—may be used. A colour filter is used to select a moderately narrow spectral region. With this arrangement irregularities of about one-twentieth of a wavelength may easily be detected. Some typical appearances are shown in fig. 9.6.

* The highest quality optical flats are plane to about one-hundredth of a wavelength.

9.13.—When it is inconvenient to place the surfaces in contact, a Fizeau type of apparatus may still be used. Since the fringes are seen by nearly parallel light, the surfaces may be widely separated. If the Fizeau arrangement is used with a glass plate as the standard optical flat, extra fringes may be seen owing to reflections between the two surfaces of the plate. These are confusing unless the plate is made wedge-shaped (so that the extra fringes are very close), or of exactly constant optical thickness.

9.14.—It is interesting to consider the production of an optical flat when no standard flat is available for test. When two pieces of glass are ground and polished on one another with a suitable rotary motion, they tend to become spherical. One is convex and the other concave. If the radius of curvature is very large, tests with contour fringes only reveal the accuracy of fit, but do not show that they are not plane. To meet this difficulty three surfaces are worked against each other, in pairs, until the fringes show that each of the three possible pairs fit. Then all three surfaces must be plane.

9.15. **Multiple-beam Fringes.**

A method of using multiple-beam interference for the examination of surfaces has recently been used by Tolansky.* His apparatus is essentially the same as that shown in fig. 9.5, but he uses a partly reflecting film of silver on both surfaces. The silver is deposited by evaporation in vacuum. Special care is taken to ensure that the light is accurately parallel, and normal to the surfaces. One of the surfaces is a good-quality optical flat, and the other is the natural surface of a crystal, such as quartz or mica. Owing to the multiple reflections between the two surfaces, the bright fringes are sharper than those obtained with the ordinary Fizeau method. " Hills " and " depressions " which are only 30 Å. in height can be detected, and the error in measuring the height is only a few Ångström units in the best cases. The molecule of mica is 20 Å. in thickness, and it is possible to detect (on the surface of mica crystals) regions which are raised above the general surface level by distances varying from 40 Å. to over 20,000 Å. These distances are always multiples of the lattice spacing.

It should be noted that the high linear discrimination is attained only in the vertical direction, and is due to the increase of effective length of path produced by multiple reflection. Some multiple-beam fringes of the type described above are shown in Plate II *i*, *j*, *k* and *l*. Fringes obtained by the ordinary Fizeau method (§ 9.11) are shown for comparison (Plate II *g* and *h*, p. 126).

* Reference 9.11.

9.16. **Testing of Mechanical Gauges.**

Accurate standards of length are required for high-precision machine-shop work. The standards of length usually consist of cylindrical pieces of steel, whose ends have been polished to be accurately flat and parallel to one another. Two such gauges may be put together by a process known as " wringing ". The surfaces to be " wrung " are carefully cleaned and pressed together with a minute film of paraffin between them. The paraffin is squeezed out by the atmospheric pressure so that the surfaces are forced into good contact. The accuracy of an instrument such as a good screw gauge may be estimated by finding what reading it gives when applied to a test block whose length is known very accurately. The test block itself is liable to wear under constant use, and it is therefore usual to retain certain standards of specially high quality, which are used only to check the ordinary working test block. The following method enables a test block to be rapidly and accurately compared with a standard.

9.17.—The surfaces of the test block are first tested for flatness by the method described in § 9.12. If they are not sufficiently flat, the block has no unique and definite length to be accurately measured.

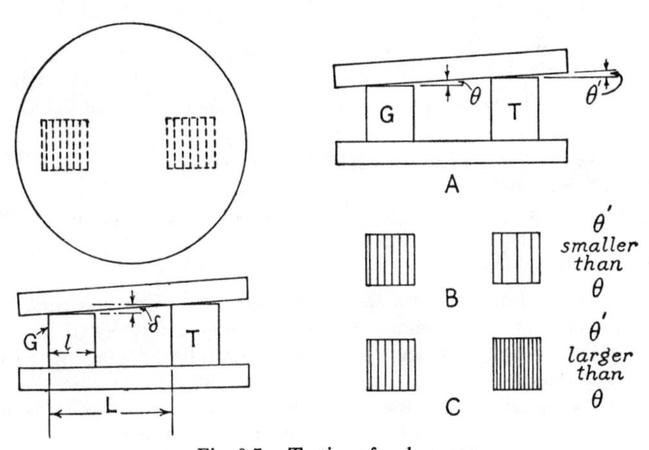

Fig. 9.7.—Testing of end-gauges

If the ends are flat, then one end of the test block and one end of the standard gauge are wrung on to an optically flat steel plate. An optically flat glass plate is placed across the top (fig. 9.7A). This flat will usually rest on one edge of each block, and there will be a wedge of air between the flat and the standard gauge. Interference fringes will be seen, and from their separation the difference of height (δh) between the gauge and the test block may be calculated.* Suppose that the number of fringes per unit length is n, and the distance between the two edges on which the upper plate rests is L, then

$$\delta h = \frac{n\lambda L}{2}. \qquad 9(1)$$

It is usual to view the fringes through a filter which passes a narrow region of the spectrum of wavelength 5100Å., or approximately 2×10^{-5} inch. If L is made equal to one inch, then one fringe per inch implies a difference of length of one-hundred-thousandth of an inch. It is not convenient to count fringes whose separation is less than 1/40 inch, and the maximum difference of length which can easily be measured by this method is therefore

$$\delta h = (40L) \times 10^{-5} = 4 \times 10^{-4}L. \qquad 9(2)$$

Thus with optical flats of $2\frac{1}{2}$-inch diameter, a difference as great as one-thousandth of an inch can be measured. Greater differences than this can be conveniently measured by micrometer screw-gauge methods.

EXAMPLES [9(i)–9(vii)]

9(i). What will be the appearance of the fringes obtained with the apparatus described in §§ 9.16 and 9.17 if the ends of the test block are flat, but not quite parallel to one another? (*b*) What changes will occur when the test block is rotated about a vertical axis?

> [The fringes will always be straight lines perpendicular to the line of greatest slope. Their separation will change when the block is rotated. See fig. 9.7B and C.]

9(ii). A standard gauge and a test block have each a one-inch square cross-section. The ends of the gauge are plane and parallel, and its length is 2 inches. The ends of the test block are plane, but not quite parallel, and its greatest length is a little less than the length of the gauge. The gauge and the test block are wrung on to an optical flat, so that their centres are 2 inches apart. An optical flat is placed across the top. In one orientation of the test block 20 fringes to the inch are seen in the film above it. When it is rotated through 180 degrees there are 5 fringes per inch. In each position the fringes above the test block are parallel to those above the gauge. What deductions concerning the test block can be derived from these observations?

* Fringes are also seen over the test block, but they do not give the difference of height directly unless the surfaces of the test block are accurately parallel [see Example 9(ii)].

What further information may be obtained by viewing the fringes over the gauge?

[There are two possible solutions which give the same number of fringes over the test block. In one case the flat rests on the same edge of the test block in both orientations. This situation gives maximum height of test block 1·99990 in., and minimum height 1·99975 in. The other solution gives heights of 1·99980 in. and 1·99970 in. The cases may be distinguished by viewing the fringes over the gauge.]

9(iii). A glass flat is placed on top of a strip of polished steel which can be bent by weights as shown in fig. 9.8*a*. The fringes obtained are as shown in fig. 9.8*b*. Give a qualitative discussion of the shape of the surface of the steel, and say generally how the results may be explained. [See answer to 9(iv).]

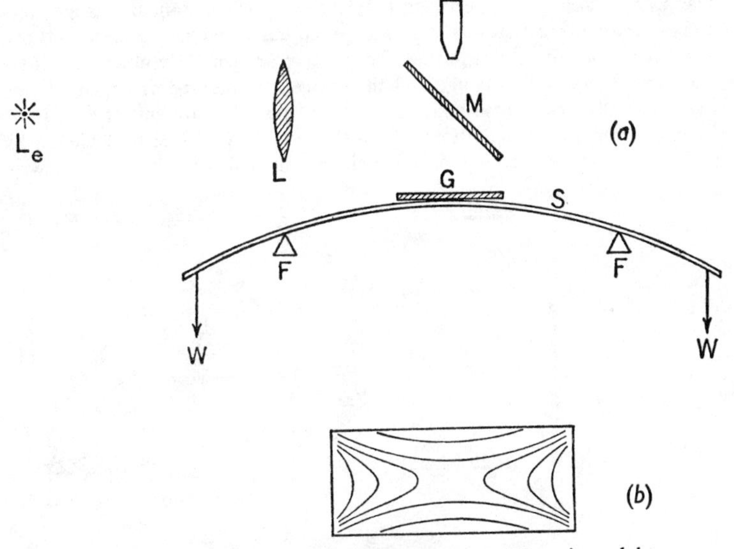

Fig. 9.8.—Cornu's method for the measurement of Young's modulus and Poisson's ratio

9(iv). Show that by means of measurements on the fringes described in Example 9(iii) it is possible to calculate Young's modulus and Poisson's ratio for the steel. Derive the appropriate formulæ.

[See CHAMPION and DAVY: *Properties of Matter* (Blackie), second edition, pp. 76–78.]

9(v). You are given two distance pieces each approximately one centimetre cube. How would you use interference methods to determine the ratio of the volumes? Give some estimate of the percentage accuracy attainable.

[0·003 per cent accuracy without taking special precautions.]

9(vi). You are given two blocks each approximately one centimetre cube, and two optical flats of known high quality. The flats are about $2\frac{1}{2}$ inches diameter.

How would you use these to test the parallelism of the surfaces of a circular piece of glass of diameter one inch, and thickness about 0·25 inch?

[First test for flatness as described in § 9.12. Then wring glass and two cubes on to one flat, and place other on cubes. Observe fringes between second flat and cubes and between second flat and glass.]

9(vii). How would you use an interference method to compare a test block of approximately one inch with a gauge of thickness 2 inches?

[Make an auxiliary one-inch block, which need not be accurate. Measure it against the one-inch block. Then wring it to the one-inch block and compare the sum with the standard two-inch gauge.]

9.18. The Double Interferometer.

The above method of comparing a test block with a standard gauge suffers from the disadvantage that the standard gauge has to be wrung on to the plane surface. It is difficult to ensure that the surfaces are perfectly clean and, if they are not, both the standard gauge and the optical flat may be damaged. Even if they are perfectly clean, some wear occurs each time. The amount of wear is very variable, but under good conditions it is of the order of 10^{-7} inch each time, which means that the standard may be appreciably affected after it has been wrung

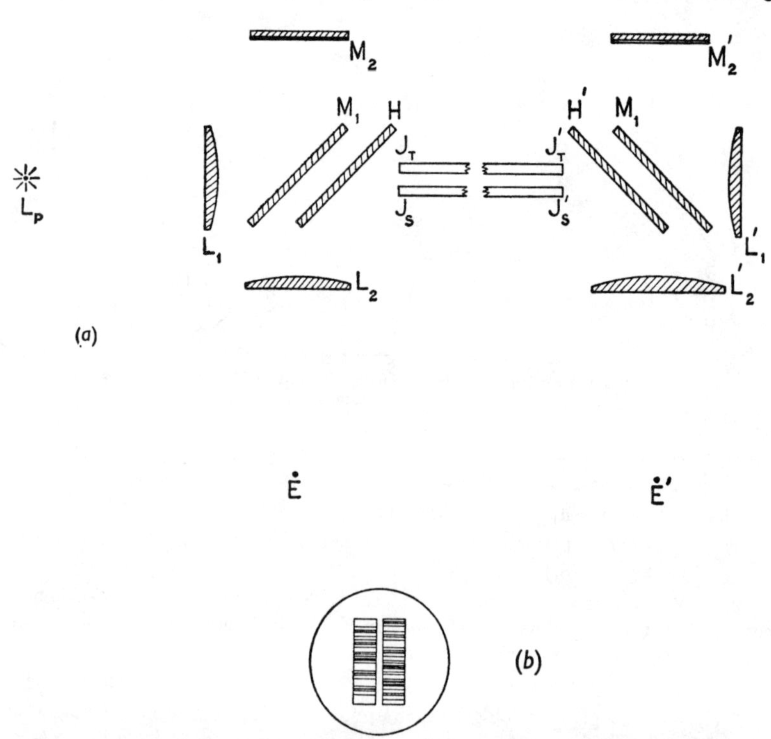

Fig. 9.9.—Double interferometer: (*a*) general arrangement, (*b*) appearance of fringes

about 50 times. This does not matter if it is being used for an annual check, but it is important if the standard is being used once a week. A modification of the Twyman–Green interferometer enables the comparison between test block and standard to be made without causing any wear on their ends. It is shown in fig. 9.9*a*. Considering the left half of the diagram, it may be seen that the arrangement is similar to that shown in fig. 9.1, except that the ends J_T and J_S of the test block and the standard gauge have replaced the prism. The fringes seen by an observer at E give a contour map of the end surfaces of the gauges. The ends are adjusted to be nearly coplanar, and their plane is tilted very slightly so that a series of parallel bands appears. White light is used and, when the compensator H is correctly adjusted, a central fringe corresponding to zero path difference can be distinguished clearly. This fringe is black, because the phase change on reflection at the steel surface differs from that at the surface of M_2. A typical appearance of the fringes is shown in fig. 9.9*b*. The relative displacement of the black fringes gives the distance by which the ends are non-coplanar. A displacement of one fringe corresponds to a distance of $\frac{1}{2}\lambda$, or about 10^{-5} inch, and displacements can fairly easily be measured to a tenth of a fringe or a millionth of an inch. The process is repeated at the other end without moving the gauges. The difference in length is then the algebraic sum of the two distances by which the corresponding ends are non-coplanar. The figure illustrates an early form of this interferometer (introduced in 1932). In a more recent modification only one source and one position of observation are used. The introduction of extra mirrors enables the fringes formed by reflection at both ends to be viewed without moving the bar. This form is more compact and more convenient.

9.19.—The methods which have been described in the preceding paragraphs are concerned with the testing of surfaces and with measurement of the distance between two flat surfaces. It is often important to test the angle between two surfaces which are not approximately parallel (e.g. to test an angle whose nominal value is 90° or 45°). The angles of a cube may be tested by wringing it on to an optical flat beside a glass cube of known high quality, and observing the interference fringes. Two cubes may be made to " test each other " by suitable alternations of the surfaces in contact with the plane. A method for testing the angles of an octagonal prism to an accuracy of 0·05 second is described by Michelson.* Interference methods are not always used for testing angles because sufficient accuracy is often obtained by observing reflected beams in an autocollimating telescope. Nevertheless, interferometers are in common use in laboratories which standardize gauges.

9.20 Measurement of Mechanical Displacements.

Interference methods have been applied to measure small mechanical displacements in the study of elasticity, and in many other

* See p. 134 of Reference 9.2.

branches of physics. One of the simplest applications of this type is due to Fizeau, who used the apparatus shown in fig. 9.10 to measure the thermal expansion of crystals of which only short specimens were available. Interference fringes are formed in the air-film between the top of the crystal and the plate P. The displacement of the fringes on heating measures the difference between the expansion of the crystal and that of the blocks which define the position of the plate. If the expansion of the material of which the blocks are made is known, the expansion of the crystal may be calculated. If the air-film is wedge-shaped, the change in the difference of length may be obtained by viewing the fringes in a microscope focused on the film, and counting the number which pass the cross-wire. An alternative and in some ways a more convenient method is to make the upper plate slightly spherical, so that Newton's rings are seen. The expansion causes fresh rings to appear or to disappear at the centre and the number of these rings may be counted.

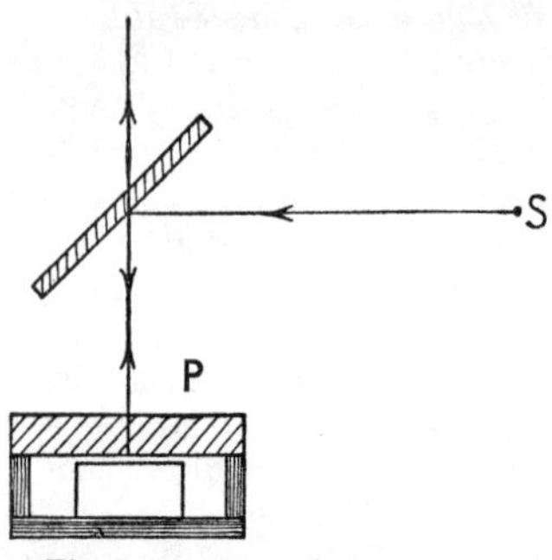

Fig. 9.10.—Fizeau's method for thermal expansion of crystals

The use of interference methods for measuring mechanical displacements is desirable only when sufficient accuracy is not obtainable by more direct means. Interference methods are unlikely to displace the mirror-and-scale method of measuring small angular displacements, because the latter is very convenient, and has sufficient accuracy and sensitivity for most purposes. The use of a high-sensitivity method of measurement in one part of an experiment is not justified unless a corresponding accuracy is obtainable in the other parts of the experiment. When interference methods are used it is necessary to be sure that the advantages of the method are not lost through irregular displacements due to temperature changes, mechanical vibrations, etc. Generally, the simpler interference methods have a rather small range, i.e. they become inapplicable if the path difference exceeds a few dozen wavelengths.

For example, the method described in § 9.16 for comparing mechanical gauges fails if the difference between the lengths of the gauges exceeds about 50 wavelengths. The range of interference methods may be extended by the use of compensators (see § 9.25), but this makes the apparatus more elaborate. Thus interference methods should be used only in conjunction with an experimental arrangement whose basic design enables full advantage to be obtained from the accuracy of the interferometer.

9.21. **Measurement of Refractive Index and of Small Differences of Index.**

Several interference methods have been devised for the measurement of the refractive indices of gases, and for small differences in refractive index of liquids and solids. The most generally convenient and accurate apparatus is the Rayleigh refractometer. Fig. 9.11*a* is

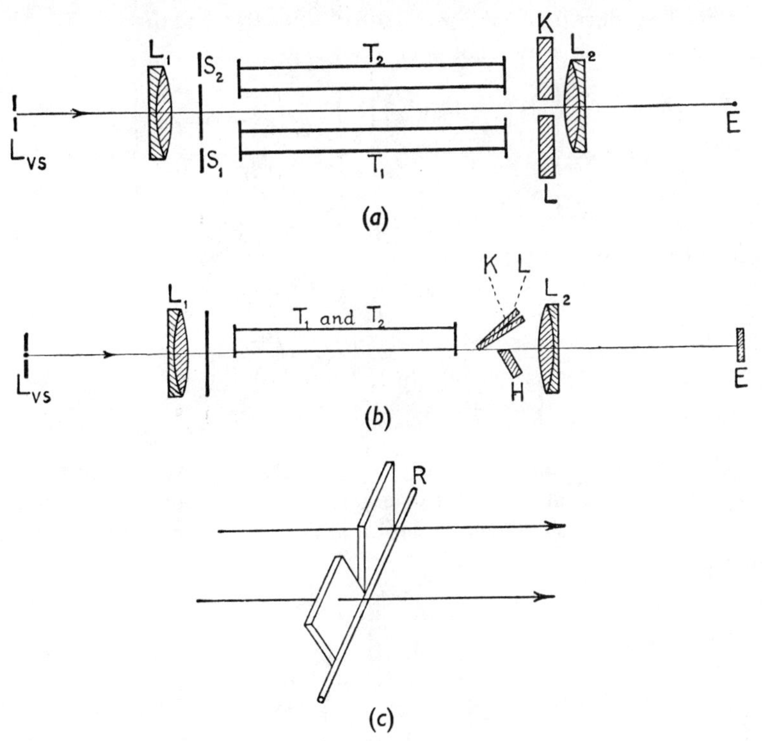

Fig. 9.11.—Rayleigh refractometer: (*a*) plan, (*b*) vertical section, (*c*) compensator

a diagram of the instrument seen from above. Light from a vertical slit source L_{vs} is collimated by the lens L_1 and passes through the tubes T_1 and T_2 and the vertical slits S_1 and S_2. Diffraction fringes are produced in the focal plane of the lens L_2 and are viewed with the aid of the eyepiece E.

Suppose that the length of each tube is t. Then, if there is a small difference $(\Delta\mu)$ between the refractive indices of the materials in the tubes, a path difference $\Delta s = t\Delta\mu$ is introduced and the fringes are displaced from the standard position obtained when $\Delta\mu = 0$. The difference of refractive index may be measured either by observing the fringe shift directly or, more conveniently, by a compensation method.

9.22.—The fringes are the diffraction fringes of a grating of two lines. If $2d$ is the width of each slit, and $2e$ is the distance between the centres of the slits, the distribution of illumination in the fringes is obtained by putting $N = 2$, $\theta_1 = 0$ and $\theta_2 = \theta$ in equation 6(55). The relative energy in a direction θ is given by

$$E(\theta) = \left\{\frac{\sin\left(\frac{2\pi}{\lambda} d \sin\theta\right)}{\frac{2\pi}{\lambda} d \sin\theta}\right\}^2 \left\{\frac{\sin\left(\frac{4\pi}{\lambda} e \sin\theta\right)}{2\sin\left(\frac{2\pi}{\lambda} e \sin\theta\right)}\right\}^2, \quad . \quad 9(3)$$

which can be put in the form

$$E(\theta) = \left\{\frac{\sin\left(\frac{2\pi}{\lambda} d \sin\theta\right)}{\frac{2\pi d}{\lambda} \sin\theta}\right\}^2 \cos^2\left(\frac{2\pi}{\lambda} e \sin\theta\right). \quad . \quad 9(4)$$

In the instrument made by Hilger and Watts, Ltd., the slits are each 2·5 millimetres wide and the separation is 11 millimetres between centres. Fig. 9.12 shows the variation of illumination in the focal plane of L_2.

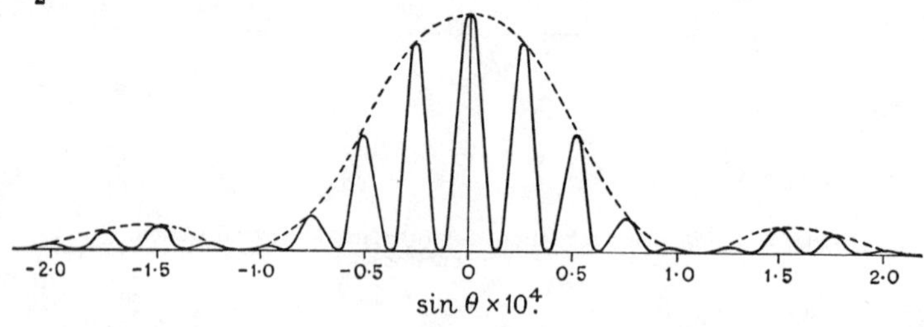

Fig. 9.12

9.23.—The instrument is reliable and convenient in operation only because a number of matters of detail have been carefully considered. The fringes in the focal plane of L_2 are extremely fine (fig. 9.12). They are viewed by means of a cylindrical lens which consists of an accurately made glass rod of diameter about 2 millimetres. This lens gives an effective magnification (in the direction of separation of the fringes) of about 150. The brightness of the field is reduced 150 times. It is still very much brighter than it would be if a spherical lens of the same magnification were used, since the spherical lens would reduce the brightness 22,500 times (i.e. 150^2 times). This is important as the total amount of light passing down the tubes is not very great. In the simple form of the instrument so far described, monochromatic light is used. The refractive index of a gas may be measured by evacuating both tubes, and then counting the number of fringes which pass a fine needle point placed in the focal plane of L_2, when gas is slowly admitted to one tube. This method places a considerable strain on the observer who is very liable to miss one of the fine fringes. Also it is not applicable to a determination of the difference of the refractive indices of liquids.

9.24.—Convenience and accuracy of observation are increased by using as fiducial mark a set of fringes similar to those due to light which passes through T_1 and T_2, instead of a cross-wire. The way in which this is done is shown in fig. 9.11*b*, which gives a side view of the instrument. Two beams pass below the tubes T_1 and T_2. An independent set of fringes is formed. By tilting the plate H these may be moved vertically so that their upper edge may be brought into coincidence with the lower edge of the fringes due to light which has passed through T_1 and T_2. In this way it is possible to take advantage of the high vernier acuity of the eye. A horizontal displacement between the two fringe systems of only one-fortieth of a fringe width can be detected, although it would not be possible to set a cross-wire on either system to an accuracy better than about one-tenth of a fringe.

9.25.—The accuracy and convenience of the measurement are improved by compensating the optical path difference. The two fringe systems are then used as a null indicator of equality of path. The compensator, in the form originally due to Jamin, is shown in fig. 9.11*c*. It consists of two glass plates set at an angle to one another, and connected through the rod R, by means of which the system may be rotated as a whole. When the rod is turned the thickness of glass in each beam is altered but, since the glass plates are inclined to one another, the change in path of one beam is slightly greater than that of the other. Thus a considerable rotation of the rod corresponds to a small shift of the fringes.

Rotation of the compensator makes the two systems of fringes slightly inclined to one another. This does not matter in the Jamin refractometer (§ 9.29) where the fringes are comparatively broad, and are viewed by an ordinary eyepiece. In the modern Rayleigh refractometer, the cylindrical lens has to be set with its axis exactly parallel to the system of fringes. The compensator must be modified so that only a negligible rotation of the fringes is introduced. In the Hilger instrument both plates are fairly thin. One is fixed, and the other rotates with the rod R. A small movement of the single plate introduces a sufficient path difference. A small and accurately reproducible rotation is produced by turning a micrometer screw which presses against a radial arm attached to the rod R.

9.26.—In order to know the refractive index for a definite wavelength it is necessary to use monochromatic light, but the central fringe obtained with monochromatic light is indistinguishable from its neighbours. The systems appear to coincide when the path difference ($t\Delta\mu$) is an integral number of wavelengths. The following method is therefore adopted.

T_1 and T_2 are evacuated (or filled with standard material) and white-light fringes are produced. The compensator is turned until the fringe systems in the upper and lower halves of the field appear to coincide. Monochromatic light is then introduced, and a small alteration is made in the setting so as to bring the two systems into coincidence. The setting of the compensator is read (on the micrometer scale). The material under test is then admitted to one tube, and the adjustment of the compensator is made again, first with white and then with monochromatic light. The final setting of the compensator is read. The difference between the two readings of the compensator gives $t\Delta\mu$ and hence $\Delta\mu$. The compensator is calibrated with monochromatic light, by using it to produce displacements of 1, 2, 3, etc., fringes and noting the corresponding readings on the scale. This calibration must be carried out with a number of different wavelengths suitably spaced through the spectrum.

9.27.—The smallest difference of path which can be detected is about $\frac{1}{40}\lambda$ and the largest which can conveniently be measured is about 200λ. The corresponding limits of index are

$$\Delta\mu_{\min} = \tfrac{1}{40}\frac{\lambda}{t} \quad \text{and} \quad \Delta\mu_{\max} = \frac{200\lambda}{t}.$$

With a tube length of 100 centimetres, the instrument will detect a difference of index of about one part in 10^8 and the maximum difference measurable is about one part in 10^4. With a tube length of one centimetre the sensitivity is one part in a million, and the maximum difference measurable is of order one part in 100. The instrument was used by Rayleigh to measure the refractivities of the rare gases helium and argon. The smaller refractivity is $3{\cdot}5 \times 10^{-5}$ at standard temperature and pressure, and is thus well within the range of the instrument.

Measurements of refractivity can also be used to detect impurities in gases, and the instrument is capable of detecting 0·01 per cent of hydrogen or 0·03 per cent of carbon monoxide in air. A modified portable form of the instrument has been used for detecting " fire-damp " in mines. The refractometer has also been widely used to measure small differences of refractive index of solutions, and hence to indicate differences in composition of solutions of chemical and biochemical interest. The final limiting factor in accuracy and convenience is set by the difficulty of temperature control. With liquids, it is often necessary to wait an hour or more to allow temperature equilibrium to be established. For this reason it is always desirable to choose the shortest tubes which give the necessary sensitivity.

9.28.—An interesting point arises when the dispersion of the material under test is different from that of the compensating plates. Suppose that one of the tubes is evacuated and the other contains material of index μ; suppose also that a difference of thickness t' of material of index μ' is introduced by the compensator. Then the centres of the fringe systems obtained with white light will be undisplaced if the extra paths introduced are traversed in equal times by two groups of waves which emerge from S_1 and S_2 respectively. This will be so if

$$\left(\mu - \lambda \frac{d\mu}{d\lambda}\right) t = \left(\mu' - \lambda \frac{d\mu'}{d\lambda}\right) t'. \qquad 9(5)$$

If

$$\frac{1}{\mu}\frac{d\mu}{d\lambda} - \frac{1}{\mu'}\frac{d\mu'}{d\lambda} = 0, \qquad 9(6)$$

then 9(5) reduces to $\mu t = \mu' t'$, and the instrument measures $\Delta\mu$ as indicated in the simple theory. If, however, the left-hand side of 9(6) exceeds $\frac{1}{2}\lambda t$, the test with white light will not bring the correct pair of fringes into coincidence. Even for smaller dispersions there is an observational difficulty in that the central fringes obtained with white light are blurred to such an extent that the central fringe cannot be identified with certainty. This difficulty may be overcome by using short tubes to determine a preliminary value of the index. This value is sufficient to show which of the fringes obtained with monochromatic light have to be brought into coincidence when the longer tubes are used.

9.29. **The Jamin Refractometer.**

An earlier type of refractometer invented by Jamin is shown on p. 280. The substance under test is placed in T_1 and the standard in T_2. Fringes due to interference between beams which have passed through two tubes are observed in the telescope T. The fringes formed are essentially fringes of superposition (§ 5.32). The glass blocks G_1 and G_2 are turned until the beams are accurately aligned. Then a slight tilt of one of the blocks about a horizontal axis gives horizontal fringes. A compensator K of the type described above (§ 9.25) may be used. The instrument can be adjusted to give wider fringes than the Rayleigh refractometer, but is less accurate because the fiduciary mark is a cross-wire instead of another system of fringes, and also

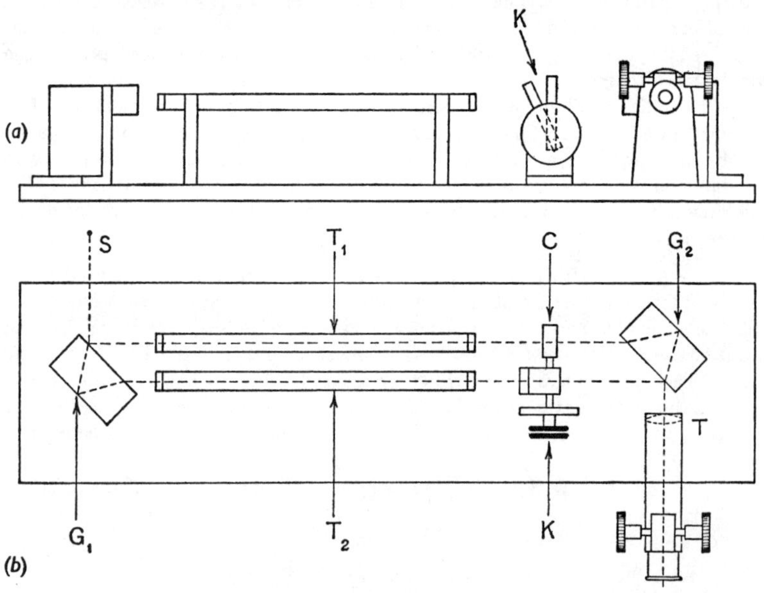

Fig. 9.13.—Jamin refractometer: (*a*) vertical section, and (*b*) plan

because errors may arise from small mechanical deformation of the parts of the instrument. Nearly all errors of this type are automatically eliminated in the Rayleigh refractometer. Although it is less satisfactory than the modern form of the Rayleigh refractometer, Jamin's instrument has, in the past, given valuable results in many different types of experiments.

EXAMPLES [9(viii)–9(xiii)]

9(viii). The refractivity of helium is $3{\cdot}5 \times 10^{-5}$ and that of argon is $2{\cdot}8 \times 10^{-4}$. Describe how you would use the Rayleigh refractometer to estimate the amount of helium in an argon/helium mixture which may contain from 50 to 100 per cent of argon. Assuming that the gas is to be used at standard temperature and pressure and that the compensator scale covers 200 fringes, calculate an appropriate length of tube. Estimate the accuracy of the determination under optimum conditions. Take $\lambda = 5000$ Å.

[The optimum length of tube is about 80 cm., since a greater length would not permit the whole range of refractivities to be covered. The accuracy of the determination is then such that a difference of 0·006 per cent of argon in the mixture can be detected.]

9(ix). In what way, if any, would your conclusions in relation to the previous problem be modified if the test samples consisted of only one cubic centimetre of gas at S.T.P?

[Since a tube of cross-sectional area much less than 0·2 cm.^{2} is impracticable, the length would have to be reduced to about 5 cm. This would reduce the sensitivity to about 0·1 per cent of argon.]

9(x). The refractivity of air is $2{\cdot}92 \times 10^{-4}$ and that of helium is $3{\cdot}5 \times 10^{-5}$. It is suspected that a specimen of helium may be contaminated with air. What is the smallest amount of contamination that could be detected by measurement of the refractive index?

[Using tubes 100 cm. long, 0·005 per cent air could be detected.]

9(xi). Discuss how the Rayleigh refractometer might be used to determine whether the refractivity of air is proportional to the density. How great a range of density could conveniently be covered? What lengths of tubes would be desirable?

[A tube one metre long would give 0·1 per cent accuracy for a pressure of about 0·05 atmosphere, and a tube 0·5 cm. long would give 200 fringes displacement with about 70 atmospheres. Hence the range is 0·05 atmosphere to 70 atmospheres.]

9(xii). The refractive index of a 4 per cent solution of a certain salt is 1·3388, and of a 6 per cent solution is 1·3418. The refractivity varies linearly between these limits. It is desired to measure the percentage strengths of solutions to within $\pm 0{\cdot}001$ per cent. Explain how the Rayleigh refractometer might be used for this purpose and estimate a suitable length of the tubes.

[Suitable length = 1 or 2 cm.]

9(xiii). A Jamin compensator consists of two plates each of thickness t set at an angle of 30 degrees to one another. Obtain an expression for the difference of path introduced when the bisector makes an angle θ with the direction of the two beams.

9.30. **Measurement of Wavelength.**

The study of the wavelengths of spectrum lines by means of interferometers forms an extension of measurements with grating spectrographs just as the measurement of distance gauges with interferometers increases the accuracy of measurements made with instruments like screw gauges. There is no difficulty in measuring the wavelengths of most spectrum lines to within 0·05 Å. (or one part in 10^5) by means of grating spectrographs. Lines which are too diffuse to be measured within this accuracy on the photograph of a grating spectrum are usually not suitable for study with an interferometer. In a similar way there is no point in trying to measure the length of an end gauge with an interferometer if measurements with a screw gauge show that the ends are seriously non-planar. If large numbers of moderately accurate measurements of wavelength are required, the spectrograph is the appropriate instrument. Interferometers are used:

(*a*) To determine the ratio between one standard wavelength and the standard metre. This comparison has been made to within one or two parts in 10^7.

(*b*) To compare wavelengths so as to be able to compile a set of a few thousand secondary wavelength standards. A large number of iron lines have been compared in this way. Spectrographic measurements may thus be referred to secondary standards of wavelength near to the unknown wavelength.

(*c*) To investigate the fine structure of spectrum lines and to measure differences of wavelength of the order of a few hundredths or thousandths of an Ångström unit.

It is convenient to defer consideration of (*a*) until the technical methods required for investigation (*b*) have been described.

9.31. **Comparison of Wavelengths by Coincidences.**

This method is similar to the well-known method of comparing the periods of two pendulums by observing the interval between times when the phases coincide. Its essentials have already been described (§ 4.11) in connection with the visibility of fringes obtained with a sodium light source and a Michelson interferometer. The Fabry-Pérot interferometer is now usually employed and fringes due to the two wavelengths are observed. The distance between the plates is varied and the separations (e_1, e_2, etc.) when the fringes show maximum clarity are noted. Suppose that λ_1 is to be determined by reference to λ_2 (a known wavelength) and that the fringe of order m for λ_1 coincides with the fringe of order $(m + n)$ for λ_2. The next coincidence will occur after p fringes of λ_1 and $(p + 1)$ of λ_2, The observations give only p and not m or n, but the ratio $\lambda_1 : \lambda_2$ may be calculated for

$$\left.\begin{aligned} 2e_1 &= m\lambda_1 = (m + n)\lambda_2, \\ 2e_2 &= (m + p)\lambda_1 = (m + n + p + 1)\lambda_2, \end{aligned}\right\} \quad . \quad . \quad 9(7)$$

and hence

$$\lambda_1 = \frac{p + 1}{p}\,\lambda_2 \quad . \quad . \quad . \quad . \quad . \quad 9(8)$$

or

$$\lambda_1 - \lambda_2 = \frac{\lambda_2\lambda_1}{2(e_2 - e_1)}. \quad . \quad . \quad . \quad . \quad . \quad 9(9)$$

Sometimes the number of fringes between coincidences is counted and 9(8) is used. Alternatively 9(9) may be applied. On the right-hand side of the equation approximate values of the wavelengths, derived from spectroscopic measurements, are used. The value of $(e_2 - e_1)$ is obtained from a micrometer screw reading. The right-hand side of 9(9) then gives a more accurate value of $(\lambda_1 - \lambda_2)$ than that originally available. It is usually necessary to observe a number of coincidences. If this is done the method can be made very accurate, but it is rather laborious especially in view of the fact that every pair of wavelengths has to be treated independently.

9.32. **Comparison of Wavelengths by Exact Fractions.**

This method, which is due to Benoît, depends on the same principle as the method of coincidences, but the principle is applied in a more subtle and elegant way. It is based on the fact that the distance between two coincidences of scales which are divided into different units is the least common multiple of the units of the different scales. Thus if three rulers are divided in units of 2, 3 and 5 inches respectively, and their zeroes aligned, coincidences between the first two occur at every sixth inch, and coincidences between all three occur at every thirtieth inch.

Suppose that we are told that the length of a certain bar is one inch greater than *one* of the lengths at which the three scales coincide, but are not told which one. If a rough measurement of the length of the bar has yielded the value (8 ± 1) feet, we can at once deduce that the length is 91 inches, because no other length would agree with both the data. In the optical problem we determine a given distance by measuring the fraction by which it exceeds an integral number of wavelengths. This measurement is made for three or more different known wavelengths, and we then know that the optical path $(2e)$ is given by

$$2e = (n_1 + f_1)\lambda_1 = (n_2 + f_2)\lambda_2 = (n_3 + f_3)\lambda_3. \quad . \quad 9(10)$$

In this equation n_1, n_2, n_3 are integers whose exact values are unknown. The fractions are known within an error of 2 or 3 per cent of a unit. If e were entirely unknown, knowledge of the fractions in equation 9(10) would only enable us to say that it had one of a certain set of very precisely defined values. Only one of these, however, agrees sufficiently well with a known approximate value of e, and we therefore know that this is the correct one.

9.33.—The method of measuring the fraction with a Fabry-Pérot etalon will now be described. Suppose that the relation between the thickness and λ_1 is given by 9(10). Then bright rings occur for angles of incidence θ_1, θ_2, etc., given by

$$\begin{aligned} n\lambda_1 &= 2e\cos\theta_1 = (n_1 + f_1)\lambda_1\cos\theta_1, \\ (n_1 - 1)\lambda_1 &= 2e\cos\theta_2 = (n_1 + f_1)\lambda_1\cos\theta_2, \\ (n_1 - m - 1)\lambda_1 &= 2e\cos\theta_m = (n_1 + f_1)\lambda_1\cos\theta_m. \end{aligned} \qquad 9(11)$$

If $\mathscr{I}_m$ is the angular *diameter* of the mth ring, we may put

$$\cos\theta_m = 1 - \tfrac{1}{8}\mathscr{I}_m^{\,2},$$

so long as θ_m is small, and we then have

$$1 - \tfrac{1}{8}\mathscr{I}_m^{\,2} = \frac{n_1 - m - 1}{n_1 + f_1} = 1 - \frac{f_1 + m + 1}{n_1 + f_1}, \qquad 9(12)$$

or

$$\mathscr{I}_m^{\,2} = \frac{8}{n_1 + f_1}(f_1 + m + 1) = \frac{f_1 + m + 1}{2e}8\lambda_1. \qquad 9(13)$$

A value of f_1 may be obtained by measuring the angular diameters of several rings for values of m from about 3 to about 7, and then plotting $\mathscr{I}_m^{\,2}$ against $(m + 1)$. The intercept * on the axis gives f_1. If this method is used it is necessary to check by direct calculation that the angles are small enough to enable the approximate value of $\cos\theta$ to be used.

9.34.—The procedure followed when f_1, f_2, f_3 have been measured may be illustrated by means of an example given by Childs.†

The fractions for a certain etalon were measured and found to be

$0{\cdot}20 \pm {\cdot}03$ for the neon line $\lambda = 6096{\cdot}163$ Å.
$0{\cdot}90 \pm {\cdot}03$ „ „ „ „ $\lambda = 5852{\cdot}488$ Å.
$0{\cdot}35 \pm {\cdot}03$ „ „ helium $\lambda = 5015{\cdot}675$ Å.

Measurement with a micrometer gave $e = 10{\cdot}040 \pm 0{\cdot}005$ mm. This implies that n_1 lies between 32,922 and 32,955. Each of these values is taken in turn and associated with the measured fraction 0·20. The corresponding values of $(n_2 + f_2)$ and $(n_3 + f_3)$ are tabulated:

* For higher accuracy the best value of f_1 is computed by the method of least squares.

† Reference 9.3.

Hypothetical order for λ = 6096·163 Å.	Corresponding order calculated for λ = 5852·488 Å.	5015·675 Å.
32,922·20	34,292·95	40,014·37
32,923·20	34,293·99	40,015·58
32,924·20	34,295·03	40,016·80
. . . .	. . . .	. . . .
32,943·20	34,314·82	40,039·90
32,944·20	34,315·87	40,041·11
32,945·20	*34,316·91*	*40,042·32*
32,946·20	34,317·95	40,043·54
. . . .	. . . .	. . . .
32,953·20	34,325·24	40,052·04
32,954·20	34,326·28	40,053·26
32,955·20	34,327·32	40,054·47

Within the range covered by the table, the only value of $(n_1 + f_1)$ which makes the three fractions correct is 32,945·20 $\pm$ 0·03. The corresponding value of e is 10·04197 $\pm$ 0·00001 mm. The nearest value of e which again gives three fractions 0·20, 0·90 and 0·35 within the limits of error, is approximately 10·049 mm. (corresponding to $n_1 = 32{,}968$). This is excluded by the "rough" measurement with the micrometer. If the initial measurement with the micrometer were less accurate it would be necessary to measure the fractions for 4 or 5 wavelengths.

9.35.—When the thickness of the etalon has been accurately determined in this way, an unknown wavelength (λ_x) can be determined by measuring the fraction (f_x). The order of interference at the centre of the ring system is about thirty to forty thousand, and the preliminary value of λ_x obtained with a grating should be correct to one in 10^5, and should give the integral part of the order without any ambiguity. If a less accurate preliminary value is available, it may be necessary to start by using a shorter etalon (say 2-millimetre separation) and to take the value obtained with this etalon as the preliminary one for use with an etalon of 10-millimetre separation.

9.36.—The method of exact fractions increases the accuracy of the determination of a wavelength from $\pm$0·05 Å. to $\pm$0·005 Å. *If a spectrum line is sharp enough,* a further increase of accuracy (to about $\pm$0·0002 Å.) may be obtained by proceeding to a 100-millimetre etalon. It will be appreciated that this accuracy is obtained only when the plates are very accurately plane and parallel. The wavelengths of a large number of lines may be measured in one experiment by focusing the ring pattern on the slit of a spectrograph with an achromatic lens. If a wide slit is used, the different spectrum lines each show a series of short arcs whose separations give the angular diameters of the rings for the corresponding wavelengths (fig. 9.14). A few known wavelengths are then used to determine the thickness of the etalon and the remainder may then be treated as unknown.*

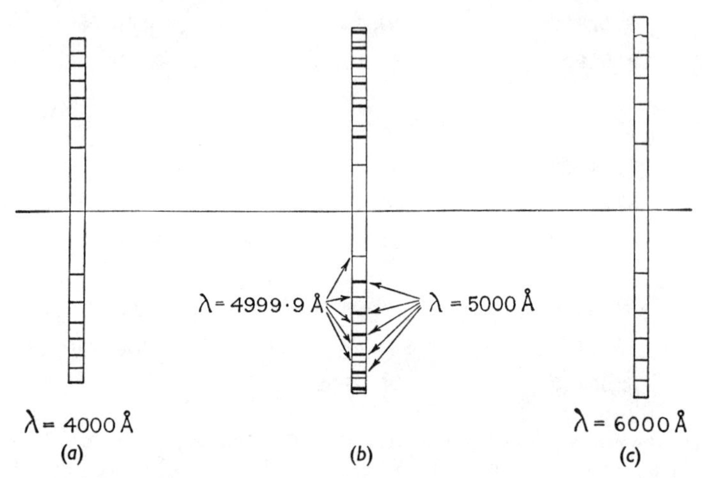

Fig. 9.14.—Diagrammatic representation of pattern obtained in spectrogram for stated wavelengths when the Fabry-Pérot rings are focused on the slit

9.37.—The usual objective of wavelength determinations is to measure the wavelength *in vacuo,* since this value forms a constant of theoretical importance, whereas the wavelength *in air* varies from day to day. If the method described in the last paragraph is used, the *ratio* of the wavelengths obtained is independent of atmospheric conditions, and is equal to the ratio of the wavelengths *in vacuo,* provided that the standards and the unknowns are all located within a fairly short interval of the spectrum. If they are not, a correction has to be made for the dispersion of air under the conditions obtaining during the experiment. It is then also necessary to make a small correction for the variation with wavelength of the phase change after reflection (§ 5.26). An account of the method by which this change is determined is given by Childs (Reference 9.3).

EXAMPLES [9(xiv)–9(xvii)]

9(xiv). An unknown wavelength of less than 6000 Å. is to be compared with a standard wavelength of exactly 6000 Å. Coincidences occur when the separation of the plates of a Fabry-Pérot interferometer are 1·5 mm., 3 mm. and 4·5 mm. Find the unknown wavelength. [5998·8 Å.]

9(xv). A certain wavelength is known to lie between 5990 Å. and 5992 Å. It gives coincidences with a standard source of 6000 Å. when the separations are

* This procedure is discussed by Meissner (Reference 9.4), whose article includes many interesting photographs of the rings.

6 mm. and 6·4 mm., but it is not certain whether these are neighbouring coincidences. Find the unknown wavelength.

[Coincidence period lies between 600 and 750. $2(e_1 - e_2) = 0{\cdot}8$ mm. = 1333 wavelengths. Thus there has been one intermediate coincidence. Exact coincidence period is 667, and $\lambda_2 = \frac{667}{668} \times 6000 = 5991$ Å.]

9(xvi). A certain wavelength is known to lie between 5000·0 Å. and 5000·1 Å. Fringes are formed with a Fabry-Pérot etalon of separation exactly 5·0000 mm. The fraction is found to be $0{\cdot}7 \pm 0{\cdot}05$. Find the wavelength to the maximum accuracy which the interferometer measurement justifies. [5000·07 Å.]

9(xvii). Three wavelengths are known to be exactly 4200 Å., 4800 Å. and 5000 Å. A certain etalon is known to have a separation of $10{\cdot}000 \pm 0{\cdot}001$ mm. The fractions are 0·20, 0·06 and 0·46 respectively. The fractions are each measured to an accuracy of $\pm 0{\cdot}05$. Find the thickness of the etalon.

[$9{\cdot}99961 \pm 0{\cdot}00002$ mm.]

(" Round " numbers have been inserted in Examples 9(xiv)–9(xvii) in order to enable the reader to practise the methods without carrying out the large amount of arithmetical work which would be needed if data were taken from real problems.)

9.38. Comparison between Optical and Mechanical Standards of Length.

At present the internationally accepted standard of length is the distance between two fine lines engraved on a platinum bar when the bar is at 0° C. This bar is called " the standard metre " or " the metre " and is kept in France. The different national standardizing laboratories, such as the National Physical Laboratory in Great Britain and the American Bureau of Standards, hold copies of this standard. These copies have been compared with the standard by the use of high-quality travelling microscopes, and the difference between each copy and the standard is known * to about $2{\cdot}5 \times 10^{-5}$ centimetre. When accurate interferometric measurements became available, it was suggested that an attempt should be made to determine the relation between the wavelengths corresponding to certain spectral lines and the standard metre. This problem was first attacked in 1892–5 by Michelson † and Benoît, who used a modification of the Michelson interferometer. They found that the ratio between the length of the standard metre and the wavelength of the red cadmium line (in air at 15° C. and 760 mm.) was 1,553,163·5. The result is accurate to about one part in two million. A new determination was made by Benoît,‡ Fabry and Pérot about fourteen years later, using a set of Fabry-Pérot etalons. The use of a

* See § 9.47. † Reference 9.5. ‡ Reference 9.6.

multiple-beam interferometric method, and various other technical improvements, enabled the accuracy to be improved to about one part in five million. The earlier determination is of considerable historical interest, but for all practical purposes it is superseded by the second determination, which we now describe.*

9.39.—*Benoît, Fabry and Pérot* used five Fabry-Pérot etalons of lengths approximately 6·25, 12·5, 25, 50 and 100 centimetres. The separators which determined the distances between the plates were V-shaped bars of Invar. Three types of measurement were involved:

(i) a determination of the length of the shortest etalon in terms of the wavelength of the cadmium red line;

(ii) an intercomparison of the etalons;

(iii) a determination of the difference between the longest etalon and a standard metre (i.e. a copy of "the metre").

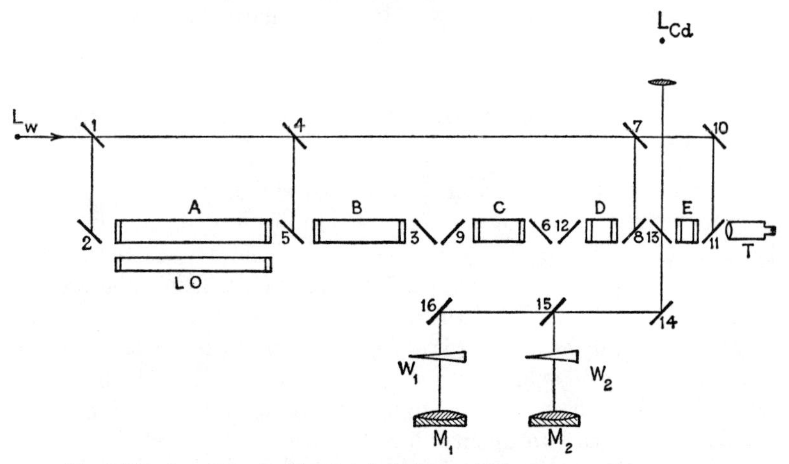

Fig. 9.15.—Apparatus used by Benoît, Fabry and Pérot for the comparison of optical and mechanical standards of length

The experimental procedure was so designed that the final series of experiments could be carried out rapidly and without disturbing the apparatus. In this way, errors due to changes of temperature and of atmospheric pressure were greatly reduced. The arrangement of apparatus is shown in fig. 9.15. L_w is a source of white light, and L_{Cd} is a source of cadmium light. Sixteen mirrors (indicated by numbers)

* A complete account of the Michelson-Benoît determination is given in Reference 9.5; see also p. 51 of Reference 9.1. The work of Benoît, Fabry and Pérot is described in Reference 9.6. Later determinations are described in References 9.7, 9.8, and 9.9.

are used to direct the light along desired paths. Arrangements are provided for inserting and removing any of these mirrors without disturbing the rest of the apparatus, or causing temperature changes. W_1 and W_2 are wedges of small angle, T is a telescope, and M_1 and M_2 are low-power microscopes focused upon W_1 and W_2. Mirror 13 is used to direct the cadmium light through the shortest etalon to the observation telescope T; mirrors 14 and 16 to direct it through W_1 to M_1; and mirrors 14 and 15 to direct it through W_2 to M_2.

9.40.—Two preliminary experiments are carried out. In the first, the length of the shortest etalon is measured in terms of the cadmium red line, by the method of exact fractions. The length so obtained may not be the exact length at the time of the final experiment, since there may be changes due to differences of temperature between the days of the preliminary and final experiments. The preliminary measurement does, however, give the length to within one wavelength, and it is necessary to re-measure only the fraction during the final experiment. In the other preliminary experiment the angles of the wedges W_1 and W_2 are determined by measuring the separation of the fringes, using cadmium light. Both of these preliminary measurements may be made with the whole apparatus *in situ*, the other mirrors being removed.

9.41.—In the final experiment, each etalon is compared with its neighbour, by using a suitable arrangement of mirrors. For example, white light may be directed through A, B, and W_1, by inserting mirrors 1, 2 and 3. The Brewster fringes (§§ 5.32 and 5.33) are observed by means of M_1, and measurement of the position of the central fringe then gives the difference between A and 2B. Extension of this procedure gives the difference between A and 16E. Observations on the circular fringes with cadmium light give the number of wavelengths of cadmium light corresponding to the length of E with an error of about 0·01 of a wavelength. The relation between A and the wavelength is then known to an only slightly lower accuracy. The final operation consists in a comparison of the distance between the plates of the etalon A and the distance between two lines engraved on an Invar bar LO. This latter length has previously been compared with the standard metre by the usual method, using travelling microscopes. The two microscopes are each set on one of the marks of the standard, and their readings noted. Without disturbing anything else, the bar LO is substituted for the standard. The algebraic sum of the distances through which the microscopes must be moved to bring them into coincidence with the marks on LO, gives the difference between LO and the standard.

9.42.—A comparison between LO and A which involved setting the travelling microscopes first on two engraved lines on LO and then on the edges of the plates of the etalon, would not be accurate because of the difference in the marks. The travelling microscopes are therefore used to measure the difference between the marks on LO and two fine lines engraved on the edges of the end-plates of A. The edges are ground, polished and silvered in order to give fine marks. The etalon A thus has an " optical length " between its mirror surfaces, and a " mechanical length " between the engraved lines. The difference is determined by an auxiliary experiment in which the plates are mounted with separators which make the distance between the marks nearly one centimetre, and then with a second set of separators which make the distance nearly two centimetres. The optical lengths of these two etalons are measured with cadmium light. Suppose that they are found to be $N_1\lambda$ and $N_2\lambda$, and that the (unknown) sum of the distances from the engraved lines to the surfaces is $X\lambda$. Three marks A, B and C are ruled on a plate so that AB and BC are approximately equal to one centimetre. The small differences

$$(N_1 + X)\lambda - \mathrm{AB} = \alpha_1,$$

$$(N_1 + X)\lambda - \mathrm{BC} = \alpha_2,$$

and

$$(N_2 + X)\lambda - \mathrm{AC} = \alpha_3,$$

are measured with travelling microscopes.

Since AB + BC = AC, we have

$$X\lambda = (N_2 - 2N_1)\lambda + \alpha_1 + \alpha_2 - \alpha_3. \quad . \quad . \quad . \quad . \quad 9(14)$$

9.43.—Thus the final result gives the relation between the standard metre and the wavelength of the cadmium red line emitted by the source L_{Cd}. The error in the interferometric measurements is a few parts in 10^7 for a single measurement but, when the average of repeated measurements is taken, the probable error is reduced to one part in 10^7. The wavelength of cadmium light in *dry* air at standard temperature and pressure is found to be $6{\cdot}438\,469\,6 \times 10^{-7}$ times the standard metre. The difference between the value found by Benoît, Fabry and Pérot and that obtained by Michelson and Benoît is four parts in 10^7. The earlier determination referred to air containing an undetermined amount of water vapour. There was also a slight uncertainty in the temperature scale. These effects are sufficient to account for the whole of the difference.

9.44. Recent Work on Standards of Length.

There have been two recent re-determinations of the relationship between the wavelength of cadmium red light and the length of the standard metre. These were made by Watanabe and Imaizumi * in Japan, and by Sears and Barrell † at the National Physical Laboratory. The results of the four determinations give the following values ‡ for the wavelength (in dry air at 15° C. and 760 millimetres):

Michelson and Benoît (1895)	$6438{\cdot}4691 \times 10^{-10}$ metre
Benoît, Fabry and Pérot (1906)	$6438{\cdot}4703 \times 10^{-10}$ metre
Watanabe and Imaizumi (1928)	$6438{\cdot}4682 \times 10^{-10}$ metre
Sears and Barrell (1934)	$6438{\cdot}4708 \times 10^{-10}$ metre
Mean	$6438{\cdot}4696 \times 10^{-10}$ metre

The maximum divergence from the mean of these four independent measurements is only 2·2 parts in 10^7. This variation is within the limits of variation of the different copies of the original "metre".

9.45.—The Japanese determination was made with apparatus essentially the same as that used by Benoît, Fabry and Pérot. The N.P.L. experiment was based on similar principles, but many improvements in technique were introduced and, in the strictly optical part of the experiment, the accuracy was greatly improved. Sears and Barrell used only three stages of intercomparison. Their subsidiary etalons were one-ninth and one-third of the length of the etalon which was compared with the standard metre.§ The comparison between etalons whose length ratio is nearly 3 : 1 was made with the Brewster fringes, using the tilt method described in § 5.32, so that the calibrated wedges were not needed. The apparatus was generally more compact (fig. 9.16*a*), and more accurate temperature control (to within 0·001° C.) was possible. The etalon separators were made of Invar with chromium-plated ends. Quartz plates were wrung on to the ends, which were polished so as to be optically flat. The joint was airtight, and the etalons could be evacuated. Adjustments for parallelism, and also to a small extent for length, could be made by means of straining wires (of Invar) joining two flanges on the ends of the tube (fig. 9.16*b*). These etalons were extremely rigid.

9.46.—An end gauge whose section is a rectangular cross could be placed inside the longest etalon (fig. 9.16*c*). The difference between

* Reference 9.7. † Reference 9.8 and 9.9.

‡ The value obtained by Michelson has been corrected, using the most accurate estimate available for the humidity and CO_2 content of the air.

§ An etalon of length approximately one-twelfth metre was used in some experiments to give an independent check.

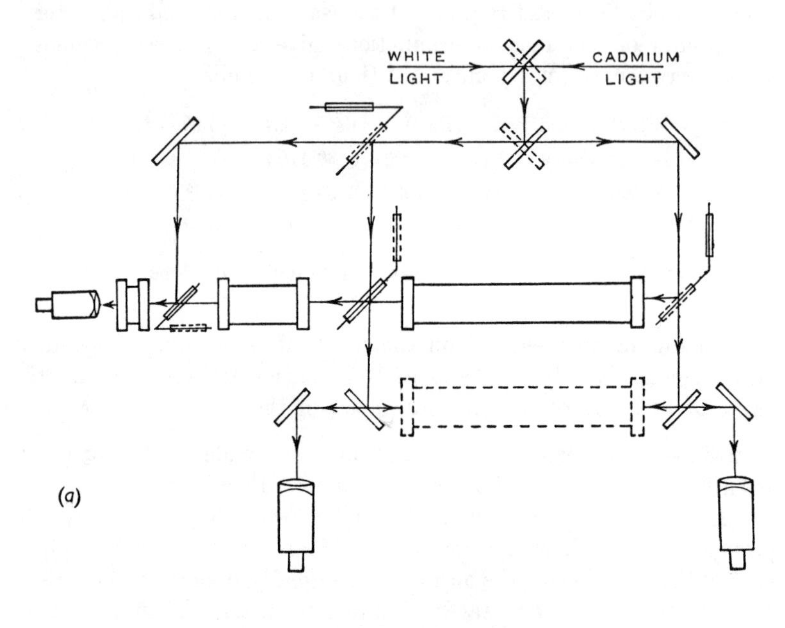

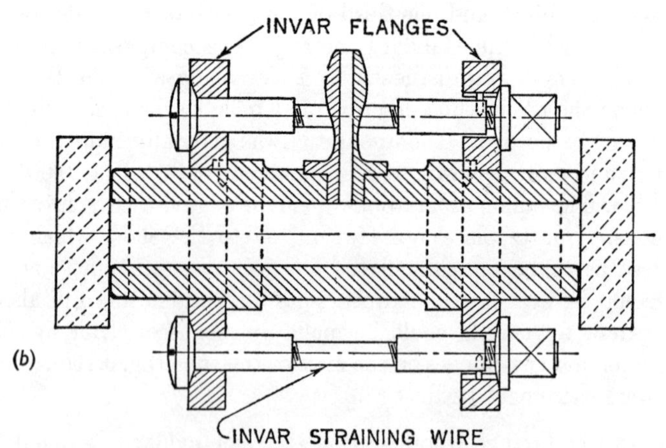

Fig. 9.16.—Apparatus used by Sears and Barrell: (*a*) general arrangement, (*b*) etalon

the length of this gauge and the distance between the plates of the etalon was determined by observing circular fringes with reflected light.

The difference of (*a*) the distance between the ends of the X-gauge, and (*b*) the distance between the lines on the standard metre, was determined by means of an intermediate composite gauge. This consisted of a bar with accurately plane, parallel ends, and of circular section. Its length was about half an inch less than that of the standard metre. Two parallel-faced blocks each half an inch in thickness, with sides respectively equal to the radius and diameter of the bar, were also used. At the middle of the half-inch face of each block, a fine line is ruled parallel to the longer edge. The two blocks are then wrung on to the two ends of

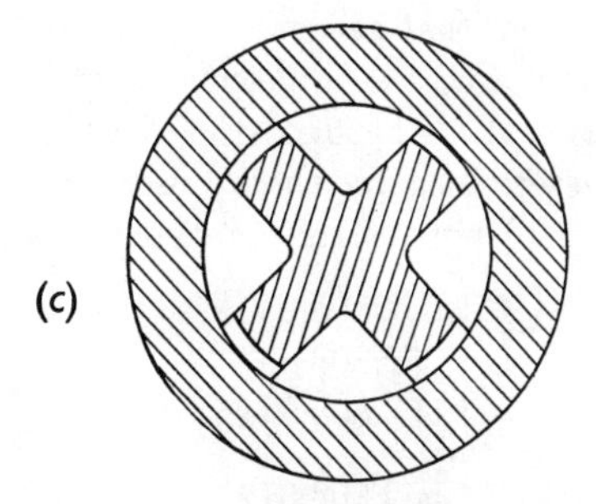

Fig. 9.16.—Apparatus used by Sears and Barrell: (*c*) X-gauge

Crown copyright reserved

the bar, so that the graduated faces are parallel to each other. A series of comparisons is then made, under microscopes, between the lengths of the composite line-gauge so formed, and the length of the standard metre, reversing the blocks between measurements so that their opposite faces are in turn in contact with the ends of the bar. The average of all these measurements gives the difference between the length of the metre and that of the end bar, plus half the sum of the thicknesses of the blocks. The end bar, with *one* block at a time wrung centrally on to its end, is then compared with the X-gauge, using an apparatus equivalent in principle to that described in § 9.18. The average of these measurements gives the difference between the length of the bar plus half the sum of the thickness of the blocks and the X-gauge. Subtracting these results gives the relation between the X-gauge and the metre. It will be noticed that the lengths of the auxiliary bar and blocks need not be known with precision since they are eliminated in the final result.

9.47.—In addition to determining the relationship between the wavelength in air and the metre, Sears and Barrell also measured the relationship between the wavelength *in vacuo* and the metre. They also made direct determinations of the relationship between the wavelengths and the standard yard. Comparison of their results gives a very accurate determination of the refractive index of air for cadmium red light. Examination of the results shows that the main uncertainty in the relation between the wavelength and the metre is in the comparisons made with the travelling microscopes. In order to define the standard metre to within one part in 10^7, it is necessary to locate the positions of the defining marks to within $0{\cdot}25 \times 10^{-4}$ millimetre or $\pm\frac{1}{2}\lambda$. The marks themselves are much wider than this. When viewed under high magnification they appear to be irregular blurred scratches. The observer has to set the crosswire on what he estimates to be the centre of the blur. There is obviously room for personal judgment, and the mean position accepted as central by one group of observers may differ appreciably from that accepted by another group—possibly of a different generation. In the experiments of Sears and Barrell the relationship between the length of the X-gauge and the wavelength *in vacuo* was determined by three observers. The values obtained (for the ratio of the length of the gauge to the wavelength) were:

	1,552,808·930
	1,552,808·897
	1,552,808·944
Mean	1,552,808·917

The standard deviation of a single determination by any one observer was only 4 parts in 10^8, and the standard deviation of the set was less than one part in 10^8. An accuracy of a similar order was obtained in the intercomparisons between etalons.*

9.48.—In view of these results it is suggested that the standard of length should be redefined in terms of a wavelength (*in vacuo*) emitted by a lamp of standard construction. If the cadmium red line † were used as standard, a lamp of a fairly simple type could be specified. Lamps produced according to the specification (within a fairly wide tolerance) would give the same wavelength to within a few parts in 10^8. Probably further study of the conditions of excitation would

* It is impossible, in the space available, to give the reader more than a general idea of the ingenuity and care used in obtaining this result. The original papers are worthy of study, both for their interesting material and also for the method of presentation. (Reference 9.8.)

† The line chosen should have no hyperfine structure (see § 9.50) and should have a small Doppler width (see § 4.25). The latter condition suggests the use of the heavy element mercury [see eqn. 4(23)] but ordinary mercury consists of several isotopes and the lines have a complicated hyperfine structure. The isotope mercury 198, prepared from gold by nuclear transmutation, is now available. It gives very narrow lines, one of which will probably be used to define an optical standard of length.

enable a standard lamp to be specified in such a way that the vacuum wavelength emitted by different specimens would be the same to within one part in 10^8, or even better. If this suggestion were adopted the " metre " would be defined to be 1,552,734·52 (or some similar number) times the wavelength of light, instead of being the distance between two marks engraved on a bar.

9.49.—Even if an optical wavelength were chosen as the ultimate standard of length it would still be necessary to have auxiliary mechanical standards. It is clearly desirable that the mechanical standards should be end gauges, not line standards.

It is of historical interest that the " Exchequer yard " of Queen Elizabeth and the original " Mètre des Archives " were end standards. When the present " Imperial standard yard " and " international prototype Metre " were introduced, line standards could be more accurately constructed than end standards, and two line standards could be more accurately compared than two end standards. Modern technique has improved the construction of the end standards, so that they can be constructed with very high accuracy, and the comparison by interferometric means now far exceeds the accuracy of comparison of line standards. The end standard is suitable for use in workshop practice and it is also suitable for direct comparison with the wavelength of light. All these reasons combine in favour of the use of end standards.

9.50. **Investigation of Hyperfine Structure.**

The use of interferometers has shown that most of the lines which appear to be single when examined in a prism or grating spectrograph, can be resolved into a number of components with separations of a few hundredths or a few thousandths of an Ångström unit. This *hyperfine structure*, as it is called, is of great interest to the theoretical physicist, who can derive important information about atomic nuclei if he is given accurate determinations of the number, intensities and separations of the components. The practical problem involved has two important aspects:

(*a*) the production of interferometers with sufficient resolving power to separate the components, and

(*b*) the production of auxiliary apparatus to effect a sufficient separation to prevent confusion owing to overlapping of orders.

There is no serious difficulty in making interferometers of the necessary resolving power. By increasing the separation of the mirrors, the Michelson or the Fabry-Pérot interferometer can be given any desired resolving power. If, however, the resolving power of an interferometer is increased merely by increasing the path difference, there is also a proportionate increase in the overlapping of orders. Thus the two problems are interconnected. If $\Delta\lambda_A$ is the largest range of wavelength which an interferometer can accept without overlapping of orders, and $\Delta\lambda_R$ is the difference of wavelength between components which are just resolved, we desire $\Delta\lambda_A/\lambda\Delta_R$ to be as large as possible.

9.51.—An echelon diffraction grating of N steps gives N beams of equal amplitude, with phases arranged in arithmetic progression. The vector diagram for calculating the resultant is given in fig. 3.2*a*. Suppose that for normal incidence and reflection the echelon gives a maximum for some wavelength λ. For this wavelength the vector diagram will give a straight line. Let us consider what would happen if the wavelength could be slowly increased. Each of the vector elements would rotate relative to its neighbour, forming a vector polygon. For the wavelength which is just resolved from λ, the vector diagram would be a completed regular polygon. Each element would have rotated through $2\pi/N$ relative to its neighbour. If the wavelength were further increased, a stage would be reached when each element had rotated through 2π relative to its neighbour. The vector diagram would be a straight line again giving a principal maximum of order one less than the order for λ. If the differences of wavelength involved were small, the ratios of the two differences of wavelength would be equal to those of the angles through which the vectors had rotated, i.e.

$$\frac{\Delta\lambda_A}{\Delta\lambda_R} = N. \qquad \qquad 9(14)$$

9.52.—In § 8.10 we saw that the resolving power of instruments of this type is proportional to the order of interference (m) multiplied by the number of beams (N). We now see that the ratio of separating power to the resolving power is determined solely by the number of beams. Similar considerations apply to a Fabry-Pérot etalon and to a Lummer-Gehrcke plate, though the problem is slightly more complicated since the amplitudes of the beams are not equal. We saw in § 8.18 that an " equivalent number " of beams of equal intensity may be calculated for the Fabry-Pérot etalon and for the Lummer plate. These " equivalent numbers " are of the order 30 to 50. The value of the ratio for these instruments is thus about the same as the value for an echelon of 40 plates. The Michelson interferometer has only two beams and is thus, from this point of view, greatly inferior. A grating, of course, gives a ratio much higher than that given by any interferometer, but gratings cannot be applied to the type of problem now under consideration, because the resolving power is not quite high enough.

Houston * has used two Fabry-Pérot etalons in series in order to obtain high resolving power together with a good separation of orders. One of the etalons had a separation of 3 millimetres and the other a separation of 9 millimetres. Every third maximum formed by the 9-millimetre etalon coincides with a maximum for the 3-millimetre etalon. Only these maxima show up strongly in the pattern produced by the compound interferometer. Houston shows that the resolving power is slightly higher than that of the 9-millimetre etalon, and the separating power is equal to that of the 3-millimetre etalon. There are two disadvantages: (*a*) a large loss of light owing to transmission through four silver films, and (*b*) the " missing " maxima are not entirely suppressed, but only made very weak. They may mask the presence of weak satellites.

9.53. *Auxiliary Apparatus.*

Sometimes it is possible to effect the separation of lines by the use of colour filters. The strongest lines of many elements can be separated in this way. Usually it is necessary to use dispersion by means of a prism or grating. When this is done, the light may be passed through the other instrument before or after the interferometer. An arrangement such as that suggested in § 9.36 (in which the interference pattern is focused on the slit of the dispersive instrument) is often used. It is also possible to place the interferometer inside the spectrograph, so that it is in the approximately parallel beam produced by the collimator. This arrangement, which has special advantages in relation to certain problems, has been used both with the Fabry-Pérot etalon and with the reflecting echelon.

9.54.—The choice of interferometer for a particular problem is determined partly by considerations of convenience of use, and of cost. The latter is a measure of the difficulty of accurate construction. In 1939, the prices of a Fabry-Pérot interferometer and of a reflecting echelon of 35 plates were each about £1000. The cost of an etalon of fixed distance was, however, only about £30, and of a Lummer-Gehrcke plate from £50 to £250 according to size.

The general tendency was to use Fabry-Pérot etalons when possible. A set of three etalons of suitable sizes (say 5, 20, and 50 millimetres) was not unduly expensive.† Such an outfit is sufficient for a wide range of problems. At one time, the Lummer-Gehrcke plate seemed to compete with the Fabry-Pérot etalon, but it is less convenient, requiring careful mounting and very good temperature control. For weak sources of light, the reflecting echelon has some advantages; but the process of making a large number of plates of identical thickness is very laborious and costly.

* Reference 9.10.

† Three separate etalons cost less than £100, but cost could be reduced to about £50 by using one pair of plates with different separators.

9.55.—It is of interest to consider what is the highest resolving power ever likely to be required. The widths of all components of a spectrum line are affected by (*a*) natural damping, (*b*) Doppler effect, and (*c*) pressure broadening. Under ordinary conditions (*b*) and (*c*) produce much larger effects than (*a*). It appears possible that by special attention to design of sources these effects may be made of the same order as (*a*). The Doppler effect may be reduced by using a discharge through a gas cooled by liquid air, or by using an atomic beam. Collisions in an atomic beam are also infrequent, so that the main cause of broadening should be natural damping. This usually implies a width of at least 0·0005 Å., indicating a maximum useful resolving power of about 10^7. A Fabry-Pérot etalon of separation 20 centimetres would give approximately this resolving power. No existing source gives lines sharp enough to require a higher resolving power.

REFERENCES

9.1. Williams: *Applications of Interferometry* (Methuen).

9.2. Michelson: *Studies in Optics* (University of Chicago Press).

9.3. Childs: *Journal of Scientific Instruments*, 1926, Vol. 3, p. 97 and p. 219.

9.4. Meissner: *Journal of the Optical Society of America*, 1941, Vol. 31, p. 405.

9.5. Michelson and Benoît: *Trav. et Mem. Bur. Int. des Poids et Mesures*, 1895, Vol. 11.

9.6. Benoît, Fabry and Pérot: *ibid*, 1913, Vol. 15.

9.7. Watanabe and Imaizumi: *Proc. Imp. Acad.* (*Tokio*), 1928, Vol. 4, pp. 3, 51.

9.8. Sears and Barrell: *Phil. Trans. Roy. Soc.*, 1932, Vol. 231, p. 75.

9.9. *Ibid.*, 1934, Vol. 233, p. 143.

9.10. Houston, W. V.: *Phys. Rev.*, 1927, Vol. 29, p. 478.

9.11. Tolansky: *Proc. Roy. Soc.*, 1945, Vol. 184, p. 41.

CHAPTER X

The Velocity of Light

10.1. Historical.

In 1676, the astronomer Römer showed that the velocity of light is finite, although it is very large compared with the velocities ordinarily occurring in laboratory experiments. After a considerable interval, experiments were carried out by Fizeau, Foucault and others during the first half of the nineteenth century. During this period it was recognized that the velocity was an interesting and important physical datum, but there was no way of relating the measured values to any physical theory. The electromagnetic theory of Maxwell, which required that the velocity of light in vacuum should be exactly equal to the ratio of certain electrical units, gave a great impetus to experimental determinations of both quantities. It was obviously important to see whether the experimental values agreed in a satisfactory way. This theory was published in 1873, and in 1879 Michelson made his first determination of the velocity of light. Measurement of the velocity was naturally correlated with attempts to find a variation under various physical conditions. The results of these experiments led to the relativity theory (1905). We shall see that, in relativity theory, the velocity of light in vacuum is regarded as one of the most important of all physical constants. Its value appears in the relation between mass and energy.

10.2. General Review of Methods.

The direct measurement of a velocity involves the determination of a distance and a transit time. Since the velocity of light is very high, it is necessary to use a large distance and, in most methods, to measure a very short time. It is not possible to observe the progress of a continuous beam of light without marking or " modulating " it in some way. Three main methods of modulation have been used: (*a*) the toothed-wheel method, (*b*) the rotating-mirror method, and (*c*) the electronic shutter. In any of these methods the transit time is derived from a measurement of the frequency of the modulator. The measurement of distance is made by standard surveying or metrological processes. The early experiments were not sufficiently accurate to detect the difference between the velocity in air and the velocity in vacuum, but, as experiments became more accurate, it became necessary to know the temperature and pressure of the air through which the beam passes. For this reason most modern experiments have been carried out in vacuum, or under conditions in which the refractivity of the air could be accurately determined.

10.3. Indirect Methods.

If certain theoretical ideas are accepted, experiments on light received from a source which is moving with respect to the observer, yield a series of values for the velocity of light. In those experiments, the velocity of light is compared with the relative velocity of source and observer. This latter velocity is then measured directly. The experiments on the Doppler-Fizeau effect, and on the aberration of light, thus form indirect determinations of the velocity. In a similar way, an experiment on the transmission of light through a medium which moves with respect to source and observer, effectively compares the velocity of light with the velocity of the medium. Such experiments are usually regarded as establishing certain points of theoretical interest rather than as determinations of the velocity of light (Chapter XI). Considered as determinations of the velocity, they are less accurate than the direct determinations.

10.4.—Once it has been established that the velocity of light, the velocity of electromagnetic waves, and the ratio of the electrical units are equal, it is legitimate to gather all the experimental evidence together, and to use it to obtain the most accurate value of this quantity. It will be shown later that the accuracy of the determination of the velocity of electromagnetic waves on wires, and of the ratio of the units, is about the same as the accuracy of direct measurements of the velocity of light; and the results are in good agreement. In the following account of the direct determinations, we shall give only an outline of four of the most interesting methods. The reader must refer to the original papers for details.*

10.5. Römer's Method.

The planet Jupiter has several satellites whose orbits are so nearly in the plane of the planet's orbit that they pass through the shadow of the planet and suffer eclipse at every revolution. The satellites revolve much more rapidly than our moon, and three of them have periods of $1\frac{3}{4}$, $3\frac{1}{2}$, and 7 days respectively. The first one passes through

* See References on p 312.

a distance equal to its diameter in $3\frac{1}{2}$ minutes. It is possible to observe the moment when an eclipse begins within a small fraction of this time. Römer made careful observations of the eclipses over an extended period of time and found irregularities ranging up to about 10 minutes. These irregularities were correlated with the variation of the distance between the earth and Jupiter, and might be due to the varying time taken for light to travel to the earth. Römer obtained a value of about $3{\cdot}5 \times 10^{10}$ centimetres per second for the velocity of light. Modern observations by the same method give a value of about $2{\cdot}98 \times 10^{10}$ centimetres per second, i.e. about 0·5 per cent less than the more accurate value obtained from terrestrial measurements. The difference of path is about $1{\cdot}5 \times 10^{13}$ centimetres and the time is about 500 seconds.

10.6. **Fizeau's Method.**

Fizeau's apparatus is shown in fig. 10.1. L_{VS} is a vertical slit source of light, and M_1 is an unsilvered piece of plane glass. (Half-silvered mirrors were not available at the time.) The light is brought to a focus at F, where it is interrupted by the toothed wheel W, which can be rotated using a clockwork mechanism actuated by a falling weight.

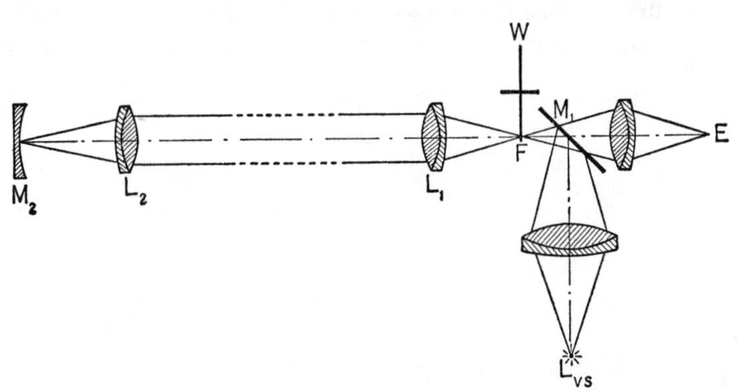

Fig. 10.1.—Fizeau's method for the velocity of light

The point F is in the focal plane of the lens L_1, from which the light goes in an approximately parallel beam to L_2, and to the concave mirror M_2. If the wheel is stationary, some of the returning light passes through M_1, and is seen by the observer through the eyepiece at E. If the speed of the wheel is gradually increased from zero, a stage will be reached when the light which passes through a given space will, on its return, be interrupted by a tooth which has moved into the appropriate position during the time of transit. At a higher speed the returning light will pass through the next space and, at a still higher speed, it will be eclipsed by the next tooth. Fizeau (1849) used a wheel with 720 teeth, and found that the first eclipse occurred when the speed was 12·6 revolutions per second. This gives a transit time of $5{\cdot}5 \times 10^{-5}$ second. The length of the double path was $1{\cdot}7266 \times 10^{6}$ centimetres, and the value obtained for the velocity is $3{\cdot}15 \times 10^{10}$ centimetres per second. The chief error in Fizeau's method is in the difficulty of observing exactly when the image is eclipsed, i.e. it is an error in the measurement of the time. Essentially the same method was used by Cornu (1874), who increased the path and was able to observe eclipses up to the thirtieth order. It was also used by Young and Forbes (1881), and by Perrotin and Prim (1903). These workers made a considerable number of technical improvements and were able to show that the velocity lies between 2·99 and $3{\cdot}01 \times 10^{10}$ centimetres per second. The method is, however, much less accurate than either of the methods which will now be described.

10.7. Rotating-mirror Method.

This method was first suggested by Wheatstone in 1834 and was first used by Foucault in 1860. The apparatus is shown in fig. 10.2.

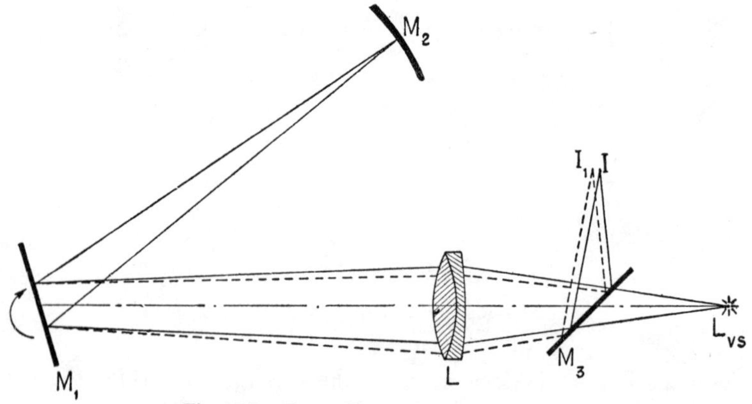

Fig. 10.2.—Foucault's rotating-mirror method

Light from the slit source L_{vs} is collimated by the lens L; it is then reflected by the mirrors M_1 and M_2, and by the unsilvered mirror M_3. The mirror M_1 can be rotated at a high speed. When it is stationary, and set at the correct angle, an image formed at I is viewed by an observer through an eyepiece. When M_1 is rotating rapidly, light falls on M_2 during only a small portion of the revolution. The light beam thus becomes intermittent. During the transit from M_1 to M_2 and back, the mirror M_1 rotates through a small angle and consequently the image is displaced slightly from I to I_1. Measurement of the distance from I to I_1 gives the angle through which M_1 turns during the transit and, if the speed of rotation is known, the transit time can be calculated.

The speed is measured by a stroboscopic method. A toothed wheel (with very small teeth) is viewed through a microscope. The wheel is intermittently illuminated (since the return beam is intermittent). If one tooth exactly replaces another between the flashes, the wheel appears to be stationary. If the wheel has n teeth, its rate of rotation is then $1/n$th that of the mirror. If the angular speed of the wheel is nearly, but not exactly, $1/n$th that of the mirror, then the wheel appears to rotate slowly and we have

$$N/n = (M - m),$$

where N is the number of rotations per second for the mirror, M is the real and m is the apparent number of rotations per second for the wheel. It is easier to measure M and m, in order to derive N, than to measure the high angular speed directly. In his final experiments, Foucault used an arrangement with five mirrors (to reflect the beam to and fro) instead of M_2. In this way he obtained a total path of 20 metres. His value of the velocity of light was $2{\cdot}98 \times 10^{10}$ centimetres per second. The distance between I and I_1 was only 0·7 millimetre and the most important source of error was in the measurement of this distance.

10.8.—The rotating-mirror method was used by Michelson (1879, 1882, 1927 and 1935) and by Newcomb (1882). Very important technical improvements were made from time to time. These included:

(*a*) rearrangement of the optical system to give a brighter image and so to permit a longer path;

(*b*) use of a compensation method with a mirror of 4, 8 or 32 sides;

(*c*) arrangements for ascertaining the refractive index of the medium so as to enable the velocity of light *in vacuo* to be calculated.

We shall give a brief description of the last two arrangements designed by Michelson, though it should be stated that many of the improvements which he incorporated were first suggested by other workers. Foucault and Newcomb both contributed to improvements of type (*a*), and Newcomb was the first to use a many-sided mirror.

10.9. *Mount Wilson Determination* (1927).

The apparatus used is shown in fig. 10.3. Light from the slit source L_{VS} is reflected from the octagonal mirror M_1 to the plane mirror M_2, and thence to the mirror M_3, which is slightly separated from M_3'. (M_3 is slightly above and M_3' is slightly below the plane of the diagram.) From M_3 the light passes to concave mirror M_4, to the concave mirror M_5 and to the plane mirror M_6. It returns, via M_5, M_4, M_3' and M_2', to the octagon, whence it goes to the eyepiece E. The right-hand portion of the apparatus was at the Mount Wilson Observatory, and the left-hand portion was about 22 miles away on Mount San Antonio. The

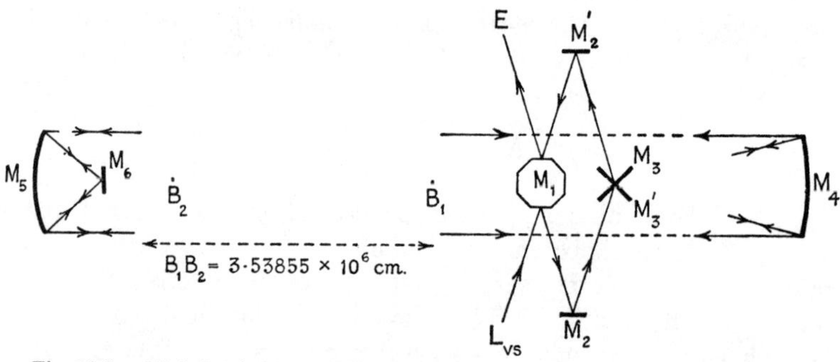

Fig. 10.3.—Michelson's Mount Wilson (1927) determination of the velocity of light

octagonal mirror makes approximately 528 turns per second, and rotates through almost exactly one-eighth of a turn during the transit of the light to and fro. If the rotation were exactly one-eighth of a turn, the image would be undisplaced by the rotation. Actually a small deviation has to be measured. The speed of the mirror was controlled and measured stroboscopically. The stroboscope was an electrically driven fork carrying a mirror. The frequency of the fork was about 132 vibrations per second, and was accurately measured by reference to a standard pendulum supplied by the U.S. Coast and Geodetic Survey. The distance between the two stations was measured by the U.S. Coast and Geodetic Survey.

Very great care was taken, and it is thought that this is the most accurate measurement, of this type, which has ever been made. The probable error of the straight-line distance (estimated from concordance of results) was only one part in 6,000,000 or about a fifth of an inch in 22 miles. The final accuracy of the distance is not so high, owing to uncertainty in the steel tape used to measure the base line, and also to the errors involved in measuring the distances from the bench marks B_1 and B_2 (fig. 10.3) to the mirrors. Michelson considered the final value for the length to be correct to within one part in a million. Dorsey* is of the opinion that Michelson misinterpreted the Bureau of Standards' certificate in regard to the length of the tape, and that the overall error in the measurement of the light path is about 1 in 200,000. Even this larger error is smaller than the errors due to (*a*) uncertainty in the measurement of the frequency of the fork (i.e. to error in the transit time), and (*b*) uncertainty in the temperature and pressure of the atmosphere over the light path (i.e. errors in the refractive index of the air, and hence in the correction required to reduce the measured velocity to a velocity in vacuum). The final result (given by Dorsey and including certain corrections not made by Michelson) is $2{\cdot}9980 \pm 0{\cdot}0002 \times 10^{10}$ centimetres per second.

10.10. *The Evacuated-tube Experiment.*

In order to avoid the error due to uncertainty in the refractive index of air, Michelson planned an experiment in which the light path was within an evacuated tube. He was unable to take part in the observations and the experiment was completed after his death by his colleagues, Pearson and Pease. The tube was a mile long, and by reflecting the light to and fro twice between two plane mirrors, a total light path of nearly 13 kilometres was obtained. A mirror with 32 sides replaced the octagonal mirror previously used, and the reflecting system was not quite the same as that shown in fig. 10.3. A number of detailed improvements in technique were incorporated, but the methods used for the measurement of distance and time were essentially the same as those used in 1927. The pressure in the tube varied from 0·5 to 5·5 millimetres in different experiments, and a correction was made for the refractive index of the residual air. The final result for the velocity in vacuum is $2{\cdot}99774 \pm 0{\cdot}00011 \times 10^{10}$ centimetres per second.

10.11. **The Kerr Cell Optical-shutter Method.**

In this method a " mark " is placed on the beam of light by passing it through a Kerr cell to which a short-wave radio-frequency voltage is applied.† In the simplest form of the apparatus the " mark " is a sinusoidal variation of the amplitude of the light transmitted. The radiation from a broadcasting station consists of a radio-frequency carrier wave modulated by frequencies in the audible range. In the light passing the Kerr cell, the radio frequency is the modulation frequency, and the frequency of the light is the carrier frequency. The

* Reference 10.9. † The theory of the Kerr cell is discussed in Chapter XVI.

method was first used by Karolus and Mittelstaedt, and has since been used by Hüttel and by Anderson (in two separate determinations). The experiments differ in technical details and, although some of these details are important, we shall describe only the second determination made by Anderson. The apparatus is shown in fig. 10.4.

Light from a mercury arc L_A passes through the modulator cell (KC), and one part is reflected by the half-silvered mirror M_1 and the plane mirror M_2, to the photocell R. The other part is transmitted by M_1 and goes to M_3, M_4, M_5 and M_6. It returns along the same path to M_1 and is reflected to the photocell. The photo-electric current pro-

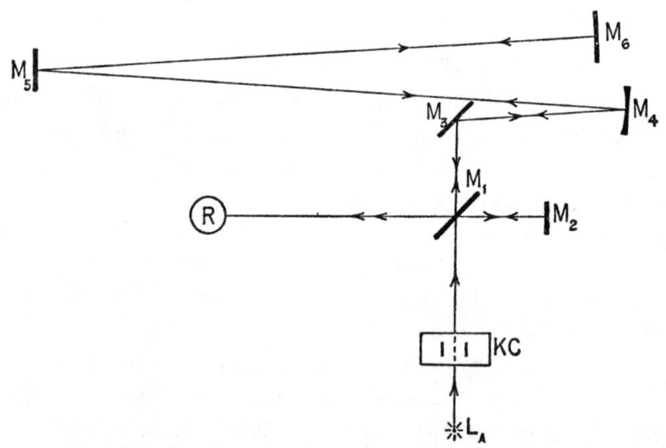

Fig. 10.4.—Electro-optical shutter method used by Anderson

duced by the joint action of the two beams is amplified by an amplifier tuned to the frequency of the oscillator. The resultant voltage is dependent on the phase relations of the light modulations. If the two beams arrive at the cell with their " modulation waves " in phase, the resultant current contains a large component whose frequency is the same as that to which the amplifier is tuned. If they arrive with the modulation waves out of phase, the total photo-electric current is the same, but the component which is amplified is small and the output current is then a minimum. The mirror M_4 could be replaced by another mirror M_4' (of different focal length) which was tilted so as to return the light direct to M_3. The following procedure was adopted.

With the mirror M_4' in place, the position of M_2 was adjusted (by a micrometer screw which moved it along accurately constructed guides) until the amplified current was a minimum. The mirror M_4

was then put in the place previously occupied by M_4', and the position of M_2 adjusted until the current was again a minimum. The time taken for the light to traverse the main path from M_4 to M_5 and M_6 and back, plus the distance between the two positions of M_2, is then equal to an integral multiple of the period of the radio-frequency oscillator. It is unnecessary to measure the distances M_1M_2, M_1M_3, M_3M_4, etc.* The length of the main path was measured by reference to the distance between two fine marks on a steel tape. The distance between the positions of M_2 is given by the difference of the readings of the micrometer screw. The overall accuracy of the measurement of path is about 1 part in 200,000. The frequency of the modulator was $19{\cdot}2 \times 10^6$ cycles per second. It was accurately controlled, and measured by the methods usually employed for radio frequencies. It was known to 1 part in a million or better. The value obtained for the velocity of light was $2{\cdot}99776 \pm {\cdot}00014 \times 10^{10}$ centimetres per second.

10.12. Discussion of Results.

It is desirable to consider whether the differences between the methods are such as to make it very unlikely that the same systematic errors enter into all the above experiments. The chief errors that have been proposed are:

(*a*) a delay in reflection, and

(*b*) an error due to abnormal reflection at the surface of a moving mirror, or abnormal reflection of a beam which is sweeping across a reflector with high velocity.

An appreciable delay on reflection should have been detected because some of the arrangements use more reflections than others. An error of type (*b*) should be revealed by a difference between the results obtained by the rotating-mirror method and the Kerr cell method. The results are, however, in good agreement. There is no theoretical reason for suspecting a delay on reflection. The theory of reflection of a rapidly moving beam has been investigated by Lorentz, who concludes that there *may* be an anomaly which is *probably* too small to be detected. It therefore appears reasonable to assume that the weighted mean of the recent accurate determinations should be regarded as the most accurate value, and that the error should be estimated from the deviations of the results. On this basis, Dorsey, after reviewing all the experimental evidence, concludes that the value

* A small correction is necessary because M_4 does not *exactly* replace M_4'.

is $2{\cdot}99773 \pm {\cdot}00001 \times 10^{10}$ centimetres per second.* It is of interest to note that the four determinations with the Kerr cell, taken alone, give this mean, and that the same value is obtained by the rotating-mirror evacuated-tube experiment. A slightly higher value is given by the Mount Wilson experiment, but this experiment is probably less accurate than those which were made later.

10.13. **Group Velocity or Wave Velocity.**

It is generally agreed that methods like that of Fizeau, in which the light beam is chopped up into sections, lead to determinations of the group velocity. It is less obvious that the rotating-mirror method measures the group velocity, but this may be shown to be so by considering the Doppler effect on the two parts of the mirror, one of which approaches, and the other of which recedes from the observer. It is, however, scarcely necessary to make this investigation since it is obvious that the rotating mirror divides the beam into trains of finite length. In a similar way the sinusoidal modulation used by Anderson in the experiments which have been described creates wave groups. Dorsey considers that the first method used by Anderson (in which the amplitude of the beam of light was constant, and the marking was done by altering the ellipticity of the polarization in a periodic way) measures the phase velocity directly. In our view, Dorsey is mistaken, because the components of the light polarized in any given plane are modulated just as much in Anderson's first method as in his second, and therefore wave groups are involved. Even if no modulator were used, it is doubtful whether any measured velocity could be other than the group velocity since the trains of waves emitted by the atoms must be represented by wave groups. The values of the velocity of light quoted above give the velocity in vacuum. The observations have been corrected for the difference between the group velocity in air and the velocity in vacuum. The difference between group velocity and phase velocity in air at standard temperature and pressure is about $0{\cdot}00006 \times 10^{10}$ centimetres per second, or 1 part in 50,000.

10.14. *Comparison of the Velocity with the Ratio of the Electrical Units, etc.*

According to the electromagnetic theory, the observed velocity of light should agree with (*a*) the observed values of the velocity of electromagnetic waves on wires, and (*b*) with the ratio of the electromagnetic units. The following are probably the most accurate values available:

(*a*) $2{\cdot}99782 \pm 0{\cdot}0003 \times 10^{10}$ cm./sec. (Mercier.)
(*b*) $2{\cdot}99784 \pm 0{\cdot}0001 \times 10^{10}$ cm./sec. (Rosa and Dorsey.)

* Birge (see Reference 10.10) assigns different weights to the experimental determinations and obtains $2{\cdot}99776 \pm {\cdot}00004 \times 10^{10}$ cm./sec.

These values * are in reasonably good agreement with the value of $2{\cdot}99773 \times 10^{10}$ centimetres per second obtained for the velocity of light.

10.15. **Recent Work.**

We now give the results of some recent measurements of the velocity of light and of radio-frequency electromagnetic waves. A review article by Essen (Reference 10.12) gives a general account of the work and includes references to the original papers. The radar apparatus used during the war included elements which, in their simplest use, estimated the distances of targets by measuring the time required for a signal to travel from the apparatus to the target and back again. The accuracy of these devices was checked by using targets whose distances could be measured by ordinary methods of surveying. At first it was assumed that the velocity of radio waves in vacuum was equal to the accepted value of c (see Reference 10.12). The velocity in air was calculated using moderately accurate values for the refractive index of air. An error due to radiation reflected or scattered from the ground was estimated and eliminated. After making these corrections it was found that the distances obtained from the radar measurements were too small by about one part in 20,000. The discrepancies could be reduced by assuming a higher value for c. After the war the radar technique was used to make systematic measurements of the velocity of radio waves. The value obtained is $299{,}792 \pm 3$ km./sec.

Essen and Gordon-Smith made a very precise measurement of the resonance frequency for a cavity resonator of accurately known dimensions. The resonator was evacuated to avoid any error due to uncertainty in the refractive index of air at radio-frequency. The phase velocity in the resonator is not equal to the velocity in free space, but the latter velocity can be calculated from their results. The value obtained is the same as that given by the radar measurements.

10.16.—Two measurements have recently been made on visible light. Bergstrand, using a modulated beam of light and a technique somewhat similar to that used in the radar measurements, obtained a result in agreement with that given by the radar method. On the other hand Houston, using a new type of electro-optical shutter, obtained the lower value previously given by the methods described in §§ 10.7–10.12. Thus all the measurements on visible light except

* The values quoted include certain small corrections suggested by Birge (Reference 10.10).

that of Bergstrand give 299,773 ± 3 km./sec. All the measurements on radio-frequencies give 299,792·5 ± 3 km./sec. The indirect measurements considered in § 10.14 give a result about half-way between these values. It thus appears that there is a discrepancy which is somewhat larger than the estimated experimental errors.* All results except those of Bergstrand suggest a difference between the velocity of light in vacuum and the velocity of radio waves in vacuum. This difference, if confirmed, would require important modifications of theory, but obviously further experimental work is needed before we can be sure that there is a real difference.

10.17. *Suggested Variation of the Velocity of Light with Time.*

It has been suggested by several writers that the velocity of light is decreasing, or possibly varying in a periodic way. It is true that the "best values" announced by the experimenters do show a general tendency to decrease from the time of the first measurements. Detailed examination shows, however, that the differences are all within the range of the experimental error obtaining at any given time. There are no systematic differences between the last six determinations. There is also good indirect evidence against the existence of any measurable variation with time. (Reference 10.10.)

The observed value of the velocity is in terms of the standard metre and the mean solar day. Let us assume that the fundamental equation $c = \nu\lambda$ is valid, where ν is the frequency of the atomic oscillator which emits light of wavelength λ. We may also assume that a vacuum is a non-dispersive medium, so that the wave velocity and group velocity in vacuum are the same. Then, if c varies, either ν or λ must vary. It is extremely improbable that ν (whose value may be calculated from constants such as e, m, h) varies, though we cannot show by direct measurement that it does not vary. Comparison of the different determinations of the relations between the wavelength of cadmium light and the metre, indicates that the ratio has not varied by more than 2 or 3 parts in 10^7 during the last fifty years (§ 9.44). It is thus very unlikely that c is varying by an amount which could be detected by direct measurement. Other indirect evidence for the constancy of the velocity is given by experiments on the ratio of the electrical units.

10.18. **Variation of Velocity with Refractive Index.**

The velocity of light in water was compared with that in air by Foucault, using the apparatus shown in fig. 10.5. The mirror M_1 revolves, and two deflected images are obtained. The ratio of the deflections is approximately inversely proportional to the ratio of the

* There is considerable difficulty in estimating the effective values of the experimental errors. The values given are upper limits for the random errors with some allowance for known sources of systematic errors.

velocities. Foucault was able to show that light travels faster in air than in water, but he gave no quantitative value for the ratio. The experiment was repeated by Michelson, who obtained the value 1·33 for the ratio of the velocity of light in air to the velocity in water. He also found the value 1·758 for the ratio of the velocity of light in air to the velocity in carbon disulphide. The ratio of the group velocities can be calculated when the refractive indices and the dispersions of the media are known. For water the calculated ratio of the group

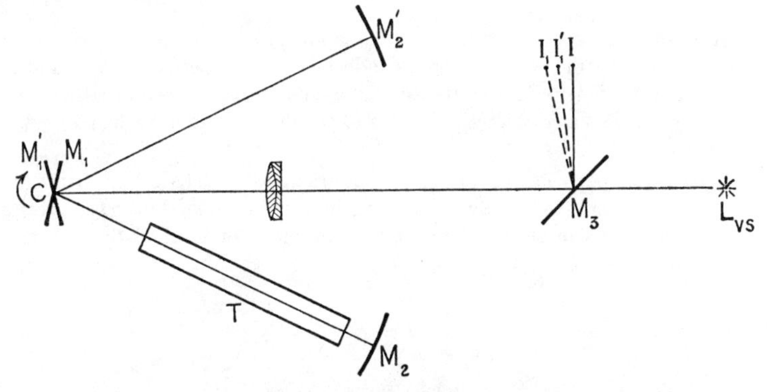

Fig. 10.5.—Comparison of velocity in air with velocity in a liquid

velocities is nearly the same as the ratio of the phase velocities (i.e. it is the inverse ratio of the indices of refraction). The observed value agrees, within experimental error, with either calculated value. For air and carbon disulphide, the calculated ratio of the phase velocities is 1·64, and of the group velocities 1·745 for the wavelength 5800 Å. The wavelength used by Michelson was not very sharply defined but was probably higher than 5500 Å. Thus his observed value (1·758) is in reasonably good agreement with the calculated ratio of the group velocities, but is definitely not in agreement with the calculated ratio of the wave velocities. Michelson was able to show that the velocity of red light in carbon disulphide is a little greater than the velocity of blue-green light, as would be expected from the difference of refractive index.

EXAMPLES [10(i)–10(iii)]

10(i). An electron is accelerated from rest in a uniform field of 100 volts per centimetre. Calculate the transit time for a path of 2 centimetres.

[$4 \cdot 75 \times 10^{-9}$ second.]

10(ii). What is the ratio of the transit time for a light path of 30 metres to the electron transit time under the conditions of the preceding example?

[20 : 1.]

10(iii). In Anderson's experiment there is an error in timing due to variation in electron transit time. Suppose that conditions are as described above, and that the variation in transit time is due to a difference in the electron path of p per cent (where p is less than 1). What value of p will cause an error of 1 in 100,000 in the velocity of light?

[$p = 0 \cdot 02$ per cent. *Note*: The conditions in Anderson's experiment are not as simple as those assumed in the example, but he states that variations in electron transit time account for most of the inaccuracy in his result.]

REFERENCES

10.1. MICHELSON: *Astrophysical Journal*, 1927, Vol. 65, p. 1.

10.2. MICHELSON, PEASE and PEARSON: *ibid.*, 1936, Vol. 82, p. 26.

10.3. MICHELSON: *Studies in Optics* (University of Chicago Press).

10.4. KAROLUS and MITTELSTAEDT: *Phys. Zeits.*, 1928, Vol. 29, p. 698.

10.5. MITTELSTAEDT: *Annalen d. Physik*, 1929, Vol. 2, p. 285.

10.6. HUTTEL: *ibid.*, 1940, Vol. 37, p. 365.

10.7. ANDERSON: *Rev. Sci. Inst.*, 1937, Vol. 8, p. 239.

10.8. ANDERSON: *J. O. S. A.*, 1941, Vol. 31, p. 187.

10.9. DORSEY: *Trans. Amer. Phil. Soc.*, 1944, Vol. XXXIV, p. 1.

10.10. BIRGE: *Reports on Progress in Physics*: 1941, Vol. VIII, p. 92.

10.11. LORENTZ: *Ach. Neerlandaises der Sci.*, 1901, Vol. 6, p. 303.

10.12. ESSEN: *Nature*, 1950, Vol. 165, p. 582.

CHAPTER XI

Relativistic Optics

11.1. Introduction.

The word " theory " is used by scientists in two ways. It may mean a co-ordinated description of the results of certain experiments, or it may mean a general method for the description of experiments. The Theory of Relativity is a " theory " in both senses of the word. It started from the results of certain experiments on the velocity of light. In order to obtain a satisfactory description of these experiments, it was necessary to reconsider the basic definitions of words like " length ", " mass " and " time ". It was then found that certain general difficulties in theoretical physics could be removed if these terms were redefined so as to give them a precise relation to the results of experiments. Concepts invented for a special purpose were seen to be useful in relation to many problems both in optics and in dynamics, and the new definitions became part of the language of theoretical physics. In this chapter we are concerned with deriving some of the main ideas of relativity from optical experiments and with applying these ideas to other optical experiments. The reader should consult textbooks on relativity for a general account of the theory and of its application to dynamical problems.*

11.2. Relative Velocity of Earth and Æther.

Suppose that the velocity of sound is measured by an observer on the ground. In still air, he will obtain a velocity which we may call V, and when there is a wind he measures a velocity $(V + v)$, where v is the component of wind velocity in the direction of propagation. The velocity of waves relative to the observer is simply the vector sum of the wind velocity and the velocity of the waves relative to the air. If the observer had no other method of measuring the wind velocity, he could determine this quantity by measuring the velocity of sound in different directions. He would find that in a certain direction the velocity of sound was a maximum, and that would give the

* References 11.1–11.5.

direction of the wind. The difference between the speed of sound up-wind and down-wind would be equal to twice the speed of the wind. It was at one time thought that this method might be used to determine the relative velocity of the earth and the æther. If light is a system of waves in the æther, and the earth is moving through the æther, we have effectively an "æther wind". Experiments on the velocity of light in different directions should determine the direction and speed of the "æther wind" or, as we more usually describe it, the velocity of the earth relative to the æther. This conclusion is valid whether we consider the æther as an elastic solid, or as the seat of the electromagnetic phenomena.

11.3.—The measurement of a possible variation of the velocity with direction is made more difficult because, in terrestrial experiments at least, we cannot measure the time for light to travel along a given path in one direction, but only the transit time to go to a certain point and to return. We shall now show that this implies that, if the velocity (v) of the earth with respect to the æther is a small fraction of the velocity (c) of light with respect to the æther, we can observe only effects of the second order, i.e. differences proportional to v^2/c^2.

Consider first the total transit time to go and to return along a path whose direction is the same as that of the relative velocity of earth and æther. If d is the length of the path, we have

$$T_1 = \frac{d}{c+v} + \frac{d}{c-v} = \frac{2dc}{c^2 - v^2}, \qquad 11(1)$$

and, when v is small compared with c,

$$T_1 = T_0(1 + v^2/c^2), \qquad 11(2)$$

where $T_0 = 2d/c$ is the time required for the transit to and fro along a similar path in a stationary æther. For a direction at right angles to the above direction, the velocity of light relative to the earth should be $(c^2 - v^2)^{1/2}$, and the transit time (to and fro) is

$$T_2 = \frac{2d}{(c^2 - v^2)^{1/2}}. \qquad 11(3)$$

When v is small compared with c,

$$T_2 = T_0\left(1 + \tfrac{1}{2}\frac{v^2}{c^2}\right). \qquad 11(4)$$

Further discussion shows that the maximum transit time (to and fro) for different directions is given by 11(2) and the minimum by 11(4). The maximum difference is thus

$$\Delta T_M = T_1 - T_2 = \tfrac{1}{2} T_0 \frac{v^2}{c^2}. \quad \ldots \ldots \quad 11(5)$$

The most direct way of finding the earth's velocity relative to the æther would consist in measuring the transit time over a certain path, whose length is defined by rigid rods, when the apparatus is placed in different orientations. The direction corresponding to the maximum transit time would be the direction of the relative velocity, and the difference between the transit times for this direction and a perpendicular direction would enable the magnitude to be calculated from equation 11(5). The smallest difference in transit time which could be detected in experiments such as those described in the preceding chapter is about one part in 100,000. This corresponds to a relative velocity of $0{\cdot}003c$ or 10^8 centimetres per second. The velocity of the earth relative to the sun is about 3×10^6 centimetres per second. An æther-earth relative velocity of this order of magnitude is thus far too small to be detected in a direct experiment. Fortunately indirect methods are much more sensitive.

11.4. The Michelson-Morley Experiment.

About 1887, Michelson and Morley used a modification of the Michelson interferometer (fig. 4.1) to investigate the problem. Suppose that the mirror M_2 (figs. 4.1 and 11.1) is set at a small angle to the reference plane R and is moved until the fringe corresponding to zero order of interference appears in the centre of the field of the telescope. We assume that the plate C exactly compensates for the path in the mirror M_1 (fig. 4.1) and we therefore omit the plate C in fig. 11.1 and regard M_1 as of zero thickness. Then the optical paths from M_1 to M_3, and from M_1 to M_2, are equal; i.e. if d_1 is the geometrical path from M_1 to M_3, and d_2 is the geometrical path from M_1 to M_2, we have

$$n_1 d_1 = n_2 d_2, \quad \ldots \ldots \ldots \quad 11(6)$$

or

$$c\left(\frac{d_1}{b_1}\right) = c\left(\frac{d_2}{b_2}\right), \quad \ldots \ldots \ldots \quad 11(7)$$

where c is the velocity of light for *a body at rest in the æther in vacuum*, and b_1 and b_2 are the velocities in the regions between M_1M_3 and M_1M_2 respectively. The terms in the brackets in 11(7) are the transit times.

Suppose that the arm M_1M_2 happens to lie in the direction of the earth-æther velocity. Then, for equal paths, the transit time M_1 to M_2 and back is greater than the corresponding transit time from M_1 to M_3 and back. In bringing the fringe of zero order to the centre of the field, we have made the transit times equal by making d_2 slightly less than d_1. If now the apparatus is rotated through a right angle, M_1M_3

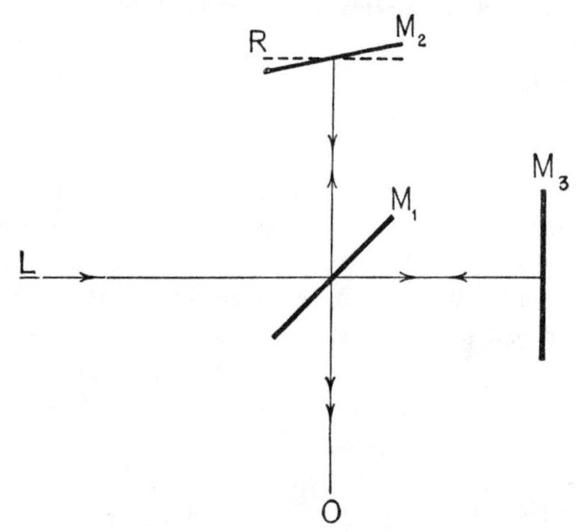

Fig. 11.1.—The Michelson-Morley experiment

will be in the direction of the æther-earth velocity and would have the longer transit time for equal paths. But if the lengths d_1 and d_2 have not altered when the apparatus was rotated, d_1 is the *longer* path, and this further increases the transit time. The difference of transit times is twice that given in equation 11(5), and this difference should produce a fringe movement corresponding to a path difference Δs, given by

$$\Delta s = cT_0 \frac{v^2}{c^2} = 2 \frac{v^2}{c^2} d, \quad . \quad . \quad . \quad . \quad . \quad 11(8)$$

where $d = \frac{1}{2}(d_1 + d_2)$.

11.5.—At the beginning of the experiment, the direction of the earth-æther velocity is unknown. The fringe of zero order is brought to the centre of the field and the apparatus is slowly rotated, observations of the position of this fringe being made at every sixteenth of a revolution. If the above theory is correct the fringes should move sideways, in a periodic way, the period being half the period of the rotation of the apparatus. When the fringes indicate a maximum transit time for M_1M_2, this arm lies along the direction of the æther-earth velocity. The magnitude of the velocity is determined by observing the amplitude of the fringe movement and applying 11(8). In fact no movement of the fringes could be observed. It was calculated that a movement corresponding to an æther-earth velocity of $3 \times 10^{-5}c$, or about 10^6 centimetres per second, could have been detected. The possibility that at the date of the original experiment the earth happened to be moving with very small velocity relative to the æther was considered. The experiment was therefore repeated at different times of the year. It always gave a null result. The result was later confirmed using more sensitive apparatus capable of detecting a velocity of 10^5 centimetres per second or $3 \times 10^{-6}c$.

11.6.—The apparatus used by Michelson and Morley is shown diagrammatically in fig. 11.2. Each of the arms of the interferometer has been lengthened by allowing reflections from corner to corner.

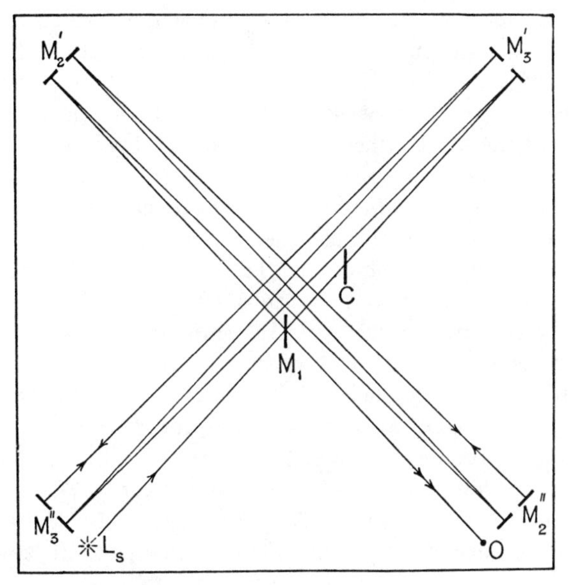

Fig. 11.2.—The Michelson-Morley experiment (multiple reflections)

(Actually there were twice as many reflections as those shown in the diagram.) This method of obtaining a long path is better than a direct increase of the length of the arms, because it is easier to secure temperature control and mechanical rigidity. The effective path was 30 metres and it was estimated that a change in path length of about $\lambda/25$ or $2{\cdot}4 \times 10^{-6}$ centimetre could have been detected. Thus the minimum detectable value of v/c, calculated from 11(8), is $2{\cdot}8 \times 10^{-5}c$. The experiment was later repeated by Joos, and by Kennedy and Illingworth.* They modified the optical arrangements so that a path difference of about $0{\cdot}001\lambda$ could be detected. No movement of the fringes was observed. Miller attempted to make the instrument more sensitive by using very large arms. This led to comparatively large random movements of the fringes owing to temperature variation, etc. He thought that there were small regular movements of the fringes superposed upon these irregular movements, but his "positive" results are generally regarded as within the limits of error.

11.7. The FitzGerald-Lorentz Contraction.

It was suggested by G. F. FitzGerald (1851–1901), and later by H. A. Lorentz, that the null result of the Michelson-Morley experiment might be due to a compensating effect, i.e. a contraction of each arm as the interferometer was turned into the direction of the æther-earth velocity. From equations 11(1) and 11(3) it may be seen that if the ratio of the length when the arm is orientated in the direction of the æther-earth velocity, to the length when orientated in the perpendicular direction is $(1 - v^2/c^2)^{1/2}$, there will be no difference in transit time, whatever the value of v may be. Lorentz gave reasons, based on the electron theory, for supposing that this contraction might be a universal property of matter. Various attempts were then made to measure v, either by detecting and measuring the FitzGerald-Lorentz contraction, or in other indirect ways. All of these experiments had one feature in common. Their objective was to show that free space is optically or electrically anisotropic. The observers looked for some difference to be produced merely by altering the orientation of their apparatus. All these experiments gave a null result. We shall see that according to relativity theory all these experiments are essentially the same experiment and all must yield a null result. It is not necessary to discuss details, but we may briefly mention the following:

(*a*) *Rayleigh* (and later *Brace*) sought to detect a photo-elastic effect produced by the FitzGerald-Lorentz contraction.

(*b*) *Rayleigh* looked for a variation in the optical rotatory power of a quartz crystal when its orientation was altered.

* References 11.7–11.9.

(*c*) *Nordmeyer* placed two sensitive thermopiles on opposite sides of a small source of light. The currents were balanced and the apparatus rotated. It was expected that when the axis of the apparatus was in the direction of the earth-æther velocity, the optical path to one thermopile would be shorter than the path to the other. The radiation received would be proportional to the inverse square of the optical paths. The balance would therefore be disturbed by rotating the apparatus so that the optical path to the other thermopile became the shorter path.

(*d*) Many electrical experiments were performed. It was suggested that the resistance of a wire should depend on its orientation because of the contraction. In a similar way the frequency of a quartz crystal vibrator might be expected to depend on its orientation. These electrical experiments were of high sensitivity.

11.8. **Special Theory of Relativity.**

It is fairly easy to explain the null result of any one of these experiments by the hypothesis of a compensating effect such as the Fitz-Gerald-Lorentz contraction, but it would obviously be unsatisfactory to explain each result by a separate and rather special hypothesis. Einstein decided to accept the experiments as a proof that the velocity of light is the same for all observers, and to re-examine the basic problems of kinematics in the light of this result. Every kinematical problem is stated, directly or indirectly, by reference to a co-ordinate system and to a clock. Suppose that a given dynamical problem is referred to a co-ordinate system x, y, z, t, and the equations of motion are set down. It is possible to express the problem in terms of another co-ordinate system x', y', z', t', if algebraic equations connecting x, y, z, t are available. Such a set of equations is called a transformation group. In the simplest case, if the two co-ordinate systems differ only in that the origin of the dashed system is situated at a distance a from the origin of the undashed system, and along the axis of x, we have

$$x' = x + a; \quad y' = y; \quad z' = z; \quad t' = t.$$

More complicated transformations refer to other geometrical changes (e.g. rotation of axes.)

11.9.—It is possible also to consider systems of axes which are moving with respect to one another. Such systems are considered in

Newtonian mechanics, and it is easy to derive the transformation group:

$$\left.\begin{aligned} x' &= x + vt, \\ y' &= y, \\ z' &= z, \\ t' &= t. \end{aligned}\right\} \quad . \quad . \quad . \quad . \quad . \quad . \quad . \quad 11(9)$$

We assume that the axes Ox and Ox' have been chosen to coincide with the direction of the relative velocity. Using this transformation, it can be shown that the laws of mechanics (expressed in Newton's Laws of Motion) are the same in the dashed and in the undashed systems, provided that the systems have no relative acceleration. For a particle moving with constant velocity U along the x-axis in the undashed system, we have

$$\left.\begin{aligned} U &= \frac{dx}{dt}, \\ U' &= \frac{dx'}{dt'} = \frac{dx}{dt} + v = U + v, \end{aligned}\right\} \quad . \quad . \quad . \quad 11(10)$$

$$\therefore \frac{dU'}{dt'} = \frac{d^2x'}{dt'^2} = \frac{d^2x}{dt^2}. \quad . \quad . \quad . \quad . \quad . \quad . \quad . \quad . \quad . \quad 11(11)$$

The velocities of the particle in the two systems are different, their relation being given by 11(10), but their accelerations are the same and laws like " force = mass × acceleration " are unaffected. Equation 11(10) appears to be in good agreement with experimental results when it is applied to ordinary velocities. For example, the equation can be used to calculate the speed of an aeroplane relative to the ground when its air-speed (i.e. speed relative to the air) and the wind velocity are known. The Michelson-Morley experiment and the associated experiments show that for light

$$c' = c, \quad . \quad . \quad . \quad . \quad . \quad . \quad . \quad . \quad 11(12)$$

and this indicates that equation 11(10) cannot apply to light.

11.10.—Starting from this fact Einstein made two hypotheses:

(1) *The principle of equivalence*—the laws of physics are independent of the motion of the co-ordinate system to which they are referred. We can never detect absolute motion of bodies through space,

but only the relative motion of one material body with respect to another.

(2) The velocity of light in every co-ordinate system has always the same value (c).

In the *special* theory of relativity (published in 1905), he applied the principle of equivalence to systems in uniform relative motion, but not to systems which are accelerated with respect to one another. In the *general* theory (published in 1915) the principle is applied to all systems. For the present, we consider only the special theory. It is important to note that in respect of systems which have relative motion, but no relative acceleration, the principle of equivalence holds in both Newtonian and in relativistic mechanics. The difference lies in the fact that the relativistic mechanics takes account of the experimental observation that the velocity of light is the same for all systems.

11.11.—Using the two fundamental hypotheses stated above, we may proceed to derive a transformation group which will replace equations 11(9). The derivation is very simple if we accept two assumptions: (a) that the equations of transformation are linear and homogeneous, and (b) that co-ordinates in directions perpendicular to the direction of the relative velocity are the same in both systems.*

We consider two systems S and S′, and choose co-ordinates such that OX and O′X′ are both in the direction of the relative velocity, We choose the origins of time and space so that $t' = 0$ when $t = 0$, and at this time O coincides with O′. We imagine a light signal emitted from the origin when $t = 0$. At time t in system S, it will have reached the sphere $x^2 + y^2 + z^2 = c^2t^2$. The same signal viewed from the system S′ will appear at time t' (corresponding to t) to have reached the sphere $x'^2 + y'^2 + z'^2 = c^2t'^2$. Thus we must have

$$(x^2 + y^2 + z^2 - c^2t^2) \equiv \gamma(x'^2 + y'^2 + z'^2 - c^2t'^2), \quad 11(13)$$

where γ is a constant. This equation is derived from the assumption that c is the same for both systems. The principle of equivalence implies that $\gamma = 1$, because, if this were not so, the "light sphere" would not be the same size in both systems. Also we have decided to assume that $y = y'$ and $z = z'$, and hence

$$x^2 - c^2t^2 \equiv x'^2 - c^2t'^2. \quad . \quad . \quad . \quad . \quad 11(14)$$

Let the relative velocity be such that O′ appears to be moving with

* For a derivation independent of these assumptions see Reference 11.5.

velocity v when viewed from S, and O appears to be moving with velocity $-v$ when viewed from S′. Then $x' = 0$ when $x = vt$, and $x = 0$ when $x' = -vt'$. Hence we must have

$$x' = k(x - vt), \qquad 11(15)$$

and

$$x = k'(x' + vt'), \qquad 11(16)$$

where k and k' are two constants which must depend on v. Inserting the value of x' given by 11(15) in 11(16), we obtain

$$t' = k\left[t - \frac{x}{v}\left(1 - \frac{1}{kk'}\right)\right]. \qquad 11(17)$$

Equations 11(15) and 11(17) give x' and t' in terms of x and t. These values may now be inserted on the right-hand side of 11(14), and the coefficients of x^2, t^2 and xt on the two sides of the identity may be equated. Three equations are obtained, and these may be regarded as simultaneous equations for k and k'. It is found that all three equations are satisfied when $k = k' = (1 - v^2/c^2)^{-1/2}$. The equations of transformation [obtained by inserting the values of k and k' in 11(15), 11(16) and 11(17)] are thus

$$\left.\begin{aligned} x' &= \frac{x - vt}{\sqrt{(1 - v^2/c^2)}}, & x &= \frac{x' + vt'}{\sqrt{(1 - v^2/c^2)}}, \\ t' &= \frac{t - \dfrac{v}{c^2}x}{\sqrt{(1 - v^2/c^2)}}, & t &= \frac{t' + \dfrac{v}{c^2}x'}{\sqrt{(1 - v^2/c^2)}}, \end{aligned}\right\} \qquad 11(18)$$

and we have also

$$y = y' \text{ and } z = z'.$$

The equations have the symmetry required by the principle of equivalence, allowing for the fact that the sign of the relative velocity measured in S′ is opposite to its sign when measured in S. It may be verified by direct substitution in 11(14) that the velocity of light is the same in both systems.

11.12. Dilation of Time and Contraction of Space.

Let us now consider four consequences which follow from equations 11(18).

(*a*) *Relativity of Simultaneity.*

From equations 11(18) we see that two events which are simultaneous in the system S, but which occur at different places, are not simultaneous in the system S′. For, if $t_1 = t_2$,

$$t_1' = \frac{1}{\alpha}\left(t_1 - \frac{v}{c^2}x_1\right) \text{ and } t_2' = \frac{1}{\alpha}\left(t_1 - \frac{v}{c^2}x_2\right), \qquad 11(19)$$

where we have written

$$\alpha = (1 - v^2/c^2)^{\frac{1}{2}}. \qquad . \quad . \quad . \quad . \quad . \quad . \quad 11(20)$$

Thus t_1' and t_2' are not equal if $x_1 \neq x_2$.

(*b*) *Dilation of Time.*

In a similar way we see that the interval between two events which occur at times t_1 and t_2, and at the same place (x_1) in system S, is not the same as the interval measured in system S′ for

$$\left.\begin{aligned} \Delta t &= t_1 - t_2, \\ \Delta t' &= t_1' - t_2' = \frac{\Delta t}{\alpha}. \end{aligned}\right\} \qquad . \quad . \quad . \quad . \quad 11(21)$$

Since α is less than unity, $\Delta t'$ is greater than Δt.

(*c*) *Contraction of Space.*

An observer who is at rest in system S may measure the length of a bar which is also at rest in that system by placing a measuring scale against the bar and reading the co-ordinates of its ends. Let the readings be x_1 and x_2. An observer in S′ can also measure the lengths of the same bar by placing a measuring scale (which is at rest in his system) against the bar, but his measuring scale will be moving past the bar. He therefore reads the co-ordinates of the ends *simultaneously*, i.e. at the same time in *his* system. The length he will measure will be

$$x_2' - x_1' = \frac{1}{\alpha}\{(x_2 - x_1) - v(t_2 - t_1)\}, \quad . \quad . \quad 11(22)$$

and he will have

$$\left.\begin{aligned} t_2 &= \frac{1}{\alpha}\left(t' + \frac{v}{c^2}x_2'\right), \\ t_1 &= \frac{1}{\alpha}\left(t' + \frac{v}{c^2}x_1'\right), \end{aligned}\right\} \qquad . \quad . \quad . \quad . \quad . \quad . \quad 11(23)$$

if he makes the observations at the same time (t') in his system. Substituting in 11(22) from 11(23), we have

$$x_2' - x_1' = \frac{1}{\alpha}\left\{(x_2 - x_1) - \frac{v^2}{\alpha c^2}(x_2' - x_1')\right\},$$

or $$\left(1+\frac{v^2}{\alpha^2 c^2}\right)\left(x_2' - x_1'\right) = \frac{1}{\alpha}\left(x_2 - x_1\right),$$

and $$x_2' - x_1' = \alpha(x_2 - x_1), \quad . \quad . \quad . \quad . \quad . \quad 11(24)$$

Since $$1 + \frac{v^2}{\alpha^2 c^2} = 1 + \frac{v^2}{c^2 - v^2} = \frac{1}{\alpha^2}.$$

Thus the length measured in system S′ is less than that measured in system S in the ratio $\alpha : 1$.

(*d*) *Addition of Velocities.*

Suppose that a particle is moving in the common direction of the x and x' axes and that its velocity as measured by an observer in the system S is U, and as measured by an observer in the system S′ is U'. Then

$$U = \frac{dx}{dt} \text{ and } U' = \frac{dx'}{dt'}.$$

$$U = \frac{d(x' + vt')}{d\left(t' + \frac{v}{c^2}x'\right)}$$

$$= \frac{\frac{d}{dt'}(x' + vt')}{\frac{d}{dt'}\left(t' + \frac{v}{c^2}x'\right)} = \frac{\frac{dx'}{dt'} + v}{1 + \frac{v}{c^2}\cdot\frac{dx'}{dt'}}.$$

Hence $$U = \frac{U' + v}{1 + \frac{vU'}{c^2}}. \quad . \quad . \quad . \quad . \quad . \quad . \quad . \quad . \quad . \quad 11(25)$$

This law of addition of velocities applies only when the two velocities are in the same direction. It is the relativistic equation corresponding to the Newtonian relation given by equation 11(10). It is in agreement with the hypothesis of the constant velocity of light because, if $U' = c$, we have

$$U = \frac{c + v}{1 + vc/c^2} = c. \quad . \quad . \quad . \quad . \quad . \quad 11(26)$$

Thus the addition of any velocity (positive or negative) to the velocity of light leaves this particular velocity unchanged.

11.13.—The conclusions of the last paragraph, that space and time measurements depend on the velocity of the observer, appear to be contrary to our " intuitive " ideas of space and time. These concepts are derived from experience, in which the observer is either at rest with respect to the objects observed, or moving with a velocity which is a small fraction of the velocity of light. The dilation of intervals and contraction of lengths are then much too small to be observed.* It is then natural to think that lengths and intervals are the same for all observers and hence to think of them as fixed objects having an existence independent of any observer. They are so conceived in Newtonian mechanics and this idea leads to equation 11(10) for the law of addition of velocities. In this law, all velocities differ according to the motion of the observer. Einstein starts from the assumption that in physics we deal only with *measured* length and *measured* time. These quantities are significant only when we have clearly defined the process of measurement.

Suppose that in the system of S, light signals are emitted at times t_1 and t_2. An observer in system S′ can note the difference between the times at which they are received. He does not regard this as the correct interval between the events, because he knows that the source has moved between the emission of the two flashes. He has to make an allowance for the difference in time of transit. According to Newtonian mechanics, this allowance must be made so that the interval measured in system S′ is the same as that measured in system S. This method of making the allowance cannot be reconciled with the constant velocity of light. According to relativity theory, the allowance must be adjusted so that the velocity of light is the same for both observers. The intervals are then unequal. We cannot have both the equality of intervals and the constant velocity of light. The Michelson-Morley experiment, and other experiments which we shall describe later, furnish very strong evidence in favour of accepting the constant velocity of light. We shall see that there is also direct evidence for the

* For an observer moving at 60 m.p.h. past the scene which he observes, α differs from unity by about one part in 10^{14}.

inequality of intervals and for the velocity addition theorem [equation 11(25)]. If we accept the inequality of intervals (and the corresponding inequality of lengths), it follows that a statement concerning a length or a time is incomplete unless the frame of reference is stated. If the given data in a problem do not all refer to the same frame, they must be converted to the same frame by means of equations 11(18)—or some equivalent process must be carried out.

11.14. Experiments in which Source and Observer are in Relative Motion.

We shall now develop equations giving the frequency and apparent direction of light emitted by a source which is moving with respect to the observer. The medium between source and observer has unit

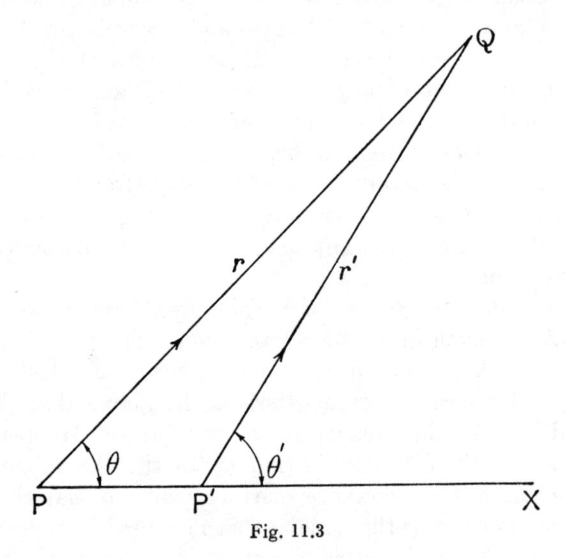

Fig. 11.3

refractive index. The observer is at rest at Q in system S, and the source is at rest in S′ (fig. 11.3). Let the waves emitted from P′ be represented by

$$\xi' = \frac{A'}{r'} \exp i\{(\omega' t' - \kappa' r') + \epsilon'\}. \quad . \quad . \quad 11(27)$$

Let the ray from P′ to the observer at Q make an angle θ' (measured in system S′) with the x' axis. Then $r' = x' \cos\theta' + y' \sin\theta'$, and we have

$$\xi' = \frac{A'}{r'} \exp i\{\omega' t' - \kappa' x' \cos \theta' - \kappa' y' \sin \theta' + \epsilon'\}. \quad 11(28)$$

To the observer in system S, the same beam is represented by

$$\xi = \frac{A}{r} \exp i\{\omega t - \kappa x \cos \theta - \kappa y \sin \theta + \epsilon\}. \quad 11(29)$$

But the expression derived by applying the transformation equations 11(18) to equation 11(28) must also represent the beam observed in the S system, and must therefore agree with 11(29) at all places and times, i.e.

$$\omega t - \kappa x \cos \theta - \kappa y \sin \theta + \epsilon$$
$$\equiv \frac{\omega'}{\alpha}\left(t - \frac{vx}{c^2}\right) - \frac{\kappa'}{\alpha}(x - vt) \cos \theta' - \kappa' y \sin \theta' + \epsilon'. \quad 11(30)$$

Equating the coefficients of x and t, and remembering that $\omega/\kappa = \omega'/\kappa' = c$, we have

$$\omega = \frac{1}{\alpha}(\omega' + \kappa' v \cos \theta'),$$

or

$$\omega = \frac{\omega'}{\alpha}\left(1 + \frac{v}{c} \cos \theta'\right); \quad . \quad . \quad . \quad 11(31)$$

and

$$\kappa \cos \theta = \frac{1}{\alpha}\left(\kappa' \cos \theta' + \frac{v\omega'}{c^2}\right),$$

$$\kappa \cos \theta = \frac{\omega'}{\alpha}\left(\frac{\cos \theta'}{c} + \frac{v}{c^2}\right),$$

$$\cos \theta = \frac{\omega'}{\omega\alpha}\left(\cos \theta' + \frac{v}{c}\right),$$

$$\cos \theta = \frac{\cos \theta' + \frac{v}{c}}{1 + \frac{v}{c} \cos \theta'}. \quad . \quad . \quad . \quad . \quad . \quad 11(32a)$$

The frequency of the light observed in the system S is given by 11(31), and in this system the position of the source is at P (fig. 11.3), the angle θ being given by 11(32*a*). It is found that both equations 11(31) and 11(32*a*) are verified by experiment (§§ 11.15, 11.16 and 11.19).

EXAMPLES [11(i)–11(v)]

11(i). Show that the transit time (to and fro) for a path which makes an angle θ to the direction of the relative velocity is given by

$$T_\theta = \frac{2d}{c^2 - v^2}\{c^2 - v^2(1 - \cos^2\theta)\}^{\frac{1}{2}}.$$

Hence show that T_1 and T_2 [equations 11(2) and 11(4)] are maximum and minimum velocities respectively.

11(ii). Show that equation 11(32a) may be written

$$\sin\theta = \frac{\alpha \sin\theta'}{1 + \frac{v}{c}\cos\theta'}. \qquad \text{11(32}b\text{)}$$

[From 11(32a) we have $(1 + \cos\theta) = (1 + \cos\theta')\left(1 + \frac{v}{c}\right)\Big/\left(1 + \frac{v}{c}\cos\theta'\right)$ and a similar expression for $(1 - \cos\theta)$. We obtain 11(32b) by putting $\sin^2\theta = (1 + \cos\theta)(1 - \cos\theta)$.]

11(iii). Show that 11(32b) implies that

$$\sin\theta' = \frac{\alpha \sin\theta}{1 - \frac{v}{c}\cos\theta}, \qquad \text{11(32}c\text{)}$$

and obtain an expression for $\cos\theta'$.

[Equation 11(32c) and a corresponding expression for $\cos\theta'$ are obtained by applying the principle of equivalence to 11(32b) and 11(32a). They may be verified algebraically.]

11(iv). Show that 11(32a) is consistent with the requirement that the coefficients of y in 11(30) must be equal.

$\left[\text{Remember that } \frac{\omega}{\kappa} = \frac{\omega'}{\kappa'} = c.\right]$

11(v). Light (emitted from a source in S′) is limited by stops which obstruct all the light except that within a small cone of solid angle $d\Omega'$ (measured in S′). Show that the solid angle measured by an observer in a system S (which is approaching the source with velocity v) is given by

$$\frac{d\Omega}{d\Omega'} = \frac{1 - v/c}{1 + v/c}. \qquad \text{11(33)}$$

[Differentiating 11(32b) we obtain

$$\cos\theta\, d\theta = \left[\frac{\alpha\cos\theta'}{1+\frac{v}{c}\cos\theta'} + \frac{v}{c}\frac{\alpha\sin^2\theta'}{(1+\frac{v}{c}\cos\theta')^2}\right]d\theta'.$$

When $\theta = \theta' = 0$, we have $\quad d\theta = \dfrac{\alpha}{1+v/c}\, d\theta',$

and 11(33) follows since $\quad \dfrac{d\Omega}{d\Omega'} = \left(\dfrac{d\theta}{d\theta'}\right)^2.$]

11.15. Radial Doppler Effect.

Equation 11(32) shows that when $\theta' = 0$ we have also $\theta = 0$. In this condition source and observer are approaching directly along the line of sight when v is positive. The relation between the frequencies is then

$$\alpha\omega = \omega'(1 + v/c), \qquad \text{11(34a)}$$

or

$$\lambda(1 + v/c) = \alpha\lambda'. \qquad \text{11(34b)}$$

This differs from the non-relativistic expression 2(58) only in respect of the factor α. Experiments of the type described in § 2.24 are not sufficiently accurate to decide between the two formulæ. The difficulty is not in the optical part of the experiment. In the early experiments on the Doppler effect, the atoms were not all moving in precisely the same direction, so that θ' [in equation 11(31)] was not known exactly, and v was known only with moderate accuracy. This makes it impossible to calculate the term in the bracket to an accuracy of better than about one part in a thousand. Also the lines are broadened owing to the variation of the component of the velocity in the line of sight.

11.16. Transverse Doppler Effect—Dilation of Time.

If a beam of atoms is observed in a line which, *according to the observer*, is at right angles to the direction of the relative velocity, then $\cos\theta = 0$, and from 11(32a) we see that $\cos\theta' = -v/c$. Inserting this value in 11(31), we have

$$\omega = \omega'\frac{(1 - v^2/c^2)}{(1 - v^2/c^2)^{1/2}} = \omega'\alpha. \qquad \text{11(35a)}$$

Hence

$$\lambda = \frac{\lambda'}{\alpha} = \frac{\lambda'}{(1 - v^2/c^2)^{1/2}}, \qquad \text{11(35b)}$$

where λ is the wavelength observed in system S, and λ' is the wavelength which would be observed by an observer at rest with respect to the source. This change is called the *transverse Doppler effect*. Comparatively recently this effect has been observed by Ives and Stilwell, and by Otting, working independently. Three technical problems have to be solved in order to attain the necessary accuracy:

(*a*) A fairly accurately collimated beam of emitting atoms must be produced.

(*b*) The experimenter must be able to set his line of observation accurately at right angles to the direction of relative motion and to know when he has achieved this setting.

(*c*) The velocity of the atoms must be fairly high (of the order $10^{-2}c$) and must be known with moderate accuracy.

Residual errors due to imperfect collimation are proportional to v/c, and their importance relative to the main effect (which is proportional to v^2/c^2) is greatly reduced by using a very high velocity. In Ives' experiment, charged atoms and molecules were focused into a parallel beam by suitably designed electrodes (working on the same principles as the well-known "electron focusing" methods). At the speeds involved, an ion may capture an electron and proceed as an excited atom without being appreciably deflected. A mirror set accurately normal to the line of observation was used, and the line of observation altered until nearly the same wavelength was given by the direct and the reflected light. When the difference (due to the small components of velocity in the line of sight) was very small, the mean could be taken as the wavelength due to an atom moving at right angles to the line of sight. Velocities ranging from about $4\times10^{-3}c$ to $7\times10^{-3}c$ were used. The variation of λ with v was that given by equation 11(35*b*). The experiment forms a satisfactory and very direct verification of the dilation of time. There is no corresponding direct verification of the contraction of space. This part of the theory is verified indirectly by the experiments described in § 11.20.

EXAMPLES 11(vi) and 11(vii)

11(vi). Calculate the difference between the Doppler shift obtained from 2(57) and that obtained from 11(34*b*) for particles whose velocities of approach are (*a*) 0·1*c*, (*b*) 0·01*c*, and (*c*) 0·003*c*. Take the wavelength of the light to be 6000 Å.

[(*a*) 27 Å., (*b*) 0·30 Å., (*c*) 0·027 Å.]

11(vii). A certain source emits light of wavelength 6000 Å., measured by an observer for whom the source is stationary. Find the displacements (*a*) for an observer who observes a transverse motion of the source, and (*b*) for an observer who observes a recession at 89° 55′ to the line of sight. In each case the velocity is $0{\cdot}0031c$.

[(*a*) 6000·03 Å. from equation 11(35*b*); (*b*) 6000·06 Å. To derive the second result use the formula $\omega' = \frac{\omega}{\alpha}\left(1 - \frac{v}{c}\cos\theta\right)$ obtained by applying the principle of equivalence to 11(31). Insert the value $v = -0{\cdot}0031c$, since the source is receding.]

11.17.—It is sometimes thought that the contraction of space implied in equation 11(24) is the contraction suggested by FitzGerald, and that this contraction is verified by the Michelson-Morley experiment. This idea is not consistent with the relativity theory because FitzGerald's hypothesis was put forward as an *alternative* to the hypothesis of a constant velocity of light. The parts of the apparatus in the Michelson-Morley experiment are not in motion relative to each other, or to the observer. Only one frame of reference is involved, and therefore there is no contraction of space or dilation of time to be considered. It has also been suggested that, because transits to and fro are involved, the Michelson-Morley experiment does not prove anything about the velocity of light *in one direction*. Essentially the experiment shows that if $c(\theta)$ is the velocity in a direction which makes an angle θ with an axis fixed with respect to the earth, then

$$\frac{1}{c(\theta)} + \frac{1}{c(\theta + \pi)} = \text{a constant independent of } \theta.$$

The repetition of the experiment at a different time of year shows that a similar relation holds when the earth has a different velocity relative to the sun, though the constants are not necessarily equal. Eddington has analysed the significance of this relation.* He concludes that a logical argument based on the Michelson-Morley experiment and on other generally accepted results shows that the velocity of light is the same in all directions and for all observers. It is not, however, strictly correct to say that the Michelson-Morley experiment *by itself* proves this proposition.

11.18. **Reflection of Light by a Moving Mirror.**

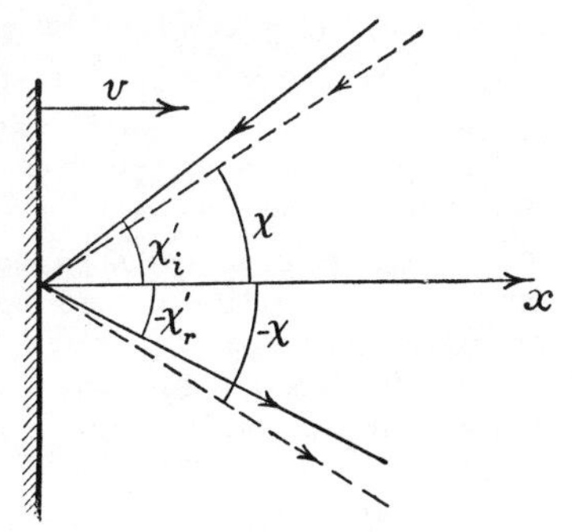

Fig. 11.4.—Reflection at moving mirror

We wish to calculate the law of reflection of light at a plane mirror which is moving (relative to the observer and the source) in the direction of its own normal (fig. 11.4). Let S be the system in which the mirror is stationary, and S′ the system of the

* Reference 11.3.

observer. Let χ be the angle of incidence measured in S. Then the angle of reflection, measured in S, will be $-\chi$, since the ordinary law of reflection is valid in S. Let the angles of incidence and reflection in S′ be χ_i' and $-\chi_r'$. Then we may apply 11(32*b*) first to obtain the relation between χ and χ_i', and again for the relation between χ and χ_r'. In the second application we must change the sign of v, because the incident beam travels towards, and the reflected beam away from, the mirror. Thus we have

$$\sin\chi = \frac{\alpha\sin\chi_i'}{1+\frac{v}{c}\cos\chi_i'} \quad \text{and} \quad \sin(-\chi) = -\frac{\alpha\sin\chi_r'}{1-\frac{v}{c}\cos\chi_r'}.$$

Thus the law of reflection is

$$\frac{\sin\chi_i'}{1+\frac{v}{c}\cos\chi_i'} = -\frac{\sin\chi_r'}{1-\frac{v}{c}\cos\chi_r'}. \quad . \quad . \quad 11(36)$$

EXAMPLES [11(viii)–11(x)]

11(viii). Show that if ω_i' and ω_r' are the circular frequencies of the incident and reflected radiation respectively, then

$$\omega_i'\left(1+\frac{v}{c}\cos\chi_i'\right) = \omega_r'\left(1-\frac{v}{c}\cos\chi_r'\right). \quad . \quad . \quad 11(37)$$

[In S, the frequency of the incident and reflected radiation is the same. Let this frequency be ω, and use 11(31) to determine the relations between ω and ω_i', and between ω and ω_r'.]

11(ix). A point source of light is placed at the focus of a lens whose focal length is f, and a plane mirror is set normal to the axis of the lens so as to give an image which coincides with the object when the mirror is stationary. Show that the displacement of the image when the mirror moves with speed v in a direction perpendicular to its normal is approximately $2vf/c$.

[Apply equation 11(32).]

11(x). Show that, in the experiment described in the preceding example, there is a change of wavelength proportional to v^2/c^2.

[Let ω be the circular frequency for an observer whose frame S is fixed relative to the mirror, and let ω_i' be the frequency measured in S′. Then 11(31) gives $\omega_i' = \omega\alpha$. For the return beam we have, approximately, $\cos\theta' = 2v/c$, and 11(31) then gives $\omega_r'(1+2v^2/c^2) = \omega\alpha = \omega_i$. The wavelength of the return beam is thus increased by $\Delta\lambda = 2\lambda v^2/c^2$.]

11.19. Aberration Experiments.

Returning to equation 11(32*a*) and to fig. 11.3, we see that an observer in the S system has to look in a direction θ in order to view an object which in the S′ system is situated in the direction θ'. Considering the simplest case when $\theta' = \frac{1}{2}\pi$, we see that the difference $(\Delta\theta)$ between θ and θ' is then given by

$$\sin(\Delta\theta) = \sin(\tfrac{1}{2}\pi - \theta) = \cos\theta = v/c. \quad . \quad . \quad 11(38)$$

The angle $\Delta\theta$ is known as the angle of aberration. Aberration was first observed by Bradley, who found an apparent difference between the angular positions of stars at different time of the year. The effect was additional to the well-known parallax effect, and was ascribed to the reversal of the earth's velocity in its orbit. Bradley used his observations to determine the velocity of light. At present it is perhaps more useful to regard the velocity of light as known, and to use the aberration effect to determine the orbital velocity of the earth.

Airy and Hoek observed the angle of aberration both (*a*) with an ordinary telescope and (*b*) with a telescope filled with water. It was expected that there would be a difference in the angle and that the difference would determine the velocity of the earth relative to the æther. No difference was observed. A rather complicated explanation of this result based on the Fresnel convection coefficient (§ 11.20) was given later. Relativity theory gives a much simpler explanation.

Suppose that, when the tube is filled with air, the observer in system S turns his telescope so that he sees a star in the centre of the field of view. *In his system*, the light from the star falling on his telescope is represented by a system of nearly plane waves whose wave normals are in the direction of the axis of his instrument. Filling the telescope tube with water makes no difference to this angular relation, and the star is seen in the same direction. The situation would, of course, be quite different if a material medium filled the whole region between source and observer.

11.20. Experiments with a Moving Medium.

In 1818 Fresnel suggested that it should be possible to determine the velocity of light in a moving medium by measuring the optical thickness of a moving plate. This experiment was carried out by Fizeau. His arrangement was, in effect, a modified form of the Rayleigh refractometer (fig. 11.5). Interference is observed at O between light which has passed from M_2 to M_1 by the upper path, returning by the lower path, and light which has passed from M_2 to M_1 by the upper path returning by the lower path. The displacement of fringes due to reversing the direction of liquid flow is measured. This displacement is proportional to the velocity of the liquid. The change of velocity of light could be calculated from the change of optical path, using 11(6), and it was found that the effective velocity of light in the moving medium was

$$b' = b + v\left(1 - \frac{1}{n^2}\right). \qquad . \quad . \quad . \quad . \quad 11(39)$$

It had been expected that the velocity would be $(b + v)$ according to the Newtonian velocity addition equation 11(10). Fizeau's result was, at first, explained by saying that the æther was convected with a moving medium, but acquired only a fraction of the velocity of the medium.

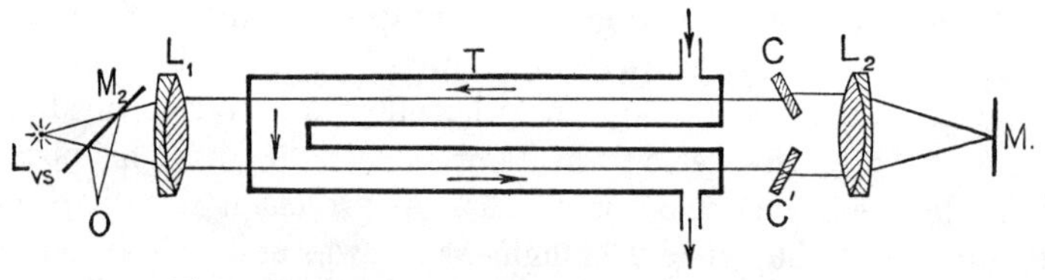

Fig. 11.5.—Transmission of light in a moving medium

The factor $(1 - 1/n^2)$ was called the *Fresnel convection coefficient.* This somewhat peculiar assumption was not derived from any satisfactory theory. In relativity the result does not require any special assumption. Applying the velocity addition equation 11(25), and writing b and b' in place of U' and U, we have

$$b' = \frac{b + v}{1 + vb/c^2} = \frac{b + v}{1 + v/nc}. \qquad . \quad . \quad . \quad 11(40)$$

When v is small compared with c, this may be written

$$b' \doteqdot (b + v)\left(1 - \frac{v}{nc}\right)$$

$$\doteqdot b + v\left(1 - \frac{1}{n^2}\right),$$

in agreement with equation 11(39).

This experiment is of importance as a direct verification of the velocity addition theorem. It also shows that it is not possible to explain the null result of the Michelson-Morley experiment by saying that the æther is completely convected with the apparatus. The medium being air, the convection coefficient would be nearly zero, whereas a convection coefficient of unity would be needed to explain the null result. Fizeau's experiment was repeated by Michelson and Morley, and by Zeeman. Part of the object of these later experiments was to investigate a small correction due to the dispersion of the medium. The fundamental result [stated in equation 11(39)] was confirmed.

11.21. General Theory of Relativity.

In 1915 Einstein extended the principle of equivalence to frames of reference which are accelerated with respect to one another. Such frames are not equivalent in Newtonian dynamics. Suppose that an observer in a closed lift finds that bodies when released have an acceleration f. Then according to Newtonian ideas he is entitled to deduce that his frame has an absolute acceleration of $(g - f)$. Einstein points out that all his observations are consistent with the view that he has been transferred to a place where the force of gravity is f instead of g. All he knows is that if he is in a gravitational field g, then his frame of reference has an acceleration $(g - f)$. He cannot determine the absolute acceleration of his frame. Einstein thus makes force and mass-acceleration equivalent in a more complete way than was contemplated in Newtonian dynamics. From this point onwards the general theory of relativity is in two parts:

(i) A general mathematical method whereby fields of force are expressed in terms of a " curved " (i.e. non-Euclidean) space-time continuum.

(ii) A theory of world structure.

11.22.—The first part is essentially the development of a general method of mathematical representation, and its application to the local gravitational fields produced by bodies like the earth and the sun. This first part of the general theory is incomplete, because it indicates the possibility of a general curvature of space-time, which cannot be determined by observations on local gravitational fields, but which may be determined by more general astronomical observations (e.g. on estimates of the number and distribution of the nebulæ). Further discussion of the general theory of relativity is outside the scope of this book, but we shall describe some optical experiments which are relevant to what we have called the first part, and one experiment which may be interpreted in terms of world structure.

11.23. Refraction of Light Rays in a Gravitational Field.

According to Newtonian theory, a material particle at a distance r from the centre of gravity of a body of mass m is subject to a gravitational acceleration m/r^2. Thus, according to non-relativistic ideas, the paths of light *corpuscles* should be curved in the neighbourhood of a large mass, but Newtonian theory does not predict any deflection of electromagnetic waves. Without reference to either corpuscular or wave concepts, the general relativistic principle of equivalence requires that light rays shall be curved by an amount corresponding to approximately twice the above force. The curvature produced by a gravitational field, such as that of the earth, is far too small to be observed.* The curvature which should be produced when a ray grazes the sun's limb is only 1·75 seconds of arc. This curvature has been measured by photographing the star field surrounding the sun during an eclipse. This photograph is compared with one of the same region of the sky taken six months earlier, at night. The outer stars in the two photographs form a co-ordinate frame of reference. Light from them does not pass near enough to the sun to suffer any deflection. The positions of stars whose light has passed near the sun may be measured on both plates, using the above co-ordinate system. It is found that the images of these stars in the photograph taken during the eclipse are slightly displaced towards the sun. From these displacements the deflection may be calculated. In 1919 the eclipse was observed at Sobral and at Principe.† Deflections of $1{\cdot}98 \pm 0{\cdot}12$ seconds and $1{\cdot}61 \pm 0{\cdot}30$ seconds were obtained. Later observations ‡ in 1923 and 1928 gave deflections of $1{\cdot}72 \pm 0{\cdot}11$ seconds and $1{\cdot}82 \pm 0{\cdot}15$ seconds. The four measurements taken together form a reasonably good verification of the effect.

11.24. **Displacement of Lines in a Gravitational Field.**

It may be shown that the wavelength of light received from an atom in a large gravitational field is greater than that from an atom which is not in a gravitational field in the ratio

$$\left(1 + \frac{m}{kr}\right) : 1, \qquad \text{11(41)}$$

where k is a constant. The value of the constant is such that the effect of the earth's gravitational field is extremely small, and the change of wavelength produced by the field at the surface of the sun is only about 2 parts in a million, or 0·01 Å. for a wavelength of 5000 Å. There is no difficulty in measuring wavelength changes of this size, but the observed wavelengths of the solar spectrum have to be corrected for Doppler effects and other effects (due to possible magnetic and electric fields and to collisions). These subsidiary effects vary from point to point across the sun's surface, and to some extent they may be eliminated by comparing the wavelengths of light from different parts

* See p. 90 of Reference 11.3. † Reference 11.3. ‡ References 11.11 and 11.12.

of the sun. When this is done, there is fairly strong evidence in favour of the existence of the gravitational-field displacement.*

Much larger displacements of the spectrum lines have been observed in the dark companion of Sirius. The large displacement is not shown in the spectrum of Sirius, and is therefore not due to a Doppler effect. If interpreted as a gravitational effect, the displacement of the spectrum lines would indicate a density of order 10^4 grammes per cubic centimetre in the star. The existence of dwarf stars with densities of this order had been predicted in theories of stellar evolution. It thus appears reasonable to regard these observations as an important verification of the general principle of equivalence.

11.25. **Interference in a Rotating System.**

Rotation is often thought to constitute a serious difficulty in relativity theory. An observer in an enclosure can detect the rotation of his system by observing the centrifugal force, or by observations on a system like a Foucault pendulum. It is suggested that he thereby detects absolute motion. According to general relativity theory he detects, not absolute motion, but motion relative to the main gravitational lines of force produced by all the masses of the universe. If all exterior masses were removed, rotational effects such as centrifugal force should disappear. In relativity theory rotational forces are represented by suitable terms in the expressions which define the curvature of space-time, and hence the gravitational fields of force. In most practical conditions, the predictions of relativistic theory in regard to rotation agree with those of non-relativistic mechanics, apart from minor corrections which are too small to be detected. We shall now describe two experiments on the interference of light. One, due to Sagnac, may be regarded as analogous to a mechanical experiment on a spinning top. The other, due to Michelson, is analogous to the Foucault pendulum experiment since it depends upon the rotation of the earth.

11.26.—The apparatus used by Sagnac is shown in fig. 11.6. The interfering beams traverse the square in two opposite directions, and the fringes produced at Q are recorded photographically. The whole apparatus including the source and camera is rotated about once a second. The fringes are found to be displaced from the positions occupied when the apparatus is at rest. According to non-relativistic theory, if the component of the velocity of one of the mirrors in the direction of the propagation of light is v, the velocity of light in one

* Reference 11.12.

direction is $(c+v)$ and in the other is $(c-v)$. If the length of path for each beam is S when the mirrors are stationary, there is a difference in transit time of

$$\frac{S}{c-v}-\frac{S}{c+v}$$

when the system is rotating.

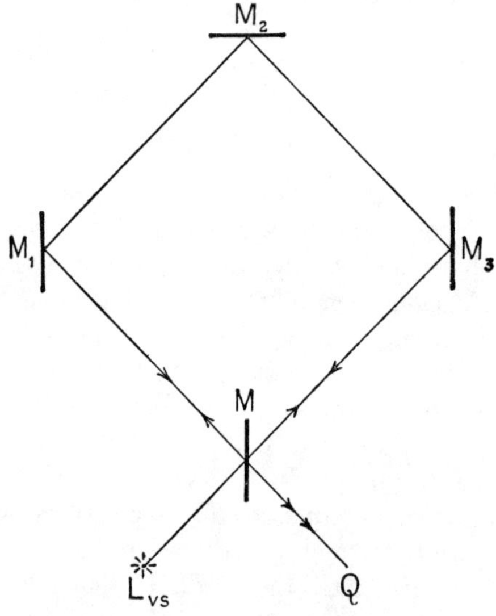

Fig. 11.6.—Sagnac's experiment

When v is small compared with c, the corresponding path difference is

$$\Delta S=\frac{2v}{c}S. \qquad 11(42)$$

The observed displacement of fringes is in agreement with this expression.

EXAMPLE 11(xi)

11(xi). Show that for a rectangular circuit the effective path difference is approximately

$$\Delta S=\frac{4A}{c}\Omega, \qquad 11(43)$$

where A is the area of the circuit and Ω is the angular velocity.

11.27.—Michelson used the rotation of the earth and was forced to use a large circuit in order to compensate for the comparatively slow rate of rotation. Moreover, he was unable to vary the rate of rotation. He therefore used two circuits, one of large area (2000 feet × 1100 feet) and one of small area. His apparatus is shown in fig. 11.7. One set of fringes is formed by interference between a beam which has traversed the main circuit (ABCD) in an anticlockwise direction, and a beam which has traversed the same circuit in a clockwise direction.

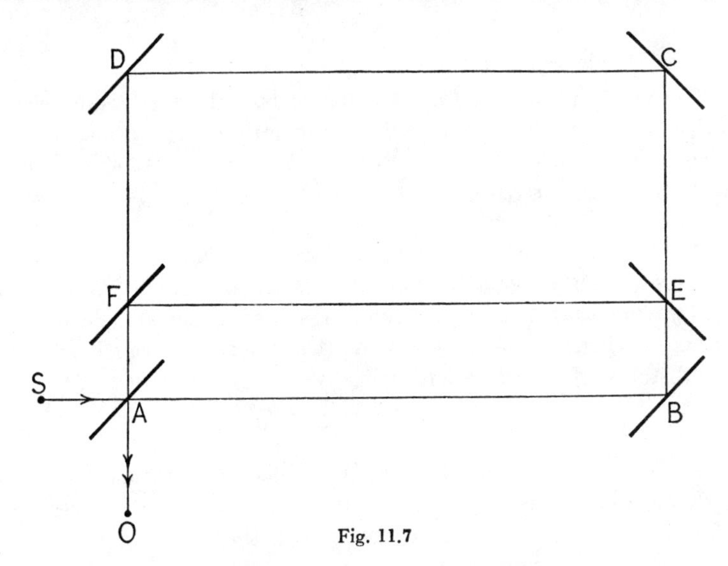

Fig. 11.7

An auxiliary set of fringes is formed by light which has traversed the small circuit ABEF in both directions. The source is a narrow slit and if images of the source corresponding to the two pairs of circuits accurately coincide, then the fringe systems should also coincide—if there were no effect due to rotation. The effect to be expected owing to the component of the earth's rotation in latitude 41° 40′, and for a wavelength of 5700 Å., was a displacement of 0·236 of a fringe width between the two systems of fringes. A displacement of 0·230 of a fringe was observed. This result was obtained only when effects due to variations of temperature and pressure had been reduced by enclosing both circuits in pipes which were partially evacuated.

11.28.—It is sometimes stated that the rotation experiments are "equally well" explained on (*a*) non-relativistic theory, (*b*) special relativity theory, (*c*) general relativity theory. This is not correct.

The " explanation " on non-relativistic theory given in § 11.25 assumes that the æther is entrained with the rotating apparatus. This, however, is quite inconsistent with the non-relativistic account of the " æther-drag " phenomenon (§ 11.20). Special relativity is not applicable to any system involving rotation, and therefore does not give any explanation. Langevin * has shown that general relativity does give a satisfactory account of the phenomenon. This is the only theory to account for these results without creating inconsistencies elsewhere.

11.29. **The Nebular Red-shift.**

The spectrum of light from a nebula shows absorption lines similar to the Fraunhofer lines in the solar spectrum, but with a displacement toward the red end of the spectrum. It is found that this displacement varies regularly with the brightness of the nebula, i.e. the fainter nebulæ give the greatest displacements. It is generally accepted that the fainter nebulæ are fainter mainly because they are the most distant, and, using other astronomical data, an expression has been derived giving the relation between brightness of the nebula and distance. If this relation is accepted, the increase of wavelength ($\Delta\lambda$) at a distance of d parsecs † is given by

$$\Delta\lambda/\lambda = 1{\cdot}7 \times 10^{-9}d. \quad . \quad . \quad . \quad . \quad 11(44)$$

This relation was discovered by E. G. Hubble, after whom it is named " Hubble's law ". It is fairly accurately verified for distances up to the order of 10^6 parsecs, i.e. to the limit of the observations.

11.30.—It is nearly certain that this phenomenon has an important bearing on the theory of light as well as on cosmological theory. Various interpretations have been proposed. The most generally accepted is that of the " expanding universe ". According to this theory, the red-shift is a radial Doppler effect. A uniform expansion of the whole universe would require that an observer at any point would observe a general recession of the nebulæ, the rate of recession being proportional to the distance from himself. If the red shift is due to a Doppler effect, we may put $\Delta\lambda/\lambda = v/c$ and equation 11(44) gives approximately

$$v = 50d, \quad . \quad . \quad . \quad . \quad . \quad . \quad . \quad 11(45)$$

where v centimetres per second is the velocity of recession for a nebula whose distance is d parsecs. The most distant nebulæ are estimated

* Reference 11.10. † 1 parsec = 3×10^{18} centimetres.

to be more than 10^6 parsecs away and to be receding with a velocity of more than 10^8 centimetres per second.

Equation 11(44) is valid only if we assume a static universe. If we assume that the distant nebulæ are receding, it is necessary to make a correction because the amount of light received from a receding body is less than the amount which would be received from a similar body which is stationary in our frame of reference. This requires a correction term proportional to d^2 in 11(44) and a consequential modification of 11(45). Thus the observations do not agree with a law of exactly uniform expansion of the universe.

11.31.—An explanation based on the idea that the effect is a gravitational shift is not satisfactory. If an observer in one nebula (A) observes a shift to the red in light coming from another nebula (B), then an observer in nebula B would observe a shift to the blue in light coming from nebula A. It would therefore be extremely probable that we should observe a red-shift in the light from some nebulæ, and a blue-shift in the light from others. The fact that we always observe a red-shift could be explained only by the assumption that our particular nebula is at, or near, a minimum of gravitational potential, and gravitational potential increases uniformly in every direction. This is highly improbable, and it is much more likely that the red-shift is due to some effect which is the same for all observers. It is desirable to point out that the interpretation in terms of an expanding universe is not the only one which satisfies this criterion. It has been suggested, for example, that a quantum may slowly lose part of its energy in transit across space. If this happened, and if h remained constant, there would be a decrease of frequency and an increase of wavelength. The increase of wavelength would be proportional to the distance of the source from the observer. Another suggestion is that the frequency of the light emitted by an atom of a given kind is slowly increasing owing to a change in one or more of the fundamental constants (h, m, e, and c). Here we are not concerned to argue the case for or against the ideas. It is, however, important to recognize that the observational data expressed in equation 11(44) are of great importance. The interpretation in terms of a radial Doppler effect, though adequate in the present state of our knowledge, may have to be abandoned when fresh experimental material becomes available.*

11.32. **Relation between Mass and Energy.**

We have seen that length and time are relative quantities, i.e. their values differ according to the movement of the frame of reference in which they are measured. It is reasonable to suppose that most derived quantities, such as momentum, energy, force, etc., will also be relative quantities, and that only certain rather special functions of length and time will be invariant (i.e. the same for all observers). It is shown in textbooks on relativity † that the general methods of

* The red-shift is given an important place in modern cosmological theory (Reference 11.17).

† Reference 11.3 or 11.5.

dealing with dynamical problems, which were invented by Hamilton on the basis of Newton's laws, may be retained if we regard the mass of a body as a function of its velocity. On this basis we have

$$m = m_0/\alpha, \qquad \ldots \ldots \quad 11(46)$$

where m_0 is the mass of the body in a frame in which it is at rest. m_0 is called the *rest mass* or the *proper mass*. m is called the *relativistic mass* or simply " the mass ". The momentum (P) and energy (E) of a body are then defined to be

$$P = mv = \frac{m_0 v}{(1 - v^2/c^2)^{1/2}}, \qquad \ldots \quad 11(47)$$

$$E = mc^2 = \frac{m_0 c^2}{(1 - v^2/c^2)^{1/2}}. \qquad \ldots \quad 11(48)$$

When v is small compared with c, the energy is given by

$$E = m_0 c^2 + \tfrac{1}{2} m_0 v^2 = E_0 + \tfrac{1}{2} m_0 v^2. \qquad \ldots \quad 11(49)$$

The energy thus consists of a constant part corresponding to the rest mass of the body, and a part corresponding to the kinetic energy.

An electrical charge e, distributed uniformly over a sphere of radius r_0, would have a potential energy e^2/r_0. Putting this energy equal to mc^2 we obtain

$$r_0 = \frac{e^2}{mc^2}. \qquad \ldots \ldots \quad 11(50)$$

The quantity r_0, so defined, is often called the "radius of the electron", even though the detailed picture of an electron as a charge distributed over a sphere is not generally accepted.

11.33.—Under the conditions of ordinary dynamical problems, with bodies moving at speeds which are small compared with that of light, the mass is effectively constant. Relativity gives the same predictions as Newtonian mechanics, because the constant term (E_0) in equation 11(49) appears on both sides of every energy equation, and is self-balancing. It is possible, however, to accelerate electrons, etc., so that they move with speeds comparable with that of light. Under these conditions a large variation of mass with velocity is obtained. Also, in the reactions of atomic nuclei it has been found that the rest mass of the final products is not always the same as that of the initial reacting material. An energy change in agreement with equation 11(48) is observed.

11.34. **Mass, Momentum and Energy of the Photon.**

From equations 11(47) and 11(48) we may derive the relation

$$E^2 = m_0{}^2c^4 + c^2P^2. \qquad \dots \qquad 11(51)$$

There is for light, however, direct experimental evidence (§ 17.21) that

$$E = cP. \qquad \dots \dots \qquad 11(52)$$

This relation is consistent with 11(49) if, and only if, the rest mass associated with light is zero. That some such assumption must be made concerning light may be seen by considering 11(47) or 11(48) alone. From these equations we see that a material particle would have to acquire an infinite momentum and an infinite energy in order to move with the velocity of light. We can, however, imagine a particle whose rest mass is extremely small, and whose velocity is extremely near to c. A suitable relation between the velocity and the rest mass will give a finite momentum and energy. The photon may be regarded as the limiting case when rest mass has tended to zero and velocity to c. In this sense, a photon may be said to have zero rest mass.

11.35.—The mass associated with a quantum of visible light is of order 10^{-33} gramme. Thus an atom whose weight is of order 10^{-24} to 10^{-23} gramme loses only a very small fraction of its energy when a quantum of visible light is emitted. The mass of the energy received per year by the earth from the sun is of the order 6×10^{10} grammes, and the total amount emitted from the sun per year is $1{\cdot}4 \times 10^{20}$ grammes. Thus, in the course of a million years' emission at this rate, the sun would lose $1{\cdot}4 \times 10^{26}$ grammes or 10^{-3} per cent of its present mass. The loss of mass by radiation is very important in connection with the process of stellar evolution. We do not know the stages of the process by which mass is converted into radiation in the stars, though there are detailed theories concerning this. It is nearly certain that some set of nuclear reactions in the core of a star is accompanied by the transformation of mass into kinetic energy, and into radiation of high frequency. Most of this energy is changed by scattering, and by absorption and re-emission into radiation of longer wavelengths as it travels towards the outside of the star. It is finally emitted in the form of ultra-violet, visible and infra-red radiation.

REFERENCES

11.1. EINSTEIN: *Relativity* (Methuen).

11.2. JOOS: *Theoretical Physics* (Blackie).

11.3. EDDINGTON: *Mathematical Theory of Relativity* (Cambridge University Press).

11.4. TOLMAN: *Relativity, Thermodynamics and Cosmology* (Oxford University Press).

11.5. MCCRAE: *Relativity Physics* (Methuen).

11.6. MICHELSON and MORLEY: *Phil. Mag.*, 1887, Vol. 24, p. 449.

11.7. KENNEDY: *Nat. Acad. Sci. Proc.*, 1926, Vol. 12, p. 621.

11.8. ILLINGWORTH: *Phys. Rev.*, 1927, Vol. 30, p. 692.

11.9. JOOS: *Annalen d. Physik*, 1930, Vol. 7, p. 385.

11.10. LANGEVIN: *Comptes Rendus Acad. Sc.*, 1921, Vol. 173, p. 831.

11.11. *Lick Observatory Bull.*, 1923, Vol. 11, p. 141, and 1928, Vol. 13, p. 130.

11.12. ST. JOHN: *Astrophysical Journal*, 1925, Vol. 67, p. 195.

11.13. ADAMS: *Proc. Nat. Acad. of Sciences* (*U.S.A.*), 1925, Vol. 11, p. 382.

11.14. HUBBLE: *The Observational Approach to Cosmology* (Oxford University Press).

11.15. IVES and STILWELL: *J.O.S.A.*, 1941, Vol. 31, p. 369.

11.16. OTTING: *Phys. Zeits.*, 1939, Vol. 40, p. 681.

11.17. MCVITTIE: *Cosmological Theory* (Methuen).

CHAPTER XII

Polarized Light

12.1. Scalar and Vector Wave-theories.

In the experiments described in Chapters II–IX, all planes which include the wave normal are equivalent. If, for example, the wavefront is a horizontal plane, anything that can be said concerning a vertical north-south plane is equally true of a vertical east-west plane. The phenomena of interference and diffraction which we have described lead naturally to a wave theory, but the quantity whose periodic fluctuations constitute the waves need not necessarily be a vector quantity. If these were the only experiments on light we should naturally tend to use a scalar wave theory because of its greater simplicity.

We now come to consider experiments in which the results depend on the relative orientation of different parts of the apparatus with respect to planes through the wave normal. These results cannot be described in terms of a scalar wave-theory, but are adequately described by a theory in which light is represented by the periodic fluctuations of a quantity which has direction as well as magnitude. The disturbance at any moment is then represented by a vector, which specifies the direction of the disturbance as well as its magnitude.* In an isotropic medium the vector is always in the wavefront, and the ray is normal to the wavefront. The result of the experiments which will now be described lead to the assumption that there are two types of light, called *polarized light* and *unpolarized light*. We find that we have to represent the former by a vector whose magnitude and orientation vary in a related way as the phase of the light alters. When the orientation remains constant and only the amplitude varies, the light is said to be *plane-polarized*. When the amplitude remains constant, but the orientation of the vector varies regularly so that the end of the representative vector moves uniformly round a circle, the light is said to be *circularly polarized light*. There is another type of

* It should be understood that this representation is entirely different from the previous " vector representation " in which the direction of the vector represented the phase.

polarized light in which both amplitude and orientation vary in such a way that the end of the representative vector moves smoothly round an ellipse. This kind is called *elliptically polarized light.*

A beam of unpolarized light may be regarded as the resultant of two beams, polarized in two different planes and having no permanent phase relation (§§ 12.17 and 12.31). The variation of the direction of the vector which represents this resultant is not related, in any regular way, to the variation of its magnitude. Most of the polarization phenomena are subject to dispersion (i.e. they vary with the wavelength of the light), and some of the effects due to dispersion are considered in §§ 12.38–12.44. In the earlier part of this chapter it is assumed that the light is monochromatic.

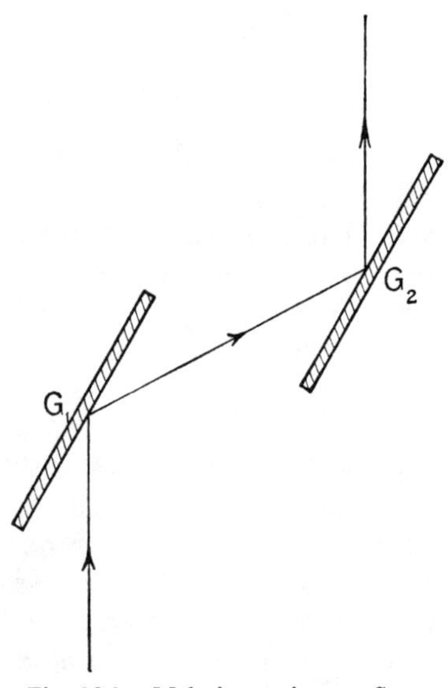

Fig. 12.1.—Malus' experiment. Successive reflections at two unsilvered mirrors (G_1 and G_2).

12.2. **The Experiment of Malus.**

Ordinary light may be plane-polarized by reflection at the unsilvered surface of a transparent medium such as glass. This phenomenon was discovered by Malus, who allowed light to be successively reflected at two such surfaces. If the relation of the two surfaces is that shown in fig. 12.1, the light is strongly reflected. If the second piece of glass G_2 is turned so as to reflect the beam out of the plane of the paper, then the amount of reflected light is greatly reduced. Thus the first reflection alters the light so that it is capable of being strongly reflected in the plane of the paper, but is not capable of being strongly reflected (by an unsilvered mirror) in a perpendicular plane. These results may be described by saying that the first mirror polarizes the light, and the second reveals that it is polarized. An arrangement which produces a beam of plane-polarized light from a beam of unpolarized light is called a *polarizer*. An arrangement which detects plane-polarized light is called an *analyser*. Any piece of apparatus which is capable of acting as a polarizer is capable of acting as an analyser and vice versa.

In the experiment of Malus the first unsilvered mirror is regarded as a polarizer, and the second as an analyser, but the experiment might equally well be performed with the light travelling in the reverse direction. When a polarizer and an analyser are oriented so as to pass the maximum amount of light, they are said to be *parallel.* When the relative orientation is such that the light emerging from the system is a minimum, they are said to be *crossed.* Two unsilvered mirrors are parallel when the reflections are in the same plane, and are crossed when the reflections are in perpendicular planes.

12.3. **Definition of the Plane of Polarization.**

The description of a parallel beam of unpolarized light is complete when the direction of propagation, the amplitude, and the frequency have been stated. To complete the description of a beam of plane-polarized light, it is necessary to give an additional datum which specifies an azimuth. It is usual to state the plane in which the beam is most strongly reflected at an unsilvered glass surface, and this plane is called the *plane of polarization.* It should be emphasized that this choice is purely a matter of convention. There is no logical reason why the plane of minimum reflection should not be called the plane of polarization, and some writers have adopted this convention. The important thing is that our definition of plane of polarization must be such that, given a beam of plane-polarized light, we can determine the plane of polarization by a simple physical test. The most simple method is to measure amounts of light reflected in different planes. At a certain stage in the elastic-solid theory of light, it was important to know whether the direction of vibration, which was usually called the direction of the "light vector", was in, or perpendicular to, the plane of polarization. For the discussions of this chapter, this question is of no significance and, for convenience, we assume that the light vector is *normal* to the plane of polarization. In Chapter XIII we shall show that, in the electromagnetic theory, a beam of plane-polarized light in an isotropic medium is represented by a magnetic vector in the plane of polarization, and an electric vector in a perpendicular plane. That is to say, we shall describe further experiments which show that the single " light vector ", which is sufficient for the description of the experiments discussed in this chapter, must be identified with the electric vector of Maxwell's theory.

12.4. **Brewster's Law.**

In § 12.2 we did not discuss the ratio of the maximum amount of light transmitted when the polarizer and analyser are parallel, to the minimum obtained when they are crossed. The variation of this quantity with the angle of incidence (θ) was investigated by Brewster, who showed that this ratio is very large for an angle θ_p given by

$$\tan \theta_p = \mu. \qquad \qquad 12(1)$$

This relation is known as *Brewster's law*, and the angle which satisfies it is known as the *polarizing angle.* A beam of light which gives zero transmission for one orientation of an analyser is said to be completely plane-polarized.

If a beam which is completely plane-polarized is added to a beam which is unpolarized, the result is said to be partially polarized. When a beam of partially polarized light is passed through an analyser, there is a maximum for one orientation and a minimum for another, but the minimum is not zero. By this test * the light reflected from an unsilvered mirror is very nearly completely polarized when $\theta = \theta_p$, and is partially polarized for any other value of θ except $\theta = 0$. It was at one time thought that the light reflected at the polarizing angle was completely plane-polarized, but detailed experiments which will be discussed later (§ 14.17) show that this is not so.

12.5. **Polarization by Transmission.**

If the light passing through a plate of glass falls upon an analyser, it is found to be partially polarized (unless $\theta = 0$). For one transmission, the degree of polarization is small. The ratio of the maximum reflection coefficient given by one orientation of the plate to the minimum (obtained with a perpendicular orientation) is largest when the light is incident at the polarizing angle, but this maximum ratio is only about 1·1 : 1. When a beam of light is incident at the polarizing angle upon a pile of parallel plates (fig. 12.2), the degree of polarization of the transmitted light increases with the number of plates. A pile of twenty-five plates gives strong polarization. If the transmitted light is reflected at the front surface of an unsilvered mirror (used as an analyser), the maximum reflection is obtained when the analyser is turned to reflect the beam out of the plane of the paper (fig. 12.2),

* We assume for the moment that auxiliary tests such as those described in § 12.28 have shown that elliptical polarization is not present.

i.e. when its orientation is at right angles to that which would give maximum reflection for a beam which has been reflected from an unsilvered surface. For this reason we say that the transmitted light is polarized in a plane perpendicular to the plane in which the reflected light is polarized.

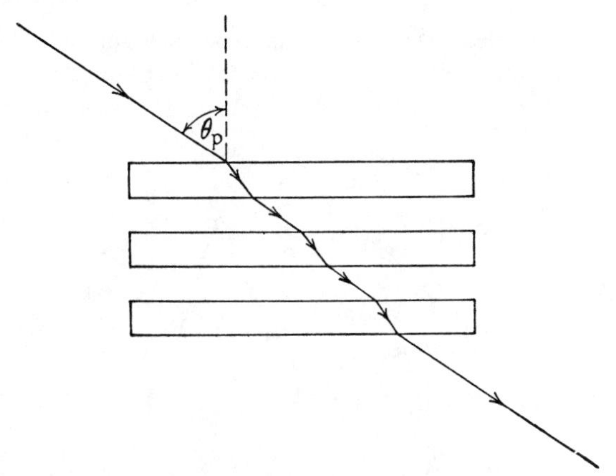

Fig. 12.2.—Polarization by transmission through a pile of plates

12.6. Double Refraction.

Crystals of the cubic system are optically isotropic. Each cubic crystal has a single refractive index, and crystalline media of this type behave like non-crystalline materials such as glass. Other crystals are optically anisotropic. The phenomena observed when a beam of unpolarized light enters a crystal of this latter kind depend on the relation between the direction of the beam and the axes of crystal symmetry. When the beam is plane-polarized, the effects also depend upon the relation between the plane of polarization and the crystal axes. In general, when a beam of light enters an anisotropic medium, it is divided into two parts which are refracted in different directions. This phenomenon is called *double refraction* or *birefringence.* From the theory which will be developed later it is possible to derive general laws of double refraction. These laws cannot be stated completely in any brief and simple way and, for our present purpose, it is sufficient to consider certain special cases.*

* In the course of the discussion in this chapter and in Chapter XVI, we shall state some of the symmetry properties of certain crystals. The reader who wishes to relate these statements to the general systematic classification of crystals is advised to consult References 12.1 and 12.2.

12.7.—Calcium carbonate crystallizes in rhombohedra to form a mineral which is called *calcite* or *Iceland spar*. The shape of the crystal is shown in fig. 12.3, in which ABCDA′B′C′D′ represent the corners of the crystal. Two of the corners A and A′ each contain three obtuse angles. The direction of a line making equal angles with each of the three edges meeting at A is called the *principal axis* of the crystal. In a perfectly formed crystal, the six faces are similar, and the line AA′ is then in the direction of the axis.* The plane ACA′C′ is called a *principal plane*. The principal axis and the principal plane are

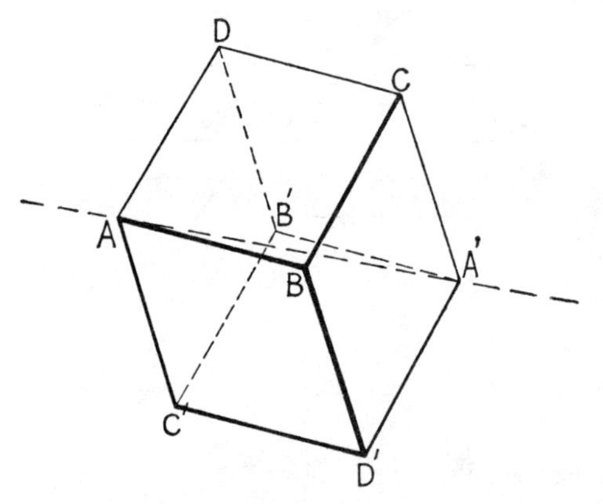

Fig. 12.3.—Regular crystal of calcite

defined by directions. Any line parallel to AA′ may be called a principal axis, and any plane parallel to ACA′C′ may be called a principal plane. It is possible to draw a line and a plane, in the directions of the principal axis and principal plane respectively, through any point in the crystal. These definitions of the principal axis and the principal plane refer to the symmetry of the crystal form, and not to the optical properties. We shall see later that there are, in general, two directions in a crystal which are of special importance in relation to the optical properties. These are called *optic axes*. The optic axes do not always coincide with any of the axes of crystal symmetry, though there is usually a fairly simple geometrical relation between the directions of the optic axes and the axes of crystal symmetry. In calcite, the two optic axes coincide, and the crystal is said to be *uniaxial*. The single optic axis also coincides with the principal axis of crystal sym-

* It should, however, be understood that the axis is defined by the angles which it makes with the appropriate faces. This definition applies to crystals which have not grown with perfect regularity, or to a broken piece of a crystal.

metry which has been defined above. These relations correspond with the fact that the laws of double refraction in calcite are not as complicated as the general laws applying to some other types of crystal.

12.8.—The following experiments may be carried out with two calcite plates which have been optically worked so that two plane faces of each plate are parallel, and so that the angle between the optic axis and the normal to the polished surfaces is the same for each plate. The exact value of this angle is not important provided that it is not too near zero or $\frac{1}{2}\pi$, i.e. the optic axis must not be in, or perpendicular to, the polished surface. For simplicity of description we shall discuss the case for which the normal to the surfaces is in a principal plane, and the optic axis is at about 45° to the surface (fig. 12.4*a*, p. 353).

A narrow beam of unpolarized light incident normally on one of the plates, is found to divide into two beams. One, called the *ordinary ray*, goes straight through; the other, called the *extraordinary ray*, is deflected on entering the crystal and emerges parallel to its original direction.* The extraordinary ray always lies in the principal plane which passes through the point at which the light enters the crystal. If the crystal is rotated about the normal to the polished surfaces, the extraordinary ray rotates with it. The separation of the two emergent beams is proportional to the thickness of the plate.

If the emergent beams are allowed to fall upon a second plate, whose optic axis is parallel to that of the first plate, two beams emerge from this second plate (fig. 12.4*b*). They are in the same plane as those emerging from the first plate, but have a greater separation. The same final result would be obtained by passing the light through one plate whose thickness is equal to the *sum* of the thicknesses of the plates. If now the second plate is rotated through an angle π, the deflections in the two plates are of opposite sign and the combination behaves like a single plate of thickness equal to the difference of the thicknesses of the two plates (fig. 12.4*c*). If the second plate is rotated through $\frac{1}{2}\pi$, there are still only two beams, but the beam which passed straight through the first plate is deflected in the second, and the beam which was deflected in the first plate is undeflected in the second. Thus the ordinary ray for the first plate behaves as an extraordinary ray in the second plate and vice versa (fig. 12.4*d*).

If the second plate is given an intermediate orientation with respect to the first, then there are four emergent beams (fig. 12.4*e*). Let

* The effect is best seen in calcite which is not of the highest optical quality, since this scatters a small amount of light, and so reveals the paths of the rays inside the crystal.

us call the ordinary beam which emerges from the first plate O_1, and the extraordinary beam E_1. Then the four images due to the beams emerging from the second plate may be called O_1O_2, O_1E_2, E_1O_2 and E_1E_2. O_1O_2 has always the same brightness as E_1E_2, and O_1E_2 has always the same brightness as E_1O_2.

If the second plate is rotated from the position in which its axis is parallel to that of the first, the following changes occur. First of all O_1E_2 and E_1O_2 are of zero brightness, and there are only two beams emerging from the second plate. As the plate is rotated O_1E_2 and E_1O_2 increase in brightness, and O_1O_2 and E_1E_2 decrease. When the plate has been rotated through an angle of $\frac{1}{4}\pi$, the four images are of equal brightness. When it has rotated through $\frac{1}{2}\pi$, O_1O_2 and E_1E_2 are of zero brightness, and there are again only two beams. As the second plate is rotated through another right angle, the same changes occur in the reverse order. When the total rotation is π, there are again only two images, and these are of equal brightness. The deflections of the images are now in the same line, but in reverse directions (fig. 12.4*c*). If the plates are of equal thickness, the images are superposed.

12.9.—These observations on double refraction, taken by themselves, give evidence of the polarization of light since the refraction in the second plate of either of the two beams which emerge from the first plate depends on the orientation of the beam with respect to the crystal axes. All the phenomena are consistent with the supposition that the ordinary and extraordinary ray are polarized in two mutually perpendicular planes. This evidence for the existence of the polarization of light is quite independent of the experiments of Malus. It did, in fact, lead Huygens to the basic idea of transverse waves more than 150 years before Malus' experiment was performed.

The experiments on polarization by reflection and by transmission have been described separately in order to show that each set, taken by itself, leads to the concept of polarized light. When this point is accepted, it is convenient to discuss them together. When the two beams emerging from a single calcite plate are examined by reflection at an unsilvered surface, it is found that the ordinary beam is polarized in the principal plane, and the extraordinary beam in a perpendicular plane. Further tests show that when a beam which has been polarized by reflection is incident normally upon a calcite plate, its behaviour depends on the relation between the plane of polarization and the principal plane of the crystal. When the plane of polarization is

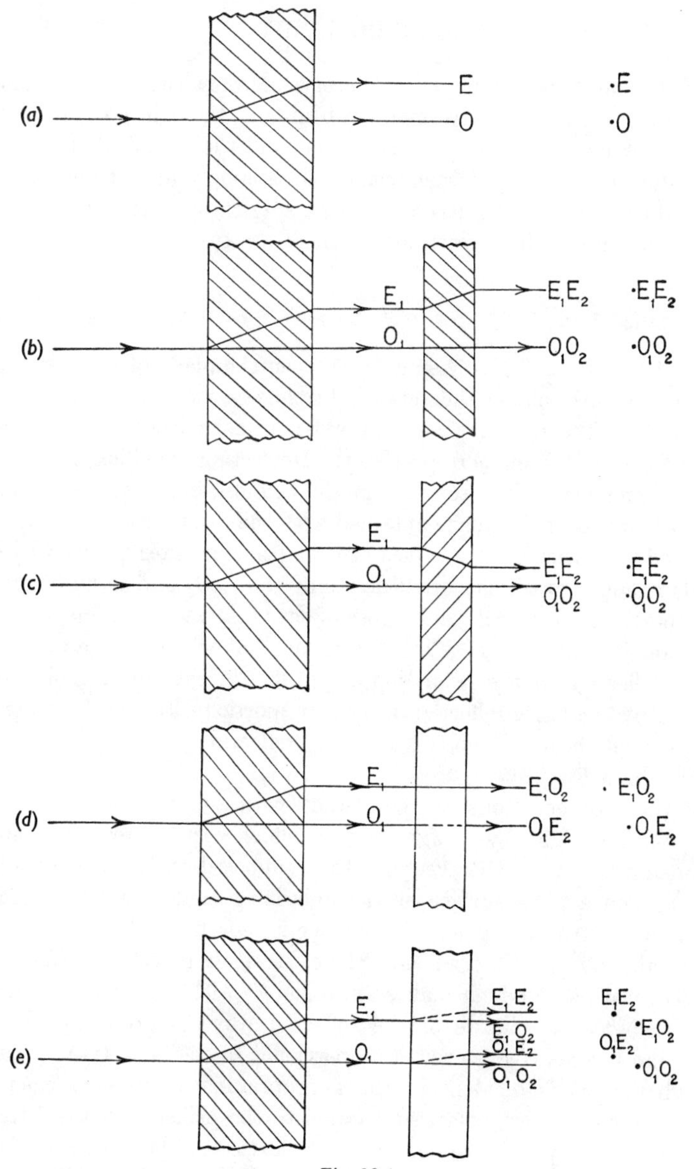

Fig. 12.4

(*a*) Refraction of light by a calcite plate.
(*b*) Refraction by two plates with the same orientation.
(*c*) Refraction by two plates, one rotated through π about the direction of the incident ray.
(*d*) Refraction by two plates, one rotated through $\frac{1}{2}\pi$.
(*e*) Refraction through two plates, one rotated through $\frac{1}{4}\pi$.

(The dots on the right-hand side show the relative positions of the spots of light received on a screen perpendicular to the direction of incidence. Dotted lines indicate rays passing out of the plane of the paper.)

parallel to the principal plane of the crystal, then there is only one image formed by an ordinary emergent beam, i.e. the light goes straight through. When the plane of polarization and the principal plane of the crystal are mutually perpendicular, there is only an extraordinary beam, which suffers a displacement in passing through the crystalline plate. For intermediate orientations there are two beams.

12.10. **Malus' Law.**

It is found that if ψ is the angle between the plane of polarization and the principal plane, the relative brightnesses of the images corresponding to the ordinary and the extraordinary beam are proportional to $\cos^2 \psi$ and $\sin^2 \psi$ respectively. Neglecting small losses due to reflection and scattering, the sum of the brightnesses is always equal to the brightness of the image obtained with the crystal removed. It is also found that when the angle between the principal axes of two calcite crystals is equal to ψ', the brightnesses of the O_1O_2 and E_1E_2 images are proportional to $\cos^2 \psi'$, and those of the O_1E_2 and E_1O_2 images are proportional to $\sin^2 \psi'$. Finally, it is found that when a beam of light is twice reflected (at the polarizing angle), the brightness of the image produced by the twice-reflected beam is proportional to $\cos^2 \psi''$, where ψ'' is the angle between the two planes of reflection. This last relation is known as *Malus' law*.

All these observations are consistent with the representation of plane-polarized light by a light vector which can be resolved into components like any other vector. It is supposed that when light is incident upon a glass surface at the polarizing angle, only the component polarized in the place of incidence is reflected. If the incident light is already polarized in this plane, it is all reflected. If it is polarized in a plane which makes an angle ψ'' with the plane of incidence, then the amplitude of the reflected wave is proportional to $\cos \psi''$, and the relative energy is proportional to $\cos^2 \psi''$. In a similar way, when light is incident in the appropriate way upon a calcite crystal, the component polarized parallel to the principal plane forms the ordinary beam, and the component polarized in the perpendicular plane forms the extraordinary beam. Detailed consideration shows that this hypothesis is consistent with all the observations on the variations in the brightness of the images. The fact that the results of calculations based on the resolution and composition of vectors are in agreement with experiment justifies the use of a vector wave representation.

12.11. Methods of producing Plane-polarized Light.

As we have seen in §12.2, plane-polarized light may be produced by reflection at the Brewsterian angle, or by transmission through a pile of plates. The former method involves great loss of light, since only a few per cent of the incident beam is reflected. The latter method does not give complete polarization and, to obtain a high degree of polarization, it is necessary to use an inconveniently large number of plates. For this reason many methods of producing plane-polarized light by means of double refraction have been devised. The simplest method is to use a thick piece of calcite, and to insert stops so as to

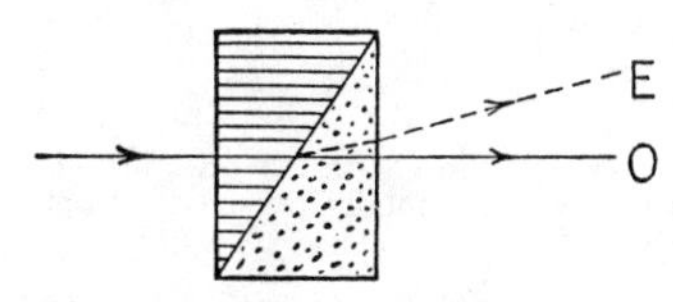

Fig. 12.5.—Rochon prism (dots indicate optic axis perpendicular to the plane of the paper).

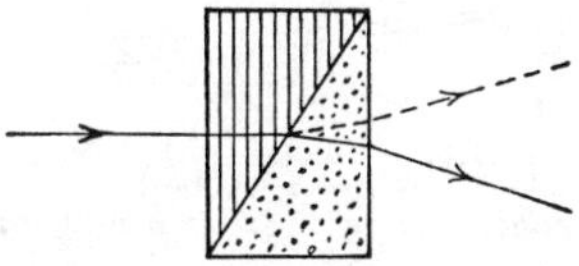

Fig. 12.6.—Wollaston prism (dots indicate optic axis perpendicular to the plane of the paper).

remove either the extraordinary or the ordinary beam. Since the separation is not very great, it is necessary to use a narrow beam. The prisms of Rochon (fig. 12.5) and of Wollaston (fig. 12.6) produce two beams of light, polarized in mutually perpendicular planes and travelling in different directions. The orientations of the optic axes are shown in the figures. It will be noticed that the Wollaston prism produces the wider separation, but the Rochon prism leaves one ray undeviated. This is an advantage in certain applications. These prisms are usually made either with quartz or with calcite. The former is more easily worked to high optical quality, but the latter provides the larger separation.

12.12. Nicol, Foucault, and Glan-Thompson Prisms.

The difference in the angles of refraction for the ordinary and extraordinary rays suggests the possibility of separating them by

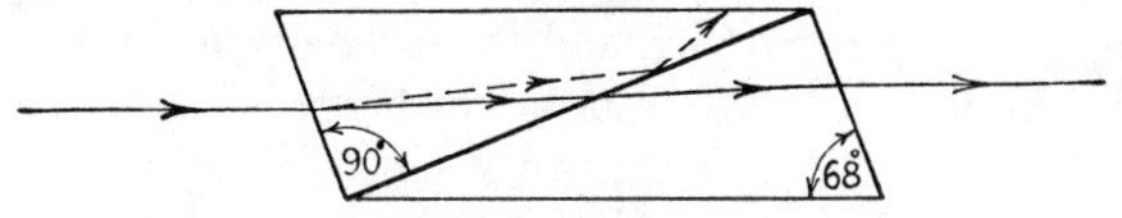

Fig. 12.7.—Nicol prism

arranging that one is totally reflected, and the other transmitted at a thin film separating two pieces of calcite. This possibility was first realized by Nicol, who cut a calcite crystal in a way shown in fig. 12.7, and cemented the two pieces together with Canada balsam. The extraordinary ray is transmitted, and the plane of polarization of the transmitted light is perpendicular to the plane of the diagram.

Calcite prisms of large size are expensive and, in order to obtain a bright source of polarized light, it is desirable to concentrate the light on the Nicol. The angular divergence of the cone which can be used is limited by the difference

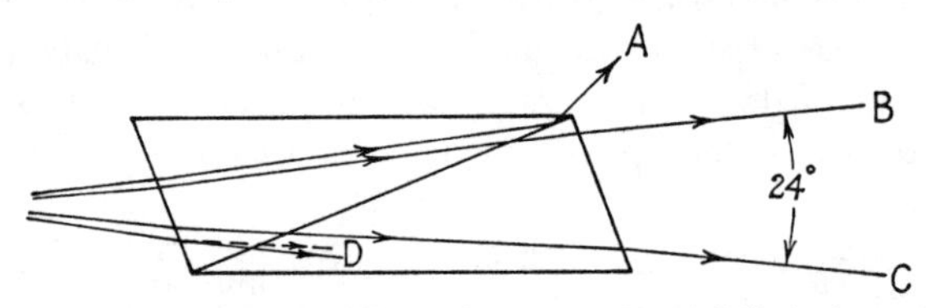

Fig. 12.8.—Angular field of Nicol prism; rays B and C limit the useful field. Outside this range either both components of ray A are totally reflected, or both components of ray D are transmitted at the interface.

between the critical angle for the ordinary and extraordinary rays. By drawing extreme rays, as shown in fig. 12.8, it is found that the maximum permissible divergence of the transmitted beam is 24°. The Nicol prism has the disadvantage that the emergent light is slightly elliptically polarized owing to a secondary effect due to the inclined end faces.

The Glan-Thompson prism (fig. 12.9) is designed to give a wider field and more perfect plane-polarization. The exact angle at which the crystal is cut varies according to the purpose for which it is designed. The prism shown in the diagram may be regarded as typical. The design of this type of prism is discussed in detail in Reference 12.3. It is not necessary that Canada balsam should be used as the separating layer, though it is most generally useful for the polarization of visible light. A prism of the Nicol type with air as the separation film was devised by Foucault for the polarization of ultra-violet radiation. Such a prism is transparent to 2300 Å.

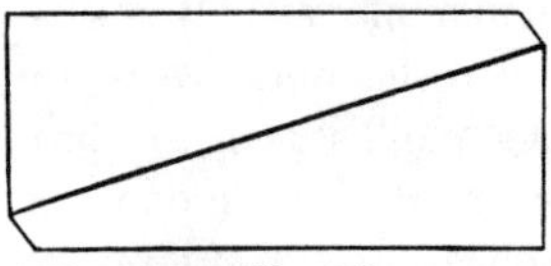

Fig. 12.9.—Glan-Thompson prism

12.13. Polarization by Absorption.

It has long been known that when unpolarized light is passed through tourmaline, the emergent light is partially plane-polarized. The crystal is doubly refracting, and the ordinary beam is much more strongly absorbed than the extraordinary beam. The natural crystal is strongly coloured and, for most wavelengths, absorbs a considerable part of the extraordinary, as well as of the ordinary beam. It is not suitable as a polarizer and its properties are chiefly of theoretical interest. A series of artificial materials, which polarize by absorption, have been developed by the Polaroid Corporation, U.S.A., following an invention by E. H. Land. The Polaroid type H film contains iodine which has been imbibed in an initially transparent plastic sheet of polyvinyl alcohol. The polarizing unit is an iodine polyvinyl alcohol complex which has been oriented by stretching the film.* This material transmits nearly 80 per cent of light polarized in one plane and less than 1 per cent of light polarized in the perpendicular plane. For wavelength 5500 Å., two pieces transmit up to 40 per cent of incident unpolarized light when parallel, and less than 0·01 per cent when crossed. At the extreme blue end of the spectrum the polarizing action is not quite so good and nearly 0·1 per cent of the incident light is transmitted. The residual light seen when a powerful source is viewed through two crossed pieces of Polaroid film is therefore blue. For an earlier type of film (known as type J) the residual light was mainly red. Large sheets of the material (up to 20 in. × 50 in.) are obtainable. Polaroid film offers the most convenient and inexpensive method of obtaining a strong source of light which is nearly completely plane-polarized. Prisms of the Glan-Thompson type are superior for use in instruments when it is important to have a very high degree of polarization for all wavelengths transmitted.

12.14. **Uses of Polarizing Devices.**

Polarizing devices are used in a variety of ways in scientific instruments, and in industry. In some experiments it is desirable to be able to reduce the effective intensity of a source of light in an accurately known ratio. This may be done by inserting a polarizer and an analyser between the source and the point of observation. The light

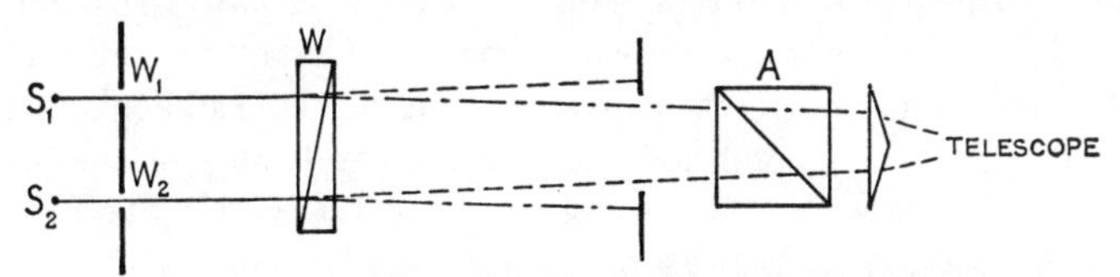

Fig. 12.10.—Polarization photometer

transmitted is proportional to $\cos^2 \theta$, where θ is the angle through which the analyser has been rotated from the parallel setting. This method is accurate and sensitive when the ratio of reduction is not less than 1 : 10 [Example 12(ii)].

A photometer based on this principle is shown in fig. 12.10. Light from the two sources S_1 and S_2 is polarized in two mutually perpen-

* The author is indebted to the Polaroid Corporation for supplying this technical information.

dicular planes by the Wollaston prism W, and the field is viewed through the analyser A which is rotated until the two halves appear equally bright. The ratio of illumination falling on the windows W_1 and W_2 is equal to $\tan^2 \theta$. If the ratio of the intensity of the weaker source to that of the stronger source is less than about 1 : 5, a filter of known transmission is placed in front of one window.

Polarizing devices are used in industry to detect strains in glassware. The glass is placed between a polarizer and an analyser which are crossed, so that the field is dark except in the strained regions which possess double refraction. Polaroid spectacles are worn to reduce glare due to sunlight reflected by the sea. In a similar way Polaroid screens are sometimes inserted in front of reading lamps to reduce reflection from glossy paper. In both of these applications the polarizing device is only partially effective because only part of the light is reflected at the polarizing angle. It has been proposed that all motor-headlights be covered with Polaroid film oriented at 45° to the vertical and that each driver be provided with film oriented at the same angle as that over his own headlights (and therefore at right angles to the film over the headlights of a car travelling in the opposite direction). He thus sees the headlights of an oncoming car only as glare-free blue discs. The advantages of this scheme are obvious. The disadvantages include a serious loss of light which it is proposed to make good by using 125-watt headlights. The overall merits of the device, and the prospects of its being generally adopted, depend on considerations which are not purely a matter of optics and which lie outside the scope of this book.

12.15. **Interaction of Beams of Plane-polarized Light.**

The conditions under which two beams of polarized light may produce interference fringes were investigated by Fresnel and Arago, whose results may be summarized as follows:

(1) Two beams of light plane-polarized in mutually perpendicular planes do not produce interference fringes under any condition.

(2) Two beams of light plane-polarized in the same plane interfere under the same conditions as two similar beams of unpolarized light, *provided that they are originally derived from the same beam of plane-polarized light, or the same component of a beam of unpolarized light.*

(3) Two beams of plane-polarized light derived *from perpendicular components of unpolarized light*, and afterwards rotated into the same plane, do not produce interference fringes under any condition.

12.16.—The generalizations stated in the preceding paragraph were derived from a rather lengthy and complicated set of experiments. The essential points are shown by the following experiments.

An apparatus similar to that used by Young in the original discovery of interference is used, and a polarizing device is placed in front of each of the slits (fig. 12.11). A pile of mica plates was actually used, but for a modern demonstration a piece of Polaroid film would be more convenient. The following results were obtained:

(i) If the polarizer P_C is omitted, and P_A and P_B are set parallel to one another, interference fringes are obtained. If they are set to polarize the two beams in mutually perpendicular planes, no interference fringes are obtained.

(ii) With the three polarizers in position, and with P_A and P_B set to polarize in parallel planes, the orientation of P_C affects the total illumination on the screen S, but not the distribution. So long as there is any illumination, fringes are obtained. In this case the interfering beams are both derived from the same polarized beam, and they are polarized in the same plane. Their plane of polarization is not the same as that for the beam from which they are derived, except when P_A, P_B and P_C are all parallel.

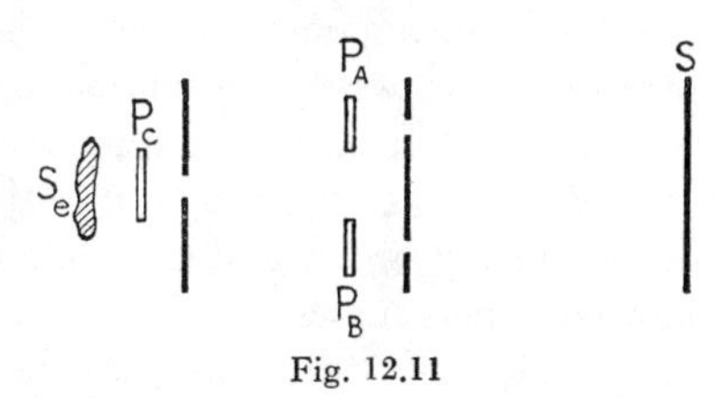

Fig. 12.11

(iii) A doubly refracting plate of calcite is placed behind each slit. The two plates are of equal thickness. Their principal planes are at right angles to one another, so that the ordinary ray passed by one plate is polarized in the same plane as the extraordinary ray passed by the other. Two sets of fringes are obtained. These are displaced to the right and left of the system obtained when the plates are removed. Further investigation shows that one set is due to the interference of the ordinary beam transmitted by the upper plate, and the extraordinary beam transmitted by the lower plate. The other set is due to the other pair of beams. The sign of the displacements indicates that, for calcite, the ordinary beam is retarded with respect to the extraordinary beam, i.e. the optical thickness of the crystal (and therefore the refractive index) is greater for the ordinary than for the extraordinary ray.

(iv) Two polarizers and a doubly refracting crystal C are placed behind the slits, as shown in fig. 12.12. The two polarizers are set with their planes perpendicular to each other and at an angle of $\frac{1}{4}\pi$ to the principal plane of the crystal. No fringes are observed. In this condition an extraordinary and an ordinary beam pass each slit. The four beams are of equal amplitude and, as in experiment (iii), there are two pairs polarized in the same plane. If, however, we consider either of these pairs, we see that its members have been derived from perpendicular components of the unpolarized source. No interference is obtained, indicating that these components are non-coherent. Similar results may be produced by inserting polarizing devices in appropriate parts of the other pieces of apparatus which are commonly used to produce interference fringes with unpolarized light. It should, however, be remembered that the polarizing devices may cause phase retardations. If these are too large, no interference will take place even if the conditions (1) and (2) stated in § 12.15 are fulfilled. Also the polarization may alter the relative amplitudes of reflected beams and thereby affect the distribution

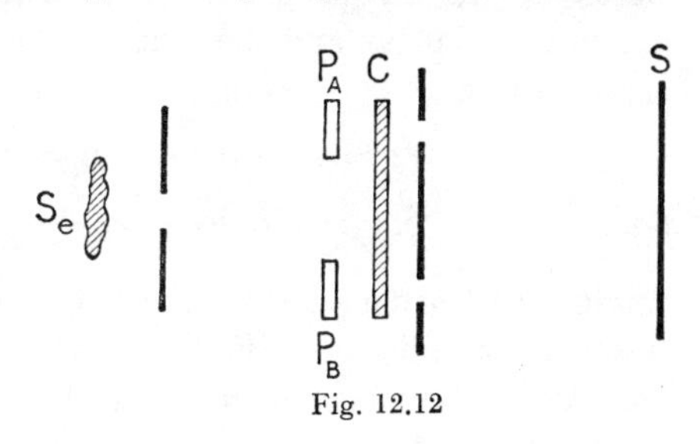

Fig. 12.12

of illumination in the interference fringes. It is even possible that one of two interfering beams may be eliminated and, if this happens, the fringes disappear.

12.17.—The vector representation of polarized light is in accord with the observation that two beams polarized in mutually perpendicular planes cannot produce interference fringes. Two vectors situated in perpendicular planes cannot annul one another, because each has no component in the plane containing the other. The vector representation would also lead us to expect that if a long train of waves polarized in a given plane is resolved into two (e.g. by the use of a Rochon prism), the beams should have a permanent phase relation. They are not able to interfere because they are situated in mutually perpendicular planes, but a component of each, polarized in an intermediate plane, may be selected. These two components interfere, provided that they are brought together without the introduction of an unduly large path difference. Since beams derived from mutually perpendicular components of unpolarized light do not interfere, it follows that, if unpolarized light is to be represented by a single transverse vector, it must be assumed that the plane of the representative vector changes in an irregular way.*

12.18. Circularly Polarized Light and Elliptically Polarized Light.

Although two disturbances in mutually perpendicular planes cannot give interference fringes, even when there is a permanent phase relation, the interaction of such vibrations does produce a type of light which has special properties of its own. Consider two disturbances, one in the xz plane and the other in the yz plane and both travelling in the direction OZ. They may be represented by

$$\xi_x = a \cos(\omega t - \kappa z) = a \cos \phi, \qquad 12(2a)$$

and

$$\xi_y = b \cos(\omega t - \kappa z + \epsilon) = b \cos(\phi + \epsilon). \qquad 12(2b)$$

These vibrations have the same frequency and velocity of propagation, but their amplitudes differ, and they have a permanent phase difference ϵ. Let us first consider the special case when $\epsilon = 0$. The resultant vibration is then represented by a vector of length $(a^2 + b^2)^{1/2}$. It lies in a plane which includes OZ and makes an angle $\chi = \tan^{-1} b/a$ with the x axis. Thus the resultant of these two disturbances is, in this special case, a linear disturbance represented by a vector in a plane intermediate between the planes of the vectors which represent the component disturbances.

* See § 12.31 for a further discussion of this point.

Now consider a second special case when $\epsilon = -\frac{1}{2}\pi$. Then we may write

$$\xi_y = b \sin(\omega t - \kappa z) = b \sin\phi. \qquad . \quad . \quad . \quad 12(3)$$

At any instant, the resultant is represented by a vector. If we plot ξ_x and ξ_y along two axes as shown in fig. 12.13, we see that, as ϕ varies, the magnitude and direction of the resultant alter. We have:

(i) when $\phi = 0$, $\quad \xi_x = a$ and $\xi_y = 0$;
(ii) when $\phi = \frac{1}{4}\pi$, $\quad \xi_x = a/\sqrt{2}$ and $\xi_y = b/\sqrt{2}$;
(iii) when $\phi = \frac{1}{2}\pi$, $\quad \xi_x = 0$ and $\xi_y = b$.

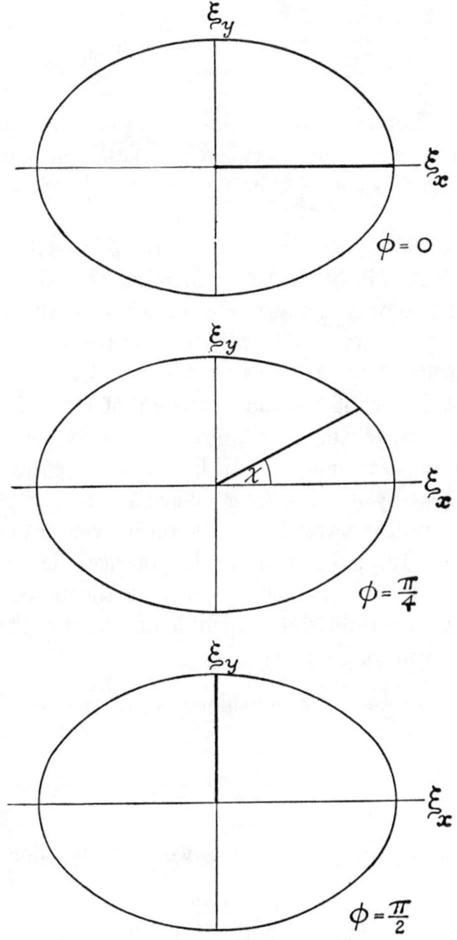

Fig. 12.13.—Vector representation of elliptically polarized light

The figure shows the resultant for these values of ϕ. It indicates that both the magnitude and orientation of the resultant vary with ϕ in a periodic way. The resultant rotates once when ϕ changes by 2π and, as it rotates, there is a gradual fluctuation in its magnitude. The changes may be expressed analytically by noting that the x and y co-ordinates of the end of the resultant are equal to ξ_x and ξ_y. Combining 12(2a) and 12(3), we see that these co-ordinates satisfy the relations

$$\frac{\xi_x{}^2}{a^2} + \frac{\xi_y{}^2}{b^2} = 1, \qquad \text{12(4)}$$

and

$$\tan \chi = \frac{\xi_y}{\xi_x} = \frac{b}{a} \tan \phi. \qquad \text{12(5)}$$

Equation 12(4) implies that the end of the resultant always lies on an ellipse. This type of light is called *elliptically polarized light.* When $a = b$, the ellipse becomes a circle, and the light is called *circularly polarized light.*

In understanding the properties of elliptically and circularly polarized light it is important to remember that ϕ may alter through a variation of t or of z. At any one place the resultant rotates and, in general, changes in magnitude with a time-period $2\pi/\omega$. The change in the magnitude of the resultant is not usually simple harmonic, but follows the variation of the radius vector of the ellipse. At any one time the direction of the resultant varies from point to point along the direction of propagation. It has a space-periodicity $\lambda = 2\pi/\kappa$. When z increases by λ, the vector makes a complete rotation, and the amplitude follows the variation of the radius vector of the ellipse given by 12(4). When the light is circularly polarized, the magnitude of the representative vector remains constant as the wave advances. The component (in any direction perpendicular to the direction of propagation) varies sinusoidally.

12.19.—So far we have considered only two special cases:

$$\text{(i)}\ \epsilon = 0$$

and

$$\text{(ii)}\ \epsilon = -\tfrac{1}{2}\pi.$$

Before considering the general case, we may consider the cases

$$\text{(iii)}\ \epsilon = \pi$$

and

$$\text{(iv)}\ \epsilon = \tfrac{1}{2}\pi.$$

It is easily verified that when $\epsilon = \pi$, the resultant is a linear vibration in a plane whose angle with the xz plane is given by $\tan \chi = -b/a$. If $\epsilon = \frac{1}{2}\pi$, we have

$$\xi_y = -b \sin (\omega t - \kappa z), \qquad \text{12(6)}$$

instead of 12(3). When 12(6) is combined with 12(2a), we again obtain equation 12(4), but equation 12(5) is replaced by

$$\tan \chi = \frac{\xi_y}{\xi_x} = -\frac{b}{a} \tan \phi. \qquad \text{12(7)}$$

Thus the end of the vector, which represents the resultant, rotates round the ellipse when ϕ increases, but in a clockwise direction. This type of elliptically polarized light is known as right-handed, or positive elliptically polarized light—the type discussed in § 12.18 being called left-handed or negative. In right-handed elliptically or circularly polarized light, the representative vector at any given point rotates clockwise when viewed by an observer who receives the beam of light.

We shall now show that when ε has some value other than one of those considered so far, the resultant is still elliptically polarized light, but the axes of the representative ellipse are not coincident with OX and OY. We need to eliminate ϕ from equations 12(2) so as to obtain a relation between ξ_x and ξ_y which is independent of ϕ. We have, from 12(2a),

$$\frac{\xi_x}{a} = \cos \phi, \qquad \text{12(8)}$$

and from 12(2b),
$$\frac{\xi_y}{b} = \cos \phi \cos \varepsilon - \sin \phi \sin \varepsilon.$$

Squaring,
$$\frac{\xi_y^2}{b^2} - 2\frac{\xi_y}{b} \cos \phi \cos \varepsilon + \cos^2 \phi \cos^2 \varepsilon = \sin^2 \phi \sin^2 \varepsilon,$$

and, substituting from 12(8), we obtain

$$\frac{\xi_y^2}{b^2} - 2\frac{\xi_x \xi_y}{ab} \cos \varepsilon + \frac{\xi_x^2}{a^2} \cos^2 \varepsilon = \left(1 - \frac{\xi_x^2}{a^2}\right) \sin^2 \varepsilon,$$

or
$$\frac{\xi_x^2}{a^2} + \frac{\xi_y^2}{b^2} - 2\frac{\xi_x \xi_y}{ab} \cos \varepsilon - \sin^2 \varepsilon = 0. \qquad \text{12(9)}$$

This is the equation of an ellipse for which one axis makes an angle θ with the x axis, where θ is given by

$$\tan 2\theta = \frac{2ab \cos \varepsilon}{a^2 - b^2}. \qquad \text{12(10)}$$

EXAMPLES [12(i)–12(vi)]

12(i). Suppose that the effective intensity of a source of light is reduced by the use of a polarizer and analyser whose relative orientation is θ. Show that the percentage error in the intensity due to an error $\Delta\theta$ in the setting is $-(200 \tan \theta)\Delta\theta$.

12(ii). Using the data of the preceding example, show that if the scale of the analyser can be read to 0·1°, an accuracy of 1 per cent is obtainable for a setting which reduces the intensity in a ratio of approximately 1 : 10.

12(iii). Find the ratio of the relative energies of the beams O_1O_2 and O_1E_2 (see § 12.8) when the relative orientation of the two crystals is θ. [$\cot^2 \theta$.]

12(iv). Show that if two coherent beams of circularly polarized light of equal amplitude (one left-handed and the other right-handed) are superposed, the resultant is plane-polarized. What determines the plane of polarization?

[The plane of polarization of the resultant contains the direction for which the components of the constituents are in phase.]

12(v). Show that the resultant of two coherent beams of elliptically polarized light is in general another beam of elliptically polarized light.

12(vi). Write down the conditions that the resultant of the preceding example shall be (*a*) plane-polarized, (*b*) circularly polarized.

[(*a*) The resultant components for any two directions of vibration must have the same phase.

(*b*) Taking any pair of axes at right angles, the resultant components must be of equal amplitude, with phases differing by $\frac{1}{2}\pi$.]

12.20. **Huygens' Wave Surface in Crystals.**

The phenomena of double refraction indicate that the velocity of propagation of a beam of light in a crystal depends on the relation between the direction of propagation and the crystal axes. Also for a given direction of propagation the velocity depends on the relation between the plane of polarization and the crystal axes. When double refraction was discovered by Bartolinus, Huygens saw that in order to apply his method of constructing rays (§ 3.8) to crystals, it was necessary to invent a special form of wave surface for crystals. Since there are two rays, it is necessary that the wave surface shall consist of two sheets. The observation that one ray always obeys both laws of refraction was taken to indicate that one sheet is a sphere. Huygens assumed, as the simplest available hypothesis, that the other sheet is an ellipsoid of revolution. He assumed that the spheroid touches the sphere either internally, as shown in fig. 12.14*a*, or externally, as shown in fig. 12.14*b*. In either case, the whole wave surface is formed by revolving the curves about the line which joins the points of contact. In the direction of this line there is only one speed of propagation, and this direction is called the *optic axis*. Huygens thought that

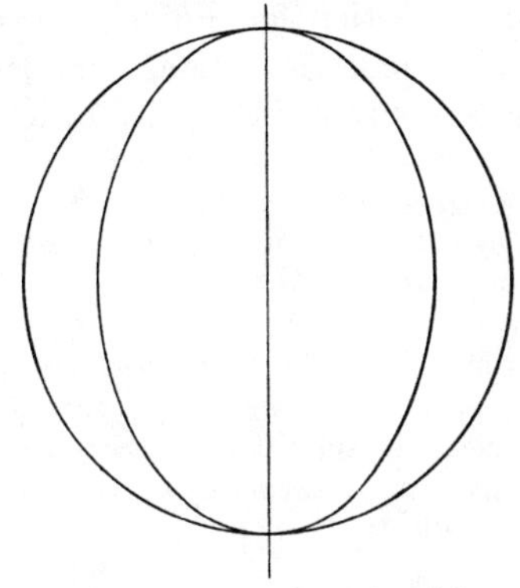

Fig. 12.14*a*.—Section of wave surface of positive uniaxial crystal

all crystals were uniaxial, and that this type of wave surface was universally applicable. Later, more extensive observations showed that the general form of the wave surface is more elaborate (fig. 16.6). It is still a surface of two sheets, but neither sheet is spherical, and the

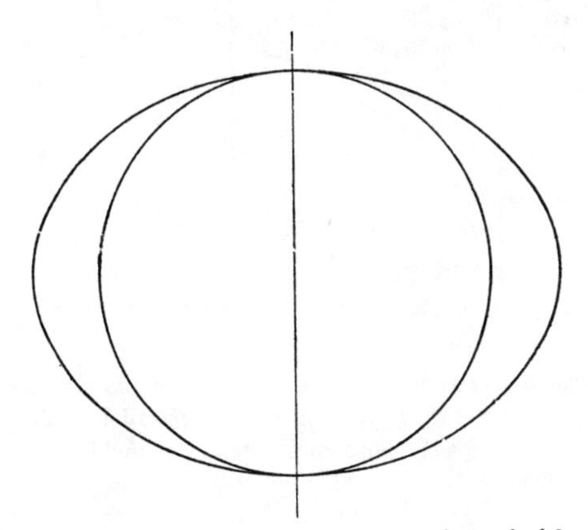

Fig. 12.14*b*.—Section of wave surface of negative uniaxial crystal

two sheets interpenetrate in a more complicated way. The most general type of crystal has two optic axes (i.e. two directions for which there is only one speed of propagation). The uniaxial crystal may be regarded as a special case in which the two axes coincide.

12.21. Verification of Huygens' Wave-surface for Uniaxial Crystals.

The form of the wave surface has been examined experimentally by Stokes, Glazebrook, and others, using modern instruments to measure the angles of refraction for different directions of incidence. The results show that for uniaxial crystals the part of the wave surface corresponding to the extraordinary ray is accurately an ellipsoid of revolution.

Earlier experiments mainly due to Malus are interesting. The following is a brief summary of these experiments.

(i) *To show that one sheet of the surface is spherical,* a composite prism is built up consisting of slices cut in different directions from a piece of calcite (fig. 12.15). There is one ordinary spectrum, the same for all slices, and a series of extraordinary spectra, most of which are deviated out of the plane of incidence. Measurements on the ordinary spectrum determine the value of the refractive index (μ_o) for any chosen wavelength.

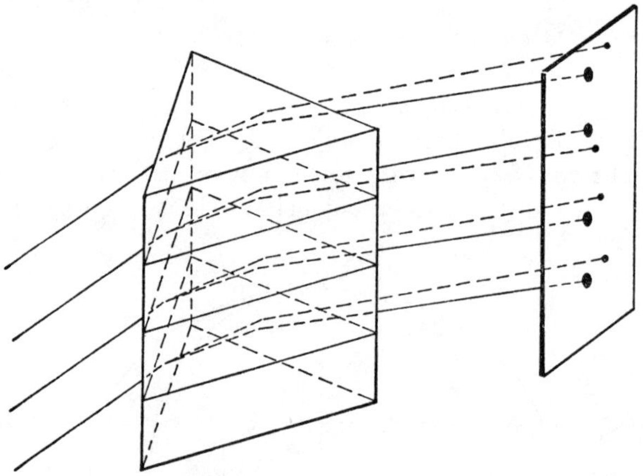

Fig. 12.15.—Refraction by composite calcite prism

(ii) *To show that one section of the other sheet is circular,* a crystal of calcite is cut so as to give a prism with its refracting edge parallel to the optic axis. Two spectra are obtained. They are both in the plane of incidence. One is polarized in the principal plane of the crystal, and the other in a perpendicular plane. For a given wavelength the index for the ordinary ray is found to be μ_o (as before), and a different index (μ_e) is found for the extraordinary ray. In this experiment the extraordinary ray obeys both the laws of refraction, and differs from the ordinary ray only in having a different index. This requires that the whole wave surface must be a surface of revolution. Fig. 12.16*a* shows Huygens' construction of the two refracted wavefronts.

Since both rays obey the sine law, measurements of the two critical angles for total reflection give μ_o and μ_e directly. These measurements may be made using a refractometer of the Abbe or Pulfrich type, with a polarizer attached. In this way, using monochromatic light, the indices may be conveniently measured with an accuracy of about 1 in 10^5 (i.e. to the fourth decimal place).

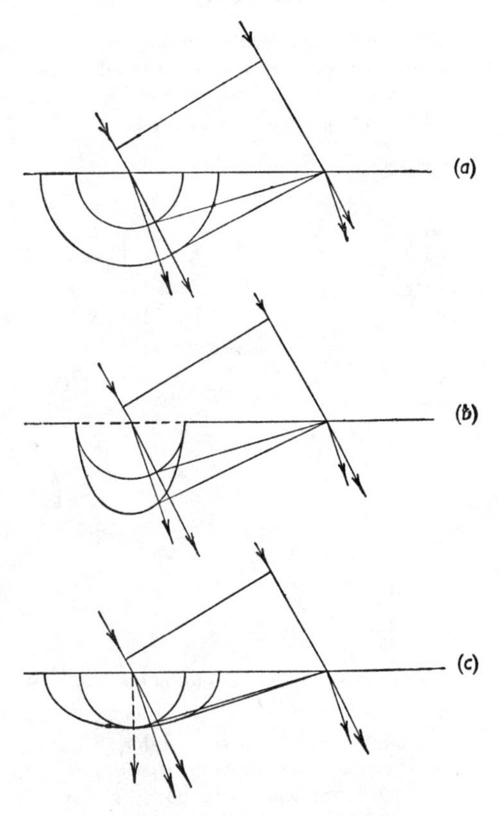

Fig. 12.16.—Huygens' construction

(*a*) Optic axis parallel to surface and perpendicular to plane of incidence.
(*b*) Optic axis parallel to surface and to plane of incidence.
(*c*) Optic axis normal to surface.

(The ellipticity of the second sheet of the wave surface has been exaggerated in order to show the construction more clearly. In calcite, the ratio of the axes of the ellipse is approximately 1·1: 1.)

(iii) *To investigate the refraction when the optic axis is parallel to the face of the crystal and to the plane of incidence.* The apparatus is shown in fig. 12.17. Two scales AB and AC are engraved on a piece of polished steel which forms a table on which the crystal is placed. The scales are observed through the telescope T, whose inclination to the vertical may be read on the scale P. The table is adjusted so that the faces of the crystal are horizontal. The crystal is turned so that the principal plane is perpendicular to AC. The line AC is then displaced in a direction perpendicular to itself. Suppose that the ordinary image of a point D on AC coincides with the extraordinary image of a point E on AB. Then, if e is the thickness of the crystal,

$$DE = e(\tan r_e - \tan r_o),$$

where r_e is the angle of refraction for the extraordinary ray, and r_o is the angle of refraction for the ordinary ray. r_o may be derived by measuring the angle of incidence on the scale P, and using the relation $\sin i = \mu_o \sin r_o$. DE may be measured and used to give r_e, when r_o and e are known. It is found that, for dif-

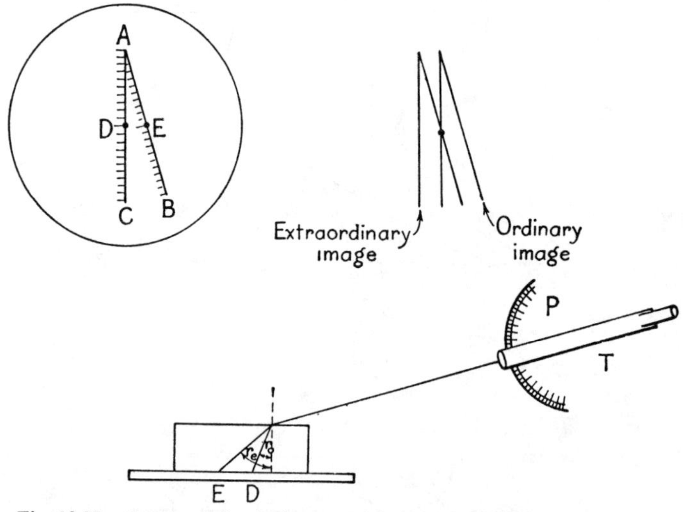

Fig. 12.17.—Apparatus for the investigation of the shape of the wave surface

ferent values of the angle i, the ratio $\tan r_e/\tan r_o$ is constant and is equal to μ_e/μ_o. Huygens' construction for this case is shown in fig. 12.16*b*. From the co-ordinate geometry of the figure it may be shown that the above relation between r_e and r_o is consistent with the assumption that the section of the two sheets of the wave surface is made up of a circle and an ellipse, which touch at two points in the way shown [see answer to Example 12(viii), p. 373].

12.22. Transmission of Polarized Light in a Thin Anisotropic Plate.

Suppose that a parallel beam of plane-polarized light is incident normally upon a thin anisotropic plate, and the relation between the plane of polarization and the crystal axis is varied by rotating the crystal about a line normal to its surface. It is found, in general, that there are two orientations in which the emergent light is plane-polarized. If the crystal does not possess the property of optical rotation (which will be considered in § 12.35), the plane of polarization of the emergent beam is the same as that of the incident beam. Two lines may be drawn on the plate (or on its mount) showing the directions in which the plane of polarization cuts the plate when the orientation is such that the emergent light is plane-polarized. The directions of these lines are called the two *privileged directions* for the given anisotropic plate. It is found that they are always perpendicular to one another.

We have seen in § 12.20 that relations of this type hold for uniaxial crystals like calcite. Further experiments show that similar results are obtained, in general, for a slice cut from any crystal in any orientation. The reason why there are two, and only two, privileged directions is that the wave surface is a surface of two sheets. In simple uniaxial crystals there is one direction (the optic axis) in which a plane-polarized beam is transmitted as a plane-polarized beam, no matter how the plane of polarization is orientated. In this direction there is only one velocity of wave propagation. A slice cut normal to the optic axis has no specially privileged directions. More complicated effects are obtained with "optically active" crystals (§ 12.37).

12.23.—The optical thickness of an anisotropic plate may be measured with an interferometer, using light plane-polarized in different planes. A simple result is obtained only when the plane of polarization is parallel to one of the privileged directions. It is found that the optical thickness for light polarized parallel to one privileged direction is greater than the optical thickness for light polarized parallel to the other privileged direction. This implies a difference in wave velocity, as we should expect from the form of Huygens' wave surface. The privileged direction corresponding to the larger velocity (and the lower refractive index) is called the *fast direction*. The other is called the *slow direction*. Let us now consider a beam of plane-polarized light, the direction of whose plane of polarization does not coincide with either of the privileged directions. We have seen above (§ 12.6) that when this is so there are two emergent beams which are parallel to one another, but displaced in a direction perpendicular to the direction of propagation. We now suppose that the anisotropic plate is so thin that the displacement sideways may be ignored. We should then expect that elliptically or circularly polarized light would be produced by the superposition of the two emergent beams which have travelled through the plate at different speeds, and have thus acquired a phase difference. We shall show, in the next paragraph, that this is so.

12.24.—Let us choose the fast and slow directions respectively as the OX and OY directions of a system of co-ordinate axes. Then OZ is the direction of propagation. The surface on which the light is incident may be taken as the XOY plane. The incident beam may then be represented by

$$\xi_0 = a \cos \omega t. \qquad \text{12(11)}$$

If the plane of polarization makes an angle ψ with OX, the components ξ_x and ξ_y (polarized parallel to OY and OX respectively) may be represented by *

$$\xi_{x0} = a \sin \psi \cos \omega t \qquad \text{12(12)}$$

and

$$\xi_{y0} = a \cos \psi \cos \omega t. \qquad \text{12(13)}$$

After the light has passed through a thickness z of the plate, the components are

$$\xi_x = a \sin \psi \cos \omega(t - z/b_1) \qquad \text{12(14)}$$

and

$$\xi_y = a \cos \psi \cos \omega(t - z/b_2), \qquad \text{12(15)}$$

where b_1 and b_2 are the fast and slow velocities of propagation.

In writing equations 12(14) and 12(15) we assume that propagation is normal to the plate. This is not strictly correct. In general, neither beam travels normally to the slice, and the two directions are not identical. We should insert $b_1 \cos \alpha_1$ for b_1, and $b_2 \cos \alpha_2$ for b_2, where α_1 and α_2 are the deviations of the two beams from the normal. Thus b_1 and b_2 are not two velocities in the same direction. In practice the values of α_1 and α_2 are fairly small and, for our present purpose, it is a sufficient approximation to assume that b_1 and b_2 are the velocities in the direction normal to the slice.

If the thickness of the crystal is e, the two components will emerge with a phase difference (ϵ_p) given by

$$\epsilon_p = \omega e \left(\frac{1}{b_2} - \frac{1}{b_1}\right) = \frac{\omega e}{c}(\mu_2 - \mu_1)$$

$$= \frac{2\pi e}{\lambda}(\mu_2 - \mu_1). \qquad \text{12(16)}$$

In this expression λ is the wavelength in air. The emergent beams are represented by

$$\xi_x = A \cos(\omega t - \kappa z - \delta) \qquad \text{12(17)}$$

and

$$\xi_y = B \cos(\omega t - \kappa z - \delta - \epsilon_p), \qquad \text{12(18)}$$

where

$$A = a \sin \psi, \ B = a \cos \psi, \qquad \text{12(19)}$$

and

$$\delta = \frac{\omega}{b_1} e. \qquad \text{12(20)}$$

* Remember that the plane of the vector represented by ξ_0 is at right angles to the plane of polarization.

We may alter the origin of time so as to eliminate δ, and 12(17) and 12(18) are then replaced by

$$\xi_x = A \cos (\omega t - \kappa z), \qquad 12(21)$$

$$\xi_y = B \cos (\omega t - \kappa z - \epsilon_p). \qquad 12(22)$$

The form of these equations is the same as that of equations 12(2*a*) and 12(2*b*). It indicates that, in general, the emergent light is elliptically polarized. Under special conditions the ellipse may reduce to a circle or a straight line.

12.25. **Quarter-wave Plate.**

An anisotropic plate for which the difference of optical thickness is a quarter of a wavelength is called a quarter-wave plate. A plate of this thickness introduces a phase difference of $\frac{1}{2}\pi$. A plate which gives a phase difference of π between the two components is called a half-wave plate, and a similar notation is applied to plates of other thicknesses. When plane-polarized light is incident upon an anisotropic plate the state of polarization in the emergent light depends upon (*a*) the difference of optical thickness, and (*b*) the relation between the plane of polarization of the light and the privileged directions of the plate. The calculations for any given condition may be made using equations 12(16), 12(19), 12(21) and 12(22). The following cases are of special importance:

(i) When plane-polarized light is incident upon a whole-wave plate, the emergent light is plane-polarized in the same plane as the incident light.

(ii) When plane-polarized light is incident upon a half-wave plate, the emergent light is plane-polarized. If the plane of polarization of the incident light makes an angle ψ with one of the privileged directions, then the plane of polarization of the emergent beam makes an angle $-\psi$ with the same direction, i.e. the plane has effectively been rotated through an angle 2ψ (fig. 12.18, p. 372).

(iii) When plane-polarized light is incident upon a quarter-wave plate, the emergent light is in general elliptically polarized. The axes of the ellipse are parallel to the privileged directions in the plate, and the ratio of the axes is given by 12(19). When the plane of polarization of the incident beam bisects the angle between the privileged directions, the light emerging from a quarter-wave plate is circularly polarized.

All the above properties depend on the *difference* of the optical thicknesses of the plate. It is of interest to consider the state of the vibration inside the crystal. From equations 12(14) and 12(15) we see that if we choose a definite value of z (i.e. if we consider a particular place in the crystal), the disturbance at that point is similar to the disturbance produced by a definite type of polarized light in an isotropic medium, i.e. the disturbance is the resultant of two vectors with

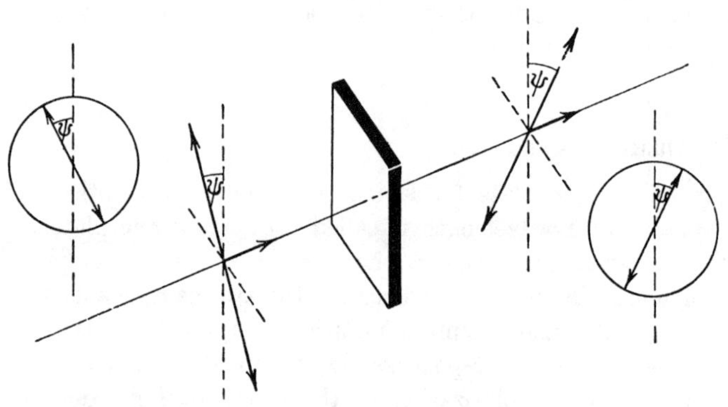

Fig. 12.18.—" Rotation " of the plane of polarization by a half-wave plate

a constant phase difference. Thus at one point the disturbance may be that of elliptically polarized light, and at another it may correspond to plane-polarized light. The *progression* of a disturbance in an anisotropic medium is not similar to that of any kind of light in an isotropic medium. In the anisotropic medium the phase difference between the components, and hence the state of the polarization, changes as the wave advances.

12.26. **Two or more Plates in Series.**

From the above discussion we should expect that the effects of transmission through two similarly oriented plates would be additive. This is found to be so. For example, if two quarter-wave plates are superimposed so that the " fast " directions coincide, the effect is exactly the same as that of a half-wave plate. If the two quarter-wave plates are superimposed so that the fast direction of one coincides with the slow direction of the other, then the total optical path is the same for both components. The plates have no effect on the state of polarization of the incident light. Generally the effects of thin plates which are superimposed so that the privileged directions coincide are algebraically additive. When the privileged directions do not coincide, the effect of each plate must be calculated successively. The resultant state of polarization produced by the first plate is calculated in the way described above. It is then regarded as a beam incident on the second plate and is resolved in two new directions (i.e. the privileged directions of the second plate).

EXAMPLES [12(vii)–12(xiv)]

12(vii). A narrow beam of light enters a crystal of monammonium phosphate at grazing incidence in a direction at right angles to the optic axis, which is parallel to the surface of the crystal. The separation of the ordinary and extraordinary beams at the opposite, parallel face of the crystal is 2·5 mm. If $\mu_o = 1{\cdot}525$ and $\mu_e = 1{\cdot}479$, calculate the thickness of the crystal. [5·1 cm.]

12(viii). Show that when an ellipse and circle touch as shown in fig. 12.14*b*, a line joining the points of contact of the tangent to the ellipse and the circle from a point on the common axis is perpendicular to this axis. Hence show that $\tan r_e/\tan r_o = \mu_e/\mu_o$.

12(ix). A uniaxial crystal is cut so that the optic axis is perpendicular to the refracting surface (fig. 12.16*c*). Show that the refracted rays and the optic axis are in the same plane, and that

$$\tan r_e = \frac{\mu_o \sin i}{\mu_e \sqrt{(\mu_e^2 - \sin^2 i)}}.$$

How would you test this relation experimentally?

12(x). Show that circularly polarized light may be produced with a three-quarter-wave plate, and also with certain plates of greater thickness.

12(xi). How may a quarter-wave plate be made to produce (*a*) right-handed and (*b*) left-handed circularly polarized light?

[Suppose the slow direction of the quarter-wave plate to be in the direction OX, and the fast direction along OY, the direction of propagation of the light being the positive direction of OZ. Then, if plane-polarized light, whose vibration direction is at $+\pi/4$ to the OX direction, is incident on the plate, the transmitted light is right-handed circularly polarized.]

12(xii). Design an arrangement to produce a vertical beam of elliptically polarized green light. The axes of the ellipse are to be in the ratio 3 : 1, and the direction of the larger axis to be north-south.

[A mercury lamp with green filter directs a beam vertically through a polarizing prism set to transmit light polarized in a plane making an angle $\cot^{-1} 3$ with the N-S direction. The light is then passed through a quarter-wave plate whose fast and slow directions are N-S and E-W respectively.]

12(xiii). How may an anisotropic plate be made to change a right-handed elliptically polarized beam of light into a left-handed beam?

[Use a half-wave plate suitably oriented. In the special case of circularly polarized light the orientation does not matter.]

12(xiv). Two plates of thicknesses e_1 and e_2 are put together so that the fast direction of one coincides with the slow direction of the other. They are of different materials. The refractive indices are μ_1 and μ_1' for the fast and slow directions respectively in plate 1, and μ_2 and μ_2' for plate 2. Given that the first plate is a whole-wave plate, that the combination acts as a quarter-wave plate, and that $e_1 = 2e_2$, derive a relation between the refractive indices.

[$5(\mu_1 - \mu_1')$ or $3(\mu_1 - \mu_1') = 2(\mu_2 - \mu_2')$.]

12.27. Analysis of Polarized Light.

The additive property may be used in the detection and analysis of circularly and elliptically polarized light. Consider, for example, a method for distinguishing between ordinary light and circularly polarized light. If a Nicol prism is placed in the beam and rotated, both these kinds of light behave in the same way. The amount of light transmitted is the same for all orientations of the analyser. If the beam is first passed through a quarter-wave plate and then through an analyser, the situation is quite different. The quarter-wave plate will change the circularly polarized light into plane-polarized light, which is extinguished at a suitable setting of the analyser. If, however, ordinary light is transmitted through a quarter-wave plate and an analyser, the amount of light transmitted is the same for all orientations. Similar methods may be used to distinguish between elliptically and plane-polarized light.

12.28.—The following procedure is suitable for making a systematic qualitative investigation of the properties of a beam of polarized light.*

Test 1.

Insert Nicol prism in the beam and rotate it.

Result (*a*).—Light extinguished for one orientation of the Nicol. Indicates that light is plane-polarized.

Result (*b*).—Amount of light transmitted is the same for all settings of the Nicol. Indicates that light is either

(i) unpolarized, or
(ii) circularly polarized, or
(iii) a mixture of circularly polarized and unpolarized light.

Apply Test 2.

* A more sensitive method of detecting a small proportion of polarized light in a mixture is described in § 12.44.

Result (*c*).—Amount of light transmitted varies as the Nicol is rotated, but is never zero. Indicates that the light is either

(i) elliptically polarized, or
(ii) a mixture of elliptically polarized and unpolarized light, or
(iii) a mixture of plane-polarized light and unpolarized light.

Apply Test 3.

Test 2.

Insert a quarter-wave plate followed by a Nicol prism. Rotate the Nicol prism.

Result (*a*).—Light extinguished for one orientation of the Nicol prism. Indicates original beam is circularly polarized.

Result (*b*).—Light unchanged by rotation of Nicol. Indicates that original beam is unpolarized.

Result (*c*).—Amount of light changed by rotation of Nicol, but minimum not zero. Indicates that original beam is a mixture of circularly polarized and unpolarized light.

Test 3.

Insert a quarter-wave plate followed by a Nicol prism. Rotate the two components independently.

Result (*a*).—Light is extinguished for one setting. Indicates that original beam is elliptically polarized, and that the axes of the ellipse are parallel to the privileged directions when the light is extinguished.

Result (*b*).—Amount of light transmitted varies when components are rotated, but minimum is not zero. Indicates that the original light is a mixture of elliptically polarized light and unpolarized light. The setting which gives the minimum transmission is one in which the privileged directions are parallel to the axes of the ellipse, and the setting of the Nicol gives the ratio of the axes. If the Nicol is parallel to one of the privileged directions, then the original beam is a mixture of plane-polarized light and unpolarized light.

12.29. Representation of Unpolarized Light.

In the above procedure of analysis we have distinguished between the following seven possibilities:

(i) unpolarized light,
(ii) plane-polarized light,
(iii) circularly polarized light,
(iv) elliptically polarized light,
(v) unpolarized light plus plane-polarized light,
(vi) unpolarized light plus circularly polarized light,
(vii) unpolarized light plus elliptically polarized light.

A theorem due to Stokes * shows that these are the only possibilities, e.g. a mixture of plane, plus elliptically, plus unpolarized light

* Reference 12.4.

is equivalent to a mixture of elliptically polarized and unpolarized light. The following brief discussion is intended only to explain the general problem involved in the mathematical representation of mixtures of light.

12.30.—In Chapter IV it was shown that a beam of light which has one definite wavelength must be represented by an infinite wave-train. An extension of the discussion shows that *a beam which has one definite wavelength must be completely polarized.* If the disturbance is not represented by a vector in one plane, it must be represented as the resultant of two vectors in two perpendicular planes. The variation of each of these vectors must be represented by an expression of simple harmonic form. The two expressions must be valid from $-\infty$ to $+\infty$. There must therefore be a permanent phase-relation between them, and the resultant is plane-polarized if the phase difference is a multiple of π. Otherwise the resultant is elliptically or circularly polarized.

Let us imagine that we have a source of light with very small damping which emits wave trains of different polarizations, so that one train lasts for a period of order 100 seconds (i.e. much longer than the trains given by laboratory sources). Suppose that the resultant of a number of these wave trains is analysed by an apparatus which displays the polarization ellipse on the screen of a cathode-ray tube. A " snapshot " of the screen will, in general, show an ellipse. An observer who watches the screen will see the parameters of the ellipse slowly change as some wave trains die out, and new ones begin. These variations will be completely irregular in the sense that no observation of the process will lead to any rational prediction of the detailed course of future alterations.

Now suppose that the wave trains become shorter and shorter. The observer will notice the irregular variations becoming more rapid and, when the average length of a wave train is of the order 10^{-2} second, or shorter,* he can observe only a general illumination covering a circular area whose boundary is the envelope of all the ellipses previously observed. The observer will see a steady pattern on the screen if, and only if, the wave trains emitted by the atoms are all polarized in the same way. When this condition is satisfied we have coherence between the components polarized in perpendicular planes, even though the wave trains are not infinitely long.†

12.31.—In representing unpolarized light we have to take into account three experimental results:

(i) The illumination produced by resolution in any plane (e.g. with a Nicol) is independent of the orientation of the plane of resolution.

(ii) This property is unaffected by previous relative retardation of any rectangular components into which the light may have been resolved (i.e. if the beam is passed first through a thin crystalline plate and then through a Nicol, the amount of light transmitted is independent of the orientation either of the plate or of the Nicol).

(iii) Interference fringes can be produced with unpolarized light.

* The wave trains which represent light are much shorter (of order 10^{-7} second or less).

† A detailed discussion of this problem has recently been given by Hurwitz (Reference 12.5).

Let a and b be the instantaneous values of the amplitudes *with respect to any pair of perpendicular axes*, and ε the phase difference of the components. Let us define A, B, C, D by the relations

$$A = \overline{a^2}, \qquad B = \overline{b^2}, \quad . \quad . \quad . \quad . \quad . \quad . \quad . \quad . \quad 12(23)$$

$$C = \overline{ab \cos \varepsilon}, \quad D = \overline{ab \sin \varepsilon}, \quad . \quad . \quad . \quad . \quad . \quad . \quad 12(24)$$

(the bars indicate mean values). Any mixture of light may be specified by giving the values of A, B, C, D. It may be shown (Reference 12.4) that necessary and sufficient conditions for satisfying (i) and (ii) above are

$$A = B,$$

and

$$C = D = 0. \quad . \quad . \quad . \quad . \quad . \quad . \quad . \quad . \quad . \quad 12(25)$$

Thus, when these conditions are satisfied, the light is unpolarized.

When $AB = C^2 + D^2$ the light is elliptically polarized. If in addition $A = B$ and $C = 0$, the light is circularly polarized or, if $D = 0$, the light is plane-polarized. Generally it is shown that for any possible values of A, B, C, D, the beam of light may be resolved into two parts, one of which satisfies the conditions for unpolarized light, and the other of which satisfies one of the above conditions for polarized light. Either of these parts may be of zero intensity, and thus the seven types of mixtures described in § 12.29 constitute all the possibilities.

12.32.—The fact that beams of unpolarized light can produce interference fringes is in agreement with the above analysis. If unpolarized light is resolved into two components, each will form its set of interference fringes separately. When the path lies entirely in isotropic media, the two sets of fringes will coincide. The *average* illumination at any point is equally divided between the components. If the path lies partly in a doubly refracting medium, then the two fringe systems may no longer be superimposed and may confuse one another.

EXAMPLES [12(xv) and 12(xvi)]

12(xv). How may a Nicol prism and a quarter-wave plate be used to distinguish between right- and left-handed circularly polarized light?

[See answer to Ex. 12(xi), p. 373.]

12(xvi). How may the procedure of § 12.28 be used to derive the proportion of unpolarized light in a mixture of right-handed circularly polarized light and unpolarized light.

[Insert a quarter-wave plate, and compare the maximum and minimum amounts of light transmitted by a rotating analyser.]

12.33. **The Babinet Compensator.**

The method described in § 12.28 is suitable for a general analysis of polarized light. More sensitive tests, some based upon colour effects which will be described later, have been devised for the detection of a small proportion of polarized light in a mixture. Special pieces of apparatus have also been invented for more accurate and convenient measurements on elliptically polarized light. The most generally useful of these is the *Babinet compensator*.* This consists of two quartz wedges cut in different directions and oriented as shown in fig. 12.19*a*. The arrangement is similar to the Rochon prism but in the compensator the angles of the wedges are so fine that there is no effective separation of the rays. The lower wedge can be moved relative to the frame (in which the upper is held) by means of a micrometer screw, so as to alter the total thickness. The difference of optical path in the compensator varies linearly from one side to the other. The central difference of path can be altered by a known amount by turning the screw. The compensator is used in conjunction with a Nicol prism or a piece of Polaroid film as an analyser. It is viewed through a low-power microscope.

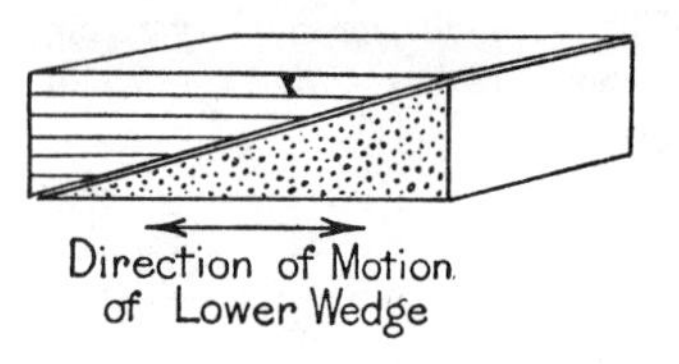

Fig. 12.19*a*.—The Babinet compensator

12.34.—The compensator is first calibrated using plane-polarized light produced by a polarizing prism. The analyser and polarizer are crossed, and the compensator is placed between them, oriented so that the plane of the analyser bisects the angle between the privileged directions for the wedges (see fig. 12.19*b*). A series of dark bands is seen in the field of the microscope. These correspond to points at which the phase difference introduced by the two wedges is equal to a multiple of 2π. For these points the emergent light is plane-polarized in the plane of the incident light, and is extinguished by the analyser. For points half-way between these points, the light is plane-polarized in a plane perpendicular to that of the incident light. For points intermediate between the positions where the emergent light is plane-polarized, it is elliptically polarized. If the analyser is rotated, the fringes at first become less clear, and then they become clear again, but the bright and dark bands are found to have changed places. The movement of the screw needed to bring successive fringes under the crosswire of the microscope (and hence to alter the phase difference at the centre by 2π) is measured. Let this distance be d. Now

* For a general description of other types of compensator, often of greater sensitivity but of more limited application, see Reference 12.6.

suppose that, still using plane-polarized light, the instrument is first adjusted so that a dark band is seen in the centre, and then the micrometer screw is turned through a distance $\frac{1}{4}d$. The phase difference produced at the centre is then equal to $(2n\pi \pm \frac{1}{2}\pi)$. For our purposes this is equivalent to a phase difference of $\frac{1}{2}\pi$. The polarizer is removed and the unknown beam of elliptically polarized light is admitted. The compensator and analyser are rotated independently

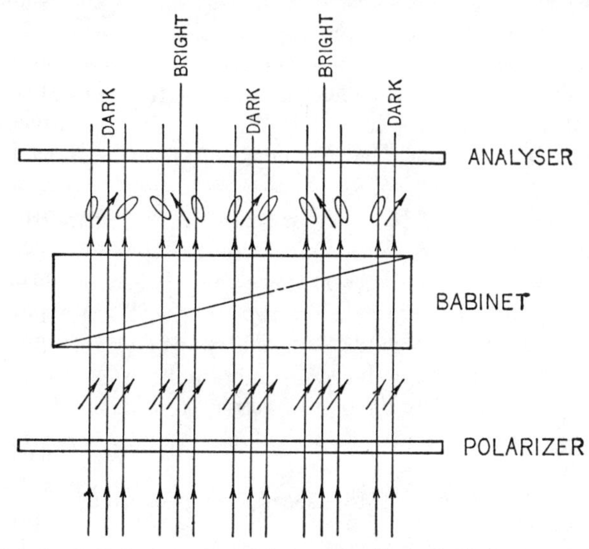

Fig. 12.19*b*.—The formation of dark bands with the Babinet compensator between crossed Nicols

until the centre of a dark band again appears under the central crosswire. The emergent light at the centre of the analyser must now be plane-polarized. Since the compensator introduces a phase difference of $\frac{1}{2}\pi$ at this point, the axes of the ellipse must be parallel to the privileged directions of the compensator. This determines the directions of the axes of the ellipse.

To determine the ratio of the axes, the analyser is rotated until the bands are as clear as possible, with the centre band dark. The analyser is now oriented perpendicular to the direction of the resultant of the two components of the elliptic vibration whose phase difference has been compensated. The tangent of the angle which the analyser now makes with the principal directions of the compensator is therefore equal to the ratio of the axes of the ellipse.

12.35. **Rotatory Polarization.**

When a beam of plane-polarized light passes through certain substances, the light remains plane-polarized but the plane of polarization is gradually rotated. This property is possessed by quartz and some other crystalline substances, and also by some liquids and vapours. The liquids include certain oils, such as turpentine, and solutions of certain substances, especially the sugars. These substances are said to be *optically active*. For solutions, the rotation is proportional to the concentration, as well as to the length of path. It is thus proportional to the number of molecules in the line of sight. This property is called *natural rotation* and must be distinguished from *magnetic rotation* (§ 16.50).

Natural rotation may be in either sense. A right-handed rotation (also called a positive rotation) is clockwise to an observer looking in the direction in which the light is travelling. Substances which give a right-handed rotation are called *dextro-rotatory*, those which give a left-handed rotation are *laevo-rotatory*. The *specific rotation* of a solution is equal to the rotation produced by a column 10 centimetres long, divided by the concentration of the active substance. The concentration is expressed in grammes of active substance per cubic centimetre of the solution.

The *molecular rotation* is the specific rotation multiplied by the molecular weight.

The following figures show the order of magnitude of the effects:

Rotation of a 10-cm. column of a solution of cane sugar (0·1 gm. per cc. of solution)	= 6·67°.
Rotation of a 10-cm. column of essence of turpentine	= −29·6°.
Rotation of a piece of quartz 1 mm. thick	= ±21·7°.

12.36.—In the vector representation, a beam of plane-polarized light may be regarded as the resultant of two equal beams of circularly polarized light, one right-handed and the other left-handed. Let the right-handed beam have components

$$\xi_{xr} = -a \cos \omega(t - z/b) = -a \cos \phi \quad . \quad . \quad 12(26)$$

and

$$\xi_{yr} = a \sin \omega(t - z/b) = a \sin \phi. \quad . \quad . \quad . \quad 12(27)$$

Let the left-handed beam have components

$$\xi_{xl} = a \cos \omega(t - z/b) = a \cos \phi \quad . \quad . \quad . \quad 12(28)$$

and

$$\xi_{yl} = a \sin \omega(t - z/b) = a \sin \phi. \quad . \quad . \quad . \quad 12(29)$$

Then the resultant is plane-polarized. The plane of polarization is the xz plane, the representative vector

$$\xi = 2a \sin \omega(t - z/b) = 2a \sin \phi \quad . \quad . \quad . \quad 12(30)$$

being in the yz plane.

The phenomenon of optical rotation may be included in the vector representation by assuming that, in optically active media, circularly polarized light is transmitted unchanged, but the velocity of left-handed circularly polarized light is not the same as that of right-handed circularly polarized light. After the light has travelled through a distance e of the optically active medium, the two components have a phase difference (ϵ_c) given by

$$\epsilon_c = \omega e \left(\frac{1}{b_r} - \frac{1}{b_l}\right) = \frac{2\pi e}{\lambda} (\mu_r - \mu_l), \quad . \quad . \quad 12(31)$$

where b_l and b_r are the speeds for the left-handed and right-handed components respectively; μ_l and μ_r are the corresponding indices and λ is the wavelength in air. The light after emerging, at $z = 0$, from the optically active material is represented by

$$\left.\begin{aligned} \xi_{xr}' &= -a \cos \phi', \\ \xi_{yr}' &= a \sin \phi', \\ \xi_{xl}' &= a \cos (\phi' + \epsilon_c), \\ \xi_{yl}' &= a \sin (\phi' + \epsilon_c), \end{aligned}\right\} \quad . \quad . \quad . \quad . \quad . \quad 12(32)$$

where

$$\phi' = \omega \left(t - \frac{e}{b_r} - \frac{z}{c}\right), \quad . \quad . \quad . \quad . \quad 12(33)$$

and the resultant has plane-polarized components in the xz and yz planes given by

$$\left.\begin{aligned} \xi_x &= -a \cos \phi' + a \cos (\phi' + \epsilon_c) = 2a \sin \tfrac{1}{2}\epsilon_c \sin (\phi' + \tfrac{1}{2}\epsilon_c), \\ \xi_y &= a \sin \phi' + a \sin (\phi' + \epsilon_c) = 2a \cos \tfrac{1}{2}\epsilon_c \sin (\phi' + \tfrac{1}{2}\epsilon_c). \end{aligned}\right\} \quad 12(34)$$

These two components are in phase, and the resultant is a vibration in a plane making an angle ψ with the yz plane, where

$$\tan \psi = \frac{\xi_x}{\xi_y} = \tan \tfrac{1}{2}\epsilon_c. \quad . \quad . \quad . \quad . \quad 12(35)$$

Thus the plane of polarization has been rotated through an angle

$$\psi = \tfrac{1}{2}\epsilon_c = \frac{\pi e}{\lambda}(\mu_r - \mu_l). \quad . \quad . \quad . \quad . \quad 12(36)$$

This angle is proportional to e, and to the difference of the refractive indices for the two circularly polarized components.

Fresnel demonstrated the above difference of refraction directly. He used right-handed and left-handed quartz prisms as shown in fig. 12.20. Owing to the difference of refractive index the light is split into two beams, one right-handed circularly polarized, and the other left-handed. This phenomenon is called *allogyric birefringence.* If the incident light is unpolarized, the two beams have no permanent phase-relation, and can never be brought to interference. If the incident light is plane-polarized, the two emergent beams are coherent. If, by the use of suitable quarter-wave plates, they are both transformed into plane-polarized light (with the same plane of polarization) they may be made to interfere.

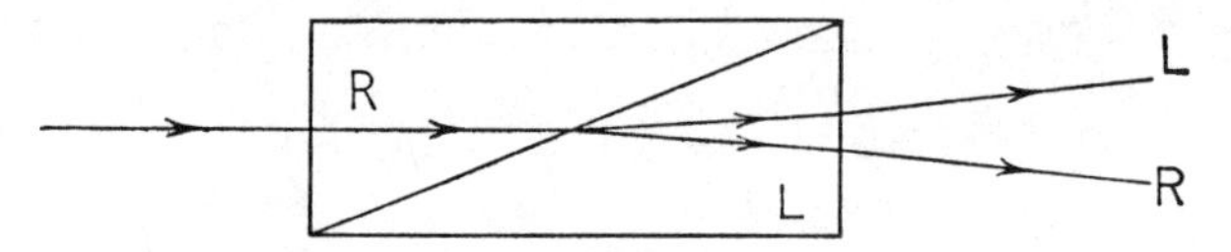

Fig. 12.20.—Fresnel's demonstration of allogyric birefringence

12.37.—Solutions of sugars and similar liquids are optically active, but isotropic. There is only allogyric birefringence and no birefringence of the ordinary kind obtained with calcite. In substances like quartz, optical activity is associated with ordinary birefringence. We have seen that a plate of inactive crystal transmits unchanged light travelling normal to the slice and polarized in one of two planes. Plane-polarized light with any orientation of the polarization is transmitted along the axis. In an active (but isotropic) substance like a sugar solution, circularly polarized light is transmitted unchanged. In quartz, circularly polarized light is transmitted unchanged when the wave normal is in the direction of the optic axis. For a direction inclined to the optic axis, only elliptically polarized light is transmitted unchanged, and for a given direction only two particular kinds of elliptically polarized light. The two ellipses are similar to one another in the orientation of the axes, and in the ratio of the minor to the major axis, but have opposite senses of rotation. The ellipses become circles for one particular direction of the wave normal (i.e. the optic axis), and shrink into lines for one other direction. In this direction (56° 10′ to the optic axis) plane-polarized light is transmitted unchanged. In quartz, the whole phenomenon is symmetrical about the single optic axis. Much more complicated effects are obtained when a biaxial crystal is also optically active.*

* Reference 12.2.

12.38. **Dispersion of Birefringence and Optical Rotation.**

All methods of producing and analysing polarized light are to some extent affected by dispersion, though in some experiments the effect is much more important than in others. The polarization of light by reflection is affected since the Brewsterian angle is a function of the index of the reflecting medium. It is not, however, a rapidly varying function of the index, and the light incident at angles near the Brewsterian angle has a very high degree of polarization (§ 14.9). Consequently, if a parallel beam of white light is incident upon a sheet of glass at the correct Brewsterian angle for a wavelength near the middle of the spectrum, the whole reflected beam is very nearly completely polarized. If it is examined with an analyser, only slight colour effects can be observed. The effects of dispersion are shown much more strongly in connection with the transmission of polarized light through thin anisotropic plates.

12.39.—Suppose that a parallel beam of white light passes through a polarizer, a plate of quartz cut perpendicular to the optic axis, and an analyser (fig. 12.21). Owing to the effects of dispersion, the rotation decreases as the wavelength increases. For certain wavelengths the rotation will be such that the light emerging from the crystal is not transmitted by the analyser. For intermediate wavelengths, some

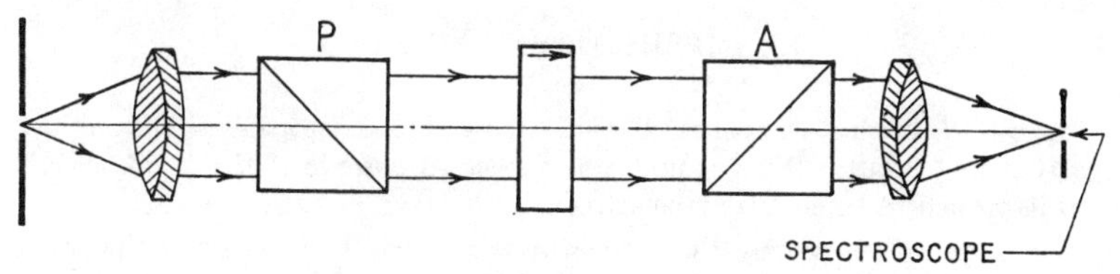

Fig. 12.21.—Apparatus for the investigation of the dispersion of optical rotation

light will be transmitted. If the light from the analyser is examined in a spectroscope, a channelled spectrum is seen. The total rotation will vary rapidly with wavelength when the plate is thick, giving a large number of close bands. When the plate is thin, there will be a small number of broad bands. The light transmitted by a fairly thin plate has a colour which depends upon its thickness, and upon the relative orientation of the analyser and polarizer. Suppose, for example, that the analyser and polarizer are crossed, and the thickness of the plate is such that a rotation of $5\pi/2$ is obtained for wavelength 5500 Å., and rotations of 2π and 3π for wavelengths 6900 Å. and

4600 Å. Then the transmitted light is green in hue.* If the analyser is rotated to the parallel position, a purple hue is obtained. When the plate is extremely thin, the total rotation is small, and its variation with wavelength is negligible. The transmitted light is then nearly white. Neither are colours obtained when the plate is very thick, because the spectrum then contains a large number of equally spaced bands, and the same fraction of each region of the spectrum is transmitted.

12.40.—The rotation (ψ) would be independent of wavelength if ϵ_c were independent of wavelength. From 12(36), we see that this would require the difference of refractive indices ($\mu_r - \mu_l$) to be proportional to the wavelength. This condition is not usually fulfilled.

Let α be the angle through which it would be necessary to rotate the polarizer *in the direction of the rotation of the optically active substance,* in order to make it parallel to the analyser. Then the amount of light transmitted is a maximum at a wavelength for which

$$\psi(\lambda) = n\pi + \alpha, \qquad \text{12(37)}$$

and a minimum when

$$\psi(\lambda) = \tfrac{1}{2}(2n + 1)\pi + \alpha. \qquad \text{12(38)}$$

EXAMPLES [12(xvii)–12(xxi)]

12(xvii). A Babinet compensator is adjusted and used according to the method suggested in § 12.34. What would you expect to observe if the "unknown" beam happened to be circularly polarized?

[In the initial setting the cross wire would be at the centre either of a bright fringe or of a dark fringe. (For the reason why there are two possibilities, see examples 12(xv) and 12(xi)). Rotation of analyser and compensator, as one unit, does not affect the fringes. Rotation of analyser alone causes the fringes to change in visibility. For four orientations they disappear and, when they reappear, the dark fringe has been replaced by a bright fringe. The clearest fringes are obtained when the analyser is at $\pi/4$ to the privileged directions of the compensator.]

12(xviii). Using a plate of quartz 4·374 cm. thick, placed between crossed Nicols, dark bands were observed at wavelengths 5990, 5510, 5130, 4820, 4560, 4340 Å. Assuming that the rotation for the sodium D line ($\lambda = 5893$ Å.) is 21·34

* Every beam of light can be regarded as the sum or difference of a certain amount of white light, plus light of a given wavelength. This wavelength defines the hue. White light, plus light of a certain hue, is complementary to white light, minus light of the same hue. For a detailed account of colour nomenclature see Reference 12.7.

degrees per millimetre, and using the above results, plot a curve showing the variation of rotation with wavelength. Show that the rotation per millimetre (ρ) is given by

$$\rho = -2{\cdot}10 + \frac{8{\cdot}14}{\lambda^2},$$

where ρ is in degrees and λ is in microns.

12(xix). Using the formula of Example 12(xviii), find the thickness of the thinnest quartz plate which would give a maximum of transmission at 5460 Å. (i) when the polarizer and analyser are crossed, and (ii) when they are at 45°.
[(i) 3·57 mm. (ii) 1·78 mm.]

12(xx). Design an arrangement to have zero transmission for 6870 Å., and maximum transmission for 6563 Å. Calculate the thickness of a suitable plate of quartz.
[5·45 cm.]

12(xxi). If a plane of polarization can be located with an accuracy of ± 10 minutes of arc, what are the corresponding limits of error in the difference between μ_r and μ_l when this difference is measured with a plate 1 cm. thick, and using light of wavelength 5000 Å?
[$\pm 4{\cdot}6 \times 10^{-8}$.]

12.41.—When a plate of anisotropic material is placed in a parallel beam of light between a polarizer and an analyser (using the apparatus shown in fig. 12.21), colour effects are obtained if the plate is thin. If white light, which has passed through a thick plate, is passed into a spectroscope, the spectrum is channelled. The amount of light of a given wavelength which is transmitted may be calculated in the following way.

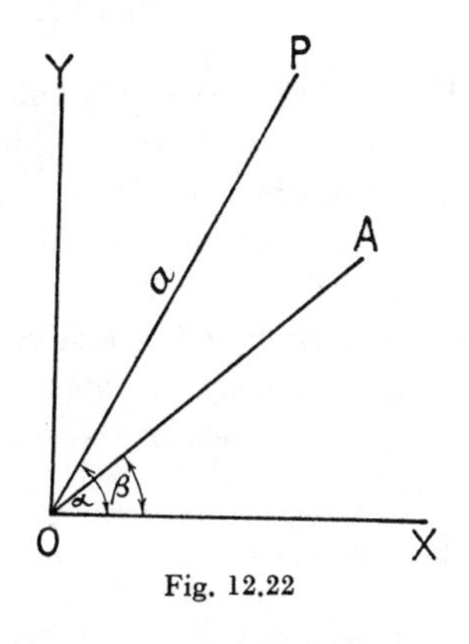

Fig. 12.22

Let OX and OY be the two privileged directions for the anisotropic plate, and let OP and OA represent the settings of the polarizer and the analyser (fig. 12.22). The light incident upon the plate has components

$$\left.\begin{aligned} &a \cos\alpha \cos\phi \quad \text{along OX,} \\ \text{and} \quad &a \sin\alpha \cos\phi \quad \text{along OY.} \end{aligned}\right\} \quad . \quad . \quad . \quad . \quad 12(39)$$

The light emerging from the plate is represented by

$$\left.\begin{aligned} &a \cos\alpha \cos\phi' \quad \text{along OX,} \\ \text{and} \quad &a \sin\alpha \cos(\phi' + \epsilon_p) \quad \text{along OY,} \end{aligned}\right\} \quad . \quad . \quad . \quad 12(40)$$

where ϕ and ϕ' are epoch angles, and ϵ_p is given by 12(16). The light transmitted by the analyser is represented by

$$\left.\begin{array}{ll} & a\cos\alpha\cos\beta\cos\phi', \\ \text{and} & a\sin\alpha\sin\beta\cos(\phi'+\epsilon_p). \end{array}\right\} \quad . \quad . \quad . \quad 12(41)$$

We may regard the light which has passed the analyser as two beams polarized in the same plane and originally derived from the same beam of plane-polarized light. They are thus in a condition to interfere. The resultant illumination is equal to the sum of the squares of the coefficients of $\cos\phi'$ and $\sin\phi'$ (see § 3.4), i.e.

$$E = (a\cos\alpha\cos\beta + a\sin\alpha\sin\beta\cos\epsilon_p)^2 + (a\sin\alpha\sin\beta\sin\epsilon_p)^2. \quad 12(42)$$

It is convenient to write 12(42) in a form in which a part involving ϵ_p is separated from a part not involving ϵ_p. This leads to

$$E = a^2\cos^2(\alpha-\beta) - a^2\sin 2\alpha\sin 2\beta\sin^2\tfrac{1}{2}\epsilon_p. \quad 12(43)$$

The first term depends only on the relative orientation of the analyser and polarizer. It is a maximum when they are parallel, and zero when they are crossed. It is known as the " white " term, since it is independent of wavelength. The second (or " colour " term) depends on ϵ_p (and hence on λ), and also on α and β. It is zero whenever either the polarizer or the analyser is parallel to one of the privileged directions. Rotation of *either* analyser *or* polarizer through a right angle (keeping the rest of the system fixed) alters the sign of this term.

12.42.—The first term on the right-hand side of equation 12(43) represents the light which would be transmitted by the analyser and polarizer if the crystal plate were removed. This light would be white and the effect of the crystal plate is to add (or, in certain relative orientations of analyser and polarizer, to subtract) a certain amount of coloured light. Addition gives one hue and subtraction gives the complementary hue.* For a plate of given material and thickness there are thus two characteristic hues. These characteristic hues are obtained only when the parallel beam is incident normally. A small tilt of the plate alters the hue of the transmitted light. A strong saturated colour is obtained when $\alpha = +\pi/4$ and $\beta = -\pi/4$. This makes the first term on the right-hand side of equation 12(43) equal to zero and gives the second term its maximum value. If the analyser and polarizer are rotated from this setting (as one unit), the colour remains saturated but the amount of light transmitted is reduced. If the polarizer is fixed ($\alpha = \pi/4$) and the analyser is rotated, the colour becomes more and more desaturated until the colour term becomes zero at $\beta = -\pi/2$ or $\beta = 0$. Further

* See footnote to § 12.39.

rotation of the analyser gives the complementary hue. The colour effects are conveniently observed with a thin plate of mica or with some kinds of transparent plastics (obtainable from general laboratory suppliers). The " chameleon " shown in Plate IV (p. 518) is made of mica of two different thicknesses. Each part gives one hue in Plate IV*e* and, by rotation of the analyser, the complementary hue in IV*f*. The channelled spectrum obtained with a thicker plate may conveniently be observed with a piece of selenite one millimetre thick.

12.43.—Owing to dispersion, a given quarter-wave plate of anisotropic material can produce a phase difference of a quarter-period for only one wavelength. In the absence of any special requirement, it is usual to make the plate exactly right for a wavelength in the yellow region of the spectrum. If a parallel beam of white light, plane-polarized in a plane which bisects the privileged directions, is incident upon such a plate, the yellow part of the emergent light is circularly polarized, and the other wavelengths are elliptically polarized. In a similar way, if plane-polarized light is incident upon a plate which is a half-wave plate for yellow light, the emergent light is not completely plane-polarized. A Nicol prism set to extinguish the yellow light will allow small amounts of neighbouring wavelengths, and larger amounts of more distant wavelengths, to pass. The emergent light contains a preponderance of blue and red, and has an easily recognized hue called the *tint of passage.*

12.44. The Biquartz.

Similar effects may occur owing to the dispersion of optical rotation. This property has been used to provide a sensitive method of locating the plane of polarization of a beam of light. Two pieces of quartz, each producing a rotation of $\frac{1}{2}\pi$ (but in opposite senses) for yellow light, are put together (fig. 12.23). This arrangement is called a *biquartz.* If plane-polarized white light is passed through a biquartz, the yellow part is rotated through a right angle in each half of the field, and is extinguished by a Nicol set parallel to the original plane of polarization. Light of other wavelengths is transmitted in similar proportions in the two halves of the field. They both show the same hue, i.e. the tint of passage. If, however, the Nicol is rotated through a small angle, one half of the field becomes more blue and the other more red. The eye is very sensitive

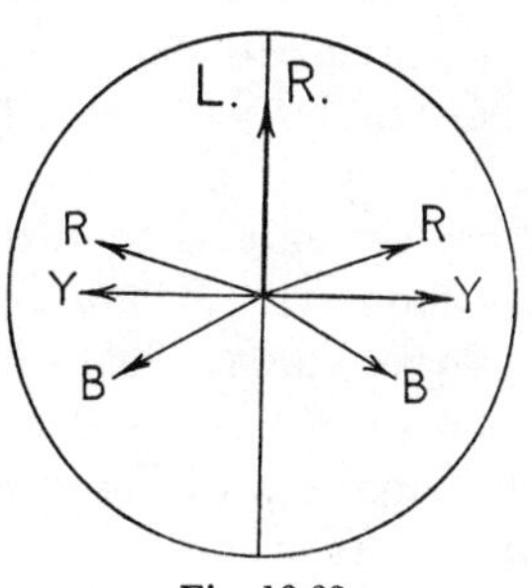

Fig. 12.23

to small differences of hue between two large uniform regions placed side by side, and the Nicol can be set with high precision. An alternative way of making the setting is to pass the light into a spectroscope, so that it produces two spectra, one above the other. There is a dark band in each spectrum. When the Nicol is in the correct setting, the bands coincide. When it is rotated slightly, one band moves to longer and the other to shorter wavelengths. The changes of hue produced by rotating the Nicol may be used to detect the presence of a small fraction of polarized light in a beam of white light.

12.45. **Saccharimetry.**

Accurate estimation of the strengths of sugar solutions can be made by measuring the specific rotation (§ 12.35). Many instruments, called *saccharimeters*, have been devised for this purpose. The simplest type of saccharimeter consists of a polarizer, a cell for containing the optically active solution, and an analyser, together with the lenses necessary to allow a parallel beam of light to pass through the solution and enter the observer's eye.

The cell is first filled with water and the analyser is turned until no light passes. The cell is then filled with the solution under test, and the analyser is rotated until the field is again dark. The angle (α) through which the analyser has been turned is read on a circular scale. Since, under practical conditions, the angle of rotation (ψ) is never greater than π, it follows that either $\psi = \alpha$ or $\psi = \pi - \alpha$. To distinguish between these possibilities, the solution is diluted to half-strength, and the new value of α is measured.

12.46.—The above arrangement is inaccurate for two reasons. In the first place, the eye is being asked to make a difficult judgment under rather unfavourable conditions. In the second place, the illumination is very nearly zero for a considerable range of setting.* Much more accurate settings can be made with the aid of a device such as the biquartz. Jellett (1817–88) devised an arrangement in which the observer views a field which is divided into two parts of equal illumination when the analyser is in the correct orientation. His arrangement was improved by Cornu, and the modern form is called the *Cornu-Jellett prism.* It consists of a Nicol (or Glan-Thompson) prism from which a wedge-shaped piece has been removed. The two halves are united so that the principal planes make a small angle δ (called the *half-shadow*

* The illumination E is proportional to $\sin^2(\alpha - \theta)$, and $dE/d\theta$ is very small when α is near to θ.

angle) with each other (fig. 12.24). If a beam of plane-polarized light is viewed through the Cornu-Jellett prism, oriented so that the plane of polarization bisects the half-shadow angle internally or externally, the two halves of the field are equally bright. The human eye is very sensitive in regard to the judgment of equality of brightness in two fairly large fields placed side by side, provided that the illumination is not too low. If the setting is made on the exterior angle of bisection, a rotation of δ from the position of equality makes one half of the field completely dark. Thus a very great change in relative brightness is caused by a small variation of angle, and the setting can be made very accurately.

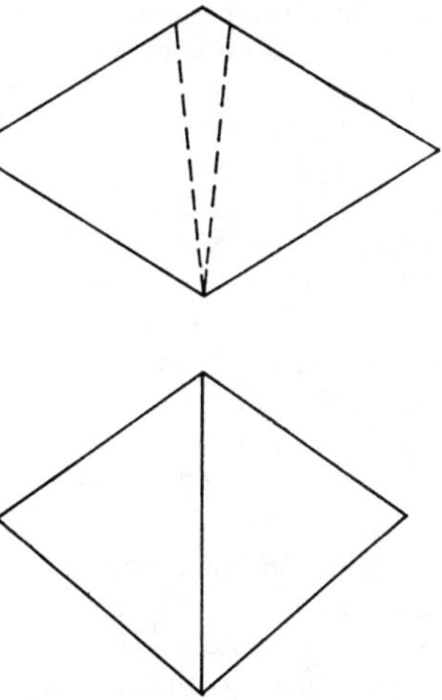

Fig. 12.24.—The Cornu-Jellett prism

12.47.—When the half-shadow method is used, there is an optimum value of the angle δ. If δ is made very small, the relative brightness varies very rapidly, but the absolute brightness of the field is very low near the point of balance [see Example 12(xxv), p. 391]. The optimum value of δ varies with the intensity of the light source, the transparency of the solution, and the wavelength of the light. In a research instrument, it is sometimes an advantage for the observer to be able to alter δ at will. In saccharimeters for routine industrial analysis by relatively unskilled observers, δ is fixed. The observations are commonly made with sodium light, using a sodium discharge lamp. This is a powerful, steady source, and is much more convenient than the older " sodium flame ". A solution containing 26 grammes of the sample per 100 millilitres is used. The length of the tube is fixed at exactly 20·00 centimetres. The scale of the instrument is divided into a hundred " sugar degrees ". The maximum reading corresponds to pure sucrose. Mixtures containing cane sugar together with other optically active substances may be analysed by measuring the rotation before and after heating the solution with hydrochloric acid. After heating, the cane sugar (which is dextro-rotatory) is changed into invert sugar, which is a mixture of two left-handed components. The specific rotation of invert sugar is $-19{\cdot}7°$. For accurate saccharimetry it is necessary to work at a standard temperature, or else to apply a correction for temperature. There is also a small correction because the rotation is not *accurately* proportional to concentration.

12.48. Light Beats.

It is well known that two sources of sound of slightly different frequency interact, so that a fixed observer hears a sound whose amplitude fluctuates with a frequency equal to the difference of the frequencies of the sources. It is difficult to produce the corresponding effect with light, because independent light sources are non-coherent. It is therefore necessary to have some means of making a small alteration in the frequency of one of two beams of light derived from a single source. One way of doing this is to reflect one beam from a moving mirror, so that the frequency is changed by the Doppler effect, and to allow the reflected beam to interfere with light received directly from the source, or by reflection at stationary mirrors. An experiment of this type may be performed with the Michelson interferometer.

Suppose an observer views a system of approximately linear fringes produced when the mirror M_2 (fig. 4.1) is at a small angle to the reference plane, and that the mirror M_2 is moved towards the observer with speed v. Then the path changes by $2v$ centimetres per second, and $2v/\lambda$ fringes will pass the observer in each second. The source of light seen in the reflected mirror moves with a speed $2v$, and the Doppler change of frequency is $\nu_0(2v/c) = 2v/\lambda$. Thus the fluctuations of illumination at a given point in the field have a frequency equal to the change of frequency introduced by the moving mirror. It will be noticed that the motion of the fringes might equally well be described in terms of a varying path difference. The position of the fringes at any instant is determined by the path difference at that instant. These alternative explanations, in terms of Doppler effect and beats, or in terms of varying path-difference, apply to many systems of moving interference fringes.

12.49.—A somewhat different method of producing two sources of slightly different frequency is due to Righi. Suppose that a beam of light is passed first through a rotating Nicol prism, and then through a quarter-wave plate and a second Nicol. Both the latter are fixed, and the plane of the second Nicol bisects the angle between the privileged directions of the quarter-wave plate. Let us suppose that at time t the Nicol makes an angle $\alpha = pt$ with the slow direction of the quarter-wave plate. The Nicol is assumed to be rotating with constant angular velocity so that p is constant. Then the components of the emergent beam, parallel to the slow and fast directions, may be represented by

$$\xi_x = a \cos \alpha \cos (\omega t - \kappa z), \qquad \text{12(44)}$$

and

$$\xi_y = -a \sin \alpha \sin (\omega t - \kappa z). \qquad \text{12(45)}$$

Each of these components makes an angle $\frac{1}{4}\pi$ with the second Nicol, which transmits a beam represented by

$$\xi = \frac{a}{\sqrt{2}}\left\{\cos \alpha \cos (\omega t - \kappa z) - \sin \alpha \sin (\omega t - \kappa z)\right\}$$

$$= \frac{a}{\sqrt{2}} \cos \left\{(\omega + p)t - \kappa z\right\}. \qquad \text{12(46)}$$

Thus the emergent beam has a frequency $(\omega + p)$. If the quarter-wave plate is turned through an angle of $\frac{1}{2}\pi$ (so that the fast and slow directions are interchanged) the sign of 12(46) is altered, and the frequency of the resultant beam is $(\omega - p)$.

12.50.—In Righi's experiment, the light is passed first through a rotating Nicol, and is then separated into two beams by a pair of Fresnel mirrors. Before being brought to interference, each beam passes through a quarter-wave plate and a fixed Nicol prism. The quarter-wave plates are placed so that the fast direction of one is parallel to the slow direction of the other. The fixed Nicol prism is placed so that its principal plane bisects the privileged directions of the plates. The fringes move so that at a fixed point the illumination fluctuates with the frequency $2p/2\pi$. This would be expected since the frequencies of the two virtual sources are $(\omega + p)/2\pi$ and $(\omega - p)/2\pi$. It will be noted that in this experiment, as in the moving-mirror experiment, a " static " explanation is possible. The illumination at a given point at any instant is determined by the instantaneous angular position of the rotating Nicol.

EXAMPLES [12(xxii)–12(xxx)]

12(xxii). Show that when analyser and polarizer are crossed, the second term in 12(43) is always positive. Hence only one hue can be obtained.

12(xxiii). With a specimen of selenite 0·913 mm. thick, and with analyser and polarizer crossed, dark bands were obtained at 6543, 6167, 5830, 5540, 5266, 5030, 4810, 4590, 4405 Å. Assuming that the difference of refractive indices for the fast and slow directions is independent of wavelength over the range covered, calculate this difference of indices.

[If successive bands occur at wavelengths $\lambda_0, \lambda_1, \ldots \lambda_k$, we may write $(n + k)\lambda_k = e(\alpha - \gamma)$, where e is the thickness, and α and γ the refractive indices of the crystal. From the slope of the straight line obtained on plotting $1/\lambda_k$ against k, the value of $(\alpha - \gamma)$ may be found. (0·012.)]

12(xxiv). The specific rotation of cane sugar is 66·5°. A solution is found to give a rotation of $t_1°$. It is then mixed with one-tenth of its own volume of hydrochloric acid, and heated. After heating the rotation is $t_2°$. Find the number of grammes of cane sugar in a litre of the original solution. (See § 12.47.)

[$1000(t_1 - 1{\cdot}1t_2)/86{\cdot}2$ gm. litre^{-1}.]

12(xxv). A Cornu-Jellett prism has a half-shadow angle δ. Find an expression for the illumination in each half of the field when the analyser receives a beam of light plane-polarized in a plane which makes an angle θ with the internal bisector of the half-shadow angle. Hence show that the contrast is varying rapidly with θ when $\theta = \frac{1}{2}\pi$, and much less rapidly when $\theta = 0$.

(Contrast = difference of illumination ÷ mean illumination.)

[$a^2 \cos^2(\theta - \frac{1}{2}\delta)$ and $a^2 \cos^2(\theta + \frac{1}{2}\delta)$.]

12(xxvi). Show that if in Righi's experiment the second Nicol is turned through a right angle, the direction of movement of the fringes is reversed.

12(xxvii). Show that a beam of light which has passed through a rotating Nicol may be represented as the resultant of two circularly polarized beams of different frequency. Show that Righi's observations may then be described by assuming that the quarter-wave plates change these beams into plane-polarized light.

12(xxviii). Design an arrangement similar to Righi's, but using fixed Nicol prisms and a rotating quarter-wave plate.

12(xxix). Discuss the production of interference fringes with circularly polarized light. Derive conditions corresponding to those for plane-polarized light (§ 12.15).

[(i) Two beams of circularly polarized light of opposite sense do not produce interference fringes under any condition.

(ii) Two beams polarized in the same sense produce interference fringes under the same conditions as two beams of unpolarized light, provided that they are originally derived from the same beam of polarized light. The beam from which they are derived may be either plane or circularly polarized.

(iii) Two beams of light, circularly polarized in the same sense, derived from perpendicular plane-polarized components of unpolarized light, do not interfere. Two beams derived from right-handed and left-handed circularly polarized components of unpolarized light also do not interfere.]

12(xxx). What are the corresponding conditions for elliptically polarized light?

[Some interference is obtained if the conditions for circularly polarized light are fulfilled. Fringes with completely dark minima are obtained only when, in addition, the ellipses are similar and similarly oriented.]

REFERENCES

12.1. Bragg, W. L.: *The Crystalline State* (Bell).

12.2. Wooster: *Crystal Physics* (Cambridge University Press).

12.3. Thompson, Silvanus P.: *Trans. Optical Convention,* 1905.

12.4. Rayleigh: *Scientific Papers,* Vol. III, pp. 140–147.

12.5. Hurwitz: *Journal of the Optical Society of America,* 1945, Vol. 35, p. 525.

12.6. Jerrard: *ibid.,* 1948, Vol. 38, p. 35.

12.7. Wright: *The Measurement of Colour* (Adam Hilger).

CHAPTER XIII

The Electromagnetic Theory

13.1. **Development of the Theory.**

The early experiments on electricity and magnetism were concerned with fields which were constant, or which varied only slowly with respect both to space and to time. These experiments led to the formulation of generalizations such as Coulomb's law, Ampère's law, and Faraday's law of induction. An experimental result connecting light with magnetism * was obtained by Faraday; he suggested that light might be electromagnetic in origin, but was not able to derive any quantitative theory. This was done later by James Clerk Maxwell. He began by summarizing the then existing laws of electromagnetism in the simple and elegant set of relations which are now known as Maxwell's equations. He extended the theory and showed that it implied the possibility of electromagnetic waves whose velocity *in vacuo* was equal to a constant c which could be derived from electrical measurements. Since this constant was equal to the observed value of the velocity of light, he suggested that light could be represented by electromagnetic waves of high frequency. He showed that this theory gave an adequate account of the reflection and refraction of light by substances such as glass, and that it could be applied to the reflection of light by metals. He began to apply it to the theory of dispersion, but his early death left others to complete the development. An important step in the verification of the theory—the generation of electromagnetic waves from an oscillating dipole—was due to Hertz.

13.2.—The advantages of the electromagnetic theory over other wave theories of light depend on a number of points of detail. The mathematical development is not very difficult, but is rather lengthy. In order that the reader may not lose sight of the objective while considering these details, the argument will now be summarized before commencing a systematic discussion and proof. We shall first define the more important electromagnetic quantities and write down Maxwell's equations. We shall assume that the equations are valid, i.e. that they form a self-consistent set whose predictions in regard to

* See § 16.50.

static or slowly varying fields are in agreement with experimental data.* From this point we proceed to show that these equations lead naturally to the hypothesis of electromagnetic waves propagated with a definite velocity, whose value is obtained from electrical measurements and is equal to the observed value of the velocity of light. We next examine the properties of these waves and find that they are transverse waves capable of representing all the states of polarization described in Chapter XII. The reflection of the waves at the boundary of two isotropic non-conducting media is examined in Chapter XIV. It is found that the theory not only gives the relations between the angles of incidence, reflection, and refraction, but also gives unambiguous predictions in relation to the amounts of light reflected and refracted, and the states of polarization. These predictions are in good agreement with experimental results. It is also found that the theory gives generally satisfactory results when applied to the propagation of light in metals and in dispersive media. The predictions here are not so precise, because they depend on hypotheses concerning molecular structure as well as upon the Maxwell equations. In Chapter XVI the theory is applied to anisotropic media and is found to give a very elegant account of the results of observations on the transmission of light in crystals. In Chapter XVII we begin the description of certain experiments on the interaction of light with atoms and molecules. We find that the electromagnetic theory in itself is not capable of representing these results, so that the quantum hypothesis must be introduced. In Chapter XVIII we show how the electromagnetic theory may formally be joined to quantum theory through the quantization of the energy and the momentum of the electromagnetic field.

13.3. **Mathematical Methods.**

Maxwell's equations and the succeeding theory are most conveniently described by the vector notation. The alternative Cartesian description is equally valid but is rather cumbersome. We shall generally use the vector method, but at the beginning the Cartesian form will also be stated, and from time to time Cartesian equations will be used where the vector method offers no appreciable mathematical advantage.†

* A systematic development of the equations is given in Chapters XII and XIII of Reference 13.1. More detailed and fundamental treatments are given in References 13.2, 13.3, and 13.4.

† All the vector theory required is given in Reference 13.1 (see especially Chapter I). A more complete account, suitable for a student specializing in mathematical physics, is given in Reference 13.5.

The following notations and conventions are used in this book:

(i) When Cartesian co-ordinates are used they are always right handed, i.e. a rotation from the positive direction of x to the positive direction of y is clockwise to an observer looking in the positive direction of z (fig. 13.1).

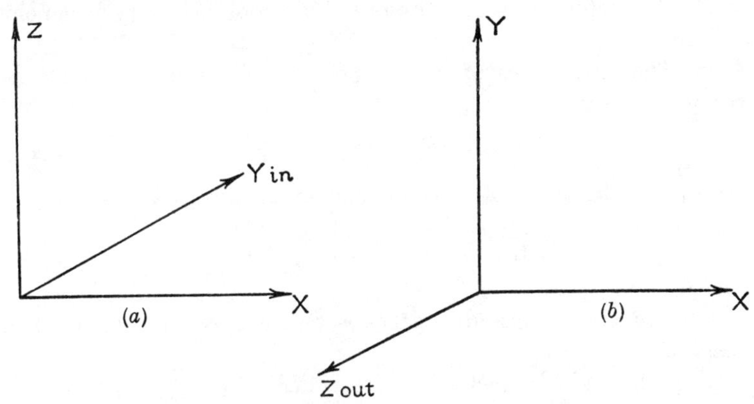

Fig. 13.1.—Right-handed axes. If X and Z are in the plane of the paper as shown in (*a*), the positive direction of Y is into the paper. If X and Y are in the plane of the paper as shown in (*b*), then the positive direction of Z is outwards from the plane of the paper.

(ii) Vectors are written in heavy type. A symbol such as **E** stands for the vector and E for the magnitude. **e** is a vector of unit magnitude in the direction of **E**. The magnitude of the amplitude of a vector which varies sinusoidally is denoted by a symbol with a suffix, e.g. we have

$$\left.\begin{aligned} \mathbf{E} &= \mathbf{e}E \\ \text{and} \qquad E &= E_0 \exp i(\omega t - \kappa z). \end{aligned}\right\} \quad . \quad . \quad . \quad . \quad 13(1)$$

(iii) The products of two vectors such as **E** and **H** are written: $\mathbf{E} . \mathbf{H}$ for the *scalar product*, whose value is $EH \cos \theta$, and $\mathbf{E} \times \mathbf{H}$ for the *vector product*, which is a vector whose magnitude is $EH \sin \theta$ and whose direction is normal to the plane containing **E** and **H**. A rotation from **E** to **H** appears clockwise to an observer looking along the direction of the vector product.

(iv) The symbols **i**, **j**, **k** are used for unit vectors in the positive directions of the axes OX, OY, and OZ respectively. The standard vector operator, called “ nabla ”, is written

$$\nabla = \mathbf{i}\frac{\partial}{\partial x} + \mathbf{j}\frac{\partial}{\partial y} + \mathbf{k}\frac{\partial}{\partial z}. \quad . \quad . \quad . \quad . \quad 13(2)$$

(v) The *gradient* of a scalar function (such as V) is written

$$\text{grad } V = \nabla V = \mathbf{i}\frac{\partial V}{\partial x} + \mathbf{j}\frac{\partial V}{\partial y} + \mathbf{k}\frac{\partial V}{\partial z}. \quad . \quad . \quad 13(3)$$

V is a scalar quantity whose value may vary from point to point; grad V is a vector whose magnitude and direction may vary from point to point.

(vi) The magnitudes of the components of a vector $\mathbf{E}$ along the axes are written E_x, E_y, E_z, so that

$$\mathbf{E} = \mathbf{i}E_x + \mathbf{j}E_y + \mathbf{k}E_z. \quad . \quad . \quad . \quad . \quad 13(4)$$

(vii) The *divergence* of a vector function $\mathbf{E}$ is defined by

$$\text{div } \mathbf{E} = \nabla . \mathbf{E} = \frac{\partial E_x}{\partial x} + \frac{\partial E_y}{\partial y} + \frac{\partial E_z}{\partial z}. \quad . \quad . \quad 13(5)$$

A vector function whose divergence is zero everywhere is said to be *solenoidal.*

(viii) The *curl* of a vector function is defined by

$$\text{curl } \mathbf{E} = \nabla \times \mathbf{E} = \mathbf{i}\left(\frac{\partial E_z}{\partial y} - \frac{\partial E_y}{\partial z}\right) + \mathbf{j}\left(\frac{\partial E_x}{\partial z} - \frac{\partial E_z}{\partial x}\right) + \mathbf{k}\left(\frac{\partial E_y}{\partial x} - \frac{\partial E_x}{\partial y}\right) \quad 13(6)$$

Note the cyclic order of permutation of letters below the line. The curl is a vector function whose x, y and z components are respectively

$$\left(\frac{\partial E_z}{\partial y} - \frac{\partial E_y}{\partial z}\right), \left(\frac{\partial E_x}{\partial z} - \frac{\partial E_z}{\partial x}\right), \text{ and } \left(\frac{\partial E_y}{\partial x} - \frac{\partial E_x}{\partial y}\right).$$

(ix) The symbol ∇^2 when applied to a scalar (V) means

$$\nabla^2 V = \nabla . \nabla V = \text{div (grad } V) = \frac{\partial^2 V}{\partial x^2} + \frac{\partial^2 V}{\partial y^2} + \frac{\partial^2 V}{\partial z^2}. \quad 13(7)$$

The result is a scalar.

(x) The sybol ∇^2 when applied to a vector ($\mathbf{E}$) means

$$\begin{aligned}\nabla^2\mathbf{E} &= \mathbf{i}\,\nabla^2 . E_x + \mathbf{j}\,\nabla^2 . E_y + \mathbf{k}\,\nabla^2 . E_z \\ &= \mathbf{i}\left(\frac{\partial^2 E_x}{\partial x^2} + \frac{\partial^2 E_x}{\partial y^2} + \frac{\partial^2 E_x}{\partial z^2}\right) \\ &\quad + \mathbf{j}\left(\frac{\partial^2 E_y}{\partial x^2} + \frac{\partial^2 E_y}{\partial y^2} + \frac{\partial^2 E_y}{\partial z^2}\right) \\ &\quad + \mathbf{k}\left(\frac{\partial^2 E_z}{\partial x^2} + \frac{\partial^2 E_z}{\partial y^2} + \frac{\partial^2 E_z}{\partial z^2}\right). \quad . \quad . \quad . \quad 13(8)\end{aligned}$$

The above operations may be repeated or applied successively. The following propositions are given for reference when required:*

(i) The divergence of the curl of any vector function is zero everywhere.

(ii) The curl of the gradient of any scalar function is zero everywhere.

(iii) The curl of a curl of a vector function (**E**) is given by

$$\nabla \times (\nabla \times \mathbf{E}) = \text{curl}\,(\text{curl}\,\mathbf{E}) = \text{grad}\,(\text{div}\,\mathbf{E}) - \nabla^2\mathbf{E}. \qquad 13(9)$$

(iv) $$\text{div}\,(\mathbf{E} \times \mathbf{H}) = \nabla \,.\, (\mathbf{E} \times \mathbf{H}) = \mathbf{H}\,\text{curl}\,\mathbf{E} - \mathbf{E}\,\text{curl}\,\mathbf{H}. \qquad 13(10)$$

13.4. Definitions of E and H.

Historically the concept of electric force was derived from observations of the mechanical forces exerted upon one another by charged bodies. In a corresponding way, the idea of magnetic force was derived from observations on small magnets. Since the concept of " magnetic pole " is somewhat artificial and unsatisfactory, it is convenient for us to regard **E** and **H** as defined by a set of ideal experiments on electrons. We observe that electrons *in vacuo* experience two kinds of force. One, which depends only on the positions of the electrons, we call the *electrostatic force.* The other, which depends also on their relative velocities, we call the *magnetic force.* The whole situation is summarized in the vector equation

$$-\mathbf{F} = e\mathbf{E} + \frac{e}{c}\,\mathbf{v} \times \mathbf{H}, \qquad . \quad . \quad . \quad . \quad 13(11)$$

where **F** is the mechanical force on a particular electron which has a velocity **v**. The magnitude of the charge e is determined by reference to Coulomb's experiment on the force between two point charges. By well-established convention the charge of the electron is negative, and this requires a negative sign on the left-hand side of 13(11). **E** and **H** are two vector fields which vary from point to point. The important constant c enters because we have, for historical reasons, certain units for **E** and **H**. These units are such that the constants of proportionality entering into the two terms on the right-hand side of 13(11) are different. The units could be altered so that c is eliminated from this equation, but it would then appear in a different way in equations 13(18), etc., below and our final result would be the same. We could

* See p. 39 ff. of Reference 13.1.

eliminate it completely only by altering our unit of length or our unit of time. This would lead to inconvenient units for length, or for time, or for both. The constant c has the dimensions of a velocity.

13.5. Definition of Charge Density and Current.

Before the molecular theory of matter, fluids were regarded as truly continuous. The density of a fluid at a given point was taken to be

$$\operatorname*{Limit}_{\Delta\tau \to 0} \left(\frac{\Delta m}{\Delta \tau}\right),$$

where Δm is the mass of a small volume $\Delta\tau$ surrounding the point of interest. When we regard the fluid as an assembly of molecules we have to admit that the density so defined does not vary continuously from point to point. Nevertheless, in hydrostatics and hydrodynamics we treat fluids as though they were continuous, and we obtain formulæ which agree with the experimental results provided that the experiments refer to volumes large enough to contain many molecules. In a similar way, we now conceive a volume distribution of charge as made up of electrons and ions, but it is still possible to use the concept of a charge density (ρ) which varies continuously from point to point. This concept is convenient and is sufficient for many purposes, though for certain applications we have to consider the detailed distribution of charge. Following the analogy of a fluid, a transport of electricity may arise from the movements of an electrical charge distribution. We may then define a vector $\mathbf{j}$, called the current density,* whose magnitude is equal to the net amount of positive charge crossing unit area of a surface (normal to the direction of flow) per second. If the velocity is $\mathbf{v}$, we then have

$$\mathbf{j} = \rho\mathbf{v}. \qquad 13(12)$$

It is found that electric charges can be transported: (*a*) through the movement of ions *in vacuo*, (*b*) through the movement of a charged conductor, and (*c*) through a conduction current in a material body. This last type of movement takes place under the action of an electric field established in the body of a conductor. According to the electron theory this transport is pictured as a steady drift superimposed upon the random motions of electrons in the conductor. The current density

* It appears, at first sight, unfortunate that $\mathbf{j}$ should stand for current vector and also for unit vector in the direction OY. In fact, the equations are such that no confusion arises.

is found to be proportional to the electric field. The conductivity of the material is defined by the relation

$$\mathbf{j} = \sigma \mathbf{E}. \qquad 13(13)$$

In isotropic materials σ is a scalar quantity which may vary from point to point in the material. Direct experimental evidence shows that the three forms of current mentioned above are equivalent in their magnetic effects. These can be calculated when the value of $\mathbf{j}$ is known.

13.6. Polarization of a Material Medium.

When a non-conductor is placed in an electric field, the distribution of the electric charges which constitute the atoms and molecules of the medium is altered so as to produce an internal field which opposes the original field. This is called *polarization* and arises from two main causes. Some substances consist of polar molecules which in the absence of a field are equivalent to a number of dipoles with random orientation. The field tends to align the dipoles parallel to its own direction. This tendency is opposed by the elastic forces in a solid, or by the randomizing action of thermal collisions in a gas or liquid, and an equilibrium condition is set up. Polarization due to rotation is ineffective at high frequencies because the movement of the dipoles is too slow to follow the variations of the field. The second cause of polarization is due to an alteration of the distribution of electric charges within the molecule which produces an induced dipole moment. This effect occurs both with polar and nonpolar molecules. Generally it is less at high frequencies than when the field is static or slowly varying, except when the high-frequency variations are in resonance with natural frequencies of the molecules. We shall consider this resonance in detail later (§ 15.18 ff.). In a dielectric, the field which would exist in the absence of a polarization is called the electric induction (**D**), the polarization field * is denoted by $4\pi\mathbf{P}$ and the resultant field is denoted by **E** (fig. 13.2). We have

$$\mathbf{E} = \mathbf{D} - 4\pi\mathbf{P}. \qquad 13(14)$$

When the medium is isotropic, the three vectors are in the same direction and for small fields E is proportional to D, so that we have

$$D = \epsilon E \qquad 13(15a)$$

and

$$\epsilon = 1 + 4\pi \frac{P}{E}, \qquad 13(15b)$$

* It may be shown by the use of Gauss's theorem that an electric moment per unit volume represented by **P** implies a field of force represented by $4\pi\mathbf{P}$ (p. 266 of Reference 13.1).

where ϵ is a scalar quantity called the dielectric constant. Similar considerations apply with regard to the magnetic properties of materials. Static fields produce a polarization of the medium which is usually interpreted as due to a diamagnetic effect (induction of fresh moments opposing the field) and a paramagnetic effect (orientation of

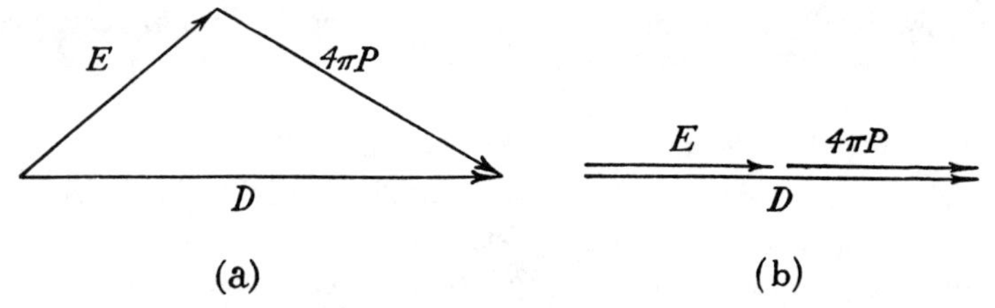

Fig. 13.2.—Relation between **E**, **D** and **P** (*a*) for an anisotropic medium, (*b*) for an isotropic medium

existing moments). The diamagnetic effect is small, and the paramagnetic effect is inoperative at optical frequencies. Thus the magnetic properties of all material media at optical frequencies are nearly the same as the properties of a vacuum, and we need not, at present, take any account of magnetic polarization.

13.7. **Maxwell's Equations.**

The experimental measurements on static and quasi-static fields in media which are non-magnetic and isotropic are summarized in the following equations:

$$\operatorname{div} \mathbf{D} = \epsilon \operatorname{div} \mathbf{E} = 4\pi\rho. \quad 13(16)$$

$$\operatorname{div} \mathbf{H} = 0. \quad 13(17)$$

$$\operatorname{curl} \mathbf{E} = -\frac{1}{c}\frac{\partial \mathbf{H}}{\partial t}. \quad 13(18)$$

$$\operatorname{curl} \mathbf{H} = \frac{4\pi}{c}\mathbf{j} + \frac{1}{c}\frac{\partial \mathbf{D}}{\partial t} \quad 13(19a)$$

$$= \frac{4\pi}{c}\sigma\mathbf{E} + \frac{\epsilon}{c}\frac{\partial \mathbf{E}}{\partial t}. \quad 13(19b)$$

In these equations electrical quantities are measured in the usual electrostatic units, and magnetic quantities in electromagnetic units. Equation 13(19*a*), without the second term on the right-hand side, is essentially Ampère's law. The second term was added by Maxwell.

He called it the *dielectric displacement-current term.* Subsequent work has shown that at low frequencies the magnetic effects of a variation of the displacement are in every way equivalent to the magnetic effects of a conduction current. The magnitude of the displacement current is

$$\mathbf{j}_d = \frac{1}{4\pi}\frac{\partial \mathbf{D}}{\partial t}. \qquad 13(20)$$

13.8. Waves in an Insulating Medium.

We shall now consider the propagation of waves in a medium which is

(*a*) uniform, so that ϵ has the same value at all points;
(*b*) isotropic, so that ϵ is independent of the direction of propagation;
(*c*) non-conducting, so that $\sigma = 0$ and therefore $\mathbf{j} = 0$;
(*d*) free from charge, so that $\rho = 0$.

With these assumptions equations 13(16)–13(19) reduce to *

$$\text{div}\,\mathbf{E} = 0. \qquad 13(21)$$

$$\text{div}\,\mathbf{H} = 0. \qquad 13(17)$$

$$-\frac{1}{c}\frac{\partial \mathbf{H}}{\partial t} = \text{curl}\,\mathbf{E}. \qquad 13(18)$$

$$\frac{\epsilon}{c}\frac{\partial \mathbf{E}}{\partial t} = \text{curl}\,\mathbf{H}. \qquad 13(22)$$

* Written in Cartesian form the equations are:

$$\frac{\partial E_x}{\partial x} + \frac{\partial E_y}{\partial y} + \frac{\partial E_z}{\partial z} = 0. \qquad 13(21a)$$

$$\frac{\partial H_x}{\partial x} + \frac{\partial H_y}{\partial y} + \frac{\partial H_z}{\partial z} = 0. \qquad 13(17a)$$

$$\left.\begin{aligned} -\frac{1}{c}\frac{\partial H_x}{\partial t} &= \left(\frac{\partial E_z}{\partial y} - \frac{\partial E_y}{\partial z}\right). \\ -\frac{1}{c}\frac{\partial H_y}{\partial t} &= \left(\frac{\partial E_x}{\partial z} - \frac{\partial E_z}{\partial x}\right). \\ -\frac{1}{c}\frac{\partial H_z}{\partial t} &= \left(\frac{\partial E_y}{\partial x} - \frac{\partial E_x}{\partial y}\right). \end{aligned}\right\} \qquad 13(18a)$$

$$\left.\begin{aligned} \frac{\epsilon}{c}\frac{\partial E_x}{\partial t} &= \left(\frac{\partial H_z}{\partial y} - \frac{\partial H_y}{\partial z}\right). \\ \frac{\epsilon}{c}\frac{\partial E_y}{\partial t} &= \left(\frac{\partial H_x}{\partial z} - \frac{\partial H_z}{\partial x}\right). \\ \frac{\epsilon}{c}\frac{\partial E_z}{\partial t} &= \left(\frac{\partial H_y}{\partial x} - \frac{\partial H_x}{\partial y}\right). \end{aligned}\right\} \qquad 13(22a)$$

We now proceed to eliminate **E** and **H** in turn from the two equations 13(18) and 13(22), using the vector theorem (iii) quoted at the end of § 13.3, namely,

$$\text{curl}\,(\text{curl}\,\mathbf{E}) = \text{grad}\,(\text{div}\,\mathbf{E}) - \nabla^2\mathbf{E} = -\frac{1}{c}\frac{\partial}{\partial t}(\text{curl}\,\mathbf{H})$$

$$= -\frac{\epsilon}{c^2}\frac{\partial^2\mathbf{E}}{\partial t^2}.$$

Hence, using 13(21), we have

$$\nabla^2\mathbf{E} = \frac{\epsilon}{c^2}\frac{\partial^2\mathbf{E}}{\partial t^2}, \qquad \text{13(23)}$$

and similarly

$$\nabla^2\mathbf{H} = \frac{\epsilon}{c^2}\frac{\partial^2\mathbf{H}}{\partial t^2}. \qquad \text{13(24)}$$

These equations are of the same form as the standard wave equation 2(38). They indicate that variations of **E** or **H** should be propagated with a velocity $c/\sqrt{\epsilon}$.

13.9. The Velocity of Light.

For a vacuum ϵ is equal to unity, and the wave velocity should be equal to the constant c. This constant may conveniently be determined by measuring the capacity of a condenser in electromagnetic and in electrostatic units. The value thus obtained is found to agree with the measured velocity of light to an accuracy of about 1 in 30,000, i.e. to within experimental error (Chapter X). If the polarization of the medium were independent of the frequency of the waves, we should expect the ratio of the velocity *in vacuo* (c) to the velocity (b) in an insulating medium such as water to be given by

$$n = \frac{c}{b} = \sqrt{\epsilon}, \qquad \text{13(25)}$$

where n is the refractive index.

Values of $\sqrt{\epsilon}$ and of n for a number of substances are shown in Table 13.1. It may be seen that, for a number of substances, n is approximately equal to $\sqrt{\epsilon}$. For others $\sqrt{\epsilon}$ is much greater than n. These latter substances derive their high dielectric constant from the orientation of polar molecules. This contribution to the polarization which is ineffective at optical frequencies is sometimes still operative at radio frequencies. For example, water gives 9·0 for $\sqrt{\epsilon}$, 8·9 for the refractive index at radio frequency, but only 1·33 for the index at optical frequencies.

TABLE 13.1

	$\sqrt{\epsilon}$	n
Air	1·000295	1·000292
Helium	1·000037	1·000035
Paraffin	1·405	1·422
Toluol	1·549	1·499
Benzol	1·511	1·501
Water	9·0	1·33
Methyl alcohol	5·7	1·34
Ethyl alcohol	5·1	1·36

13.10. Properties of Electromagnetic Waves.

Let us now consider a plane wave propagated in the positive direction of z. Any of the components may be represented by an expression of the type

$$E_x = E_{0x} \exp i(\omega t - \kappa z). \quad . \quad . \quad . \quad . \quad 13(26)$$

Since the wave fronts are parallel to the xy plane, all these expressions are independent of x and y. We now derive the following properties:*

(i) *The waves are transverse.*

Since $\mathbf{E}$ depends only on z, we have

$$\frac{\partial E_x}{\partial x} = 0; \quad \frac{\partial E_y}{\partial y} = 0; \quad \frac{\partial E_z}{\partial x} = 0; \text{ and } \frac{\partial E_z}{\partial y} = 0. \qquad 13(27)$$

Taking the first two of these equations with 13(21a), we have

$$\frac{\partial E_z}{\partial z} = 0. \quad . \quad . \quad . \quad . \quad . \quad . \quad 13(28)$$

This relation, together with the latter part of 13(27), shows that E_z does not depend on x, y, or z.

We are not interested in an electric field which is the same at all points. We therefore put E_z equal to zero with the understanding that E_x, E_y, E_z, etc., stand for parts of the fields which are propagated. Thus the wavefield has no component in the direction of propagation, and $\mathbf{E}$ is perpendicular to the direction of propagation. It may similarly be shown that $\mathbf{H}$ is perpendicular to the direction of propagation, i.e. the waves are transverse.

* These properties require reconsideration when we come to deal with anisotropic media and with absorbing media.

(ii) *The electric and magnetic vectors are mutually perpendicular.*

Let us now choose the direction of the electric vector as the x axis. Then $E_y = 0$ everywhere. Writing out part of equation 13(18a) we have

$$-\frac{1}{c}\frac{\partial H_x}{\partial t} = \left(\frac{\partial E_z}{\partial y} - \frac{\partial E_y}{\partial z}\right) = 0. \quad . \quad . \quad . \quad 13(29)$$

The first term in the bracket vanishes because E_z is independent of y [equation 13(27)], and the second because $E_y = 0$ everywhere. Since we understand that H_x stands for the fluctuating part of the magnetic field, 13(29) implies that $H_x = 0$. Thus **H** is perpendicular to **E**.

(iii) *The magnetic and electric vectors are in phase.*

Let us assume that E_x is given by 13(26) and H_y by

$$H_y = H_{0y} \exp i(\omega t - \kappa z). \quad . \quad . \quad . \quad 13(30)$$

We now find the ratio of H_{0y} to E_{0x}. If this ratio is real and positive, then the vectors are in phase; if it is real and negative, there is a phase difference of π. Any phase difference which is not a multiple of π will be represented by a complex ratio between the amplitudes of the vectors (§ 2.26 and compare § 15.9). Applying equation 13(18) we have

$$-\frac{1}{c}\frac{\partial H_y}{\partial t} = \frac{\partial E_x}{\partial z} - \frac{\partial E_z}{\partial x}, \quad . \quad . \quad . \quad . \quad 13(31)$$

and since E_z is zero, this implies that

$$\frac{\omega}{c} H_{0y} = \kappa E_{0x}, \quad . \quad . \quad . \quad . \quad . \quad . \quad . \quad . \quad . \quad 13(32)$$

i.e.

$$H_{0y} = \frac{c}{b} E_{0x} = n E_{0x} = \sqrt{\epsilon} E_{0x}, \quad . \quad . \quad 13(33)$$

since $\omega/\kappa = b = c/\sqrt{\epsilon}$. The ratio is real and positive.

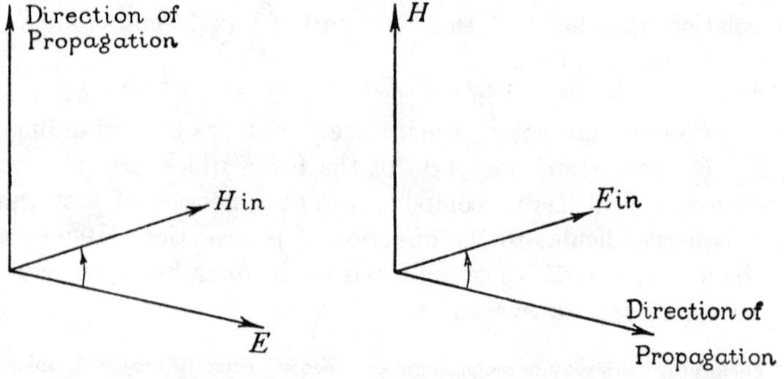

Fig. 13.3.—Relation between signs of **E**, **H** and direction of propagation

The two vectors **E** and **H** are in phase. When **E** has its maximum value (directed along the positive direction of x), **H** has also a maximum value (directed along the positive direction of y). *To an observer looking in the direction of propagation, a rotation from the direction of* **E** *to the direction of* **H** *is clockwise* (fig. 13.3).

EXAMPLES [13(i)–13(vii)]

13(i) Show that the scalar products **n** . **H**, **n** . **E**, and **E** . **H** (where **n** is the wave normal) are all zero.

13(ii). Show that if axes x', y', z' are chosen so that z' is the direction of propagation, but **E** and **H** are not in the directions of OX′ and OY′, we have

$$E_{x'}H_{x'} + E_{y'}H_{y'} = 0. \qquad 13(34)$$

13(iii). Show that the following are solutions of Maxwell's equations:

(*a*)

$$\left.\begin{aligned} E_x &= Ae^{i\phi}; & E_y &= 0; & E_z &= 0. \\ H_x &= 0; & H_y &= \sqrt{\varepsilon}Ae^{i\phi}; & H_z &= 0. \end{aligned}\right\} \qquad 13(35)$$

(*b*)

$$\left.\begin{aligned} E_x &= 0; & E_y &= Ae^{i\phi}; & E_z &= 0. \\ H_x &= -\sqrt{\varepsilon}Ae^{i\phi}; & H_y &= 0; & H_z &= 0. \end{aligned}\right\} \qquad 13(36)$$

$\phi = \omega t - \kappa z$ and A is a constant.

[13(35) is the solution we have obtained above, in simplified notation. 13(36) is the same solution with the X and Y axes rotated through a right angle.]

13(iv). Show that the following is a solution of the equations corresponding to a wave travelling in the negative direction of z:

$$\left.\begin{aligned} E_x &= A\exp i(\omega t + \kappa z); & E_y &= 0; & E_z &= 0. \\ H_x &= 0; \quad H_y = -\sqrt{\varepsilon}A\exp i(\omega t + \kappa z); & & & H_z &= 0. \end{aligned}\right\} \qquad 13(37)$$

Write down the solution analogous to 13(36).

$$[E_x = 0; \quad E_y = A\exp i(\omega t + \kappa z); \quad E_z = 0.$$
$$H_x = \sqrt{\varepsilon}A\exp i(\omega t + \kappa z); \; H_y = 0; \; H_z = 0.]$$

13(v). Using the Cartesian forms 13(21*a*), etc., derive the wave equation 13(23).

13(vi). Show that the system defined by 13(38) on p. 407 is a solution of the wave equation.

13.11. **Superposition of Electromagnetic Waves.**

It is an experimental observation that under static or quasi-static conditions the resultant electromagnetic field due to two systems of charges is obtained by adding the fields due to each system taken by itself, except when the fields are very strong or when ferromagnetic media are involved. Thus under all relevant conditions electromagnetic waves should obey the principle of superposition, and the velocity should be independent of the amplitude. If this were not so, the electromagnetic theory would be unable to give a simple explanation of interference, diffraction, and polarization. Given this condition, the electromagnetic theory can take over the main structure of the general wave theory as developed by Fresnel, Rayleigh, Kirchhoff, etc. The details of the theory of diffraction of electromagnetic waves require consideration. This problem has been examined by several writers.* It may appear at first sight that the electromagnetic theory has made the wave theory of light more difficult by introducing two vectors instead of one. This extra complication seems to be unnecessary from the point of view of optical theory, though it is required to satisfy the electromagnetic equations. As against this view, it should be understood that the components of the electromagnetic wave are not independent. The relation between **E** and **H** is such that, when we are given either of them (together with the constants of the medium and the direction of propagation), we can derive the other. The relation between them is, however, dependent on the medium and this relation yields the boundary conditions which are just sufficient to give a complete theory of reflection and refraction. Also the relation between **E** and **H** is different in anisotropic media and this contributes to a satisfactory theory of the propagation of light in these media. Thus the electromagnetic wave has a sufficient, but not an excessive, " flexibility " for the representation of light.

13.12. Representation of Polarized Light.

The wave described by equations 13(26) and 13(33), or more simply by equation 13(35), represents plane-polarized light. The electric vector is normal to the direction of propagation and remains in the same plane as the wave advances. The magnetic vector behaves in a similar way. It remains in the YZ plane, whereas the electric vector is in the XZ plane. For reasons which will appear later (§ 14.12), we find that the plane of the magnetic vector is the plane of polarization defined in § 12.3. If we accept the principle of superposition, we may form the concept of circularly and elliptically polarized light by combining beams which are plane-polarized in two mutually perpendicular planes, and

* Reference 13.6.

which have a constant phase difference. Elliptically polarized light may be represented by the following system:

$$\left.\begin{aligned} E_x &= A\cos(\omega t - \kappa z), & H_x &= -\sqrt{\epsilon}\,B\sin(\omega t - \kappa z),\\ E_y &= B\sin(\omega t - \kappa z), & H_y &= \sqrt{\epsilon}\,A\cos(\omega t - \kappa z),\\ E_z &= 0. & H_z &= 0. \end{aligned}\right\} \qquad 13(38)$$

If $A = B$, 13(38) represents circularly polarized light. If either A or $B = 0$, it represents plane-polarized light. The representation of unpolarized light is obtained (as in § 12.17) by imagining the superposition of beams which have no permanent phase relation.

13.13. Energy of the Electromagnetic Field.

The electromagnetic field has an energy W, given by *

$$\frac{\delta W}{\delta\tau} = \frac{\mathbf{D}\,.\,\mathbf{E}}{8\pi} + \frac{\mathbf{H}^2}{8\pi}, \qquad . \quad . \quad . \quad . \quad 13(39a)$$

where δW is the energy in an element of volume $\delta\tau$. This expression includes the effect of electrical polarization, but we have omitted the effect of magnetic polarization since it is negligible at high frequencies.

In vacuum, $\mathbf{D} = \mathbf{E}$ and the first term on the right-hand side of 13(39a) becomes equal to $E^2/8\pi$. For the wave represented by 13(26) the mean value of E^2 is $\frac{1}{2}E_0{}^2$ and the mean value of H^2 is $\frac{1}{2}H_0{}^2$. But in vacuum $H_0 = E_0$ [see 13(33)] and therefore

$$\frac{\delta W}{\delta\tau} = \frac{E_0{}^2}{8\pi} = \frac{H_0{}^2}{8\pi}. \qquad . \quad . \quad . \quad . \quad 13(39b)$$

13.14. Poynting's Theorem.

It was shown in 1884 by J. H. Poynting (1852–1914) that when there are both an electric and a magnetic field at the same point, there is in general a flow of the field energy. This flow is completely described by a vector (now known as the *Poynting vector*) whose direction gives the direction of flow, and whose magnitude is the amount of energy crossing unit area (normal to the direction of flow) per second. We shall now show that this vector ($\mathbf{G}$) is equal to $c/4\pi$ times the vector product of $\mathbf{E}$ and $\mathbf{H}$, i.e

$$\mathbf{G} = \frac{c}{4\pi}\,\mathbf{E}\times\mathbf{H}. \qquad . \quad . \quad . \quad . \quad . \quad 13(40)$$

It thus vanishes when $\mathbf{E}$ and $\mathbf{H}$ are in the same direction.

* Reference 13.1.

For a non-dispersive medium we have

$$\frac{\partial W}{\partial t} = \frac{1}{8\pi}\frac{\partial}{\partial t}\int (\epsilon \mathbf{E}^2 + \mathbf{H}^2)\, d\tau \quad . \quad . \quad . \quad 13(41)$$

$$= \frac{1}{4\pi}\int \left(\epsilon \mathbf{E}\frac{\partial \mathbf{E}}{\partial t} + \mathbf{H}\frac{\partial \mathbf{H}}{\partial t}\right) d\tau, \quad . \quad . \quad 13(42)$$

where we assume that ϵ is not dependent on ω, which is true for a non-dispersive medium. Hence using Maxwell's equations 13(18) and 13(19), we have

$$\frac{\partial W}{\partial t} = \frac{c}{4\pi}\int \left[\mathbf{E}\,.\left\{\nabla \times \mathbf{H} - \frac{4\pi}{c}\mathbf{j}\right\} - \{\mathbf{H}\,.\,\nabla \times \mathbf{E}\}\right] d\tau, \quad 13(43)$$

and using 13(10)

$$\frac{\partial W}{\partial t} = -\frac{c}{4\pi}\int \nabla\,.\,(\mathbf{E}\times\mathbf{H})\, d\tau - \int \mathbf{E}\,.\,\mathbf{j}\, d\tau. \quad . \quad 13(44)$$

The second term represents the resistive dissipation of energy (Joule's law). It is zero for a non-conducting medium. We then have

$$\frac{\partial W}{\partial t} + \int \operatorname{div} \mathbf{G}\, d\tau = 0, \quad . \quad . \quad . \quad . \quad 13(45a)$$

where $\mathbf{G}$ is given by 13(40).

One of the fundamental equations of hydrodynamics * is

$$\frac{\partial \rho_F}{\partial t} + \operatorname{div} \rho_F \mathbf{V} = 0, \quad . \quad . \quad . \quad . \quad 13(45b)$$

where ρ_F is the density of the fluid and $\mathbf{V}$ is the local velocity of flow. This equation applies when there are no sources or sinks, and it is effectively a statement that the total quantity of the liquid is constant. In a similar way, if $\rho_E = \partial W/\partial \tau$ is the density of the electromagnetic field energy and $\mathbf{U}$ is a velocity defined by $\rho_E U = G$, equation 13(45a) becomes equivalent to 13(45b). In this form it may be regarded as a statement of the conservation of field energy and $\mathbf{U}$ may be called the velocity. The mean rate at which the energy is propagated is therefore the mean value of G/ρ_E, i.e. we have

$$\overline{U} = \frac{2c\,.\,\mathbf{E}\times\mathbf{H}}{\epsilon \mathbf{E}^2 + \mathbf{H}^2}, \quad . \quad . \quad . \quad . \quad . \quad 13(46)$$

* See p. 185 of Reference 13.1.

the mean being taken over a time which is long compared with the period of the wave.

If **E** and **H** are given by 13(35), we have

$$\overline{U} = \frac{\frac{1}{2}c\sqrt{\epsilon}A^2}{\frac{1}{2}\epsilon A^2} = \frac{c}{\sqrt{\epsilon}} = b. \quad . \quad . \quad . \quad 13(47)$$

Thus *in a non-dispersive medium* the rate of flow of energy is equal to the phase velocity b. From the definition of the Poynting vector, we see that the direction of flow is the direction of propagation.

13.15. Momentum of the Electromagnetic Waves.

It is easy to show, in a qualitative way, that electromagnetic waves must possess momentum in the direction of propagation. Suppose that a beam falls upon a conductor which partially absorbs it. The electric vector starts an electric current which is at right angles to the magnetic vector. There is, therefore, an electromagnetic force on the conductor. Since the electric field and the magnetic field change sign together, this force is always in the same direction—the direction of propagation. The detailed calculation of the magnitude of the force, allowing for the fact that the current in the conductor is not exactly in phase with the wave, is fairly complicated. The most satisfactory general method is to appeal to relativity. According to relativistic principles, the electromagnetic energy W implies the presence of a mass W/c^2 [equation 11(48)], and a momentum W/c must also be present [equation 11(52)]. We shall refer to this matter later in connection with experiments on light pressure. It will also be shown later that a beam of circularly polarized electromagnetic waves possesses angular momentum about the direction of propagation (§ 17.23).

EXAMPLES [13(viii)–13(x)]

13(viii). Does 13(38) represent a right-handed or a left-handed polarized wave? [Left-handed.]

13(ix). Find the energy density and the rate of flux of energy in an elliptically polarized wave represented by 13(38).

$$\left[\frac{\delta W}{\delta \tau} = \frac{\varepsilon}{8\pi}(A^2 + B^2). \quad G = b\frac{\partial W}{\partial \tau}.\right]$$

13(x). Give a general non-mathematical discussion of the propagation of energy in spherical electromagnetic waves (i.e. revise the arguments of Chapter II in terms of the electromagnetic theory).

REFERENCES

13.1. Joos: *Theoretical Physics* (Blackie).

13.2. Jeans: *Theory of Electricity and Magnetism* (Cambridge University Press).

13.3. Maxwell: *Electromagnetic Theory* (Oxford University Press).

13.4. Thomson, J. J.: *Elements of Electricity and Magnetism* (Cambridge University Press.)

13.5. Rutherford: *Vector Methods* (Oliver and Boyd).

13.6. Baker and Copson: *The Mathematical Theory of Huygens' Principle* (Oxford University Press).

APPENDIX XIII A

Representation of the Electromagnetic Field by Potentials

1. In Chapter XIII, the electromagnetic field is represented by two vectors **E** and **H** which satisfy equations 13(16) to 13(19). For some purposes it is convenient to express **E** and **H** in terms of a scalar potential (ϕ) and a vector potential (**A**). Since the div (curl) of any function is zero and since div $\mathbf{H} = 0$, we may put

$$\mathbf{H} = \text{curl}\,\mathbf{A}. \qquad \text{13(48)}$$

Since curl (grad) of any function is zero, we may write, in place of 13(18),

$$\mathbf{E} + \frac{1}{c}\dot{\mathbf{A}} = -\text{grad}\,\phi, \qquad \text{13(49)}$$

as may be verified by applying the operation " curl " to 13(49).

Suppose that a vector function $\mathbf{A}_0$ and a scalar function ϕ_0 satisfy the above equations, and that Φ is another scalar function, then the functions

$$\mathbf{A} = \mathbf{A}_0 - \text{grad}\,\Phi \qquad \text{13(50)}$$

and

$$\phi = \phi_0 + \frac{1}{c}\dot{\Phi} \qquad \text{13(51)}$$

also satisfy 13(48) and 13(49), since curl (grad Φ) = 0. Thus 13(48) and 13(49) do not completely determine **A** and ϕ. We may introduce any function Φ, which can be differentiated, just as we may add one or more constants of integration to the solution of a differential equation. We choose Φ in the following way in order to simplify subsequent equations. Let

$$\nabla^2\Phi - \frac{1}{c^2}\ddot{\Phi} = \text{div}\,\mathbf{A}_0 + \frac{1}{c}\dot{\phi}_0. \qquad \text{13(52)}$$

Taking the divergence of each side of 13(50) and adding $(1/c)\,\partial/\partial t$ times the corresponding side of 13(51), we obtain

$$\operatorname{div}\mathbf{A} + \frac{1}{c}\dot{\phi} = \operatorname{div}\mathbf{A}_0 - \operatorname{div}(\operatorname{grad}\Phi) + \frac{1}{c}\dot{\phi}_0 + \frac{1}{c^2}\ddot{\Phi}, \quad . \quad 13(53)$$

and, using 13(52) and 13(7), $$\operatorname{div}\mathbf{A} + \frac{1}{c}\dot{\phi} = 0. \quad . \quad . \quad . \quad . \quad . \quad . \quad . \quad 13(54)$$

When $\varepsilon = 1$, equations 13(19) and 13(12) give

$$\operatorname{curl}\mathbf{H} = \frac{1}{c}\dot{\mathbf{E}} + \frac{4\pi}{c}\rho\mathbf{v} \quad . \quad . \quad . \quad . \quad . \quad . \quad . \quad 13(55)$$

and 13(16) gives $$\operatorname{div}\mathbf{E} = 4\pi\rho. \quad . \quad . \quad . \quad . \quad . \quad . \quad . \quad . \quad 13(56)$$

Substituting for **H** and **E** from 13(48) and 13(49), and using 13(9) and 13(54), we obtain

$$\nabla^2\mathbf{A} = \frac{1}{c^2}\ddot{\mathbf{A}} - \frac{4\pi}{c}\rho\mathbf{v} \quad . \quad . \quad . \quad . \quad . \quad . \quad . \quad 13(57)$$

and $$\nabla^2\phi = \frac{1}{c^2}\ddot{\phi} - 4\pi\rho. \quad . \quad . \quad . \quad . \quad . \quad . \quad . \quad 13(58)$$

In free space (when $\rho = 0$) the potentials each satisfy the wave equation, and variations of the potentials are propagated with velocity c. Even when 13(52) is satisfied there is still some choice of Φ, and for free space we may choose Φ so that $\phi = 0$. We then have (for the parts of **E** and **H** which are propagated)

$$\mathbf{H} = \operatorname{curl}\mathbf{A} \quad . \quad . \quad . \quad . \quad 13(59)$$

and $$\mathbf{E} = -\frac{1}{c}\dot{\mathbf{A}} \quad . \quad . \quad . \quad . \quad 13(60)$$

in free space.

2. Analysis of the Electromagnetic Field.

We saw in § 3.20 that, when a beam of light is incident upon a partially reflecting surface, both progressive waves and standing waves are present. Now consider the radiation in a cube with partially reflecting walls. The values of the fields are the same (but not necessarily zero) on any pair of opposite sides of the cube. The side L is assumed to be large in comparison with any wavelength in which we are interested. Suppose that q_s is a function of t (but not of x, y, z) which satisfies the relation

$$\ddot{q}_s + \omega_s^2 q_s = 0, \quad . \quad . \quad . \quad . \quad . \quad . \quad . \quad . \quad 13(61)$$

and that $\mathbf{A}_s$ is a vector function of x, y, z, but not of t, which satisfies the relations

$$\nabla^2\mathbf{A}_s + \frac{\omega_s^2}{c^2}\mathbf{A}_s = 0 \quad . \quad . \quad . \quad . \quad . \quad . \quad . \quad 13(62)$$

and $$\operatorname{div}\mathbf{A}_s = 0. \quad . \quad . \quad . \quad . \quad . \quad . \quad . \quad 13(63)$$

Then if $$\mathbf{A} = \sum_s q_s\mathbf{A}_s, \quad . \quad . \quad . \quad . \quad . \quad . \quad . \quad . \quad 13(64)$$

the wave equation will be satisfied by **A**. A possible solution of 13(62) is

$$\mathbf{A}_s = \mathbf{A}_{0s} \sin (\kappa_{sx}x + \kappa_{sy}y + \kappa_{sz}z), \qquad 13(65)$$

where

$$\kappa_{sx}{}^2 + \kappa_{sy}{}^2 + \kappa_{sz}{}^2 = \frac{\omega_s{}^2}{c^2}, \qquad 13(66)$$

and, in order to satisfy the boundary condition,

$$\kappa_{sx} = \frac{2\pi}{L} n_x, \quad \kappa_{sy} = \frac{2\pi}{L} n_y, \quad \kappa_{sz} = \frac{2\pi}{L} n_z, \qquad 13(67)$$

where n_x, n_y, n_z are integers.

Thus the field may be analysed into a number of waves propagated in different directions, and the time-varying part has been separated from the space-varying part of each wave. For any direction of propagation, there are two possible directions of polarization. In 13(61) to 13(64) these are included by giving two values of s to each direction. In 13(65) it must be understood that the vectors $\mathbf{A}_{0s}$ appear in pairs which have the same magnitude but mutually perpendicular directions. It is, of course, possible to use cosines or complex exponentials in place of 13(65), but the form given above is the most simple for our purposes.

3. Number of Standing Waves between ω and $(\omega + d\omega)$.

Suppose that three co-ordinate axes κ_x, κ_y, κ_z are set up and that each point which corresponds to a solution of 13(65) is marked with a dot. The number of dots per unit volume (for a volume of regular shape which is large enough to contain many dots) is $L^3/8\pi^3$. The number between spheres of radii κ and $(\kappa + d\kappa)$ is $(4\pi\kappa^2 d\kappa)L^3/(8\pi^3)$. Since $c\kappa = \omega$, the number between ω and $(\omega + d\omega)$ is

$$\frac{\omega^2\, d\omega}{2\pi^2 c^3} L^3. \qquad 13(68)$$

We have shown that **A** represents the electromagnetic field and it follows that there are two types of polarized wave, so that the total number of vibrations between ω and $(\omega + d\omega)$ is $\omega^2\, d\omega/(\pi^2 c^3)$ per unit volume of the cube, or, between ν and $(\nu + d\nu)$,

$$\frac{8\pi\nu^2\, d\nu}{c^3}. \qquad 13(69)$$

It may be shown that, when all the linear dimensions are large compared with the wavelengths, the number is independent of the shape of the volume.

APPENDIX XIII B

Radiation from a Dipole

1.—In electrostatics, a dipole consists of two equal and opposite charges placed at a small distance from one another. The moment is equal to the magnitude of one of the charges multiplied by their separation. It is positive when the positive charge is on the positive side of the common centre taken as origin. We now consider an oscillating dipole whose moment is represented by

$$\mathbf{M}(t) = \mathbf{M}_0 e^{i\omega t}. \quad \quad 13(70)$$

The physical moment is the real part of $\mathbf{M}$. The amplitude of oscillation of electrons which emit light is not infinitesimal but, for the present, we consider an idealized dipole for which the separation of the charges is always infinitesimal

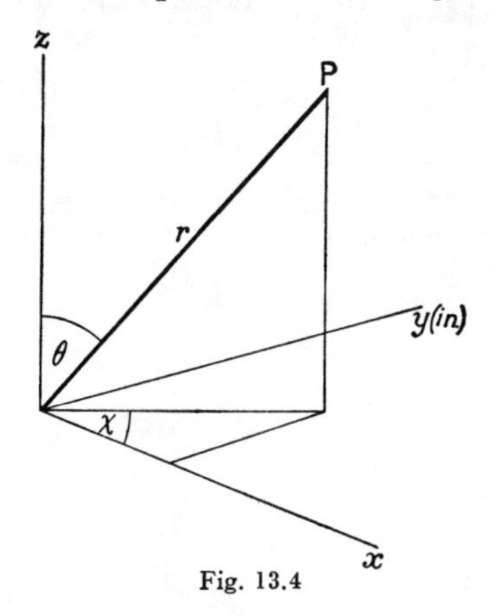

Fig. 13.4

though the moment is finite. It is convenient to use Cartesian co-ordinates in one part of the calculation and spherical polar co-ordinates (r, θ, χ) in another (see fig. 13.4). We take the axis of the dipole as the z axis. The field must be symmetrical about this axis and therefore all components are independent of χ. Consider the vector

$$\mathbf{\Pi}(t) = \mathbf{M}_{-}/r, \quad \quad 13(71)$$

where $\mathbf{M}_{-}$ is the value of $\mathbf{M}$ at time $(t - r/c)$. The vector $\mathbf{\Pi}$ (known as the Hertzian vector) is a solution of the wave equation.

2.—We require to find two vectors $\mathbf{E}$ and $\mathbf{H}$ which represent the field of the dipole. We know that the solution must satisfy the following tests:

(i) When r is small, the field $\mathbf{E}$ must reduce to the field calculated for a doublet in electrostatics. Similarly $\mathbf{H}$ must reduce to the field for a current element $\dot{\mathbf{M}} = i\omega\mathbf{M}_0$.

(ii) The field at P at time t should depend on the state of the dipole at time $t - r/c$.

(iii) At large distances from the dipole, the fluctuating part of the field should represent spherical electromagnetic waves with **E** and **H** perpendicular to each other and to **r**. The amplitudes should be inversely proportional to r.

Let us take as a trial solution

$$\mathbf{A} = \frac{1}{c}\dot{\mathbf{\Pi}} \quad . \quad . \quad . \quad . \quad . \quad . \quad . \quad . \quad . \quad . \quad . \qquad 13(72)$$

and

$$\phi = -\text{div}\,\mathbf{\Pi}. \quad . \quad . \quad . \quad . \quad . \quad . \quad . \quad . \quad . \qquad 13(73)$$

3.—We have $\Pi_x = \Pi_y = 0$, so that $A_x = A_y = 0$,

and

$$A_z = \frac{1}{c}\dot{\Pi}_z = \frac{i\omega}{rc}M_-, \quad . \quad . \quad . \quad . \quad . \quad . \quad . \quad . \quad . \qquad 13(74)$$

$$\phi = -\frac{\partial \Pi_z}{\partial z} = -\frac{\partial \Pi_z}{\partial r}\cdot\frac{\partial r}{\partial z}$$

$$= -M_0 \cos\theta\, e^{i\omega t}\frac{\partial}{\partial r}\left(\frac{e^{-i\omega r/c}}{r}\right). \quad . \quad . \quad . \quad . \qquad 13(75)$$

Since $A_x = A_y = 0$ everywhere, their derivatives are also zero, and 13(48) implies $H_z = 0$. Therefore **H** at P lies in the plane through P normal to Oz.

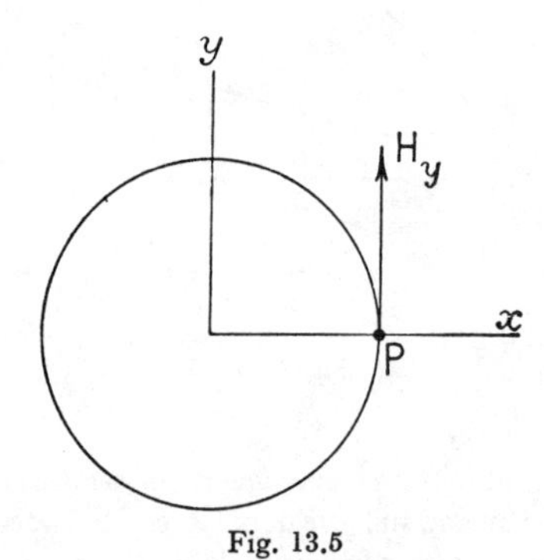

Fig. 13.5

Suppose, for the moment, that the y co-ordinate of P is zero as shown in fig. 13.5. Since A_z is the same at all points on the circle through P, it follows that $\partial A_z/\partial y$ at P $= 0$, $(H_x)_P = 0$ and

$$H_\chi = (H_y)_{\text{at P}} = -\left(\frac{\partial A_z}{\partial x}\right)_P$$

$$= \frac{i\omega}{c}M_0 e^{i\omega t}\sin\theta\,\frac{\partial}{\partial r}\left(\frac{e^{-i\omega r/c}}{r}\right)$$

$$= -\frac{M_-}{r}\frac{\omega^2}{c^2}\sin\theta\left(1 + \frac{c}{i\omega r}\right). \quad . \quad . \quad . \quad . \qquad 13(76a)$$

We have shown that, at P, **H** is tangential to the circle and has the magnitude given by 13(76). Symmetry requires that **H** is always tangential to the circle and has the above magnitude, i.e. the above expression gives $H = H_\chi$ even when the y co-ordinate of P is not necessarily zero.

When r is large compared with c/ω (i.e. with the wavelength), we have

$$H_\chi = -\frac{M_-}{r}\frac{\omega^2}{c^2}\sin\theta. \qquad \text{13(76}b\text{)}$$

From 13(49) we obtain

$$\left.\begin{aligned} E_r &= \frac{M_-}{r}\cos\theta\left(\frac{2i\omega}{rc}+\frac{2}{r^2}\right). \\ E_\theta &= -\frac{M_-}{r}\frac{\omega^2}{c^2}\sin\theta\left(1+\frac{c}{i\omega r}-\frac{c^2}{\omega^2r^2}\right). \\ E_\chi &= 0. \end{aligned}\right\} \qquad \text{13(77}a\text{)}$$

When r is large compared with c/ω, the component E_r becomes negligible compared with the component E_θ, which reduces to

$$E_\theta = -\frac{M_-}{r}\frac{\omega^2}{c^2}\sin\theta. \qquad \text{13(77}b\text{)}$$

4.—When r is large, H_χ [see 13(76*b*)] is the only component of **H** and similarly E_θ is the only component of **E**. The fields given by 13(76*b*) and 13(77*b*) represent transverse spherical electromagnetic waves, so that conditions (ii) and (iii) stated above are satisfied. The fields obtained from 13(76*a*) and 13(77*a*), by including only the highest power of $1/r$, which is the dominant term when r is small, satisfy condition (i). We shall therefore accept 13(76) and 13(77) as the appropriate solutions of the equations.*

5.—The rate of propagation of energy is obtained by calculating the Poynting vector (see § 13.14). The value is

$$G = \frac{c}{4\pi}\frac{M_-^2}{r^2}\frac{\omega^4}{c^4}\sin^2\theta. \qquad \text{13(78)}$$

The time average of M_-^2 is $\frac{1}{2}M_0^2$, so that the average rate at which energy crosses unit area normal to the direction θ is

$$G_A = \frac{c}{8\pi}\frac{M_0^2}{r^2}\frac{\omega^4}{c^4}\sin^2\theta. \qquad \text{13(79)}$$

Since the mean value of $\sin^2\theta$ over a sphere is $\frac{2}{3}$, the total energy emitted per unit time is

$$-\frac{\partial W}{\partial t} = \frac{M_0^2\omega^4}{3c^3} = \frac{16\pi^4M_0^2\nu^4}{3c^3}. \qquad \text{13(80)}$$

* For a further discussion on the singularity at the origin, etc., the reader should consult References 18.4 and 6.2.

6.—Let us now consider the radiation from an electron which is moving along the x axis so as to satisfy the equation

$$\ddot{x} + \gamma\dot{x} + \omega^2 x = 0, \qquad 13(81)$$

with a damping constant γ which is very small compared with ω. The solution is

$$x = x_0 e^{-\gamma t/2} e^{i\omega t}. \qquad 13(82)$$

The energy at time t is

$$W = W_0 e^{-\gamma t}, \qquad 13(83)$$

where

$$W_0 = \tfrac{1}{2}\omega^2 m x_0^2. \qquad 13(84)$$

The rate of loss of energy is γW, which is approximately γW_0 when t is small. The vibrating electron is equivalent to an oscillating dipole (plus a static charge which does not contribute to the radiation field). The equivalent value of M_0 is ex_0. Substituting in 13(84), we obtain for the rate of loss of energy

$$-\frac{\partial W}{\partial t} = \gamma W_0 = \tfrac{1}{2}\gamma \frac{\omega^2}{e^2} m M_0^2. \qquad 13(85)$$

This agrees with 13(80) if

$$\gamma = \frac{2e^2}{3mc^3}\omega^2. \qquad 13(86)$$

The time τ_0 for the energy to fall to $1/e$ of the original value is given by

$$\tau_0 = \frac{1}{\gamma} = \frac{3}{2}\frac{mc^3}{e^2\omega^2}. \qquad 13(87)$$

7.—In the above discussion there is an inconsistency. We assumed in equation 13(70) that the amplitude of the dipole was constant, and we are now considering an electron which continually loses energy so that the equivalent dipole amplitude is decreasing. The emission of radiation reacts on the motion of the electron and is equivalent to a damping force. To take account of this, we should compare 13(85), not with 13(80), but with an expression obtained by substituting $M = M_0 e^{-\gamma t/2} e^{i\omega t}$ for 13(70) and then calculating the fields. The procedure we adopted is satisfactory only because the damping is small. In cases of practical importance γ is of the order $10^{-6}\omega$ or less. It is therefore permissible to ignore the reaction of the radiation on the oscillator when calculating the rate of radiation.

8.—If we have a distribution of charges which is electrically neutral as a whole, the equivalent moment is

$$\mathbf{M}_0 = \sum e\mathbf{r}_0. \qquad 13(88)$$

If the displacements are varying sinusoidally, we have

$$\mathbf{M} = \sum e\mathbf{r}_0 e^{i\omega t}. \qquad 13(89)$$

The mean square of the magnitude of the moment is

$$\begin{aligned}\overline{M^2} &= \tfrac{1}{2}\sum e r_0^2 \qquad 13(90)\\ &= \tfrac{1}{2}\sum e(x_0^2 + y_0^2 + z_0^2).\end{aligned}$$

When the system is spherically symmetrical this becomes

$$\overline{M^2} = \tfrac{3}{2} \sum e x_0^2. \qquad 13(91)$$

It is important to note the difference between the formulæ obtained when x_0 is a co-ordinate chosen to be parallel to the axis of a dipole, and when x_0 is a co-ordinate fixed in relation to the apparatus with the dipole oriented at random. In Chapter XV we consider dipoles created by the action of the field. In an isotropic medium these dipoles are necessarily parallel to the electric vector of the field by which they are induced.

9.—Since the electric vector of the emitted radiation is proportional to the equivalent moment, it follows that the light emitted by an electron oscillator is represented by

$$E = E_0 e^{-\gamma t/2} e^{i\omega t}. \qquad 13(92)$$

This is the typical damped harmonic wave. We have discussed the equivalent energy distribution in Appendix IV B (equation 4(92), p. 111].

10. Scattering by Free Electrons.

Suppose that an electromagnetic wave represented by the expression

$$E = E_0 e^{i\omega t}$$

falls upon a free electron (charge e). Then, neglecting the forces due to the magnetic field and to the radiation damping, the equation of motion is

$$m\ddot{x} = eE_0 e^{i\omega t}. \qquad 13(93)$$

If the origins of x and t have been chosen so that $x = 0$ when $t = 0$, a solution is

$$x = -\frac{eE_0}{m\omega^2} e^{i\omega t}. \qquad 13(94)$$

Since e is negative, this equation implies that the electron undergoes simple harmonic motion with the frequency and phase of the incident wave. This electron is equivalent to an oscillating dipole for which

$$M_0 = -\frac{e^2}{m\omega^2} E_0. \qquad 13(95)$$

Applying 13(79) and 13(80) we see that the energy scattered in a direction θ is proportional to $\sin^2\theta$ and that the total energy scattered per unit time is

$$\frac{e^4 E_0^2}{3m^2c^3}. \qquad 13(96)$$

Equation 13(39*b*) gives the energy density. The energy crossing unit area in unit time is therefore $cE_0^2/8\pi$ and the *fraction* scattered in unit length of path is

$$k = \frac{8\pi}{3} \frac{e^4}{m^2c^4}, \qquad 13(97)$$

and, using 11(50),

$$k = \frac{8\pi}{3} r_0^2. \qquad . \quad . \quad . \quad . \quad . \quad . \quad . \quad . \quad . \quad . \qquad 13(98)$$

Note that this quantity is a universal constant independent of the frequency of the incident light. It has the dimensions of an area because we have taken the ratio of the total light scattered by one electron to the energy incident per unit area.

11. Scattering by Bound Electrons.

When an electromagnetic wave is incident upon a bound electron whose free vibration is represented by 13(81), the equation of motion may be written

$$\ddot{x} + \gamma_s \dot{x} + \omega_s^2 x = \frac{e}{m} E_0 e^{i\omega t}, \qquad . \quad . \quad . \quad . \quad . \qquad 13(99)$$

the suffixes having been inserted to distinguish the constants of the free motion. The complete solution of 13(99) consists of two parts: (*a*) a vibration of frequency ω_s which is quickly damped out, and (*b*) a steady vibration of frequency ω. We are interested only in the latter and, to investigate it, we put $x = x_0 e^{i\omega t}$ in 13(99) and obtain

$$x_0 = \frac{e}{m} \frac{E_0}{\omega^2{}_s - \omega^2 + i\gamma_s \omega}. \qquad . \quad . \quad . \quad . \quad . \quad . \qquad 13(100)$$

This gives an equivalent dipole whose moment is complex and it is easily shown that we must substitute the square of the modulus for M_0^2 in 13(80). We then follow the procedure of § 10 of this Appendix and obtain

$$k' = \frac{8\pi}{3} \frac{e^4}{m^2 c^4} \frac{\omega^4}{(\omega_s^2 - \omega^2)^2 + \omega^2 \gamma_s^2}. \qquad . \quad . \quad . \qquad 13(101)$$

In dispersion theory we are often interested in a situation where there are N scattering centres per unit volume, each with an oscillator strength equal to f_s times the oscillator strength given by electromagnetic theory for one electron. We then have (neglecting damping)

$$k'' = \frac{8\pi}{3} \frac{Ne^4}{m^2 c^4} f_s^2 \frac{\omega^4}{(\omega_s^2 - \omega^2)^2}. \qquad . \quad . \quad . \quad . \quad . \quad . \qquad 13(102)$$

Since M_0 appears to the second power in 13(80), f_s appears to the second power in 13(102).

12. Multipole Radiation.

It sometimes occurs that the moment M given by 13(89) is nearly zero, and the rate of radiation is then extremely small. It then becomes necessary to take account of other small effects. A vibrating electron is not exactly equivalent to a mathematical dipole because there is always some finite amplitude. The retarded time $(t - r/c)$ is to be measured from the actual position of the electron at any time and not, as we have assumed, from the mean position. This is equivalent to saying that we must allow for the fact that light takes a finite time to travel a distance equal to the amplitude of motion. This higher-order effect leads to expressions for an additional radiation which may be expressed in higher powers of the displacement, i.e. the rate of radiation is proportional to $ax_0^2 + bx_0^4 + \ldots$ etc. Since x_0 is always very small, the second term becomes important only when the coefficient of the first vanishes, and so on. This higher-order radiation may be regarded as due to a pair of dipoles placed very close to one another and out of phase by π. The first term then disappears by interference. An arrangement of this type is known as a " quadripole ". Some weak lines are due to quadripole radiation. Part of the x_0^4 term may be attributed to a magnetic dipole (see Reference 18.4, p. 619).

CHAPTER XIV

The Electromagnetic Theory of Reflection and Refraction

14.1. Boundary Conditions.

The field equations of electromagnetism apply to the whole of one medium and, with different constants, to a second medium. A complete theory must give an account of what happens at the boundary of two media. Mathematically, we require a set of boundary conditions which specify the way in which two solutions of the equation, each valid on one side of the boundary, are to fit together at the boundary. Since, at optical frequencies, all media are effectively non-magnetic, we shall expect magnetic lines of force to cross the boundary without any discontinuity. We also know, from experiments on electrostatics, that the tangential component of electric force and the normal component of electric displacement are continuous at a boundary.* If the boundary between medium 1 and medium 2 is parallel to the XY plane, we have

$$\left.\begin{aligned} E_{1x} &= E_{2x} \qquad & E_{1y} &= E_{2y} \\ H_{1x} &= H_{2x} \qquad & H_{1y} &= H_{2y} \\ D_{1z} &= D_{2z} \qquad & H_{1z} &= H_{2z} \end{aligned}\right\} \quad . \quad . \quad . \quad 14(1)$$

where E_{1x}, etc., refer to a point near the boundary but in medium 1, and E_{2x}, etc., refer to a corresponding point in medium 2. These conditions are not independent, for, if the first two are satisfied, the relation between H_{1z} and H_{2z} follows from the fact that the field equation

$$\frac{\partial E_y}{\partial x} - \frac{\partial E_x}{\partial y} = -\frac{1}{c}\frac{\partial H_z}{\partial t} \quad . \quad . \quad . \quad . \quad 14(2)$$

must be true in each medium. Similarly the relation between D_{1z} and D_{2z} is implied by the second pair of equations. Thus there are only

* A quantity is said to be *continuous* when the values at two neighbouring points, one on each side of the boundary, are equal.

four independent relations, and we shall generally use the first four of equations 14(1) as boundary conditions.

14.2. Laws of Reflection and Refraction.

All real materials are dispersive, and any treatment of reflection and refraction which ignores this fact must be slightly artificial. Nevertheless, it is reasonable to assume that, *for a given frequency*, the dielectric constant has a definite value in each medium. We assume that the *effective* value for each medium is equal to n^2 [equation 13(25)]. Using this value we apply the boundary conditions. Let us choose a

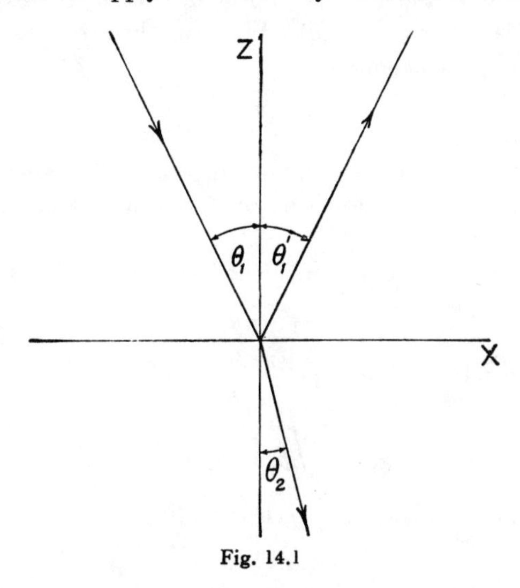

Fig. 14.1

system of co-ordinate axes so that the XY plane is the surface of separation, and XZ is the plane of incidence (fig. 14.1). Let θ_1 be the angle of incidence. We assume that the incident beam is plane-polarized, but do not at present specify in which plane it is polarized. Let the y components of the electric vectors in the incident, reflected and refracted beams be given as follows:

Incident beam $E_{1y} = A_{1y} \exp i\{\omega_1 t - \kappa_1(l_1 x + m_1 y + n_1 z)\}$ 14(3)

Reflected beam $E_{1y}' = A_{1y}' \exp i\{\omega_1' t - \kappa_1'(l_1' x + m_1' y + n_1' z)\}$ 14(4)

Refracted beam $E_{2y} = A_{2y} \exp i\{\omega_2 t - \kappa_2(l_2 x + m_2 y + n_2 z)\}$. 14(5)

In these equations A_{1y} is assumed to be real and positive. If the reflected or refracted beam has a phase difference of π from the incident beam, we shall expect A_{1y}' or A_{2y} respectively to be real and negative. A change of phase other than π will be indicated by a complex value of the corresponding amplitude (§ 2.27). Owing to our choice of axes

$$l_1 = \sin\theta_1;\ m_1 = 0;\ \text{and}\ n_1 = -\cos\theta_1. \qquad 14(6)$$

14.3.—The second of equations 14(1) implies that

$$E_{1y} + E_{1y}' = E_{2y} \quad . \quad . \quad . \quad . \quad . \quad 14(7)$$

when $z = 0$ for all values of x, y, and t. This implies that the same coefficients of x, y and t must appear in each of the exponentials of equations 14(3), 14(4) and 14(5). We have these results:

(*a*) Equating coefficients of t,

$$\omega_1 = \omega_1' = \omega_2, \quad . \quad . \quad . \quad . \quad . \quad . \quad . \quad 14(8)$$

so that the reflected and refracted beams have the same frequency as the incident beam. Henceforth we omit the suffix and write ω for the circular frequency.

(*b*) Since the reflected and incident beams are in the same medium,

$$\kappa_1' = \kappa_1. \quad . \quad . \quad . \quad . \quad . \quad . \quad . \quad . \quad 14(9)$$

(*c*) For the refracted wave,

$$\frac{\kappa_2}{\kappa_1} = \frac{\omega}{b_2}\bigg/\frac{\omega}{b_1} = \frac{b_1}{b_2} = n_{12}, \quad . \quad . \quad . \quad . \quad 14(10)$$

(*d*) Equating coefficients of y, using 14(6),

$$0 = m_1' = m_2, \quad . \quad . \quad . \quad . \quad . \quad 14(11)$$

so that the three beams are in the plane containing the normal to the surface and the incident beam.

(*e*) Equating coefficients of x, and writing

$$l_1' = \sin\theta_1'\ \text{and}\ l_2 = \sin\theta_2, \quad . \quad . \quad . \quad 14(12)$$

we have

$$\sin\theta_1 = \sin\theta_1', \quad . \quad . \quad . \quad . \quad . \quad 14(13)$$

$$\kappa_1 \sin\theta_1 = \kappa_2 \sin\theta_2, \quad . \quad . \quad . \quad . \quad 14(14)$$

or

$$\sin\theta_1 = n_{12} \sin\theta_2. \quad . \quad . \quad . \quad . \quad 14(15)$$

The electromagnetic waves thus obey all the experimental laws of reflection and refraction at a surface separating two isotropic media.

14.4.—It is satisfactory that they do so, but this fact does not give any support to the electromagnetic theory as opposed to other wave theories of light.

In Chapter III we showed that these relations could be derived from any self-consistent wave-theory. They are not dependent upon any special form of the boundary conditions, but only upon the facts that there are boundary conditions, and that there are different velocities of propagation on the two sides of the boundary. It will be noted that we have, so far, used only one of the four boundary conditions. The derivation of reflection coefficients for different angles of incidence and states of polarization involves the full boundary conditions, and thus forms an important test of the electromagnetic theory.

14.5.—We now wish to calculate the fraction of the incident energy which is reflected for different angles of incidence, and for different planes of polarization. We are chiefly interested in the components of the vectors parallel and perpendicular to the plane of incidence, rather than in components resolved along the co-ordinate axes. We must use the latter, however, in applying the boundary conditions.

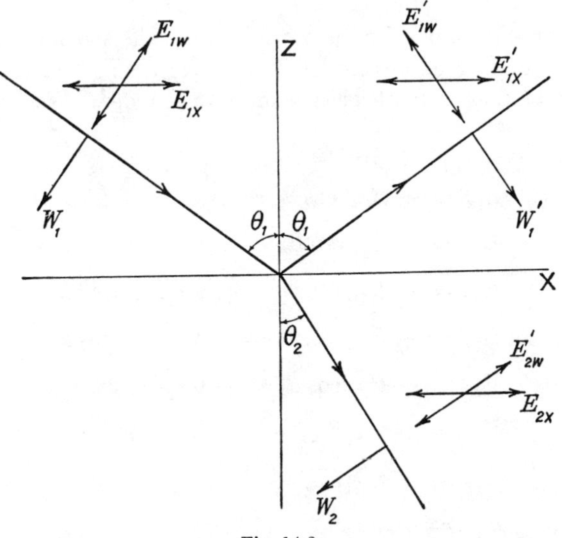

Fig. 14.2

The axis OY perpendicular to the plane of incidence serves for all three beams, and we introduce new axes w_1, w_1', and w_2. Each of these is in the plane of incidence, and they are perpendicular to the directions of propagation for the incident, reflected and refracted beams respectively. The positive directions of w_1, w_1', and w_2 are shown in fig. 14.2. They are such that an observer looking in the direction of propagation would make a clockwise rotation from the positive direction of w to the positive direction of y, i.e. w, y, and the direction of propagation, taken in that order, form a right-handed set of axes. The incident beam may be specified by giving

$$E_{1w} = A_{1w} \exp i\{\omega t - \kappa_1(x \sin \theta_1 - z \cos \theta_1)\} \quad . \quad 14(16)$$

$$E_{1v} = A_{1v} \exp i\{\omega t - \kappa_1(x \sin \theta_1 - z \cos \theta_1)\}. \quad 14(17)$$

The reflected and refracted beams are specified in a similar way. Since the direction of propagation of the reflected wave has a component in the positive direction of z, $n_1' = + \cos \theta_1$ [see 14(4) and 14(6)]. The problem is to deduce the amplitudes A_{1w}', A_{1v}', A_{2w} and A_{2v}, when A_{1w} and A_{1v} are given. We have, from 13(33)*,

$$H_{1w} = -\sqrt{\epsilon_1} \,.\, E_{1v}, \quad . \quad . \quad . \quad . \quad 14(18)$$

$$H_{1v} = +\sqrt{\epsilon_1} \,.\, E_{1w}, \quad . \quad . \quad . \quad . \quad 14(19)$$

and similar expressions for the other beams. The signs are the same in the corresponding expressions for H_{1w}', H_{2w}, etc., because our three sets of local axes have all been made right-handed. We have also

$$E_{1x} = -E_{1w} \cos \theta_1, \quad 14(20)$$

$$E_{1x}' = +E_{1w}' \cos \theta_1, \quad 14(21)$$

$$E_{2x} = -E_{2w} \cos \theta_2, \quad 14(22)$$

$$H_{1x} = -H_{1w} \cos \theta_1 = +\sqrt{\epsilon_1} E_{1v} \cos \theta_1, \quad 14(23)$$

$$H_{1x}' = +H_{1w}' \cos \theta_1 = -\sqrt{\epsilon_1} E_{1v}' \cos \theta_1, \quad 14(24)$$

$$H_{2x} = -H_{2w} \cos \theta_2 = +\sqrt{\epsilon_2} E_{2v} \cos \theta_2, \quad 14(25)$$

$$H_{1v} = +\sqrt{\epsilon_1} E_{1w}, \quad 14(26)$$

$$H_{1v}' = +\sqrt{\epsilon_1} E_{1w}', \quad 14(27)$$

$$H_{2v} = +\sqrt{\epsilon_2} E_{2w}. \quad 14(28)$$

Each of equations 14(20) to 14(28) contains an exponential factor on either side [see e.g. 14(16) and 14(17)]. Since all these factors are equal when $z = 0$, we may write the boundary conditions as relations between the amplitudes, instead of relations between the vectors.

14.6.—Applying the boundary condition for E_v, we have

$$A_{1v} + A_{1v}' = A_{2v}. \quad . \quad . \quad . \quad . \quad . \quad 14(29)$$

* See also Examples 13(iii) and 13(iv) on p. 405.

Applying the condition for H_x, and using 14(23), 14(24) and 14(25),

$$\sqrt{\epsilon_1}(A_{1v} - A_{1v}') \cos \theta_1 = \sqrt{\epsilon_2} A_{2v} \cos \theta_2, \qquad 14(30a)$$

$$(A_{1v} - A_{1v}') \cos \theta_1 = n_{12} A_{2v} \cos \theta_2. \qquad 14(30b)$$

Similarly the condition for E_x gives

$$(A_{1w} - A_{1w}') \cos \theta_1 = A_{2w} \cos \theta_2; \qquad 14(31)$$

and the condition for H_y, using 14(26), 14(27) and 14(28), gives

$$\sqrt{\epsilon_1}(A_{1w} + A_{1w}') = \sqrt{\epsilon_2} A_{2w}. \qquad 14(32)$$

Hence

$$(A_{1w} + A_{1w}') = n_{12} A_{2w}. \qquad 14(33)$$

Multiplying 14(33) by $\cos \theta_2$, and 14(31) by n_{12}, and subtracting, we have

$$A_{1w}' = A_{1w} \frac{n_{12} \cos \theta_1 - \cos \theta_2}{n_{12} \cos \theta_1 + \cos \theta_2}. \qquad 14(34a)$$

Using the sine law of refraction [equation 14(15)],

$$A_{1w}' = A_{1w} \frac{\sin 2\theta_1 - \sin 2\theta_2}{\sin 2\theta_1 + \sin 2\theta_2}, \qquad 14(34b)$$

or

$$A_{1w}' = A_{1w} \frac{\tan (\theta_1 - \theta_2)}{\tan (\theta_1 + \theta_2)}. \qquad 14(34c)$$

In a similar way, we obtain from 14(29) and 14(30b),

$$A_{1v}' = -A_{1v} \frac{n_{12} \cos \theta_2 - \cos \theta_1}{n_{12} \cos \theta_2 + \cos \theta_1}. \qquad 14(35a)$$

Using Snell's law,

$$A_{1v}' = -A_{1v} \frac{\sin (\theta_1 - \theta_2)}{\sin (\theta_1 + \theta_2)}. \qquad 14(35b)$$

Substituting from 14(34a) in 14(33), and simplifying,

$$A_{2w} = A_{1w} \frac{2 \sin \theta_2 \cos \theta_1}{\sin (\theta_1 + \theta_2) \cos (\theta_1 - \theta_2)}. \qquad 14(36)$$

Using 14(35a) and 14(29),

$$A_{2v} = A_{1v} \frac{2 \sin \theta_2 \cos \theta_1}{\sin (\theta_1 + \theta_2)}. \qquad 14(37)$$

14.7.—Equations 14(34) to 14(37) are indeterminate for normal incidence, but equations 14(29), etc., may easily be solved directly to give

$$A_{1w}' = A_{1w}\frac{n_{12}-1}{n_{12}+1}, \quad . \quad . \quad . \quad . \quad . \quad 14(38)$$

$$A_{1v}' = -A_{1v}\frac{n_{12}-1}{n_{12}+1}, \quad . \quad . \quad . \quad . \quad 14(39)$$

$$A_{2w} = A_{1w}\frac{2}{n_{12}+1}, \quad . \quad . \quad . \quad . \quad . \quad 14(40)$$

$$A_{2v} = A_{1v}\frac{2}{n_{12}+1}. \quad . \quad . \quad . \quad . \quad . \quad 14(41)$$

Equations 14(34) to 14(37) are known as Fresnel's relations. They were originally obtained by Fresnel using the elastic-solid theory of light. It should be understood that the elastic-solid theory was never able to give a completely consistent and satisfactory account of reflection and refraction, even when the assumptions concerning the boundary conditions were skilfully adjusted to produce the desired results. The electromagnetic theory does not introduce any *special* assumptions for this purpose. It uses the standard boundary conditions which are derived from the results of experiments on electricity and magnetism.

14.8. **Reflection Coefficients.**

The energy densities for the incident and reflected beams are proportional to the squares of the amplitudes. The reflection coefficients

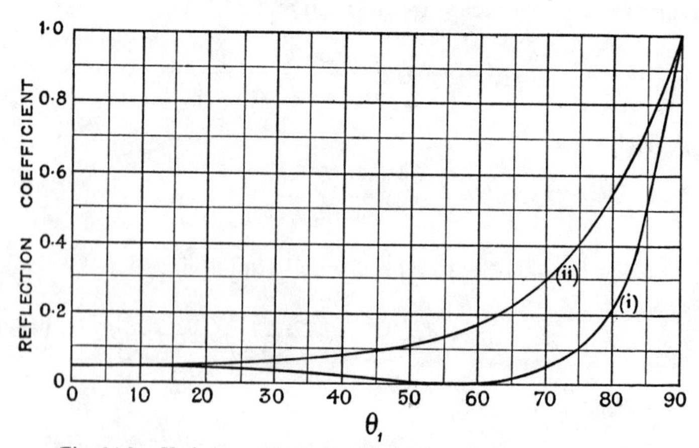

Fig. 14.3.—Variation with angle of incidence of reflection coefficient for components of E: (i) in the plane of incidence (ρ_w), and (ii) perpendicular to the plane of incidence (ρ_y). $n_{12} = 1{\cdot}5$.

may thus be obtained directly from 14(34), 14(35), 14(38), and 14(39). For example, we have for the component of the electric vector in the plane of incidence,

$$\rho_w = \left(\frac{A_{1w}'}{A_{1w}}\right)^2 = \frac{\tan^2(\theta_1 - \theta_2)}{\tan^2(\theta_1 + \theta_2)}. \quad . \quad . \quad . \quad 14(42)$$

The reflection coefficients for each component, and for different angles of incidence, are shown in fig. 14.3 for $n_{12} = 1{\cdot}5$. From the figure, or directly from equation 14(42), we see that when $\theta_1 + \theta_2 = \frac{1}{2}\pi$, the component of the electric vector in the plane of incidence is always zero, no matter what the condition of the incident light may be. Thus the phenomenon of polarization by reflection at the Brewsterian angle is included in the electromagnetic description of light waves if we assume that the *magnetic vector is in the plane of polarization.* This assumption enables us to give a consistent account of other experimental results (§ 14.13).

14.9. **Degree of Polarization.**

In a mixture of plane-polarized and unpolarized light,* the degree of polarization may be defined as

$$Q = \frac{A_{\max}{}^2 - A_{\min}{}^2}{A_{\max}{}^2 + A_{\min}{}^2}, \quad . \quad . \quad . \quad . \quad 14(43)$$

where $A_{\max}$ is the maximum amplitude of the electric vector for different orientations with respect to the direction of propagation, and $A_{\min}$ is the minimum amplitude. The plane of polarization includes the direction of $A_{\min}$. When unpolarized light is reflected at the boundary of two insulating media, the reflected light is partially or wholly polarized in the plane of incidence. The refracted light is always partially polarized in a plane perpendicular to the plane of incidence except in the limiting case of normal incidence. The degree of polarization for the reflected light may be calculated by substituting A_v' and A_w' for $A_{\max}$ and $A_{\min}$ respectively in 14(43). Similarly for calculations on the refracted light A_{2w} must be substituted for $A_{\max}$ and A_{2v} for $A_{\min}$.

Calculations on this basis show that complete polarization is obtained only by reflection at the polarizing angle. At this angle, reflection of the component of the electric vector in the plane of incidence is zero, but reflection of the com-

* Note that this definition does not apply when elliptically or circularly polarized light is present.

ponent perpendicular to the plane of incidence is not complete. Thus the transmitted light always contains some of each component, and we never get complete polarization by transmission. Repeated refractions, either from a dense to a light medium or vice versa, produce a continually increasing degree of polarization in a plane perpendicular to the plane of incidence. Similarly, repeated reflections at an angle other than the polarizing angle produce a continually increasing degree of polarization in the plane of incidence.

14.10. **Rotation of the Plane of Polarization.**

When the incident light is plane-polarized, reflections and refractions cannot increase the *degree* of polarization. They do, however, rotate the plane of polarization. The rotation is towards the plane of incidence for reflection and away from the plane of incidence for refraction.

14.11. **Change of Phase on Reflection.**

From equation 14(35*b*) we see that when reflection takes place in the less dense medium ($\theta_1 > \theta_2$), the component of the electric vector perpendicular to the plane of incidence changes sign. A_{1w}' has the same sign as A_{1w} when the angle of incidence is less than the polarizing angle [equation 14(34*c*)]. The directions of A_{1w}' and A_{1w} are not the same, and the fact that the signs are the same implies that A_{1x}' is opposite in sign to A_{1x}, and that A_{1z}' has the same sign as A_{1z} (fig. 14.2). Thus any component of the electric vector *parallel* to the reflecting surface changes sign when the angle of incidence is less than the polarizing angle. The component *normal* to the reflecting surface does not change sign. From 14(18) we see that the sign of H_{1w} is linked to that of E_{1y}, and must change with it. This implies that H_z changes sign, but H_x does not. Similarly 14(19) shows that the sign of H_y must follow that of E_w, i.e. it does not change. A little consideration will show that these sign changes of the components of **H** are those required to make **E**, **H** and the direction of propagation for the reflected beam form a right-handed set of axes as required by the fundamental equations (§ 13.10).

When the reflection takes place at an angle greater than the polarizing angle, $\tan(\theta_1 + \theta_2)$ is negative and A_w' has a sign opposite to that of A_w. If the angle of incidence is increased, the component of the electric vector in the plane of incidence decreases to zero, and then reappears with the opposite sign. In this sense there is a change of phase at the polarizing angle.

EXAMPLES [14(i)–14(iv)]

14(i). Writing χ, χ_1', and χ_2 for the angles between the planes of polarization of the incident, reflected and refracted light, and the plane of incidence, show that

$$|\tan \chi_1'| = \tan \chi \frac{\cos(\theta_1 + \theta_2)}{\cos(\theta_1 - \theta_2)}, \quad . \quad . \quad . \quad . \quad 14(44)$$

and

$$|\tan \chi_2| = \tan \chi . \sec(\theta_1 - \theta_2). \quad . \quad . \quad . \quad . \quad 14(45)$$

Hence verify the statement in §14.10.

14(ii). Show that when unpolarized light is incident upon a single plate, the degree of polarization of the transmitted light is greatest near grazing incidence.

14(iii). Taking the case of light polarized in the plane of incidence, show that energy is conserved, i.e. all the energy in an incident beam is found in the reflected and refracted beams. [In estimating the energy in the refracted beam, remember that (*a*) the energy density is proportional to εE^2, (*b*) the cross-section of the beam is altered by refraction, and (*c*) the velocity is altered.]

14(iv). Discuss the phase relations between components of the vectors which exist when a beam of light incident at a small angle to the normal is reflected in the denser medium.

14.12. **Stationary Waves.**

Electromagnetic waves reflected at a plane surface may interfere with the incident waves. Since the latter are usually appreciably stronger than the former, the resultant disturbance consists of a stationary wave, plus a system of progressive waves (§ 3.21). In the following discussion we ignore the progressive part, and consider the disturbance which would be produced if the two beams were of equal amplitude. We first consider the case where the incidence is nearly normal and $n_{12} > 1$. The two components of the electric vector in the plane of the surface then constitute nearly the whole of the vector, and both change sign. This implies that the electric vectors are in opposite directions at the surface of reflection. This must therefore be a node for the electric vector. Other nodes will be situated $\lambda/2$, λ, $3\lambda/2$, etc., from the surface. As shown in the preceding paragraph, the two components of the magnetic vector parallel to the surface do not change sign. The two magnetic vectors are therefore in the same direction at the surface. The antinodes of the magnetic vector thus coincide with the nodes of the electric vector in the standing wave. Wiener's experiment with fluorescent material indicates that, when there is strong reflection, a node is formed at the surface. Thus the maximum fluorescent effect coincides with the maximum of the electric vector.*

14.13.—Let us now consider a beam of light for which $\theta_1 = 45°$. The w' and w directions are mutually perpendicular, and components of the light in the plane of incidence cannot interfere. Using the criterion for plane of polarization given in § 12.3, it is found that stationary waves are formed when light incident at 45° is polarized in the plane of incidence, but not when it is polarized perpendicular to the plane of incidence. If we assume (as required in § 14.8) that the magnetic vector is in the plane of polarization, this implies that the photochemical action is directly related to the electric vector, rather than to the magnetic vector.

On elementary considerations, we should expect that the atoms and molecules would be directly affected by the electric vector which is able to polarize them directly. In view of the above observations, it is sometimes said that the electric vector is the " light vector ". This statement should not be interpreted to imply that the magnetic vector is of minor importance in optics. The electric and magnetic vectors are two aspects of our description of the electromagnetic field, and are linked in such a way that either implies the other.

14.14.—The above qualitative discussion may be supplemented by the following treatment.

Consider an incident ray polarized in the plane of incidence. We have, for the incident beam,

$$E_{1w} = E_{1x} = E_{1z} = 0; \qquad 14(46)$$

$$E_{1y} = A_{1y} \exp i\,\{\omega t - \kappa_1(x \sin \theta_1 - z \cos \theta_1)\}. \qquad 14(47)$$

For the reflected beam,

$$E_{1w}' = E_{1x}' = E_{1z}' = 0; \qquad 14(48)$$

$$E_{1y}' = A_{1y}' \exp i\,\{\omega t - \kappa_1(x \sin \theta_1 + z \cos \theta_1)\}. \qquad 14(49)$$

Put $\chi = \omega t - \kappa_1 x \sin \theta_1$; $\eta = \kappa_1 z \cos \theta_1$ and $f = -A_{1y}'/A_{1y}$. Then the resultant electric vector is

$$E_{1y} + E_{1y}' = e^{i\chi}(A_{1y}e^{i\eta} + A_{1y}'e^{-i\eta}). \qquad 14(50)$$

Equation 14(35b) shows that f is positive when $n_{12} > 1$. Then

$$E_{1y} + E_{1y}' = e^{i\chi}A_{1y}\{(1 - f)e^{i\eta} + f(e^{i\eta} - e^{-i\eta})\}. \qquad 14(51)$$

The second term in 14(51) represents the stationary part of the wave (see § 3.20). Writing E_y for this term we have

$$E_y = 2ifA_{1y} \sin (\kappa_1 z \cos \theta_1) \exp i(\omega t - \kappa_1 x \sin \theta_1). \qquad 14(52)$$

* In this discussion, we have assumed that phase changes on reflection at a metal surface follow the same rule as that found for an insulator. It will later be shown that this is justified under the conditions in which we are interested [equation 15(18)].

The corresponding parts E_w, E_z, E_x are all zero. For the magnetic vector $H_y = 0$ and

$$H_{1w} = -\sqrt{\varepsilon}\,E_{1y}, \quad . \quad . \quad . \quad . \quad . \quad . \quad . \quad 14(53)$$

$$H_{1w}' = -\sqrt{\varepsilon}\,E_{1y}'. \quad . \quad . \quad . \quad . \quad . \quad . \quad . \quad 14(54)$$

The component of H in the x direction is

$$-H_{1w}\cos\theta_1 + H_{1w}'\cos\theta_1 = \sqrt{\varepsilon}\cos\theta_1(E_{1y} - E_{1y}'). \quad . \quad 14(55)$$

The procedure which led to 14(52) gives, for the stationary part:

$$H_x = 2f\sqrt{\varepsilon}\cos\theta_1 A_{1y}\cos(\kappa_1 z\cos\theta_1)\exp i(\omega t - \kappa_1 x\sin\theta_1), \qquad 14(56)$$

and similarly

$$H_z = 2if\sqrt{\varepsilon}\sin\theta_1 A_{1y}\sin(\kappa_1 z\cos\theta_1)\exp i(\omega t - \kappa_1 x\sin\theta_1). \quad 14(57)$$

When the incidence is nearly normal, the nodes of **E** and **H** are formed as stated in § 14.13. Also, when $\theta = 45°$, H_x and H_z are out of phase, so that no ordinary standing waves are formed by the magnetic vector.

EXAMPLES [14(v) and 14(vi)]

14(v). Show that the relation between **E** and **H** in the standing waves formed at normal incidence is in accord with the fundamental equations 13(18) and 13(22).

14(vi). Discuss the interference of the incident and reflected light when the incident light is polarized perpendicular to the plane of incidence. Obtain formulæ similar to those given in § 14.14. [It is simplest to start with the magnetic disturbance.]

14.15. Total Reflection.

When a beam of light is incident on a boundary from the denser side * at an angle of incidence whose sine is greater than n_{12}, the sine of the angle of refraction [given by 14(15)] is greater than unity. If we apply the usual relations the cosine is imaginary, for we have

$$\sin\theta_2 = \frac{1}{n_{12}}\sin\theta_1, \quad . \quad . \quad . \quad . \quad . \quad . \quad . \quad 14(58)$$

and

$$\cos\theta_2 = \sqrt{(1 - \sin^2\theta_2)},$$

i.e.

$$\cos\theta_2 = \frac{i}{n_{12}}\sqrt{(\sin^2\theta_1 - n_{12}^2)}. \quad . \quad . \quad . \quad 14(59)$$

An interesting result, which is verified by experiment, is obtained by inserting the value from 14(59) in equations 14(34) to 14(37). We have, from 14(34a),

$$A_{1w}' = A_{1w}\frac{n_{12}\cos\theta_1 - (i/n_{12})\sqrt{(\sin^2\theta_1 - n_{12}^2)}}{n_{12}\cos\theta_1 + (i/n_{12})\sqrt{(\sin^2\theta_1 - n_{12}^2)}}. \qquad 14(60)$$

* The denser medium 1 is assumed to be on the "negative Z" side of the OXY plane (fig. 14.5).

The complex quantity $(p - iq)/(p + iq)$ is equal to $e^{-2i\delta}$, where $\tan \delta = q/p$. Hence we may write

$$A_{1w}' = A_{1w} e^{-2i\delta_w}, \qquad 14(61)$$

and
$$E_{1w}' = A_{1w} \exp i\{\omega t - \kappa_1(x \sin\theta_1 + z \cos\theta_1) - 2\delta_w\}, \qquad 14(62)$$

where

$$\tan \delta_w = \frac{\sqrt{(\sin^2 \theta_1 - n_{12}^2)}}{n_{12}^2 \cos \theta_1}. \qquad 14(63)$$

Similarly, it may be shown that

$$E_{1y}' = A_{1y} \exp i\{\omega t - \kappa_1(x \sin\theta_1 + z \cos\theta_1) - 2\delta_y\}, \qquad 14(64)$$

where

$$\tan \delta_y = \frac{\sqrt{(\sin^2 \theta_1 - n_{12}^2)}}{\cos \theta_1} = n_{12}^2 \tan \delta_w. \qquad 14(65)$$

Thus the reflected wave has the same energy as the incident wave. The component of the electric vector in the plane of incidence is retarded by $2\delta_w$, and the perpendicular component is retarded by $2\delta_y$ with respect to the incident wave. If the incident wave is plane-polarized, the reflected wave is elliptically polarized. The phase difference (δ) between the components of the electric vector in and perpendicular to the plane of incidence is given by

$$\tan \tfrac{1}{2}\delta = \tan(\delta_w - \delta_y) = \frac{\tan \delta_w - \tan \delta_y}{1 + \tan \delta_w \tan \delta_y}$$

$$= \frac{\tan \delta_w (1 - n_{12}^2)}{1 + n_{12}^2 \tan^2 \delta_w}$$

$$= \frac{\cos \theta_1 \sqrt{(\sin^2 \theta_1 - n_{12}^2)}}{\sin^2 \theta_1}. \qquad 14(66)$$

Total reflection provides a very useful method of producing elliptically polarized light of any desired type. By using one or more internal reflections, any desired phase difference from 0 to $\frac{1}{2}\pi$ can be produced. Circularly polarized light is conveniently produced by two internal reflections at an angle * of about 53° in a glass whose refractive index is 1·5, i.e. $n_{12} = 0{\cdot}667$. The incident light must be plane-polarized in a plane which makes an angle of $\frac{1}{4}\pi$ with the plane

* There are two possible angles—see Examples 14(vii) and 14(viii).

of incidence. A glass block of the shape shown in fig. 14.4 allows the two internal reflections at the correct angle *. It is known as ***Fresnel's rhomb.***

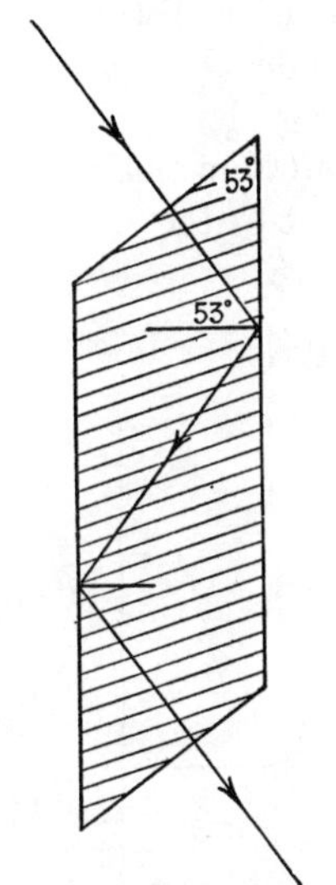

Fig. 14.4.—Fresnel's rhomb

EXAMPLES [14(vii)–14(x)]

14(vii). Show that, when the index is large enough, there are two angles of incidence which may be used to give a phase difference of $\frac{1}{4}\pi$ between the components polarized in and perpendicular to the plane of incidence. Find the angles when $\mu = 1{\cdot}6$. [58·5° and 42·5°.]

14(viii). Using the data of the previous example, show that the rate of variation of δ with the angle of incidence is greater when the angle is smaller. (Hence it is desirable to use the larger angle in constructing a rhomb for producing circularly polarized light.)

14(ix). Show that when the rotation from the plane of polarization in the incident beam to the plane of incidence is clockwise (viewed by an observer looking in the direction of propagation), the light emerging from a Fresnel rhomb is right-handed circularly polarized light.

14(x). Write down equations corresponding to 14(63) and 14(65) for the magnetic vector. Show that its changes of phase are such that it is always perpendicular to the electric vector.

* Small strains in the surface of the glass (due to polishing) have an important effect on the ellipticity of the reflected light. It is necessary to reduce these strains by annealing, and then to adjust the angles of a rhomb by trial and error in order to obtain circularly polarized light (see § 14.18).

14.16. **Disturbance in the Second Medium.**

Since the incident and reflected waves do not have a phase difference of π, there must be a disturbance in the second medium in order to satisfy the boundary conditions. The equations describing this disturbance are obtained by inserting the values for $\sin\theta_2$ and $\cos\theta_2$ from 14(59) and 14(60) in the expression for the refracted wave. Considering the component of the electric vector perpendicular to the plane of incidence, we have

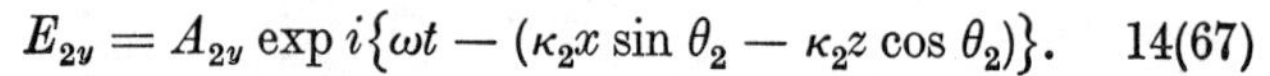

$$E_{2y} = A_{2y} \exp i\{\omega t - (\kappa_2 x \sin\theta_2 - \kappa_2 z \cos\theta_2)\}. \quad 14(67)$$

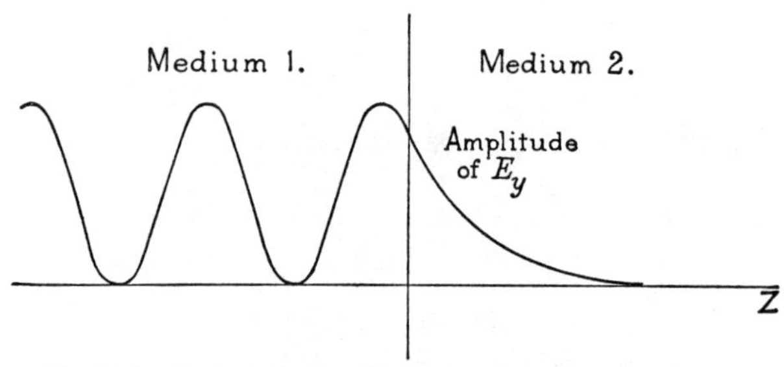

Fig. 14.5.—Variation of *amplitude* of stationary wave with z for incidence at an angle greater than the critical angle. Note that there is no discontinuity of the amplitude at the boundary.

A_{2y} is found to be complex and, more important, the factor multiplying z is found to be real. We have, from 14(29) and 14(64),

$$\begin{aligned} A_{2y} &= A_{1y}\{1 + e^{-2i\delta_y}\} \\ &= 2A_{1y}e^{-i\delta_y}\cos\delta_y, \quad . \ . \ . \ . \ . \quad 14(68) \end{aligned}$$

and hence

$$E_{2y} = 2A_{1y}\cos\delta_y \exp\{-\kappa_1 z\sqrt{(\sin^2\theta_1 - n_{12}^2)} + i(\omega t - \kappa_1 x \sin\theta_1 - \delta_y)\}. \quad 14(69)$$

The disturbance in the second medium is thus of a rather peculiar type, since it is periodic in x, but not in z. It is best considered in conjunction with the total disturbance in the first medium. Using 14(47) and 14(65), we have

$$E_{1y} + E_{1y}' = 2A_{1y}\cos(\kappa_1 z\cos\theta_1 + \delta_y)\exp i(\omega t - \kappa_1 x \sin\theta_1 - \delta_y). \quad 14(70)$$

Thus, in medium 1, there is a stationary wave of the type considered in § 14.14 (with $f = 1$). This wave is continued into the second medium for a short distance. By putting $z = 0$ in 14(69) and 14(70), we see that the amplitude of the stationary wave does not change discontinuously at the boundary. If it did change at the boundary, the continuity of E_y would fail. The amplitude of the stationary wave is periodic with z in medium 1, but decays exponentially with z in medium 2 (fig. 14.5).

The above discussion applies to a beam of indefinite extent in space and in time. A complete treatment of a finite beam must take into account conditions at the time when the beam is first incident on the surface. It is also necessary to consider in detail the conditions at the edges of a beam of finite spatial extension. In the steady state, the Poynting vector is parallel to the boundary, and there is no net flow of energy across the boundary, except at the edges of the beam. There is a flow of energy parallel to the boundary in each medium. For a more detailed treatment of the theory, Reference 14.1 may be consulted.

14.17. Experimental Test of the Theory of Reflection and Refraction.

The results derived above have been tested in the following ways:

(1) measurement of the reflection coefficient for each component for different angles of incidence;

(2) measurement of the state of polarization of the beam reflected in a less dense medium;

(3) measurement of the phase differences between the two components when total reflection occurs;

(4) verification of the disturbance in the second medium when total reflection occurs.

Measurements of the first type were made by Rayleigh,* and by Murphy.† The results are in good general agreement with the values given by equations 14(34) and 14(35). As a test of the theory, measurements of reflection coefficient are less sensitive than measurements of polarization. The results obtained by the latter method appeared at first to disagree with the theory. With most surfaces, the light reflected at the Brewsterian angle is elliptically polarized. The ellipse is usually a slender one, but the light is certainly not plane-polarized as required by the theory. Further study has shown that when freshly formed liquid surfaces are used, the measurements are much nearer to the theoretical prediction. A slight residual ellipticity may be assigned to the effect of surface tension on the arrangement of molecules near the surface. The study of the polarization of light reflected at the Brewsterian angle has been made an interesting method of investigating the properties of surface films.‡

* Reference 14.2. † Reference 14.3.

‡ For a general summary of the experiments, see p. 1582 of Reference 14.1. Recent work is described in Reference 14.4.

14.18.—The phase difference between the components of a totally reflected beam has been studied by Kynast,* using the Babinet compensator (§ 12.33). He used an accurately parallel beam of monochromatic light ($\lambda = 5461$ Å.). He found that for glass of high refractive index, and for quartz, the phase differences are greater than those predicted by the theory. The discrepancy increases as the angle of incidence increases. These results have been confirmed by Volke.† It has been suggested that polishing may cause the production of a doubly refracting film on the surface, and that this film may cause an increase in the phase difference. There is no independent evidence in favour of the presence of such a film, and there is evidence from electron diffraction that films formed by polishing are generally amorphous. Recent experimental work with liquid surfaces gives the phase difference predicted by 14(66).

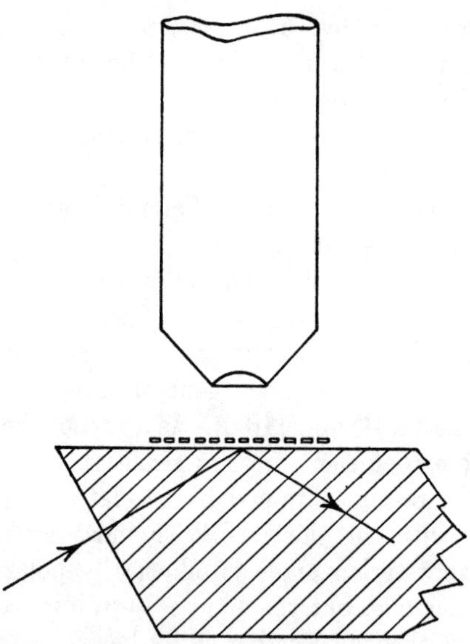

Fig. 14.6.—Arrangement for viewing light scattered from fine particles on the upper side of a surface when light is incident at an angle greater than the critical angle on the lower side.

14.19.—Various attempts have been made to study the disturbance in the second medium. In one method, fine particles of carbon are deposited on the surface of a glass prism at which a beam of light is totally reflected (fig. 14.6). The particles, observed in a microscope, are seen to be illuminated. It is claimed that the light entering the

* Reference 14.5. † Reference 14.6.

microscope is scattered from the "radiation" specified by equation 14(69). The experiment has been criticized on the ground that the particles disturb the boundary conditions in their own neighbourhood, so that the reflection is not total reflection. Another type of experiment on the penetration of light into the second medium is more interesting, since quantitative measurements are possible. Newton observed that reflection is not quite complete when light is incident (at an angle greater than the critical angle) upon a thin film of the less

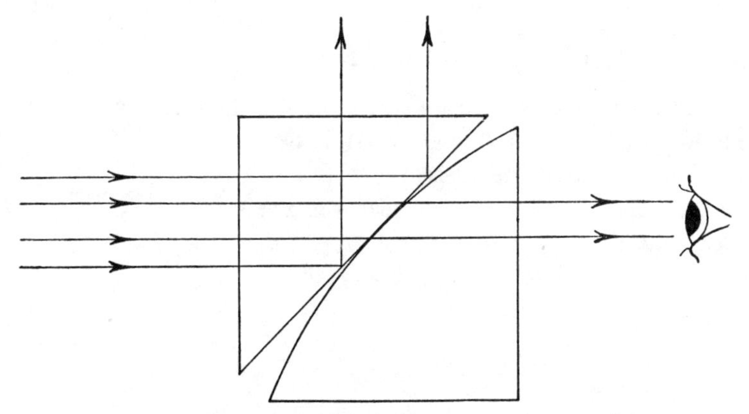

Fig. 14.7.—Penetration of light incident at an angle greater than the critical angle through a thin film

dense medium. A convenient apparatus is shown in fig. 14.7. If light is incident from the left, an observer looking from the right sees a bright central disc surrounded by a more faintly illuminated ring. The central disc corresponds to the area of contact, and the ring to a region where the thickness of the air is very small. The amount of light transmitted at a given thickness depends on the wavelength, and on the angle of incidence. Some transmitted light can be observed when two glass surfaces are separated by an air film 3λ thick, or by a water film of over 5λ thick. The flow of energy through the thin film has been studied theoretically by Eichenwald.*

14.20.—No experiment can reveal the disturbance in the second medium without drawing energy. Thus the conditions at the boundary must be modified to some extent, and the theory given in § 14.16 does not apply exactly. It has been suggested that this implies that the disturbance in the second medium can never be observed. Nevertheless it is possible, in principle, to draw different small amounts of energy and to extrapolate from a small modification of the boundary

* See p. 1587 of Reference 14.1.

conditions to no modification at all. This is essentially what is done when the transmission of light through thin films of different thicknesses is measured. It is also suggested that the disturbance in the second medium is not transverse, and is therefore not light and cannot be observed as light. This objection is based on a misunderstanding. Stationary light waves are not transverse in the ordinary sense, as is shown by the equations of § 14.14. The peculiarity of the disturbance in the second medium is entirely in the rapidity with which the amplitude decreases with z, so that there is no periodicity in the z direction.

REFERENCES

14.1. Müller-Pouillets: *Lehrbuch der Physik,* Vol. II, Chapter 28 (Vieweg).

14.2. Rayleigh: *Scientific Papers,* Vol. III, p. 496.

14.3. Murphy: *Ann. d. Phys.,* 1896, Vol. 57, p. 593.

14.4. Bouhet: *Annales de Physique,* 1931, Vol. XV, p. 1.

14.5. Kynast: *Ann. d. Phys.,* 1907, Vol. 22, p. 726.

14.6. Volke: *ibid.,* 1910, Vol. 31, p. 609.

CHAPTER XV

The Electromagnetic Theory of Absorption and Dispersion

15.1.—We have shown in Chapters XIII and XIV that the electromagnetic theory can give a satisfactory account of the transmission of light in transparent media, and of the reflection and refraction of light at the boundaries of such media. No medium, except a vacuum, is *perfectly* transparent for any region of the spectrum and all material media show strong absorption in some region of the electromagnetic spectrum. For example, quartz, which is nearly transparent for the visible region, shows very strong selective absorption for certain wavelengths in the infra-red. Absorption is a general property; the transparent medium is a limiting case.

15.2.—Metals form one important class of absorbing media. Their absorption is so strong that no measurable amount of visible light is transmitted through a film of metal which is more than a few wavelengths thick.* Our chief practical interest is not in the propagation of light in metals but in the associated strong reflection of light by metals. Metals have strong absorption for a wide range of wavelengths including all the visible spectrum, but they also have selective absorption, and the colour of metals like gold and copper is due to selective absorption. Other metals show selective absorption for wavelengths outside the visible region.

15.3.—Most insulating media have much lower absorption than metals in the visible part of the spectrum except for certain narrow regions of selective absorption. In these regions, insulating media may show absorption and reflection which are comparable with metallic absorption and reflection. Insulating media which are nearly transparent in the visible region begin to absorb at some wavelength in the ultra-violet, e.g. ordinary glasses begin to absorb at about 3500 Å., quartz begins to absorb strongly at about 1900 Å. and fluorite at about 1300 Å. For most materials this general absorption increases to a

* A film of aluminium one wavelength thick transmits less than 1 per cent of the light incident upon it.

maximum * at about 1000 Å. Below this wavelength the absorption shows a general tendency to decrease as the X-ray region is approached. Selective absorption of electromagnetic waves by insulating media is associated with wide variations in the refractive index and with anomalous dispersion.

15.4.—From the above general review of the experimental observations, we see that the field is an extensive one. It becomes important to deal with the difficulties one by one, rather than all at once. For this reason we shall start with the theory of propagation of monochromatic waves in an absorbing medium and shall omit dispersion—although later developments will reveal an intimate relation between absorption, dispersion and scattering. We begin by distinguishing, from the experimental side, between absorption and scattering. Both these processes remove energy from a parallel beam of light. Scattering changes the direction of propagation but does not directly affect the amount of radiant energy. In true absorption the energy of the radiation decreases and usually the temperature of the medium is raised.† This suggests that true absorption is associated with some kind of dissipative force. We are familiar with the dissipation of electrical energy in metals due to the resistance, and it is natural to relate the absorption of light to the resistivity of a metal. In insulators the presence of dissipative forces is not so obvious. We assume that insulators contain elastically bound electrons. These electrons behave like simple harmonic oscillators except when they are subject to forces which are nearly in resonance with their natural periods. When this happens they behave like weakly damped oscillators, and a transformation of radiant energy into other forms occurs through the operation of the damping forces. The electromagnetic theory is thus able to describe the chief phenomena of dispersion and absorption of insulating media in terms of a theory of damped oscillators, but the calculation of the natural frequencies and the damping coefficients for the oscillators is not within the scope of the electromagnetic theory.

15.5. Transmission of Light in an Absorbing Medium.

Consider a parallel beam of monochromatic light propagated in the positive direction of z. Let $L(z)$ be the relative energy and let L_0 be the value of $L(z)$ when $z = 0$. We assume that the beam has entered

* At this wavelength a film of cellophane one wavelength thick transmits about 20 per cent of the incident radiation.

† It is also possible that the absorbed energy may cause chemical or electrical action.

the medium through a plane boundary perpendicular to the direction of propagation. Experimental observations on the transmission in a homogeneous medium which absorbs, but does not scatter, the light are summarized in Lambert's law, which may be written

$$L(z) = L_0 e^{-2\alpha z}. \quad \quad 15(1)$$

The constant 2α is called the absorption coefficient. If $A(z)$ is the amplitude of one of the vectors, then

$$A(z) = A_0 e^{-\alpha z} \quad \quad 15(2)$$

and

$$\frac{dA}{dz} = -\alpha A, \quad \quad 15(3)$$

or

$$\frac{d(\log A)}{dz} = -\alpha. \quad \quad 15(4)$$

The wave may be represented by the expression

$$E_x = A_0 \exp\{-\alpha z + i(\omega t - \kappa z)\}. \quad \quad 15(5)$$

It is convenient to introduce an auxiliary symbol $\varkappa$ equal to α/κ, so that 15(5) may be written

$$E_x = A_0 \exp i\{\omega t - \kappa(1 - i\varkappa)z\}, \quad \quad 15(6)$$

or

$$E_x = A_0 \exp i\omega\left\{t - \frac{n}{c}(1 - i\varkappa)z\right\}, \quad \quad 15(7)$$

where

$$n = \frac{\kappa c}{\omega}. \quad \quad 15(8)$$

Equation 15(7) may be written in the form

$$E_x = A_0 \exp i\omega\left\{t - \frac{\boldsymbol{n}}{c} z\right\}, \quad \quad 15(9)$$

where

$$\boldsymbol{n} = n(1 - i\varkappa). \quad \quad 15(10)$$

Equation 15(9) is formally similar to the expression for a wave in a transparent medium and, for this reason, $\boldsymbol{n}$ is often called " the complex index ". This name is a little unfortunate because, as is shown in Appendix XV A, the real part of $\boldsymbol{n}$ is only indirectly related to angles of refraction. We shall call $\varkappa$ the *extinction coefficient.*

EXAMPLES [15(i)–15(iv)]

15(i). Find the change in log A when the wave advances through a distance equal to one vacuum wavelength in an absorbing medium. [$-2\pi n\varkappa$.]

15(ii). Show by substitution that 15(9) satisfies the equation

$$\frac{\partial^2 E_x}{\partial x^2} + \frac{\partial^2 E_x}{\partial y^2} + \frac{\partial^2 E_x}{\partial z^2} = \frac{\boldsymbol{n}^2}{c^2}\frac{\partial^2 E_x}{\partial t^2}. \quad . \quad . \quad . \quad . \quad 15(11)$$

15(iii). Show that, in an absorbing medium, **E** and **H** are mutually perpendicular. [Follow the procedure of § 13.10.]

15(iv). Show that, in an absorbing medium, **H** lags behind **E** by a phase angle γ such that $\tan\gamma = \varkappa$. [Refer to § 13.10. If **E** is directed along OX and **H** along OY, we have $H_y = \boldsymbol{n}E_x$ {see 13(33)}; using 15(10) we obtain the phase angle between the two vectors by writing $H_y = n(1 - i\varkappa)E_x = n(1 + \varkappa^2)^{1/2}E_x e^{i\gamma}$].

15.6. Reflection of Light by an Absorbing Medium.

Consider a parallel beam of light incident from vacuum * on to the plane boundary of an absorbing medium. The boundary conditions stated in § 14.1 may be formally satisfied by a treatment similar to that used in §§ 14.2 and 14.3, i.e. we write

$$E_{1v} = A_{1v} \exp i\omega \left\{t - \frac{1}{c}(x \sin\theta_1 - z\cos\theta_1)\right\}, \quad 15(12a)$$

$$E_{1v}' = A_{1v}' \exp i\omega \left\{t - \frac{1}{c}(x \sin\theta_1 + z\cos\theta_1)\right\}, \quad 15(12b)$$

$$E_{2v} = A_{2v} \exp i\omega \left\{t - \frac{\boldsymbol{n}}{c}(x \sin\theta_2 - z\cos\theta_2)\right\}. \quad 15(12c)$$

In order that equation 14(7) may be satisfied, it is necessary that

$$\sin\theta_2 = \frac{1}{\boldsymbol{n}}\sin\theta_1, \quad . \quad . \quad . \quad . \quad . \quad . \quad 15(13)$$

and that A_{1v}', A_{2v}, etc., have the values obtained by substituting $\boldsymbol{n}$ for n in equations 14(34) to 14(42). The angle θ_2 is complex except for normal incidence ($\theta_1 = 0$). It is shown in Appendix XV A that there is a real angle of refraction and that the theory gives a reasonable description of certain special properties of the refracted wave. For the present we assume that the value of $\sin\theta_2$ given by 15(13) may

* Or from air, since the index of refraction for air at ordinary pressures is very near to unity.

be used in equation 14(34), etc., to calculate A_{1v}' and A_{1w}'. We therefore put

$$\cos \theta_2 = \frac{1}{\boldsymbol{n}} \sqrt{(\boldsymbol{n}^2 - \sin^2 \theta_1)}, \quad . \quad . \quad . \quad 15(14)$$

and

$$\tan \theta_2 = \frac{\sin \theta_1}{\sqrt{(\boldsymbol{n}^2 - \sin^2 \theta_1)}} \quad . \quad . \quad . \quad . \quad 15(15)$$

15.7. Reflection at Normal Incidence.

The reflecting power (ρ) is defined to be the ratio of the reflected to the incident *energy* for normal incidence. Putting $\boldsymbol{n}$ for n in 14(39) or 14(40) (and remembering that the energy is the product of the complex amplitude with its conjugate) we have

$$\rho = \frac{A_{1w}'A_{1w}'^*}{A_{1w}^2} = \frac{A_{1v}'A_{1v}'^*}{A_{1v}^2} = \frac{\boldsymbol{n}-1}{\boldsymbol{n}+1}\frac{\boldsymbol{n}^*-1}{\boldsymbol{n}^*+1} \quad 15(16)$$

$$= \frac{(n-1)^2 + n^2\varkappa^2}{(n+1)^2 + n^2\varkappa^2}. \quad . \quad . \quad . \quad . \quad . \quad . \quad . \quad . \quad . \quad 15(17)$$

From 15(17) we see that, when $n\varkappa$ is large compared with $(n + 1)$, the reflecting power is nearly unity. Strong reflection at normal incidence is associated with strong absorption. This result is in agreement with experimental observations. Light which enters a metal is strongly absorbed but most of the light which is incident upon a metal surface is reflected. Substances like solid iodine, which absorb strongly, also show strong reflection and have a " metallic appearance ". Mercury vapour strongly absorbs radiation of wavelength 2537 Å. (§ 15.31). It also reflects this radiation very strongly.

EXAMPLE 15(v)

Show that the change of phase on reflection at normal incidence is given by

$$\tan \delta_n = \frac{-2n\varkappa}{n^2 - 1 + n^2\varkappa^2}. \quad . \quad . \quad . \quad . \quad . \quad 15(18)$$

[To obtain the change of phase the complex quantity A_{1v}'/A_{1v} must be put in the form $e^{i\gamma}$ where γ is real. The phase angle is $\tan^{-1}\gamma$ (see § 2.27).]

Note that $\tan \delta_n \to 0$ when $\varkappa \to 0$, and also when $\varkappa$ is large compared with unity. $\tan \delta_n = 0$ is in agreement with either $\delta_n = 0$ or $\delta_n = \pi$. We have seen that when $\varkappa = 0$, δ_n is zero when $n < 1$, and π when $n > 1$ (see §§ 5.10 and 14.11). Using 14(39) directly and putting $\boldsymbol{n} = -in\varkappa$, i.e. neglecting n in comparison with $n\varkappa$, we see that when $\varkappa \to \infty$, $A_{1v}'/A_{1v} \to -1$ and the phase difference is π.

15.8. **Reflection at Oblique Incidence.**

The amount and the state of polarization of the reflected light at oblique incidence may be calculated by putting $\boldsymbol{n}$ for n in 14(34) and 14(36). The detailed calculation is very lengthy. The reader should consult the references at the end of this chapter * for details of the calculation and for certain approximate formulæ which are useful.

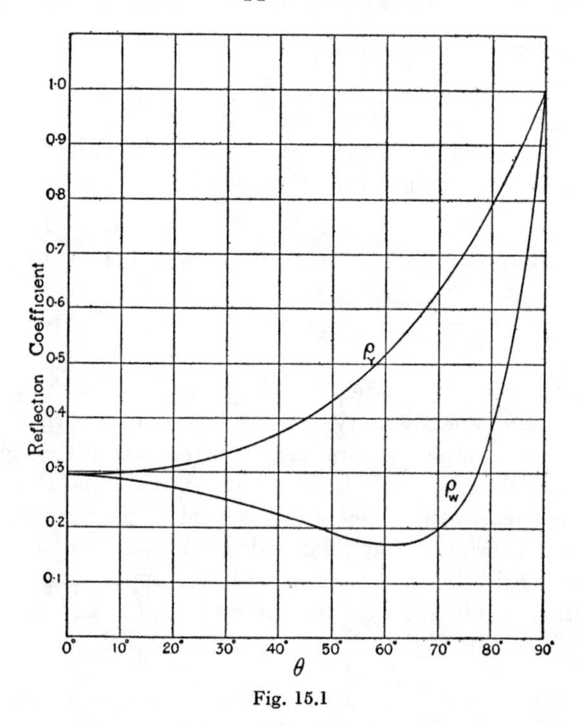

Fig. 15.1

The following equations are valid when $n^2 + n^2\varkappa^2$ is large compared with unity:

$$\rho_w = \frac{A_{1w}'A_{1w}'^*}{A_{1w}^2} = \frac{n^2(1+\varkappa^2)\cos^2\theta_1 - 2n\cos\theta_1 + 1}{n^2(1+\varkappa^2)\cos^2\theta_1 + 2n\cos\theta_1 + 1}, \qquad 15(19a)$$

$$\rho_v = \frac{A_{1v}'A_{1v}'^*}{A_{1v}^2} = \frac{n^2(1+\varkappa^2) - 2n\cos\theta_1 + \cos^2\theta_1}{n^2(1+\varkappa^2) + 2n\cos\theta_1 + \cos^2\theta_1}. \qquad 15(19b)$$

Fig. 15.1 shows the way in which the ratios ρ_w and ρ_v depend on the

* See especially p. 1595 of Reference 15.1; Chapter VI of Reference 15.2; and p. 242 ff. of Reference 15.3.

angle of incidence when $n = 1{\cdot}5$ and $\varkappa = 1{\cdot}00$, i.e. for a typical strongly absorbing medium. This figure may be compared with fig. 14.3, which shows corresponding functions for a transparent medium. Both for transparent and for absorbing media the component of the electric vector reaches a minimum for one particular angle of incidence. This minimum is zero for transparent media but is not zero for absorbing media, i.e. there is no angle of incidence which gives complete polarization by reflection.

15.9.—When the incident light is plane-polarized, the light reflected from an absorbing medium is elliptically polarized. We now consider the polarization with special reference to one particular condition of incidence which is of practical importance. We start with the exact formulæ and derive equations 15(20) to 15(24). The approximation $(n^2 + n^2\varkappa) \gg \sin^2 \theta_1$ is introduced to obtain the final result. Let a be the ratio of the amplitude of the component of the electric vector in the plane of incidence to the amplitude of the perpendicular component, and let δ be the phase difference. Then from 14(34c) and 14(35b)

$$ae^{i\delta} = \frac{A_{1w}'}{A_{1v}'} = -\frac{A_{1w}}{A_{1v}} \cdot \frac{\cos(\theta_1 + \theta_2)}{\cos(\theta_1 - \theta_2)}, \quad . \quad . \quad 15(20)$$

Now consider the case when the incident light is plane-polarized at 45° to the plane of incidence so that $A_{1w} = A_{1v}$. We have

$$ae^{i\delta} = -\frac{\cos(\theta_1 + \theta_2)}{\cos(\theta_1 - \theta_2)} = \frac{\tan\theta_1 \tan\theta_2 - 1}{\tan\theta_1 \tan\theta_2 + 1}, \quad . \quad 15(21)$$

$$\frac{1 + ae^{i\delta}}{1 - ae^{i\delta}} = \tan\theta_1 \tan\theta_2, \quad . \quad . \quad . \quad . \quad . \quad . \quad . \quad . \quad . \quad 15(22)$$

and, eliminating θ_2 with the aid of 15(15),

$$\frac{1 + ae^{i\delta}}{1 - ae^{i\delta}} = \frac{\tan\theta_1 \sin\theta_1}{\sqrt{(\boldsymbol{n}^2 - \sin^2\theta_1)}}. \quad . \quad . \quad . \quad 15(23)$$

This expression gives the amount and polarization of the reflected light when $\boldsymbol{n}$ and the angle of incidence are given.

15.10. **Principal Angle of Incidence.**

At normal incidence and at grazing incidence (i.e. $\theta_1 = 0$ and $\theta_1 = \frac{1}{2}\pi$) the angle $\delta = 0$ or π, and the reflected light is plane-polarized.

In general δ has an arbitrary value. The reflected light is elliptically polarized and the axes of the ellipse have no special relation to the plane of incidence (fig. 15.2). There is, however, one angle of incidence for which $\delta = \frac{1}{2}\pi$. The reflected light is then elliptically polarized with one axis of the ellipse in the plane of incidence (fig. 15.3). This angle of incidence, which we shall denote by Θ_1, is called the *principal angle of incidence.* Since $e^{i\delta} = i$ when $\delta = \frac{1}{2}\pi$, we have, from 15(23),

$$\frac{1+ia}{1-ia} = \frac{\tan \Theta_1 \sin \Theta_1}{\sqrt{(\boldsymbol{n}^2 - \sin^2 \Theta_1)}}. \quad . \quad . \quad . \qquad 15(24)$$

15.11. **Principal Azimuth.**

Whatever the angle of incidence, it is possible to change the elliptically polarized reflected light into plane-polarized light by inserting a crystal plate of suitable thickness and orientation. A procedure using the Babinet compensator described in § 12.33 is convenient. When this has been done, the plane of polarization of the reflected

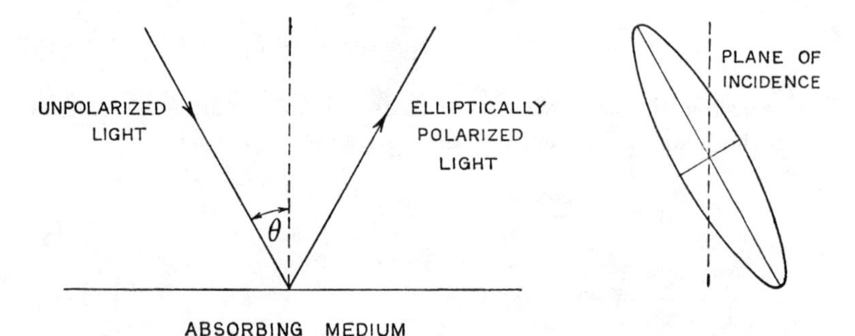

Fig. 15.2.—Reflection of light at the boundary of an absorbing medium. The diagram on the right shows the ellipse of polarization as viewed by an observer who receives the reflected light.

light may be determined with a Nicol prism. The angle between the plane of incidence and the plane of polarization of the reflected light (when compensated by introducing a phase difference $-\delta$) is called the azimuth.* The *principal azimuth* (Ψ) corresponds to the principal angle of incidence (figs. 15.2 and 15.3). Since, for this angle, $\delta = \frac{1}{2}\pi$, the compensator may be a quarter-wave plate with its slow direction

* Some writers define the complementary angle (i.e. the angle between the plane of the electric vector and the plane of incidence) as the azimuth.

in the plane of incidence. We have, for incidence at the principal angle,

$$a = \tan \Psi. \quad \text{15(25)}$$

For most cases of practical importance $(n^2 + n^2\varkappa^2) \gg \sin^2 \theta_1$. The denominator of 15(24) may then be put equal to $\boldsymbol{n}$, and we have

$$\tan \Theta_1 \sin \Theta_1 = \boldsymbol{n} \frac{1 + i \tan \Psi}{1 - i \tan \Psi},$$

$$\tan \Theta_1 \sin \Theta_1 = \boldsymbol{n} \frac{\cos \Psi + i \sin \Psi}{\cos \Psi - i \sin \Psi} = \boldsymbol{n} e^{2i\Psi}. \quad \text{15(26)}$$

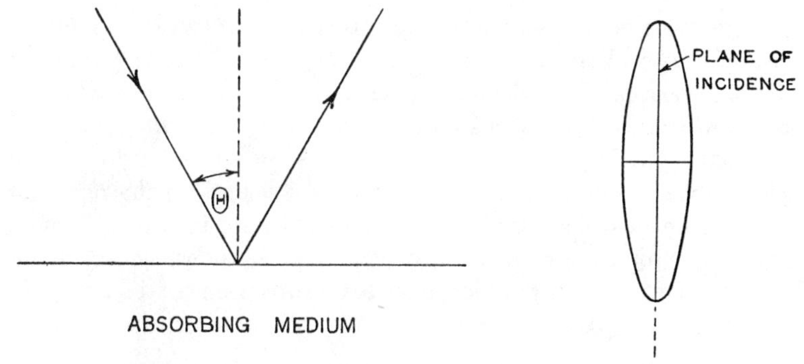

Fig. 15.3.—Reflection of light at the principal angle of incidence

Since the quantity on the left is real, the imaginary part of the right-hand side must vanish, i.e.

$$\varkappa = \tan 2\Psi. \quad \text{15(27)}$$

Inserting this value in 15(26) we obtain

$$n \cos 2\Psi + n \sin 2\Psi \tan 2\Psi = \tan \Theta_1 \sin \Theta_1,$$

i.e.

$$n = \tan \Theta_1 \sin \Theta_1 \cos 2\Psi. \quad \text{15(28)}$$

15.12. **Comparison of Theory and Experiment.**

A convenient test of the agreement between theory and experiment is obtained by measuring the angles Θ_1 and Ψ, using 15(27) and 15(28) to calculate n and $\varkappa$, and then using 15(17) to obtain ρ. The value so calculated may be compared with the results of direct measurements of ρ. Some experimental results are shown in Table 15.1 (p. 448).

TABLE 15.1

Metal	n	$\varkappa$	ρ (calculated from n and $\varkappa$)	ρ (direct measurement)
Copper	0·64	4·08	73·2	89·0
Gold	0·366	7·70	85·1	88·2
Nickel	1·79	1·86	62·0	65·9
Platinum	2·06	2·06	70·1	66·3
Silver	0·181	20·2	95·3	93·5
Steel	2·41	1·38	57·5	58·4

The agreement of the last two columns is as good as can reasonably be expected. The observations of n and $\varkappa$ are due to Drude,* who did not use a very precisely defined wavelength. The measurements of ρ are due to Hagen and Rubens † who used a completely different set of specimens. The large discrepancy in the case of copper is due to the rapid change of reflection coefficient with wavelength in the region of the spectrum in which Drude's measurements were made. Moreover, the optical properties of a metal depend to a considerable extent on the method of polishing which affects the arrangement of atoms in the surface layer. Exact agreement cannot therefore be expected.

It is possible to measure $\varkappa$ and n by experiments on thin metal films and metal prisms of very small angle. The measurements of n are not accurate and the properties of these thin films cannot be expected to be the same as those of the metal in bulk. It is therefore not possible to test the theory by comparing these values with those obtained in reflection measurements.

EXAMPLE 15(vi)

Show that when $\varkappa$ approaches zero the principal angle approaches the polarizing (or Brewsterian) angle.

[Do not use 15(26), which is valid only when $(n^2 + n^2\varkappa^2) \gg 1$. Use 15(24). If $\boldsymbol{n}$ is real $a = 0$ and the reflected light is plane-polarized in the plane of incidence. We then have $n^2 - \sin^2 \Theta_1 = \tan^2 \Theta_1 \sin^2 \Theta_1$, i.e. $n = \tan \Theta_1$.]

15.13. Optical Constants of Metals.

Eventually we are going to consider a metal as an assembly of atoms and of free electrons. As a preliminary stage we now consider it as a continuous medium which has a conductivity which is large,

* Reference 15.4. † Reference 15.5.

and a dielectric constant which is not very different from unity. We apply Maxwell's method to this medium and show that the equations yield a complex index. We may then compare the calculated values of n and $\varkappa$ with those obtained experimentally by the methods described in the preceding paragraphs.

15.14.—Let us return to Maxwell's equations 13(16) to 13(19) and write 13(19) in the form

$$\operatorname{curl} \mathbf{H} = \left(\frac{4\pi}{c}\sigma + \frac{\epsilon}{c}\frac{\partial}{\partial t}\right)\mathbf{E}. \qquad 15(29)$$

Hence
$$\nabla^2 \mathbf{E} = \frac{1}{c}\frac{\partial}{\partial t}\left(\frac{4\pi\sigma}{c} + \frac{\epsilon}{c}\frac{\partial}{\partial t}\right)\mathbf{E},$$

i.e.
$$\nabla^2 \mathbf{E} = \frac{\epsilon}{c^2}\frac{\partial^2 \mathbf{E}}{\partial t^2} + \frac{4\pi\sigma}{c^2}\frac{\partial \mathbf{E}}{\partial t}. \qquad 15(30)$$

Let us try the solution

$$E_x = A \exp i\omega\left(t - \frac{\boldsymbol{n}}{c}z\right). \qquad 15(31)$$

We do not assume that $\boldsymbol{n}$ is necessarily complex but only that it may possibly be so. Differentiating 15(31) and substituting in 15(30), we obtain

$$\boldsymbol{n}^2 = \epsilon\left(1 - i\frac{4\pi\sigma}{\epsilon\omega}\right). \qquad 15(32)$$

The solution of the equations thus requires the complex index given by 15(32). Equating real and imaginary parts, we obtain

$$n^2(1 - \varkappa^2) = \epsilon, \qquad 15(33a)$$

$$n^2\varkappa = \frac{2\pi}{\omega}\sigma = \sigma T = \frac{\sigma\lambda}{c}, \qquad 15(33b)$$

where T is the period of vibration and λ is the wavelength *in vacuo.*

Hence
$$2n^2 = \sqrt{(\epsilon^2 + 4\sigma^2T^2)} + \epsilon, \qquad 15(34a)$$

$$2n^2\varkappa^2 = \sqrt{(\epsilon^2 + 4\sigma^2T^2)} - \epsilon. \qquad 15(34b)$$

Multiplying these equations, we obtain

$$n^2\varkappa = \sigma T. \qquad 15(35)$$

When ϵ is small compared with σT, equations 15(34) reduce to

$$n = n\varkappa = \sqrt{(\sigma T)}. \quad . \quad . \quad . \quad . \quad . \quad 15(36)$$

If σT is large compared with unity, the reflecting power [equation 15(17)] becomes

$$\rho = 1 - \frac{2}{\sqrt{(\sigma T)}} = 1 - 2\sqrt{\frac{c}{\sigma\lambda}}. \quad . \quad . \quad 15(37)$$

15.15.—The values of n and $\varkappa$ calculated from the electrical constants do not even approximately agree with the results obtained by measurements made with visible light. For all the metals listed in Table 15.1 $\varkappa$ is greater than 1. This would require ϵ to be negative if 15(33*a*) were correct. The variation of $n^2\varkappa$ with wavelength or from one metal to another is not at all in agreement with equation 15(33*b*). These equations are based upon equations which are adequate at low frequencies and even at radio frequencies (10^8 cycles per second). They fail at optical frequencies (10^{15} cycles per second). We shall later consider the optical problem in more detail. In the meantime it is interesting to see whether equations 15(33) to 15(37) are valid in the frequency range 10^{13} cycles per second to $1{\cdot}5 \times 10^{14}$ cycles per second, i.e. in the infra-red.

15.16.—Measurements of the reflecting powers of metals in the range $1{\cdot}5 \times 10^{14}$ to about 10^{13} cycles per second (i.e. $\lambda = 2\,\mu$ to about $30\,\mu$) were made by Hagen and Rubens * and other workers. These experiments show that the reflecting power of most metals increases rapidly with wavelength and approaches the value given by 15(37). The reflecting powers of most metals are so high that direct measurements are not suitable for revealing the difference between one metal and another, nor for investigating the variation of reflecting power with temperature. It is more accurate to measure the emissivity E which, according to Kirchhoff's law, is related to the reflecting power by the equation

$$E = \frac{A}{B} = 1 - \rho, \quad . \quad . \quad . \quad . \quad . \quad 15(38)$$

where A is the energy emitted by the metal surface and B is the energy emitted by an equal area of a black body at the same temperature.

* Reference 15.5.

The apparatus is shown in fig. 15.4. A narrow range of frequency is selected by passing the radiation from the body B through an aperture and then reflecting it three times from the fluorite surfaces F_1, F_2, F_3. The reflection of fluorite is small except for a few narrow regions

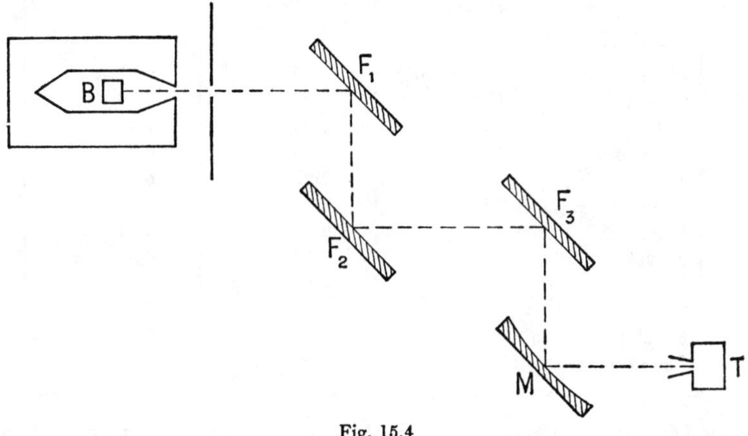

Fig. 15.4

of the spectrum and only one of these (a region near 25·5 μ) has appreciable energy after three reflections. This radiation is focused by the metal mirror (M) on to the thermopile (T). Radiation is obtained either (*a*) from a small hole in the side of the furnace, which gives black-body radiation, or (*b*) from a piece of metal placed in the furnace. The ratio gives an accurate value * of E.

15.17.—The results of these measurements are shown in Table 15.2. Equation 15(37) predicts that the product $(1 - \rho)\sqrt{\sigma}$ should be constant. The last column of the Table shows that this law is obeyed to a good approximation by many metals. Bismuth is a striking exception. The mean value of the constant † for the metals listed (excluding bismuth) is $6{\cdot}96 \times 10^6$. The value calculated from 15(37) for $\lambda = 25{\cdot}5\ \mu$ is $6{\cdot}86 \times 10^6$. The variation of reflecting power with temperature agrees satisfactorily with that calculated from the variation of conductivity. The results of measurements at different

* We have described these experiments because of their great historical importance. For a discussion of the technique of modern methods of infra-red measurements, the reader should consult Reference 15.6. The theory of the reflection from crystals like fluorite will be discussed in § 15.32.

† When σ is in e.s.u.

TABLE 15.2

Metal	$(1-\rho)$ for temperature 170° C. and wavelength 25·5 μ		$(1-\rho)\sigma^{\frac{1}{2}}$
	Observed	Calculated from 15(37)	
Silver	$1{\cdot}13 \times 10^{-2}$	$1{\cdot}15 \times 10^{-2}$	$6{\cdot}71 \times 10^{6}$
Copper	1·17	1·27	6·19
Gold	1·56	1·39	7·69
Aluminium	1·97	1·60	8·46
Zinc	2·27	2·27	6·87
Cadmium	2·55	2·53	6·92
Platinum	2·82	2·96	6·53
Nickel	3·20	3·16	6·96
Tin	3·27	3·23	6·95
Steel	3·66	3·99	6·28
Mercury	7·66	7·55	6·96
Bismuth	25·6	10·09	17·4

wavelengths are shown in Table 15.3, which gives the mean observed values of the constant and the calculated values.

TABLE 15.3

VALUES OF $(1-\rho)\sqrt{\sigma} \times 10^{-6}$

λ	4 μ	8 μ	12 μ	25·5 μ
Observed mean value	18·4	12·3	10·4	6·96
Calculated from 15(37)	17·3	12·2	9·96	6·86

The agreement is very good down to a wavelength of 8 μ. It is less good at 4 μ and rapidly becomes worse as the optical region is approached. The general results of these experiments show that the Maxwell theory in its simplest form is adequate up to frequencies of the order 6×10^{13} cycles per second (i.e. down to wavelengths of about 5 μ).

15.18. **Dispersion Theory. Dielectric Media.**

We shall now derive equations giving the variation of $\boldsymbol{n}$ with frequency for dielectric media. We shall find that, in general, $\boldsymbol{n}$ is complex and that both the real and the imaginary parts are functions of the frequency. There is, in general, absorption which varies with the frequency and reaches very high values, sometimes comparable with metallic absorption, at the centres of absorption lines. Apart from these narrow ranges of frequency, the absorption of dielectrics is very much less than that of metals. Even substances like coal, which we call " opaque ", appear nearly transparent when in slices 10 μ to 50 μ thick,* although, as we have seen above, metallic films 1 μ thick transmit less than 1 per cent of the incident light. The absorption of pure water in the visible region is so small that layers several metres thick have to be used to obtain accurate measurements of absorption.

15.19.—The fundamental problem in calculating $\boldsymbol{n}$ is to obtain the relation between the polarization of the dielectric and the electric vector of the incident light wave, i.e. we want P/E as a function of ω. The value of P/E is connected with ϵ and $\boldsymbol{n}$ by the equation †

$$4\pi\frac{P}{E} = \epsilon - 1 = \boldsymbol{n}^2 - 1. \quad . \quad . \quad . \quad . \quad 15(39)$$

In accordance with the discussion of § 2.27, we shall interpret the complex ratio of P/E as indicating a phase difference between the vectors. In the molecular theory of dielectrics, the polarization field is regarded as the resultant field of a large number of dipoles. The local field is assumed to vary rapidly over distances comparable with the intermolecular distance. $\mathbf{P}$ gives an average value of the molecular field, the average being taken over a volume which is large enough to contain many molecules, but small in comparison with smallest specimens for which we measure ϵ or any similar property. In our present calculation, it is important that a volume $(0{\cdot}1\lambda)^3$ is in fact large enough to contain many molecules. The vector $\mathbf{E}$ does not vary much over a distance comparable with the intermolecular distance, and we are justified in using an average value of P in equation 15(39) to calculate ϵ and $\boldsymbol{n}$. In calculating the force acting on a single electron, we are concerned with the local force due to the molecular dipoles, which we shall denote by $\mathbf{X}$.

15.20.—Suppose that there are $\mathcal{N}$ oscillators per unit volume and that each one may be represented by an electron which is controlled by an elastic restoring force and a small damping force. The latter is

* Slices of this thickness are used for microscopic examination of the structure of coal.

† § 13.6 and equation 13(15b).

assumed to be proportional to the velocity. The equation of motion is

$$m\ddot{\mathbf{r}} + g\dot{\mathbf{r}} + k\mathbf{r} = \mathbf{X}e + \mathbf{E}e. \quad . \quad . \quad . \quad 15(40)$$

m is the electronic mass and e is the charge, g and k are constants whose values should—eventually—be calculated from molecular theory, $\mathbf{r}$ is the displacement from a mean position. We may divide the field $\mathbf{X}$ (due to the dipoles) into two parts: (*a*) the contribution due to the dipoles in a small sphere (whose radius is of the order $0{\cdot}1\lambda$) surrounding the electron, and (*b*) the contribution due to the rest of the medium. The electric force $\mathbf{E}$, and hence the strengths of the induced dipoles, do not vary appreciably over the sphere. In an isotropic medium the dipoles in the sphere are arranged so that their field at the centre is zero.* The field due to the material outside the sphere may be shown † to be $(4\pi/3)\mathbf{P}$. This force may be regarded as due to surface charges on the cavity left when the sphere is removed. In an isotropic medium $\mathbf{E}$, $\mathbf{P}$, $\mathbf{X}$ and the displacement $\mathbf{r}$ are all in the same direction. We may therefore use the magnitudes of the vectors and rewrite equation 15(40)

$$m\ddot{r} + g\dot{r} + kr = \left(E + \frac{4\pi}{3}P\right)e. \quad . \quad . \quad 15(41)$$

The introduction of the term $(4\pi/3)P$ instead of P is known as the Lorentz-Lorenz correction after H. A. Lorentz and L. Lorenz.

15.21.—Consider first the free motion of the electron oscillator in the absence of a field. Putting E (and therefore P) equal to zero, we obtain an equation whose solution is

$$r = r_0 \exp\left(-\tfrac{1}{2}\gamma t + i\omega_0 t\right), \quad . \quad . \quad . \quad 15(42)$$

where

$$\gamma = \frac{g}{m} \quad \text{and} \quad \omega_0^2 = \frac{k}{m} - \frac{g^2}{4m^2}. \quad . \quad . \quad . \quad 15(43)$$

When the damping is small, we may put $\omega_0^2 = k/m$, i.e. we assume that the free period is equal to that of an undamped oscillator. Using these constants, 15(41) may be written

$$\ddot{r} + \gamma\dot{r} + \omega_0^2 r = \left(E + \frac{4\pi}{3}P\right)\frac{e}{m}. \quad . \quad . \quad 15(44)$$

* In an amorphous medium the molecules are arranged perfectly irregularly, and in a cubic crystal they are arranged symmetrically about the central point. In either case the resultant field at the centre is zero provided **E** does not vary appreciably from one side to the other.

† See p. 270 of Reference 15.7.

Now the polarization P is equal to the dipole moment per unit volume, so that *

$$P = \mathcal{N}er. \qquad 15(45)$$

Multiplying each side of 15(44) by $\mathcal{N}e$ and using 15(45), we obtain

$$\ddot{P} + \gamma\dot{P} + \omega_0^2 P = \left(E + \frac{4\pi}{3}P\right)\frac{\mathcal{N}e^2}{m}. \qquad 15(46)$$

We assume that P has the same frequency as E, but not necessarily the same phase, and write

$$E = E_0 e^{i\omega t}, \qquad 15(47)$$

$$P = \alpha E_0 e^{i\omega t}, \qquad 15(48)$$

where α is a constant which is in general complex. Substituting from 15(47) and 15(48) in 15(46), we obtain

$$\alpha(-\omega^2 + i\gamma\omega + \omega_0^2) = \frac{\mathcal{N}e^2}{m}\left(1 + \frac{4\pi}{3}\alpha\right), \qquad 15(49)$$

and hence, using 15(39),

$$\boldsymbol{n}^2 - 1 = 4\pi\alpha = 4\pi\frac{P}{E} = \frac{4\pi\mathcal{N}e^2}{m}\,\frac{1}{\omega_0^2 - \omega^2 - \dfrac{4\pi}{3}\dfrac{\mathcal{N}e^2}{m} + i\gamma\omega}. \qquad 15(50)$$

From 15(50) we obtain

$$\frac{\boldsymbol{n}^2 - 1}{\boldsymbol{n}^2 + 2} = \frac{4\pi\alpha}{4\pi\alpha + 3} = \frac{4\pi}{3}\frac{\mathcal{N}e^2}{m}\,\frac{1}{\omega_0^2 - \omega^2 + i\gamma\omega}. \qquad 15(51)$$

15.22.—Now consider a medium containing in unit volume N molecules each with f_1 oscillators whose constants are ω_1 and γ_1, f_2 whose constants are ω_2 and γ_2, and so on. Then P in 15(45) is now equal to $\sum_s P_s$, where

$$P_s = Nef_s r_s = \alpha_s E_0 e^{i\omega t} \qquad 15(52a)$$

and

$$\alpha = P/E = \sum_s \alpha_s. \qquad 15(52b)$$

On the left of 15(44) r is replaced by r_s, but P (*not* P_s) remains on the right because the polarization due to all types of oscillators acts on any one electron. Equation 15(49) is replaced by

$$\alpha_s = \frac{Ne^2}{m}\left(1 + \frac{4\pi}{3}\alpha\right)\frac{f_s}{\omega_s^2 - \omega^2 + i\gamma_s\omega}.$$

Summing the effects of all oscillators we obtain, instead of 15(51),

$$\frac{\boldsymbol{n}^2 - 1}{\boldsymbol{n}^2 + 2} = \frac{4\pi}{3}\frac{Ne^2}{m}\sum_s \frac{f_s}{\omega_s^2 - \omega^2 + i\gamma_s\omega}. \qquad 15(53)$$

* See p. 267 of Reference 15.7.

Let us define a new constant given by

$$\omega_s'^2 = \omega_s^2 - \frac{4\pi}{3}\frac{Ne^2}{m}f_s \quad . \quad . \quad . \quad . \quad . \quad 15(54)$$

Then 15(50) is replaced by

$$\boldsymbol{n}^2 - 1 = 4\pi\frac{Ne^2}{m}\sum_s \frac{f_s}{\omega_s'^2 - \omega^2 + i\gamma_s\omega}. \quad . \quad 15(55)$$

Note that ω_s' appears in 15(55) and ω_s in 15(53). The following expression for the dielectric constant (measured with static or slowly varying fields) is obtained by putting $\omega = 0$ in 15(55):

$$\epsilon - 1 = 4\pi\frac{Ne^2}{m}\sum_s \frac{f_s}{\omega_s'^2}. \quad . \quad . \quad . \quad . \quad 15(56)$$

On classical theory we should expect the f's to be integers, but it is found that the values which have to be inserted in 15(55) in order to agree with experiment are nearly always less than unity, and are sometimes small compared with unity (see Table 18.1, p. 601). This requires a change in the interpretation of the f's. We now take the electron oscillator considered in §§ 15.18–21 as an oscillator of unit strength. Each molecule is then regarded as containing one oscillator of strength f_1, one of strength f_2 and so on. The strength of an oscillator is proportional to its dipole moment (see Appendix XIII B, p. 413).

This change of interpretation makes no difference to the equations of dispersion [15(58) etc.] because f_s and N enter only as the product Nf_s. In the theory of scattering (§§ 15.41–46) the actual number of oscillators is important, and it is necessary to assume one oscillator of each type per molecule in order to obtain formulæ in agreement with experiment. The concept of an oscillator whose strength is expressed as a fraction of the strength of the classical theory oscillator plays an important part in the discussion of the relation between the classical and quantum theories of dispersion. It affects also a group of related problems, including the emission, absorption, and scattering of light.

15.23.—Equation 15(55) may be regarded as the fundamental equation of classical dispersion theory. Slightly different forms are obtained by inserting approximations, or by minor changes in the basic assumptions. Some writers assume that the internal field is equal to $4\pi aP$, where a is a constant. They usually put $a = \frac{1}{3}$ at the end of the calculation and we have preferred to use this value from the beginning. In classical electromagnetic theory an accelerated electron emits radiation and suffers a damping force corresponding to the loss of energy. It is thus possible to calculate a value of γ for an oscillator whose natural period is ω_s (see Appendix XIII B). It is found that

$$\gamma_s = \frac{2\pi}{3}\frac{e^2}{mc^3}\omega_s^2. \quad . \quad . \quad . \quad . \quad . \quad . \quad . \quad 15(57)$$

The values of γ derived from this formula do *not* agree with those derived from absorption measurements. The classical radiation damping force is thus inadequate to account for the observations. The calculation of the frequencies and of the damping constants by quantum mechanics is discussed in Chapter XIX.

15.24. **Dispersion in Regions of Small Absorption.**

Let us consider the dispersion in regions of the spectrum for which the magnitude of $\omega_s'^2 - \omega^2$ is large compared with $\gamma_s\omega$. Then the absorption may be neglected and the real index is given by modifying 15(55). We have

$$n^2 - 1 = 4\pi \frac{Ne^2}{m} \sum_s \frac{f_s}{\omega_s'^2 - \omega^2}, \quad . \quad . \quad . \quad 15(58)$$

and this equation may be written

$$n^2 = 1 + \frac{Ne^2}{\pi c^2 m} \sum_s \frac{f_s \lambda^2 \lambda_s'^2}{\lambda^2 - \lambda_s'^2}, \quad . \quad . \quad . \quad 15(59)$$

where

$$\lambda_s' = \frac{2\pi c}{\omega_s'}.$$

If we put

$$a_s = \frac{Ne^2}{\pi c^2 m} f_s \lambda_s'^2, \quad . \quad . \quad . \quad . \quad . \quad . \quad . \quad 15(60)$$

we obtain

$$n^2 = 1 + \sum_s \frac{a_s \lambda^2}{\lambda^2 - \lambda_s'^2}. \quad . \quad . \quad . \quad . \quad 15(61)$$

This is known as Sellmeier's dispersion formula. It may be written

$$n^2 = c_0 + \sum_s \frac{c_s}{\lambda^2 - \lambda_s'^2}, \quad . \quad . \quad . \quad . \quad 15(62)$$

where

$$c_0 = 1 + \sum_s a_s,$$

$$c_s = a_s \lambda_s'^2.$$

15.25. **Dispersion of Gases in Regions remote from Absorption Lines.**

For a gas N is small enough to make ω_s' nearly equal to ω_s and, when n is only a little greater than 1, we may put $n^2 - 1 = 2(n - 1)$. Then 15(58) reduces to

$$n - 1 = 2\pi \frac{Ne^2}{m} \sum_s \frac{f_s}{\omega_s^2 - \omega^2}, \quad . \quad . \quad . \quad . \quad 15(63)$$

indicating that, at low pressures, the refractivity for a given gas should be proportional to N, i.e. to the density. This result is confirmed by experiment. It has also been shown that for the rare gases, the empirical dispersion curve is fitted by equation 15(63) if only one term is used. This would appear to indicate that only one type of oscillator is concerned, and we should then expect that the value of λ_s found experimentally would agree with the value obtained from the wavelength of some dominant line in the spectrum. It is found, however, that the wavelengths of all the lines in the principal series are greater than the value indicated by the dispersion formula. This discrepancy finds no immediate explanation in the classical dispersion theory.* As would be expected, the value of $n^2 - 1$ obtained by putting $\omega = 0$ in the dispersion formula agrees well with the experimental value of $\varepsilon - 1$.

15.26. **Molecular Refractivity.**

Let us now return to equation 15(53) and insert the value

$$N = N_0 \frac{\rho}{M}, \qquad 15(64)$$

where N_0 is Avogadro's number ($6{\cdot}02 \times 10^{23}$), M is the molecular weight and ρ the density. For a region where the absorption may be neglected the " molecular refractivity " $[n]$ is given by

$$[n] = \frac{n^2 - 1}{n^2 + 2} \frac{M}{\rho} = \frac{4\pi}{3} \frac{N_0 e^2}{m} \sum_s \frac{f_s}{\omega_s'^2 - \omega^2}. \qquad 15(65)$$

From 15(65) we should expect that the refractivity of a mixture of substances which do not interact would be given by the formula

$$\frac{n^2 - 1}{n^2 + 2} \frac{\Delta}{\rho} = \frac{n_1^2 - 1}{n_1^2 + 2} \frac{\Delta_1}{\rho_1} + \frac{n_2^2 - 1}{n_2^2 + 2} \frac{\Delta_2}{\rho_2} + \ldots, \qquad 15(66)$$

where Δ grammes is the mass of a mixture which contains Δ_1 grammes of a substance whose density is ρ_1 and whose index is n_1, etc. This rule is confirmed by many experimental results. When a substance is formed by the combination of atoms, molecules, or molecular ions, it is reasonable to expect a dispersion formula containing terms derived from the constituents and some extra terms which belong to the electrons which form the linkage. This second rather more vague rule is also confirmed, and it is found that in certain groups of organic compounds the refractivity can be calculated by assigning a refractivity to a given type of bond and adding this to the contributions derived from the components.† There are other classes of compounds in which

* See § 19.11 for the quantum theory discussion of this matter.

† Reference 15.8.

the rearrangement of electrons resulting from the combination has modified the structure of the units so much that little trace of the original refractivities remains.

15.27. Region of Absorption.

To deal with a region near to one of the natural frequencies of the oscillators we return to equation 15(55) and write it in the form

$$\mathbf{n}^2 = n_0^{\,2} + 4\pi \frac{Ne^2}{m} \cdot \frac{f_s}{\omega_s'^2 - \omega^2 + i\gamma_s\omega}. \quad . \quad 15(67)$$

In writing this equation we assume that the oscillators other than the sth oscillator contribute nothing to the absorption and a constant amount to the real part of the complex index. This real part is included in n_0. An expression of this type may be expected to hold for a narrow range of frequency near the frequency of the sth oscillator. Such a formula is useful in dealing with the typical narrow absorption lines of vapours like sodium or mercury at low pressure. For small ranges round ω_s' we may put

$$\omega_s'^2 - \omega^2 = 2\omega_s'(\omega_s' - \omega)$$

and 15(67) then becomes

$$\mathbf{n}^2 - n_0^{\,2} = \frac{4\pi Ne^2}{m\omega_s'} \frac{f_s}{2(\omega_s' - \omega) + i\gamma_s}, \quad . \quad . \quad 15(68)$$

since ω is approximately equal to ω_s'.

Equating real and imaginary parts we obtain

$$n^2\varkappa = \frac{2\pi Ne^2}{m\omega_s'} \frac{f_s\gamma_s}{4(\omega_s' - \omega)^2 + \gamma_s^{\,2}}, \qquad 15(69a)$$

$$n^2 - n^2\varkappa^2 - n_0^{\,2} = \frac{8\pi Ne^2}{m\omega_s'} \frac{f_s(\omega_s' - \omega)}{4(\omega_s' - \omega)^2 + \gamma_s^{\,2}}. \qquad 15(69b)$$

If $(n - n_0)$ and $n\varkappa$ are both small compared with n_0, the left-hand side of 15(69b) is approximately equal to $2n_0(n - n_0)$ and we have

$$n = n_0 + \frac{4\pi Ne^2}{n_0 m\omega_s'} \frac{f_s(\omega_s' - \omega)}{4(\omega_s' - \omega)^2 + \gamma_s^{\,2}}. \quad . \quad . \quad 15(70)$$

In fig. 15.5 we show the variation of n and of $n\varkappa$ with ω for the case when $\omega_s' = 7{\cdot}4 \times 10^{15}$ sec.$^{-1}$; $\gamma_s = 3{\cdot}3 \times 10^{11}$ sec.$^{-1}$ and $N_s f_s = 9{\cdot}2 \times 10^{16}$ cm.$^{-3}$ We see that n varies between 1·03 and 0·97. This is typical for a region of absorption in a gas at fairly low pressure.* In the figure we show the variation of $n\varkappa$ although 15(69a) gives $n^2\varkappa$. Since n varies very little, the variation of $n\varkappa$ is nearly all due to variation of $\varkappa$, and there is little difference between $n\varkappa$ and $n^2\varkappa$. Note that in the region of absorption n decreases as ω increases. This is the so-called " anomalous " dispersion. The absorption coefficient 2α (§ 15.5) is equal to $2\omega n\varkappa/c$, i.e. we have

$$2\alpha = \frac{4\pi Ne^2\omega}{mcn\omega_s'}\frac{f_s\gamma_s}{4(\omega_s' - \omega)^2 + \gamma_s^2}. \quad . \quad . \quad 15(71)$$

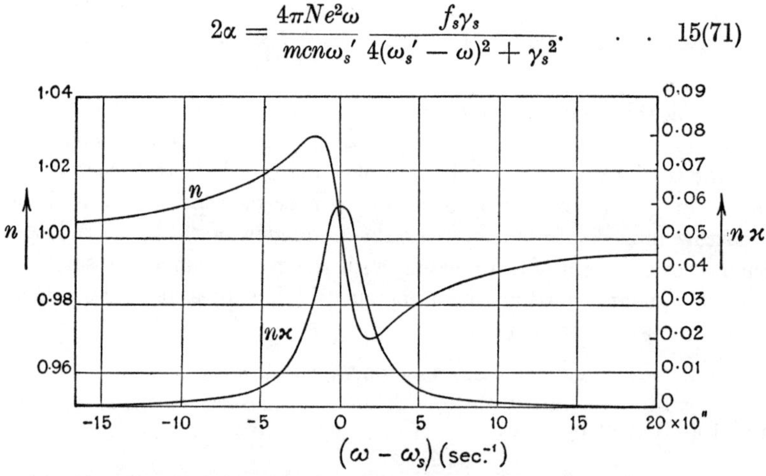

Fig. 15.5.—Variation of n and $n\varkappa$ in the region of a weak absorption line

Under many conditions of practical importance the line is so narrow that we may put $\omega = \omega_s'$ in the numerator, and so weak that we may put $n = 1$ in the denominator; thus we obtain

$$(2\alpha)_{\max} = \frac{4\pi Ne^2 f_s}{mc\gamma_s}. \quad . \quad . \quad . \quad . \quad 15(72)$$

and

$$\int_0^\infty 2\alpha\, d\omega = \frac{2\pi^2 Ne^2 f_s}{mc}. \quad . \quad . \quad . \quad . \quad 15(73a)$$

Note that limits 0 to ∞ for ω imply effective limits $-\infty$ to ∞ for $\omega_s' - \omega$ since ω_s/γ_s is large.

$$\int_0^\infty 2\alpha\, d\nu = \frac{\pi Ne^2 f_s}{mc}. \quad . \quad . \quad . \quad . \quad . \quad 15(73b)$$

15.28.—The maximum absorption is directly proportional to f_s and inversely to γ_s, and the half-width is equal to γ_s. The integral is

* In the regions remote from absorption lines, the indices of gases differ from unity by only a few parts in a thousand at one atmosphere pressure.

independent of γ_s and proportional to f_s. The conditions stated above for the validity of 15(70) imply that 15(71) and 15(72) are valid only in relation to weak absorption lines. For strong lines more complicated formulæ are derived by solving 15(67) to a higher degree of approximation. The absorption maximum does not then coincide exactly with ω_s'. Neither $(n - n_0)$ nor $n\kappa$ are symmetrical functions of $(\omega_s' - \omega)$ when the absorption is very strong.

15.29. **Measurement of the f-value.**

The determination of the value of f_s associated with a given natural frequency is now of considerable theoretical interest (§ 17.40). If the absorption coefficient of a gas or vapour at known pressure can be measured as a function of the wavelength, it is possible to calculate f_s from 15(73). In practice, absorption lines are broadened by the Doppler effect and by collisions (§ 4.25). When only the Doppler effect is operative the absorption is the same as that which would be given by a number of stationary atoms having slightly different values of ω_s but the same value of f_s. The integrated absorption coefficient for each of these atoms is the same, and therefore the value of f_s calculated by taking the integrated absorption coefficient for the broadened line is the same as that given by any one of the narrow lines. Investigation has shown that broadening of the 2537 Å. line of mercury by collisions with inactive gases (such as argon) has the same effect as an increase of γ_s when the pressure is fairly low (up to about a quarter of an atmosphere). When the pressure is increased to several atmospheres the value of f_s is increased by a few per cent. Measurements on the absorption of lines subject to pressure broadening may be used to give the value of f_s by measuring $\int_0^\infty 2\alpha\, d\nu$ for different pressures and extrapolating to zero pressure.

15.30.—Measurements on the dispersion for wavelengths near to the region of strong absorption may also be used to derive f-values. The most elegant and accurate method is the "hook" method of Roschdestvenski.* A Jamin interferometer is used and the fringes are projected to the slit of a spectrograph so that in the absence of the compensating plate the spectrum appears as shown in fig. 15.6a. With a suitably adjusted plate, the appearance is that of fig. 15.6b. The f-value can be calculated from a measurement of the difference between the wavelengths at the loops of the hook.

* Reference 15.9.

We assume that the hooks occur at points sufficiently far from the centre of the absorption line to make 15(59) rather than 15(70) the appropriate approximation for the refractive index, but sufficiently near to enable us to use only one term in the summation on the right-hand side of 15(59). This latter assumption is not justified when absorption lines are fairly close together, e.g. in the D lines of sodium. It is then necessary to make a correction to the formula given below [equation 15(75)]. Neglecting this correction we derive from 15(59)

$$n = 1 + \tfrac{1}{2}\,\frac{Ne^2}{\pi c^2 m}\cdot\frac{f_s\lambda^2\lambda_s'^2}{\lambda^2 - \lambda_s'^2}, \quad . \quad . \quad . \quad . \quad 15(74)$$

and hence

$$n - 1 = \frac{Ne^2}{4\pi c^2 m}\cdot\frac{f_s\lambda_s'^3}{\lambda - \lambda_s'}. \quad . \quad . \quad . \quad . \quad . \quad . \quad 15(75)$$

Let l be the length of the column of gas, and let l' and n' be the thickness and refractive index of the compensating plate. Each fringe corresponds to a definite phase difference between the two beams, i.e. to a definite order of interference.

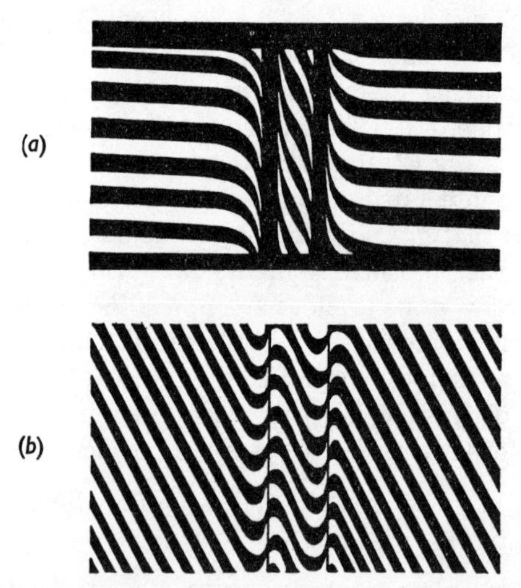

Fig. 15.6.—Appearance of fringes (*a*) without compensating plate, (*b*) with compensating plate suitably adjusted. The two sodium D lines are shown

Consider the fringe of order k. The path difference is made up of three parts: (i) a part $(n - 1)l$ due to the gas, (ii) a part $(n' - 1)l'$ due to the compensator, and (iii) a part due to inequalities in the end-plates of the tubes and the blocks in which the reflections occur. In the absence of the gas and the compensator, the fringes are uniformly spaced along the slit. We may therefore take the distance (y) along the slit (at a given place in the spectrum) to be proportional to (iii). We may therefore write

$$cy - (n - 1)l + (n' - 1)l' = k\lambda. \quad . \quad . \quad . \quad . \quad 15(76)$$

Usually the value of y corresponding to zero path difference (in the absence of the gas and the compensator) is off the slit, but this does not matter since we are concerned only with differences of y. The loops of the hooks are points at which y is a maximum or a minimum. From 15(76) we may find $dy/d\lambda$ and hence the values of λ corresponding to the loops. Calling these values λ_+ and λ_-, we have

$$f_s = \frac{\pi mc^2}{Ne^2\lambda_s'^3}\frac{K(\lambda_+ - \lambda_-)^2}{l}, \qquad 15(77)$$

where

$$K = k - l'\frac{dn'}{d\lambda}. \qquad 15(78)$$

In deriving 15(77) we assume that $dn'/d\lambda$ is constant over the region between the loops. K is found by removing the gas. The positions of the fringes are then given by

$$cy + (n' - 1)l' = k\lambda, \qquad 15(79)$$

and, if y is kept fixed, k and λ vary in such a way that

$$K\,\Delta\lambda = -\lambda\,\Delta k, \text{ or } K = -\lambda\frac{dk}{d\lambda}. \qquad 15(80)$$

K is thus obtained by counting the number of fringes (crossing a horizontal line placed over the spectrum) in a given wavelength interval.

15.31. **Absorption in Liquids and Solids.**

Molecules in a liquid are continually in collision with their neighbours. We should therefore expect their absorption and dispersion to be similar to that produced by gas molecules at very high pressure, i.e. we should expect the absorption lines to be very much broadened

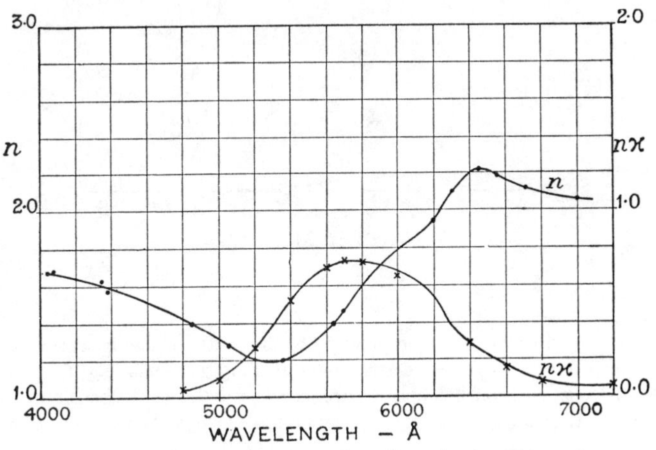

Fig. 15.7.—Absorption and anomalous dispersion in solid cyanin

and the dispersion curve to be affected in a similar way. The absorption and dispersion curves given by liquids may be regarded as due to a pronounced distortion of a series of absorption " lines " which are broadened so that they overlap. Coloured glasses and similar amorphous solids may usually be regarded as supercooled solutions of materials which absorb light. The absorption is due to molecules which are placed at varying distances from, and are therefore subject to varying amounts of distortion by, their neighbours. Similar effects are obtained with crystalline solids (fig. 15.7). In these solids the atoms are placed at regular distances from one another and there arises the possibility of another effect, i.e. sharply defined regions of absorption, characteristic of certain groupings. The most interesting absorption of this type usually occurs in the infra-red region of the spectrum.

15.32. The " Reststrahlen ".

We have seen from § 15.7 and equation 15(17) that a high absorption coefficient should imply a strong reflecting power. This effect has been shown to exist in gases by R. W. Wood, who found that mercury vapour strongly reflects

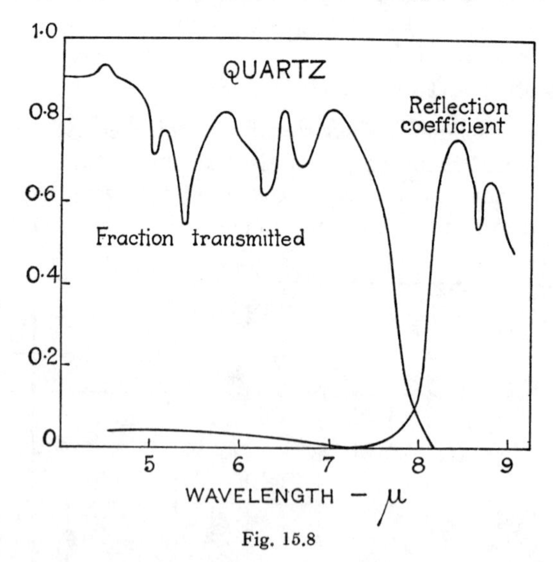

Fig. 15.8

light of wavelength 2537 Å., corresponding to the centre of an absorption line. It has also been found with certain strongly absorbing dyestuffs, e.g. fuchsin. For crystals, the phenomenon has been investigated in detail by Rubens and Nichols and later by other workers.* A typical reflection curve is shown in fig. 15.8. The

* Reference 15.2.

material has a very high reflection coefficient for certain sharply defined spectral regions. By using thin plates and thin films it is possible to show that for some of these regions the absorption and dispersion are related to the reflection in the way predicted by theory. In other regions the absorption is so high that it is difficult to measure the dispersion, though the resonance frequency may be determined by the reflection measurements. The relation between the resonant frequencies and the molecular structure will be considered later. The selective reflection may be employed to isolate radiation of certain frequencies, using the arrangement shown in fig. 15.4. The *residual rays* or *reststrahlen* produced by three or four successive reflections from the same kind of crystal contain a small number of well-defined frequencies. One of these may be separated from the others by passing the reflected radiation through an appropriate filter.

15.33. Dispersion Formulæ for Metals.

In §§ 15.13 to 15.17 we obtained formulæ giving the complex index for a metal as a function of the circular frequency. The metal was regarded as a continuous medium with a dielectric constant (ϵ) and a conductivity (σ). The formulæ obtained agree with experimental measurements of the reflecting power of metals in the infra-red, but are not in good agreement with measurements of n and $\varkappa$ in the visible region of the spectrum. We shall now discuss attempts to improve the theory by introducing the concept of the metal as an assembly of atoms which are fixed, and of electrons which move freely between the atoms. This idea was developed soon after the discovery of the electron by Lorentz, Drude and others. Let us first consider it in its application to the conduction of electricity for steady fields. Suppose that there are N atoms per unit volume and f_e *free* electrons per atom. We shall expect f_e to be of the order of magnitude of unity. In the absence of an applied field, the electrons have random motions. When a field (E) is applied, each free electron has an acceleration Ee/m. A drift velocity is superposed on the random motion. The extra velocity of a particular electron increases until it makes a collision with an atom. A collision destroys the extra component of velocity in the direction of the field and restores random motion. If the time between collisions is 2τ, then the velocity component (due to the field) is $2\tau Ee/m$ immediately before a collision. The time-average velocity is $\tau Ee/m$. The current is $Nf_e e$ times this velocity, i.e. it is $Nf_e e^2\tau E/m$. Thus the theory predicts that the current should be proportional to the field and gives for the constant of proportionality

$$\sigma = \frac{\text{current}}{\text{field}} = \frac{Nf_e e^2\tau}{m}. \qquad \qquad 15(81)$$

The constant τ is called the relaxation time. It is reasonable to expect that when the period (T) of an incident electromagnetic wave is *large* compared with τ, the collisions will have the same effect as for a steady field, i.e. the effective conductivity will be the same as for a steady field. We therefore expect 15(81) to be valid at low frequencies but to fail when T is of the same order as τ or is less than τ.

15.34.—Let us now approach the problem in a rather different way. We regard the metal as an assembly of electrons, some elastically bound like the electrons in insulators and some which are free. We apply the theory of dielectric dispersion to the bound electrons and include the free electrons by introducing a term corresponding to electrons whose "natural frequency" is zero. We also make one important change. We assume * that, owing to the presence of the free electrons, the local field in a metal is effectively the same as E instead of being equal to $[E + (4\pi/3)P]$. Equation 15(50) becomes, on substituting Nf_e for $\mathcal{N}$,

$$\boldsymbol{n}^2 - 1 = \frac{4\pi N f_e e^2}{m} \frac{1}{i\gamma_e\omega - \omega^2} \quad . \quad . \quad . \quad 15(82)$$

when we consider only the free electrons. The term $(4\pi/3)(\mathcal{N}e^2/m)$ in the denominator has disappeared because we make the local field equal to E. The complete equation contains terms due to free electrons and terms due to bound electrons, i.e. we have

$$\boldsymbol{n}^2 - 1 = \frac{4\pi N e^2}{m}\left[\frac{f_e}{i\gamma_e\omega - \omega^2} + \Sigma \frac{f_s}{\omega_s{}^2 - \omega^2 + i\gamma_s\omega}\right]. \quad 15(83)$$

15.35.—Let us now consider the dispersion at low frequencies and, for the present, neglect the effect of bound electrons. When ω is small compared with γ_e, equation 15(82) becomes

$$\boldsymbol{n}^2 - 1 = \frac{4\pi N f_e e^2}{i m \gamma_e \omega}. \quad . \quad . \quad . \quad . \quad 15(84a)$$

Also when we put $\epsilon = 1$ (in accordance with our decision to neglect the effect of bound electrons) in 15(32), we obtain

$$\boldsymbol{n}^2 - 1 = \frac{4\pi\sigma}{i\omega}. \quad . \quad . \quad . \quad . \quad . \quad 15(84b)$$

* This assumption is not accepted by all writers but its discussion is outside the scope of this book.

We note that these two equations agree if

$$\gamma_e = \frac{Nf_e e^2}{m\sigma}, \qquad . \quad . \quad . \quad . \quad . \quad 15(85)$$

or, using 15(81), if

$$\gamma_e = \frac{1}{\tau}. \qquad . \quad . \quad . \quad . \quad . \quad . \quad 15(86)$$

To understand this relation, we observe that the dispersion theory introduces an undetermined damping constant γ_e to represent some dissipative action whose nature is unknown. In the discussion leading to 15(81) we assume that the relation between current and field is the same as that for steady fields. We know that under these conditions the resistance of the metal gives rise to the Joule heating effect. Equations 15(84*a*) and 15(84*b*) imply that if we give γ_e a certain value, then the dissipation of energy at low frequencies will be σE^2, i.e. our dispersion theory will, in the limit of low frequency, agree with ordinary equations—including the Joule heating effect. The correct value satisfies 15(85). In the simple electron theory of metallic conduction, the collisions of electrons with atoms is the method by which electrical energy is changed into heat. In order that the rate of change of electrical energy into heat may agree with the observed rate, τ must have the value given in 15(81), and hence the relation between τ and γ_e is that given by 15(86).

15.36.—In the preceding paragraph we applied the dispersion equation 15(82) to very low frequencies in order to discover what relation must be satisfied in order that it may agree with observations at these frequencies. Let us now consider the other extreme case and assume that, for certain high frequencies, γ_e $(= 1/\tau)$ may be neglected in comparison with ω, i.e. we assume that the period of the incident wave is so short that the electrons make enormous numbers of vibrations to and fro between collisions. In the limit, none of the energy of the high-frequency field is transformed into heat. We cannot expect this very simple picture to give *exactly* correct results under any conditions, but it does give one very important prediction which is in accord with observations. If we put $\gamma_e = 0$ in 15(82), we obtain

$$\boldsymbol{n}^2 = 1 - \frac{4\pi Nf_e e^2}{m\omega^2}. \qquad . \quad . \quad . \quad . \quad 15(87)$$

15.37.—According to this equation, there should be a transition wavelength for which $\boldsymbol{n} = 0$. For shorter wavelengths $\boldsymbol{n}$ is real but less than unity. The metal is transparent and n is less than unity, so total reflection occurs for angles of incidence above a certain critical angle. On the long wavelength side of the transition wavelength, $\mathbf{n}$ is a pure imaginary. No periodic wave can be propagated in the metal. The disturbance dies away exponentially from the boundary. The radiation is totally reflected at *all* angles of incidence, including normal

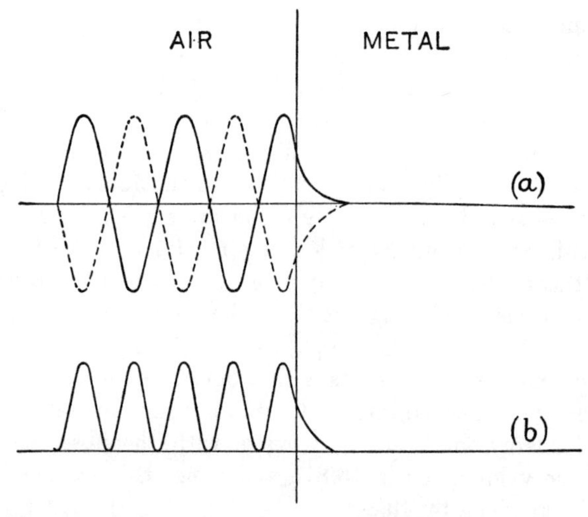

Fig. 15.9.—Reflection of light incident normally upon a metal surface

(*a*) shows the displacement at a certain moment (full line) and at half a period later (dotted line). (*b*) shows the variation of energy

incidence (fig. 15.9, and compare fig. 14.5). The reflection coefficient for normal incidence changes suddenly from nearly zero to unity at the transition wavelength. This extreme transition is not observed, but a very sudden change is obtained with the alkali metals (fig. 15.10). Table 15.4 shows the relation between the critical wavelengths observed and the values calculated from 15(87), assuming that $f_e = 1$.

TABLE 15.4

Metal	Cs	Rb	K	Na	Li
λ_0 in Å. (obs.)	4400	3600	3150	2100	2050
λ_0 in Å. (cal.)	3600	3200	2900	2100	1500

These wavelengths are far removed from those at which 15(37) and 15(84) are valid.

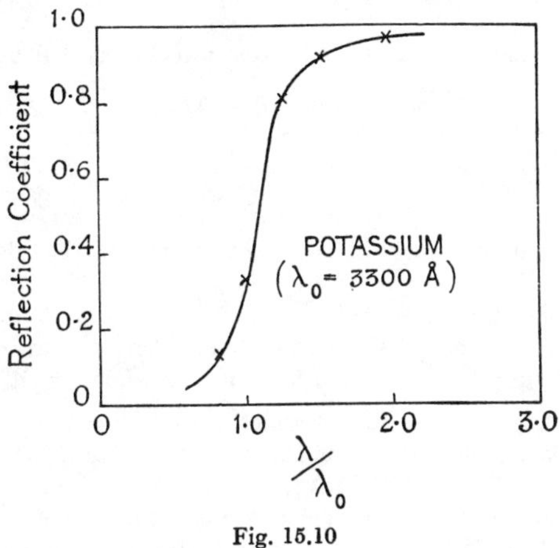

Fig. 15.10

EXAMPLE 15(vii)

Calculate the values of f_e which correspond to the observed values of transition wavelength given in Table 15.4.

[0·68 for Cs: 0·79 for Rb: 0·85 for K: 1·00 for Na: 0.54 for Li. Note that these values are all of order unity.*]

15.38.—In § 15.35 we derived an approximation to 15(82) which is valid at very long wavelengths, and in §§ 15.36 and 15.37 we derived another approximation which should apply at very short wavelengths. Let us now return to the original equation 15(82) and insert $1/\tau$ for γ_e to obtain

$$\boldsymbol{n}^2 - 1 = \left(\frac{4\pi N f_e e^2}{m\omega}\right) \Big/ \left(\frac{i}{\tau} - \omega\right). \quad . \quad . \quad 15(88)$$

Or, separating real and imaginary parts,

$$n^2(1 - \varkappa^2) = 1 - \frac{4\pi N f_e e^2}{m}\left(\omega^2 + \frac{1}{\tau^2}\right)^{-1}, \quad . \quad 15(89a)$$

$$n^2\varkappa = \frac{2\pi N f_e e^2}{m\omega\tau}\left(\omega^2 + \frac{1}{\tau^2}\right)^{-1}. \quad . \quad . \quad 15(89b)$$

* This calculation takes no account of the effect of polarization of the atomic cores. When this is allowed for (see p. 122 of Reference 15.10) the values of f_e become: 0·85 for Cs: 0·94 for Rb: 0·97 for K: 1·1 for Na: 0·55 for Li.

EXAMPLES [15(viii) and 15(ix)]

15(viii). Show that, if λ_0 is the transition wavelength defined in §15.37, then

$$n^2(1 - \varkappa^2) = 1 - (\lambda/\lambda_0)^2, \quad \ldots \quad 15(90a)$$

$$2\pi n^2\varkappa = \frac{\lambda^3}{2c\tau\lambda_0^2}, \quad \ldots \quad 15(90b)$$

is an approximation to 15(89). Under what conditions is this approximation valid?

[It is valid when we neglect γ_e^2 in comparison with ω^2. Note that this is a less drastic assumption than that used in § 15.36 where we put $\gamma_e = 0$ at the beginning, i.e. this approximation should be valid at moderately short wavelengths.]

15(ix). If λ_τ is the wavelength for which $\omega = 2\pi/\tau = 2\pi\gamma_e$, calculate λ_τ for (*a*) silver at 0° C., (*b*) platinum at 0° C., (*c*) platinum at 1000° C. Take $N = 1{\cdot}475 \times 10^{22}$ and $\sigma = 6{\cdot}13 \times 10^{17}$ for Ag; $N = 1{\cdot}667 \times 10^{22}$ and $\sigma = 0{\cdot}821 \times 10^{17}$ for Pt at 0° C., and $N = 1{\cdot}534 \times 10^{22}$ and $\sigma = 0{\cdot}191 \times 10^{17}$ for Pt at 1000° C. Take $f_e = 1$ for all cases. σ is given in e.s.u. [(*a*) 49 μ, (*b*) 5·8 μ, (*c*) 1·5 μ.]

15.39.—Equation 15(89) is the exact formula for the contribution of free electrons to metallic absorption and dispersion. It represents the observed results for mercury in the visible spectrum very well

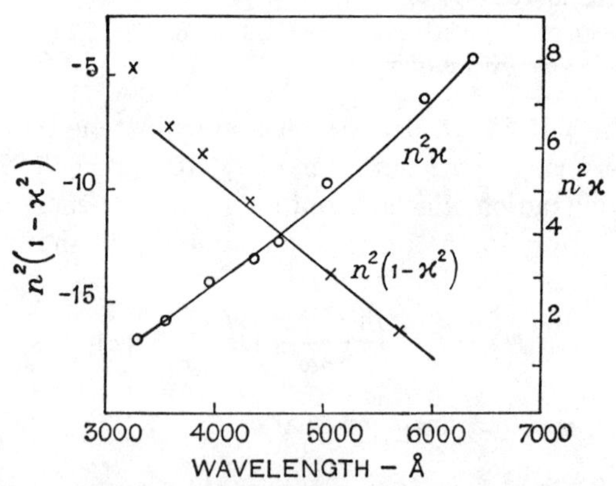

Fig. 15.11.—Absorption data for liquid mercury. Full lines calculated from equation 15(89). Points represent experimental results

(fig. 15.11). For most other metals the absorption calculated from 15(89) is considerably smaller than the observed values. Also the variation with temperature is not in accord with the temperature coefficient of σ. This suggests that the main absorption in the visible region by metals like silver and copper is due to bound electrons. It has been found possible to represent the observations for most metals by means of equation 15(83), taking four terms in the summation term as well as a term for free electrons. This gives three adjustable constants (f_s, ω_s and γ_s) for each bound electron, and two for the free electrons, i.e. 14 constants in all to be adjusted for each metal. It is not surprising that a fair agreement with observations is obtained. A real test of agreement between theory and experiment can be made only when the theory can predict the natural frequencies, etc., for the bound electrons.

15.40.—One other defect of the simple free-electron theory should be mentioned. Even in the infra-red, where the effect of bound electrons may be neglected, there are certain difficulties. We should expect that in the far infra-red 15(84) would be adequate, and that as we approach λ_τ (i.e. $c\tau$) it would be necessary to use 15(89); on the short-wave side of λ_τ the approximation 15(90) would become adequate. Qualitatively this appears to be satisfactory, but the value of λ_τ and hence of γ_e ($=1/\tau$) obtained from the measurements in the near infra-red are lower than those predicted by the theory [equation 15(85)]. It is possible to avoid this discrepancy by saying that the conductivity in the metal surface is lower than that of the metal in bulk, or that the number of free electrons per atom is not nearly unity. Every suggestion of this type removes the local difficulty at the expense of creating a fresh difficulty somewhere else. For example, if we assume a special conductivity for the surface layer, the agreement between theory and experiment discussed in § 15.17 is invalidated. The difficulty is a fundamental one and is inherent in the classical electron theory. Thus, while emphasizing the success of that theory in dealing with results in the far infra-red and in describing the observations at the transition wavelength, we must recognize that it is not adequate to deal with the general problem of dispersion in metals.

15.41.—We shall consider the general quantum theory of dispersion later, but, as we do not intend to return to the dispersion of metals, it is convenient to state some of the more important results here. The quantum theory indicates that a metal should behave as an assembly of free electrons and of atoms which possess bound electrons. It accepts the general form of 15(83) as the dispersion equation but recognizes the possibility of certain special types of absorption (e.g. photo-electric absorption and internal transitions between the Brillouin zones). In the quantum theory the number of electrons which are effectively free is not necessarily the same at all frequencies. This gives at least the possibility of removing the difficulty stated in the preceding paragraph. *In principle*, the quantum theory is able to predict values for the constants f_s, γ_s, etc. In practice, the calculation is extremely difficult, but some progress has been made. No irreconcilable difficulty has appeared but a detailed quantitative comparison between theory and experiment is not possible.*

15.42. The Relation between Dispersion and Molecular Scattering.

The interactions of light with material media fall into two classes:

(*a*) those which depend on the mean number of atoms per unit volume (the mean being taken over a volume large enough to contain many atoms but whose linear dimensions are small compared with λ); and

(*b*) those which depend on local deviations from the mean.

Reflection, refraction and dispersion belong to class (*a*). They can be described in terms of relations between the vectors **P**, **E**, **D**, etc., which, by definition, represent mean values.† The scattering of light is the most important phenomenon of class (*b*). Calculations based on the vectors **P**, **E**, **D** are useful to show the continuity between the electromagnetic theory of light and the electromagnetic theory of static and slowly varying fields. These calculations cannot predict the existence of scattering since the local deviations are excluded at the beginning. We shall now consider a method of calculation in which we start with atoms as scattering centres and obtain equations giving the amount of light scattered. Then, at a later stage, we take an average resultant so as to obtain some of the fundamental equations connected with phenomena of class (*a*). It is important to note that we are now introducing a new mathematical procedure and not a new set of physical assumptions. If we take averages at the beginning we obtain equations for phenomena of class (*a*) in the most simple way, but we exclude class (*b*) from consideration. We are now to deal with class (*b*) before introducing the averaging process which gives the equations for class (*a*).

15.43.—Suppose a parallel beam of light is incident normally from vacuum upon the surface of a material medium (fig. 15.12). The incident beam is represented by

$$E_x = A \exp i(\omega t - \kappa z), \quad . \quad . \quad . \quad . \quad 15(91)$$

where $\omega/\kappa = c$. The direction of **E** is parallel to OX. The medium is now regarded as a system of electrical charges which form radiating dipoles under the action of the light.‡ We assume that the beam

* Reference 15.10. † See § 13.5.

‡ The properties of these dipoles are discussed in Appendix XIII B and we shall quote the appropriate results.

represented by 15(91) continues to advance through the medium but is modified by the spherical waves emitted by the dipoles. For the present the calculation is confined to transparent isotropic media. We assume that the dipoles are irregularly distributed in space and that their axes are parallel to the electric vector of the field by which they are created. We also assume that

(*a*) the amplitude at Q is proportional to A and to $\sin\chi$;

(*b*) there is a constant difference of phase (δ) between the phase of the incident wave at P and the phase of the emitted wave.

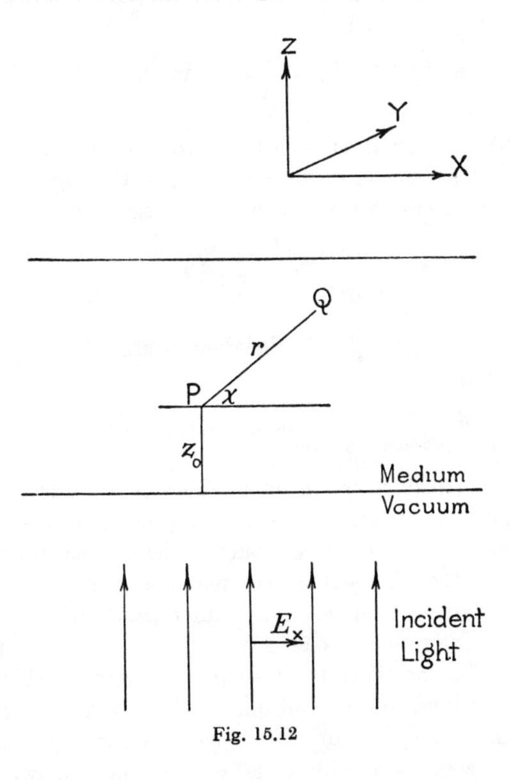

Fig. 15.12

These assumptions are reasonable on any wave theory. They are justified, in their application to electromagnetic waves, in Appendix XIII B. The wave emitted by one of the dipoles at P may then be represented at Q by

$$E(r, \chi) = \frac{\beta A}{r} \sin\chi \exp i(\omega t - \kappa z_0 - \kappa r - \delta), \qquad 15(92)$$

where β is a real constant, z_0 is the co-ordinate of the particular dipole, r is the distance PQ and χ is the angle between OX (the direction of the axis of the dipole) and PQ.

15.44.—Now consider the light scattered in some direction for which χ is not zero or π. Since the dipoles are irregularly distributed, the term $\kappa(z_0 + r)$ in the exponential introduces an irregularity of phase, and the waves from different dipoles are non-coherent. If there are N dipoles per unit volume, the energy (L_0) scattered in the direction χ per unit cross-section of the incident beam per unit solid angle and per unit path is

$$L_0 = \frac{\beta^2 A^2}{r^2} N \sin^2 \chi, \quad . \quad . \quad . \quad . \quad 15(93)$$

provided that N is small enough for us to neglect secondary scattering processes (i.e. the scattering by one atom of the radiation emitted by another). The energy scattered in all directions is *

$$k = \frac{8\pi}{3} \beta^2 N \quad . \quad . \quad . \quad . \quad . \quad . \quad . \quad 15(94)$$

per unit energy density in the incident beam. k is called the coefficient of scattering.

15.45. **Relation between k and μ.**

Now let us consider the effect of the waves scattered from a layer between z and $z + \Delta z$ inside the material medium, but fairly near to its surface, at some point Q′ in the path of the incident beam (fig. 15.13). Let us assume that Δz is small compared with λ but that there is a fairly large number of dipoles in a volume $(\Delta z)^3$. Then we may apply to the waves from the dipoles a calculation similar to that used in §§ 7.1–7.6 to " reconstruct the wavefront " from the Huygens wavelets. We may divide the wavefront into zones, and we find that the resultant amplitude due to all wavelets is equal to half that due to the first zone whose area is $\pi(z' - z_0)\lambda$ and which contains $\pi(z' - z_0)\lambda N \Delta z$ dipoles. In accordance with the discussion of § 7.3, we must multiply by $1/\pi$ and insert a phase difference $\frac{1}{2}\pi$. We also put $z' - z_0 = r$ since we are considering the forward direction. The resultant obtained from 15(92) using these assumptions is

$$E' = \beta \lambda N A \, \Delta z \exp i(\omega t - \kappa z' - \delta - \tfrac{1}{2}\pi). \quad . \quad 15(95)$$

* Appendix XIII B.

If δ is small, the phase of this wave is $\frac{1}{2}\pi$ behind that of the incident wave [represented by 15(91)] at Q. If the scattered wave is weak compared with the main wave, the phase of the resultant lags be-

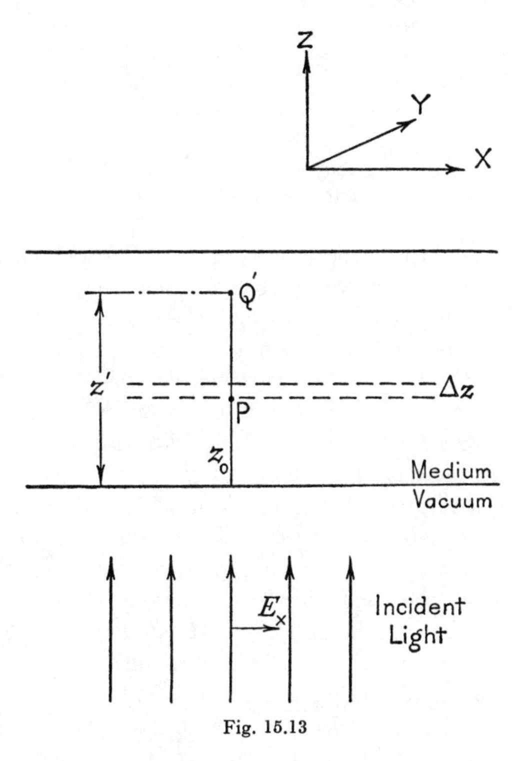

Fig. 15.13

hind that of the main wave by the angle $\beta\lambda N\,\Delta z$. But we know that the retardation due to a layer Δz of index n must be $\kappa(n-1)\,\Delta z$, and hence we have

$$2\pi(n-1) = \beta\lambda^2 N, \qquad . \quad . \quad . \quad . \quad 15(96)$$

or, using 15(94),

$$k = \frac{32\pi^3}{3\lambda^4 N}(n-1)^2. \qquad . \quad . \quad . \quad . \quad 15(97)$$

EXAMPLES [15(x) and 15(xi)]

15(x). Calculate the value of k (for $\lambda = 5000$ Å.),

(a) for air at S.T.P.,

(b) for He at 100 atmospheres, given that $(n - 1)$ for air is $2{\cdot}93 \times 10^{-4}$, and for He is $3{\cdot}6 \times 10^{-5}$ at S.T.P., and the number of molecules per c.c. at S.T.P. is $2{\cdot}7 \times 10^{19}$. What are the dimensions of k?

[(a) $1{\cdot}7 \times 10^{-7}$, (b) $2{\cdot}5 \times 10^{-9}$. $[k] = L^{-1}$.]

15(xi). Show that, for a given density, the amount of light scattered per unit volume of a gas is proportional to the volume.

15.46.—The relation 15(97) was derived by the elder Rayleigh who showed that it was valid both for the electromagnetic theory and for certain forms of elastic-solid theory. He also showed that the brightness of the blue sky was of the order of magnitude to be expected if the light is due to molecular scattering in the atmosphere. It is not necessary to take account of scattering by dust particles. The factor λ^{-4} in the expression for k accounts for the blue colour. The expression has also been verified by laboratory experiments on gases. The apparatus used is shown in fig. 15.14. A parallel beam of light is allowed to pass into the tube T. A series of diaphragms (each slightly larger than the preceding one) is used to prevent light scattered from the edges of the apparatus from emerging from the window W. For the same reason the horns H_1 and H_2 are used. If the surface of one of these horns is smooth, any light entering is reflected backwards and forwards along the horn, and there is only a very small return beam. This design provides a " dead-black " background against which the scattered light is observed. It is necessary to take these precautions and also to filter the gas very carefully to remove dust particles, because light scattered from pure gases is a very small fraction of the incident light [Example 15(x)]. Note that since N and $(n - 1)$ in 15(97) are both proportional to the density of the gas, it follows that k should be directly proportional to the density. It is possible to measure the polarization of the scattered light by allowing it to pass through a double image prism P and then estimating the relative energies of the two images. Observations of this type have been made by the younger Rayleigh, by Cabannes, by Raman and other workers.* It is found that the scattered light from the rare gases is at least 99·5 per cent polarized but that other gases show a much less complete polarization (e.g. 96 per cent for N_2 and 90

* See p. 384 of Reference 15.2 for a table of results and references to original papers.

per cent for CO_2). This result would be expected if the molecule possesses an intrinsic dipole moment. The defect of polarization may be correlated with other effects which depend on the polarity of the molecule.*

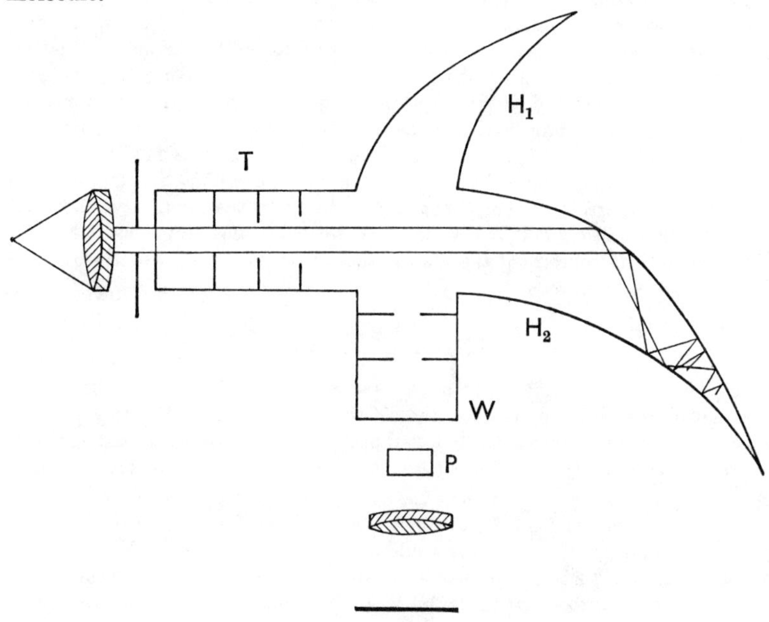

Fig. 15.14

15.47.—The results of Appendix XIII B enable us to calculate the constant k of 15(97). Substituting the value from 13(102) we obtain

$$n - 1 = 2\pi \frac{Ne^2}{m} \frac{f_s}{\omega_s^{\,2} - \omega^2},$$

which agrees with 15(63). The method of calculation which we introduced in § 15.42 is thus capable of giving the fundamental equations of dispersion theory.

It must be admitted that we have had to walk delicately in order to obtain this result. On the one hand the number of scattering centres in a volume λ^3 must be fairly large so that we can take the average in § 15.44. On the other hand, to avoid very lengthy mathematics, we have treated only the case when $(n - 1)$ is small. We have also not specified closely what we mean by the statement that the atoms are distributed at random. There is a range of gas pressures from about 10^{-3} to 10^3 atmospheres for which our approximations are justified. The relation

* Reference 15.2, p. 383.

15(97) between refractive index and scattering does not apply when the distance between the molecules becomes of the same order as the molecular diameter. The molecules are not then " distributed at random " even in an amorphous solid, because their separations cluster round a mean distance. The amount of scattered light then becomes very much less than the amount calculated from 15(97). It may be shown experimentally that the scattered light from good crystals is very small indeed, and the scattering decreases with temperature because the lattice becomes more nearly perfect when the thermal motions are reduced. It is fairly easy to see that once the effective velocity of propagation is given by a relation such as 15(96), other properties such as the laws of reflection and refraction and the amount of scattered light can be deduced. The type of calculation we have just discussed can be extended to absorbing media by assuming that the scattered wave has an appreciable component in opposition to the main wave, i.e. that δ in 15(92) is *not* negligible. It may also be extended to anisotropic media by suitable assumptions concerning the distribution of the scattering centres and the form of the scattered wave.*

EXAMPLE 15(xii)

If the atoms are sufficiently closely packed the scattered waves should be coherent in a direction opposite to that of the incident wave. Use 15(95) to calculate the amount of light in the reflected beam from a parallel-sided slab of thickness t and show that your result is consistent with the Fresnel relations (§ 14.7).

[The return scattered wave is the same as the forward wave for a thickness Δz. Integrating from $z = 0$ to t we find that the return wave is of zero amplitude if $2t = n\lambda$ (as would be expected in view of the phase change at one surface) and has a maximum amplitude of $(n - 1)A$ which agrees with equation 14(39) if we remember that $(n - 1)$ is small so that $(n + 1)$ is approximately equal to 2, and also that there are equal reflections from two surfaces.]

15.48. **Other Types of Scattering.**

Two types of scattering, due to atoms and molecules, are not included in the above discussion. They are (*a*) resonance radiation from a gas, and (*b*) the Raman effect. The resonance radiation, which is described in detail in § 17.14 is, under practical conditions, non-coherent with the primary beam (§ 19.13). The Raman radiation has not the same wavelength as the incident beam and cannot therefore be coherent with it, though it is possible that the Raman radiation from one atom may be coherent with that from another. In addition to scattering from atoms and molecules we have scattering from larger particles. The name " Tyndall effect " is given to this type of scattering. J. J. Thomson demonstrated (originally in connection with X-ray scattering) that when the size of the particle is large compared with λ, the scattering is independent of λ. More detailed theory is needed to deal with scattering by particles whose diameter is of the same order as λ. It is generally found that for a given kind of particle the variation of scattering with λ follows a curve of the general form shown in fig. 15.15, i.e. there is a flat portion where $\lambda \ll r$, a maximum for a wavelength near r, and a fall

* References 15.2 and 15.11.

(proportional to λ^{-4}) when $\lambda \gg r$. The theory of scattering by particles of the same order as, and larger than, the wavelength is of importance in connection with the transmission of light through haze and through colloidal solutions. Media which show large scattering of this type are called "turbid" media. The turbidity of photographic emulsion is an important factor limiting the quality of photo-

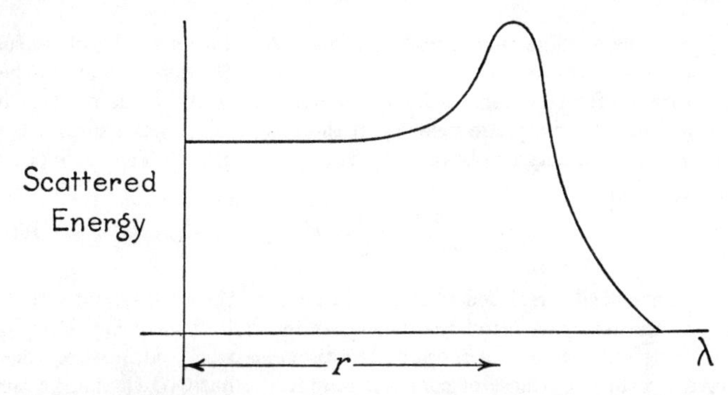

Fig. 15.15.—Typical relation between energy scattered and wavelength for particles of radius r. This is intended to show qualitatively the shape of the curve and does not represent the properties of any particular substance.

graphic images. Practical calculations on the transmission of radiation through turbid media are very complicated when it becomes necessary to take account of multiple scattering. The size of particles in a slight haze is of the same order as the wavelength of the light. The penetration by infra-red radiation is much better than the penetration by visible light. Consequently in a fairly clear atmosphere (visibility 5 to 10 miles) the quality of infra-red photographs of distant objects is much better than that of photographs by visible light. The particles present in a fog are much larger than the wavelengths of the near infra-red and for them the scattering in the near infra-red is nearly the same as that for visible light. The "penetration of fog" by infra-red is thus little better than the penetration by visible light.

REFERENCES

15.1. MÜLLER-POUILLETS: *Lehrbuch der Physik*, II, *Optik* (Vieweg).

15.2. BORN: *Optik* (Springer).

15.3 *Handbuch der Physik*, 1928, Vol. XX (Springer).

15.4. DRUDE: *Wied. Ann.*, 1890, Vol. 39, p. 481.

15.5. HAGEN and RUBENS: *Ann. der Phys.*, 1903, Vol. 11, p. 873.

15.6. SUTHERLAND: *Infra-red and Raman Spectra* (Methuen).

15.7. JOOS: *Theoretical Physics* (Blackie).

15.8. DENBIGH: *Trans. Far. Soc.*, 1940, Vol. 36, p. 936.

15.9. KORFF and BREIT: *Rev. Mod. Phys.*, 1932, Vol. 4, p. 482.

15.10. MOTT and JONES: *The Theory of the Properties of Metals and Alloys* (Oxford University Press, 1936).

15.11. DARWIN: *Trans. Camb. Phil. Soc.*, Vol. XXIII, p. 137.

APPENDIX XV A

The Refracted Wave in an Absorbing Medium

Let $P_1P_2P_3$ be a plane wave incident from a vacuum on to the plane surface of an absorbing medium. Let θ_1 be the angle of incidence, and, as before, take the plane xz for the plane of incidence and xy for the plane of the surface (fig. 15.16). Let b be the phase velocity in the medium. A wave surface $Q_1Q_2Q_3$ of the refracted wave may be obtained by Huygens' method. The angle of refraction θ_2 is given by

$$\frac{\sin\theta_1}{\sin\theta_2}=\frac{c}{b}=n'. \qquad 15(98)$$

This angle is essentially real and n' is a real number. We shall see later that it is not equal to the real part (n) of the " complex index ". The surface $Q_1Q_2Q_3$ is a surface of constant phase. In most of our previous work, surfaces of constant phase have also been surfaces of constant amplitude, but $Q_1Q_2Q_3$ is not a surface

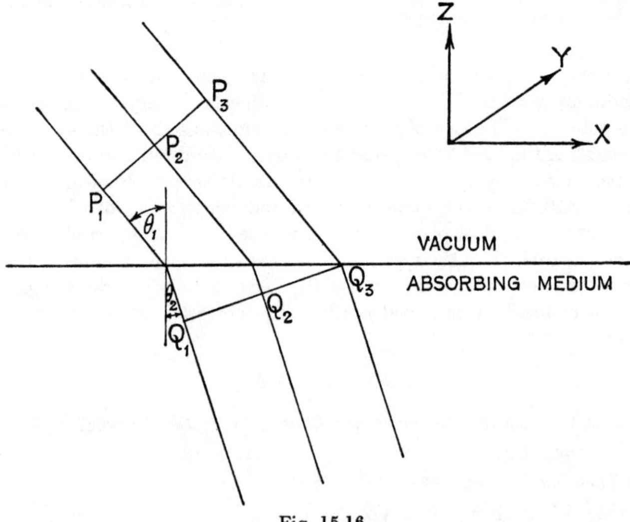

Fig. 15.16

of constant amplitude. The wave at Q_3 has just entered the medium and has suffered no absorption. That which has reached Q_1 has advanced a considerable distance in the absorbing medium and has been attenuated. It is reasonable to assume, as a working hypothesis, that the attenuation at different points is determined by the length of path (measured along the ray) in the absorbing medium. This implies that the surfaces of constant amplitude are parallel to the surface of the medium, i.e. that the amplitude at any point is proportional to $\exp(-n'\varkappa' z)$, where $\varkappa'$ is an extinction coefficient. We know that $\varkappa'$ will be related to $\varkappa$, but we cannot assume that it is equal to $\varkappa$ because the wave is not incident normally on the surface of separation.* We may then write

$$E_{2y} = A_{2y} \exp i\omega \left\{ t - \frac{n'}{c} (x \sin \theta_2 - z \cos \theta_2) - \frac{in'\varkappa' z}{c} \right\}. \qquad 15(99)$$

This expression must satisfy the wave equation 15(11), i.e. we must have

$$n'^2 \sin^2 \theta_2 + n'^2 (\cos \theta_2 - i\varkappa')^2 = \mathbf{n}^2. \quad . \quad . \quad . \quad 15(100)$$

Equating real and imaginary parts, we obtain

$$n'^2(1 - \varkappa'^2) = n^2(1 - \varkappa^2), \quad . \quad . \quad . \quad . \quad . \quad . \quad 15(101)$$

$$n'^2 \varkappa' \cos \theta_2 = n^2 \varkappa. \quad . \quad . \quad . \quad . \quad . \quad . \quad . \quad 15(102)$$

Equations 15(98), 15(101) and 15(102) may be regarded as three simultaneous equations from which n', $\varkappa'$ and θ_2 can be calculated when n and $\varkappa$ are given. Thus an expression whose form is that of 15(99) can satisfy the wave equation. Since 15(98) is satisfied, the boundary condition is also satisfied. From the three equations 15(98), 15(101) and 15(102) we see that n' depends on θ_1, i.e. Snell's law is not obeyed. The phase velocity in the absorbing medium depends on the angle between the surfaces of constant phase and the surfaces of constant amplitude. Hence the phase velocity varies with the direction of propagation although the medium is isotropic. It is, in principle, possible to measure θ_2, and hence n', using a metal prism of small angle. The method is very inaccurate because the angles involved are very small.

* See definition of $\varkappa$ in § 15.5. Note that the last term in 15(99) is negative because the wave is travelling in the negative direction of z.

CHAPTER XVI

Anisotropic Media

16.1. **Optical and Electrical Anisotropy.**

In Chapter XII we showed that the main properties of polarized light, including many of the phenomena associated with its transmission through crystals, could be described in terms of a general theory of transverse waves. Since the electromagnetic theory of light is a theory of transverse waves, it can take over most of the discussion of that chapter with only a few verbal changes. In this chapter we consider the propagation of light in anisotropic media in a more detailed way, and show formally that the electromagnetic theory can account for a wider range of observed phenomena. The equations of Maxwell are applied to media which are assumed to be electrically anisotropic and hence we deduce the equations of optical anisotropy. The advantages of the electromagnetic theory over an elastic-solid transverse wave theory are:

(*a*) the electromagnetic theory gives a more detailed account of the observed phenomena of natural optical anisotropy;

(*b*) it is able to include in its treatment anisotropy created by mechanical strains or by electric and magnetic fields;

(*c*) it is able to relate electrical anisotropy, and hence optical anisotropy, to molecular theories of matter, i.e. to the arrangements of atoms and molecules in crystal lattices, and in some cases to anisotropic properties of the molecules themselves.

16.2.—In the treatment of isotropic media the displacement **D** is related to the electric vector **E** by the relation $\mathbf{D} = \epsilon\mathbf{E}$, where ϵ is a scalar quantity, **D** and **E** being in the same direction. It is found that when an electric field **E** is established in an optically anisotropic medium, the direction of **D** does not in general coincide with that of **E**. The relation between **D** and **E** may be expressed by writing

$$\left.\begin{aligned} D_x &= \epsilon_{xx}E_x + \epsilon_{xy}E_y + \epsilon_{xz}E_z, \\ D_y &= \epsilon_{yx}E_x + \epsilon_{yy}E_y + \epsilon_{yz}E_z, \\ D_z &= \epsilon_{zx}E_x + \epsilon_{zy}E_y + \epsilon_{zz}E_z. \end{aligned}\right\} \quad . \quad . \quad . \quad . \quad 16(1)$$

We summarize 16(1) by writing

$$\mathbf{D} = \epsilon\mathbf{E}. \quad . \quad . \quad . \quad . \quad . \quad . \quad . \quad . \quad 16(2)$$

The quantity ϵ now belongs to the class of mathematical variables called tensors. Multiplication by a tensor changes the direction of a vector as well as its magnitude. The tensor ϵ has nine components. The form of equation 16(1) preserves the linear relation between **D** and **E**, and so satisfies the principle of superposition. In accordance with our treatment of isotropic media, we assume that the magnetic permeability is always effectively unity at optical frequencies, and thus do not assign any part of optical anisotropy to the magnetic anisotropy observed when static fields are applied to crystals. We must also expect that the effective values of ϵ_{xx}, etc., will vary with frequency, so that, in order to relate optical anisotropy in a quantitative way to measurements of anisotropy in static or slowly varying electric fields, it will be necessary to use the full dispersion theory. The variation of ϵ_{xx}, etc., with frequency is in general such that the ratios $\epsilon_{xx}/\epsilon_{yy}$, etc., and therefore the ratios of the refractive indices, are themselves functions of the frequency. Thus both the magnitude of the birefringence and the direction of the optic axes are, in general, functions of the frequency. For the present we ignore the complications due to dispersion and assume that we are dealing with a range of frequencies for which ϵ_{xx}, etc., are independent of frequency.

16.3.—It is shown in Reference 16.3 that we may retain equations 13(39) and 13(42) for the energy of the electromagnetic field and for the flux of energy, provided we assume that the tensor ϵ is symmetrical, i.e. that

$$\epsilon_{xy} = \epsilon_{yx}, \text{ etc.} \quad . \quad . \quad . \quad . \quad . \quad . \quad . \quad 16(3)$$

We make this assumption and have, for the energy,

$$W = V + T = \frac{1}{8\pi}\mathbf{E}\,.\,\mathbf{D} + \frac{1}{8\pi}\mathbf{H}^2, \quad . \quad . \quad 16(4)$$

or

$$W = \frac{1}{8\pi}\sum_{x,y,z} E_x\epsilon_{xy}E_y + \frac{1}{8\pi}\mathbf{H}^2. \quad . \quad . \quad . \quad . \quad 16(5)$$

The symmetrical tensor has *six* independent components. The electrical part of the energy is now given by

$$8\pi V = \mathbf{E} \cdot \mathbf{D} = \epsilon_{xx}E_x^2 + \epsilon_{yy}E_y^2 + \epsilon_{zz}E_z^2 + 2\epsilon_{xy}E_xE_y + 2\epsilon_{yz}E_yE_z + 2\epsilon_{zx}E_zE_x. \quad . \quad 16(6)$$

16.4.—If we put $x = E_x$, $y = E_y$, and $z = E_z$ in 16(6), we have the equation of an ellipsoid surface which represents the properties of the symmetrical tensor * ϵ. If the quadric is referred to co-ordinate axes which are in the same directions as its own principal axes, the components ϵ_{xy}, ϵ_{yz} and ϵ_{zx} vanish and 16(6) reduces to

$$8\pi V = \mathbf{E} \cdot \mathbf{D} = \epsilon_x E_x^2 + \epsilon_y E_y^2 + \epsilon_z E_z^2. \quad . \quad . \quad 16(7)$$

In 16(7) we write ϵ_x instead of ϵ_{xx} because only one suffix is required. It must be emphasized that ϵ_x, ϵ_y and ϵ_z are *not* the components of a vector. Of the six independent components of ϵ, three now determine the axes of the quadric in relation to some physically identifiable system

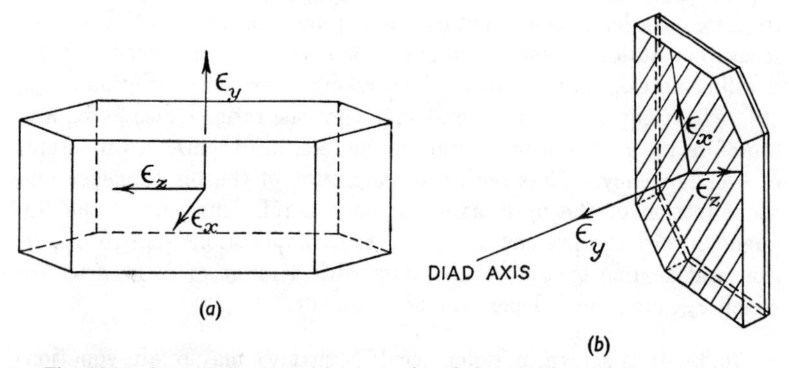

Fig. 16.1.—Axes of the ϵ quadric shown in relation to (*a*) an orthorhombic crystal, (*b*) a monoclinic crystal.
In (*a*) the quadric axes are parallel to the crystal axis. In (*b*) one of the axes of the quadric is parallel to the diad axis of the crystal.

of co-ordinates, e.g. certain axes of crystal symmetry (fig. 16.1). The other three define the principal axes of the quadric (i.e. ϵ_x is equal to the reciprocal of the square of one axis). We call ϵ_x, ϵ_y, and ϵ_z the "principal dielectric constants". In a similar way we shall later use the symbols b_x, b_y, and b_z for principal velocities, and n_x, n_y, and n_z for principal refractive indices.† In the remainder of this chapter it is

* See p. 349 of Reference 16.1. † The quadric is sometimes called the "index ellipsoid".

assumed that the principal axes of the quadric are used as axes of co-ordinates unless otherwise stated. Since the material is homogeneous, any point may be regarded as an origin of co-ordinates. As explained in § 12.7 an axis is, for our present purposes, only a method of defining a direction, and any parallel line is equivalent.

16.5. The Ray in an Anisotropic Medium.

If we define the ray as the line of flow of the energy (§ 7.27), then the Poynting vector 13(40) defines the ray as it does in a medium which is isotropic. Since we are considering a homogeneous medium, the rays must be straight lines under the limiting conditions to which ray optics is applicable. The vector $\mathbf{G}$ is, by definition, perpendicular to $\mathbf{E}$ and to $\mathbf{H}$, so these vectors are still perpendicular to the ray. Since $\mathbf{D}$ and $\mathbf{E}$ are not in general collinear, the ray is not necessarily perpendicular to $\mathbf{D}$. In Chapter VII we showed that the ray could be regarded either (*a*) as the locus of the centres of successive Fresnel zones, or (*b*) as the path defined by Fermat's principle. It may be shown that, even in an anisotropic medium, the ray defined by 13(40) satisfies both these alternative definitions. We shall see that the ray velocity is, in general, neither equal in magnitude to the wave velocity, nor has it the same direction. We define $\mathbf{s}$ to be a unit vector in the direction of the wave normal and $\boldsymbol{\rho}$ to be a unit vector in the direction of the ray, and we take α to be the angle between the wave normal and the ray. $\mathbf{d}$, $\mathbf{e}$, and $\mathbf{h}$ are unit vectors in the directions of $\mathbf{D}$, $\mathbf{E}$, and $\mathbf{H}$ respectively.

16.6. Propagation of Plane Waves.

We shall now obtain an expression for the velocity of a plane wave in the direction $\mathbf{s}$ in an anisotropic material. Let us assume that Maxwell's equations 13(16) to 13(19) remain valid, so that we may write

$$c \operatorname{curl} \mathbf{E} = -\dot{\mathbf{H}} \quad \ldots \ldots \quad 16(8)$$

and

$$c \operatorname{curl} \mathbf{H} = \dot{\mathbf{D}} \quad \ldots \ldots \quad 16(9)$$

$$\operatorname{div} \mathbf{D} = \operatorname{div} \mathbf{H} = 0, \quad \ldots \ldots \quad 16(10)$$

but div $\mathbf{E}$ is *not* zero.

Using 16(2),

$$\left.\begin{aligned} c^2 \operatorname{curl} (\operatorname{curl} \mathbf{E}) &= -\epsilon \ddot{\mathbf{E}}, \\ \text{or} \qquad \nabla^2 \mathbf{E} - \operatorname{grad} (\operatorname{div} \mathbf{E}) &= \frac{\epsilon}{c^2} \ddot{\mathbf{E}}. \end{aligned}\right\} \quad \ldots \quad 16(11a)$$

The second term on the left is not zero as for isotropic media, and we proceed to solve these equations in the following way. One component of 16(11a) may be written

$$\frac{\partial^2 E_x}{\partial x^2}+\frac{\partial^2 E_x}{\partial y^2}+\frac{\partial^2 E_x}{\partial z^2}-\frac{\partial}{\partial x}\left(\frac{\partial E_x}{\partial x}+\frac{\partial E_y}{\partial y}+\frac{\partial E_z}{\partial z}\right)=\frac{\epsilon_x}{c^2}\frac{\partial^2 E_x}{\partial t^2}. \quad 16(11b)$$

Let us now take as a trial solution the plane wave

$$\mathbf{D}=a\mathbf{d}\exp i\omega\left(t-\frac{\mathbf{r}\,.\,\mathbf{s}}{b}\right)=a\mathbf{d}\exp i\phi, \quad . \quad 16(12a)$$

where a is an amplitude, $\mathbf{r}$ is the radius vector drawn from the origin to the point (x, y, z), and b is the magnitude of a phase velocity whose direction is that of $\mathbf{s}$. In Cartesian co-ordinates we write

$$\epsilon_x E_x=D_x=a\,d_x\exp i\omega\left(t-\frac{s_x x+s_y y+s_z z}{b}\right), \quad 16(12b)$$

and two similar equations.

Substituting D_x/ϵ_x for E_x in 16(11b), we obtain

$$\frac{D_x}{\epsilon_x b^2}-\frac{s_x}{b^2}\left(\frac{s_x D_x}{\epsilon_x}+\frac{s_y D_y}{\epsilon_y}+\frac{s_z D_z}{\epsilon_z}\right)=\frac{D_x}{c^2}, \quad . \quad 16(13)$$

and two similar equations.

Put $$\frac{c^2}{\epsilon_x}=b_x^2,\quad \frac{c^2}{\epsilon_y}=b_y^2,\quad \frac{c^2}{\epsilon_z}=b_z^2; \quad . \quad . \quad . \quad . \quad 16(14)$$

and $$P^2=\sum_x b_x^2 s_x D_x=c^2\sum_x s_x E_x=c^2\mathbf{E}\,.\,\mathbf{s}. \quad . \quad 16(15)$$

b_x, b_y, and b_z are called the *principal phase velocities. Note that they are constants of a given crystal and not the components of the phase velocity.*

On multiplying 16(13) by b^2c^2 and simplifying, we obtain

$$D_x=-\frac{s_x}{b^2-b_x^2}P^2. \quad . \quad . \quad . \quad . \quad 16(16a)$$

In a similar way we find

$$D_y=-\frac{s_y}{b^2-b_y^2}P^2, \quad . \quad . \quad . \quad . \quad 16(16b)$$

$$D_z=-\frac{s_z}{b^2-b_z^2}P^2. \quad . \quad . \quad . \quad . \quad 16(16c)$$

From 16(12a)

$$\text{div}\,\mathbf{D} = -\frac{i\omega}{b}(s_xD_x + s_yD_y + s_zD_z)$$

$$= -\frac{i\omega}{b}(\mathbf{s}\,.\,\mathbf{D}). \qquad . \quad . \quad . \quad . \quad . \quad . \quad . \quad 16(17)$$

And since div $\mathbf{D}$ is zero, we have

$$\frac{s_x^{\,2}}{b^2 - b_x^{\,2}} + \frac{s_y^{\,2}}{b^2 - b_y^{\,2}} + \frac{s_z^{\,2}}{b^2 - b_z^{\,2}} = 0. \quad . \quad . \quad . \quad 16(18)$$

Thus the plane wave defined by 16(12) is a solution of Maxwell's equations and the phase velocity in the direction of $\mathbf{s}$ is given by 16(18). Equation 16(18) is known as *Fresnel's equation*. Since it is a quadratic in b^2, there are generally two possible values of the phase velocity for any given direction of the wave normal. For certain special directions there is only one solution of 16(18), i.e. there is only one phase velocity [Example 16(iv)]. These directions are called *optic axes*.

EXAMPLES [16(i)–16(iv)]

16(i). What are the values of the phase velocity when the wave normal is parallel to OZ? [b_x and b_y.]

16(ii). Obtain expressions for d_x, d_y, d_z; i.e. obtain the direction of $\mathbf{D}$ corresponding to a given direction of the wave normal $\mathbf{s}$.

$$\left[d_x = -\frac{s_x}{b^2 - b_x^{\,2}}\,Q^2, \text{ etc.,} \qquad . \quad . \quad . \quad . \quad . \quad 16(19)\right.$$

$$\left.\text{where } Q^2 = \sum b_x^{\,2} s_x d_x.\right]$$

16(iii). If

$$n = c/b; \quad n_x = c/b_x = \varepsilon_x^{\,1/2}, \text{ etc.,} \qquad . \quad . \quad . \quad . \quad 16(20)$$

deduce an expression for n in terms of s_x, n_x, etc.

$$\left[\sum_x \frac{s_x^{\,2}}{1/n^2 - 1/n_x^{\,2}} = 0.\right]$$

(16iv). Assuming that the axes of co-ordinates have been chosen so that

$$\varepsilon_x < \varepsilon_y < \varepsilon_z, \qquad . \quad . \quad . \quad . \quad . \quad . \quad . \quad . \quad 16(21a)$$

and hence

$$b_x > b_y > b_z, \qquad . \quad . \quad . \quad . \quad . \quad . \quad . \quad . \quad 16(21b)$$

show that when

$$\left.\begin{aligned} s_x^2 &= \frac{b_x^2 - b_y^2}{b_x^2 - b_z^2}, \\ s_y &= 0, \\ s_z^2 &= \frac{b_y^2 - b_z^2}{b_x^2 - b_z^2}, \end{aligned}\right\} \quad \ldots\ldots \quad 16(22)$$

there is only one possible phase velocity.

[Treat 16(18) as a quadratic in b^2 and show that when 16(22) is satisfied, the discriminant is zero.]

16.7. Angular Relations between D, E, H, s, and ρ.

For a plane wave in an isotropic medium, **D**, **E**, and **H** have the same phase over any plane which is perpendicular to the wave normal. We now complete the specification of our trial solution by assuming that **E** and **H** are given by expressions containing the same phase

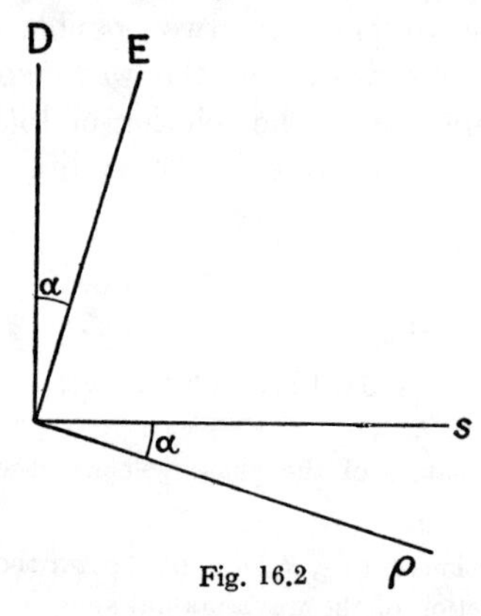

Fig. 16.2

factor $e^{i\phi}$ as **D** [see 16(12*a*)]. We shall show that this is satisfactory provided that the vectors have certain angular relations. The operator $\partial/\partial t$ *applied to one of these expressions* is equivalent to multiplication by $i\omega$, and the operator $\partial/\partial x$ to multiplication by $-\dfrac{i\omega}{b}s_x$. Hence we have

$$\dot{\mathbf{D}} = i\omega D, \quad \ldots\ldots \quad 16(23)$$

and

$$\text{curl}\,\mathbf{E} = \frac{i\omega}{b}\,\mathbf{E}\times\mathbf{s}, \quad \ldots\ldots \quad 16(24)$$

$$\text{div}\,\mathbf{E} = -\frac{i\omega}{b}\,(\mathbf{E}\,.\,\mathbf{s}), \quad \ldots\ldots \quad 16(25)$$

and similar expressions may be obtained for curl **H**, etc.*

* 16(24) may be verified by writing out a component in Cartesian co-ordinates.

We then have [using 16(8) and 16(9)]

$$n(\mathbf{E} \times \mathbf{s}) = -\mathbf{H} \quad . \quad . \quad . \quad . \quad . \quad 16(26)$$

and

$$n(\mathbf{H} \times \mathbf{s}) = \mathbf{D}, \quad . \quad . \quad . \quad . \quad . \quad 16(27)$$

where n is a scalar with the value defined by 16(20).

Thus **H** is normal to the plane defined by **E** and **s** and also to **D**. The direction of $\boldsymbol{\rho}$ is, by definition, normal to **E** and to **H**. Thus **E**, **D**, **s**, and $\boldsymbol{\rho}$ are coplanar, and **H** is normal to them. Also, since **D** is

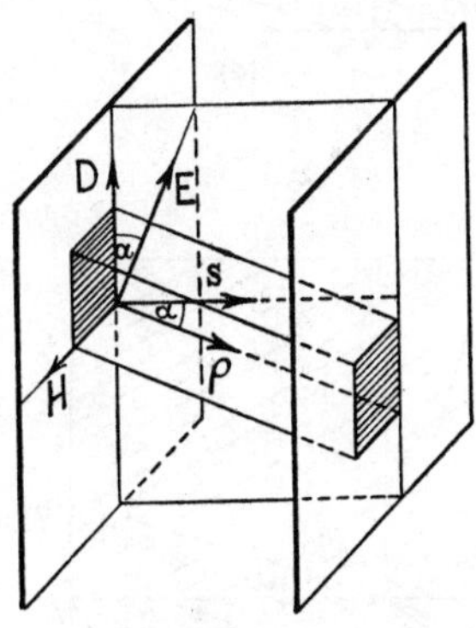

Fig. 16.3

normal to **s**, and **E** to $\boldsymbol{\rho}$, the angle between **D** and **E** is equal to α (the angle between **s** and $\boldsymbol{\rho}$). These angular relations are shown in figs. 16.2 and 16.3. Note that **E** has the same value at all points on the plane normal to **s** although it does not lie in that plane.

16.8. The Two Possible Directions of D for a Given Wave-normal are mutually Perpendicular.

Let b' and b'' be the two solutions of 16(18) corresponding to a given direction of **s**. Then, by inserting values successively in 16(19), we obtain two directions **d′** and **d″** for the vector **D**. We have

$$\mathbf{d}' \cdot \mathbf{d}'' = Q^4 \sum_{x,y,z} \frac{s_x^{\,2}}{(b'^2 - b_x^{\,2})(b''^2 - b_x^{\,2})}$$

$$= \frac{Q^4}{(b''^2 - b'^2)} \sum_{x,y,z} \left[\frac{s_x^{\,2}}{b'^2 - b_x^{\,2}} - \frac{s_x^{\,2}}{b''^2 - b_x^{\,2}}\right] = 0,$$

since b' and b'' both satisfy 16(18).

Thus the two possible directions for **D** are mutually perpendicular (fig. 16.4*a*).

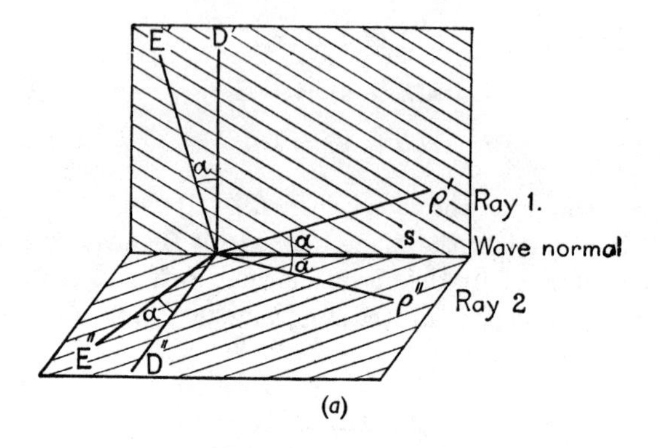

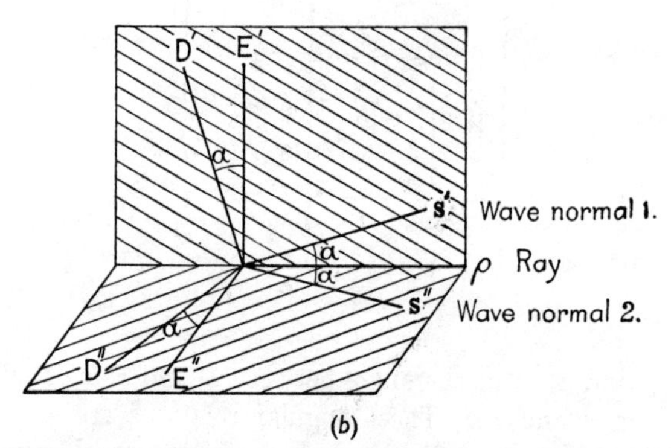

Fig. 16.4.—(*a*) Showing the two possible directions for **ρ**, **D** and **E** corresponding to a given wave-normal **s**. **ρ**′, **D**′, **E**′ lie in one plane; **ρ**″, **D**″, **E**″ lie in another plane. These planes are perpendicular and intersect in **s**.

(*b*) Showing the two possible directions for **s**, **E**, and **D** corresponding to a given ray **ρ**. **s**′, **E**′, **D**′ and **s**″, **E**″, **D**″ lie in perpendicular planes intersecting in **ρ**.

EXAMPLES [16(v)–16(viii)]

16(v). Show that for any given direction of **D** there is only one possible phase velocity.

[The equations 16(18) and 16(19) give

$$\sum_{x,y,z} (b^2 - b_x^2)\, d_x^2 = 0$$

and hence, since **d** is a unit vector,

$$b^2 = \sum_{x,y,z} b_x^2\, d_x^2. \qquad 16(28).]$$

16(vi). Show that

$$Q^{-4} = \Sigma\left(\frac{s_x}{b^2 - b_x^{\,2}}\right)^2. \qquad 16(29)$$

[Use 16(16) to find the sum of the squares of the components of the unit vector **d**.]

16(vii). Find an expression for d_x/e_x.

[$D_x = D \,.\, d_x$, where D is the magnitude of **D**. Hence $d_x/e_x = (E/D)\,\varepsilon_x$.]

16(viii). Show that the components of **e** are

$$e_x = -\frac{b_x^{\,2} s_x}{b^2 - b_x^{\,2}} M^2, \qquad 16(30)$$

where

$$M^{-4} = \Sigma\left(\frac{b_x^{\,2} s_x}{b^2 - b_x^{\,2}}\right)^2. \qquad 16(31)$$

[Use 16(16) to obtain the components of **E**. The components of **e** are proportional to those of **E**, and the sum is unity.]

16.9. Rate of Transport of Energy. Ray Velocity.

Suppose that a certain value ϕ of the phase is found on the plane AB (fig. 16.5) at time $t = 0$ and on the plane A′B′ at a later time t, the distance (AA′) between the planes being bt. Consider the region bounded by a cylinder of rays of which MM′ and NN′ lie in the plane

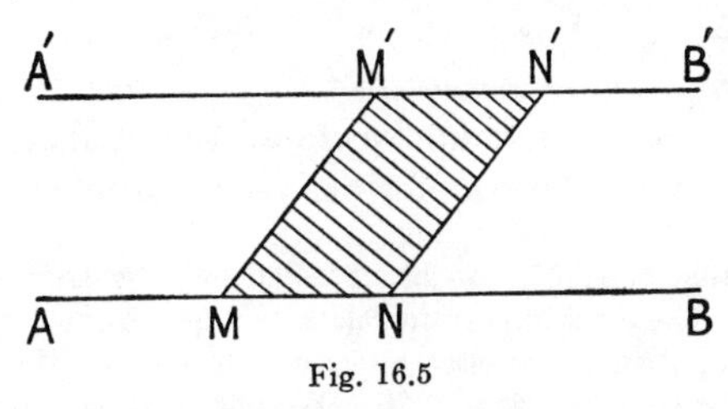

Fig. 16.5

of the paper. Let the area of cross-section be A. Suppose that there is a minute variation of amplitude which forms a " mark " on the wave. This " mark " enters the lower surface when $t = 0$, and emerges from the upper surface at time t. During this time an amount of energy AGt will have entered the lower surface and an equal amount will have left through the upper surface.* The energy in the space at any time is $WAbt$. The amount entering and leaving is the same as if the whole energy moved forward with a velocity g given by

$$g = \frac{G}{W}. \qquad 16(32)$$

* G and W are defined on p. 407.

This velocity is called the *ray velocity*. From the figure we have

$$g \cos \alpha = b. \qquad \qquad 16(33)$$

The ratio of c to the ray velocity g is sometimes called the ray index n_g. It obeys the equation

$$n_g = \frac{c}{g} = n \cos \alpha. \qquad \qquad 16(34)$$

16.10. **Properties of the Ray.**

It is possible to derive expressions involving the *ray* velocity and direction corresponding to those obtained for the *wave* velocity and direction. The method used is similar to that given in § 16.6. There is such correspondence between the results that the following rule is convenient.

Let the variables be written in two rows.

$$\left.\begin{array}{ccccccccccccc} \mathbf{E}, & \mathbf{D}, & \mathbf{s}, & \boldsymbol{\rho}, & b, & n, & \epsilon_x, & \epsilon_y, & \epsilon_z, & b_x, & b_y, & b_z, & c, \\ \mathbf{D}, & \mathbf{E}, & -\boldsymbol{\rho}, & -\mathbf{s}, & \frac{1}{g}, & \frac{1}{n_g}, & \frac{1}{\epsilon_x}, & \frac{1}{\epsilon_y}, & \frac{1}{\epsilon_z}, & \frac{1}{b_x}, & \frac{1}{b_y}, & \frac{1}{b_z}, & \frac{1}{c}. \end{array}\right\} \qquad 16(35)$$

Then any relation which is valid for members of one row remains valid when all the corresponding members of the other row are substituted.

If we show that this rule applies to all the fundamental equations, then it will necessarily be true for derived equations.

A complete proof of 16(35) can be obtained only by analysis of the quadric associated with ϵ. The most important relations required are the angular relations (see fig. 16.2, p. 488) which we have already deduced, and the vector equations 16(36*b*) and its analogous relation 16(39). We obtain 16(36*b*) by first eliminating **H** from 16(26) and 16(27), giving

$$\mathbf{D} = -n^2(\mathbf{E} \times \mathbf{s}) \times \mathbf{s}. \qquad \qquad 16(36a)$$

The vector **D** is now expressed in terms of a vector in the direction of **E** and a vector in the direction of **s**. Resolving the right-hand side of 16(36*a*), we obtain

$$\mathbf{D} = n^2\{\mathbf{E} - \mathbf{s}(\mathbf{E} \, . \, \mathbf{s})\}. \qquad \qquad 16(36b)$$

Forming the scalar product of each side of 16(36*b*) with $\boldsymbol{\rho}$ we obtain

$$\mathbf{D} \, . \, \boldsymbol{\rho} = -n^2(\mathbf{E} \, . \, \mathbf{s}) \cos \alpha \qquad \qquad 16(37)$$

(remembering that $\mathbf{E} \, . \, \boldsymbol{\rho} = 0$). Since **D**, **E**, and $\boldsymbol{\rho}$ are coplanar, there must be a linear relation of the type

$$\boldsymbol{\rho} = l\mathbf{D} + m\mathbf{E}, \qquad \qquad 16(38)$$

where l and m are scalar constants. Forming the scalar products first with $\boldsymbol{\rho}$ and then with $\mathbf{s}$, we obtain

$$l\mathbf{D} . \boldsymbol{\rho} = 1 \text{ and } m\mathbf{E} . \mathbf{s} = \boldsymbol{\rho} . \mathbf{s} = \cos \alpha,$$

so that 16(38) may be written

$$\mathbf{E} = \frac{1}{m}(\boldsymbol{\rho} - l\mathbf{D}) = \frac{\mathbf{E} . \mathbf{s}}{\cos \alpha}\left(\boldsymbol{\rho} - \frac{\mathbf{D}}{\mathbf{D} . \boldsymbol{\rho}}\right).$$

Substituting for $\mathbf{E}$ (on the right-hand side) from 16(37) we obtain

$$\mathbf{E} = -\frac{\mathbf{D} . \boldsymbol{\rho}}{n^2 \cos^2 \alpha}\left(\boldsymbol{\rho} - \frac{\mathbf{D}}{\mathbf{D} . \boldsymbol{\rho}}\right)$$

$$= \frac{1}{n_g^{\,2}}[\mathbf{D} - \boldsymbol{\rho} . (\mathbf{D} . \boldsymbol{\rho})], \qquad 16(39)$$

which is the relation analogous to 16(36b).

16.11.—In order to find the direction of $\mathbf{E}$ when the direction of the ray is given we apply 16(35) to 16(18) and obtain

$$\Sigma \frac{\rho_x^{\,2}}{\left(\frac{1}{b_x^{\,2}} - \frac{1}{g^2}\right)} = 0$$

or

$$\Sigma \frac{\rho_x^{\,2} b_x^{\,2}}{(g^2 - b_x^{\,2})} = 0. \qquad 16(40)$$

This relation gives the ray velocity when the direction of the ray is known. From 16(16) and 16(15) we have

$$D_x = -\frac{s_x c^2}{b^2 - b_x^{\,2}} \mathbf{E} . \mathbf{s},$$

and applying 16(35) we obtain

$$E_x = \frac{g^2 \rho_x b_x^{\,2}}{c^2(g^2 - b_x^{\,2})} \mathbf{D} . \boldsymbol{\rho}. \qquad 16(41)$$

In general there are two directions of $\mathbf{E}$ obtained by inserting the two possible ray velocities [obtained from 16(40)] in 16(41). It may be shown, by a discussion similar to that of § 16.8, that these directions are mutually perpendicular. It may also be shown that when the direction of $\mathbf{E}$ is given, both the direction and magnitude of the ray velocity are completely determined [cf. Example 16(v), p. 490]. The relation between the two directions of $\mathbf{E}$ and a ray is shown in fig. 16.4b.

EXAMPLES [16(ix)–16(xi)]

16(ix). Show that 16(40) may be obtained by using 16(41) and the fact that **E** is perpendicular to $\boldsymbol{\rho}$.

16(x). Show that, when the ray is along one of the principal axes, the ray and the phase velocities are equal. Find the velocities for a ray directed along the x-axis. [b_y and b_z.]

16(xi). Show that when the direction of $\boldsymbol{\rho}$ is given by

$$\left.\begin{aligned} \rho_x^2 &= \frac{b_z^2}{b_x^2}\left(\frac{b_x^2 - b_y^2}{b_x^2 - b_z^2}\right), \\ \rho_y &= 0, \\ \rho_z^2 &= \frac{b_x^2}{b_z^2}\left(\frac{b_y^2 - b_z^2}{b_x^2 - b_z^2}\right), \end{aligned}\right\} \qquad 16(42)$$

the two ray velocities are equal.

These directions are known as the axes of single ray velocity.

16.12. *The Angle between the Ray and the Wave Normal.*

From the geometrical relations of fig. 16.2 (p. 488) we have

$$\sin\alpha = \mathbf{e}\,.\,\mathbf{s} \text{ and } \cos\alpha = \mathbf{e}\,.\,\mathbf{d}. \qquad 16(43)$$

The components of $\mathbf{d}$ are given by 16(19) and those of $\mathbf{e}$ are *proportional* to d_x/ε_x, etc., i.e. to $b_x^2\,d_x$, etc. [Note that $d_x \neq \varepsilon_x e_x$, see Example 16(vii), p. 491.]

We have therefore

$$\tan\alpha = \frac{Q^2\sum\dfrac{-s_x^2 b_x^2}{b^2 - b_x^2}}{Q^4\sum\dfrac{s_x^2 b_x^2}{(b^2 - b_x^2)^2}}, \qquad 16(44)$$

since the constant of proportionality cancels.

Using 16(18), we obtain

$$\sum\frac{-s_x^2 b_x^2}{b^2 - b_x^2} = \sum s_x^2 - \sum\frac{s_x^2 b^2}{b^2 - b_x^2} = 1, \qquad 16(45a)$$

and similarly, using 16(18) and 16(29),

$$\sum\frac{s_x^2 b_x^2}{(b^2 - b_x^2)^2} = \sum\frac{s_x^2}{b^2 - b_x^2}\left(\frac{b^2}{b^2 - b_x^2} - 1\right) = b^2 Q^{-4}, \qquad 16(45b)$$

so that

$$\tan\alpha = \frac{Q^2}{b^2}. \qquad 16(46)$$

EXAMPLES [16(xii) and 16(xiii)]

16(xii). Use 16(19) and 16(30) to verify 16(46) by showing that

$$\cos\alpha = \frac{b^2}{(b^4 + Q^4)^{\frac{1}{2}}}. \qquad 16(47)$$

16(xiii). Show that

$$Q^4 = b^2(g^2 - b^2). \qquad 16(48)$$

[Use 16(33) and 16(47).]

16.13. Direction of the Ray.

It is sometimes desirable to be able to calculate the direction of the ray (i.e. the components of $\boldsymbol{\rho}$) when the direction of the wave normal is given. Since $b_x^2 D_x = c^2 E_x$, we have [from 16(16) and 16(41)]

$$\frac{s_x c^2}{b^2 - b_x^2}\,\mathbf{E}\,.\,\mathbf{s} = -\frac{g^2 \rho_x}{g^2 - b_x^2}\,\mathbf{D}\,.\,\boldsymbol{\rho}, \qquad 16(49)$$

and, using 16(37) with 16(33),

$$\frac{s_x b}{b^2 - b_x^2} = \frac{g\rho_x}{g^2 - b_x^2}, \qquad 16(50)$$

or

$$\rho_x = s_x \cos\alpha \left(\frac{g^2 - b_x^2}{b^2 - b_x^2}\right). \qquad 16(51)$$

On eliminating g and using 16(33),

$$\rho_x = s_x \cos\alpha \left(1 + \frac{b^2 \tan^2\alpha}{b^2 - b_x^2}\right)$$

$$= \frac{s_x}{(Q^4 + b^4)^{\frac{1}{2}}}\left(b^2 + \frac{Q^4}{b^2 - b_x^2}\right). \qquad 16(52)$$

EXAMPLE 16(xiv)

Show that

$$\Sigma \frac{\rho_x s_x}{g^2 - b_x^2} = 0. \qquad 16(53)$$

[Use 16(50) and 16(18).]

16.14. The Wave Surface or Ray Surface.

Consider the plane wave shown in fig. 16.3. The distance between two wave surfaces measured along the ray is d_r, and measured along the wave normal is $d_n = d_r \cos\alpha$. The phase difference δ is given by

$$\delta = \frac{2\pi}{\lambda}\,d_n = \frac{\omega}{b}\,d_n = \frac{\omega}{g}\,d_r, \qquad 16(54)$$

i.e. the phase difference between two points on a ray is proportional to the distance divided by the ray velocity. Now consider a series of lines radiating from a point O (fig. 16.6*a*). Along each line mark off two points, such as R_1' and R_1'', so that OR_1' and OR_1'' are equal to $g_1't$ and $g_1''t$ where g_1' and g_1'' are the two ray velocities for the direction $OR_1'R_1''$. The locus of the points R′ and R″ is a surface (of two sheets) called the ray surface (fig. 16.6*b*).

Consider, for the moment, one sheet only. For any ray from O to the surface the phase difference has the same value (ωt), so the *ray surface* is a surface of constant phase for waves spreading from a point

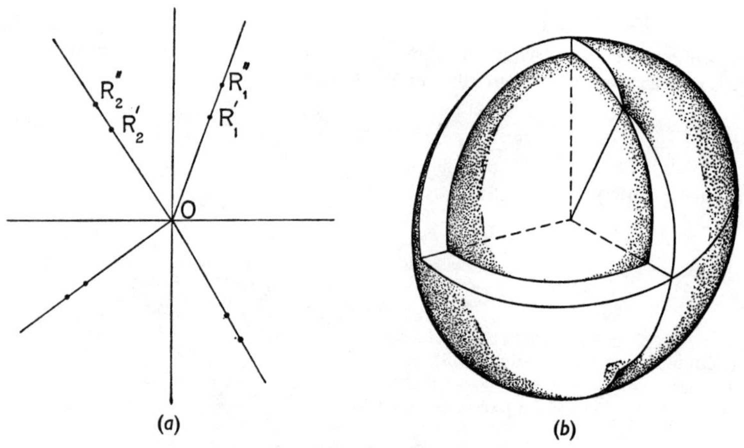

Fig. 16.6.—(*a*) The construction of the wave surface (or ray surface). (*b*) The wave surface for a biaxial crystal

source at O. It is a *wave surface* in the sense in which we defined the term in Chapter III and used it in Chapter XII. If we consider a plane wave passing through O at time $t = 0$, its position at time t must be such that it intersects the sheet of the ray surface which we are now considering once and only once, i.e. it is tangential to the ray surface. Since this is true for all possible directions of the wave normal, the ray surface is the envelope of the positions of all these plane waves at time t. This property is sometimes used as a definition of a wave surface in anisotropic media. When we take account of the fact that in any direction there are two possible ray velocities, the above argument is slightly more complicated, but the important conclusion is unchanged. We have to associate each plane wave with the sheet of the ray surface to which, by reason of its direction of polarization, it belongs. With this provision, the identification of the ray surface with the wave surface remains valid. Each plane wave is tangential to its own sheet of the ray surface, though it may cut the other sheet. In the course of the ensuing discussion it will be shown that the identity of the ray surface and the wave surface may be proved analytically. The surface may be drawn for any time after $t = 0$, but when no time is specified it is assumed that $t = 1$.

16.15.—The equation of the surface for $t = 1$ may be obtained by putting r for g and x/r for ρ_x, etc., in 16(40). We have (after multiplying by r^2)

$$\Sigma \frac{x^2 b_x^2}{r^2 - b_x^2} = 0. \qquad 16(55a)$$

This equation may be written

$$\Sigma\, x^2 b_x^2 (r^2 - b_y^2)(r^2 - b_z^2) = 0, \qquad 16(55b)$$

or

$$r^2(b_x^2x^2 + b_y^2y^2 + b_z^2z^2) - b_x^2(b_y^2 + b_z^2)x^2 - b_y^2(b_z^2 + b_x^2)y^2$$
$$-b_z^2(b_x^2 + b_y^2)z^2 + b_x^2b_y^2b_z^2 = 0. \qquad 16(55c)$$

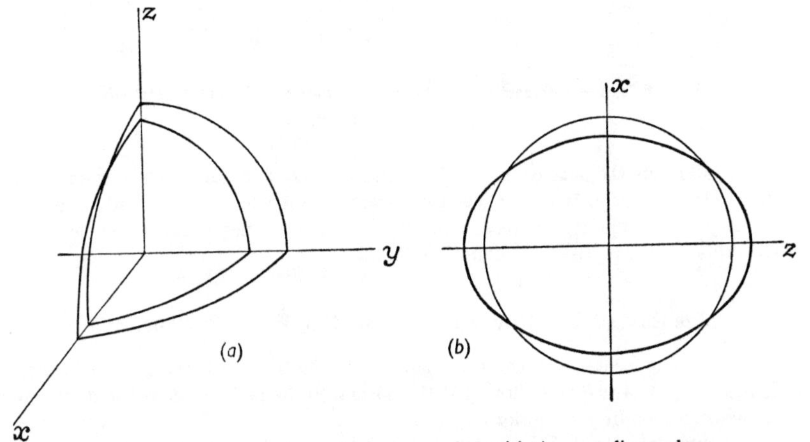

Fig. 16.7*a*, *b*.—Intersections of the wave surface with the co-ordinate planes (see also fig. 16.7*c*, *d* on p. 498)

This equation is of the fourth degree. Its intersections with the co-ordinate planes are shown in fig. 16.7. The intersection with $y = 0$ is given by

$$(r^2 - b_y^2)(x^2b_x^2 + z^2b_z^2 - b_x^2b_z^2) = 0. \qquad 16(56)$$

The curve of section is of two parts:

(*a*) the circle $\quad x^2 + z^2 = r^2 = b_y^2, \qquad 16(57a)$

(*b*) the ellipse $\quad b_x^2x^2 + b_z^2z^2 - b_x^2b_z^2 = 0. \qquad 16(57b)$

The ellipse and the circle intersect at points which satisfy a relation obtained by multiplying 16(57*b*) by b_y^2 and substituting from 16(57*a*) in the last term.

$$\frac{z^2}{x^2} = \frac{b_x^2(b_y^2 - b_z^2)}{b_z^2(b_x^2 - b_y^2)}. \qquad \text{16(58)}$$

Since $b_x > b_y > b_z$, there are four real points of intersection. It is easily seen that the co-ordinate plane $x = 0$ intersects the surface in a circle and an ellipse, but that these do not intersect. The circle encloses the ellipse. Similarly the plane

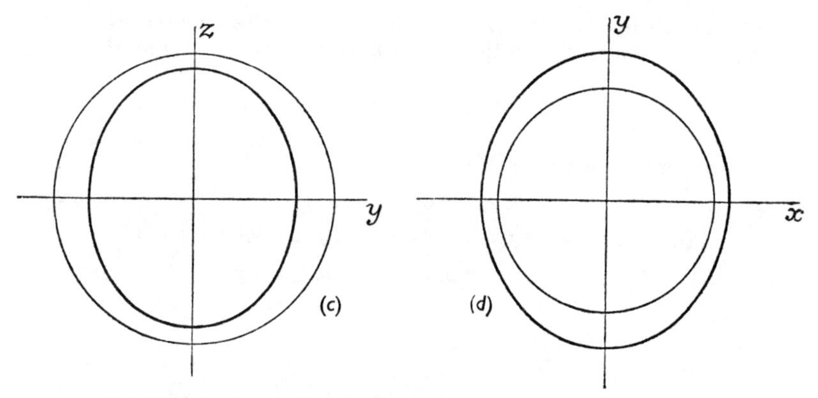

Fig. 16.7*c*, *d*.—Intersections of the wave surface for a biaxial crystal with the *xy* and *yz* planes

$z = 0$ intersects the surface in an ellipse and a circle, but the ellipse now encloses the circle. Equation 16(58) defines two directions for which there is only one ray velocity. The equation agrees with the direction cosines obtained earlier [see Example 16(xi), p. 494, and equation 16(42)].

16.16. Identity of the Ray Surface and the Wave Surface.

In order to prove that the wave surface is identical with the ray surface as defined in § 16.14, we now show that the normal to the ray surface is the direction of the corresponding wave normal.

A convenient form of the expression for the ray surface is obtained by writing 16(55) as

$$\Sigma \frac{x^2 b_x^2}{r^2 - b_x^2} = \Sigma \left(\frac{x^2 r^2}{r^2 - b_x^2} - x^2\right) = 0,$$

whence

$$\Sigma \left\{\frac{x^2}{r^2 - b_x^2} - 1\right\} = 0 \qquad \text{16(59)}$$

since $\Sigma x^2 = r^2$. If we put the left-hand side equal to F, then the direction cosines of the normal to the surface at the point x, y, z are proportional to the values of $\partial F/\partial x$, $\partial F/\partial y$, and $\partial F/\partial z$ at the point.

We have
$$\frac{\partial F}{\partial x} = 2x \left[\frac{1}{r^2 - b_x^2} - \Sigma \frac{x^2}{(r^2 - b_x^2)^2}\right].$$

Putting $r = g$ and $x = \rho_x g$ (see § 16.15),
$$\frac{\partial F}{\partial x} = 2g\rho_x \left[\frac{1}{g^2 - b_x^2} - \Sigma \frac{\rho_x^2 g^2}{(g^2 - b_x^2)^2}\right]$$
$$= 2g\rho_x \left[\frac{1}{g^2 - b_x^2} - \Sigma \frac{s_x^2 b^2}{(b^2 - b_x^2)^2}\right] \qquad \text{by 16(50).}$$

Using 16(45*b*) and 16(48),
$$\frac{\partial F}{\partial x} = 2g\rho_x \left[\frac{1}{g^2 - b_x^2} - \frac{1}{g^2 - b^2}\right]$$
$$= \frac{-2g\rho_x}{(g^2 - b^2)} \left[\frac{b^2 - b_x^2}{g^2 - b_x^2}\right]. \qquad \text{16(60)}$$

Using 16(50) again, this becomes
$$\frac{\partial F}{\partial x} = \frac{2bs_x}{b^2 - g^2}, \qquad \text{16(61)}$$

i.e. $\partial F/\partial x$, etc., are proportional to the direction cosines of the wave normal and the direction of the normal to the surface is the direction of the wave normal.

16.17. The Normal Surface.

We may define a second surface by considering a number of lines radiating from O and marking off lengths equal to corresponding phase velocities. The surface so obtained is called the normal surface. It is not as important in practice as the wave surface, but it does offer a convenient way of finding the directions of the optic axes (i.e. axes of single phase velocity). Putting r for b and x/r for s_x in 16(18), we have
$$\Sigma\, x^2(r^2 - b_y^2)\,(r^2 - b_z^2) = 0. \qquad \text{16(62)}$$

Exactly as in § 16(15) we find that it intersects each co-ordinate plane in an oval and a circle. In the $x = 0$ plane the circle encloses the oval, in the $z = 0$ plane the oval encloses the circle. In the $y = 0$ plane the two curves intersect in points whose co-ordinates satisfy the relation
$$\frac{z^2}{x^2} = \frac{b_y^2 - b_z^2}{b_x^2 - b_y^2}. \qquad \text{16(63)}$$

These define the directions of single wave velocity in agreement with 16(22).

16.18. Difference of the Two Phase Velocities for a Given Direction of the Wave Normal.

We shall now show that, if b' and b'' are the phase velocities corresponding to a wave normal which makes angles χ_1 and χ_2 with the two optic axes,
$$b'^2 - b''^2 = (b_x^2 - b_z^2) \sin\chi_1 \sin\chi_2. \qquad \text{16(64)}$$

When the difference between b_x and b_z is small, we have approximately

$$b' + b'' \simeq b_x + b_z,$$

and 16(64) reduces to

$$b' - b'' = (b_x - b_z) \sin \chi_1 \sin \chi_2. \quad \quad 16(65)$$

Equation 16(18) may be written

$$\sum s_x^2(b^2 - b_y^2)(b^2 - b_z^2) = 0$$

or

$$b^4 - b^2 \sum s_x^2(b_y^2 + b_z^2) + \sum s_x^2 b_y^2 b_z^2 = 0.$$

The square of the difference of the roots of the equation is equal to the discriminant, i.e.

$$(b'^2 - b''^2)^2 = \{\sum s_x^2(b_y^2 + b_z^2)\}^2 - 4 \sum s_x^2 b_y^2 b_z^2. \quad \quad 16(66a)$$

Put $A = s_x^2(b_y^2 - b_z^2)$, $B = s_y^2(b_z^2 - b_x^2)$ and $C = s_z^2(b_x^2 - b_y^2)$,

then

$$\begin{aligned}(b'^2 - b''^2)^2 &= A^2 + B^2 + C^2 - 2AB - 2BC - 2CA \\ &= (A - B + C)^2 - 4AC. \quad \quad 16(66b)\end{aligned}$$

Now, from 16(22), we have

$$\cos \chi_1 = s_x \sqrt{\left(\frac{b_x^2 - b_y^2}{b_x^2 - b_z^2}\right)} + s_z \sqrt{\left(\frac{b_y^2 - b_z^2}{b_x^2 - b_z^2}\right)}$$

and

$$\cos \chi_2 = -s_x \sqrt{\left(\frac{b_x^2 - b_y^2}{b_x^2 - b_z^2}\right)} + s_z \sqrt{\left(\frac{b_y^2 - b_z^2}{b_x^2 - b_z^2}\right)}.$$

Putting $\sin^2 \chi_1 = 1 - \cos^2 \chi_1 = s_x^2 + s_y^2 + s_z^2 - \cos^2 \chi_1$, we obtain

$$\begin{aligned}(b_x^2 - b_z^2) \sin^2 \chi_1 = (b_x^2 - b_z^2)(s_x^2 + s_y^2 + s_z^2) - s_x^2(b_x^2 - b_y^2) \\ - s_z^2(b_y^2 - b_z^2) - s_x s_z \sqrt{\{(b_x^2 - b_y^2)(b_y^2 - b_z^2)\}},\end{aligned}$$

i.e.

$$(b_x^2 - b_z^2) \sin^2 \chi_1 = A - B + C + 2\sqrt{(AC)}.$$

Similarly

$$(b_x^2 - b_z^2) \sin^2 \chi_2 = A - B + C - 2\sqrt{(AC)},$$

so that

$$(b_x^2 - b_z^2)^2 \sin^2 \chi_1 \sin^2 \chi_2 = (A - B + C)^2 - 4AC, \quad \quad 16(67)$$

and 16(64) is obtained by comparing 16(66b) and 16(67).

16.19. The Wave Surface in Uniaxial Crystals.

If we put

$$b_x = b_y = b_o, \text{ and } b_z = b_e, \quad \quad 16(68)$$

then 16(55b) becomes

$$(r^2 - b_o^2)\{b_o^2(x^2 + y^2) + b_e^2 z^2 - b_o^2 b_e^2\} = 0. \quad \quad 16(69)$$

This is a surface of two sheets:

(*a*) the sphere of radius b_o,

(*b*) the spheroid formed by rotating an ellipse of semi-axes b_o and b_e about the direction of the latter axis, which is the OZ direction.

The sphere and the spheroid touch as shown in fig. 16.8. If b_o is greater than b_e, the sphere includes the spheroid. The surface is identical with that obtained experimentally for a positive uniaxial crystal. Similarly, if we make $b_o < b_e$, we obtain the wave surface for a negative uniaxial crystal. The electromagnetic theory thus gives a

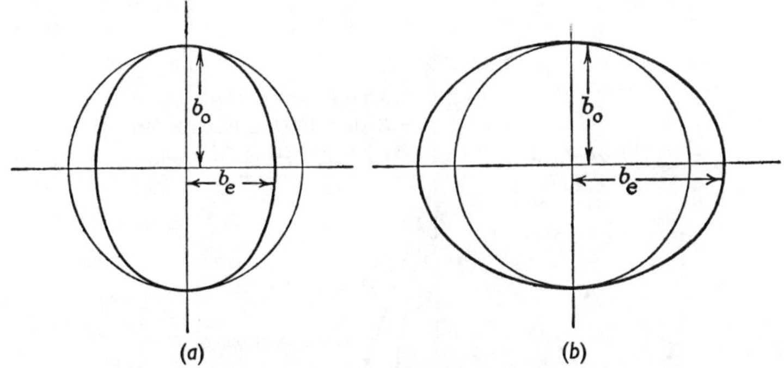

Fig. 16.8.—Sections of the wave surfaces of a uniaxial material: (*a*) positive—$b_o > b_e$. (*b*) negative—$b_o < b_e$

satisfactory account of all the results which lead to these forms of the wave surface in uniaxial crystals. The uniaxial crystal is to be regarded as a limiting case in which the axes of the biaxial crystal are indefinitely near to one another. The continuity of the relation between the biaxial and uniaxial crystals is shown by sodium sulphate. This substance is biaxial at room temperature, but, as the temperature is raised, the angle between the optic axes decreases. For violet light the axes coincide at about 40° C. and the substance is uniaxial at that temperature. The axes diverge and the crystal again becomes biaxial at higher temperatures.

16.20. Double Refraction.

We now consider a plane wave incident upon the plane surface of an anisotropic medium from a vacuum. Let us choose axes so that the plane $x = 0$ coincides with the boundary, and so that the plane $z = 0$ is the plane of incidence, i.e. the plane containing the wave normal of the incident beam and the normal to the surface. In order to make the discussion general, we do not assume that these axes necessarily coincide with the directions of the principal velocities in the crystal. Let θ_1 be the angle of incidence (fig. 16.9). We assume that there are two refracted beams and that their *wave normals* (not the rays) make angles θ_2' and θ_2'' with the normal to the surface. There will also be a reflected beam, which need not concern us at present. The incident wave will contain a factor

$$\exp i\omega \left\{t - \frac{1}{c}(x \cos\theta_1 + y \sin\theta_1)\right\}.$$

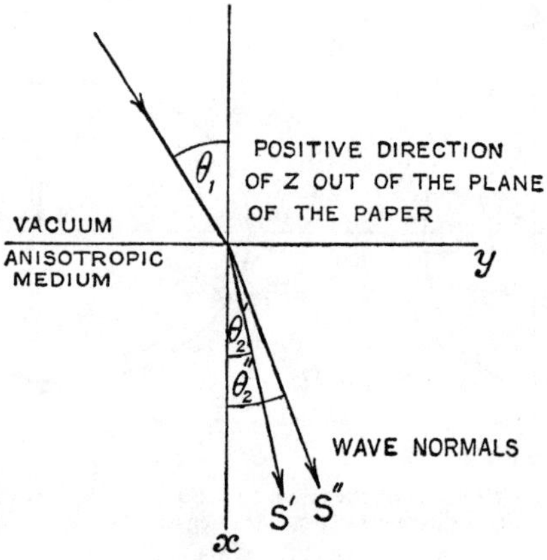

POSITIVE DIRECTION OF Z OUT OF THE PLANE OF THE PAPER

VACUUM

ANISOTROPIC MEDIUM

WAVE NORMALS

Fig. 16.9

The refracted waves will contain factors

$$\exp i\omega \left\{t - \frac{1}{b'}(s_x' x + s_y' y + s_z' z)\right\}$$

and

$$\exp i\omega \left\{t - \frac{1}{b''}(s_x'' x + s_y'' y + s_z'' z)\right\},$$

where $\mathbf{s}'$ and $\mathbf{s}''$ are unit vectors in the direction of the wave normals of the two refracted beams. If the usual boundary conditions are to be fulfilled (§ 14.2), the three exponentials must be equal for all values of y and z when $x = 0$, i.e. we must have

$$s_z' = 0 \quad \text{and} \quad s_z'' = 0, \qquad 16(70)$$

$$\frac{\sin\theta_1}{c} = \frac{s_y'}{b'} = \frac{s_y''}{b''}, \qquad 16(71a)$$

or

$$\frac{\sin\theta_1}{c} = \frac{\sin\theta_2'}{b'} = \frac{\sin\theta_2''}{b''}. \qquad 16(71b)$$

Equation 16(70) implies that the wave normals of the two refracted beams are in the plane of incidence, and 16(71) gives the relation between directions and velocities. Application of Huygens' construction shows that, when a ray is incident upon the surface of an anisotropic medium *from the inside,* there are in general two *reflected* rays. There are also two critical angles for total reflection. This property is used in the Nicol prism and in the measurement of the principal refractive indices (§ 16.44).

16.21. **Double Refraction in Uniaxial Crystals.**

In a uniaxial crystal one of the refracted waves is associated with the spherical sheet of the wave surface and for this wave the ray coincides with the wave normal.* Thus this *ray* obeys both laws of refraction and is correctly described as the *ordinary ray.* Its direction can be determined by putting $b' = b_o$ in 16(71). The phase velocity of the second wave (corresponding to the extraordinary ray) varies with the direction, and thus 16(71) gives the direction of the refracted wave normal, but only by a process of trial and error. In practice the direction of the second wave normal is obtained by Huygens' construction (as shown in § 12.21) or by an equivalent analytical method. This process gives also the direction of the ray which is the radius vector from the origin to the point where the tangent plane touches the spheroid. Let us now define the principal plane to be the plane containing the incident wave normal and the optic axis, i.e. the z axis if we use the notation of equation 16(68). The direction of $\mathbf{D}$ for the ordinary wave cannot readily be obtained from 16(19), because the result is indeterminate when $Q = 0$. It may be shown † indirectly that the direction of $\mathbf{D}$ for the ordinary ray is normal to the principal plane. In view of the discussion in § 12.3 this implies that the ordinary ray is polarized in the principal plane. Similarly it may be shown that the extraordinary ray is polarized perpendicular to the principal plane. These results agree with experimental data.

16.22. **Refraction in Biaxial Crystals.**

We now wish to discuss the refraction of a beam of light which is incident upon a plane surface of a biaxial crystal. It is convenient to start by considering some special cases.

* This is almost certainly true by symmetry. Analytically we see that it is so because 16(29) shows that $Q = 0$ when $b = b_y$, and hence $\alpha = 0$ from equation 16(46).

† See p. 353 of Reference 16.1.

(*a*) *The plane of incidence contains two of the principal axes and is therefore normal to the other* (fig. 16.10*a*, *b*, *c*).

Since the wave surface is symmetrical about the plane of incidence, both refracted rays are in this plane. The direction of one ray (OR_2) is obtained by drawing a tangent from P to the elliptical section of the wave surface. The direction of this wave normal is given by

$$\frac{\sin \theta_1}{c} = \frac{\sin \theta_2''}{b''}, \qquad \qquad 16(71c)$$

where b'' is a solution of Fresnel's equation [16(18)], but the ratio of sines is not constant since b'' is a function of θ_2''. The vector **D** is in the plane of incidence since both the wave normal (ON_2) and the ray (OR_2) are in that plane and they do not coincide, i.e. this ray is

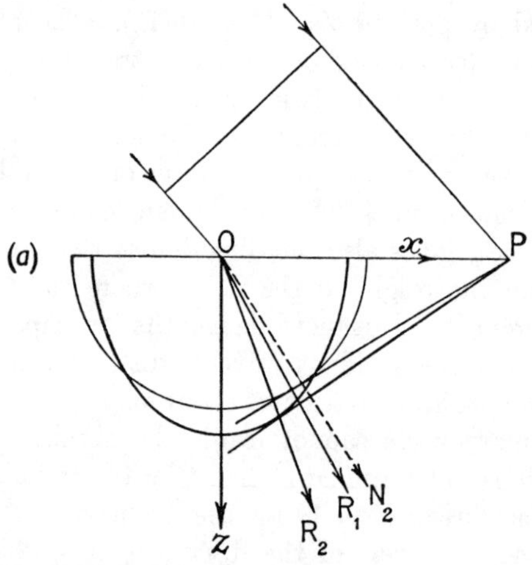

Fig. 16.10*a*.—Refraction at the surface of a biaxial crystal

polarized in a plane normal to the plane of incidence. The other ray OR_1 is obtained by drawing the tangent from P to the circular section of the wave surface. The ray coincides with the wave normal, so that they do not jointly define the plane of **D**, but, by using the tensor ellipsoid, it may be shown that the direction of **D** is normal to the plane of incidence, i.e. this ray is polarized in the plane of incidence. For this ray the ratio of the sines is constant. It is equal to n_x for the plane shown in fig. 16.10*a*; to n_y for the plane shown in fig. 16.10*b*; and to n_z for the plane shown in fig. 16.10*c*.

(*b*) *The incident ray is normal to the surface of separation* which is set at an arbitrary angle to the principal axes.

In this case there is only one wave normal since 16(71*b*) gives $\sin \theta_2' = \sin \theta_2'' = 0$. The two refracted rays are not collinear with

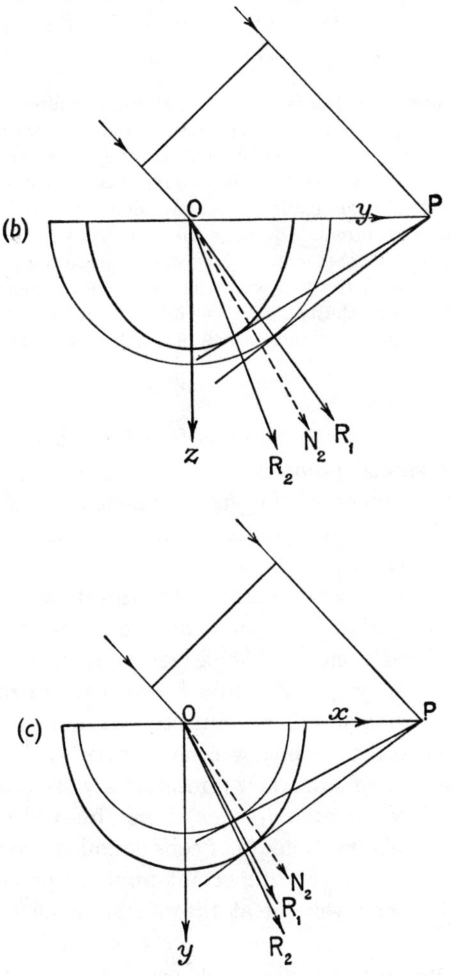

Fig. 16.10*b*, *c*.—Refraction at the surface of a biaxial crystal

the incident ray, and the three rays (one incident and two refracted) are not coplanar. Since they belong to the same wave normal, the two refracted rays are polarized in mutually perpendicular planes (§ 16.8).

(c) *The general case.*

In general there are two rays and two wave normals. All four directions can be obtained by drawing tangent planes to the two sheets of the wave surface. The normals to the wave surface at the points of contact then give the wave normals, and the lines drawn from O to the points of contact give the rays. Each ray and wave normal then determines a plane of polarization.

16.23.—It may be shown (Reference 16.3) that the two directions of **D** corresponding to a given wave-normal are the internal and external bisectors of the angle between two planes, each of which contains the wave normal and one of the optic axes. This does not show that the two refracted rays in a biaxial crystal are polarized in mutually perpendicular planes, since they do not in general belong to the same wave-normal. Owing to the fact that the difference between the greatest velocity (b_x) and the smallest (b_z) is always small compared with either velocity, the angle between the two wave normals is never more than a small fraction of a radian. Thus the two rays belong to neighbouring wave normals and are polarized in planes which are nearly perpendicular to one another.

16.24. **Conical Refraction.**

Hamilton noticed that the wave surface for a biaxial crystal has the following two special properties:

(a) An infinite number of tangent planes may be drawn to the surface at any one of the four points at which it is intersected by one of the axes of single ray velocity.

(b) A single tangent plane normal to the direction of single phase velocity touches the surface not in one or two points but in a circle.

Hamilton predicted from (a) that in certain circumstances a single ray within a crystal may give rise to a hollow cone of rays on emergence, and from (b) that, in other circumstances, a single ray entering a crystal may give rise to a hollow cone of rays within the crystal. These phenomena, which are known respectively as *external conical refraction* and *internal conical refraction*, were observed by H. Lloyd. Their prediction and observation are of considerable historical interest, although they are not now very important from the practical point of view, nor have they any very great theoretical significance. To the student their most important aspect is perhaps that they form a test of his understanding of the relation between ray and wave normal.

16.25. **External Conical Refraction.**

Lloyd's first experiment is illustrated in fig. 16.11. A hollow cone of rays is incident upon the crystal. The rays inside the slice are confined to a very small range of directions near to the axis of single ray velocity by suitably placed pinholes (H_1 and H_2). The directions of the axis of the incident cone and its semi-angle have been chosen so that one refracted ray corresponding to each wave normal is in the direction of single ray velocity. When the narrow beam of rays in the crystal reaches the second surface of the crystal, it will be

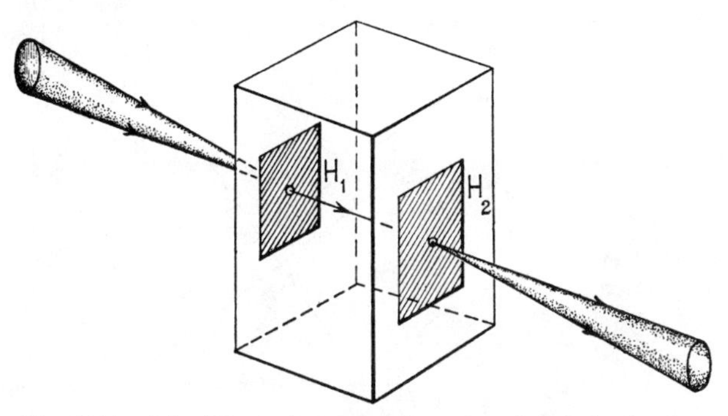

Fig. 16.11.—Lloyd's experiment showing external conical refraction

refracted into a cone of rays similar to the one from which it was derived. Each ray in the original beam will, in fact, emerge parallel to its original direction. The emergent cone of rays forms a ring on a suitably placed screen, and the radius of the ring is proportional to the distance of the screen from the crystal.

16.26. **Internal Conical Refraction.**

The second special property of the wave surface, i.e. (*b*) in § 16.24, implies that there is a common normal for all points of contact. This common normal is the optic axis. The axis of single ray velocity lies within the circle, but does not intersect it quite centrally. The shape of the wave surface in the region of the singular directions may be likened to that of the crater of a volcano. If a ray is directed on to the surface of a biaxial crystal in a certain direction, there is only one refracted wave normal, and the direction of this wave normal coincides with the optic axis. That there is only one refracted normal is expressed geometrically by the fact that, for this direction of incidence, only one tangent plane can be drawn from P to the wave surface. The algebraic expression of the same experimental background is that for this direction the two solutions of Fresnel's equation [16(18)] coincide, so that only one angle θ_0 satisfies equation 16(71). When the refracted wave normal coincides with the optic axis, the vector **D** may have any direction provided that it is perpendicular to the wave normal. Application of 16(52) then shows that each possible direction of **D** corresponds to a possible ray. These rays form a hollow cone. One member (OA) of this cone of rays is drawn to the circular section C of the wave surface (fig. 16.12), and this ray coincides with its wave normal, i.e. with the optic axis.

16.27.—From the above discussion we may expect that a ray incident upon the crystal in a suitable direction will give rise to a hollow cone of rays within the crystal. Each member of this cone will be refracted back to the direction of the incident ray upon emergence. The emergent light should, therefore, form a hollow cylinder which may be received upon a suitably placed screen S as a circular ring of light (fig. 16.13). The diameter of this circle should be independent of the distance of the screen from the crystal. In the early experiments of Lloyd, a

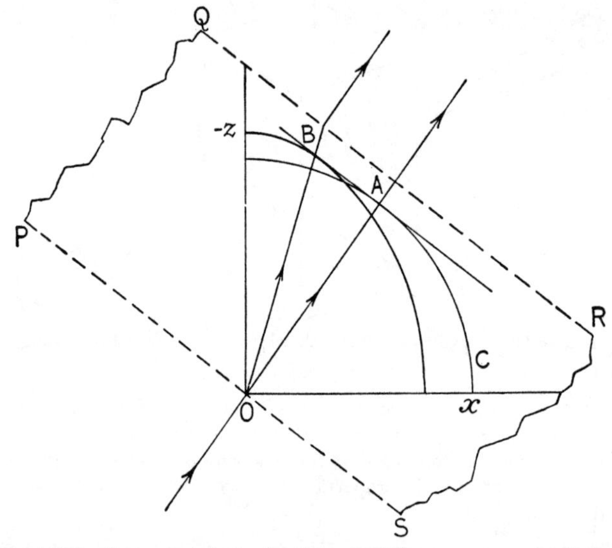

Fig. 16.12.—Internal conical refraction. PQRS represents a crystal slice with plane faces PS and QR perpendicular to the wave surface which touches the surface at A in a circle.

single ring of light was obtained. Later observations by Poggendorf and by Haidinger, using a better technique, show that the ring of light is divided into two parts by a fine dark line. This line occupies the position calculated for the ring corresponding to internal conical refraction, and its presence shows that the above discussion is incomplete. The following treatment is due to Voigt.

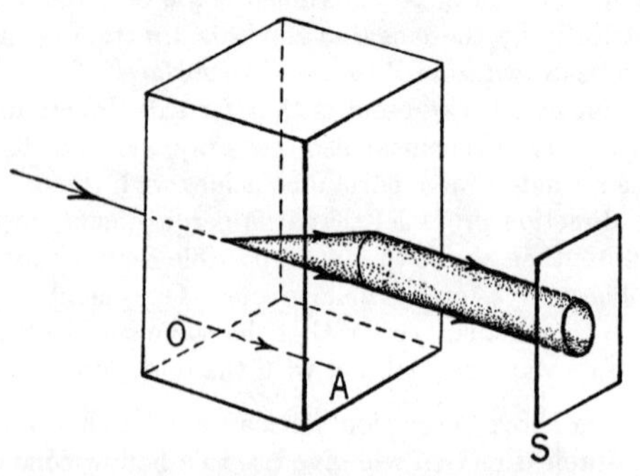

Fig. 16.13.—Internal conical refraction

16.28.—Even if effects due to diffraction could be neglected, no experimental arrangement would produce an ideal mathematical pencil of light. Let us therefore consider a small solid cone of rays incident upon the crystal so that the axis of the cone is *near* the direction corresponding to internal conical refraction. The small, but finite, semi-angle of the cone is denoted by γ, and the angle between its axis and the direction of conical refraction by β. Let us suppose that at first β is much larger than γ. Then there is no conical refraction in the sense of the preceding discussion. There are two small cones of rays within the crystal, and the emergent beam also consists of two small cones of rays which give spots of light on the screen. The two small cones of rays within the crystal correspond to two small areas on the wave surface. If the axis of the small cone of incident rays is allowed to approach the direction of internal conical refraction, the two small areas on the wave surface approach the rim of the " crater "—one from

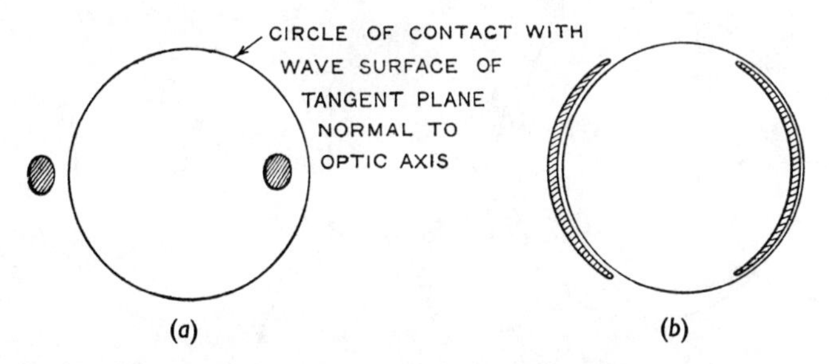

Fig. 16.14.—Showing the formation of the double ring of light observed in internal conical refraction: (*a*) β much greater than γ, (*b*) β only slightly greater than γ.

inside and the other from outside (fig. 16.14*a*). When β is only slightly greater than γ, the small areas begin to assume the shapes shown in fig. 16.14*b*. When $\beta = 0$, the two spots have each spread into a circular ring whose width is proportional to γ. The two rings are still separated by a thin dark line. Let us now divide our small finite bundles of rays into a set of infinitesimal hollow cones and consider the one lying between γ and $(\gamma + d\gamma)$. The amount of light energy in this cone will be proportional to $2\pi\gamma\, d\gamma$, and this light will go to form two rings of light, one on either side of the ring corresponding to conical refraction. The width of these rings will be proportional to $d\gamma$, and their area to $2\pi R\, d\gamma$, where R is the radius of the ring corresponding to conical refraction; thus the light per unit area will be proportional to γ/R, so it will fall to zero as γ itself falls to zero. Hence the illumination is zero along the line corresponding to $\gamma = 0$, i.e. to a ray coincident with the direction of conical refraction. It rises approximately linearly as we go away from the ring corresponding to conical refraction, until we come to the edge of the rings defined by γ.

16.29.—From the preceding paragraph we can see that the dark ring is present because the amount of light making an infinitesimal angle $d\gamma$ with the optic axis is proportional to $(d\gamma)^2$, whereas the amount in an infinitesimal range of angle between γ and $(\gamma + d\gamma)$ is proportional to $d\gamma$. The first amount is infinitesimal compared with the second. So long as one of the " spots " on the wave surface corresponding to a narrow cone of rays is clear of the rim of the crater, its area is proportional to the semi-angle of the cone. When it reaches the rim, one dimension is determined by the radius of the rim (and is finite), whereas the other is infinitesimal, i.e. the area over which a given amount of light is spread becomes of the first order of small quantities, whereas the amount of light remains of the second order. Thus the illumination (light per unit area) becomes infinitesimal.

16.30.—The direction of **D** for any ray may be obtained from the condition that **D** is coplanar with the ray and the wave normal. As explained at the end of § 16.26, one member of the internal cone of rays coincides with the wave normal, i.e. with the optic axis. The directions of **D** for the others are shown in fig. 16.15*a*. These directions may be verified experimentally, remembering that the plane of

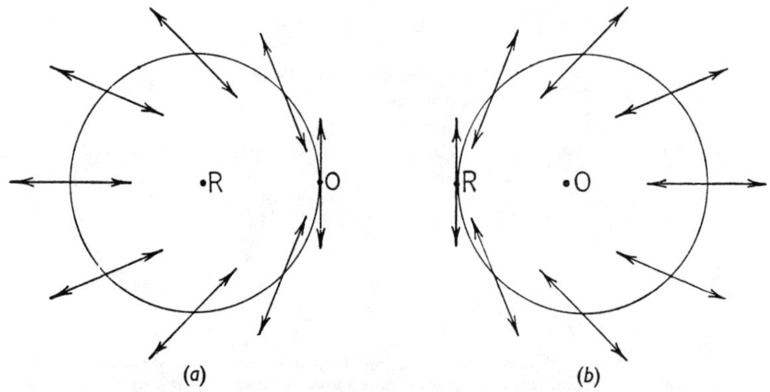

Fig. 16.15.—Directions of **D** for rays in (*a*) internal, (*b*) external conical refraction

polarization is always normal to the plane of **D**. In a similar way, when we consider external conical refraction, there is an internal cone of wave normals. One wave normal (the one normal to the circular section of the wave surface) coincides with its ray, i.e. with the axis of single ray velocity. The directions of **D** for other wave normals are shown in fig. 16.15*b*. Each emergent ray is polarized in the same way as the wave normal from which it is derived.

16.31. Transmission of Convergent Plane-polarized Light through a Thin Crystal Slice.

In § 12.41 ff. we considered the colours produced when a parallel beam of light was passed first through a polarizer, then normally through a thin crystal slice, and finally through an analyser using the the arrangement shown in fig. 12.21. It would be possible to study this phenomenon in greater detail by rotating the slice through moderately small angles (about lines perpendicular to the direction of light) in order to see how the colours change. It is, however, more convenient to view all directions of transmission at once, using the apparatus shown in fig. 16.16. A beam of light is made approximately parallel by the lens L_1; it is passed through a polarizer and then rendered fairly strongly convergent by the lens L_2. After passing the crystal slice X, it is received by the lenses L_3 and L_4. The analyser is placed between these lenses in a position where the light is approximately parallel. Taken together, these lenses form a telescopic system. The rays which reach a given point on the screen S have passed through different points of the crystal, but all in the same direction. If a knife edge is moved across the crystal, its image does not appear on the screen S, but the illumination on all parts of the screen is reduced in proportion to the area of the crystal which is obscured. Since

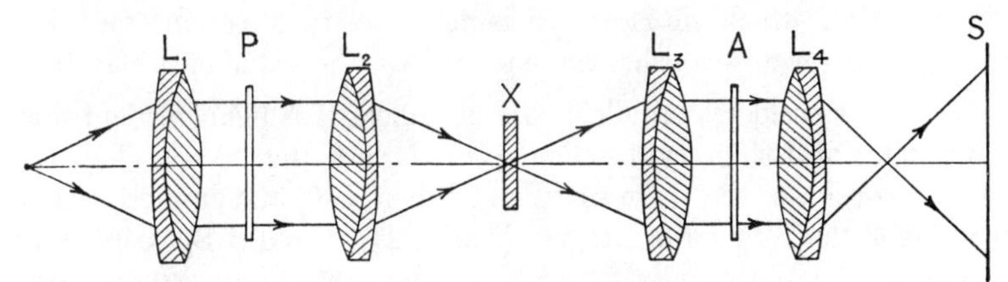

Fig. 16.16.—Apparatus for showing isochromatic lines *

all parts of the crystal contribute to the light at any one point, clear patterns are obtained only when all the part of the crystal which is being used is free from inclusions, etc., and when it is of nearly uniform thickness. The picture on the screen S is not an image of the crystal, but a representation in which each point gives the colour and amount of light transmitted through the slice in a given direction. To obtain a complete review of the whole phenomenon, pictures of this type for slices of different thicknesses cut in several different directions relative to the crystal axes are required. Examples of the results are shown in Plate IV, p. 518. The typical pattern consists of a set of lines along which the colour is constant (*isochromatic lines or rings*) and another set of lines which are uncoloured (*achromatic lines*). These latter are usually much less well marked than the former and are sometimes called *brushes*.

16.32.—In § 12.41 we obtained the following relation for the amount of light transmitted when the incident beam is parallel and normal to the slice:

$$E = a^2\{\cos^2(\alpha - \beta) - \sin 2\alpha \sin 2\beta \sin^2 \tfrac{1}{2}\delta\} \quad . \quad 16(72)$$

[equation 12(43)].

* The positions shown for P and A are suitable when Polaroid film is used. Nicol prisms would be placed at the left-hand and right-hand positions where the beams are narrow.

In this equation α and β depend on the orientations of the polarizer and the analyser relative to the slice. δ is the phase difference between two beams, each plane-polarized in one of the two possible directions for transmission normal to the slice. It is given by

$$\delta = \kappa(\mu'' - \mu')d, \quad . \quad . \quad . \quad . \quad . \quad 16(73)$$

where κ is the wavelength constant in air, and d is the geometrical length of the path in the crystal. We shall show that the isochromatic lines are lines joining points corresponding to directions of transmission for which δ is constant. The achromatic lines correspond to directions for which $\sin 2\alpha \sin 2\beta$ is zero, i.e. to directions for which the " colour term " vanishes no matter what the value of δ may be.

16.33.—Consider a parallel beam of unpolarized light incident upon the surface of a thin crystal slice at an angle θ_1 (fig. 16.17). Let A_1B_1 be the position of the wave surface at time $t = 0$, and suppose that in the crystal the light splits into two beams, A_2A_3'' and B_2B_3'' being wave

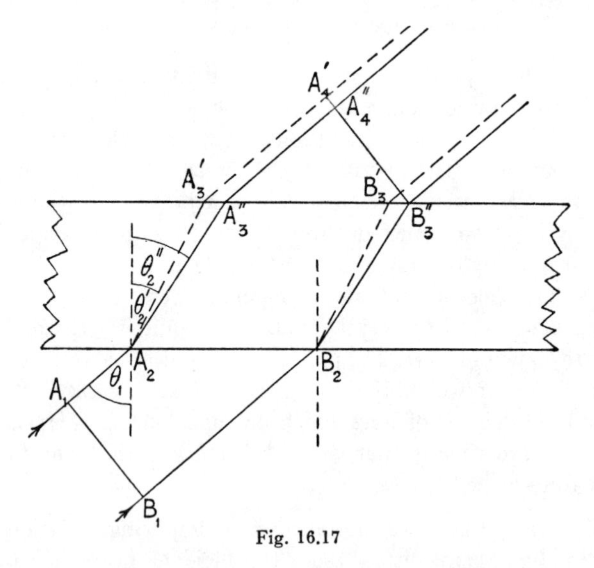

Fig. 16.17

normals for one beam and A_2A_3' and B_2B_3' for the other. $A_3''A_4''$ and $A_3'A_4'$ represent two of the emergent wave normals. On either side of the crystal the wave normals and the rays coincide, but within the crystal they are distinct. It is important to note that A_2A_3'', etc., are not rays but wave normals. In general, the rays would be out of the plane of the paper, but the wave normals must lie in the plane of incidence, which is the plane of the diagram. All the emergent wave normals are parallel to the incident wave normals (and therefore they are parallel to each other), and the wave surfaces are parallel after transmission. The emergent waves may be collected by a lens, and the components which pass through the analyser may be brought to interference at a point in its focal plane. The phase difference at the focal point will be the same as the phase difference on the plane $B_3''A_4'$ (§ 8.1). It is shown in § 5.13 that when a beam of light passes through an inclined transparent plate of thickness e, the transmitted wave suffers a phase delay δ given by

$$\delta = \kappa\mu e \cos\theta_2,$$

where θ_2 is the angle between the wave normal and the normal to the surface of the plate, and κ is the wavelength constant in air. It follows that, for a plate of anisotropic material, the phase difference between the two emergent rays is

$$\delta = \kappa e(\mu'' \cos\theta_2'' - \mu' \cos\theta_2'). \qquad . \quad . \quad 16(74)$$

Equation 16(74) expresses the fact that the phase difference between the two emergent beams depends on:

(*a*) the difference in the refractive index, and

(*b*) a small difference in the length of path in the crystal, since the wave normals make slightly different angles with the normal to the surface of the crystal.

We shall now show that, when the birefringence is small, the second effect may be neglected in comparison with the first. We then have

$$\delta = \frac{\kappa e}{\cos\theta_2}(\mu'' - \mu'), \qquad . \quad . \quad . \quad . \quad 16(75)$$

where $\cos\theta_2$ is the mean of $\cos\theta_2'$ and $\cos\theta_2''$.

When the birefringence is small, we may write

$$\delta = \kappa e(\mu'' - \mu')\frac{d}{d\mu}(\mu\cos\theta_2)$$

$$= \kappa e(\mu'' - \mu')\left\{\cos\theta_2 - \mu\sin\theta_2\frac{d\theta_2}{d\mu}\right\}. \qquad . \quad . \quad . \quad 16(76)$$

By differentiating the relation

$$\mu\sin\theta_2 = \sin\theta_1,$$

we obtain
$$\sin \theta_2 + \mu \cos \theta_2 \frac{d\theta_2}{d\mu} = 0, \qquad . \quad . \quad . \quad . \quad . \quad 16(77)$$

and 16(75) is obtained by eliminating $d\theta_2/d\mu$ between 16(76) and 16(77).

16.34. The Isochromatic Surfaces.

Let us now temporarily consider not a crystal slice but an anisotropic medium of unlimited extent. We take a point O and draw a

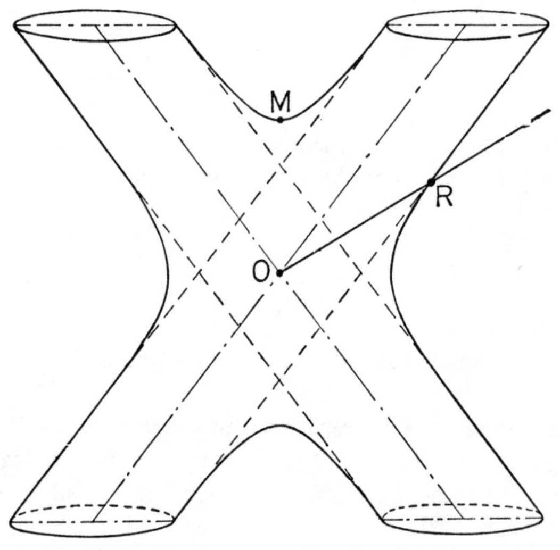

Fig. 16.18.—Isochromatic surface for a biaxial crystal

wave normal OR in a given direction (fig. 16.18). Suppose that two plane waves corresponding to this wave normal have the same phase when they pass through O. Then the phase difference when they pass through R is *

$$\delta = \kappa(n'' - n')r \qquad . \quad . \quad . \quad . \quad . \quad 16(78)$$

where OR $= r$.

A surface may be defined as the locus of all points for which δ has some chosen value (say δ_1), i.e. by drawing in each direction a radius vector whose length is inversely proportional to the difference of the indices. A family of surfaces is obtained by taking different values of the parameter δ. These surfaces are called the *isochromatic* surfaces. If the directions are defined by stating the angle which the wave

* Note that n (index to vacuum) is nearly equal to μ (index to air).

normal makes with the optic axes, then from 16(65) we obtain, as an approximation,*

$$\kappa r(\mu_z - \mu_x) \sin \chi_1 \sin \chi_2 = \delta_1 \quad . \quad . \quad . \quad 16(79)$$

for the equation of one of the isochromatic surfaces.

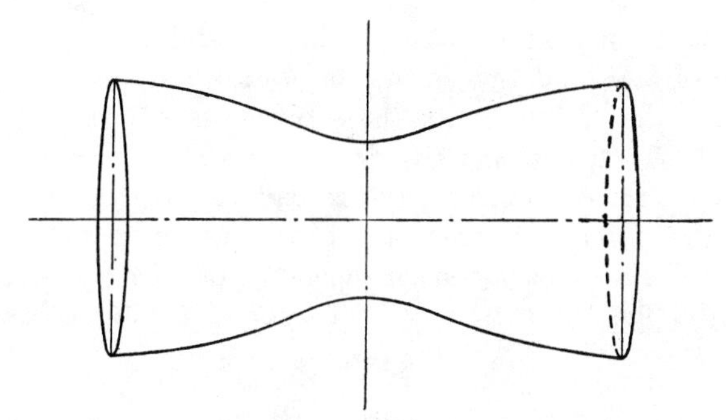

Fig. 16.19.—Isochromatic surface for a uniaxial crystal

The form of a typical isochromatic surface for a biaxial crystal is shown in fig. 16.18 and the reduced form for a uniaxial crystal is shown in fig. 16.19.

The equation of the isochromatic surface in Cartesian co-ordinates may be shown to be

$$[\Sigma\, x^2(\mu_y^2 + \mu_z^2) - \delta^2/\kappa^2]^2 = 4r^2 \,\Sigma\, x^2\mu_y^2\mu_z^2. \qquad 16(80)$$

We have, from equation 16(78),

$$\delta = \kappa r(n'' - n')$$
$$\doteqdot \kappa r(n_g'' - n_g'), \quad . \quad . \quad . \quad . \quad . \quad . \quad . \quad 16(81a)$$

since the angles between the rays and wave normals are small.

The difference between the ray indices n_g'' and n_g' may be obtained from equation 16(40), which may be written

$$\Sigma \frac{\rho_x^2}{n_g^2 - n_x^2} = 0.$$

From this equation, which is quadratic in n_g^2, we obtain

$$n_g'^2 + n_g''^2 = \Sigma\, \rho_x^2(n_y^2 + n_z^2), \quad . \quad . \quad . \quad . \quad 16(81b)$$
$$n_g'^2 n_g''^2 = \Sigma\, \rho_x^2 n_y^2 n_z^2.$$

* In this approximation we assume $b'b'' = b_x b_z$, and we also replace indices to vacuum (n_x, n_z) by μ_x and μ_z (indices to air).

Now, from 16(81a), $\quad [n_g'^2 + n_g''^2 - \delta^2/\kappa^2 r^2]^2 = 4n_g'^2 n_g''^2,$

whence, substituting from 16(81a) and writing $\rho_x = x/r$, etc., and replacing n_x, etc., by μ_x, etc., we obtain 16(80).

16.35. The Isochromatic Lines.

Let O be a point on one face of a thin crystal slice and let the isochromatic surface be drawn in the correct orientation, i.e. so that the directions of the optic axes for the surface are the same as those for the crystal. Then any line OR drawn from O to a point where the isochromatic surface intersects the second face of the crystal represents a wave normal for which the phase difference is δ_1. The projection of the directions of the corresponding rays *outside the crystal* on the screen S (fig. 16.20) by means of the lens L produces a series of

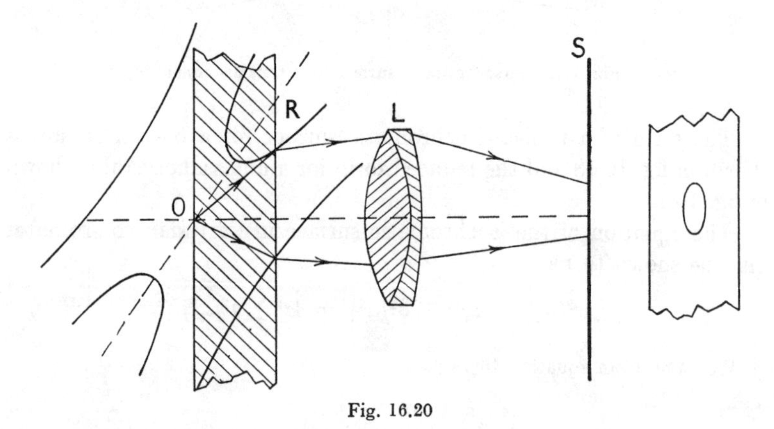

Fig. 16.20

lines. These lines are called the *isochromatic lines*, since for all points on one of these lines the " colour term " in equation 16(72) has the same magnitude, though not necessarily the same sign (§ 12.42). Only one hue, or its complement occurs because, along one line, the phase difference of the two interfering beams is the same for a given wavelength. If monochromatic light is used we obtain bright lines for $\delta_1 = 2m\pi$ and dark rings for $\delta_1 = (2m + 1)\pi$, where m is an integer. The isochromatic lines must not be regarded as a magnified picture of the intersections of the surface of the crystal with the isochromatic surface. The above procedure is only a way of defining directions and, for the purposes of the above argument, any point on the first surface of the crystal may be taken to be O. Note that the isochromatic lines are not even directly related to the directions of the wave normals in the crystal. Since θ_2 is a simple function of θ_1, the isochromatic lines have the same general appearance as the intersections of a plane with the isochromatic surface, and it is sufficient to discuss these intersections when we desire only a general qualitative discussion. From the shapes of the isochromatic surfaces (figs. 16.18 and 16.19) it may be seen that the isochromatic lines can assume a wide variety of different forms. We shall discuss a few of the more important cases.

16.36. **Uniaxial Crystal cut perpendicular to the Optic Axis.**

For a uniaxial crystal the isochromatic surface 16(79) reduces to

$$\kappa r(\mu_x - \mu_z) \sin^2 \chi = \delta_1, \quad . \quad . \quad . \quad . \quad 16(82)$$

where χ is the angle between the wave normal and the optic axis. Consider a crystal section normal to the optic axis. Obviously the pattern has circular symmetry. In the notation of fig. 16.17, $\chi = \theta_2$ and $\sin \theta_1 = \mu \sin \chi$. μ is the mean index of refraction for the two wave normals. The radius (R) of the isochromatic ring corresponding to a phase difference of $2m\pi$ is given by $R = f \sin \theta_1 = \mu f \sin \chi_m$, where f is a constant (proportional to the focal length of L_4 in fig. 16.16), and $\sin \chi_m$ is obtained by substituting $2\pi m$ for δ_1 in 16(82),

i.e. $$R = Am^{1/2},$$

where $$A^2 = \frac{\lambda \mu^2 f^2}{r(\mu_x - \mu_z)}. \quad . \quad . \quad . \quad . \quad . \quad 16(83)$$

A is independent of θ_1 if we neglect the slow variation of μ with direction. The radii of the rings are thus proportional to the square roots of the natural numbers, so that their general appearance is like Newton's rings. For a given value of m, the radius is a function of the wavelength, partly because λ appears directly in 16(83), and partly because the indices depend upon the wavelength. The rings are coloured and, for high values of m, the rings of different colours overlap so that they cannot be seen clearly (see Plate IV*a*, *b*).

16.37. **Achromatic Lines (Uniaxial Crystal).**

Each point in the ring system corresponds to a single direction of incidence upon the crystal. All the points on a radial line correspond to directions of incidence (and of transmission) which lie in a certain plane containing the optic axis. From the discussion of § 16.21, it follows that the two beams in the crystal are polarized in and perpendicular to this plane. The polarizer will be set to transmit light polarized in a plane corresponding to a certain radius. For any point on this radius there is only one beam in the crystal, and no interference can take place, i.e. at all these points the colour term in 16(72) is zero because $\alpha = 0$ and $\sin 2\alpha = 0$. Similarly, for a perpendicular radius, $\alpha = \pi/2$ and again $\sin 2\alpha = 0$. A grey cross is thus superimposed upon the ring pattern. A second grey cross is similarly related to the analyser. The brightness in these parts of the pattern is proportional to $\cos^2 (\alpha - \beta)$. It may be shown that the colour term, which vanishes at points in one of these grey crosses, is small for points fairly near the crosses. Thus the crosses are not sharply defined and are called " brushes ". When $\alpha = \beta$ the crosses coincide to give a single white cross (Plate IV*b*). Similarly, when the analyser and polarizer are crossed, there is a single black cross (Plate IV*a*). When $\sin 2\alpha$ or $\sin 2\beta$ passes through zero, it changes sign. Thus the hue of a given ring changes to its complement as we pass one of the grey crosses and the rings appear broken. Since both $\sin 2\alpha$ and $\sin 2\beta$ change sign on passing the black or the white cross, there is no change in the sign of the colour term and the rings are unbroken.

16.38. **Biaxial Crystal cut perpendicular to a Bisector of the Optic Axes.**

Sections of the isochromatic surface for a biaxial crystal (fig. 16.18) by planes normal to the bisector of the optic axes and at different distances from O are shown

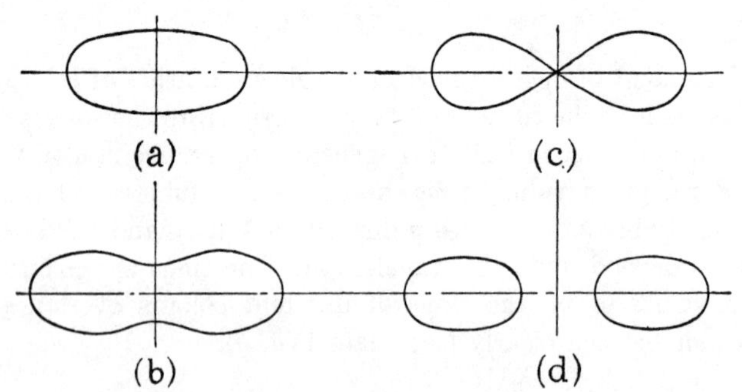

Fig. 16.21.—Sections normal to the bisector of the optic axes of the isochromatic surface for a biaxial crystal

in fig. 16.21. The shape of the section depends on the relation of the plane to the saddle in the isochromatic surface. When the plane is below M (fig. 16.18), then the section is like 16.21*a* or *b*. When it touches M, we obtain fig. 16.21*c*; and when

(*a*) to (*d*). Patterns obtained with crystals in convergent plane-polarized light.

(*a*) Uniaxial crystal cut normal to the axis, polarizer and analyser crossed.

(*b*) Uniaxial crystal cut normal to the axis, polarizer and analyser parallel.

(*c*) Biaxial crystal cut perpendicular to bisector of the axes (polarizer and analyser crossed).

(*d*) Uniaxial and optically active crystal (quartz) cut perpendicular to the axes (polarizer and analyser crossed).

(*e*) and (*f*) " Chameleon " made of mica of different thicknesses viewed in parallel plane-polarized light. The sign of the colour term is reversed by rotating the analyser.

PLATE IV

This plate is reproduced in full color on the inside back cover.

it is above M, we obtain curves like fig. 16.21*d*. For the moment let us consider light of one wavelength. For a thin slice of crystal even the isochromatic surface corresponding to $\delta = 2\pi$ (i.e. $m = 1$) is cut beneath the saddle point, and the isochromatic lines are all like those in fig. 16.21*a* or *b*. For a thicker slice, a complete family of curves is obtained (Plate IV*c*).

16.39. **Achromatic Lines (Biaxial Crystal cut normal to Bisector of Axes).**

We shall show that the achromatic lines are branches of rectangular hyperbolas. In general there are two hyperbolas, one with asymptotes parallel and perpendicular to the direction of the polarizer, and the other similarly related to the plane of the analyser. When the analyser and polarizer are parallel, we obtain a single white hyperbola, and when they are perpendicular, a single black hyperbola. When the direction of either polarizer or analyser is parallel or perpendicular to the plane of the axes, the corresponding hyperbola coincides with its asymptotes so that we obtain a grey cross. When both the polarizer and the analyser are parallel or perpendicular to the plane of the axes, a white cross is obtained when they are parallel to each other and a black cross when they are perpendicular to each other (Plate IV*c*).

16.40.—Let R (fig. 16.22) represent a point on the achromatic lines due to the polarizer, and let OX and OY be directions parallel and perpendicular to the plane of the analyser. Let P′ and P″ be the points corresponding to the optic

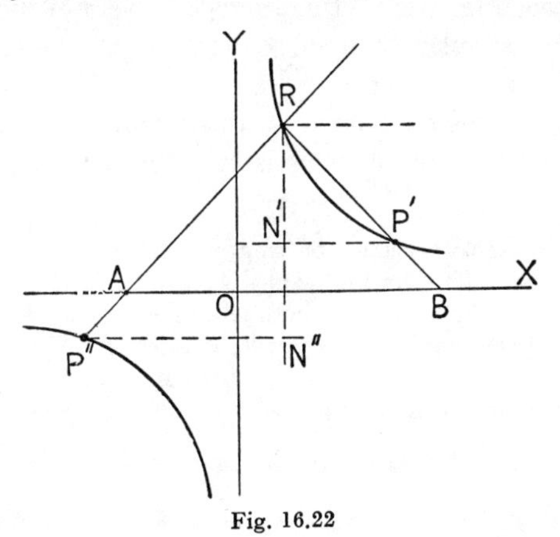

Fig. 16.22

axes. The directions of vibration corresponding to R are the internal and external bisectors of the angle between P′R and P″R. The achromatic line is the locus of a point R which moves so that these bisectors are parallel to OX and to OY. If the co-ordinates of R are x and y, and those of P′ are X and Y, those of P″ will be $-X$ and $-Y$. The triangle RAB is isosceles and

$$\tan \mathrm{RBO} = \tan \mathrm{RP'N'} = \frac{y - Y}{X - x}.$$

Similarly, $\tan \mathrm{RP''N''} = \dfrac{y+Y}{X+x}$,

and hence $xy = XY = \text{constant}$,

i.e. the locus of R is a rectangular hyperbola whose asymptotes are parallel and perpendicular to the plane of the polarizer.

16.41. **Transmission of Circularly Polarized Light.**

If a suitably oriented quarter-wave plate is inserted between P and L_2 (fig. 16.16), the light incident upon the crystal is circularly polarized. The emergent beam may be analysed either with a linear analyser (e.g. a Nicol) or with a circular analyser consisting of a quarter-wave plate followed by a linear analyser set to bisect the privileged directions of the quarter-wave plate. The linear analyser resolves the beam which it receives into plane-polarized components, and transmits the component in one plane. In a corresponding way the circular analyser may be regarded as resolving the incident beam into right-handed and left-handed circularly polarized components, of which one is transmitted. It may be shown that when the linear analyser is used, both isochromatic and achromatic lines are obtained, though the patterns are simpler than those described above (§§ 16.35–16.40). When the circular analyser is used, there are no achromatic lines. The isochromatic lines are similar to those previously described. This separation of the two sets of lines is sometimes an experimental convenience.

16.42.—It is easier to consider the transmission of circularly polarized light through a crystal plate from first principles than to use 16(72). Consider a beam of circularly polarized light incident normally upon a thin crystal slice. As in § 12.41, let OX and OY represent the privileged directions of vibration. Then the incident beam may have components $\xi_x = a \cos \omega t$ and $\xi_y = a \sin \omega t$. The emergent beam will have components

$$a \cos \omega t \quad \text{and} \quad a \sin (\omega t + \delta), \qquad 16(84)$$

and the components passing a linear analyser set at an angle β to OX are

$$\left.\begin{aligned} &a \cos \beta \cos \omega t, \\ \text{and} \quad &a \sin \beta \sin (\omega t + \delta). \end{aligned}\right\} \qquad 16(85)$$

The amount of light transmitted by the analyser is given by

$$E = a^2(1 + \sin 2\beta \sin \delta). \qquad 16(86)$$

There is a white term and a colour term. The value of β determines whether the latter is positive, zero, or negative. With convergent light there is only one set of achromatic lines. By moving the origin of time we may write the components of 16(84) in the form

$$x = a\cos(\omega t - \tfrac{1}{2}\delta) \quad \text{and} \quad y = a\sin(\omega t + \tfrac{1}{2}\delta). \qquad 16(87)$$

These may be resolved into

$$x_1 = a\cos\omega t\cos\tfrac{1}{2}\delta \quad \text{and} \quad y_1 = a\sin\omega t\cos\tfrac{1}{2}\delta$$

and

$$x_2 = a\sin\omega t\sin\tfrac{1}{2}\delta \quad \text{and} \quad y_2 = a\cos\omega t\sin\tfrac{1}{2}\delta,$$

i.e. into a left-handed and a right-handed circularly polarized component. If the circular analyser is set to transmit right-handed circularly polarized light, we have

$$E = a^2\sin^2\tfrac{1}{2}\delta, \qquad 16(88)$$

and, if it is set to transmit left-handed circularly polarized light, we have

$$E = a^2\cos^2\tfrac{1}{2}\delta = a^2(1 - \sin^2\tfrac{1}{2}\delta). \qquad 16(89)$$

Applying this result to convergent beams, we see that in either case there are no achromatic lines. The isochromatic rings depend on $\sin^2\frac{1}{2}\delta$ and are thus similar to those obtained with plane-polarized light. The left-handed circular analyser gives colours which are complementary to those obtained with the right-handed circular analyser.

16.43. Optically active Crystals in Convergent Polarized Light.

The most general type of crystal is both anisotropic and optically active. In convergent polarized light these crystals produce characteristic patterns of isochromatic and achromatic lines. The detailed calculation of these patterns involves no new principles but the calculations are rather lengthy. It is necessary to apply the methods of the preceding paragraphs to crystals in which, in a given direction, two elliptically polarized beams of light can be transmitted unchanged. Plate IV*d* (p. 518) shows the pattern obtained when convergent plane-polarized light is passed through a quartz plate cut normal to the axis and the emerging vibration is passed through an analyser set perpendicular to the polarizer. The absence of the cross in the centre leads to the assumption that, in a uniaxial optically active crystal, the two sheets of the wave surface do not touch. The one sheet encloses the other completely. The phenomena occurring in optically active biaxial crystals are still more complicated, and experimental data on most of these crystals are confined to measurements of the principal indices and of optical rotatory power for transmission in the directions of the axes. In general, the rotatory power for transmission along one axis is not equal to the rotatory power for transmission along the other.*

16.44. Measurement of the Optical Constants of Crystalline Media.

The relations between the optic axis (or axes) and the axis (or axes) of crystal symmetry may be determined by cutting several slices from

* See p. 154 of Reference 16.2.

a crystal and examining them in convergent polarized light. Usually one of the slices will be sufficiently near to the optic axis of a uniaxial crystal to show the centre of the ring pattern, and hence give approximately the direction of the axis. The direction may then be obtained more accurately by cutting a slice normal to this direction. If the rings are not quite central when the centre of the convergent beam of light is normal to the slice, then the small rotation needed to make them central enables the angle between the normal to the slice and the optic axis to be calculated. When the optic axis of a uniaxial crystal is known, the values of μ_o and μ_e may be obtained by measurements of the appropriate critical angles (§ 12.21). The dispersion of the birefringence may be examined by measuring the critical angle for different wavelengths, or by forming a pair of spectra (as suggested in § 12.21), or by measurements on the channelled spectrum (described in §§ 12.41–42). Similarly, observations with monochromatic convergent polarized light may be used to determine the plane of the axes of a biaxial crystal. Suppose that a section has been cut normal to one of the bisectors of the axes. For the axes used earlier in this chapter, this direction must be either OX or OZ, and we take it to be OZ. By measuring the critical angle corresponding to the circular section of the wave surface, with the incident ray in the plane of the axes (i.e. the XZ plane), we obtain μ_y directly. By measuring the corresponding angle for a ray incident in the YZ plane, we obtain μ_x. The other index may be obtained directly by cutting another section so as to be able to use a ray incident in the XY plane. It may also be obtained by measurements on the second ray in the XZ plane. When the values of the three indices are known, the angle between the optic axes may be calculated. The result may be confirmed by measurement of the angular separation of the " eyes " of the lemniscate pattern (Plate IV*c*, p. 518). This measurement is usually made by rotating the crystal so as to bring the two " eyes " in turn into the centre of the field of view. This reading gives the " apparent " angle between the axes, i.e. the angle outside the crystal between the beams of light which, inside the crystal, have normals in the directions of the axes. The true angle may be calculated from the apparent angle if μ_y is known [Example 16(xviii), p. 536]. The dispersion in biaxial crystals may be obtained by measuring the indices with different kinds of monochromatic light in turn. The angle between the principal (optical) directions and the crystallographic axes is the same for all wavelengths. Each principal index varies independently with wavelength so that the ratios of the principal indices vary with wavelength. Hence the angle between the optic axes varies with wavelength. It is even possible that the angle may be zero for one wavelength and not for others, so that the crystal is uniaxial for one wavelength and biaxial for others.

16.45. Relation of Optical Anisotropy to Crystal Structure.

By measuring the optical constants of a large number of crystals, it is possible to deduce the following empirical relations between optical and crystallographic properties:

(*a*) Crystals of the cubic system are always isotropic; the index ellipsoid is a sphere.

(*b*) Crystals which have a single principal axis (trigonal, tetragonal, or hexagonal) are uniaxial; the optic axis coincides with the principal axis of crystal symmetry; the index ellipsoid is a spheroid.

(*c*) Crystals which have three axes of different lengths (orthorhombic, monoclinic, and triclinic) are biaxial; the ellipsoid has three unequal axes. When there are three diad axes, the principal axes of the ellipsoid coincide with these axes. When there is one diad axis (monoclinic crystals), then one of the principal axes of the ellipsoid coincides with this axis. When there is no diad axis (triclinic system), the principal axes of the ellipsoid may have any arbitrary relation to the crystal axes.

These results are summarized in Table 16.1 which is due to Born.*

TABLE 16.1

Crystallographic symmetry	Triclinic	Monoclinic	Orthorhombic	Trigonal Tetragonal Hexagonal	Cubic
Orientation of ε surface	Unrelated to crystal axes and depending on wavelength.	One axis parallel to 2-fold crystal axis; the directions of the other two depend on wavelength.	Three axes fixed and independent of wavelength.	One axis parallel to 3-, 4- or 6-fold crystal axis; the other two unrelated to crystal axes.	All orientations are equivalent.
ε surface	Triaxial ellipsoid			Spheroid	Sphere
Optically	Biaxial			Uniaxial	Isotropic

* See p. 232 of Reference 16.3.

16.46. **Dispersion.**

The above relations are associated with certain limitations in regard to the dispersion properties of crystals. In uniaxial crystals the optic axis, since it coincides with a crystallographic axis, has the same direction for all wavelengths. The values of μ_o and μ_e are quite free and $(\mu_o - \mu_e)$ is not in general proportional to the mean index. It is possible for a crystal to be positive uniaxial for one wavelength and negative uniaxial for another. In the orthorhombic system the axes of the indicatrix must be in the same direction for all wavelengths (since they coincide with the crystallographic axes). The optic axes are situated in the plane which contains the largest and the smallest velocity. In the preceding discussion of this chapter, we have assumed $b_x > b_y > b_z$, and the axes are then in the xz plane. If this relation is true for all wavelengths, then the optic axes are always in the xz plane and b_y is a common bisector for them all. This is known as *normal dispersion.* It is possible, however, that for some wavelengths $b_x > b_z > b_y$, so that the axes are in the xy plane, or $b_y > b_x > b_z$, giving axes in the yz plane. Thus, in this type of crystal, the axes for different wavelengths may lie in any one of three mutually perpendicular planes. If the axes for two wavelengths lie in one plane, then they have a common bisector. In the monoclinic system of crystals, one axis of the indicatrix is fixed, and the other two must be in a plane perpendicular to this axis. They may rotate through any angle provided they remain perpendicular to one another. Two cases arise:

(i) If the fixed axis is the bisector of the optic axes for one wavelength, then the optic axes for other wavelengths may lie in different planes but have a common bisector (inclined dispersion).

(ii) The optic axes may be in the plane perpendicular to the principal crystallographic axis for all wavelengths. In this case, there will be no common bisector.

If the relation between the magnitudes of the principal velocities varies in the extreme way suggested as possible for orthorhombic crystals, it is possible that the axes for some wavelengths will be situated as in case (i) and those for other wavelengths as in case (ii). This extreme variation is, however, unlikely to occur with monoclinic crystals. If the principal crystal axis is associated with the largest velocity for one wavelength, it is usually associated with this velocity for all wavelengths. In triclinic crystals no limitation of this type exists and the dispersion of the axes is very complicated. It is possible to recognize the different types of dispersion described above by inspecting the lemniscate figures. Observation of the type of dispersion, together with the apparent angle between the axes, forms a method of identifying certain crystals in mineralogical specimens. By a suitable modification of the apparatus shown in fig. 16.16, the observations may be made on very small specimens.

16.47. **Relation to Molecular Structure.**

The calculation of the indices for crystalline media forms a development of the dispersion theory described in Chapter XV. Many problems discussed in that chapter reappear—for example, the refractivity of the bond between two atoms or ions. It is possible to approach the problem from the two viewpoints described in Chapter XV, i.e. to consider the relation between **D** and **E** when a high-frequency field is applied, or to consider the crystal as a set of scattering centres —the scattered radiation being mainly coherent with the incident radiation. As in isotropic substances, there is a relation between dispersion and absorption, strong anisotropic dispersion being accompanied by anisotropic absorption. The essentially new problem introduced is to account for the anisotropy of the optical properties of the crystal in terms of the arrangement of atoms in the crystal lattice or by the regular arrangement of molecules which are electrically anisotropic. The molecules may be electrically anisotropic individually, or they may owe their anisotropy to the polarizing effects of interactions with their neighbours. Wooster * has given an empirical classification of the optical properties of non-molecular crystals. This shows that strong double refraction tends to occur when:

(*a*) the atoms or molecules are arranged in parallel layers (e.g. calcite or PbO);

(*b*) one ion or group is strongly planar (CO_3, NO_3, etc., form planar ions; this is an additional reason for the strong birefringence of calcite);

(*c*) the system forms a chain lattice (e.g. NaN_3).

Weak double refraction occurs when the general arrangement of the crystal is a three-dimensional lattice with no strong geometrical anisotropy (e.g. quartz) and also with compounds containing ions like SO_4, which are themselves nearly isotropic.

16.48. Calculations of Birefringence from Crystal Structure.

Direct calculations from the theory of crystal lattices have been made by Ewald, Born, etc. These calculations are very difficult and they do not give all the indices directly. Instead, they provide certain relations between the different optical properties. The fact that these relations were in accord with experiment gave an important verification of the theory of crystal structure developed by Born and his school. Another approach is due to W. L. Bragg who uses the ionic refractivities (§ 15.26) as the basic data. He then shows how certain arrangements of molecules are likely to make the ratio of the polarization **P** (and therefore of **D**) to **E** anisotropic. For example, if the atoms of a molecular group lie in one plane, the resultant polarization is less when **E** is perpendicular to the plane of the group than when it is

* See p. 177 of Reference 16.2.

in the plane. This is because when all the atoms are polarized in their own plane the field of each dipole tends to reduce the polarization of its neighbours (fig. 16.23).

Results of calculations by this method for calcite and for several other substances are in good agreement with experiment.* For these

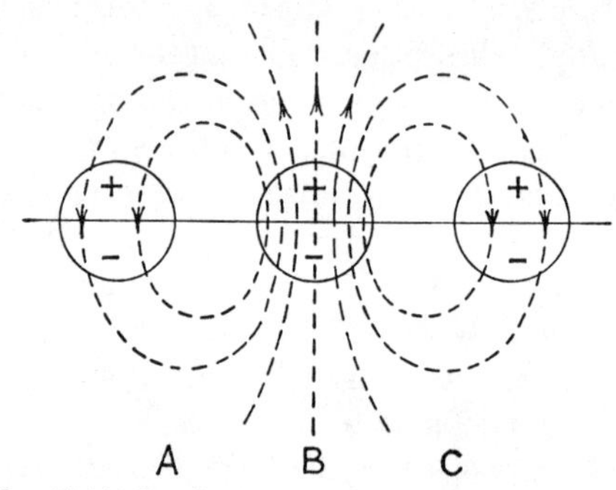

Fig. 16.23.—Field of a dipole at B. The dotted lines show the directions of the lines of force due to the dipole. The field of the dipole opposes the main field at A and C.

substances it is not necessary to assume that the ions themselves are electrically anisotropic. In general, the ions and molecules must be assumed to be anisotropic, and the anisotropy at optical frequencies bears no very direct and simple relation to the anisotropy at low frequencies.

16.49. **Optical Activity.**

It is not difficult to make a formal theory of optical activity (i.e. of a difference of velocity between left-handed and right-handed circularly polarized light). We may formally postulate that when the wave normal is in the OZ direction, D_x depends not only on E_x but also upon $\partial E_y/\partial z$, i.e. we write

$$D_x = \varepsilon E_x - i\gamma E_y; \quad D_y = \varepsilon E_y + i\gamma E_x; \quad \text{and } D_z = \varepsilon E_z,$$

where ε and γ are constants. It may then be shown† that the index for right-handed circularly polarized light is $n_R = \sqrt{\varepsilon} + \gamma/\sqrt{\varepsilon}$ and that for left-handed circularly polarized light is $n_L = \sqrt{\varepsilon} - \gamma/\sqrt{\varepsilon}$. This formal proof cannot, by itself, give very much satisfaction, since it is based on what appears to be a highly artificial assumption. The real problem is to relate this assumption to molecular theory. In isotropic substances such as solutions, the optical activity must be related to the structure of the molecule itself. It has been shown that molecules which possess " mirror symmetry " (i.e. which have two distinct forms, one being the mirror image of the other) should possess this type of property. The more

* See p. 185 of Reference 16.2. † See p. 411 of Reference 16.3.

difficult problem of relating the properties of an optically active crystal like quartz to the structure of the individual molecule or to the arrangement of the molecules in a "helical" lattice has also been solved in principle. The full calculation of the optical constants of a crystal which is both strongly active and strongly birefringent is naturally very difficult.

16.50. **Faraday Effect.**

In 1845 Faraday investigated the transmission of plane-polarized light through a block of glass in the presence of a magnetic field. He found that, when the direction of transmission has a component along the lines of force, there is a rotation of the plane of polarization. When H_p is the component of the field in the direction of transmission and l is the length of path, the angle of rotation θ_r is given by

$$\theta_r = CH_pl, \qquad \qquad 16(90)$$

where C is a constant.

This phenomenon is known as the *Faraday effect*, and the constant is called Verdet's constant. If C is positive, the direction of rotation is the same as that of a current which could produce the magnetic field. The sign, *viewed by an observer who looks along the field*, is independent of the direction of transmission of the light. This means that, when light is transmitted to and fro along a path in a magnetic field, there is a double rotation.* When the effect is very weak, it may be amplified by reflecting the beam several times backwards and forwards in the field. The effect has been measured in a large number of substances including liquids and gases. There is a universal positive effect associated with diamagnetism, and also a negative effect associated with para- and ferro-magnetism. Particularly strong effects are found when light is passed through thin films of iron, nickel, or cobalt, as would be expected in view of the very strong internal fields. The following rotations, obtained with a field of 10^4 oersteds, show the order of magnitude of the effect:

Water	2° 10′ for a 1-cm. path,	
Quartz †	2° 46′ „ „ „	
O_2	0·06′ „ „ „	(at S.T.P.)
Fe	130° for a 10^{-3} cm. path.	

The effect also varies rapidly in the neighbourhood of an absorption frequency. The theory is discussed on p. 353 of Reference 16.3. A detailed account of the Faraday effect is given in pp. 2119–2182 of Reference 16.1.

* The natural rotation when a beam is sent to and fro is zero (see § 12.35).

† Additional to the natural rotation for quartz.

16.51. The Kerr Effect.

In 1876 J. Kerr showed that many isotropic substances, when placed in an electric field, behave like a uniaxial crystal with the optic axis in the direction of the lines of force. If n is the index of the substance in the absence of a field, and n_p and n_s are the indices for directions of **D** parallel and perpendicular to the field, it is shown that

$$(a) \qquad (n_p - n_s) = \lambda B E^2 \quad \text{(Kerr's law)} \qquad . \quad . \quad . \quad . \quad 16(91)$$

$$\text{and } (b) \qquad (n_p - n) = 2(n_s - n) \quad \text{(Havelock's law)}. \qquad . \quad 16(92)$$

The constant B is known as Kerr's constant.* At high fields small deviations from Kerr's law have been observed (p. 137 of Reference 16.4) and Havelock's law is not universally obeyed (particularly when there is strong dispersion), but any satisfactory theory must include these laws as first approximations. The order of magnitude of the Kerr effect is shown by the following data:

	Substance	Kerr constant for λ = 5890 Å.
Gases	CO_2 (at S.T.P.)	$0{\cdot}25 \times 10^{-10}$
	N_2 ,,	$0{\cdot}36 \times 10^{-10}$
	CH_3Cl ,,	$8{\cdot}7 \times 10^{-10}$
Liquids	H_2O	$4{\cdot}7 \times 10^{-7}$
	CS_2	$3{\cdot}2 \times 10^{-7}$
Solids	$CHCl_3$	$-3{\cdot}46 \times 10^{-7}$
	Glasses	$2{\cdot}9 \times 10^{-9}$ to $1{\cdot}5 \times 10^{-8}$

Since the effect is proportional to E^2, its sign is independent of the sign of the field. Most substances behave in the field as positive uniaxial crystals (i.e. have positive Kerr coefficients), but a few behave as negative crystals.

16.52.—The Kerr constant may be measured by passing a beam of plane-polarized light between two condenser plates and compensating the resulting ellipticity by a Babinet compensator or by one of the more sensitive compensators described in Reference 12.6 (p. 392).

The term *Kerr cell* is usually applied to the condenser when it is intended for investigation of the electro-optical effect. The apparatus is shown in fig. 16.24*a*. Since the effect is proportional to E^2, and the constant is small, it is desirable to use as high a value of E as possible.

* In 16(91) E is measured in absolute e.s.u. (300 volts = 1 e.s.u.).

The distortion of field at the ends of the plates may be calculated or may be eliminated by using plates of different lengths. When the substance to be tested has some conductivity, the apparatus shown in fig. 16.24*b* is used. The light is passed through two Kerr cells in series. The directions of the fields are mutually perpendicular. The two

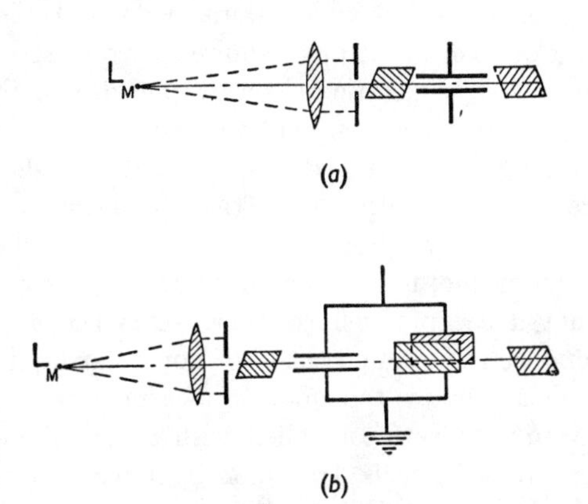

Fig. 16.24.—(*a*) Apparatus for the investigation of the Kerr effect. (*b*) Arrangement of apparatus for substances possessing appreciable conductivity.

Nicols are crossed, so that no light is transmitted in the absence of the field. The distances between the plates in one condenser are varied until, on applying a voltage of short duration, no transmission of light is observed. We then have

$$\frac{B_1}{B_2} = \frac{a_1{}^2 l_2}{a_2{}^2 l_1}, \quad . \quad . \quad . \quad . \quad . \quad . \quad . \quad 16(93)$$

where a_1, l_1 are the separation of the plates and the length of path for the substance whose Kerr constant is B_1, and a_2, l_2 are the corresponding quantities for the second substance whose constant is B_2. In this way the Kerr constant for a substance which cannot sustain a permanent field is determined in terms of the constant for an insulator. The above methods give $n_p - n_s$ but, in order to verify Havelock's law, it is necessary to measure $(n_p - n)$ and $(n_s - n)$. This may be done by means of a Jamin type of interferometer. The substance under test is placed in both arms of the interferometer and the readings of the compensator are obtained: (*a*) when no field is applied, and (*b*) when field is applied to the substance in one arm. An analyser is used to separate the light polarized parallel to the field from that polarized perpendicular, and thus the two differences $(n_p - n)$ and $(n_s - n)$ can be obtained.

16.53. The Kerr Electro-optical Shutter.

In his early experiments, which were made with glass, Kerr observed that the optical anisotropy required several seconds to reach full value when a steady electric field was applied, and that there was a corresponding delay when the field was removed. Later work has confirmed this long delay in many solids and has shown that, for viscous liquids which are also polar, there is a measurable, though very much smaller, delay. For example, it has been shown that the delay for undecyl alcohol is between 10^{-8} and 10^{-9} second. The delay is obviously not a sharply defined time, since the approach to an equilibrium condition is asymptotic. For non-polar liquids with small molecules, the delay time is too small to measure and is probably less than 10^{-11} second. A Kerr cell, filled with one of these liquids and placed between crossed Nicols, transmits light only when the electric field is applied. It is usual to orient the Nicols so that their principal planes are at 45° to the direction of the applied field. By using high-frequency fields it is possible to modulate a beam of light just as a beam of high-frequency radio waves can be modulated with an audio frequency. The modulation frequency can be made as high as 10^9 cycles per second without difficulty and frequencies up to 10^{10} cycles per second are possible. It is possible to apply pulses of high voltage and of short duration to the Kerr cell. In this way a light beam can be " chopped " into a series of pieces, each a centimetre or so long, separated by intervals of darkness. The arrangement of a Kerr cell between two crossed Nicols is known as the Kerr optical shutter. Its effect is the same as that of a mechanical shutter, but the frequency can be made enormously higher than that of a mechanical shutter. The Kerr shutter has many interesting technical applications.* Its use in connection with the accurate measurement of the velocity of light is described in Chapter X.

16.54. The Theory of the Kerr Effect.

In the Lorentz-Lorenz theory of dispersion in *isotropic* media, it is assumed that the electric action on a given electron due to a light wave may be calculated by imagining the electron to be placed at the centre of a small *spherical* cavity,

* Reference 16.5.

the net effect of the matter removed from the cavity being zero (see § 15.20). Havelock assumed that an external electric field makes this cavity elliptical, and from this assumption he deduced the relations

$$n_p - n = K \frac{(n^2 - 1)^2}{n} E^2, \qquad 16(94a)$$

and

$$n - n_s = \frac{K}{2} \frac{(n^2 - 1)^2}{n} E^2. \qquad 16(94b)$$

From these equations he obtained 16(91), 16(92), and also the relation

$$B\lambda = K'(n^2 - 1). \qquad 16(95)$$

This last relation gives the variation of B with the ordinary dispersion of the medium and is in agreement with a good deal of experimental evidence. A more detailed theory of the effect is due to Born and Langevin. They consider the effect to be due to (*a*) orientation of polar molecules, and (*b*) creation of electric moments in non-polar molecules and alteration of existing moments in polar molecules. The orientation effect is naturally important in polar liquids and in gases. It takes an appreciable time, of course, to alter the orientation of molecules in highly viscous liquids, and this accounts for the relaxation effects.

16.55. **Cotton-Mouton Effect.**

A magneto-optical effect closely analogous to the Kerr electro-optical double refraction was discovered by Cotton and Mouton in 1905.* The difference of indices $(n_p - n_s)$ varies with the square of the field as does the electro-optical effect, but the Cotton-Mouton constant is small compared with the corresponding Kerr constant. The following are some typical values for pure liquids:

Water ..	$-1{\cdot}1 \times 10^{-14}$	Benzene ..	$+75 \times 10^{-14}$
Acetone ..	$+4{\cdot}1 \times 10^{-14}$	Chloroform ..	-660×10^{-14}

It is necessary to use very strong magnetic fields to measure this constant. The technical difficulty is increased because, if the direction of transmission of the light is not perpendicular to the lines of force during the whole transit through the field, there is a rotation of the plane of polarization due to the Faraday effect. This comparatively large effect has to be distinguished from the small ellipticity of the transmitted light due to the Cotton-Mouton double refraction. The magnetic double refraction in gases and in amorphous solids is generally too small to measure. Certain interesting effects have been observed in crystals, but most of the investigations have been made on liquids such as those listed above. The theory is very similar to that of the Kerr effect. It is probable that the relations 16(94) and 16(95) apply, but the experimental data are not sufficient to provide accurate verification. The most important theoretical interest is in the wide range of values of the constant and their interpretation in terms of the relation between magnetic anisotropy and optical anisotropy (see §§ 16.50 and 16.51).

* There was an early observation of the effect by Kerr himself in 1901 but the first systematic investigation is due to Cotton and Mouton (see p. 360 of Reference 16.3).

16.56. **Photo-elasticity.**

In 1816 Brewster discovered that transparent isotropic materials become optically anisotropic when subject to mechanical stress. The effect may be investigated by means of the apparatus shown in fig. 16.25*a*. Light from the small source L_p is rendered nearly parallel by the lens L_1 and passes through the polarizer P, the specimen O, and the analyser A, to the lens L_2. *This lens forms an image of* O *upon the screen* S. The specimen is usually in the form of a thin plate.

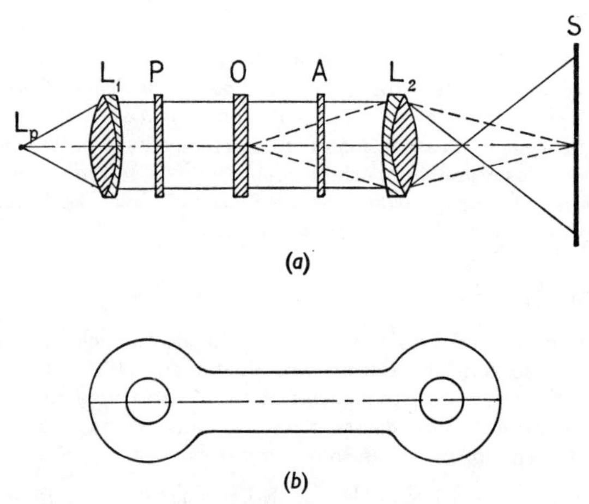

Fig. 16.25.—Demonstration of photo-elastic effect. The dotted lines show image formation by " rays " of wider angle than those actually used

The image of O is formed by narrow bundles of rays which are all nearly normal to the strip. It is convenient to cross the analyser and polarizer so that no light is transmitted when the specimen is free from stress. When the specimen is stressed (either by compression or extension) the light is restored. By using specimens of simple shape, such as that shown in fig. 16.25*b*, it is possible to investigate the relation between the stress and the double refraction. It is found that:

(*a*) The privileged directions of **D** for the stressed specimen (considered as if it were a crystal slice) are along the directions of the principal stress.

(*b*) If n_P and n_Q are the indices of refraction for directions of **D** parallel to the principal stresses P and Q at any point, then

$$n_P - n_Q = C(Q - P). \qquad 16(96)$$

In the central part of the specimen shown in fig. 16.25*b*, there is only one stress (P) which may be calculated from the load and cross-section. If we use monochromatic light and apply a gradually increasing load, the transmitted light reaches a maximum when $(n_P - n_Q)d$ is equal to $\lambda/2$; it falls to zero when $(n_P - n_Q)d$ is equal to λ, and so on. In this way the *relative stress-optical constant* (C) may be measured. To verify equation 16(96) completely it is necessary to subject a specimen to variable shearing force. The equation is found to be valid even beyond the elastic limit (Reference 16.6). The stress optical coefficient has the dimensions of the reciprocal of a stress. For glasses the coefficient lies between 10^{-13} and 10^{-12} square centimetre per dyne, but plastics such as Xylonite have coefficients in the range 10^{-12} to 10^{-11} square centimetre per dyne.

16.57.—The photo-elastic effect is used by engineers to investigate the stresses in structures where calculation would be very laborious. A model of the structure is made in a suitable plastic, and suitable stresses are applied gradually. When white light is used, *isochromatic lines* correspond to points on the model at which $(P - Q)$ has a constant value. Uncoloured lines correspond to points at which the directions of the principal stresses are parallel to the directions of the analyser and polarizer.* These are known as *isoclinic lines* since they are the loci of points at which the principal stresses make the same angle to an external system of co-ordinates. The isoclinic lines cross at *isotropic* points, i.e. when the *difference* of the principal stresses vanishes.† The photo-elastic method gives directly $(P - Q)$. The separate stresses may be obtained by using the fact that there is only one principal stress at a free boundary and applying a method of integration. This method is not very accurate and it is often more convenient to measure the lateral contraction [which is proportional to $(P + Q)$] at different points on the model by means of a mechanical extensometer.‡ Even when quantitative measurements are not made, the photo-elastic method can be used to show by inspection whether any point of a given structure is excessively stressed. The directions of the principal stresses at a given point can also be determined by rotating the specimen (keeping analyser and polarizer fixed) until the

* This may be seen from equation 16(72), remembering that the analyser and polarizer are crossed, and that the principal stresses are perpendicular to one another.

† Note that although the systems of coloured and uncoloured lines are generally similar to the patterns obtained with crystals and convergent polarized light, the mode of formation is entirely different. Here we are using nearly parallel light and each point on the screen corresponds to a point on the specimen.

‡ Reference 16.6.

isoclinics pass through the point. On the theoretical side, it is easy to see that an anisotropic mechanical deformation of a solid should cause optical anisotropy, but the calculation of the constant C from molecular structure is very difficult.

16.58. **Anisotropy in Liquids.**

It is found that liquids may show optical anisotropy when certain molecules—particularly long chain molecules—are present. Such molecules have a tendency to align themselves with their axes all pointing in the same direction. In a stationary liquid this tendency is opposed by the random thermal motions. Despite this opposition, local domains may be formed which show crystalline properties. These domains are very unstable except when the liquid is in layers whose thickness is of the same order as the length of the chain. In such thin films stable " liquid crystals " may be formed. Anisotropy is found in bulk liquid when there is lamellar flow. If adjacent layers of liquid have a relative velocity, the long chain molecules tend to lie across the velocity gradient, and optical anisotropy results. This phenomenon is of assistance in the investigation of the flow of fluids past obstacles. Measurements of the double refraction give the direction and magnitude of the velocity gradient at any point. Regions of turbulent flow are indicated by the absence of double refraction.

16.59. **General Conclusion.**

In this chapter we have considered natural and induced optical anisotropy from three aspects:

(*a*) in relation to the development of the electromagnetic theory of light;

(*b*) in relation to crystal and molecular structure;

(*c*) as the basis of methods for the practical investigation of stresses, velocity gradients, etc.

Our treatment has necessarily been somewhat condensed and in a number of places it has been possible to give only an indication of the main lines of certain developments.* We may now draw certain general conclusions.

In the first place, the electromagnetic theory can include in its scheme the main features of all the experimental results on natural and induced optical anisotropy. No theory which was " simpler "

* As an indication of the complexity of the subject, it may be stated that Reference 16.4 lists nearly 200 scientific papers on the Kerr effect and the Cotton-Mouton effect

would be adequate. An elastic-solid theory (with one vibration vector) must have great difficulty in relation to natural anisotropy, and cannot even begin the theory of the electro- and magneto-optical effects. The subject of optical anisotropy as a whole brings out the full power of the electromagnetic theory. It is not profitable to pursue calculations beyond a certain point because very laborious computations are not justified unless there is a specific point of practical or theoretical significance to be elucidated.

16.60.—Our basic knowledge of the structure of crystals is due to studies of X-ray diffraction and electron diffraction, and not to the study of optical anisotropy. The main optical properties of many crystals were known long before X-rays were used, but it was the latter which caused the great advance in our knowledge of the detailed arrangement of atoms in crystals and of the structure of large molecules. While recognizing that the optical study of crystals is only a secondary method, it is still a very important one. X-ray analysis may lead to more than one possible structure, and study of the optical properties may resolve the ambiguity. Also calculation of the optical constants when the structure of a crystal has been determined by X-rays forms an important check upon the correctness of the assumed structure. It is probable that a good deal more information about molecular structure will be obtained from detailed studies of the electro-optical and magneto-optical effects. The wider use of these methods has been restrained partly by technical difficulties, and partly by the difficulty of theory and calculation. Recent advances in methods for the production and measurement of high voltages (particularly at high frequencies) have reduced the technical difficulties, and the use of calculating machines should remove some difficulties of calculation. There still remains, however, the difficulty that, in general, the results of a single optical measurement depend on several molecular constants. Any one constant can be obtained only by taking differences (and not usually first differences) from several optical results. It is very difficult to get accurate determinations of molecular constants this way. On the other hand, if the constants have been determined independently, it is very useful to see whether they lead to the observed values for the electro-optical and magneto-optical constants.

EXAMPLES [16(xv)–16(xxi)]

16(xv). Discuss the validity of the approximation 16(75) for a beam of light incident at an angle of 45° upon a plate of calcite, 3 mm. thick, cut so that the optic axis is parallel to the surface, and the plane of incidence is perpendicular to the optic axis. Calcite is uniaxial. $\mu_x = \mu_y = 1{\cdot}658$, $\mu_z = 1{\cdot}486$ for $\lambda = 5893$ Å.

[16(75) gives $\delta = 0{\cdot}191\kappa e$ and 16(74) gives $\delta = 0{\cdot}193\kappa e$.]

16(xvi). A beam of convergent circularly polarized light is passed through a crystal and then through a linear analyser. Show that rotation of the analyser through $\pi/2$ changes all hues into their complements.

16(xvii). Convergent circularly polarized light is incident upon a plate of a unaxial crystal, cut normal to the optic axis. Show that the isochromatic lines obtained with a linear analyser are displaced by a quarter of an order from those obtained when the incident light is plane-polarized. Show that the sign of this displacement changes at an achromatic line.

16(xviii). Find the relation between the true and apparent angles between the optic axes. [If $2\theta_1 =$ the true angle and $2\theta_2 =$ the apparent angle, $\sin\theta_2 = \mu_y \sin\theta_1$.]

16(xix). Light is passed through a polarizer, a Kerr cell, and an analyser The polarizer and the analyser are crossed. Show that, for a weak field, the maximum transmission of light is obtained when the plane of the polarizer makes an angle of $\pi/4$ with the applied field. [Use equation 16(72).]

16(xx). A beam of light of circular frequency ω is passed through a Kerr shutter while an electric field of circular frequency p is applied to the plate. Obtain an expression for the profile of the transmitted light wave. Assume that the birefringence produced by the applied field is always so weak that the phase difference $(n_p - n_s)l\kappa$ is very small compared with 2π.

[The phase difference δ is proportional to $E_a^2\cos^2 pt$, where E_a is the amplitude of the applied field. When the Nicols are crossed, and at 45° to the applied field, the amount of light transmitted is obtained by putting $\alpha = -\beta = \frac{1}{4}\pi$ in 16(72), and is proportional to $\sin^2\frac{1}{2}\delta$, i.e. when δ is small, the amplitude is proportional to δ. Hence the light beam is represented by $\xi = AE_a^2\cos^2 pt\cos\omega t$, where A is a constant and the profile is $\xi_0 = AE_a^2\cos^2 pt$.]

16(xxi). Using the data of the previous example, make an analysis (into three components of different frequencies) of the light transmitted by the Kerr shutter.

$$[\xi = \tfrac{1}{2}AE_a^2[\cos\omega t + \tfrac{1}{2}\cos(\omega + 2p)t + \tfrac{1}{2}\cos(\omega - 2p)t].]$$

REFERENCES

16.1. Joos: *Theoretical Physics* (Blackie).

16.2. Wooster: *Crystal Physics* (Cambridge University Press).

16.3. Born: *Optik* (Springer, Berlin).

16.4. Beams: *Rev. Mod. Phys.*, 1932, Vol. 4, p. 133.

16.5. Lawrence and Beams: *Phys. Rev.*, 1928, Vol. 32, p. 478.

16.6. Filon: *Photoelasticity for Engineers* (Cambridge University Press).

CHAPTER XVII

The Interaction of Radiation and Matter

17.1.—In the classical wave theory, the absorption of light is described as a process in which electrons are accelerated by the electromagnetic field which represents light. The energy thus given to the electrons may be transformed into thermal energy by collisions with neighbouring atoms or molecules. The rate at which matter can absorb energy from the electromagnetic field is low, except when the frequency of the field is near to certain natural frequencies of the electrons. Absorption is regarded as a continuous process and there is no lower limit to the amount of energy which an atom or an electron can absorb. We shall now describe some experiments whose results do not fit into this picture. They appear to require the hypothesis that the absorption and emission of light by an atom or molecule are discontinuous processes in which the amount of energy exchanged is always the same for radiation of a given frequency. This fixed amount of energy is called the *quantum*. For different frequencies the size of the quantum is proportional to the frequency, i.e.

$$W = h\nu, \qquad \qquad 17(1)$$

where h is a universal constant known as Planck's constant. The value of h is $6{\cdot}62 \times 10^{-27}$ erg second.

Planck's original theory was concerned only with the interaction of radiation and matter. He hoped that it would involve only minor modifications of the classical electromagnetic theory of light and the classical electron theory of matter. This expectation was not fulfilled and it became necessary to modify the theory of light to include the fact that energy of the radiation field which represents a beam of light can change only by an integral number of quanta. It is also necessary to assume that an atom can change its energy only by discrete amounts, and hence that atoms can exist only in certain states separated by finite differences of energy. Nothing in the classical laws of electromagnetism, which are based on experiments with static fields or alternating fields of low frequency, would lead us to expect this effect.

The quantum theory thus becomes both a theory of radiation and a theory of matter instead of being merely a special hypothesis concerning their interaction. The theory is a connected whole and we cannot deal with a section of the theory concerning radiation without referring extensively to atomic structure. In this chapter we discuss the experimental basis of the quantum theory, giving special prominence to experiments on light and on electromagnetic radiation of shorter wavelength. The order is chosen for convenience of exposition and does not follow the historical order.

17.2. **The Photo-electric Effect.**

It is found that electrons are ejected from the surfaces of metals by light and by radiation of shorter wavelength (X-rays and γ-rays). If the radiation is able to penetrate the substance, electrons in the interior may be removed from their equilibrium positions. In this paragraph we are concerned with the emission of electrons from surfaces, and we shall call this the photo-electric effect, although, strictly,

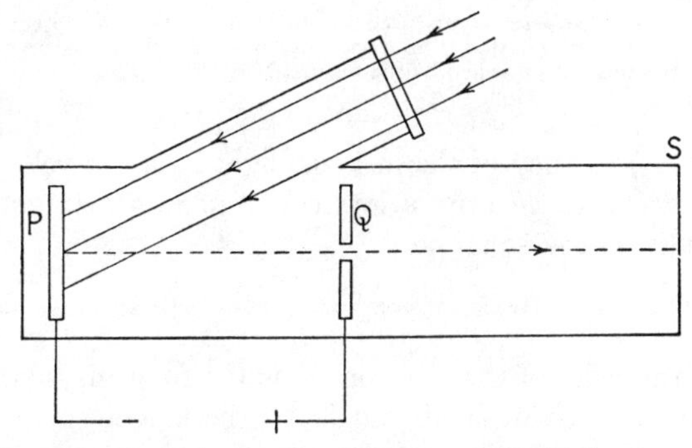

Fig. 17.1.—Photo-electric effect. Diagram of simplified apparatus

it should be called the *surface* photo-electric effect. The number and velocities of electrons emitted have been measured for different metals and for different wavelengths. Fig. 17.1 shows in a diagrammatic way a simple apparatus. Fig. 17.2 shows a more elaborate experiment in which the surface of the metal is freshly cut *in vacuo* immediately before the measurements are made. The results of the experiments in which monochromatic light is incident normally upon the surface of a metal may be summarized as follows:

(i) For every metal there is a maximum wavelength (implying a *minimum* frequency) above which no electrons are emitted, however strong the radiation. The maximum wavelength is called the *photo-electric threshold.* It depends on the metal and also upon the crystal structure, cleanliness, etc., of the surface layer. Some typical values are given in Table 17.1.

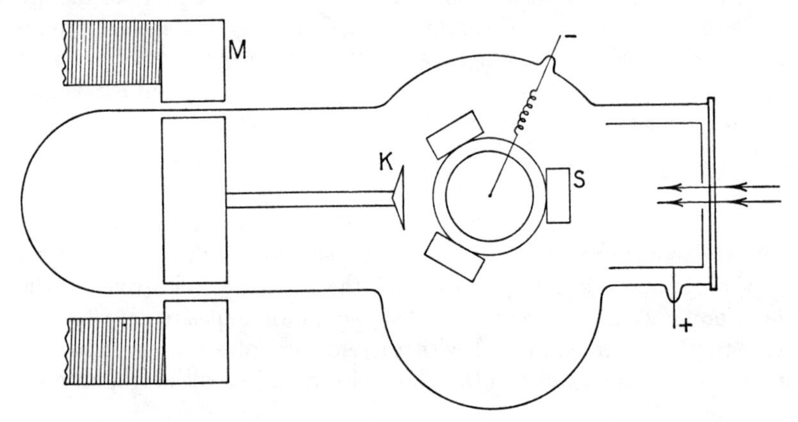

Fig. 17.2.—Millikan's apparatus used to obtain data which verify equation 17(2). The photo-electrons are emitted from a sodium surface S. This has been freshly cut by the knife K which is controlled by the magnet M.

(ii) If the frequency of the incident light is ν, the velocities of the emitted electrons range from a low value up to a maximum velocity v_m which satisfies the equation

$$W_m = \tfrac{1}{2}mv_m{}^2 = h(\nu - \nu_0), \quad . \quad . \quad . \quad . \quad 17(2)$$

where m is the mass of the electron, ν_0 is the frequency corresponding to the photo-electric threshold, and W_m is the kinetic energy (fig. 17.3).

(iii) The maximum velocity of the emitted electrons is independent

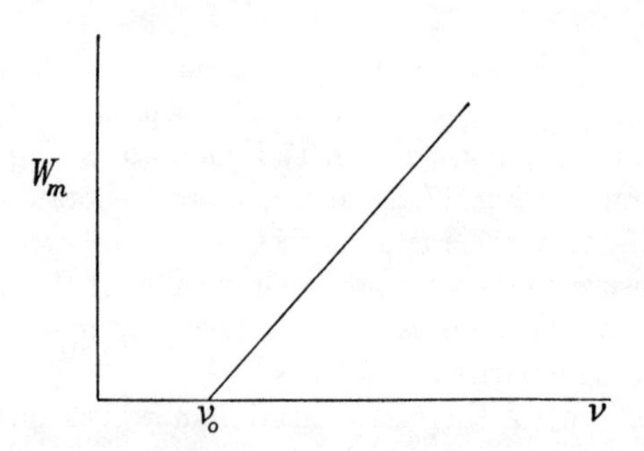

Fig. 17.3.—Variation of maximum energy of photo-electrons with frequency

of the energy of the incident beam of light when its frequency is constant.

(iv) The number of emitted electrons is accurately proportional to the incident energy (for any one frequency).

TABLE 17.1

Photo-electric thresholds.*			
Metal	λ_0 (Å.)	ν_0 (sec.$^{-1}$)	$V = 300h\nu_0/e$ (volts)
Ag	3,250	$9{\cdot}22 \times 10^{14}$	3·82
Bi	2,980	10·1	4·18
Cd	3,140	9·55	3·95
Pb	2,980	10·1	4·18
Pt	2,570	11·7	4·84
W	2,300	13·0	5·38
† Cs on Ag	~ 12,000	~ 2·5	~ 1·0

17.3.—The classical theory does not predict, in any simple way, either the existence of a photo-electric threshold or the relation 17(2) giving the maximum velocity of the emitted electrons. Planck's quantum hypothesis does give a fairly simple picture, provided that we assume that a certain minimum energy ($W = h\nu_0$) is lost by an electron in escaping from the surface.‡ According to Planck's hypothesis, if an electron receives any energy at all from light of frequency ν it must receive an amount $h\nu$. If the electron does not escape immediately, the energy gained is always lost by collision before a second absorption can take place. Therefore the electron cannot escape unless the energy gained in a single process exceeds W, i.e. unless $h\nu$ is greater than W. This implies $\nu > \nu_0$. Thus there is a minimum frequency (and maximum wavelength) for release of electrons. If $\nu > \nu_0$ then the *maximum* energy of escape is given by 17(2). The electron may lose some energy by interaction with an atom, and emerge with much less energy than this maximum. The maximum energy of any one electron

* The values given apply to fairly clean surfaces. To show the range of variation for surfaces which are not clean, we may mention that platinum gives values from 2840 Å. to 2780 Å. for surfaces which have not been "outgassed", and 2570 Å. for an outgassed surface.

† Composite surface of Cs, CsO, and Ag used in sensitive photo-cells.

‡ Many later experimental results have justified this assumption of a work function which is characteristic of the metal and of the state of the surface.

should depend on ν but not on the energy density of the radiation. The number of quanta absorbed from a beam of light (of one frequency) is proportional to the energy of the beam. The number of electrons which absorb energy is proportional to the number of incident quanta, and hence to the energy of the incident beam. Thus all four of the experimental observations described in § 17.2 are in accord with the hypothesis.

17.4. The Line Spectra of Atoms.

During the nineteenth century the wavelengths of very large numbers of lines were measured with an accuracy of about one part in a hundred thousand. This compared favourably with the accuracy of other contemporary physical measurements, and many attempts were made to find empirical relations between the wavelengths. These relations might be expected to provide a foundation for a theory of spectra related to theories of atomic structure. No important progress was made until it was realized that the empirical laws of spectrum analysis are all expressed more simply in terms of wave numbers (or reciprocal wavelengths) rather than wavelengths. Even so, the general analysis of spectra is very complicated. For our purposes it is sufficient to state certain principles of spectrum analysis, and these may be shown by considering the spectrum of the hydrogen atom. It is found that the spectrum of the hydrogen atom can be analysed into the following series:

(*a*) *The Lyman Series.*

This series of lines is in the far ultra-violet region of the spectrum. The wave numbers are given by

$$\frac{1}{\lambda}=\frac{\nu}{c}=R\left(\frac{1}{1^2}-\frac{1}{n^2}\right), \qquad 17(3)$$

where R is a constant whose numerical value is 109,677·6 cm.$^{-1}$, and n is an integer greater than 1.

(*b*) *The Balmer Series.*

This series is in the visible spectrum and the near ultra-violet. The wave numbers are given by

$$\frac{1}{\lambda}=\frac{\nu}{c}=R\left(\frac{1}{2^2}-\frac{1}{n^2}\right), \qquad 17(4)$$

where n is an integer greater than 2. This series appears as a set of absorption lines in many stellar spectra. It may be obtained in emission by a discharge in hydrogen at low pressure. More than 40 lines have been observed.

(*c*) *The Paschen Series.*

This series is in the near infra-red. The wave numbers are given by

$$\frac{1}{\lambda} = \frac{\nu}{c} = R\left(\frac{1}{3^2} - \frac{1}{n^2}\right), \quad . \quad . \quad . \quad . \quad 17(5)$$

where n is an integer greater than 3.

(*d*) *The Brackett Series.*

This series is in the far infra-red. The wave numbers are given by

$$\frac{1}{\lambda} = \frac{\nu}{c} = R\left(\frac{1}{4^2} - \frac{1}{n^2}\right), \quad . \quad . \quad . \quad . \quad 17(6)$$

where n is an integer greater than 4.

The arrangement of these series is shown in fig. 17.4.

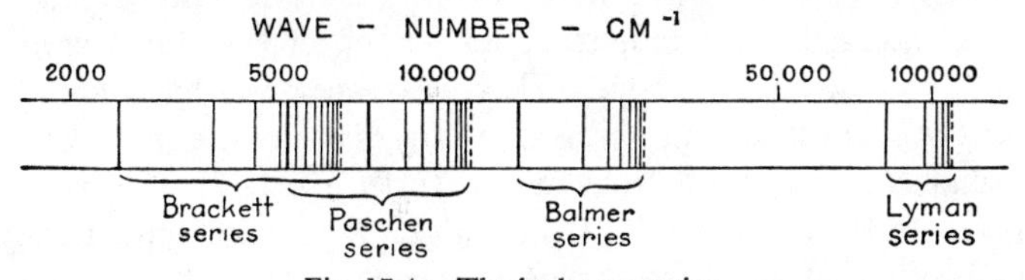

Fig. 17.4.—The hydrogen series

17.5.—These formulæ suggest that the wave numbers of all these series may be expressed as differences of a set of wave numbers which are known as spectroscopic " terms ". The value of the nth term for the hydrogen atom is R/n^2. This analysis of spectra, first into series and then into terms, can be carried out for the optical spectra of atoms other than hydrogen and also for X-ray spectra. Sometimes the lines of different series are not so clearly separated as in the case of hydrogen, and they then have to be grouped together by picking out those lines which are strong under a given type of excitation, or by the appearance of the lines (e.g. " sharp " series or " diffuse " series). The general formulæ are not so simple as 17(4), e.g. for X-ray spectra a typical term formula gives for the nth term the value $R(Z - \alpha)^2/n^2$, where Z is the atomic number and α is a constant. Even though the formulæ for large atoms become complicated, the formulæ for terms are always much more simple than any analysis based on wavelengths or wave numbers of lines. We shall next see how the terms can be related to atomic theory. The use of terms was first suggested by Ritz (1908), who called it the " combination principle ". The constant R is called " Rydberg's constant " after R. Rydberg, who, together with Ritz, analysed many spectra into terms. The extension to X-ray spectra is due to Moseley (1913).

17.6. **The Rutherford-Bohr Atom.**

In 1912 the work of Rutherford and his pupils suggested that the atom should be regarded as a central nucleus surrounded by a number of electrons revolving in orbits like planets round a sun. There was one fundamental difficulty in this theory. Calculation based on the classical electromagnetic theory showed that the electrons should quickly radiate all their energy and spiral in towards the nucleus. All the energy would pass into the radiation field. Bohr suggested that, inasmuch as Planck had already shown that the classical theory did not adequately describe the interaction between radiation and matter, it was reasonable to assume that there were certain *stationary states* in which an atom did not radiate at all. Each of the states is characterized by a definite energy. They are separated by finite energy differences. Emission and absorption of radiation take place when the atom makes a transition from one state to another. If a transition from a state W_2 to a state W_1 (when $W_2 > W_1$) is accompanied by the emission of radiation of frequency ν_{21}, then

$$h\nu_{21} = W_2 - W_1. \qquad \qquad 17(7)$$

This scheme provides an obvious foundation for the Ritz combination principle. We suppose that each term corresponds to a stationary state. To give numerical agreement with the observations, it is necessary that, for hydrogen,

$$W_n = -\frac{Rhc}{n^2}, \qquad \qquad 17(8)$$

where W_n is the energy of the nth state.* Bohr made a second application of quantum ideas by assuming that, in the stationary states,

* The energy of an electron revolving round a nucleus consists of two parts: the potential energy and the kinetic energy. The zero of potential energy is chosen to correspond to an electron at an infinite distance from the nucleus. This makes the total energy always negative.

the angular momentum of the electron is an integral multiple of $h/2\pi$. Using this hypothesis, he was able to obtain equation 17(8) and to calculate R in terms of the fundamental constants; his result was in good agreement with the value obtained from spectroscopic observations.

17.7. The Stationary State.

The advances in atomic theory which Bohr initiated, and to which he has himself contributed much, soon made the early forms of his theory obsolete. His method of calculating energies did not always give values in agreement with experiment and is now superseded by the quantum mechanics. The idea of an analysis of spectra by terms is more important than the detailed formulæ for term values. In a similar way, the concept of quantum states is more important than any process of calculating the energy values, though any complete theory must include calculations of energy values which agree with experimental observations. The idea of quantum states of matter remains as a fundamental hypothesis in each new form of atomic theory. It is so closely in relation with a wide range of observations that it must appear, directly or indirectly, in any theory which is to describe these observations. We have, so far, mentioned only the analysis of spectra. In the following paragraphs we shall consider some of the other experiments which support the hypothesis of stationary states. Most of these are experiments on gases. They show the processes of interaction between radiation and matter more clearly than experiments on solids because, for many purposes, the atoms of a gas act independently and do not affect one another.

17.8. The Relation between Absorption and Emission Spectra of Gases

Under ordinary laboratory conditions every line in the absorpti spectrum of a monatomic gas is found in the emission spectrum exc ed by an electric discharge at a pressure of about 0·1 millimetre. O the other hand, there are many lines in the emission spectrum whic are not found in the absorption spectrum. If the temperature of e gas is raised a few hundred degrees, most elements show more bsorption lines. These observations may be understood by comb ng the hypothesis of stationary states with a general physical law st stated by Boltzmann. In its application to the present problem, B tzmann's theorem states that if a large number of identical at ms are in equilibrium with one another and with their surroundings at a temperature T, then

$$n_1 : n_2 : n_3 = e^{-W_1/kT} : e^{-W_2/kT} : e^{-W_3/kT}, \qquad . \quad 17(9a)$$

where n_1 is the number of atoms in State 1 of energy W_1, n_2 is the number in State 2 of energy W_2, and so on, and k is a universal constant known as Boltzmann's constant. Its value is $1{\cdot}38 \times 10^{-16}$ erg/deg.

So far we have considered states as characterized solely by their energy. It sometimes happens that atoms in states which have equal energy can be distinguished in other ways, e.g. by the splitting of spectrum lines in a magnetic field (see § 19.14). States which can be distinguished in any way are separate and distinct states. If State 2 and State 2′ have the same energy W_2, then 17(9) gives the number of the atoms in *each* state, and the total number of atoms with energy W_2 is $2n_2$. More generally, suppose that g_1 states have energy W_1, and that N_1 is the *total* number of atoms with energy W_1; g_2 states have energy W_2 and N_2 is the total number of atoms with energy W_2, etc., then

$$N_1 : N_2 : N_3 = g_1 e^{-W_1/kT} : g_2 e^{-W_2/kT} : g_3 e^{-W_3/kT}. \qquad . \quad 17(9b)$$

Using the values of W_1, W_2, etc., obtained from the analysis of spectral terms, it is easily calculated that, at room temperature, most of the atoms are in the stationary state of lowest energy [Example

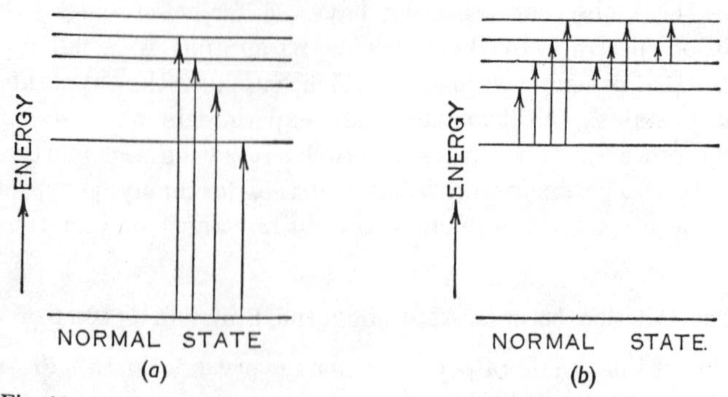

Fig. 17.5.—(*a*) Transitions from the normal state of an atom. The horizontal lines represent different states of the atom and are plotted on a vertical energy scale, the normal state being the one of lowest energy. (*b*) Transitions between excited states.

17(v), p. 578]. This state is often called the *normal state,* and other states (which are reached only when energy is supplied) are called *excited states*. We should therefore expect that, at room temperature, the atoms can absorb only quanta of those frequencies which correspond to transitions based on the lowest state, i.e. to those shown in fig. 17.5*a*.

If some atoms are raised by electronic collision to high excited states, then all the lines indicated in fig. 17.5*a* and many more can be seen in absorption (fig. 17.5*b*). If the temperature is raised, more atoms pass into higher states so that more absorption lines appear. In addition to the effect of temperature on absorption spectra, it is also found that fresh absorption lines appear if a gas is made to glow feebly by means of a weak electric discharge. These extra lines are due to some of the atoms being transferred to higher states by collision with electrons.

17.9.—The change of the absorption spectrum with temperature is the basis of Saha's theory of absorption lines in stellar spectra. According to earlier theories, the strength of an absorption line in the spectrum of a given star is determined by the abundance of the corresponding element in the stellar atmosphere. We now see that it depends on the number of atoms *in the appropriate stationary state.* This number depends both on the abundance of the element and (through Boltzmann's law) on the temperature of the stellar atmosphere. According to Saha's calculations, the temperature effect is very important, and the presence of certain absorption lines in a stellar spectrum and the absence of others is to be regarded, in the first place, as an indication of the temperature of the outer layers of the star.

17.10. **Excitation of Spectra by Slow Electrons.**

The apparatus shown in fig. 17.6 may be used to pass electrons of known velocity into a gas. Electrons from the hairpin filament F are accelerated by a potential V volts between F and the grid G, and then pass into the field-free space between G and A. The pressure is low so that the chance of an electron colliding with an atom in the short path between F and G is very small. The number of electrons is maintained constant by adjusting the filament current, and the energy of the electrons entering the space between G and A is gradually increased from a low value. It is found that, at first, no light is emitted but that, as the energy is increased, a stage is reached at which a single line is emitted. This line corresponds to a transition from the first excited state back to the normal state. It is found that the *critical potential* (i.e. the minimum potential for emission) is connected with the frequency by the relation

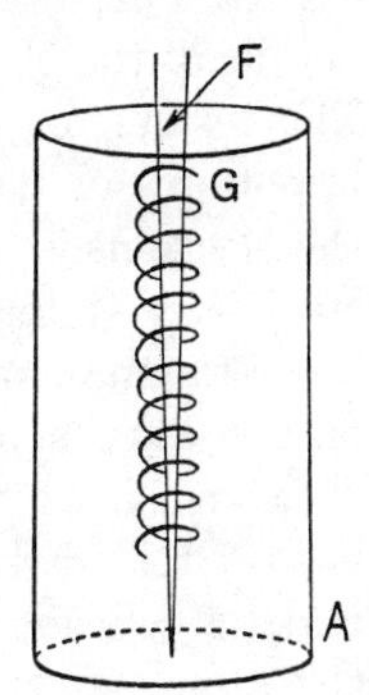

Fig. 17.6.—Excitation of light by slow electrons.

$$Ve \doteqdot h\nu. \qquad \qquad 17(10)$$

This relation is only approximately verified because the electrons have

a small amount of energy when they leave the filament. This amount is not quite the same for all electrons, so the energy of the electrons cannot be accurately measured. The results are in agreement with the relation within the accuracy with which this energy can be estimated. As the potential is raised, more lines appear, indicating that higher states are excited. The differences between the energies of the fourth and fifth excited states, the fifth and sixth, etc., are all small and, owing to the spread in electron energies, they all appear to be excited simultaneously.

17.11. **Critical Potentials.**

The experiments described in the preceding section show that an electronic collision may increase the internal energy of an atom, provided that the energy of the electron exceeds a certain minimum value. Another approach to this problem is obtained by studying the loss of energy by the electron. Many experiments of this type have been carried out. The reader should consult Reference 17.2 or 17.3 for details of the electrical measurements which need not concern us here. The results form a most detailed and convincing verification of the hypothesis of stationary states. It is found that, when the energy of the electrons is too small to raise the atom from the first to the second stationary state, the collision is similar to that between elastic bodies. Since the mass of the atom is large compared with that of the electron, only a very small interchange of kinetic energy occurs. When the electron has an energy slightly greater than a minimum critical value, it may still make an elastic collision with the atom, but it may also make a collision in which it loses exactly the energy required to raise the atom to the second state. Electrons of higher energy may lose different amounts of energy, corresponding to different degrees of excitation. If the energy of the incident electron is high enough, the atom may be ionized. That critical potential which corresponds to ionization is called the *ionization potential.* There is a series of critical potentials corresponding to the different transitions. In every case the energy differences between states obtained from the electrical measurements are the same (within the error of the electrical measurements) as the more accurate values obtained by the analysis of spectra.

17.12. **Series Limit Absorption in Monatomic Gases.**

The absorption spectrum of sodium vapour in the near ultra-violet is shown in Plate II*b* (p. 126). It consists of a series of lines which gradually close up to a series limit [Plate V*e* (p. 636)]. Beyond this limit there is continuous absorption. Any one of the lines is due to the absorption of quanta whose energy is just sufficient to transfer the atoms from the normal state to one of the excited states. Quanta which have appreciably more energy than that required for a given transition (but not enough for the next transition) cannot be absorbed, because there is no way of disposing of the excess energy. If the extra energy were given to the atom as kinetic energy, the law of conservation of momentum could not be satisfied. If, however, the energy of the quantum is sufficient to ionize the atom, it is possible for an excess energy to be divided between the ion and the electron, in such a way that both energy and momentum relations are satisfied. In working out the equations of energy and momentum, it is necessary to take account of the fact that the light possesses a small momentum as well as considerable energy (§ 17.19). The existence of continuous absorption on the short-wavelength side of the series-limit is in accord with the theory of stationary states, provided that the frequency ν_L of the series limit is connected with the ionization potential by the relation

$$V_i e = h\nu_L, \qquad 17(11)$$

where V_i is the ionization potential in absolute electrostatic units. This relation has been verified experimentally.

Absorption spectra of the type shown in Plate II*b* are not obtained for many elements. Sometimes this is because the series-limit (calculated from the ionization potential) should lie in the far ultra-violet. Comparatively little work has been done in this region of the spectrum because of technical difficulties. Other elements do not give this type of absorption spectrum because their vapours are not monatomic. Molecules (in the gaseous state, or in solids or liquids) may give continuous spectra, but the theory is rather complicated and we shall not discuss it here. Hydrogen is diatomic at room temperature. Hydrogen atoms, in the normal state, would be expected to have a series limit, with associated continuous absorption, at 912 Å. This absorption has recently been observed. Stellar spectra show a series-limit at 3647 Å. and continuous absorption on the short-wavelength side. This frequency corresponds to an ionization potential of 3·4 volts, i.e. to the amount of energy necessary to free an electron from the second stationary state (i.e. the first excited state) of hydrogen. We have already seen (§ 17.8) that we should expect to find excited atoms in a gas at the high temperature of a stellar atmosphere.

17.13. **Photo-electric Effect in Gases.**

If the above account of series-limit absorption in gases is correct, we should expect to find that the wavelength of the series-limit agrees with that of the photo-electric threshold for the gas. This relation has been confirmed for a number of elements. We should also expect that there would be one photo-electron emitted for every absorbed quantum (on the short-wave side of the series-limit). From this consideration it is possible to deduce an equation connecting the absorption coefficient and the number of electrons emitted when a vapour is irradiated under certain standard conditions. This relation has been experimentally verified.

17.14. Resonance Radiation.

The spectrum of the mercury arc contains a strong line of wavelength approximately 2537 Å. In 1905, R. W. Wood showed that light of this wavelength is strongly absorbed by mercury vapour at low pressures. The energy absorbed from a directed beam is re-emitted as radiation of the original wavelength (fig. 17.7). The re-emitted radiation, which is known as *resonance radiation*, is not confined to the direction of the original beam, and is approximately equally strong in

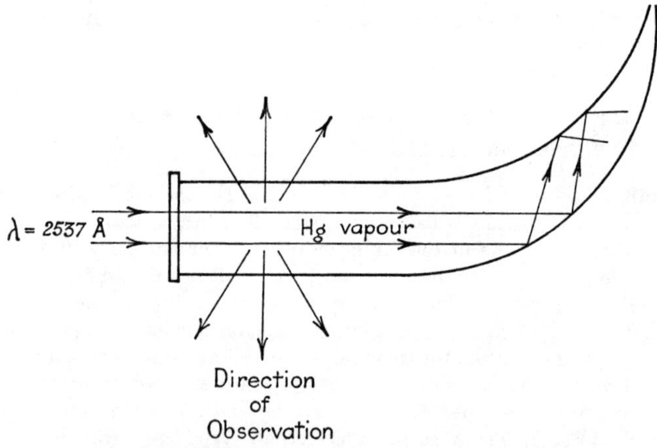

Fig. 17.7.—Apparatus for study of resonance radiation

all directions. This strong absorption is associated with anomalous dispersion (and with a strong reflection coefficient) in the way predicted by the electromagnetic theory of light and the Lorentz theory of the electron (Chapter XV). The quantum theory is able to give a simple account of the absorption and re-emission by assuming that absorption of radiation of wavelength 2537 Å. raises the atom from the normal state of lowest energy to the next state and that, at low pressure, the atom very quickly makes a transition back to the normal state re-emitting the radiation. It is also found that sodium absorbs and re-emits as resonance radiation the two well-known yellow lines at wavelengths 5890 Å. and 5896 Å. These also are associated with anomalous dispersion in the predicted way.

Rayleigh (R. J. Strutt) showed that, when sodium vapour is illuminated with one of the fairly close pair of lines whose wavelengths are approximately 3303 Å., the re-emitted radiation is only partly of the same wavelength as the incident radiation. Both the yellow lines

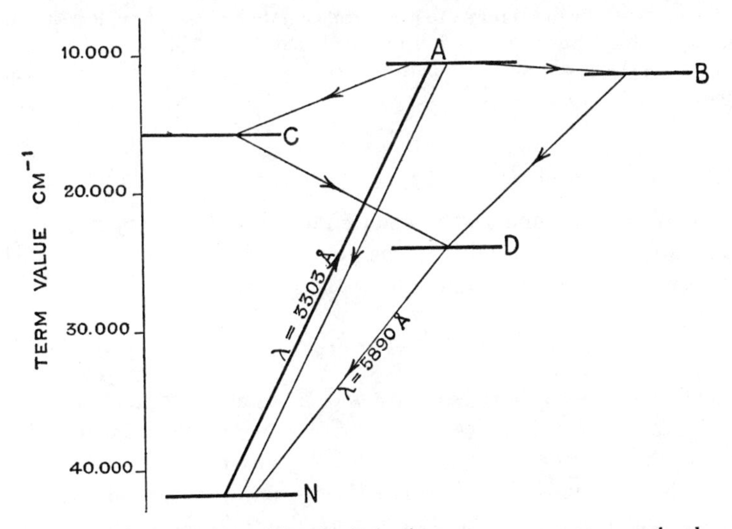

Fig. 17.8.—Possible transitions in the sodium atom consequent upon the absorption of a quantum of wavelength 3303 Å. The return to the normal state (N) may occur in the following ways:

(1) A — N, 3303-Å. line re-emitted.
(2) A — B, B — D } Infra-red lines.
D — N, 5890-Å. line.
(3) A — C, C — D } Infra-red lines.
D — N, 5890-Å. line.

For reasons discussed in Chapter XIX, certain transitions, e.g. A to D, do not occur.

are also emitted. This observation cannot be described as a classical resonance phenomenon but follows immediately from the idea of a set of stationary states. Suppose that sodium has a series of stationary states whose energy differences are shown in fig. 17.8. Absorption of 3303 Å. raises the atom to one of the higher excited states. From this state it may return to the normal state directly, emitting 3303 Å., or it may also return via intermediate levels, emitting lines in the infra-red which are not observed, and one of the yellow lines.

Absorption and re-emission with change of wavelength as well as direction is given by solids, liquids, and gases. It is known as *fluorescence*. It was studied, in the nineteenth century, by a number of workers including Stokes, who discovered an empirical law that the wavelength of the re-emitted radiation is never less than that of the incident radiation. If the absorbing atom is initially in the lowest state, and if after the absorption it can lose, but not gain, energy, the re-emitted quantum must be of lower frequency (i.e. longer wavelength) than the quantum absorbed. Thus the quantum theory includes Stokes' law but also suggests that in special circumstances there may be exceptions. If the absorbing atom is not in the lowest state and is raised to a still higher state by absorption, it may return directly to the lower state. It will then emit radiation of higher frequency than that of the absorbed radiation. This is found experimentally under certain conditions.

17.15. **Collisions of Second Type.**

R. W. Wood found that resonance radiation of mercury is quenched when a foreign gas is added to the vessel containing the vapour. This effect was later shown to be due to the process

$$\mathrm{Hg}^* + \mathrm{M} \rightarrow \mathrm{Hg} + \mathrm{M}^*, \qquad 17(12)$$

where * indicates an excited state and M stands for a molecule of the foreign gas. In this process some of the energy of excitation of the mercury is transferred to the molecule (raising it to an excited state), the balance being turned into kinetic energy. This type of collision in which some energy of excitation is turned into kinetic energy is called the " second type ", to distinguish it from the collision of the first type in which the kinetic energy (usually of an electron) is used to excite the atom. Not every collision between an atom in an excited state and another atom is a collision of the second type. Some are ordinary elastic collisions. A reaction such as the one described in 17(12) is often followed by

$$\mathrm{M}^* \rightarrow \mathrm{M} + h\nu, \qquad 17(13)$$

i.e. the molecule M emits its own resonance radiation, although the incident radiation has quite a different wavelength. This *sensitized fluorescence* could not readily be explained on classical theories of light and of matter. For a description of experiments on sensitized fluorescence see p. 499 of Reference 17.3. It should be stated that quenching of resonance radiation is not always associated with sensitized fluorescence. When mercury resonance radiation is quenched by hydrogen, the molecule is dissociated. There is no fluorescence, but the presence of atomic hydrogen may be demonstrated by chemical methods. When the resonance radiation is quenched by oxygen, some of the mercury combines with the activated oxygen.

17.16. The Einstein Photochemical Law.

Einstein put forward the hypothesis that the primary process in any photochemical reaction is the absorption of a quantum. Therefore, under the simplest conditions, the number of reacting stoichio-chemical units is equal to the number of quanta absorbed. This generalization is called the *Einstein photochemical law.* This law scarcely ever applies directly to a *completed* photochemical reaction, because chain reactions tend to increase the number of reacting units, and also because there are various ways in which a molecule which has absorbed a quantum can lose its energy before it has time to enter into the reaction. Under a fairly wide range of conditions, the number of reacting units is proportional to the number of quanta absorbed, and it is then usual to measure the quantum efficiency, i.e. the ratio of the number of reacting units to the number of quanta absorbed. Values of the quantum efficiency up to 30,000 are observed when chain reactions are operative, and values much less than unity are obtained when collisions of the second type or other deactivating processes are effective. Nevertheless, all discussion of photochemical reactions is now based on the assumption that Einstein's law applies to the primary photochemical process.

17.17. Einstein Theory of Photons.

So far we have considered only the energy exchanges involved in the interaction of radiation and matter. Relativity principles require us to associate mass with the energy of radiation, and it is reasonable to suppose that an exchange of momentum may also take place. In the succeeding paragraphs we shall describe experiments which show that a beam of light exerts a pressure upon a piece of matter which absorbs, reflects, or refracts it, i.e. radiation is able to exchange momentum with matter. We shall describe also certain experiments which show that the exchange of momentum between free electrons and radiation is very similar to the exchange which occurs when two particles collide. These experiments suggest that a beam of light should be considered as an assembly of " units ", each of which possesses energy (W), momentum (p), and mass (m), given by

$$W = h\nu; \quad p = \frac{h\nu}{c}; \quad \text{and } m = \frac{h\nu}{c^2}. \qquad . \quad . \quad 17(14)$$

The units thus possess many of the properties of particles. This general picture was first suggested by Einstein in 1906, although the most important experimental evidence came later. The units are now called *photons*. In some ways the photon theory appears to constitute a return to the corpuscular theory of radiation, but the photons are not supposed to have *all* the properties of material particles. Under the conditions of ordinary experiments, the number of material particles, such as electrons, is fixed. When an atom receives an extra electron it becomes negatively charged, and must emit one and only one electron in order to return to the normal state. We have seen that when an atom absorbs a quantum of frequency ν (i.e. when it receives a photon) it may give up its energy in a collision, emitting no photon, or it may emit two photons of frequencies ν' and ν'', if $\nu' + \nu'' = \nu$. Thus photons do not have the permanent " identity " which we normally associate with material particles. Moreover, the spreading of light by diffraction shows that the energy of a photon cannot be permanently concentrated in a small volume like the energy of a material particle.

17.18. **The Inverse Photo-electric Effect.**

Consider the apparatus shown diagrammatically in fig. 17.9. A beam of electrons is accelerated by a voltage V (of order 100,000 volts) and strikes the plate P_1. X-rays are emitted and some of the radiation

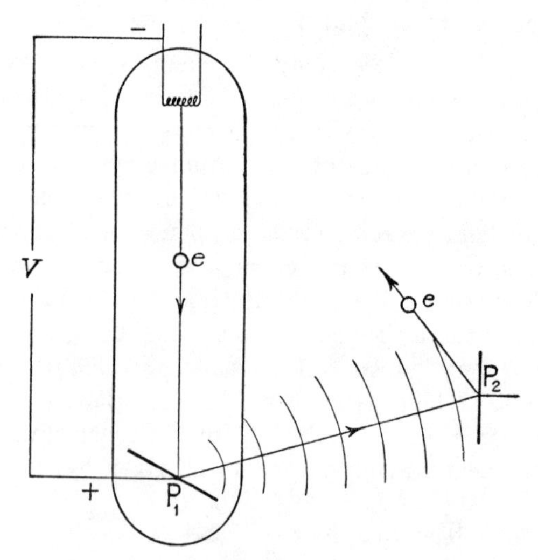

Fig. 17.9.—The inverse photo-electric effect

falls on the plate P_2, which emits photo-electrons with energy approximately V. The energy of one of the photo-electrons is independent of the distance of P_2 from P_1. We may describe these results in the following way. Some of the electrons which strike P_1 cause the emission of photons of frequency ν (where $h\nu = Ve$). Some of these photons fall on P_2 and give up their energy to electrons in the surface of the metal. These electrons are emitted with energy approximately * Ve. If the number of electrons striking P_1 is small, it is possible to show that nearly the whole energy of a single electron can be transferred from P_1 to a single electron in P_2. These results are not compatible with any theory which assumes that the X-ray energy produced when a single electron strikes P_1 spreads in the way predicted by a simple wave theory. They are most easily described in terms of localized concentrations of energy.

An apparatus which is capable of detecting individual photons has been developed using this effect. A fine wire is stretched along the axis of a metal cylinder (fig. 17.10). A potential of a few hundred volts is applied between the wire and the cylinder, the latter being positive. A small window is provided through which light can reach the wire. If the voltage is increased beyond a certain value, a discharge occurs. The voltage is adjusted to be a little below this critical value.

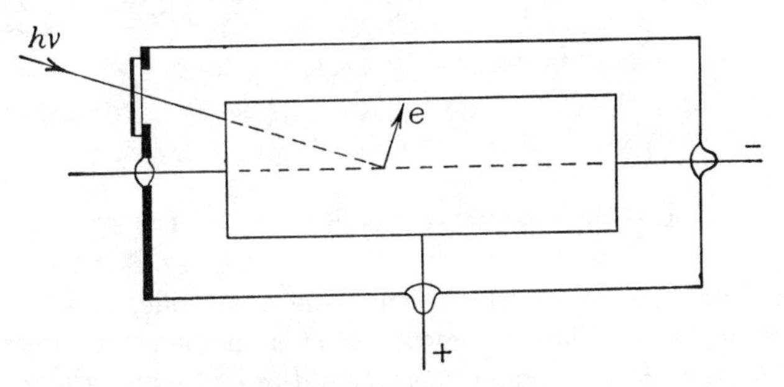

Fig. 17.10.—The photon counter

If a photon enters the apparatus, it may release a photo-electron from the wire. This electron produces others through ionization by collision, and there is a small pulse of current. This pulse can be amplified and used to operate a mechanical counter. In this way individual photons can be detected and numbers of photons counted. It should be stated that the device does not give a pulse for every photon which enters the window, but only for those which release photo-electrons in or near the wire.

17.19. Light Pressure.

The measurement of light pressure is difficult because, under normal experimental conditions, the pressure is only 10^{-5} to 10^{-6} dynes

* Approximately because we are neglecting the work function and other effects which are relatively small when V is large.

per square centimetre. The true radiation pressure can be measured only when a certain thermal action known as the *radiometer effect* has been eliminated. A simple apparatus is shown in fig. 17.11. Light from two powerful sources S_1 and S_2 falls upon the vanes A_1 and A_2 which are at the ends of a rod suspended by a thin quartz fibre. Rotation of the system may be measured by the usual mirror-and-scale method, and the torsional coefficient of the fibre may be obtained by measuring the period of oscillation. If the radiation heats the surface vane, then the molecules of gas which strike one side will be moving

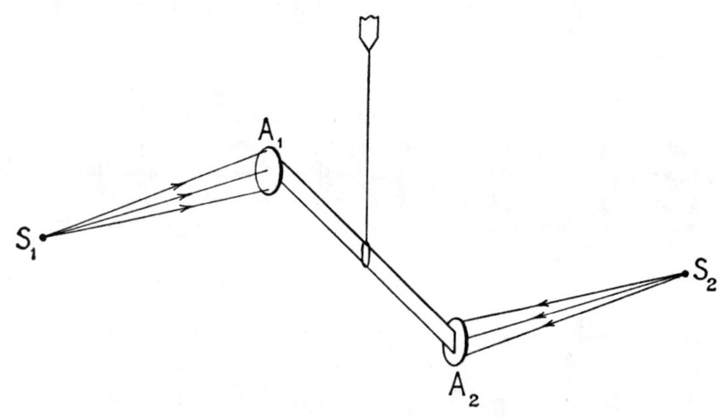

Fig. 17.11.—Apparatus for measurement of light pressure

more rapidly than those which strike the other side, and there will be a net pressure. This radiometer effect is in the same direction as the light pressure and under most conditions is many times larger. The radiometer effect depends on pressure, etc., and, by choosing suitable experimental conditions, it may be reduced so that the light pressure can be measured. In some of the experiments * the radiometer effect was greatly reduced by enclosing the reflecting or absorbing surface in thin glass cells as shown in fig. 17.12. One side of the inner surfaces is blackened and the other is silvered. The radiometer effect at the inner surfaces produces no turning moment, and the effect at the outer surfaces (which reflect very little light) is small. It is then possible to measure the light pressure within an error of a few per cent. The pressure p, exerted by a parallel beam incident normally on a body which completely absorbs it, is found to be given by

$$p = \rho_p, \qquad \qquad 17(15)$$

* Reference 17.4.

where ρ_p is the energy per unit volume of the incident radiation. The pressure on a surface whose reflection coefficient is r is found to be greater in the ratio $(1 + r) : 1$.

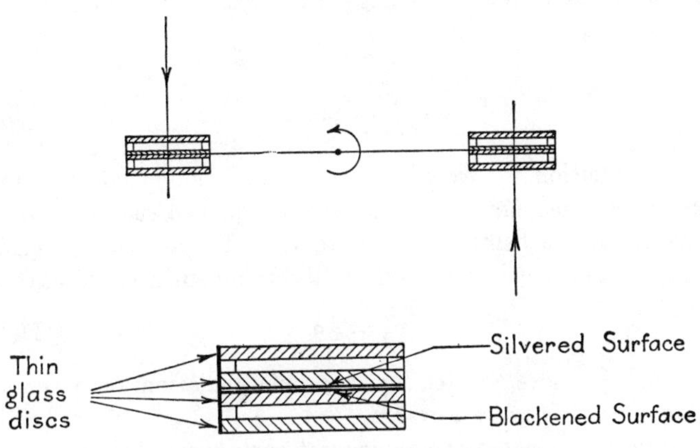

Fig. 17.12.—Details showing special cells

17.20.—We shall now use this result to calculate the pressure due to isotropic radiation.

Consider a nearly parallel beam of light lying within a small solid angle $d\Omega$. Let the energy per unit volume whose direction lies within the solid angle be $\rho_\theta \, d\Omega$. The light is incident upon a surface S (fig.

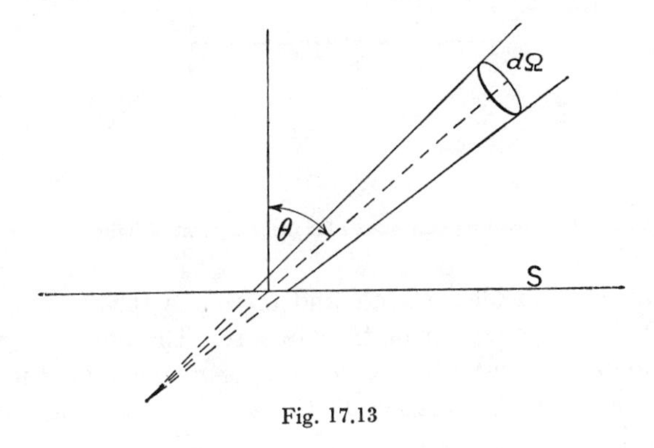

Fig. 17.13

17.13) and the central direction of the pencil makes an angle θ with the normal to S. If the beam is of unit cross-section, the area irradiated is $1/\cos\theta$. The normal component of the force exerted on the surface is $\rho_\theta \cos\theta \, d\Omega$, and the normal pressure (i.e. force on unit area) is

$\rho_\theta \cos^2\theta \, d\Omega$. The normal pressure for unit energy density is $\cos^2\theta \, d\Omega$. Similarly there is a tangential component $\sin\theta \cos\theta \, d\Omega$ per unit energy density. If radiation is incident upon a surface over a wide solid angle, the normal pressure per unit energy density is

$$\frac{\int \rho_\theta \cos^2\theta \, d\Omega}{\int \rho_\theta \, d\Omega}. \qquad \qquad 17(16)$$

When the radiation is *isotropic* (i.e. when ρ_θ is independent of θ), the normal pressure per unit energy density is equal to the mean value of $\cos^2\theta$ taken over a hemisphere, i.e. to 1/3. For an energy density ρ of isotropic radiation incident on an absorbing surface, we have

$$p_n = \tfrac{1}{3}\rho. \qquad \qquad 17(17)$$

The normal pressure excited by isotropic radiation on a perfectly reflecting surface is twice as large.

Poynting investigated the tangential component due to a directed beam, using the apparatus shown in fig. 17.14. A beam of light is directed upon a vane whose surface is perpendicular to the supporting

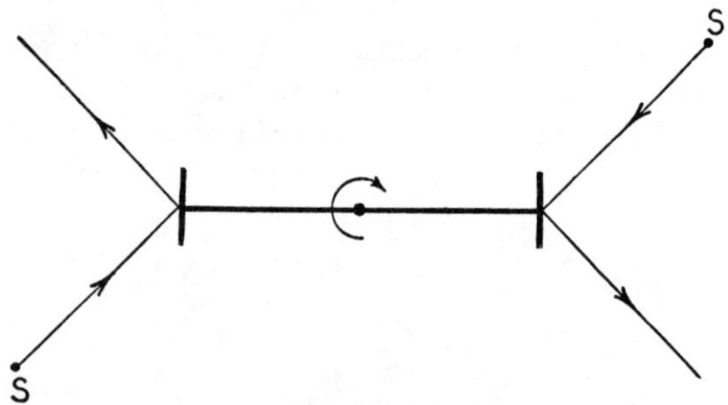

Fig. 17.14.—Poynting's apparatus for measurement of light pressure

rod. The angle of incidence is 45° and there is a turning moment due to the tangential component of the pressure. The normal component and the radiometer action exert no turning moment on the system. The results agree with equation 17(16) within an accuracy of about ten per cent.

Although, under the conditions of laboratory experiments, the pressure of radiation is very small, it is very important in stars. The high radiation density in a star gives a large pressure tending to drive the matter outwards from the centre. This opposes the gravitational force which tends to make the star collapse.

In a star the emission and absorption of radiation cause a rapid transfer of mass from one part of the star to another. Radiation coming from the centre and absorbed in the outer layers possesses less angular momentum than the matter which absorbs it. This tends to reduce the speed of rotation and increases the effective viscosity of the gas.

17.21.—Let us now consider the radiation pressure of a parallel beam of light, incident normally on an absorbing body, from the point of view of the Einstein photon theory. Suppose that the light is of frequency ν and that there are N quanta per unit volume. Then we must have

$$\rho_p = Nh\nu. \qquad 17(18)$$

Since all quanta included in a cylinder of volume c cubic centimetres are incident upon unit area of the surface in one second, the pressure p is given by

$$p = NcP, \qquad 17(19)$$

where P is the momentum of one photon. Combining 17(18) and 17(19) with 17(15), we see that the experiments imply that

$$P = \frac{h\nu}{c} = \frac{h}{\lambda}. \qquad 17(20)$$

The results obtained for the pressure on reflecting surfaces and for isotropic radiation are in agreement with the above hypothesis.

17.22. Wave Theory of Light Pressure.

A beam of electromagnetic radiation sets up currents in the surface of a metal which partly absorbs and partly reflects the radiation. The action of the magnetic field of the radiation on these currents should be manifested as a force on the conductor which is carrying the currents. In a similar way the radiation sets up displacement currents in a dielectric medium which absorbs or reflects the light, and again the action of the magnetic field on the currents constitutes a " pressure ". It is possible to show by direct calculation that the magnitude of the pressure predicted is in agreement with experimental results.* The following indirect proof, due to Larmor, enables the result to be obtained more easily.

It is easiest to apply the argument to a plane wave which is incident normally on a surface which gives total reflection. The wave train is assumed to be very long compared with the wavelength, but not infinitely long. Suppose that the

* Reference 17.5.

reflector moves towards the waves with a velocity v which is small compared with c. Let the incident and reflected waves be represented by

$$\xi = a \cos(\omega t - \kappa x), \qquad 17(21a)$$

$$\xi' = a' \cos(\omega' t + \kappa' x). \qquad 17(21b)$$

The position of the reflector at time t is given by $x = -vt$, and at this point we must have $\xi + \xi' \equiv 0$. This requires that $a = -a'$. Also since

$$\frac{\omega}{\kappa} = \frac{\omega'}{\kappa'} = c, \qquad 17(22)$$

then

$$\omega'(c - v) = \omega(c + v). \qquad 17(23)$$

If the energy density for waves of different frequency but the same amplitude is proportional to $\dot{\xi}^2$ and hence to ω^2, we have

$$\frac{\rho'}{\rho} = \left(\frac{c + v}{c - v}\right)^2, \qquad 17(24)$$

where ρ and ρ' are the energy densities in the incident and reflected beams. The reflected train contains the same number of waves as, and is therefore shorter than, the incident train in the ratio $(c - v)/(c + v)$, so that the total energy of the reflected wave is greater in the ratio $(c + v)/(c - v)$. When v is small, the excess energy in the reflected beam is $2v/c$ times the energy falling upon the surface. In unit time the incident energy is ρc per unit area, and the excess energy is $2v\rho$. This is exactly equal to the work which the mirror would have to do if it were pushing against a pressure 2ρ due to the light. (Note that the pressure is independent of v, so that we may apply the result to a stationary surface.)

17.23. **The Compton Effect.**

We have seen that the macroscopic experiments on light pressure may be satisfactorily included either in a wave theory or in a photon theory. The photon theory of light pressure, being very similar to the kinetic theory of gas pressure, is easier to visualize, and a very simple calculation leads to the correct value for the light pressure. Nevertheless, the wave-theory treatment does form an integral part of wave theory, and does not require any special hypothesis. The experiments we shall now describe point more definitely to a photon theory.

It was shown by A. H. Compton in 1923 that when a beam of X-rays is scattered, part of the radiation scattered through an angle θ has a wavelength greater than that of the incident radiation by an amount

$$\Delta\lambda = \frac{h}{m_0 c}(1 - \cos\theta), \qquad 17(25)$$

where m_0 is the mass of the electron. It is shown in Appendix XVII A

(p. 579) that this relation may be derived by applying the laws of conservation of momentum and energy to a collision between a photon and an electron, provided we assume that the momentum of the photon is given by 17(20). Compton's original experiments deal with average effects due to large numbers of collisions. They cannot, therefore, give direct evidence concerning the change of momentum in a single collision. The existence of single collisions of the type considered is shown by the Wilson chamber photographs of " fish-tail " tracks. It was also shown by Geiger and Bothe * that the scattered photon and the scattered electron appear simultaneously. Collisions of photons of visible light with electrons have not been demonstrated directly, because the energies associated with the forces which hold the electrons in the atoms are so large that the light photon must effectively collide with the whole atom. The energies of the X-ray photons are so large, however, that the binding energies of the outer electrons in an atom are negligibly small by comparison.

It is possible to obtain the change of wavelength from a purely wave-theory by assuming that the scattering is a double process in which the light is absorbed and is then emitted by the moving electron. The change of wavelength is then ascribed to the Doppler effect.

17.24. **Angular Momentum associated with Circularly Polarized Light.**

It has been shown experimentally that a beam of circularly polarized light exerts a torque upon a half-wave plate which reverses the sense of the rotation (i.e. changes left-hand circularly polarized light into right-hand circularly polarized light, or vice versa). The apparatus is shown in fig. 17.15. H is a half-wave plate suspended by a fine quartz fibre which passes through a hole in a fixed quarter-wave plate M. A beam of circularly polarized light passes up through H, and is reflected by M. The sense of the circular polarization is changed first in H (causing a torque on the suspended system), then in M (which it traverses twice), and a third time in H (causing a second torque on H which reinforces the first). In the diagram the direction and polarization of the incident light are shown on the left, and of the reflected light on the right, although both beams cover the whole plate. The beam of light is periodically interrupted in such a way as to produce a set of impulses which are in resonance with the natural period of the suspended system. The resulting oscillation of the system is observed by reflecting a weak beam of light from the mirror m. The

* Reference 17.6.

results obtained may be regarded as showing that a beam of circularly polarized light can transfer angular momentum Ω per second per unit area when it is absorbed, where

$$2\pi\,\Omega = \rho\lambda. \qquad \qquad 17(26)$$

The experimental difficulties involved in detecting and measuring the effect are very formidable. The torque obtained is of order 10^{-11}

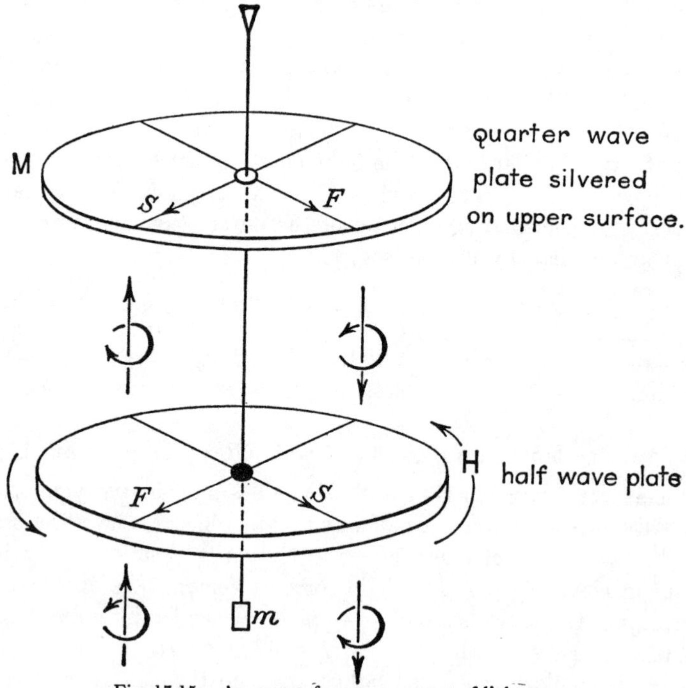

Fig. 17.15.—Apparatus for measurement of light torque

dyne centimetre. If the beam of light does not pass through the apparatus in a symmetrical way, larger torques may be produced by radiometer effect or even by the light pressure. By varying various parameters it is possible to show that these spurious effects have been eliminated.* The final result verifies equation 17(26) within an accuracy of ± 10 per cent.

17.25.—Let us assume that a beam of circularly polarized light is represented by photons, each of which possesses angular momentum $h/2\pi$. If there are N

* Reference 17.7 should be consulted for details of this experiment.

quanta per unit volume, the number crossing unit area per second is Nc, and the total angular momentum is $\Omega = Nhc/2\pi$. The energy per unit volume $\rho = Nh\nu$, so that

$$2\pi\Omega = \frac{\rho}{\nu}c = \rho\lambda. \qquad \text{17(27)}$$

If left-hand circularly polarized light is represented by photons whose angular momentum is of opposite sign to right-hand circularly polarized light, then a double torque will be obtained when left-hand circularly polarized light is changed into right-hand circularly polarized light.

It is possible to show from a wave-theory treatment that a beam of circularly polarized light should exert a torque in the circumstances of the above experiment. The most simple wave-theory treatment connects the effect with diffraction at the edges of a beam of circularly polarized light (or at the edge of the half-wave plate if it is smaller than the beam).

17.26.—In the experiments described in § 17.19 the pressure of light on a piece of matter containing many atoms is measured. In a similar way the experiments described in § 17.24 give the light torque on a vane containing many atoms. We have seen (in §§ 17.21, 17.22, and 17.25) that the wave theory and the photon theory are equally successful in dealing with macroscopic experiments of this type. The experiments on the Compton effect (and particularly the fish-tail tracks) refer more directly to processes of interaction between radiation and single atoms. These experiments are more easily understood in terms of photons. The corresponding evidence on the exchange of angular momentum between radiation and single atoms exists, but is less direct. Atoms oriented in a magnetic field are found to emit circularly polarized light. The Gerlach and Stern experiment on the behaviour of atoms in a magnetic field shows, in principle, that those atoms which emit circularly polarized light suffer a change of angular momentum (equal to $h/2\pi$ in the simplest case). Again the photon theory is superior. The whole process is most simply described by assuming that each atomic transition is accompanied by the emission of a photon, and that in each individual interaction between radiation and matter the conservation laws for energy, momentum, and angular momentum are obeyed.

17.27. **Temperature Radiation.**

It is found experimentally that an enclosure which forms part of a body maintained at a uniform temperature always contains radiation. This radiation quickly reaches an equilibrium value for the total energy density, and an equilibrium for the distribution of energy with wavelength. All the properties of this equilibrium state depend only upon the temperature, and are independent of the material which forms the walls of the enclosure. The radiation which is in equilibrium in a cavity whose walls have a uniform temperature is called the *temperature radiation.*

The temperature radiation is investigated by allowing a very small fraction to escape through a small hole in the wall of the cavity. This fraction may be allowed to fall directly upon a thermopile when the total energy is to be measured. Alternatively, the radiation in a given range of wavelength may be isolated by a mono-

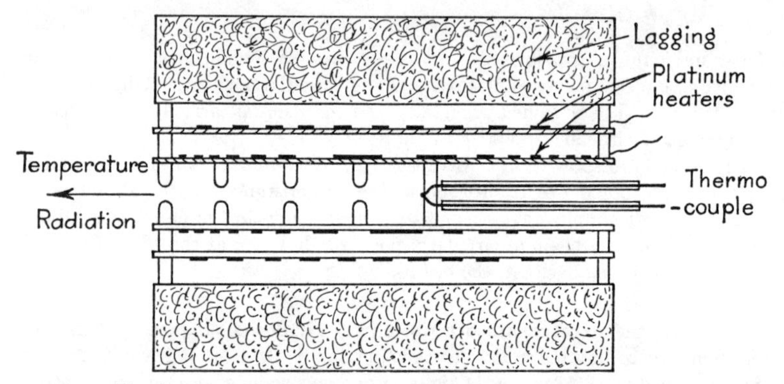

Fig. 17.16.—Investigation of variation of energy with wavelength

chromator, and then passed to a sensitive measuring instrument. The apparatus used by Coblentz for the investigation of the distribution of energy with wavelength is shown in fig. 17.16. Reference 17.8 should be consulted for details of the experimental method.

It is found that, if E_a is the total energy (for all wavelengths) emitted in unit time from unit area of the wall of the cavity, then

$$E_a = \sigma T^4, \quad \ldots \ldots \quad 17(28)$$

where T is the absolute temperature. This energy is emitted into a solid angle 2π. Equation 17(28) is known as Stefan's law, and the constant σ is called Stefan's constant.

The distribution of energy with wavelength is described by a function $\rho(\lambda)$, such that the energy in the wavelength range $(\lambda + \frac{1}{2}d\lambda)$ to $(\lambda - \frac{1}{2}d\lambda)$ is $\rho(\lambda)d\lambda$. Fig. 17.17 shows the results obtained from measurements of the radiation at 900° K., 1000° K. and 1100° K. It is found that $\rho(\lambda)$ is a maximum for a certain wavelength λ_m, and that the maximum wavelengths for different temperatures are connected by the relation

$$\lambda_m T = \text{const.} \quad \ldots \ldots \quad 17(29)$$

The maximum shifts towards the ultra-violet as the temperature rises, but at any wavelength a higher temperature gives a higher value of $\rho(\lambda)$. In 1893, W. Wien showed that the distribution of energy with

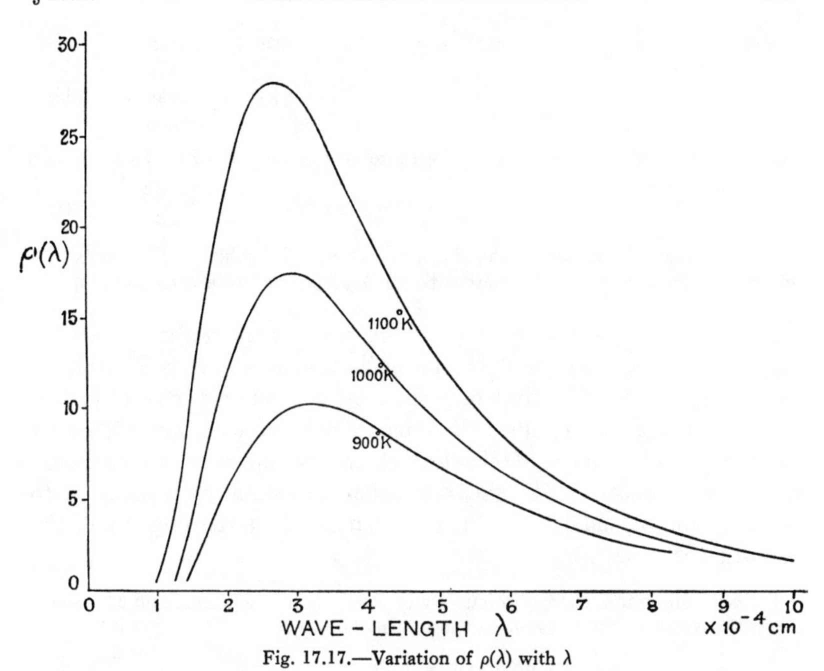

Fig. 17.17.—Variation of $\rho(\lambda)$ with λ

wavelength obeyed the following law (known as Wien's displacement law),

$$\rho(\lambda) = \lambda^{-5} F\left(\frac{1}{\lambda T}\right). \qquad 17(30)$$

For the present we may regard 17(28), 17(30) and 17(31) as empirical relations derived by inspection of the experimental results. The form of F is considered later.

EXAMPLE 17(i)

By (*a*) integrating and (*b*) differentiating 17(30), show that both 17(28) and 17(29) may be derived if 17(30) is assumed to be correct.

[(*a*) Integrating, we have

$$\rho = \int_0^\infty \rho(\lambda)\, d\lambda = \int_0^\infty \lambda^{-5} F\left(\frac{1}{\lambda T}\right) d\lambda. \qquad 17(31)$$

Putting $x = \lambda T$, we have (for one fixed temperature)

$$\rho = T^4 \int_0^\infty x^{-5} F\left(\frac{1}{x}\right) dx, \qquad 17(32)$$

i.e. ρ is proportional to T^4.

(*b*) To obtain λ_m we put $\partial\rho(\lambda)/\partial\lambda = 0$, and we have

$$-\frac{5}{\lambda_m}F\left(\frac{1}{\lambda_m T}\right) - \frac{1}{\lambda_m{}^2 T}F'\left(\frac{1}{\lambda_m T}\right) = 0, \quad . \quad . \quad . \quad 17(33)$$

where F' implies the derivative of F with respect to $y = 1/(\lambda T)$. This equation may be written

$$yF'(y) + 5F(y) = 0. \quad . \quad . \quad . \quad . \quad . \quad . \quad . \quad 17(34)$$

We now have an equation in the single variable y. Its solution, if there is one, gives a value of y in terms of constants, i.e. $\lambda_m T$ is equal to a constant.]

17.28.—Since the properties of the temperature radiation are independent of the material of the enclosure, it seems reasonable to seek a theory which involves only the most general properties of matter and of radiation. In 1884 Boltzmann derived equation 17(28) by applying the methods of classical thermodynamics to an enclosure containing radiation. In this discussion he assumed, in addition to the first and second laws of thermodynamics, that the pressure of the radiation is equal to $\frac{1}{3}\rho$.

If dS represents a change in entropy of the system as a whole, and dW represents the corresponding change in the internal energy,

$$T\,dS = dW + p\,dV. \quad . \quad . \quad . \quad . \quad . \quad . \quad . \quad . \quad 17(35)$$

Now

$$dW = V\,d\rho + \rho\,dV, \quad . \quad . \quad . \quad . \quad . \quad . \quad . \quad . \quad 17(36)$$

so that

$$T\,dS = V\left(\frac{\partial\rho}{\partial T}\right)_V dT + \frac{4}{3}\rho\,dV.$$

Hence

$$\left(\frac{\partial S}{\partial V}\right)_T = \frac{4}{3}\frac{\rho}{T} \text{ and } \left(\frac{\partial S}{\partial T}\right)_V = \frac{V}{T}\left(\frac{\partial\rho}{\partial T}\right)_V. \quad . \quad . \quad . \quad 17(37)$$

Differentiating the left-hand equation with respect to T and the right-hand equation with respect to V, we obtain

$$\frac{\partial^2 S}{\partial V\,\partial T} = \frac{\partial}{\partial T}\left(\frac{4}{3}\frac{\rho}{T}\right) = \frac{\partial}{\partial V}\left(\frac{V}{T}\frac{\partial\rho}{\partial T}\right),$$

i.e.

$$\frac{4}{3}\left(\frac{1}{T}\frac{\partial\rho}{\partial T} - \frac{\rho}{T^2}\right) = \frac{1}{T}\frac{\partial\rho}{\partial T}.$$

When the volume is constant,

$$\frac{d\rho}{\rho} = 4\frac{dT}{T}, \quad . \quad . \quad . \quad . \quad . \quad . \quad . \quad . \quad 17(38)$$

so that ρ is proportional to T^4.

17.29.—Wien * deduced equation 17(30), employing the methods of thermodynamics. He used the formulæ for the Doppler effect and

* See p. 327 of Reference 17.1.

for reflection from a moving mirror given in Chapter XI. Up to this point the theory of temperature radiation had been derived by applying the methods of thermodynamics to a system obeying laws which could be directly verified by experiments on light pressure, Doppler effect, etc. In order to go further and to derive the form of the function F in equation 17(30), it is necessary to introduce some hypothesis which cannot be directly referred to experiments which were available at the end of the nineteenth century. In the light of our present knowledge, it is desirable to introduce the quantum hypothesis, which is verified by the experiments described in §§ 17.2–17.19. Before doing this, let us consider an attempt to find this function F, using the methods of classical physics.

17.30.—Rayleigh and Jeans assumed that the general principles of mechanics, which are summarized in Hamilton's equations, were applicable to the radiation field. They assumed that radiation behaves, in its energy relations, like a mechanical system which possesses a certain number of degrees of freedom. This assumption appeared reasonable since the Hamiltonian method had been successfully applied to derive expressions for currents in inductive circuits and to many other problems in electromagnetism. They assumed that the amount of energy associated with any one degree of freedom can vary continuously. The problem of assemblies of systems with very large numbers of degrees of freedom had been analysed in detail by Gibbs and by Boltzmann, who showed that the average energy associated with each degree of freedom is kT, where T is the absolute temperature and k is a constant now known as Boltzmann's constant. Rayleigh and Jeans considered the equilibrium of radiation in a volume V, whose linear dimensions are very large compared with λ_m. We have seen * that the number of standing waves with wavelengths between $(\lambda + \frac{1}{2}d\lambda)$ and $(\lambda - \frac{1}{2}d\lambda)$ is equal to $(8\pi/\lambda^4)d\lambda$. If it is assumed that each possible mode of vibration (i.e. each possible standing wave) constitutes a degree of freedom, we obtain at once

$$\rho(\lambda) = \frac{8\pi}{\lambda^4} kT. \quad . \quad . \quad . \quad . \quad . \quad 17(39)$$

17.31.—This expression agrees approximately with the observations at the long-wavelength end of the spectrum but, instead of giving a maximum energy for one wavelength, the value of $\rho(\lambda)$ in-

* Appendix XIII A (p. 412).

creases without limit as λ decreases. Moreover, the integral $\int_0^\infty \rho(\lambda)\,d\lambda$ is not finite. This result may be deduced by very simple arguments. If it be assumed that "free space" has an infinite number of degrees of freedom for electromagnetic vibrations, and that each degree of freedom has an equal finite amount of energy, then the total energy of the radiation is infinite. Since material systems possess only a finite number of degrees of freedom, equilibrium between matter and radiation means that *all* the energy must be in the radiation field. This difficulty is fundamental and no minor amendment of the classical theory will provide a way of escape from this conclusion.

17.32. **Planck's Law.**

Let us now introduce the quantum hypothesis which was, in fact, first invented in order to provide a solution of this difficulty. Let us assume that the possible energies of a mode of vibration of frequency ν are $0, \epsilon_\nu, 2\epsilon_\nu, 3\epsilon_\nu, \ldots$, where $\epsilon_\nu = h\nu$. Then, according to Boltzmann's theorem, the probabilities of a given mode of vibration possessing energies $0, \epsilon_\nu, 2\epsilon_\nu, \ldots$ are in the ratios

$$1 : e^{-\epsilon_\nu/kT} : e^{-2\epsilon_\nu/kT} : \ldots,$$

so that the most probable mean energy is

$$\bar{\epsilon}_\nu = \frac{\sum_{n=0}^{\infty} n\epsilon_\nu e^{-n\epsilon_\nu/kT}}{\sum_{n=0}^{\infty} e^{-n\epsilon_\nu/kT}} \qquad 17(40)$$

$$= -\epsilon_\nu \frac{d}{dz}\left[\sum_{n=0}^{\infty} e^{-nz}\right] \Big/ \sum_{n=0}^{\infty} e^{-nz},$$

where $$z = \epsilon_\nu/kT;$$

thus $$\bar{\epsilon}_\nu = \frac{\epsilon_\nu}{e^{\epsilon_\nu/kT} - 1}. \qquad 17(41)$$

We insert this value of $\bar{\epsilon}_\nu$ in 17(39) instead of kT and obtain

$$\rho(\lambda) = \frac{8\pi}{\lambda^4} \frac{\epsilon_\nu}{e^{\epsilon_\nu/kT} - 1} \qquad 17(42)$$

or $$\rho(\lambda) = \frac{8\pi}{\lambda^5} hc \frac{1}{e^{hc/\lambda kT} - 1}, \qquad 17(43a)$$

putting $\epsilon_\nu = hc/\lambda$. This equation may be written

$$\rho(\lambda) = \frac{c_1}{\lambda^5}\frac{1}{e^{c_2/\lambda T} - 1}, \quad \text{. . . . . 17(43b)}$$

where
$$c_1 = 8\pi hc \quad \text{and} \quad c_2 = \frac{hc}{k}. \quad \text{. . . 17(44)}$$

If the energy corresponding to frequencies between $(\nu + \frac{1}{2}d\nu)$ and $(\nu - \frac{1}{2}d\nu)$ is $\rho(\nu)\,d\nu$, and if these frequency limits correspond to the wavelength range $(\lambda - \frac{1}{2}d\lambda)$ to $(\lambda + \frac{1}{2}d\lambda)$, we have

$$\rho(\nu)d\nu = -\rho(\lambda)d\lambda;$$

and since
$$\lambda\nu = c, \quad \nu d\lambda + \lambda d\nu = 0,$$

so that
$$\rho(\nu) = \frac{\lambda}{\nu}\,\rho(\lambda) = \frac{8\pi h\nu^3}{c^3}\frac{1}{e^{h\nu/kT} - 1}. \quad \text{. . 17(45)}$$

Equations 17(43) and 17(45) are alternative forms of Planck's law. These equations agree with Wien's law [17(30)], and hence with 17(28) and 17(29). They are also in satisfactory agreement with the direct measurements of the variation of energy with wavelength and frequency.

17.33.—Planck's constant h and Boltzmann's constant k can both be deduced from measurements on the temperature radiation. A convenient way is to measure the total energy (for all wavelengths) and the wavelength corresponding to the maximum energy at one and the same temperature. The solution of equation 17(34) yields *

$$\lambda_m T = \frac{c_2}{4{\cdot}9651} = \frac{hc}{4{\cdot}9651k}. \quad \text{. . . 17(46)}$$

* To obtain this formula put $F(y) = \dfrac{1}{e^{c_2 y} - 1}$ in 17(34).

We have
$$F'(y) = -c_2 e^{c_2 y}/(e^{c_2 y} - 1)^2$$

and hence 17(34) becomes
$$yc_2 e^{c_2 y} - 5(e^{c_2 y} - 1) = 0.$$

This equation may be solved by successive approximation (Newton's method) to yield 17(46).

Integrating * 17(43*b*), and comparing the result with 17(28), we obtain

$$\sigma = \frac{2\pi^5 k^4}{15c^2h^3}. \qquad . \quad . \quad . \quad . \quad . \quad . \quad . \qquad 17(47)$$

Thus when σ and $\lambda_m T$ have been measured, 17(46) and 17(47) form two equations in which h and k are the only unknowns.

17.34.—In the above derivation of Planck's law we have applied the quantum condition to a set of electromagnetic vibrations. It should be possible to obtain the same result by representing the temperature radiation as a " gas " in which the molecules are photons, and applying the methods of statistical mechanics. If we assume that the photons possess all the properties we associate with the material particles, we only repeat the calculations which Maxwell made for gas molecules and must obtain his result. If, however, we introduce the assumptions:

(*a*) that the number of photons is not fixed (although the total energy is fixed), and

(*b*) that the photons are not distinguishable individuals,

then we obtain Planck's law for the distribution of energy with frequency (see §§ 18.34–18.39).

17.35. **Principle of Detailed Balance.**

Let us now consider an assembly of atoms and radiation in thermodynamic equilibrium at a common temperature. It is observed that the radiation is independent of the kind of atom and depends only on the temperature. Planck's law gives the relation between $\rho(\lambda)$ and T. In a similar way, the distribution of the atoms among different stationary states depends on the temperature, and this relation is summarized in Boltzmann's law. Planck's law is obtained by applying the methods of statistical mechanics to the radiation—taking the quantum hypothesis into account. Boltzmann's law is obtained by applying the same principles to an assembly of atoms, and the quantum

* To integrate put $u = \frac{c_2}{\lambda T}$, and we obtain

$$\int_0^\infty \rho(\lambda)\,d\lambda = \frac{8\pi hcT^4}{c_2^4}\int_0^\infty \frac{u^3\,du}{e^u - 1}.$$

The value of the integral may be shown to be $\pi^4/15$.

Hence
$$\int_0^\infty \rho(\lambda)\,d\lambda = \frac{8\pi hcT^4}{c_2^4}\,\frac{\pi^4}{15} = \frac{8\pi^5k^4T^4}{15h^3c^3},$$

so that
$$\sigma = \frac{c\rho}{4T^4} = \frac{2\pi^5k^4}{15h^3c^2}.$$

[See Example 17(ix).]

hypothesis is introduced when we assume that there are discrete "stationary" states. The relation between statistical mechanics and thermodynamics is such that any departure from Planck's law or Boltzmann's law (apart from the small fluctuations predicted by statistical mechanics) would imply a breach of the second law of thermodynamics. It is clear that in an assembly of atoms and molecules, equilibrium is dynamic rather than static. All the time some atoms are absorbing radiation and some atoms are emitting radiation. These absorptions and emissions are accompanied by changes in the stationary states of the atoms, so that the conservation laws are satisfied in each individual process. The distribution laws of Planck and Boltzmann are maintained by the compensating action of the different processes.

17.36.—It is found that the equilibrium is maintained if, and only if, each individual interaction process is balanced by its own reverse process. For example, if an atom with stationary states 1, 2, 3, etc., is present, certain atoms will in unit time be transferred from state 1 to state 3 by the absorption of radiation of frequency ν_{13}. This process must be balanced by the emission of an equal number of quanta of frequency ν_{13} leading to a transfer of an equal number of atoms from state 3 to state 1. The emission of quanta of frequency ν_{12} followed by quanta of frequency ν_{23} would maintain the balance for the atoms but would destroy the equilibrium for the radiation. The latter could be put right by bringing in further compensating processes, but when this is done it is found that a fresh unbalance has been created elsewhere. The best that can be done by assuming "indirect" balancing is to create an equilibrium which might hold good for one temperature, pressure, etc., but which would be destroyed by an infinitesimal change in one of these physical parameters. The necessity of satisfying this "principle of detailed balance" is now generally accepted.

17.37. The Einstein Coefficients.

Let us consider the transference of a given atom to and from a state 1 and a state 2 by means of the absorption and emission of radiation of frequency ν_{12}. Let the state 2 be the state of higher energy. Let N_1 be the number of atoms per unit volume in state 1 when the system is in thermodynamic equilibrium at temperature T. Then the number per unit volume in state 2 is given by

$$N_2 = N_1 \exp\left(-\frac{h\nu_{12}}{kT}\right). \quad \ldots \quad 17(48)$$

Let $\rho(\nu)\,d\nu$ be the energy of isotropic radiation in the range $(\nu_{12} + \frac{1}{2}d\nu)$ to $(\nu_{12} - \frac{1}{2}d\nu)$. Then the number of atoms transferred from state 1 to state 2 per unit time by the absorption of radiation will be proportional to $\rho(\nu)N_1$, provided that the absorption line is so narrow that $\rho(\nu)$ is nearly constant over the spectral range for which there is appreciable absorption. Call the coefficient of proportionality B_{12}. In unit time a certain fraction of the atoms in state 2 will be transferred to state 1 by spontaneous emission. These transitions are not dependent on the presence of the radiation. Their number is proportional to N_2 and we call the constant of proportionality A_{21}. If these were the only processes operative, the principle of detailed reversibility would give

$$\rho(\nu)B_{12}N_1 = A_{21}N_2, \qquad 17(49)$$

which, taken with 17(48), gives

$$\rho(\nu) = \frac{A_{21}}{B_{12}} \exp\left(-\frac{h\nu_{12}}{kT}\right). \qquad 17(50)$$

The variation of radiation density with temperature given by 17(50) is not in accordance with experimental measurements which support Planck's law [equation 17(45)]. We may remove the discrepancy by assuming that, in addition to the spontaneous emissions which are independent of the presence of the radiation, there are also some induced emissions whose number is proportional to the density of radiation. The coefficient of proportionality is called B_{21}. We then have

$$\rho(\nu)B_{12}N_1 = A_{21}N_2 + \rho(\nu)B_{21}N_2 \qquad 17(51)$$

instead of 17(49), and this leads to

$$\rho(\nu) = \frac{A_{21}}{B_{12}\exp\left(\frac{h\nu_{12}}{kT}\right) - B_{21}}. \qquad 17(52)$$

This expression is identical with 17(45) if, and only if,

$$B_{12} = B_{21} \qquad 17(53a)$$

and

$$A_{21} = \frac{8\pi h\nu^3}{c^3} B_{21}. \qquad 17(53b)$$

The coefficients A_{21}, B_{12} and B_{21} are known as the Einstein coefficients, the above theory being due to Einstein. The quantity A_{21} is also called the "transition probability".* For densities of radiation which are normally available in optical experiments on absorption and emission, the second term on the right-hand side of 17(51) is very small compared with the first. It must, of course, be taken into account in deriving the radiation laws.

17.38. The Life of an Excited Atom.

When one of the Einstein coefficients is known, equations 17(53*a*) and 17(53*b*) enable us to determine the others. It is convenient to regard the coefficient of spontaneous emission A_{12} as the primary one, and to consider how it may be measured experimentally or calculated from atomic theory. Let us now consider the following experiment.

A vessel containing sodium vapour at low pressure is irradiated with light from a sodium lamp for a time, and the source is then removed. During the period of irradiation, light is absorbed by sodium atoms in the normal state and is re-emitted as resonance radiation. After the irradiation has been in progress a short time, an equilibrium is reached in which the number of atoms in the upper state (state 2) is constant, because the number of quanta absorbed per second is equal to the number re-emitted. When the radiation is cut off, the emissions go on, and since there are only a very few absorptions (due to the temperature radiation) the number of atoms in state 2 decreases rapidly. The number of atoms leaving the upper state in any short time dt is equal to

$$dN_2(t) = -A_{21}N_2(t)\,dt, \qquad \qquad 17(54)$$

where we write $N_2(t)$ for the number of atoms in the upper state in order to emphasize that this number is a function of the time. From 17(54) we see that the decrease of the number of atoms in the upper state is exponential, i.e. it follows the same law as the law of radioactive decay. We have

$$N_2 = (N_2)_0 e^{-A_{21}t} = (N_2)_0 e^{-t/\tau}. \qquad \qquad 17(55)$$

The constant τ introduced in 17(55) is called the *life of the excited state.* In a time τ the number of atoms in the upper state decreases by a

* Note that the above discussion and the relations 17(53) apply to individual states. When several states have the same energy, these relations must be applied to each state separately (see § 17.8 and Example 17(x), p. 578).

factor e. When there is only one possible transition (accompanied by the emission of radiation) from a given state, we have

$$\frac{1}{\tau} = A_{21}, \qquad \qquad 17(56a)$$

where τ is the life of state 2. When the atom is in a state from which it can emit several different lines, we have

$$\frac{1}{\tau_n} = \sum_m A_{nm}, \qquad \qquad 17(56b)$$

i.e. τ depends on the sum of the transition probabilities for all possible transitions.

17.39. Direct Measurement of τ.

The value of the constant τ for the transition corresponding to the emission of one of the yellow lines of sodium is of the order 10^{-8} second. A direct determination has been made using the apparatus shown in fig. 17.18. A vessel V containing sodium vapour is irradiated

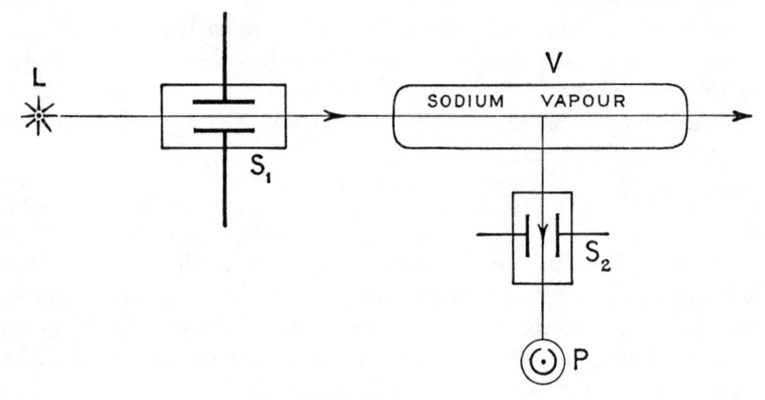

Fig. 17.18.—Measurement of τ

intermittently by means of a Kerr electro-optical shutter (S_1). This produces flashes of light of less than 10^{-8} second. The emitted light passes through a second shutter (S_2) and is measured by the photocell P. By appropriate electrical circuits it may be arranged that S_2 passes the light emitted by the vapour during a certain small interval which is subsequent to the time of irradiation, and the time which elapses between the two intervals can be varied. Since the amount of light emitted is proportional to the number of atoms in the upper state, this measurement enables the rate of decrease of this number to be calculated. An alternative method of measuring τ is due to Wien, who allowed a fast-moving beam of excited hydrogen atoms (fig. 2.7) to emerge from a hollow cathode. Wien measured the Doppler shift of the lines in order to determine the velocity. The variation in brightness along the length of the beam then gives an estimate of τ. Neither of these methods of measuring τ is very accurate, but it is important to have this very direct evidence of the decay of atoms in the upper state.

17.40. Relation between f-value and τ.

In the classical theory of dispersion a gas is regarded as an assembly of dipole oscillators which do not interact with one another. Each oscillator can emit, absorb, or scatter electromagnetic waves. The amount of absorption or emission is proportional to the number of oscillators, and it is assumed that there are f oscillators of a given type per molecule. The emission or absorption is thus proportional to the number of molecules per cubic centimetre (N) and to the value of f. In the quantum theory, emission and absorption are regarded as due to transitions between stationary states. The amount of emission or absorption is proportional to N, and to a constant (A or B). There must therefore be a relation between the "f-value" for a given absorption line and the corresponding Einstein B coefficient. Once this relation is established, equation 17(53b) gives a relation between f and τ (which is related to the Einstein constant A).

The constant f is defined by concepts which have no place in the quantum theory, at any rate so far as our present discussion has gone. Similarly, the concept of a life of an excited state has no place in classical theory. That there is a relation between these two quantities suggests that the two theories may have more common ground than has appeared so far. We shall consider the theoretical importance of the relation in Chapters XVIII and XIX. For the present let us regard it as a formal relation which may be submitted to experimental test.

17.41.—Consider a parallel beam of light of unit cross-sectional area passing normally through a thin layer of gas of thickness Δz (fig. 17.19). Suppose that the pressure is low enough to make the velocity of propagation nearly equal to c. Let the energy per unit volume in the layer be $\rho(\nu)\,d\nu$. Then the amount of energy passing through per second is $c\rho(\nu)\,d\nu$. The energy absorbed per second is

$$c\,\Delta z\int_0^\infty 2\alpha\rho(\nu)\,d\nu,$$

where 2α is the absorption coefficient (§ 15.5). From the definition of B_{12} (§ 17.37) the number of quanta absorbed per second is $B_{12}\rho(\nu)N_1\,\Delta z$ and the energy absorbed per second is $B_{12}\rho(\nu)N_1h\nu\,\Delta z$. Equating the

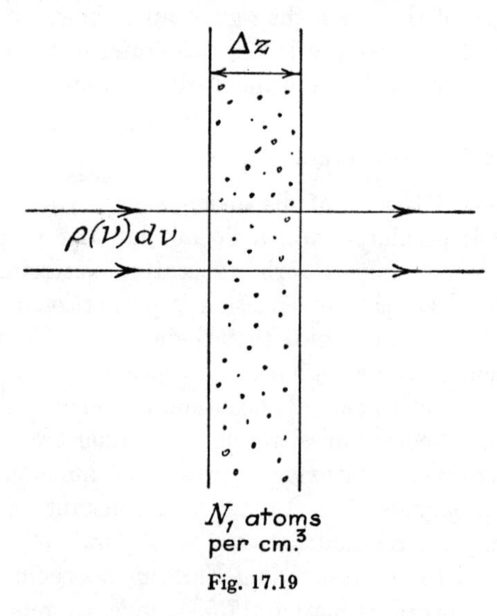

Fig. 17.19

two expressions for the energy absorbed per second and assuming that $\rho(\nu)$ is nearly constant over the absorption line (i.e. over the range of ν for which α differs appreciably from zero), we have

$$B_{12}=\frac{2c}{N_1h\nu}\int_0^\infty \alpha\,d\nu. \quad . \quad . \quad . \quad . \quad 17(57)$$

Using 15(73*b*) we have

$$B_{12}=\frac{\pi e^2}{mh\nu}f. \quad . \quad . \quad . \quad . \quad . \quad . \quad . \quad . \quad 17(58)$$

From 17(53*b*) and 17(57) we obtain

$$\frac{1}{\tau}=A_{21}=\frac{16\pi\nu^2}{N_1c^2}\int_0^\infty \alpha\,d\nu \quad . \quad . \quad . \quad 17(59)$$

and

$$\frac{1}{\tau}=A_{21}=\frac{8\pi^2\nu^2e^2}{mc^3}f. \quad . \quad . \quad . \quad . \quad . \quad . \quad 17(60)$$

The value of f for a given spectrum line may be derived from measurements on absorption or on anomalous dispersion. The constant τ may be measured in one of the ways described in § 17.39. Thus 17(59) may be checked by comparing values of f obtained from measurements of absorption or anomalous dispersion with values calculated * from measurements of τ. For example, for the mercury 2537 Å. line we have $f = 0{\cdot}027$ from absorption, and $f = 0{\cdot}0285$ from measurements of τ (Wien's method). For the sodium D lines (taken together) we have $f = 0{\cdot}975$ from anomalous dispersion, $f = 1{\cdot}05$ from direct measurements of τ, and $f = 0{\cdot}97$ from measurements of line absorption. This agreement is satisfactory in view of the difficulty of the measurement of τ.

17.42. **Relation between Natural Line Breadth and Transition Probability.**

In Chapter IV we saw that the frequency distribution in a beam of light is related to the effective length of the wave train. In elementary expositions of the quantum theory it is sometimes assumed that each atom emits its radiation instantaneously. If this were so, we should not have sharp spectral lines at all. It is therefore necessary to assume that the amplitude of each atomic oscillator decreases with time and is proportional to $e^{-A_{21}t/2}$. This makes the amount of light proportional to $e^{-A_{21}t}$, and is in agreement with experiment. Using the result obtained in Appendix IV B (p. 111), we obtain

$$f(\nu) = \frac{A_{21}}{4\pi^2} \frac{1}{(\nu - \nu_0)^2 + \frac{1}{4}A_{21}{}^2} \quad . \quad . \quad . \quad . \quad . \quad 17(61)$$

for the frequency distribution in a line. The half-width of the line is $A_{21} = 1/\tau$. It thus appears that the lifetime of the excited state determines the *natural* width of the line as well as the integrated absorption. This calculation takes no account of Doppler effect and collisions between atoms (see § 4.25–4.27).

17.43. **Calculation of Transition Probabilities.**

We have seen that the constant A_{21} is related to the lifetime of the excited state, to the f-value and to the natural width of the spectral line emitted when the atom makes a transition. The strength of the line determines the dispersion and, as we shall see later, is connected with a number of other optical properties (magnetic rotation, scattering, etc.). This constant is of great practical importance, since its value is needed for calculations on the transference of energy in the stars and in our own atmosphere. It must also play an important part in calculations on spectra emitted from discharges. Any satisfactory atomic theory must provide a method of calculating the transition probabilities as well as the energies of the stationary states. Neither

* In general it is necessary to use equation 19(69), p. 656, rather than equation 17(60).

the classical electromagnetic theory nor the early quantum theory of the atom is equipped to tackle this problem. The classical electromagnetic theory provides a method for making calculations on the relation between absorption, dispersion, etc., but the strength of the line (f-value) appears always as a constant to be determined by experiment. Calculation of transitions on the early Bohr theory consisted essentially of attempts to find a logical method of modifying some of the calculations of the classical theory. The new quantum mechanics, which we shall discuss in the next chapter, provides a general method of calculation.

EXAMPLES [17(ii)–17(x)]

17(ii). Show that the energy corresponding to the photo-electric threshold for cadmium (see Table 17.1, p. 541) is $6{\cdot}3 \times 10^{-12}$ ergs per electron released.

17(iii). Find the maximum energy of an electron released by light of wavelength 2000 Å. from a surface whose threshold is 3000 Å. [$3{\cdot}3 \times 10^{-12}$ ergs.]

17(iv). Find the wavelengths of (*a*) the first line (i.e. the one of longest wavelength) of the Balmer series, (*b*) the 40th line, and (*c*) the 41st line.

[(*a*) 6564·7 Å., (*b*) 3655·3 Å., (*c*) 3655·0 Å. Note that these are wavelengths *in vacuo*. The usual tabulated values are for wavelengths measured in air. Note that $n = 42$ in eqn. 17(4) corresponds to the 40th line.]

17(v). Show that the difference between the reciprocal of λ for successive lines of any one of the hydrogen series is approximately $2R/n^3$ when n is large.

17(vi). The lowest critical potential of caesium is 1·448 volts. Find the ratio of the number of atoms in the normal state to the number in the first excited state at temperatures of 27° C., 327° C., and 1727° C., using equation 17(9*b*) and taking $g_1 = 2$, $g_2 = 6$. [(i) $6{\cdot}7 \times 10^{23}$, (ii) $4{\cdot}7 \times 10^{11}$, (iii) $1{\cdot}5 \times 10^{3}$.]

17(vii). What is the tangential component of the pressure of isotropic radiation? [Zero.]

17(viii). Show that the dimensions are the same on both sides of 17(15).

[Remembering that ρ represents energy per unit volume, the dimensions on each side are $ML^{-1}T^{-2}$.]

17(ix). Show that if ρ is the energy density of isotropic radiation inside the enclosure, then

$$\rho = \frac{4}{c} E_a = \frac{4\sigma}{c} T^4.$$

[The density ρ applies to a solid angle 4π. The emission is into an angle 2π.]

17(x). Show that if g_1 states have energy W_1, and g_2 have energy W_2 (see § 17.8, small type, and § 17.37), then

$$g_1 B_{12} = g_2 B_{21},$$

and that 17(53*b*) is still valid. [Use equation 17(9*b*).]

REFERENCES

17.1. BORN: *Atomic Physics* (Blackie).
17.2. ANDRADE: *Structure of the Atom* (Bell).
17.3. RUARK and UREY: *Atoms, Molecules and Quanta* (McGraw Hill).
17.4. NICHOLS and HULL: *Phys. Rev.*, 1901, Vol. 13, p. 293.
17.5. MAXWELL: *Treatise on Electricity and Magnetism*, Vol. II, p. 440.
17.6. GEIGER AND BOTHE: *Z. Physik.*, 1925, Bd. 32, p. 639.
17.7. BETH: *Phys. Rev.*, 1936, Vol. 50, p. 115.
17.8. *Dictionary of Applied Physics*, Vol. IV, p. 541.

APPENDIX XVII A

THEORY OF THE COMPTON EFFECT

Suppose that a photon of frequency ν is scattered by an electron of rest mass m_0. Take a frame of reference in which the electron is at rest before the collision. Suppose that in this frame the scattered photon has frequency ν' and the

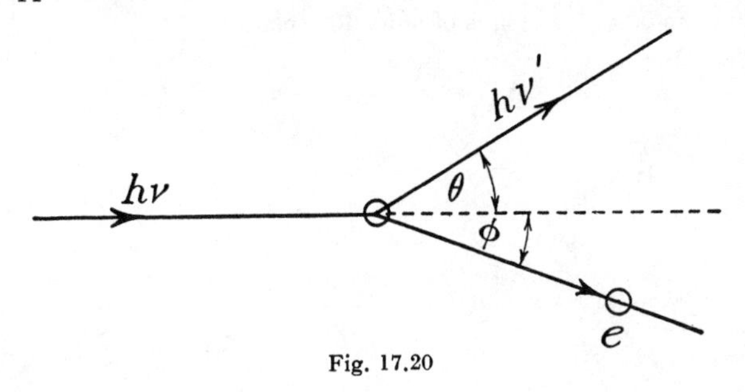

Fig. 17.20

scattered electron has speed v, the directions of motion making angles θ and ϕ with the original direction of the radiation (fig. 17.20). Then

(*a*) equating the relativistic energies,

$$h\nu + m_0c^2 = W_p + h\nu', \qquad 17(62)$$

where W_p is the relativistic energy (mc^2) of the electron after collision;

(*b*) equating components of momentum (and putting P_p for the momentum of the electron after collision),

$$\frac{h\nu}{c} = \frac{h\nu'}{c}\cos\theta + P_p\cos\phi, \qquad 17(63a)$$

$$\frac{h\nu'}{c}\sin\theta = P_p\sin\phi. \qquad 17(63b)$$

We have the general relation [see equation 11(49)]

$$c^2P_p{}^2 = W_p{}^2 - m_0{}^2c^4 = (W_p - m_0c^2)(W_p + m_0c^2). \quad . \quad 17(64)$$

Hence, using 17(62),

$$c^2P_p{}^2 = h(\nu - \nu')[h(\nu - \nu') + 2m_0c^2], \quad . \quad . \quad . \quad 17(65)$$

and from 17(63*a*) and 17(63*b*)

$$c^2P_p{}^2 = h^2[(\nu - \nu' \cos\theta)^2 + \nu'^2 \sin^2\theta]. \quad . \quad . \quad . \quad 17(66)$$

Equating the right-hand sides of 17(65) and 17(66)

$$(\nu - \nu')[h(\nu - \nu') + 2m_0c^2] = h[\nu^2 - 2\nu\nu' \cos\theta + \nu'^2]$$

or

$$2(\nu - \nu')m_0c^2 = 2h\nu\nu'(1 - \cos\theta).$$

Hence

$$\frac{\nu - \nu'}{\nu\nu'} = \frac{h(1 - \cos\theta)}{m_0c^2}$$

or

$$\lambda' - \lambda = \frac{h}{m_0c}(1 - \cos\theta). \quad . \quad . \quad . \quad . \quad . \quad 17(67)$$

In the above discussion we neglect the forces which bind the electron to the atom, because the binding energies of outer electrons are only a few volts, and the energy of an X-ray photon is of order 10^5 volts.

CHAPTER XVIII

Quantum Theory of Radiation

18.1. **Relation between Theory of Radiation and Theory of Atomic Structure.**

In the preceding chapter we saw that the quantum theory began as a theory of the *interaction* between radiation and matter, but that quantum principles came to be applied both to the radiation itself, in the Einstein theory of photons, and to the atoms in Bohr's theory of stationary states. The connecting link, Bohr's frequency condition [equation 17(7)], is a direct application of the law of conservation of energy. There is also good experimental evidence that the conservation laws apply to individual processes involving the interaction of energy and matter. Every important change in the theory of radiation will involve a corresponding change in the theory of atomic structure. Any satisfactory quantum theory must apply both to radiation and to matter. Up to 1924 it was thought that the similarity between radiation and matter was chiefly a matter of energy relations. Radiation theory was admitted to be difficult because light showed wave properties in some experiments and particle properties in others, but the behaviour of atoms and electrons was described by ordinary dynamical theories of particles. In 1924 L. de Broglie suggested that there is a wave associated with every material particle and that the relation between the relativistic momentum (p) of the particle and the associated wavelength (λ) is

$$\lambda = \frac{h}{p}. \qquad \qquad 18(1)$$

This relation is the same as the relation between the wavelength and the momentum of the photon.* The waves postulated by de Broglie are not electromagnetic waves. Their special properties will be considered later. For the present discussion only general wave properties (wavelength, periodicity, etc.) are relevant.

* See equation 17(20).

18.2. **Diffraction of Electrons.**

Insertion of numerical values shows that the wavelength associated with an electron of energy 1 volt is about 12 Å., whereas that associated with an electron of 40,000 volts is about 0·06 Å. The order of magnitude is thus comparable with the order of X-ray wavelengths rather than with that of the wavelength of visible light. In 1927 G. P. Thomson obtained diffraction patterns by passing a beam of electrons of about 15,000 volts energy through very thin metal foils. About the same time Davisson and Germer obtained evidence of the diffraction of slow electrons by reflection from single crystals of nickel. Two photographs of the diffraction of a beam of electrons which has passed through thin foil are shown in Plate V*h*, *i* (p. 636). Some time after the original discovery of electron diffraction, experiments more closely parallel to the usual experiments on optical diffraction were carried out with electrons, and typical diffraction patterns were obtained. Plate V*f*, *g* also shows the diffraction of a beam of electrons at the edge of a small crystal. It may be seen that one picture differs from that obtained with light [Plate III*a*, p. 214] only in that in the electron picture the diffraction fringes are somewhat sharper despite the high magnification. Thomson showed that the electron wavelength was inversely proportional to the momentum and, using the X-ray values for the spacing of atoms in his thin metal films, he was able to deduce the constant of proportionality. It was found to be equal to h to within the limit set by experimental error. All subsequent work on electron diffraction, including the direct measurement of wavelengths with the ruled gratings, has confirmed the validity of equation 18(1). It has also been shown that protons, atoms of He, and molecules of H_2 can also be diffracted under suitable conditions. It is now generally accepted that 18(1) is valid for material particles.

EXAMPLES [18(i)–18(v)]

18(i). Insert numerical values in 18(1) and calculate the electron wavelengths for velocities 10^5 cm./sec., 10^8 cm./sec., and 10^9 cm./sec.

[7280 Å.; 7·28 Å.; 0.728 Å.]

18(ii). Show that the wavelength associated with an electron which has been accelerated by a field of V volts is

$$\lambda = h\sqrt{\frac{150}{m_0 eV}}, \qquad \text{18(2a)}$$

or approximately $$\lambda = 12{\cdot}3/\sqrt{V}, \qquad \text{18(2}b\text{)}$$

when λ is in Ångström units and the velocity is so small that the change of mass with velocity may be neglected.

18(iii). Calculate from 18(2*b*) the electron wavelengths for electrons accelerated by fields of 3 volts and 30,000 volts. [7·1 Å.; 0·071 Å.]

18(iv). Derive an expression for the error in wavelength due to neglecting the relativity correction. For what velocity is this error 10 per cent of the true wavelength? [$1{\cdot}25 \times 10^{10}$ cm./sec.]

18(v). Assuming that 18(1) applies to protons, calculate the wavelength associated with a proton of energy 10^5 electron volts. [0·00093 Å.]

18.3.—To the classical physicist, wave and particle theories were alternative ways of representing the transfer of energy from one place to another. At the end of the nineteenth century it appeared that the particle concept applied to atoms, electrons, and molecules, as well as to larger pieces of matter which were composed of molecules. On the other hand the wave theory was very successful in representing the results of experiments on light. By 1927 it was clear that light had some " particle properties ", and that matter had some " wave properties ". The results of *some* experiments on light were more easily represented by means of a particle picture, and the results of *some* experiments on matter were more easily represented by a wave theory. Yet the original evidence which had led to the wave theory of light and the particle theory of matter remained valid, and had to be included with the newer results in any complete and satisfactory representation. It became necessary to invent a new theory in which wave and particle ideas would be complementary, not alternative. A complete solution of this problem should apply both to radiation and to matter.

18.4. **The Uncertainty Principle—Particle Dynamics.**

Any satisfactory physical theory must give an account of actual experiments. Relativity theory is essentially an analysis of the relation between experimental data concerning length and time measurements, and the symbols which appear in the equations of mathematical physics. It shows that paradoxes may be avoided by discussing theory in terms of measured quantities whose relations are determined by experiment. The uncertainty principle which we shall now describe is an extension of this process to the measurement of dynamical variables for particles of atomic dimensions.

In expositions of thermodynamics it is often convenient to discuss idealized experiments made with pistons of zero friction, cylinders of zero heat capacity, etc. We know that these conditions can never be realized in any real apparatus, yet the results derived from the discussion of these " ideal " experiments are valid because the departures of real substances from the ideal properties do not affect the argument or, in the case of certain inequalities, they actually reinforce it. We show that A is greater than B under the ideal condition, and also that any departure from the ideal will increase the difference. The inequality, proved to hold for ideal conditions, is valid generally.

Let us now consider a series of ideal experiments designed to verify the laws of Newtonian mechanics. Let us assume that we have at our disposal apparatus which is capable of focusing radiation or material waves of any wavelength without any of the errors of geometrical optics, i.e. that we can use a microscope with X-rays or γ-rays, or electrons of any desired wavelength. Suppose that some elementary dynamical problem, such as the motion of an electron in the absence of a field of force, is to be investigated. We desire to show that there is no field of force by testing the validity of the equation

$$\frac{d^2q}{dt^2} = 0, \quad \ldots \ldots \ldots \quad 18(3)$$

where q is a co-ordinate specifying the position of the particle at time t.

The obvious procedure is to measure the position and momentum of the electron at some time $t = t_0$, in order to obtain two " initial conditions " which can be inserted in the solution of 18(3). From this solution we can then calculate the position and momentum at some later time t_1 and see if the calculation agrees with the result of an observation made at that time. Suppose that we observe the particle with light of wavelength λ. The discussion of Chapter VIII shows that the diffraction of the wave sets a limit to the accuracy of a position measurement such that

$$\Delta q \sim \frac{\lambda}{2 \sin \theta}, \quad \ldots \ldots \ldots \quad 18(4)$$

where Δq is the probable error in our measurement of q, and θ is the semi-angle of the cone of rays accepted by the microscope (fig. 18.1). The sign $\sim$ means " at least of the order of magnitude of ". The experiment of Compton (§ 17.23) shows that the interaction between the photon and the electron involves an exchange of momentum. We may assume that the momenta of the particle and of the photon were known exactly before their interaction, but our knowledge of the momentum of the electron *after the interaction* depends on the accuracy with which we can estimate the magnitude and direction of the momentum exchanged during the interaction. We know that the photon enters the microscope, and therefore we know its direction of motion (after the interaction) within an angle 2θ. Any attempt to determine along what path the photon entered the microscope (by inserting stops or in any other way) reduces the effective aperture of the microscope and hence increases Δq. Thus, if the aperture is 2θ the component of the momentum of the photon in the plane perpendicular to the axis of the microscope (which is the plane in which q is measured) is uncertain by an amount

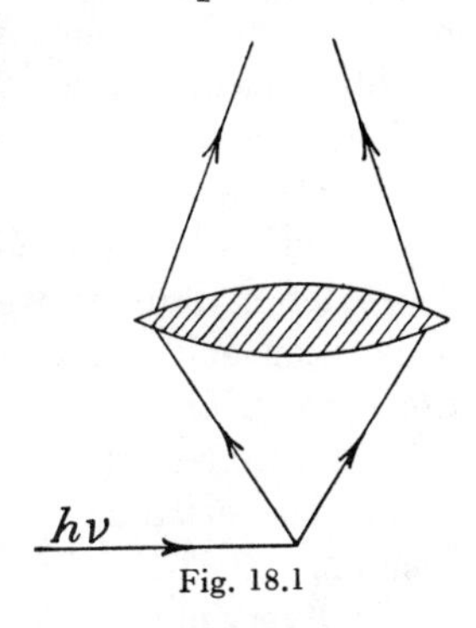

Fig. 18.1

$$\Delta p \sim \frac{2h\nu}{c} \sin \theta. \qquad 18(5)$$

The momentum of the particle after the interaction is uncertain by Δp. Combining 18(4) and 18(5), we have

$$\Delta p \, \Delta q \sim \frac{\lambda}{2 \sin \theta} \frac{2h\nu}{c} \sin \theta,$$

i.e.

$$\Delta p \, \Delta q \sim h. \qquad 18(6)$$

18(6) represents the smallest error product which can result. Any deviation from the ideal conditions (e.g. lack of precise knowledge of the momentum of the photon before interaction) can only increase one of the errors. Note that the momentum of the particle may be measured accurately in one experiment and its position in another. The second experiment will always affect the momentum in such a way that the uncertainty product complies with 18(6). We cannot obtain exact information about the two initial conditions *at the same time.* It is precisely this information which is required to enable us to make accurate predictions about the future position of the particle.

18.5.—This result is not dependent on the details of the method of measurement, such as the wavelength used or the angle of the cone of rays. The analysis of a quite different kind of experiment leads to the relation 18(6). It has the same sort of generality as the second law of thermodynamics, which is in no way dependent on the details of any one experiment which exemplifies it. Many attempts have been made to find systems which infringe the second law, but detailed analysis always shows that they are invalid, because some compensating effect has been overlooked. In a similar way Heisenberg has been able to show that all proposed methods * of making measurements with a higher accuracy than that permitted by 18(6) are invalid. The relation 18(6) is a general relation applying to all measurements of position and momentum, because it derives from three generalizations which are themselves directly based upon experiment. These generalizations are:

(*a*) The accuracy of all spatial measurements is limited by diffraction.

(*b*) Every observation involves a finite exchange of energy between the object and the measuring system.

(*c*) The relation between the wavelength (which determines the error of a spatial measurement) and the momentum of the associated particle (which determines the momentum error in the measurement) is given by 18(1). This is true whether the measurement be made with light or with material particles.

18.6. **The Uncertainty Principle—Wave Theory.**

The above discussion can also be applied to measurements on waves. Suppose that a theoretical calculation has shown that a certain radio transmitter should emit plane waves of a given frequency travelling with velocity c. We know that such waves are represented by an expression of the type

$$\xi = A \cos \phi = A \cos (\omega t - \kappa x + \epsilon). \qquad . \quad . \quad 18(7)$$

We saw in Chapter II that A and ϵ in this expression may be regarded as arbitrary constants to be determined from the initial conditions. If we make suitable measurements on the wave at a given time at a point near the transmitter, we can calculate the displacement for a distant point at another time. The theory may be tested by seeing whether the calculated value correctly predicts the result of a second measurement. In an ideal experiment we may determine the initial conditions by allowing the radiation to fall upon a mirror placed at $x = 0$ and measuring the pressure it exerts upon the mirror. This pressure will fluctuate during a period of the wave motion, but at a

* Reference 18.1.

certain time ($t = 0$) we may observe that it is a maximum, so that we know that when $x = 0$ and $t = 0$, ϕ is also zero and $\epsilon = 0$. By measuring the maximum pressure we determine A. Hence knowing A and ϵ, we can calculate the value of ξ at any other time and place from 18(7).

Now consider the accuracy of these measurements. We can measure the pressure only by measuring the momentum given to the mirror by the radiation. The discussion of the preceding paragraphs shows that we can never know the position and momentum of the mirror to an accuracy better than that given by 18(6). We have determined the phase of the wave at a point whose position is uncertain by Δx, so that the phase at a given point *in our co-ordinate system* is uncertain by an amount of order $\kappa\,\Delta x$. If Δp is the uncertainty in the momentum of the mirror the total momentum of the wave is uncertain by Δp and, since its frequency is known to be exactly $\omega/2\pi$, the number of quanta in the wave is uncertain by ΔN, where

$$\Delta N \frac{h}{c}\frac{\omega}{2\pi} = \Delta p. \qquad 18(8)$$

Hence we have

$$\Delta N\,\Delta\phi = \frac{c}{h}\frac{2\pi}{\omega}\,\kappa\,\Delta p\,\Delta x$$

$$\sim 2\pi. \qquad 18(9)$$

For a given frequency the number of quanta is proportional to the energy, which is proportional to A^2. Thus the uncertainty in N implies an uncertainty in A, and we cannot know both A and ϕ with complete accuracy. The relation 18(9) differs from 18(6) by the factor h, because of the way we have chosen to define our variables. It is of the same generality and, like 18(6), it cannot be evaded by any recourse to ingenious devices.

18.7. Energy-time Relations.

Suppose that we wish to know with high precision the time at which a photon is emitted from a source. We may place an ideal shutter in front of the source and open it for a short time Δt. The time of emission will then be known within an error equal to Δt. From the discussion of wave trains of finite length in Chapter IV and Appendix IV B, we know that when the length of a wave train is $c\,\Delta t$, there is a distribution of frequency covering a range $\Delta\nu = 1/\Delta t$; so the energy

(W) of a photon in the beam is uncertain by $\Delta W = h\,\Delta\nu = h/\Delta t$. We thus have

$$\Delta W\,\Delta t \sim h. \qquad \text{18(10)}$$

This relation, like 18(6) and 18(9), is generally valid. An interesting and important example arises in connection with the natural widths of spectral lines. Suppose we attempt to measure the energy difference between the normal state and one of the higher stationary states of an atom by measuring the wavelength of the line emitted during transition between the states. The energy will be subject to an error $\Delta W = h\,\Delta\nu$, where $\Delta\nu$ is the half-value width of the line. The time at which a given atom was in this state is uncertain by a quantity Δt which is of the same order as τ, the life of the state. But it was shown in § 17.42 that $\Delta\nu = 1/\tau$, and hence $\Delta W\,\Delta t \sim h$, and 18(10) is satisfied. Any experiment designed to discover whether energy is conserved when light is emitted, absorbed or scattered will be subject to this error. At best we may be able to say that energy is conserved within the accuracy of measurement.

18.8. **Indeterminism.**

The classical theory of dynamics is a method of exact calculation. The details of its application to the dynamics of particles differ from the details of its application to wave motion, but the essentials are the same for both. It is assumed that certain differential equations are exactly correct. The solutions of these equations contain two arbitrary constants for each degree of freedom. If we assume that these constants can be exactly determined by measurements made at some given time, then the behaviour of the system at any later or earlier time can be calculated exactly. The uncertainty relations show that the arbitrary constants can never be known exactly, even when " ideal " apparatus is available, so that it is impossible to apply either classical particle dynamics or classical wave theory in such a way as to obtain the exact solution of any physical problem. When we are dealing with large pieces of matter such as we ordinarily handle in the laboratory, the errors involved in using the classical method are small compared with the ordinary experimental errors, and we can detect no deviation from classical dynamics. We may neglect the transfer of momentum from the measuring apparatus to the object under observation when the amount transferred is small in relation to the general errors of observation. In calculations involving small numbers of electrons and single photons, the momentum transferred from the measuring system to the object is either the whole momentum involved, or is an appreciable fraction of the whole momentum; the uncertainty relation then becomes of primary importance. The Heisenberg analysis shows that, in order to represent the results of physical measurements, we need a theory which is essentially different *both* from the classical particle dynamics and from the classical wave theory. These theories are mathematical methods of using precise data to calculate exact predictions. They both deal with variables which, for mathematical purposes, are assumed to be precisely known. The actual experimental errors are taken to be accidental and of no logical significance. Once we accept the three experimental results stated at the end of § 18.5, we admit the presence of uncertainties which are not accidental. In order to attain logical consistency, it is necessary to insert these uncertainties into the foundation of the theory. The new theory must be a calculation of probable values of certain variables from probable values—not of precise values from precise values. We shall now consider a theory of this type.

18.9. **The Wave Mechanics.**

As an introduction to the new theory let us discuss the relation between geometrical optics and wave optics. In the former, light is assumed to travel along certain lines called " rays ". The positions of the rays can be calculated by applying the laws of reflection and refraction which are summarized in Fermat's relation

$$\delta \int \frac{ds}{\lambda} = 0. \qquad \qquad 18(11)$$

The wave theory does not differ from geometrical optics in predicting that light will travel along a different set of lines. It denies that light energy travels along lines at all. According to wave optics, light is diffracted, and any attempt to confine it to precise lines by means of slits and diaphragms must be unsuccessful. The wave theory predicts that nearly all the energy travels very near to certain lines, provided that the wavelength is small in relation to the dimensions of the apparatus. These lines are found to coincide with the rays of geometrical optics. Thus all these results which can be correctly calculated by geometrical optics are also predictions of the wave theory.

18.10.—According to classical mechanics, particles move along lines. The position of these lines may be calculated from Newton's laws of motion, which we may now regard as analogous to the laws of reflection and refraction in geometrical optics. Hamilton showed that the laws of classical mechanics could be summarized in the variational relation

$$\delta \int L\,dt = 0, \qquad \text{18(12}a\text{)}$$

where L is the difference of the kinetic and potential energy. It was shown by Maupertuis that, for self-contained systems which obey the conservation laws, this relation may be replaced by

$$\delta \int p\,ds = 0, \qquad \text{18(12}b\text{)}$$

where p is the momentum. Experiment has shown that the wavelength of material particles is inversely proportional to the momentum. Inserting the experimental relation 18(1) into 18(11) we obtain 18(12*b*). Thus the trajectories of material particles calculated from 18(12*b*) are exactly the same as " rays " calculated from 18(11). The rays of geometrical optics are rectilinear when the refractive index is constant, and are curved when the medium is heterogeneous. Similarly, the paths of particles are straight lines in free space, but become curved when the particle enters a field of force. The close analogy between equation 18(12*b*) and equation 18(11) suggests that there may be a wave equation for mechanics and that this equation will be similar in some ways to the wave equation for optics. In the next paragraph this analogy is used as a guide to obtain the fundamental equation of wave mechanics.

18.11.—The wave equation of optics may be written

$$\nabla^2 \Psi = \frac{n^2}{c^2}\frac{\partial^2 \Psi}{\partial t^2}. \qquad \text{18(13)}$$

If we put

$$\Psi = \psi e^{i\omega t}, \qquad \text{18(14)}$$

where ψ is a function of x, y, z, but not of t, we obtain

$$\nabla^2 \psi + \frac{n^2}{c^2}\,\omega^2 \psi = 0 \qquad \text{18(15}a\text{)}$$

or

$$\nabla^2 \psi + n^2 \kappa_0{}^2 \psi = 0, \qquad \text{18(15}b\text{)}$$

where κ_0 is the wavelength constant *in vacuo*. The wave theory [based on 18(13) and 18(15)] leads to rays [determined by 18(11)] when the approximations of geometrical optics are valid. We now wish to find an equation, analogous to 18(15), from which we may deduce 18(12*b*) by introducing corresponding approximations. In order to determine the constants of this equation, we have the following requirements:

(*a*) The new equation must be dimensionally correct.

(*b*) The wavelength of a particle in free space must be equal to h/p.

(*c*) Since $n\kappa_0$ is proportional to $1/\lambda$, it follows that $n\kappa_0$ must be proportional to p.

We shall now use these requirements to help to find the new equation, and then make it the starting point of the new theory.

From (*c*) we have

$$n^2 \propto m^2v^2 \propto m(W - V), \quad . \quad . \quad . \quad 18(16)$$

where W is the total energy and V is the potential energy.

The following equation, proposed by Schrödinger, satisfies these conditions:

$$\nabla^2\psi + \frac{8\pi^2}{h^2} m(W - V)\psi = 0. \quad . \quad . \quad . \quad 18(17)$$

This equation is of the same mathematical form as 18(15).

When $V = 0$, the equation reduces to

$$\nabla^2\psi + \frac{4\pi^2}{h^2} p^2\psi = 0. \quad . \quad . \quad . \quad . \quad 18(18)$$

This corresponds to a wavelength h/p for a free particle. The equation 18(17) implies

$$n^2 = (W - V)/W, \quad . \quad . \quad . \quad . \quad 18(19)$$

provided that $\kappa_0 = 2\pi mv/h = 2\pi p/h$.

We must not expect that the ψ-waves will have all the properties of optical waves. For the immediate problem this is not of great importance. The constant in the second term of 18(17) is proportional to $m(W - V)$, and this guarantees that condition (*c*) is satisfied.

18.12.—The relation between the wave equation in a heterogeneous medium (n varying from point to point) and certain problems in classical mechanics may be discussed in the following way. Suppose that

$$\psi = Ae^{i\phi} = A \exp in\kappa_0(lx + my + nz). \quad . \quad . \quad . \quad 18(20)$$

If A and n are constants, then ψ represents a plane wave in a *homogeneous* medium. Now let us suppose that n varies from point to point, but so slowly that the variation in a distance of one wavelength is small. Then A becomes a slowly varying function of x, y, z, and the derivatives of A are small compared with κ_0. We then have

$$\frac{\partial \psi}{\partial x} = \left(\frac{\partial A}{\partial x} + iA\frac{\partial \phi}{\partial x}\right)e^{i\phi}$$

or, approximately, since $\partial\phi/\partial x$ is of the same order of magnitude as κ_0,

$$\frac{\partial \psi}{\partial x} = iA\frac{\partial \phi}{\partial x}e^{i\phi}$$

and, to the same approximation,

$$\frac{\partial^2 \psi}{\partial x^2} = -A\left(\frac{\partial \phi}{\partial x}\right)^2 e^{i\phi}.$$

Equation 18(15b) then becomes

$$\left(\frac{\partial \phi}{\partial x}\right)^2 + \left(\frac{\partial \phi}{\partial y}\right)^2 + \left(\frac{\partial \phi}{\partial z}\right)^2 = n^2\kappa_0^2. \quad . \quad . \quad . \quad . \quad . \quad 18(21)$$

This is a fundamental equation in ray optics. In isotropic (but not necessarily homogeneous) media, rays may be taken to be orthogonal trajectories of the family of surfaces $\phi = c$. For certain problems in dynamics, it is usual to introduce a function S (known as Jacobi's function) defined by

$$\frac{\partial S}{\partial x} = p_x, \text{ etc.} \quad . \quad . \quad . \quad . \quad . \quad . \quad . \quad . \quad . \quad 18(22)$$

For a single particle we have

$$\left(\frac{\partial S}{\partial x}\right)^2 + \left(\frac{\partial S}{\partial y}\right)^2 + \left(\frac{\partial S}{\partial z}\right)^2 = 2m(W - V). \quad . \quad . \quad . \quad 18(23)$$

This equation is identical with 18(21) if we substitute for n from 18(19) and identify the phase ϕ of the wave with $h/2\pi$ times Jacobi's function S from dynamics. We see that:

(i) The fundamental equation 18(21) of ray optics in a heterogeneous medium can be deduced from the wave equation.

(ii) This equation is directly analogous to the equation 18(23) which applies to a particle moving in a field which varies from point to point.

Thus the equation of motion of such a particle can be deduced from 18(17) by inserting the approximations appropriate to geometrical optics.*

18.13.—Equation 18(17) is generally known as the *wave equation* or as *Schrödinger's equation*. It will be noticed that it is not a complete

* For a detailed discussion of the relation between wave and ray optics on the one hand, and wave and classical mechanics on the other, see Reference 18.2. Note that in the references the constant $k_0 = 2\pi/h$ and is not equal to our κ_0; there is also a corresponding difference in the definition of n. The method used in this book is intended to give the most direct analogy between mechanics and optics.

wave equation, since it does not contain the time. It is analogous to the reduced wave equation of optics which gives the variation of amplitude from one place to another. The quantity ψ is generally known as the *wave function* although it should strictly be called a wave amplitude function.

We shall discuss the physical interpretation of the wave functions later. We now wish to show how the wave equation may be used to calculate the energies of stationary states. In doing this we assume that the only solutions of the equation which have physical significance are continuous, single-valued, and finite at all points in the space considered, and on the boundary. This assumption is formally a new postulate, but one which is inherently very reasonable. When this condition is applied to the wave equation, it is found that under most conditions only certain values of the energy are permitted. The equation gives satisfactory solutions only when the constant W has one of these values. The solutions so obtained are called " proper functions " (*eigenfunctions*) and the associated values of W are called " proper values " (*eigenvalues*).

The situation is similar to the boundary problems which arise with sound waves and light waves. Vibrations of any frequency may be transmitted along a string of unlimited length. The mechanical properties of the material determine the speed of propagation, but not the frequency. When boundary conditions are imposed (e.g. if we assume that two points on the string are fixed so that the displacement is zero at these points), then the system vibrates with its fundamental frequency, or with one or more of the harmonics. In a similar way, the electromagnetic theory allows us to calculate the velocity of light in a vacuum. Any frequency is possible until we impose a boundary condition. In a space of finite size, standing waves are set up and these have frequencies determined by the boundary conditions (Chapter III and Appendix XIII A).

18.14. Application of the Wave Equation to Free Particles.

Let us consider a free particle under no forces. We put $V = 0$ in the wave equation [18(17)] and we obtain

$$\nabla^2\psi = -\frac{8\pi^2}{h^2} mW\psi = -\frac{4\pi^2}{h^2} m^2v^2\psi. \quad . \quad . \quad 18(24)$$

Consider the solution appropriate to plane waves travelling along the x axis. We have

$$\psi = Ae^{i\kappa x} + Be^{-i\kappa x}, \qquad \ldots \ldots \quad 18(25)$$

where
$$\kappa^2 = \frac{8\pi^2}{h^2} mW = \frac{4\pi^2}{h^2} m^2v^2. \qquad \ldots . \quad 18(26)$$

We see that as long as κ is real (W positive) the solution satisfies the condition. Thus any positive value of W is permitted. Since W is equal to the kinetic energy ($\frac{1}{2}mv^2$), κ is equal to $2\pi mv/h$, and the associated wavelength is that given by 18(1). Equation 18(25) corresponds to plane waves travelling in the direction of the axis. There is no restriction on the wavelength or the velocity of the particle.

18.15. **Particle in a Box.**

Let us now consider a particle in a box of finite size. Suppose that the x co-ordinate of any point in the box is between 0 and a_1, the y co-ordinate between 0 and a_2 and the z co-ordinate between 0 and a_3. The value of ψ must be zero whenever $x \geqslant a_1$ or $y \geqslant a_2$ or $z \geqslant a_3$. These are the conditions which we should apply to an optical wave function to find the frequencies of the standing waves in the box.* The wave functions which satisfy the boundary conditions are

$$\psi = \sin\left(2\pi \frac{n_1}{a_1} x\right) \sin\left(2\pi \frac{n_2}{a_2} y\right) \sin\left(2\pi \frac{n_3}{a_3} z\right), \qquad 18(27)$$

where n_1, n_2, n_3 are any integers.

The associated values of the wavelength constants are

$$\kappa_x = \frac{2\pi n_1}{a_1}, \quad \kappa_y = \frac{2\pi n_2}{a_2}, \quad \kappa_z = \frac{2\pi n_3}{a_3}, \qquad . \quad 18(28)$$

leading to
$$\kappa^2 = 4\pi^2\left(\frac{n_1^2}{a_1^2} + \frac{n_2^2}{a_2^2} + \frac{n_3^2}{a_3^2}\right). \qquad \ldots \quad 18(29)$$

The associated energies are

$$W = \frac{h^2}{2m}\left(\frac{n_1^2}{a_1^2} + \frac{n_2^2}{a_2^2} + \frac{n_3^2}{a_3^2}\right). \qquad \ldots \quad 18(30)$$

Thus, in a box, the particle can have one of a certain selected set of energies. Since the energy is all kinetic, this implies that only certain velocities are possible. The above discussion applies to any number

* In Appendix XIII A we have considered waves in a *cubical* box.

of particles which do not exert forces upon one another. It thus applies to photons generally and to material particles when the effects of gravitational or electrical forces are neglected.

18.16. The Simple Harmonic Oscillator.

Let us now consider a simple harmonic oscillator as a dynamical system for which the potential energy is proportional to the square of the displacement of the particle from an equilibrium position. We have (see § 2.5)

$$V = \tfrac{1}{2}\omega_0{}^2 m x^2. \qquad \text{18(31)}$$

If we insert this expression for V in the wave equation, we obtain

$$\frac{d^2\psi}{dx^2} + \frac{8\pi^2 m}{h^2}(W - \tfrac{1}{2}\omega_0{}^2 m x^2)\psi = 0. \qquad \text{18(32)}$$

To simplify the notation, we put

$$\alpha = \frac{2\pi m\omega_0}{h} \quad \text{and} \quad \beta = \frac{8\pi^2}{h^2} mW, \qquad \text{18(33)}$$

so that

$$\frac{d^2\psi}{dx^2} + (\beta - \alpha^2 x^2)\psi = 0. \qquad \text{18(34)}$$

We wish to find solutions of 18(34) which are single-valued, finite, and continuous throughout all space ($x = -\infty$ to $x = +\infty$). We expect that such solutions will be obtained only for certain values of the parameter β, and we wish to find these proper values. The calculation is given below (in small type). It shows that the simple harmonic oscillator has discrete energy states and that the energies of successive states differ by $h\nu$. The lowest state has energy $\frac{1}{2}h\nu_0$.

We can write down one solution by inspection, namely

$$\psi = a_0 e^{-\alpha x^2/2}. \qquad \text{18(35a)}$$

Differentiation shows that this is a solution if β has the value $\beta_0 = \alpha$. The solution satisfies all the necessary conditions, and β_0 is one of the proper values. In order to find others, we try the expression

$$\psi = v e^{-\alpha x^2/2}, \qquad \text{18(35b)}$$

where v is some function of x to be determined. Inserting this expression for ψ in 18(34) we obtain

$$\frac{d^2 v}{dx^2} - 2\alpha x\frac{dv}{dx} + (\beta - \alpha)v = 0. \qquad \text{18(36)}$$

We may further simplify the notation by introducing a new variable $\xi = x\sqrt{\alpha}$ and replacing $v(x)$ by $H(\xi)$, to which it is equal. Equation 18(36) becomes

$$\frac{d^2H}{d\xi^2} - 2\xi\frac{dH}{d\xi} + \left(\frac{\beta}{\alpha} - 1\right)H = 0. \qquad 18(37)$$

We now take as a trial solution

$$H(\xi) = a_0 + a_1\xi + \ldots + a_n\xi^n + \ldots$$

$$= \sum_{n=0}^{\infty} a_n\xi^n. \qquad 18(38)$$

We insert this value in 18(37) and obtain on the left-hand side a power series, each of whose coefficients must vanish if 18(38) is a solution. Equating the coefficient of ξ^n to zero, we have

$$(n + 1)(n + 2)a_{n+2} + (\beta/\alpha - 1 - 2n)a_n = 0. \qquad 18(39)$$

or

$$\frac{a_{n+2}}{a_n} = \frac{(2n - \beta/\alpha + 1)}{(n + 1)(n + 2)}. \qquad 18(40)$$

A power series whose coefficients satisfy these relations is a solution of 18(37), and by replacing ξ by x and using 18(35), we obtain a solution of the wave equation. This solution is not necessarily an acceptable solution. For large values of n, the ratio a_{n+2}/a_n tends to the value $2/n$. For large values of n, the ratio of successive terms in the expansion

$$e^{\xi^2} = 1 + \xi^2 + \frac{\xi^4}{2!} + \ldots + \frac{\xi^n}{(n/2)!} + \ldots \qquad 18(41)$$

tends to the same value. When ξ is very large, $H(\xi)$ in general tends to increase in the same proportion as e^{ξ^2}, because only the terms containing very high powers of ξ are then important. Now, from 18(35) we know that the wave function is proportional to $e^{-\xi^2/2}H(\xi)$ and hence, for large values of ξ, to $e^{-\xi^2/2} . e^{\xi^2}$, i.e. to $e^{\xi^2/2}$. Since $e^{\xi^2/2}$ tends to infinity, the solution defined by 18(40) does not in general provide an acceptable solution of the wave equation. If, however, β has the value

$$\beta = \alpha(2n + 1), \qquad 18(42)$$

the numerator of 18(40) vanishes, and if n is even, the even-numbered terms of the power series [18(38)] terminate at the nth term. If, in addition, we have $a_1 = 0$, the odd-numbered terms are all zero. The series then terminates at the nth term and, although $H(\xi)$ diverges for large values of ξ, the wave function ψ becomes zero for large values of x, because the exponential $e^{-\alpha x^2/2}$ in 18(35) decreases more rapidly than any finite power of n increases. In a similar way, if n is odd and if $a_0 = 0$, the series also terminates at the nth term, and acceptable solutions are obtained when β has one of the values defined by 18(42). To satisfy 18(42) we must have

$$\frac{8\pi^2}{h^2} mW = \left(\frac{2\pi m\omega_0}{h}\right)(2n + 1),$$

i.e.

$$W = \frac{h}{2\pi}(n + \tfrac{1}{2})\omega_0 = (n + \tfrac{1}{2})h\nu_0. \qquad 18(43)$$

18.17. Rules of Interpretation.

A new theory is essentially a new language designed to describe a set of experimental observations. The inventor of a new artificial language is entitled to define the rules of its grammar, and his rules need not be the same as those of any existing language, provided that they are logical and self-consistent. In the same way, the statement of a new theory must include rules by which the symbols which it uses are to be related to experimental observations. It must have a procedure by which it provides a set of numbers which can be compared with experimental results. The wave mechanics provides one such set in its calculation of the energies of the stationary states by means of the proper values of the constant in the wave equation. In the preceding paragraphs we have applied it to three important cases—the free particle, the particle in a box, and the simple harmonic oscillator—and it has yielded results which are in agreement with a great deal of experimental evidence. The further application to atomic and molecular energy levels is outside the scope of this book. It may be sufficient to state that very large numbers of such calculations have been made, and none of the results is in conflict with experimental data.

18.18. Completion of the Wave Equation.

Another series of numbers suitable for comparison with experiment may be obtained by calculations based on the wave functions. Before considering these, we need to complete the wave equation by reintroducing the time. To do this we use the assumption, originally due to de Broglie, that the relation between the energy of a particle and the frequency for a material particle is the same as the corresponding relation for a photon; i.e. we have

$$W = mc^2 = h\nu, \qquad \qquad 18(44)$$

where m is the *relativistic* mass, not the rest mass. We then have

$$\omega = \frac{2\pi}{h} W, \qquad \qquad 18(45)$$

and hence we may write

$$\Psi = \psi \exp\left(\frac{2\pi i}{h} Wt\right) \qquad \qquad 18(46)$$

and

$$\frac{\partial \Psi}{\partial t} = \frac{2\pi i}{h} W\psi \exp\left(\frac{2\pi i}{h} Wt\right), \qquad \qquad 18(47)$$

since ψ is independent of the time. From 18(46) we may also derive

$$\nabla^2\Psi = \nabla^2\psi \exp\left(\frac{2\pi i}{h} Wt\right). \quad . \quad . \quad . \quad 18(48)$$

Substituting from 18(47) and 18(48) into 18(17), and re-arranging terms, we obtain

$$-\frac{h^2}{8\pi^2 m}\nabla^2\Psi + V\Psi = \frac{h}{2\pi i}\frac{\partial\Psi}{\partial t} = -\frac{ih}{2\pi}\frac{\partial\Psi}{\partial t}. \qquad 18(49)$$

We may put

$$\Psi^* = \psi^* \exp\left(-\frac{2\pi i}{h} Wt\right) \quad . \quad . \quad . \quad 18(50)$$

and hence

$$\Psi\Psi^* = \psi\psi^*. \quad . \quad . \quad . \quad . \quad . \quad 18(51)$$

We may note that the equation which has been obtained by reintroducing the time is not the same as the wave equation [18(13)] from which we started. Equation 18(49) is strictly speaking not a wave equation. It is a diffusion equation which possesses certain wave properties because the diffusion coefficient is imaginary. The fact that the equation is not a true wave equation does not matter. The important question is whether the solutions (manipulated according to certain rules to be stated in §§ 18.19 ff.) yield numbers which agree with those obtained from experiments. We shall see that by this criterion the theory is very successful.

18.19.—Let us suppose that the wave equation corresponding to a certain physical problem (such as the harmonic oscillator or the hydrogen atom) has been solved. The proper values have been found and these lead to energies W_1, W_2, . . . for the stationary states. Corresponding to them, we have found a series of proper functions Ψ_1, Ψ_2, ... We note that the form of the equation 18(49) is such that if any solution is multiplied by a constant, it is still a solution. Moreover, solutions are linearly additive, so that

$$\Psi = c_1\Psi_1 + c_2\Psi_2 + c_3\Psi_3 + \ldots, \quad . \quad . \quad 18(52)$$

where c_1, c_2, etc., are constants, is also a solution. Let us now consider the solutions belonging to individual stationary states (ψ_1, ψ_2, etc.), and assume that each solution has been normalized to unity, i.e. that each solution has been multiplied by a suitable constant, so that

$$\int \Psi_n \Psi_n^* \, d\tau = \int \psi_n \psi_n^* \, d\tau = 1, \quad . \quad . \quad 18(53)$$

where the integral is taken over the whole space.

18.20.—We can now make a square array or *matrix* of products of the solutions taken in pairs as follows:

$$\left.\begin{matrix} \Psi_1\Psi_1^* & \Psi_1\Psi_2^* & \Psi_1\Psi_3^* & \Psi_1\Psi_4^* & \cdots \\ \Psi_2\Psi_1^* & \Psi_2\Psi_2^* & \Psi_2\Psi_3^* & \Psi_2\Psi_4^* & \cdots \\ \Psi_3\Psi_1^* & \Psi_3\Psi_2^* & \Psi_3\Psi_3^* & \Psi_3\Psi_4^* & \cdots \\ \Psi_4\Psi_1^* & \Psi_4\Psi_2^* & \Psi_4\Psi_3^* & \Psi_4\Psi_4^* & \cdots \\ \cdots & \cdots & \cdots & \cdots & \cdots \\ \cdots & \cdots & \cdots & \cdots & \cdots \end{matrix}\right\}. \qquad 18(54)$$

or, substituting for each Ψ or Ψ^* from 18(46) or 18(50),

$$\begin{matrix} \psi_1\psi_1^* & \psi_1\psi_2^* \exp\frac{2\pi i}{h}(W_1-W_2)t & \psi_1\psi_3^* \exp\frac{2\pi i}{h}(W_1-W_3)t \ldots \\ \psi_2\psi_1^* \exp\frac{2\pi i}{h}(W_2-W_1)t & \psi_2\psi_2^* & \psi_2\psi_3^* \exp\frac{2\pi i}{h}(W_2-W_3)t \ldots \\ \ldots & \ldots & \ldots \quad \ldots \\ \ldots & \ldots & \ldots \quad \ldots \end{matrix} \qquad 18(55)$$

We note that the diagonal members of the second matrix are all independent of the time. Also they each depend on only one proper function. This suggests that they represent some property of the corresponding stationary state. Let us assume that $\Psi_1\Psi_1^*\,d\tau$ $(=\psi_1\psi_1^*\,d\tau)$ represents the probability of finding the particle in the small volume element $d\tau$, and hence that $e\Psi_1\Psi_1^*$ represents an effective charge density in the volume. The normalized ψ functions are suitable for representing a probability in this way because relation 18(53) makes the total probability of finding the particle at some point in the space equal to unity. Calculations based on this assumption have been made to predict the results of experiments on electron scattering, crystal structure, etc., and have been found to yield results in agreement with experiment.

18.21. Transition Probabilities.

If $e\Psi_n\Psi_n^*$ represents an electron density, it is reasonable to assume that $e\Psi_n x\Psi_n^*\,d\tau$ represents the x component of an electric dipole moment, and that the total component of an electric dipole moment in the direction of the co-ordinate x is

$$(M_{nn})_x = \int e\Psi_n x\Psi_n^*\,d\tau = \int e\psi_n x\psi_n^*\,d\tau. \qquad 18(56)$$

This form has been used to calculate the dipole moment of certain molecules. The results of calculation which include this assumption agree with measurements of the dielectric constant. Let us now consider dipole moments of the type

$$(M_{nm})_x = e\int (\Psi_n x \Psi_m^* + \Psi_m x \Psi_n^*)\, d\tau. \qquad 18(57)$$

We have

$$(M_{nm})_x = e\int \psi_n x \psi_m^* \exp (2\pi i \nu_{nm} t)\, d\tau + e\int \psi_m x \psi_n^* \exp (-2\pi i \nu_{nm} t)\, d\tau. \qquad 18(58)$$

This expression may be regarded as the x component of a dipole oscillator of frequency $\nu_{nm} = (W_n - W_m)/h$. This moment is real, because the second term in 18(58) is the complex conjugate of the first.

If

$$x_{mn} = \int \psi_n x \psi_m^*\, d\tau = \int \psi_m^* x \psi_n\, d\tau \qquad 18(59)$$

(note that $x_{nm} = x_{mn}^*$),

$$(M_{nm})_x = e x_{mn} \exp (2\pi i \nu_{nm} t) + e x_{mn}^* \exp (-2\pi i \nu_{nm} t); \qquad 18(60a)$$

and if x_{mn} is real,

$$(M_{nm})_x = 2 e x_{mn} \cos 2\pi \nu_{nm} t. \qquad 18(60b)$$

The mean value of $(M_{nm})_x^2$ is $2e^2 x_{mn}^2$. Similar expressions may be obtained for the y and z components. According to classical electromagnetic theory [Appendix XIII B, equation 13(80)], the energy radiated per second by the dipole whose moment is given by 18(60b) is

$$\frac{64}{3} \frac{\pi^4 e^2}{c^3} \nu_{nm}^4 (x_{mn}^2 + y_{mn}^2 + z_{mn}^2) = \frac{64}{3} \frac{\pi^4 e^2}{c^3} \nu_{nm}^4 r_{mn}^2. \qquad 18(61)$$

It may be shown that a similar expression holds if the moduli of x_{mn}, y_{mn} and z_{mn} are inserted in 18(61), when x_{mn}, etc., are complex. The Einstein coefficient A_{nm} is equal to the rate of radiation divided by $h\nu_{nm}$. Thus we have

$$\frac{1}{\tau} = A_{nm} = \frac{64}{3} \frac{\pi^4 e^2 \nu_{nm}^3}{hc^3} (\mathbf{r}_{mn} \mathbf{r}_{mn}^*), \qquad 18(62a)$$

i.e. the rate is that appropriate to an oscillator whose moment is

$$\mathbf{M}_{nm} = 2e\mathbf{r}_{mn} \cos 2\pi \nu_{nm} t, \qquad 18(62b)$$

where

$$\mathbf{r}_{mn} = \int \psi_n \mathbf{r} \psi_m^*\, d\tau. \qquad 18(62c)$$

Using 17(53), we have

$$B_{nm} = B_{mn} = \frac{8\pi^3 e^2}{3h^2} \mathbf{r}_{mn}\mathbf{r}_{mn}^*. \quad . \quad . \quad . \quad 18(63)$$

18.22. Relation between Classical and Quantum Oscillators.

The classical theory of dispersion includes the concept of dipole oscillators which can absorb, emit or scatter light. The f-value is the number of oscillators per atom or molecule. The quantum theory associates an " oscillator " with each possible transition, and it is usual to assign to each transition an oscillator strength or f-value. This quantity is defined to be the number of classical oscillators per atom which would absorb the same amount of radiation from a parallel beam of light [for which $\rho(\nu)$ is nearly constant over the region of the absorption line]. The relation between f and B is given by equation 17(58) and, using 18(63), we obtain the following relation between f and the matrix elements $\mathbf{r}_{mn}$:

$$f = \frac{8\pi^2 m\nu_{nm}}{3h} \mathbf{r}_{mn}\mathbf{r}_{mn}^*. \quad . \quad . \quad . \quad . \quad 18(64)$$

Calculations of f based on this equation agree reasonably well with observed values (Table 18.1).

TABLE 18.1

EXPERIMENTAL AND THEORETICAL f-VALUES

Element	Spectrum line	Experimental f-values	Theoretical f-values
Na	5893	0·9755	0·9796
Na	3303	0·01403	0·01426
Na	2853	0·00205	0·00221
Na	2680	0·00063	0·00073
Li	3233	0·00549	0·00551
Li	2741	0·00478	0·00471
Li	2563	0·00314	0·00253

On classical theory it would be reasonable to expect that f would be a small *integer*. In fact the results of experiments show that f is usually less than unity, and is often a small fraction. This result is not surprising in quantum theory, because there is a general proof that the *sum* of the f-values for all transitions from a single state is unity. This "f-sum rule" is not valid for all states, but it applies under a very wide range of conditions.†

18.23.—An atom may absorb radiation from a parallel beam but emits radiation in all directions. In classical dispersion theory the electron is free to move in any direction in response to the field. The oscillator of quantum theory is a directed oscillator with an axis whose direction is determined by the properties of the wave functions. If a parallel beam of plane-polarized light falls upon a directed oscillator, the dipole moment induced is proportional to $\cos\theta$, where θ is the angle between the direction of polarization and the axis of the oscillator. The amount of absorption is proportional to the square of the induced moment, i.e. to $\cos^2\theta$. When a parallel beam falls upon atoms oriented at random, the absorption is proportional to the mean value of $\cos^2\theta$ taken over a sphere. This implies that the absorption is one-third as large as it would be for an atom whose dipole moment is given by 18(62b). It has the value appropriate to an oscillator whose moment is given by 18(60b), i.e. to the component of M_{nm} in the direction of polarization. This factor of one-third has been taken into account in equations 17(58) and 18(64). The rate of radiation is that appropriate to an oscillator whose moment is given by 18(62b). We have shown in Appendix XIII B that a classical oscillator loses a fraction γ_0 of its energy per second, where γ_0 is given by 13(86). In considering absorption from a parallel beam we associate an oscillator of strength f with a given transition. In considering emission we have to associate an oscillator of strength $3f$ with the same transition. The rate of radiation becomes

$$3f\gamma_0 h\nu_{nm} = \frac{64}{3}\frac{\pi^4 e^2 \nu_{nm}^{\ 4}}{c^3}\mathbf{r}_{mn}\mathbf{r}_{mn}^* \quad . \quad . \quad . \quad . \quad 18(65)$$

in agreement with 18(61). This relation may be expressed in a slightly different way. Let τ_0 be the "life" of a classical oscillator of unit strength, i.e. the time required for its energy to fall to $1/e$ of the original value. Then $\tau_0 = 1/\gamma_0$ and the quantum theory τ is given by‡

$$\tau = \frac{f}{3}\tau_0. \quad . \quad . \quad . \quad . \quad . \quad . \quad . \quad . \quad . \quad 18(66)$$

It is possible to "make a picture" by saying that the quantum theory associates three oscillators with moments proportional to x_{mn}, y_{mn}, z_{mn} with a given transition. All three are effective for emission. Only one is effective for absorption from a parallel beam of plane-polarized light. This analogy helps to build up the quantum theory of scattering and of dispersion, but it should not be pressed too far.

Since the product $\psi_n x \psi_n^*$ is independent of the time, the corresponding rate of radiation is zero. We therefore conclude that, in the lowest stationary state, the atom does not radiate. This removes—at least formally—a difficulty inherent in the older quantum theory which viewed the atom as a dynamical system in which the electron revolved round the nucleus as a planet revolves round the sun. According to this theory the electron is always accelerated towards the

† See p. 169 of Reference 19.3.

‡ When more than one pair of states is involved, equation 19(67c) must be used.

nucleus, so that it should radiate—and radiate so fast that stationary states could not exist long enough to be observed. On the present theory, the point electron is replaced by a charged cloud and, when the atom is in the lowest stationary state, the total radiation from all elements of this cloud is zero.

18.24.—The method which we have outlined in the preceding paragraphs is consistent with our original assumptions concerning the interpretation of the ψ functions (§ 18.20). At a later stage we shall see that it is possible to give a more detailed account of the radiation process (§ 19.1). For the present let us accept it as a rule of calculation and see if it yields correct results when applied to some optical problems. We may begin by considering the transition probabilities for the simple harmonic oscillator. From 18(35), 18(37), and 18(42) we see that the proper functions are of the form

$$\psi_n = e^{-\xi^2/2} N_n H_n(\xi), \qquad \text{18(67)}$$

where H_n is the solution of the equation

$$\frac{d^2H_n}{d\xi^2} - 2\xi\frac{dH_n}{d\xi} + 2nH_n = 0, \qquad \text{18(68)}$$

and N_n is a constant chosen to normalize the ψ functions. Equation 18(68) is obtained by inserting in 18(37) the proper value of β given by 18(42). Since the problem is a one-dimensional one, the $\mathbf{r}$ of 18(62) is parallel to x and to ξ. From equation 18(61) we see that the transition probabilities are proportional to the square of x_{mn}, where

$$x_{mn} = \int_{-\infty}^{+\infty} \psi_n x \psi_m{}^* \, dx = \frac{N_n N_m}{\alpha} \int_{-\infty}^{+\infty} H_n H_m e^{-\xi^2} \xi \, d\xi, \qquad \text{18(69)}$$

since the wave functions are real.

The factor $1/\alpha$ enters through the change of the variable to be integrated. From the known mathematical properties of these functions (§ 11*c* of Reference 18.3), we find that

$$x_{n,\,n+1} = \sqrt{\left(\frac{n+1}{2\alpha}\right)} = \sqrt{\frac{(n+1)h}{4\pi m_0\omega_0}}, \qquad \text{18(70)}$$

$$x_{n,\,n-1} = \sqrt{\left(\frac{n}{2\alpha}\right)} = \sqrt{\frac{nh}{4\pi m_0\omega_0}}, \qquad \text{18(71)}$$

and that $x_{nm} = 0$ when $m \neq (n \pm 1)$. Thus the transition probabilities are zero except for transitions to adjacent states. Transitions whose probabilities are zero are said to be *forbidden*, others are said to be *permitted*. The probabilities for transitions to adjacent states are obtained by substituting from 18(70) and 18(71) into 18(62). Substituting from 18(71) into 18(62) gives directly the probability for spontaneous transitions from state n to state $(n - 1)$. There are no spontaneous transitions from state n to state $(n + 1)$. Equation 18(70) gives the Einstein A for transitions from state $(n + 1)$ to state n, and, by means of the relations 17(57) and 17(53), it gives the probability of induced transitions from state n to state $(n + 1)$ (i.e. the Einstein B's) and the coefficient for the absorption of light. No atomic system behaves *exactly* like a simple harmonic oscillator but, as an approximation, a diatomic molecule may be pictured as two masses attached by a spring. For small vibrations the restoring force is proportional to the displacement but for larger displacements the force is no longer proportional to the displacement. The system therefore behaves like a simple harmonic oscillator for small displacements. It is found that there is a series of stationary states with nearly constant energy differences between successive states in the region corresponding to small amplitudes of vibration. It is also found that, in this region, transitions take place only between adjacent vibrational states. The absolute values of the transition probability can be obtained by measuring the absorption coefficient and the result is in agreement with the calculation. More elaborate calculations are needed to take account of the deviations from the linear law of force and displacement, and also to include effects due to rotational energy. The results of these calculations agree with the observed values over the wider range of conditions to which they apply (see Chapter X of Reference 18.3).

18.25. **Quantization of the Electromagnetic Field.**

We have seen that the quantum mechanics provides a method of calculation which, when applied to the simple harmonic oscillator, gives both the energies of the stationary states and the probabilities of transitions between the states. The results obtained from these calculations, and from similar calculations on the frequency and amount of radiation emitted from atoms and molecules, are in good agreement with experimental data. We may therefore regard this most important aspect of the interaction between radiation and matter as satisfactorily brought within the theory and made amenable to calculation. We now wish to see what account the theory can give of the radiation itself. How can it simultaneously describe the wave and particle properties of light? How can the radiation field be quantized and yet retain its characteristic wave properties? In order to apply the quantum mechanics, we have to regard the radiation as a mechanical system. We have seen that Rayleigh and Jeans were able to derive the classical radiation laws in this way. We now wish to obtain an expression for the energy of the radiation in the form which Hamilton derived for mechanical systems. When this has been done it is found that the expression for the energy is the same as that for a system of simple harmonic oscillators, and we are able to apply the quantum theory of harmonic oscillators to the radiation field.

It is convenient to analyse the field energy by an extension of the method used in Appendix XIII A. We represent the field of radiation enclosed in a cube of side L by the vector $\mathbf{A}$ which satisfies 13(59) and 13(60). The scalar potential ϕ has been made equal to zero. We put

$$\mathbf{A} = \Sigma\, q_s \mathbf{A}_s, \qquad 18(72)$$

where q_s is a function of t (but not of position) which satisfies the relation

$$\ddot{q}_s + \omega_s^2 q_s = 0, \qquad 18(73)$$

and $\mathbf{A}_s$ is a vector function of position (but not of t) which satisfies 13(62) and 13(63). We choose for $\mathbf{A}_s$ the expression appropriate for standing waves:

$$\mathbf{A}_s = \mathbf{e}_s c(8\pi L^{-3})^{1/2} \sin(\boldsymbol{\varkappa}_s \cdot \mathbf{r}), \qquad 18(74)$$

where $\boldsymbol{\varkappa}_s$ is a vector of magnitude $\kappa_s = \omega_s/c$ whose direction is that of the wave normal. The components of $\boldsymbol{\varkappa}_s$ satisfy 13(67). $\mathbf{e}_s$ is a unit vector specifying the plane of polarization. Equation 18(74) is of the same form as 13(65); the multiplying factor has been chosen for reasons which will appear later. Using 13(60) we have

$$\mathbf{E} = -\frac{1}{c}\dot{\mathbf{A}} = -\frac{1}{c}\Sigma\, \dot{q}_s \mathbf{A}_s, \qquad 18(75)$$

and hence $\quad \mathbf{E} = \Sigma\, \mathbf{E}_s,$

where $\quad \mathbf{E}_s = -\mathbf{e}_s \dot{q}_s (8\pi L^{-3})^{1/2} \sin(\boldsymbol{\varkappa}_s \cdot \mathbf{r}). \qquad 18(76)$

Applying the method of § 16.7 to 13(59) and 18(74), we have

$$\mathbf{H}_s = c q_s (8\pi L^{-3})^{1/2} (\boldsymbol{\varkappa}_s \times \mathbf{e}_s) \cos(\boldsymbol{\varkappa}_s \cdot \mathbf{r}). \qquad 18(77)$$

The energy W may be written

$$W = \frac{1}{8\pi}\int (E^2 + H^2)\, d\tau, \quad . \quad . \quad . \quad 18(78)$$

the integral being taken over the cube. We then have

$$E^2 = (\sum_s E_s)^2 = \sum_s E_s^2 + \sum_s \sum_r^{s \neq r} E_s E_r. \quad . \quad . \quad 18(79)$$

Inserting the value of E_s from 18(76) and using § 2 of p. 104, we obtain

$$\int E_s^2\, d\tau = 4\pi \dot{q}_s^2 \quad . \quad . \quad . \quad . \quad . \quad . \quad . \quad . \quad 18(80)$$

and, similarly,

$$\int E_s E_r\, d\tau = 0, \text{ when } s \neq r. \quad . \quad . \quad . \quad 18(81)$$

Similarly, remembering that $\mathbf{e}_s$ is a unit vector perpendicular to $\boldsymbol{\varkappa}_s$, so that the magnitude of $\boldsymbol{\varkappa}_s \times \mathbf{e}_s$ is κ_s, we have

$$\int H_s^2\, d\tau = 4\pi c^2 \kappa_s^2 q_s^2 = 4\pi \omega_s^2 q_s^2. \quad . \quad . \quad 18(82)$$

Again, the integral of cross products $H_s H_r$ is zero. Hence

$$W = \tfrac{1}{2}\sum (\dot{q}_s^2 + \omega_s^2 q_s^2). \quad . \quad . \quad . \quad . \quad . \quad 18(83)$$

The expression for the energy is thus of the same form as that for a set of simple harmonic oscillators, each of unit mass. According to the method of Hamilton, we put $\dot{q} = p$, and we have

$$\mathscr{H} = \tfrac{1}{2}\sum (p_s^2 + \omega_s^2 q_s^2) = \sum \mathscr{H}_s. \quad . \quad . \quad 18(84a)$$

and

$$\frac{\partial \mathscr{H}_s}{\partial q_s} = -\dot{p}_s, \quad \frac{\partial \mathscr{H}_s}{\partial p_s} = \dot{q}_s, \quad . \quad . \quad . \quad 18(84b)$$

since q_s is a solution of 18(73).

18.26.—The above analysis of the radiation field applies in detail only to standing waves. In the quantum theory we are often interested in progressive waves or in a mixture of standing waves and progressive waves. For these we replace 18(72) by the more general form

$$\mathbf{A} = \sum (q_s \mathbf{A}_s + q_s^* \mathbf{A}_s^*). \quad . \quad . \quad . \quad . \quad . \quad 18(85)$$

$\mathbf{A}$, $\mathbf{E}$, and $\mathbf{H}$ are real but q_s, $\mathbf{A}_s$, $\mathbf{E}_s$, $\mathbf{H}_s$ are complex quantities.

Let us choose q_s, which is a solution of 18(73), to be proportional to $\exp i\omega_s t$ and let us make $\mathbf{A}_s$, which is a solution of 13(62), proportional to $\exp(-\boldsymbol{i}\boldsymbol{\varkappa}_s \mathbf{r})$. Then one pair of terms in the summation on the right-hand side of 18(85) is proportional to $\cos(\omega_s t - \boldsymbol{\kappa}_s \cdot \mathbf{r})$ and this represents a purely progressive wave. As before, the constant of proportionality is adjusted to obtain an expression for the energy in a suitable form and we write

$$\mathbf{A}_s = \mathbf{e}_s c(4\pi L^{-3})^{1/2} \exp(-i\boldsymbol{\kappa}_s \cdot \mathbf{r}), \quad . \quad . \quad . \quad . \quad 18(86)$$

so that, instead of 18(76) and 18(77), we obtain

$$\mathbf{E}_s = -\mathbf{e}_s \dot{q}_s (4\pi L^{-3})^{1/2} \exp(-i\boldsymbol{\kappa}_s \cdot \mathbf{r}) \quad . \quad . \quad . \quad 18(87)$$

and

$$\mathbf{H}_s = c q_s (4\pi L^{-3})^{1/2} (\boldsymbol{\kappa}_s \times \mathbf{e}_s) \exp(-i\boldsymbol{\kappa}_s \cdot \mathbf{r}). \quad . \quad . \quad 18(88)$$

E^2 is now equal to $(\Sigma\, E_s + E_s^*)^2$ and terms like E_s^2, $E_s E_r^*$ and $E_s E_r$ vanish on integration. We then have

$$\int E^2\, d\tau = \Sigma\, 2\int E_s E_s^*\, d\tau = 8\pi \,\Sigma\, \omega_s^2 q_s q_s^*, \quad . \quad . \quad 18(89)$$

and. since the magnitude of $(\boldsymbol{\kappa}_s \times \mathbf{e}_s)$ is $\kappa_s = \omega_s/c$, $\int H^2\, d\tau$ has an equal value and

$$W = 2\,\Sigma\, \omega_s^2 q_s q_s^*. \quad . \quad . \quad . \quad . \quad . \quad . \quad 18(90)$$

The co-ordinates q_s in 18(85) do not obey Hamilton's equations 18(83) and 18(84), but if we put $q_s' = q_s + q_s^*$ we obtain co-ordinates for the progressive wave which satisfy these equations.

18.27.—The above analysis of the electromagnetic field energy as the sum of a series of terms each having the same form as the expression for the energy of a simple harmonic oscillator is based on Maxwell's electromagnetic theory. We now introduce the quantum theory by assuming that what has been proved concerning the quantum mechanics of a linear harmonic oscillator applies to these oscillators (or modes of vibration) which represent the radiation field, i.e. the q_s variables obey a wave equation.

Since the equivalent oscillators have unit mass,

$$\frac{\partial^2 \phi}{\partial q^2} + \frac{8\pi^2}{h^2}(W - \tfrac{1}{2}\omega^2 q^2)\phi = 0. \quad . \quad . \quad . \quad 18(91)$$

We insert ϕ instead of ψ because we shall later need to distinguish the wave equation for the radiation field from the equation for an atom. The equation implies that each mode possesses a series of stationary states. The energy difference between successive states is $h\nu$, and the only permitted transitions are those between adjacent states, i.e.

those which involve emission or absorption of energy $h\nu$. Thus Planck's original postulate that a single quantum is interchanged in each emission or absorption has been incorporated into the theory without abandoning the wave concept.

18.28.—The energy of the radiation field is the sum of the energies of the individual oscillators. It has a stationary state corresponding to every possible combination of stationary states of the oscillators. Its lowest state will be that corresponding to all the oscillators being in their lowest states. If each oscillator has energy $\frac{1}{2}h\nu$, and there is an infinite number of oscillators, then the lowest possible energy of the field would be infinite. This conclusion, which seems at first sight absurd, need not occasion any serious difficulty. The energy involved in almost any problem in dynamics can be made infinite by a suitable choice of the zero for potential energy. We are concerned with changes of energy, and an infinite energy which can never become available does not generally affect the result of a calculation, though it may make the calculation difficult. In the particular case at issue it is found, as a result of detailed considerations,* that we are not forced to accept the view that the lowest state of energy is $\frac{1}{2}h\nu$. It is possible to choose our variables in such a way that the energy of the lowest state is zero. The distribution of radiation energy with frequency then follows Planck's law [17(43)].

In early attempts to apply the quantum theory to the electromagnetic field, it was sometimes thought that only one quantum of a given frequency and plane of polarization could be present in a closed space, i.e. that each oscillator possessed only two stationary states of energies, zero and $h\nu$. The quantum mechanics which we are now considering predicts that each mode of vibration may have energy $nh\nu$. The earlier theory was in conflict with the principle of superposition. When two equal beams of radiation of the same frequency illuminate the same space, the resulting radiation has the same frequency and, taking the space as a whole, twice as much energy as one of the beams. If each original beam has one quantum of energy $h\nu$, the resultant beam must have two. The idea that the field could have only one quantum of a given frequency is directly in conflict with experimental data on radio waves, whose frequency is 10^9 times lower than that of light. The energy in one quantum is extremely low. We know that it is possible to produce beams with sharply defined frequency and with considerable energy. These must be represented by stationary states in which n is very large—large enough to make the product $nh\nu$ large even when ν is relatively small.

* See p. 60 of Reference 18.4.

18.29. Analysis of the Momentum of the Field.

In a beam of progressive electromagnetic waves, the energy crossing unit area in unit time is c times the energy per unit volume. The momentum crossing unit area in unit time is equal to the energy per unit volume [equation 17(15)]. The momentum per unit volume is thus equal to $1/c$ times the energy per unit volume and to $1/c^2$ times the energy crossing unit area in unit time. Thus if $\mathbf{P}$ is the total momentum and $\mathbf{G}$ is the Poynting vector, then

$$\mathbf{P} = \frac{1}{c^2}\int \mathbf{G}\, d\tau = \frac{1}{4\pi c}\int [\mathbf{E} \times \mathbf{H}]\, d\tau. \quad . \quad . \quad 18(92)$$

Using equations 18(87) and 18(88), the momentum of progressive waves may be expressed in the form $\mathbf{P} = \sum \mathbf{P}_s$, where

$$4\pi c\mathbf{P}_s = \int (\mathbf{E}_s + \mathbf{E}_s^*)(\mathbf{H}_s + \mathbf{H}_s^*)\, d\tau, \quad . \quad . \quad 18(93)$$

since $\mathbf{E}_s$ is perpendicular to $\mathbf{H}_s$.

Referring to the discussion in § 18.26, we see that the only non-zero terms on the right-hand side of 18(93) are $\int (E_s H_s^* + E_s^* H_s)\, d\tau$. Substituting from 18(87) and 18(88) we obtain

$$cP_s = 2\omega_s^2 q_s q_s^*. \quad . \quad . \quad . \quad . \quad . \quad 18(94)$$

Hence P_s is $1/c$ times the energy of the mode of vibration associated with $\mathbf{A}_s$ and $\mathbf{A}_s^*$. It may be shown that $\mathbf{P}_s$ is in the same direction as $\boldsymbol{\varkappa}_s$ so that the direction of the momentum is normal to the wavefront. Thus, if the energy associated with a particular mode of vibration is $n_s h\nu_s$, then the momentum associated is $n_s h\nu_s/c$. The mode behaves like a set of n particles each with energy $h\nu$ and momentum $h\nu/c$ (directed along the wave normal), and any single process of emission or absorption will alter n by unity. We may describe this change by saying that a new particle will appear or a particle will disappear. The quantized wave behaves in its energy and momentum relations like a set of n photons (with the properties originally postulated by Einstein). It still retains the ordinary wave properties of interference and diffraction.

The momentum of the photon representing a standing wave calculated from 18(76) and 18(77) is zero. This is in accordance with the experimental fact that the pressure of standing waves on the box which contains them is a purely hydrostatic pressure and does not transfer any momentum to the box. In wave theory, a system of standing waves may be regarded as the superposition of two sets of progressive waves travelling in opposite directions. When the energy associated with any mode of vibration is large compared with $h\nu$, a similar analysis may be made in quantum theory. The standing wave may then be represented by a group of photons, each possessing momentum, but with the total momentum for the group equal to zero.

18.30. **Indeterminism in Relation to the Wave Equation.**

In Chapters IV and VI we considered the theory of wave groups and it was shown that a pure simple-harmonic plane wave must fill the whole of space and time. If the wave is restricted in the direction of propagation, then it ceases to be of one frequency, and there is a frequency distribution. If it is restricted in the perpendicular direction, the plane wave is diffracted and becomes a group of waves travelling in slightly different directions. We have seen that, according to quantum mechanics, the momentum is proportional to the frequency (and hence to the wave number κ). Thus the relations we discussed in Chapters IV and VI may now be regarded as aspects of the uncertainty relation. The probability of an atom at a given point absorbing a quantum is proportional to the square of the amplitude * of the wave (ξ_0) at that point. In a similar way the square of the amplitude $a(\kappa)$ is proportional to the probability that any one of the photons represented by the given system of waves will have the wave number κ, and the corresponding values of frequency and momentum. If ξ_0 differs from zero only for a comparatively small range of a co-ordinate measured in the direction of propagation, the corresponding wave train will be short. This implies that the total momentum (which is the same as the co-ordinate of momentum in the direction of propagation) will be ill defined. In a similar way, if the wave is restricted in a direction perpendicular to the direction of propagation (e.g. by passing it through a narrow slit) the co-ordinate in that direction becomes well defined. At the same time the direction of propagation of waves which have passed the slit is ill defined, and hence the component of the momentum in the direction perpendicular to propagation becomes ill defined. In each situation any method whereby a co-ordinate specifying the space extension of the wave is determined or controlled leads inevitably to an uncertainty in the corresponding component of momentum. Also, if a beam of light is represented by a short train of quantized electromagnetic waves, the time when the light energy passes a given point can be measured accurately but,

* Or the square of the modulus if the amplitude is complex.

because the train is short, there is a large uncertainty in the frequency, and hence in the energy. This is in agreement with 18(10) and indirectly with 18(9).

It is possible to discuss the uncertainty principle in relation to photons by assuming that ξ_0^2 represents the "probability of finding the photon at a given point". The discussion then becomes closely parallel with the corresponding discussion for material particles. This way of discussing the problem is convenient for some purposes but leads to serious difficulties. It carries the undesirable suggestion that the photon can, in general, be regarded as a particle localized at a point. In quantum mechanics, a photon is a quantized wave. Every photon in a system of standing waves in a box extends over the whole volume of the box and is equally present at all points. It behaves as a particle *only* in that it can deliver up all its energy and momentum to a single atom. We therefore do not use the term "position of the photon".

18.31.—The discussion of the preceding paragraph shows that qualitatively the uncertainty principle is included in the theory of quantized waves. This is so because the theory is a wave theory and yet gives the relations between momentum and wavelength and the interaction energy required by the relations $\lambda = h/p$ and $W = h\nu$. Thus the three points mentioned at the end of § 18.5 are included. It remains to use the quantum theory to formulate the uncertainty principle in a more precise way than that given by 18(6) and 18(10).

Consider the function $g(p)$ defined by

$$g(p) = \frac{1}{h}\int_{-\infty}^{+\infty} \psi(q) \exp\left(-\frac{2\pi i}{h} pq\right) dq. \quad . \quad . \quad . \quad 18(95)$$

Apart from multiplicative constants which are inserted to take account of normalization, this equation gives a relation between $g(p)$ and $\psi(q)$ which is similar to the one between $a(\kappa)$ and $\xi(x)$ given by 4(60). By Fourier's theorem, 18(95) implies

$$\psi(q) = \int_{-\infty}^{+\infty} g(p) \exp\left(\frac{2\pi i}{h} pq\right) dp. \quad . \quad . \quad . \quad . \quad 18(96)$$

It will be in agreement with our previous discussion if we assume that $hg(p)g^*(p)dp$ is the probability that the momentum of the photon is between p and $(p + dp)$, the factor h being required for normalization. For example, let us suppose that a certain measurement has determined the co-ordinate q with a Gaussian distribution of error; then

$$\psi(q)\psi^*(q) = e^{-(q/q_1)^2}, \quad . \quad . \quad . \quad . \quad . \quad . \quad . \quad 18(97)$$

where the most probable value of q is chosen as origin and $q_1/\sqrt{2}$ is the root-mean-square error. This implies that

$$\psi(q) = \exp(-\tfrac{1}{2}q^2/q_1^2)\,e^{iQ}, \qquad 18(98)$$

where Q is a suitable function of q. In the simplest case we may take it to be proportional to q, i.e. we write

$$\psi(q) = \exp\left[-\frac{q^2}{2q_1^2} + \frac{2\pi i}{h}p_0 q\right], \qquad 18(99)$$

where p_0 is a constant having the dimensions of a momentum. Inserting this function for ψ in 18(95), we obtain

$$g(p) = \frac{\sqrt{2\pi}}{h}q_1 \exp\left[-\frac{(p-p_0)^2}{2p_1^2}\right]. \qquad 18(100)$$

The calculation is exactly similar to that by which 4(98) is derived (see Appendix IV B, p. 112). $p_1/\sqrt{2}$ is the R.M.S. error of p and we have

$$\frac{p_1}{\sqrt{2}}\cdot\frac{q_1}{\sqrt{2}} = \frac{h}{4\pi}. \qquad 18(101)$$

Thus the product of the R.M.S. errors is of the order of h. Further calculation shows that when $\psi(q)\psi^*(q)$ has any other form than a Gaussian error function,† the product of the uncertainties is larger than that given by 18(101). Thus we have generally

$$\Delta p\,\Delta q \geqslant \frac{h}{4\pi}, \qquad 18(102)$$

where Δp and Δq are the R.M.S. errors.

18.32. The Quantum State.

In classical dynamics the result of a measurement gives the exact values of two quantities which are used in the form of initial conditions as the basis for further calculation. We know that actual experiments give only probabilities, and we have seen that the wave equation is of such a form that it can use probable values as its initial conditions and derive probable values as a result of its calculation. It is not necessary that the result of a measurement should lead to probable values of the position and momentum. By suitable transformations the wave equation can use probable values of any two variables which are suitable for the calculations of generalized dynamics. Let us suppose that we know the general properties of a system. The wave equation has been solved, so that the energies of the stationary states and the corresponding wave functions are known. Usually

† For proof of the inequality, see Reference 18.5.

the result of an experiment will tell us that a system is in a certain state specified by a solution of the wave equation of the form of 18(52), i.e. it will determine the coefficients c_1, c_2, etc., of this equation, c_1^2 being the probability that the system is in the first stationary state, c_2^2 that it is in the second, and so on. In this case the two variables involved are the energy and the time. When the time becomes long compared with any of the natural periods involved in the problem, it may happen that the system has a very high probability of being in one particular stationary state. One of the coefficients is nearly unity, and the rest are near zero. This is the exceptional case. As a rule the states determined as the result of experiment are formed by the superposition of stationary states.

18.33. **Quantum Statistics.**

In the classical kinetic theory, a gas is regarded as an assembly of particles, each of which has a definite position and momentum. Statistical methods are used to determine the most probable distribution in space and the most probable distribution of momentum. The corresponding calculation in quantum theory is concerned not with possible positions and momenta of particles, but with possible states of the assembly. The calculation is essentially a counting of states which are physically distinguishable. It is therefore essential to know which of the states which can be formed by superposition of stationary states are physically distinguishable.

We know that photons are themselves physically indistinguishable since two waves of the same frequency and phase travelling in the same direction cannot be distinguished by any experiment. Further, if we have two photons a and b of the same frequency travelling in directions A and B respectively, the state of the system as a whole is physically indistinguishable from one in which photon b travels in direction A, and photon a in direction B. We have no means of attaching to photons labels which have physical significance. In a similar way we know that electrons are physically indistinguishable. These experimental facts must appear in the theory.

18.34.—Consider the ψ functions for two similar particles. First of all imagine that they are in two different boxes, and that q is a co-ordinate in one box and q' in the other. Then $\psi_1(q)\psi_1^*(q)\,dq$ is the probability that particle 1 is in the range q to $(q + dq)$ and $\psi_2(q')\psi_2^*(q')\,dq'$ is the probability that 2 is in the range q' to $(q' + dq')$. The probability that both these events occur simultaneously is $\psi_1(q)\psi_1^*(q)\psi_2(q')\psi_2^*(q')\,dq\,dq'$, since the events do not influence one another. We could have derived this value for the joint probability by forming a wave function for the system as a whole, this function being

$$\psi_{12}(qq') = \psi_1\psi_2. \qquad 18(103)$$

This method may be extended to give the wave function for an assembly of photons in the same space, *provided that the interaction between them is negligible.*

Let us now consider a number of particles which, for the moment, we call particle 1, particle 2, etc., and let the wave functions for the stationary states be ψ_a, ψ_b, etc. Then, when particle 1 is in state a, particle 2 in state b, etc., the wave function for the whole system is

$$\psi'_{abc\ldots} = {}_1\psi_a\, {}_2\psi_b\, {}_3\psi_c \cdots \qquad 18(104)$$

If the first and second particles change places, we obtain a second solution of the wave equation, namely

$$\psi''_{abc\ldots} = {}_2\psi_a\, {}_1\psi_b\, {}_3\psi_c \cdots \qquad 18(105)$$

These two wave functions may be superposed in the way discussed in § 18.19 to form a wave function

$$\psi(q) = c'\,{}_1\psi_a\, {}_2\psi_b\, {}_3\psi_c + c''\,{}_2\psi_a\, {}_1\psi_b\, {}_3\psi_c \cdots \qquad 18(106)$$

The most general form of $\psi(q)$ will be

$$\psi(q) = \sum_P c_P\, {}_1\psi_a\, {}_2\psi_b\, {}_3\psi_c \cdots, \qquad 18(107)$$

where the symbol $\sum_P$ indicates that the summation is to be taken over all possible permutations of the left-hand subscript, each permutation having its own coefficient.

18.35.—We now have to introduce the assumption that nothing which we can observe changes when one particle is exchanged with another. This means that $\psi\,\psi^*$ must be unchanged by a permutation which merely expresses the fact that two particles have changed places. This implies that either ψ does not change at all (*symmetrical case*), or that ψ remains the same in magnitude but changes sign (*antisymmetrical case*). For the symmetrical case the coefficients are all equal; and for the antisymmetrical case they are equal in magnitude but alternate in sign. It is easily shown that, when there is no interaction, the transitions between an antisymmetrical and a symmetrical state have zero transition probability. This is confirmed by experimental evidence from atomic spectra. Thus an assembly must exist in either symmetrical or antisymmetrical states, but not sometimes in one and sometimes in the other.

Consider two particles among an assembly which is represented by an antisymmetrical wave function. Taking any positive term in the summation of equation 18(107), there exists a negative term differing only in that the two particles in which we are interested have been exchanged. If the two particles have the same quantum numbers (and hence the same wave functions), each positive term will be equal in magnitude to the corresponding negative term, and the wave function will vanish. Thus we see that antisymmetrical states of an assembly for which two particles have the same quantum numbers do not exist. On the other hand, two particles in a symmetrical state can have the same quantum numbers. The experimental data summarized in the Pauli principle, which states that two electrons in an atom or molecule never have the same set of quantum numbers, imply that the electrons, in addition to being indistinguishable, are able to exist only in antisymmetrical states. On the other hand, we know that large numbers of photons can exist simultaneously in the same state, and hence we deduce that the states are symmetrical. We have also deduced that, since the coefficients in 18(107) are all equal, all accessible states, i.e. all symmetrical states, are equally probable.

18.36.—Using the results that photons are indistinguishable in the sense defined in § 18.33, and that any number may exist in a given quantum state, we may apply ordinary statistical methods to determine the distribution of energy among photons of different frequencies in a state of temperature equilibrium. Consider the radiation in a volume V; the number of states corresponding to energies between $h\nu$ and $h(\nu + \Delta\nu)$ given by 13(69) is

$$g_s = \frac{8\pi\nu^2 V}{c^3}\Delta\nu. \qquad 18(108)$$

Let us suppose that there are n_s photons in this energy range. These have to be distributed among the g_s states. If the photons and the states were distinguishable, the total number of possible arrangements would be *

$$g_s(g_s + n_s - 1)! \qquad 18(109)$$

The photons are indistinguishable, and the states are indistinguishable, so the total number of arrangements is

$$\frac{g_s(g_s + n_s - 1)!}{g_s!\, n_s!} = \frac{(g_s + n_s - 1)!}{(g_s - 1)!\, n_s!}. \qquad 18(110)$$

To take a simple example, suppose there are 100 states and 5000 photons. An arrangement in which 100 photons are found in each of 50 states, and none in the other states, is different from one in which 100 photons are found in each of 49 states, 50 photons are in each of 2 states, and no photons in the remaining states; but with either

* This permutation formula may be deduced by the following argument given in p. 208 of Reference 18.5. Denote the cells by z_1, z_2, etc. Denote the photons by a_1, a_2, ... Write down a series of letters, e.g. $z_1a_1a_2z_2z_3a_3a_4a_5 \ldots$, with the understanding that the photons standing between two z's are in the cell to the left. We obtain all arrangements by first choosing a z to set at the left (which can be done in g_s ways) and then choosing z's or a's to fill the remaining $(g_s + n_s - 1)$ places [which can be done in $(g_s + n_s - 1)!$ ways].

arrangement it does not matter which states are the ones with 100 photons, nor does it matter which photons go to make up a group of 100.

18.37.—Using 18(110) we obtain for the probability of a given distribution among all ranges of frequency the expression

$$\prod_s \frac{(g_s + n_s - 1)!}{(g_s - 1)!\, n_s!}. \quad \ldots \ldots \quad 18(111)$$

Using the usual method of statistical theory in which the maximum probability is obtained by the method of undetermined multipliers, the most probable value of n_s is found to be

$$n_s = \frac{g_s}{e^{h\nu/kT} - 1} \quad \ldots \ldots \quad 18(112)$$

for the number of photons in the range ν to $(\nu + d\nu)$. Hence, using the value of g_s given by 18(108), and putting $\rho(\nu) = n_s h\nu$, we obtain

$$\rho(\nu) = \frac{8\pi h\nu^3}{c^3} \frac{1}{e^{h\nu/kT} - 1} \quad \ldots \quad 18(113)$$

for the energy per unit volume and unit frequency range. The statistical theory applied to photons, assumed to be indistinguishable particles of which any number can be in one state, leads to the radiation law, which we have found to be in accord with experiment.

18.38.—It may seem remarkable that Planck's law can be obtained in two ways so very different as those given in §§ 17.32 and 18.37. Actually both these deductions are in accord with quantum mechanics and with the general principles of statistical mechanics. In § 17.32 we considered the partition of energy among the different modes of electromagnetic vibration, i.e. the different stationary waves which could be formed in the enclosure. These modes of vibration are physically distinguishable. Both the frequency and direction of propagation are available to identify one of the modes. Since the statistical elements are distinguishable we may apply the Maxwell-Boltzmann method of statistics. This we did in using Boltzmann's theorem. In § 18.37 we dealt with photons which, according to statistical mechanics, are indistinguishable. By modifying the statistical calculation to take account of the indistinguishability, we obtain the same result. This problem is one of many problems in quantum mechanics which can be started either from a wave or from a particle standpoint. The two methods yield the same result precisely, because the theory is internally consistent and does take account effectively of both the wave and the particle properties of light.

18.39.—Both the deductions of Planck's law mentioned in the preceding paragraph may be derived by applying quantum mechanics to the radiation. A third deduction can be given by applying quantum mechanics in detail * to the processes of emission and absorption by mechanical systems (Appendix XIX A) In this way it is possible to compute the relations 17(53*a*) and 17(53*b*) between the Einstein coefficients. Using the principle of detailed balance, it is then possible to use the relations between the coefficients to deduce Planck's law, instead of using Planck's law to derive the relations. The fact that this is possible shows that the quantum mechanics is self-consistent in its description of the emission and absorption of radiation. The same result is obtained whether we start the calculation from the side of the radiation or from that of the atom.

18.40. **The Relation between Photons and Material Particles.**

In this chapter we have stressed the similarity between the behaviour of photons and the behaviour of material particles. The quantum mechanics applies the same fundamental quantum ideas—originally due to Planck and Einstein—both to matter and to the electromagnetic field. The results of its calculations for the two problems show the degree of similarity required by the experimental results. It thus provides a formal description of matter and radiation, both in their wave and particle aspects. The similarity between the behaviour of matter and radiation is so striking that the reader may feel that there are no important differences—apart from differences of scale, such as those due to the fact that the electron wavelengths usually encountered in practical problems are much shorter than optical wavelengths. It is therefore important to discuss the relation between matter and radiation in such a way as to bring out the differences as well as the similarities, and to show that the theory includes these differences.

18.41.—The three relations

$$W = mc^2 = h\nu, \qquad 18(114)$$

$$p\lambda = h, \qquad 18(115)$$

$$W^2 = m_0{}^2c^4 + c^2p^2, \qquad 18(116)$$

apply both to electromagnetic radiation and to material particles, provided that we assume that a photon is a particle of zero rest mass. In 18(16) and in the subsequent discussion the *relativistic* energy is involved. The relation

$$\nu\lambda = b, \qquad 18(117)$$

* Reference 18.4.

where b is the phase velocity, applies to ψ waves as well as to electromagnetic waves. It leads to the result

$$b = \frac{c^2}{v} \quad \ldots\ldots\ldots \quad 18(118)$$

for the phase velocity of the ψ waves associated with a particle of velocity v. It is easily shown that the group velocity of the ψ waves associated with the free particle is v, i.e. it is, as we should expect, the actual observed velocity. The photon thus appears as a particle whose phase velocity is equal to the group velocity (*in vacuo*) and hence to the particle velocity. In these relations, which apply both to radiation and to matter, the photon appears as a limiting case.

18.42.—Despite these similarities, the classical theory of radiation is a wave theory, and the classical theory of matter is a particle theory. This is more than a historical accident. A particle must be able to change its energy in order to be observed, and there must be a corresponding change of momentum in order to obey the conservation laws. Under usual laboratory conditions, material particles change their energy by changing their velocity, photons by changing their mass (i.e. their frequency). Very high-energy material particles whose velocities are near to c change their energy by changing their mass, and are very like high-energy photons. They are little deflected by electric, magnetic or gravitational fields and have a very high power of penetrating matter. At low energies, on the contrary, the differences become very marked because the energy of a material particle tends to a minimum value determined by the rest mass—whereas the photon energy can become indefinitely near to zero. Thus, at low energies, the material particle remains an easily observable unit. It is indeed more easily observable at low energies than at high energies, because the velocity is conveniently low and changes of velocity can be observed in the course of experiment. At low energies the photon becomes more and more difficult to observe, because one photon is unable to affect our apparatus, and also because it is still moving with velocity c. At high energies the " track " of a photon may be rendered visible in a Wilson chamber by noting the points at which electrons have been accelerated by it. No corresponding experiment is possible at low frequencies, and it is not possible to attach any meaning to the phrases " position of the photon " or " probability of the position of the photon ". The reader should note that the q's introduced in equation 18(73), etc., are dynamical co-ordinates specifying the radiation field, but they do not stand for the "position of a photon at time t". The theory of radiation at low frequencies is thus a theory of fluctuating fields, and passes naturally into a static field theory when the frequency (and therefore energy of the photon) tends to zero.* Thus the natural limiting theory for matter is a particle theory, and the limiting theory for radiation is a wave theory.

REFERENCES

18.1. HEISENBERG: *The Physical Principles of the Quantum Theory* (University of Chicago Press).

18.2. DE BROGLIE, L.: *Wave Mechanics* (Methuen).

18.3. PAULING and WILSON: *Introduction to Quantum Mechanics* (McGraw-Hill).

18.4. HEITLER: *Quantum Theory of Radiation* (Oxford University Press).

18.5. BORN: *Atomic Physics* (Blackie).

* It can be shown (p. 64 of Reference 18.4) that when the wave functions are symmetrical, the theory passes easily to a field theory as $\nu \to 0$. With antisymmetrical functions there is no obvious passage to a classical field theory.

CHAPTER XIX

Interaction Processes in Relation to Quantum Mechanics

19.1. Absorption and Emission of Radiation.

In § 18.21 an expression for the rate of emission of light by excited atoms was derived using the assumption that an excited atom is equivalent to an oscillating dipole whose moment is given by 18(57). This calculation gives the Einstein coefficient A_{nm} directly, and the coefficients B_{nm} (for induced emission) and B_{mn} (for absorption) may be obtained by applying equations 17(53). We shall now discuss a theory of the *processes* of emission and absorption.

This detailed theory of the exchange of energy between radiation and matter is due to Dirac. He applies the fundamental principles of quantum mechanics to a system consisting of an atom placed in a field of electromagnetic radiation. The energy of this system consists of three parts: (*a*) the energy ($\mathscr{H}_p$) which the atom would possess in the absence of the radiation, (*b*) the energy ($\mathscr{H}_r$) of the radiation, and (*c*) an interaction energy ($\mathscr{H}_i$). In principle the wave equation can be solved, and the eigenvalues and transition probabilities can be calculated. In practice the equations are too difficult for exact solutions, and it is necessary to resort to approximations which use the fact that $\mathscr{H}_i$ is small compared with $(\mathscr{H}_p + \mathscr{H}_r)$. Even with these approximations the mathematics is fairly difficult. In the following paragraphs we give an outline of the process and state the results. A mathematical account of some of the more important parts of the theory is given in Appendix XIX A.

19.2.—Consider a system of atoms and radiation in a box and, for the moment, let us disregard the interaction energy. The wave equation will be

$$(\mathscr{H}_p + \mathscr{H}_r)\Psi = -\frac{ih}{2\pi}\frac{\partial \Psi}{\partial t}, \qquad \qquad 19(1)$$

i.e. the left-hand side is the sum of all terms in the corresponding equation for the atoms and all terms in the wave equation for the radiation. Let us assume that the wave functions for the atoms have been found by the usual methods of approximate computation. The wave functions for the radiation have been discussed in § 18.24 ff., where it is shown that they are similar to those for a system of simple harmonic oscillators. If the dimensions of the box are very large compared with the wavelengths in which we are interested, the frequencies of the oscillators are so close that they form an effectively continuous spectrum. An energy $nh\nu$ (where n is an integer) is associated with each " oscillator ", i.e. with each quantized wave. It has been shown in § 18.34 that any wave function for a system of two parts (which do not interact) is the product of a wave function for the first system and a wave function for the second system. Thus the solutions of 19(1) are obtained by combining solutions for the atoms alone with solutions for the radiation alone.

19.3.—The interaction energy (regarded as a small " perturbation ") is then taken into account, and it is found that the system (atom plus radiation) undergoes transitions in which the atom changes from one stationary state to another, and at the same time the appropriate quantized wave changes its energy by one quantum, so that (in the system as a whole) energy is conserved within the accuracy of measurement (§ 18.7). Only processes which comply with this condition have an appreciable transition probability. The theory further shows that the probability of absorption of radiation leading to a transition from a state m to a state n is proportional to $U_{mn}U_{mn}{}^*$, where

$$U_{mn} = \int \psi_m{}^* \mathscr{H}_i \psi_n \, d\tau. \quad . \quad . \quad . \quad . \quad . \quad 19(2)$$

The calculation of the energy of interaction between the electrons in the atoms and the high-frequency electromagnetic field is complicated, but it may be shown that the expression given by elementary electrostatics is correct to a first approximation. This expression is

$$\mathscr{H}_i = E \sum_j e_j r_j, \quad . \quad . \quad . \quad . \quad . \quad . \quad 19(3)$$

where e_j is the charge and r_j the displacement of one of the particles, i.e. $\mathscr{H}_i$ is proportional to the dipole moment. Since from 19(2) and 19(3) $U_{mn}U_{mn}{}^*$ is proportional to E^2, the theory predicts that the transition probability is proportional to the energy density of the radiation. The fact that the interaction energy is proportional to the dipole moment means that $U_{mn}U_{mn}{}^*$ is proportional to $M_{mn}M_{mn}{}^*$ (§ 18.21). In order to find the constant of proportionality it is necessary to integrate the probability of transitions for all possible orientations of the atom relative to the electric vector of the radiation field. It is also necessary to integrate over a range of energies corresponding to the finite width of the absorption line. When this is done, it is found that the Dirac theory gives an expression for the transition probability which is equivalent to 18(61).

19.3.—It may appear at first sight that the detailed theory is only a rather roundabout way of obtaining a result which is given directly by the method of § 18.21. It should, however, be understood that § 18.21 starts from an assumption based on an analogy with the classical theory of a Hertzian dipole. It was necessary to use this analogy before the quantum mechanics had been developed in a form suitable for application to the radiation field. Dirac's theory of absorption of radiation is an essential part of a formal quantum theory which includes both matter and radiation. By means of the Dirac theory it is possible to calculate each of the Einstein coefficients separately, and to show that the relations 17(53) are satisfied. It is also possible to obtain further important results by extensions of the Dirac method to Rayleigh scattering, dispersion theory, etc.

19.4. **Selection Rules.**

From the experimental side the accurate measurement of the relative strengths of lines in a spectrum is technically difficult, especially when there is an appreciable wavelength difference. When this is so, it is not possible to assume that the sensitivity of a selective receptor, such as the eye, photocell, or a photographic plate, is the same for both lines. Until comparatively recently the available nonselective receptors, such as thermopiles, were very insensitive. Thus, in the early days of the quantum theory, there were few measurements of absolute transition probabilities, or even of the relative values when the lines involved were not members of close multiplets. It was, however, well known that certain transitions occurred with a high probability of order 10^8 per second, whereas transitions between certain other states either did not occur at all or else occurred with very much lower probability (e.g. 10^2 per second).

The spectroscopists gradually drew up a classification of terms. This classification started as an empirical one rather like early botanical classifications. Then, as the theory progressed, the classification was referred to the quantum numbers of the states. Transitions were divided into " permitted " transitions and " forbidden " transitions.

Selection rules, which made it possible to predict whether a transition would occur with high probability or not, were invented. At an early stage the quantum theory was able to give some account of these rules by showing that the functions U_{mn} are necessarily zero for certain types of transitions. For example, one important empirical rule is that a quantum number J can change only by 0 or ± 1. The theory shows that all transition probabilities which do not obey this rule vanish (to first order). It also shows that the permitted transitions involve changes in the angular momentum of the atom by 0 or $\pm h/2\pi$. It is found that the field can change its angular momentum by these amounts, and the empirical " ΔJ selection rule " is then seen to be an expression of the law of conservation of angular momentum. Other selection rules can be referred to symmetry conditions.

19.5. **Forbidden Lines.**

As the number and range of spectroscopic observations increased, it became clear that many lines which are theoretically forbidden (i.e. for which the function U_{mn} is zero) are actually observed. Usually these lines are rather weak when observed in the laboratory. Most of them are observed only in emission, but some (e.g. the mercury line $\lambda = 2270$ Å.) are observed in absorption at fairly low pressures. Many lines which are observed in the laboratory with very great difficulty are quite prominent in the spectra of nebulæ or of the upper atmosphere. The auroral green line ($\lambda = 5577$ Å.) is now known to be a forbidden line of oxygen. The reason why lines of this type appear in auroræ or nebulæ is explained in the following way.

Suppose that *all* the downward radiative transitions from a given state have extremely low probability. Then the " radiative life " of the state will be very long. If atoms are transferred to this state by collisions with atoms, or with electrons, or by radiative transition from a higher state and then left undisturbed, they remain in this state for relatively long periods (10^{-2} second or more). Such states are said to be *metastable.* In an ordinary discharge tube, atoms in a metastable state lose their energy by collision with other atoms, or with the wall of the tube. In the nebulæ and in the upper atmosphere pressures are very low, and the mean time between collisions is very long, so the atoms have time to radiate, even though the radiative transition probability is very low. In collisions with rare gases, the non-radiative transition processes have a very low probability. They also have low probability on the wall of a discharge tube which has been treated in certain ways. Thus by carefully adjusting experimental conditions, it is possible to obtain in the laboratory forbidden lines which originate on metastable states.

19.6. Theory of Forbidden Lines.

Since many hundreds of " forbidden " lines have now been observed, it is important to see whether they can be included in the theory. It is found that they fall into two important classes. In the first class the emission is due to some secondary effect which distorts the atom, so that the appropriate dipole moment (M_{mn}) is not quite zero. For example, an external electric field may cause the emission of lines which, in the absence of the electric field, would have zero transition probability. The interaction with the electric field enables angular momentum to be conserved, even when ΔJ is not 0 or ± 1. In a similar way, internal interaction between the magnetic moments of the electrons and the intrinsic moments of the electrons causes the emission or absorption of some lines, and interaction between the magnetic moment of the nucleus and the orbits causes the emission or absorption of other lines. Under suitable conditions the formation of extremely weakly bound (and short-lived) molecules may cause the emission or absorption of atomic lines which would normally be called forbidden. In all these cases a recalculation of the transition probability, taking account of some secondary effect, shows that U_{mn} is not zero. The radiation emitted is ordinary electric dipole radiation, corresponding to a transition whose probability is zero in a first approximation, but not zero when the secondary effect is included. The origin of these lines may be verified by observing the alteration of intensity when some experimental condition is varied. For example, the intensity of a line due to an external electric field should depend on the strength of the field; the intensity of a line due to " temporary molecules " should depend on temperature, since the number of these molecules is sensitive to change of temperature.

19.7. Multipole Radiation.

There are certain forbidden lines whose presence cannot be explained by any process of the type we have just discussed. These lines may be included in the theory by developing the expressions given by Dirac to a higher order of approximation. The essential difference is that in the first approximation it is assumed that the wavelength of the radiation is so large compared with the diameter of the atom that phase differences between light from different parts of the atom are negligible, i.e. the time for light to cross the atom is taken as zero.

A similar assumption is usually made in the classical theory of a Hertzian oscillator.* If the resulting dipole radiation is accurately zero, it is necessary to use a higher approximation. Higher-order terms involving products of the type $(ex^2)^2$ instead of $(ex)^2$ then have to be considered. This second-order term was at first called "electric quadripole radiation", but it is now recognized that part of it should be called "magnetic dipole radiation". When the higher approximation is calculated, it is found that the transition probability for a typical electric quadripole radiation is less than that for a typical dipole in the ratio $(a/\lambda)^2$, where a is the atomic radius and λ is the wavelength of light. Since λ is of order 6000 Å. and a is of order 2 Å., the factor is of the order 10^{-7}, which is in accord with observed transition probabilities of 10^7 to 10^9 per second for dipole radiation, and 0·1 to 10^2 per second for quadripole radiation. It is, of course, possible that the second-order matrix element may also vanish, in which case the quadripole radiation is also "forbidden". Thus, quadripole lines have their own selection rules. The Zeeman effect (§ 19.14) for quadripole lines is different from that for dipole lines. By these and similar tests, the auroral line 5577 Å. is shown to be an electric quadripole line.

The transition probabilities for magnetic dipole radiation in the optical region are less than those for the corresponding electric dipole radiation in the ratio $10^{-5}:1$. Magnetic dipole radiation should, therefore, be stronger than electric quadripole radiation. It is found that, in many cases, the matrix elements corresponding to magnetic dipole radiation vanish, so that very few lines of this type are actually observed. Some lines are jointly due to electric quadripole and magnetic dipole radiation. In principle, terms of higher order than the second (electric octopole, etc.) should exist, but the factor $(a/\lambda)^4$, which is of order 10^{-14}, appears in the expression for the transition probabilities, so that it is very unlikely that they will be observed in the optical region.†

19.8. Interactions between Radiation and Matter involving two Quanta per Process.

Absorption of light takes place when the incident radiation fulfils the Bohr frequency condition within the accuracy required by the uncertainty principle. In each individual process of absorption, one quantum of energy is transferred from the radiation field to the atom. One interaction energy (U_{mn} for dipole radiation) occurs to the second power in the expression for the transition probability. This expres-

* See Appendix XIII B (p. 413).

† For further treatment of forbidden lines and multipole radiation see Reference 19.1.

sion * therefore contains the factor e^2. We now consider what processes may occur when an atom interacts with radiation whose frequency is relatively far from that of an absorption line. On the experimental side we know that the light may be scattered without change of wavelength (Rayleigh scattering) or with change of wavelength (Raman effect).† Also the presence of the atoms causes an effective change in the phase velocity of the light, and this velocity depends on the frequency, i.e. we have dispersion. We have shown in § 15.45 that the change of velocity is closely related to the scattering of light.

19.9.—The Dirac theory of the interaction of radiation and matter gives the following general account of the scattering of light. When the frequency is remote from that of an absorption line, a quantum of energy may still pass from the radiation field to an atom, which makes a transition to another stationary state called " the intermediate state ". The atom may then make a transition from the intermediate state back to the initial state, emitting radiation of the same wavelength as the incident radiation, but not entirely in the same direction (in the simplest case the incident wave is plane and the emitted wave is spherical). This is the fundamental process giving rise to Rayleigh scattering and to dispersion. Under suitable conditions the atom may make a transition from the intermediate state to a final state which is different from the initial state. The emitted light will then differ from the incident light in frequency as well as in direction, as is observed in the Raman effect.‡ In these processes the total energy and momentum at the end is the same as at the beginning. As a rule, energy and momentum are *not* conserved in the intermediate stages. These are not observable and, indeed, their introduction is essentially part of the mathematical process of calculation.

19.10.—The probability of a process of this type occurring is proportional to the product of probabilities of the transition from the initial to the intermediate state, and of the transition from the intermediate to the final state, so that the probability for a scattering process always includes e^4 as a factor. The transition probability for a Raman process involves the square of the product of two interaction

* The same power of e occurs in expressions for the probabilities of quadripole radiation.

† See § 19.21.

‡ With X-rays, a change of wavelength associated with scattering by " free " electrons (Compton effect) is also observed (see § 17.23). This also is a two-quantum process.

energies. In the Rayleigh scattering, one interaction energy is involved twice, and thus appears to the fourth power.* For extremely short wavelengths (γ-rays) transitions involving *three* quanta may be observed, but they are of no importance in the optical region.

19.11. **Dispersion Theory in Quantum Mechanics.**

The relation between refractive index and Rayleigh scattering given in 15(97), and the main lines of the analysis by which it was derived, is accepted in quantum mechanics. Thus, when the probability of two-stage transitions in which one of the radiation oscillators (corresponding to a given direction) loses a quantum, and another (corresponding to the same frequency but a different direction) gains a quantum, has been calculated, the dispersion formula follows. We obtain (when the effect of damping is neglected)

$$n - 1 = \frac{2\pi e^2 N}{m} \sum_s \frac{f_s}{\omega_s^2 - \omega^2}, \qquad . \quad . \quad . \quad 19(4a)$$

where ω is the circular frequency of the incident light and ω_s a natural frequency of the atom. This equation is the same as equation 15(63) which was derived from the classical electromagnetic theory. In both theories the formula is valid when ω is not too near to ω_s. While the equation is the same in both theories, there is a difference of interpretation. As Kramers first pointed out, in classical theory the values of ω_s are those corresponding to absorption lines. In quantum theory the values of ω_s are proportional to the energy differences for possible transitions. If an atom is in an excited state, the possibility of downward transitions must be taken into account. The terms corresponding to these transitions are negative, so that the formula becomes

$$n - 1 = \frac{2\pi e^2 N}{m} \left[\sum_a \frac{f_a}{\omega_a^2 - \omega^2} - \sum_e \frac{f_e}{\omega_e^2 - \omega^2} \right], \qquad 19(4b)$$

where f_a, ω_a stand for upward transitions and f_e, ω_e stand for downward transitions. According to quantum theory there should be added also a term to allow for transitions which ionize the atom, i.e. transitions to a continuous state. This term and a second additional term (considered in § 20 of Appendix XIX A, p. 660) contribute very little to optical dispersion. For most substances at room temperature the atoms are nearly all in the normal state, so that the quantum theory

* In the final dispersion formula it appears only to the second power, because the scattering coefficient is proportional to $(n - 1)^2$ [see equation 15(97)].

formula becomes identical with the classical one. Ladenburg and his collaborators have measured the dispersion of electrically excited gases, and have shown that at high current densities the negative terms in the formula become important.* Their apparatus is shown in fig. 19.1.

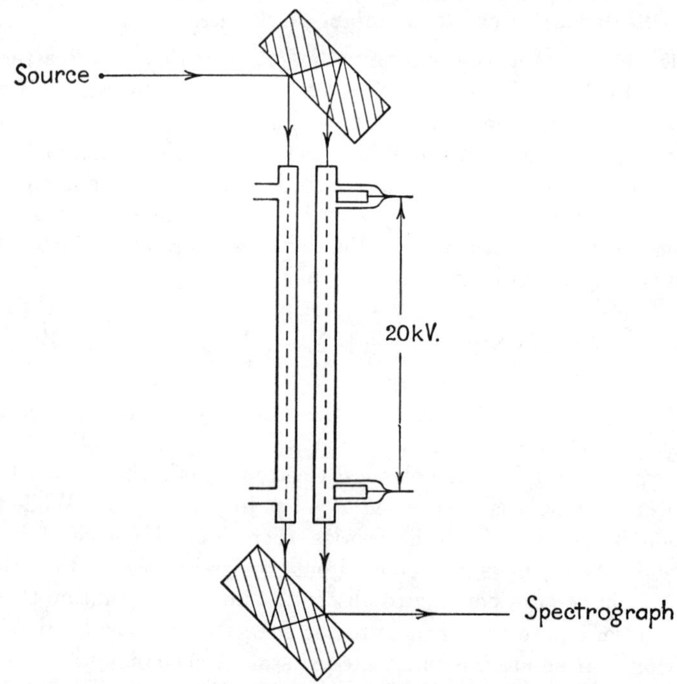

Fig. 19.1.—Apparatus for investigation of dispersion in electrically excited gases, using a Jamin refractometer (cf. fig. 9.13)

19.12. **Calculated and Observed *f*-Values.**

The quantum theory of dispersion provides a method of calculating the f-value. It also gives a relation between an f-value (f_s) and the associated damping coefficient [γ_s in equation 15(53)]. As explained above, the absolute measurement of f-values was, at one time, rather difficult, and the calculation by lengthy methods of successive approximation was very laborious. During the past twenty years, ways of measuring f-values have been improved and new ones have been devised. The increasing use of calculating machines has greatly reduced the labour of calculation and considerable numbers of f-values

* Reference 19.4.

have been calculated. It is thus now possible to compare theory with experiment and a good agreement is obtained (Table 18.1, § 18.22). The theory gives certain general rules for the sum of the f-values for certain groups of lines. The simplest one is that the sum of the f-values for all transitions starting from a single state (including the transitions leading to the photo-ionization) is unity. This rule is fairly well verified for transitions of the valence electrons of sodium and lithium.

19.13. **Resonance Radiation.**

In § 17.14 we saw that an atom of a gas (e.g. mercury vapour) could absorb radiation and make a transition to an excited state. The excited atom could then emit radiation and return to the normal state. This radiation was called resonance radiation. In the quantum theory this phenomenon is regarded as two separate processes—first the absorption, then the re-emission. The transition probabilities for the processes are calculated independently. The radiation absorbed from a parallel beam is re-emitted in all directions, yet we do not speak of this as a Rayleigh scattering process. The essential difference is that the re-emitted radiation is non-coherent with the incident beam, and cannot combine with the incident wave in the way discussed in § 15.45. It thus appears that we apply one theory (the two-stage process with the intermediate state) to obtain the dispersion equation [19(4)], and another theory (two successive processes—absorption followed by re-emission) when we consider " resonance ". There is no real inconsistency. In dispersion theory we want a variation of the index n with frequency. We consider the frequency as given *exactly*. This implies irradiation with very homogeneous light. In order to measure the variation of n with ω as we pass through an absorption line (i.e. to obtain a curve such as fig. 15.5), we must use radiation which is confined to a frequency region which is narrow compared with the absorption line. This implies a steady illumination with a pure sine wave of constant amplitude for a period which is long compared with the natural life of the excited state. The single two-stage process then operates even in the absorption line, and the scattered radiation is coherent with the incident wave. When the gas is irradiated with light whose frequency distribution covers the absorption line, the main effect is the double process (resonance radiation). The re-emitted radiation is not coherent. The exploration of effects in the centre and at the edges of an absorption line presents some interesting problems. In making calculations on the absorption, scattering, etc., in or very near an absorption line, it is not possible to use approximate formulæ such as 19(4).

19.14. **Zeeman Effect.**

It has long been known that when a source of light is placed in a magnetic field, each line in the spectrum splits into a number of components. The wavelength differences are of order 1 Å. for 20,000 oersteds. The light emitted is usually polarized, and the polarization depends on the relation between the direction of observation and the direction of the field. There are corresponding effects on the absorption spectrum and on the dispersion in the neighbourhood of the absorption lines. The first theory of this effect was based on classical electromagnetism and on the Lorentz theory of the electron. It was shown that electrons tend to rotate about the lines of force in a magnetic field under the action of the forces described by 13(11). An electron, which in the absence of the field would undergo simple harmonic motion along a line, will move in a rosette-shaped orbit in the field

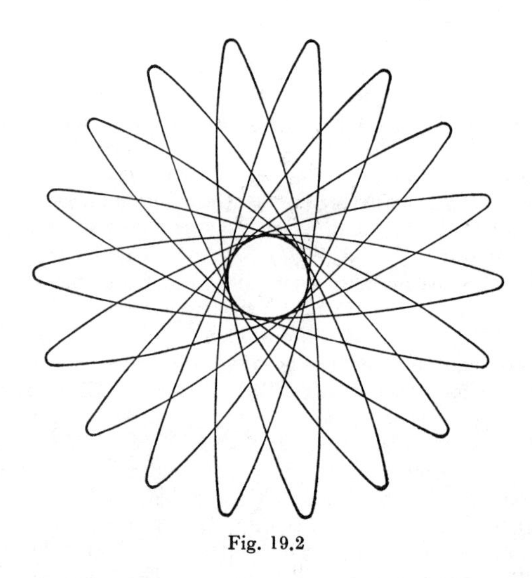

Fig. 19.2

(fig. 19.2). There will then be two frequencies involved, the original frequency and the frequency of rotation. The classical theory predicts the emission of three lines: one corresponding to the original frequency, one to the sum, and one to the difference of the frequencies. In weak fields, the simple effects predicted by the classical theory are obtained with only a few types of spectra. All spectra show the simple arrangement of components when the field is very strong. This is known as the Paschen-Bach effect.

19.15.—The quantum theory considers the problem of the atom in a magnetic field in a rather different way. New terms have to be added to the wave equation to take account of the energy of interaction between the atom and the field. As a rule these terms are sufficiently small to be treated as perturbations. It is found that there are more eigenvalues of the wave equation when the perturbation is included, i.e. there is a larger number of stationary states, and consequently a larger number of spectral terms. In the early quantum theory it was said that the states observed in the absence of the field were " degenerate ", and that in the field they split into a number of " non-degenerate " states. In this book we recognize only single states (see § 17.8). In the absence of a field, several states may have the same energy, and these states may differ in energy when the field is applied. The number of states is thus the same with or without the field.

19.16. Vector Model.

The light emitted in the magnetic field is due to transitions between the stationary states of the perturbed atom. The theory gives correctly the number of components, the relative intensities, the polarization, and the variation with field strength.* This part of the quantum theory is part of the theory of the atom and is outside the scope of this book. One important result of the calculation may be stated. It is found that, under a wide range of conditions (though not under all conditions), an atom behaves as though it were a magnetic top, i.e. as though it possessed a magnetic moment and mechanical angular momentum. The vector which defines the magnetic moment is not, in general, collinear with the vector which defines the angular momentum. The magnitude of the angular momentum can be calculated by taking the *vector* sum of:

(*a*) An angular momentum which, on classical theory, is due to motion of the electrons around the nucleus of the atom.

(*b*) An intrinsic moment of the electron (called " electronic spin ").

(*c*) An intrinsic moment of the nucleus (called " nuclear spin ").

The *magnetic* moment of the atom as a whole is obtained from the vector sum of three similar terms:

(*a*) The magnetic moment due to the orbit.

(*b*) An intrinsic moment of the electron.

(*c*) An intrinsic moment of the nucleus.

In the early years of the quantum theory the wave numbers of spectral terms were calculated without taking account of spin, and many puzzling differences between theory and experiment were obtained. Then the " spin " was introduced as an *ad hoc* hypothesis.

* For intermediate field strengths, calculations are sometimes very difficult and only approximations can be obtained.

Finally Dirac showed that the spin property can be derived by applying the methods of quantum mechanics to a relativistic wave equation of an electron.

It is found that, in very weak fields, the top precesses about the lines of force, but that the relative orientations of the three angular momentum vectors (*a*), (*b*), (*c*) remain unaffected. In very strong fields, the internal coupling of the different vectors breaks down and each electron precesses independently, so that we have an approach to classical conditions. The intermediate condition is obviously very complicated. The relation between the magnetic moment and the angular momentum of the atom can be investigated directly by means of an experiment due to Gerlach and Stern. A beam of atoms is passed through a strong non-homogeneous field. According to quantum theory, the atoms in a given state should orient themselves in one of a small number of angles to the field. From the deflections observed, the relation between angular momentum and magnetic moment can be measured. Although the information obtained is not very detailed, it is important to have this direct determination of the relation.

19.17. Stark Effect.

It is found that the spectrum lines are broadened and split when the source of light is in a strong electric field. This is known as the Stark effect. According to classical theory, the natural frequency of a dipole oscillator is unaffected by a steady electric field. The quantum theory treats the interaction energy as a perturbation, and the frequency, intensity, polarization, etc., are calculated as for the Zeeman effect. It predicts correctly that the separations in moderate fields are usually proportional to the square of the field strength. The calculations are more difficult than the corresponding calculations for Zeeman effect, and the results are less useful in the classification of spectra. On the experimental side, the technique of maintaining and measuring a strong electric field across a light source is difficult. For these reasons less work has been done on the Stark effect than on the Zeeman effect. Nevertheless, there is sufficient to provide an important additional check on the theory of the emission and absorption of light.

19.18. The Polarization of Resonance Radiation.

It is found that resonance radiation is partially polarized. The polarization depends upon the polarization of the incident light, and upon the direction of observation. If a magnetic field is present, it also depends upon the direction and magnitude of the field. The polarization can be calculated when all the data are known. The calculation is fairly complicated. In preliminary calculations of the Zeeman effect it is possible to neglect the nuclear moment. In the early calculations on the polarization of resonance radiation the nuclear moments (which were then imperfectly understood) were not included, and the results obtained did not give a satisfactory agreement with experiment. The present theory is in reasonably good agreement with experiment.

We shall now discuss one effect which leads to a determination of the life of the excited atom and thus provides a datum which can be compared with values obtained from dispersion measurements and absorption lines. Suppose that mercury vapour is illuminated with polarized radiation of wavelength 2537 Å., so that this resonance line

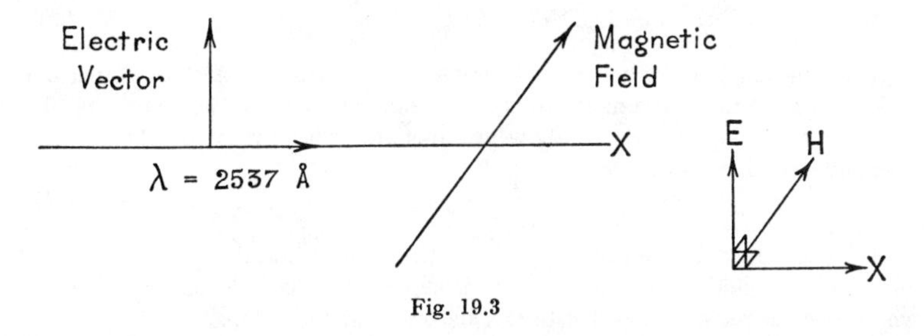

Fig. 19.3

is excited and that a magnetic field is applied perpendicular to the electric vector and to the direction of the illuminating beam (fig. 19.3). It is then found that the emitted light is partially polarized and there is a plane of maximum polarization of the emitted light. As the field is increased from zero, two effects are observed. In the first place the plane of maximum polarization rotates and, secondly, the percentage of polarization decreases. The theory gives the following account of these effects.

In an extremely weak field the emitting atoms are oriented in a certain way and precess very slowly, so that they all emit their radiation before they have turned through any appreciable angle. At higher fields, the majority of the atoms precess through a certain angle before emitting radiation, and this angle determines the plane of maximum polarization. This angle is, however, only a mean angle, as atoms rotate for different times (and hence through different angles) after excitation before emitting radiation. Thus there is a depolarizing effect. This description of the phenomenon is confirmed by an experiment of Fermi and Rasetti, who applied rapidly alternating magnetic fields. They showed that when the frequency is much higher than that of the atomic precession (so that the atoms merely oscillate through a small angle), the polarization is the same as that obtained in the absence of the field.

19.19.—The relation between the life of the excited atom and quantities which can be determined by experiment is given by the following calculation.

The light emitted at any time t comes from atoms which have been excited at various earlier times. Consider a particular group of N_0 atoms which were excited at time $t = 0$. Let τ be the mean life of the excited atom and let A_{nm} (equal to $1/\tau$) be the transition probability [see eqn. 18(62a)]. The number of atoms which are still excited at time t is

$$N = N_0 \exp(-A_{nm}t). \quad \quad 19(5)$$

The energy emitted between t and $(t + dt)$ is

$$dL = h\nu A_{nm} N\, dt = h\nu A_{nm} N_0 \exp(-A_{nm}t)\, dt. \quad \quad 19(6)$$

Let $g\omega_H$ be the angular velocity of precession for a magnetic "top" when the ratio of the resultant magnetic to the resultant mechanical moment is g. Then the angle ϕ, through which the above group of atoms have rotated between excitation and emission, is

$$\phi = g\omega_H t. \quad \quad 19(7)$$

Consider the light emitted in the positive direction of Z and suppose that, in the absence of the field, the electric vector* is along OX. Then the energy of the light whose electric vector is parallel to OX becomes, in the field,

$$\begin{aligned} dL_x &= h\nu N_0 A_{nm} \exp(-A_{nm}t) \cos^2\phi\, dt \\ &= B \cos^2\phi\, dt. \end{aligned} \quad \quad 19(8)$$

The component with electric vector in a direction which makes an angle θ with OX is

$$dL_p = B \cos^2(\theta - \phi)\, dt, \quad \quad 19(9a)$$

and the component with **E** perpendicular to this direction is

$$dL_s = B \sin^2(\theta - \phi)\, dt. \quad \quad 19(9b)$$

After substituting for ϕ from 19(7), and for B from 19(8), and integrating, we have

$$P_\theta = \frac{L_p - L_s}{L_p + L_s} = \frac{a}{a^2 + 4}(a \cos 2\theta + 2 \sin 2\theta), \quad \quad 19(10)$$

where P_θ is the degree of polarization for the direction θ and $1/a = g\omega_H\tau$. This ratio has a maximum value when

$$\tan 2\theta = 2g\omega_H\tau, \quad \quad 19(11)$$

and the maximum value is given by

$$P_m^2 = \frac{1}{1 + 4g^2\omega_H^2\tau^2} \quad \quad 19(12)$$

* For mercury 2537 Å., the electric vector of the resonance radiation is nearly in the same direction as that of the incident light when $H = 0$.

for the case when the radiation is completely polarized in the absence of the field. When H is small and θ is small, the angle of maximum polarization is $g\omega_H\tau$, i.e. the angle through which the atom rotates during its average life in the excited state. When H is large, P_m tends to zero. Most of the atoms have then rotated through angles large compared with 2π before emitting any appreciable amount of radiation. The value of τ may be deduced from the experimental measurements of θ and P_m using 19(11) and 19(12).

19.20. Effect of Foreign Gas on Resonance Radiation.

We have seen in § 17.15 that resonance radiation may be " quenched " by the presence of a foreign gas. This effect is ascribed to collisions of the second type. If every collision (within the atomic radius derived by applying kinetic theory to measurements of viscosity, etc.) were a collision of the second type, then experiments on this effect would lead to a value of τ. If r is the ratio of the energy of resonance radiation emitted in the presence of a foreign gas (at a pressure which makes the time between collisions equal to T) to the energy emitted in the absence of the gas, then we should expect

$$r = e^{-T/\tau}. \qquad \qquad 19(13)$$

Hence, knowing r and T, we could calculate τ. The values of τ obtained by this method differ widely; this shows that not every kinetic-theory collision is a collision of the second type, so that we cannot determine τ in this way. We may then assume that we know τ from other measurements (§ 17.41) and use 19(13) to determine the frequency of collisions of the second type. It is found that with excited mercury and oxygen, there are more collisions of the second type than the number calculated, by kinetic theory, for the total number of collisions. We must assume that the effective collision-radius of the excited mercury atom for collisions of the second type with oxygen is larger than the kinetic-theory radius of the normal atom. On the other hand, the collisions of the second type between mercury and helium are very rare (certainly less than one per cent of the kinetic-theory collisions). The interaction between the excited atom and the atom or molecule of the foreign gas has been investigated theoretically by the methods of quantum mechanics. It is found that the probability of the transfer of energy of excitation from one atom to another is high only when the energies of excitation for the two atoms are nearly equal. This is in accordance with experimental results. The excitation energy for mercury 2537 is 4·9 electron volts; oxygen has an excitation energy of 4·86 electron volts, and the nearest value for helium is 19·75 electron volts.

19.21. Raman Effect.

In 1923 Smekal suggested that energy might be interchanged between radiation and matter during the scattering of light. The atom or molecule would change from state n to state l, and the frequency of the quantum would increase or decrease by ν_{nl} so as to conserve energy. In 1928 C. V. Raman discovered that scattered radiation of this type is obtained with solids and liquids, and with gases under high pressure.* The displaced radiation had not been observed previously because it is very weak, and the probability of scattering without change of wavelength is much higher (Rayleigh scattering). Apparatus for the study of the Raman effect is shown in fig. 19.4. Typical Raman spectra are shown in Plate V*a*, *b*, *c*, *d*. The frequency differences involved in some of the Raman transitions are quite large, so that the difference of wavelength between the Raman radiation and light scattered by the Rayleigh process is several hundred Ångström units. With others the corresponding difference is only a few Ångström units. The line due to Rayleigh scattering is always heavily over-exposed when the exposure is sufficient to show the Raman lines. It is therefore necessary to use a spectrograph of moderately good dispersion as well as of high light-gathering power. It is also very desirable to use a monochromatic, or nearly monochromatic, source in order to prevent the Raman lines being masked by Rayleigh scattering of lines (other than the exciting line) in the source.

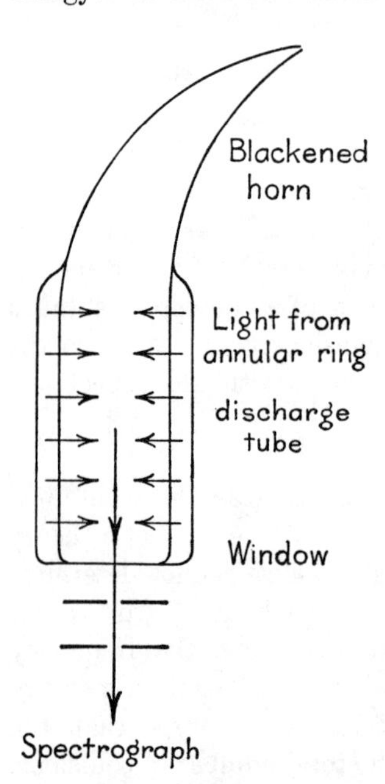

Fig. 19.4.—Apparatus for observation of Raman spectra. A mercury-vapour discharge between the walls of the double-walled quartz tube acts as light source. The scattered radiation leaves through the window and is examined with a spectrograph.

19.22.—When all the atoms or molecules are in the normal state, the only possible Raman transitions are those in which the scattered light is of lower frequency than the incident light. These are called the Stokes lines. This name is given because they obey the general

* Landsberg and Mandelstamm discovered the effect independently in the same year.

PLATE V

(*a*), (*b*), (c), (*d*) Raman effect. (*a*) Source only, (*b*) cyclo-hexane, (*c*) paraxylene, (*d*) carbon tetrachloride. Note that both Stokes and anti-Stokes lines appear in (*c*) and (*d*).

(*e*) Microphotometer trace of sodium absorption spectrum.

(*f*), (*g*) Patterns due to Fresnel diffraction of electrons by very small obstacles.

(*h*) Patterns due to Fraunhofer diffraction of electrons by many small crystals.

(*i*) Pattern due to Fraunhofer diffraction o electrons by a single crystal.

4047 Å.

PLATE V

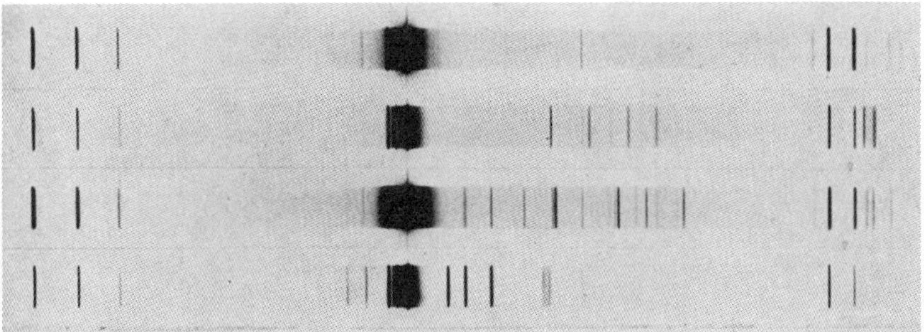

(a)

(b)

(c)

(d)

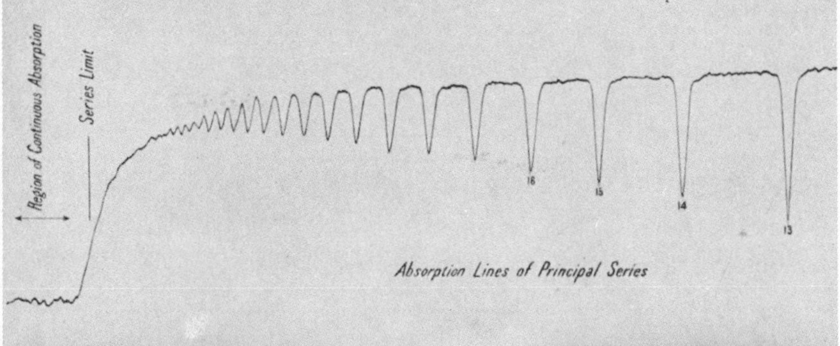

Region of Continuous Absorption

Series Limit

16

15

14

13

Absorption Lines of Principal Series

(e)

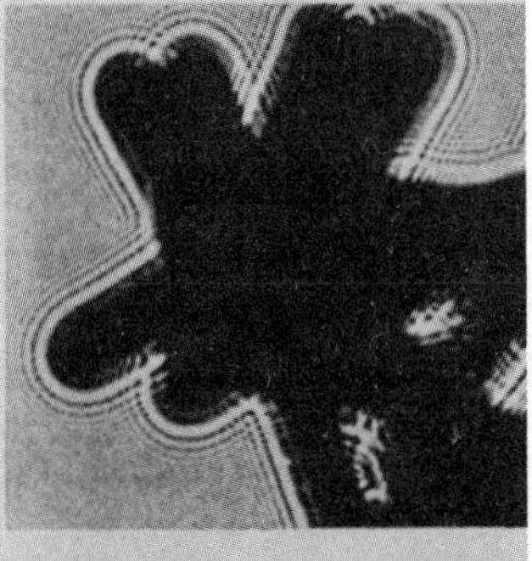

(g)

)

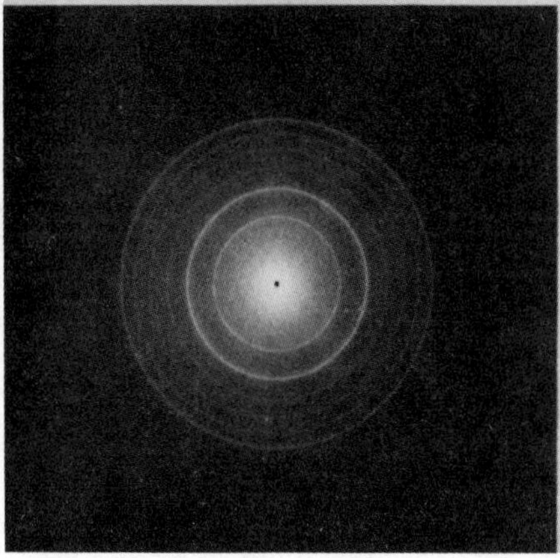

(i)

law, due to G. G. Stokes, that the fluorescent radiation is not of higher frequency than the incident light. Even at room temperature, substances like water have some molecules in states other than the normal state, and Raman lines of frequency higher than that of the incident light are obtained (Plate V*d*). The strength of these "anti-Stokes lines" increases when the temperature is raised, because more molecules of the scattering substance pass into states of higher energy. The energy exchanged between radiation and matter in the Raman process is equal to the difference of energy between two stationary states of the scattering atom or molecule. If direct transition between these states could be produced, the resulting emission or absorption line would usually lie in the far infra-red region. It is found that, in some cases, a line can be observed in the infra-red spectrum, and a Raman line with the corresponding frequency difference is also found. On the other hand, many Raman lines are seen when the corresponding infra-red lines are forbidden. Some infra-red lines are observed and the corresponding Raman lines do not appear. The quantum-mechanical theory is able to account for this result in a satisfactory way. The transition probability for direct transition between two states, n and l, depends on the interaction energy U_{nl} [equation 19(2)]. The Raman process is a two-quantum process and involves two interaction energies U_{nm} and U_{ml}, where m is the " intermediate level " mentioned in § 19.9. The Raman process thus depends on two interaction energies, neither of which is the same as that involved in the calculation of the probability of a direct transition. Thus the selection rules for a Raman process are quite different from the corresponding selection rules for the direct transition. The Raman process takes place if there is an intermediate level * which combines both with the level n and with the level l, even though these states do not combine with one another. The Raman effect is of great theoretical interest in that it forms an important confirmation of the general theory of the interaction of radiation and matter. It is also of importance as a tool in exploring molecular structure. It enables the energies of levels which would not otherwise be accessible to be measured. The relative strengths of the different Raman lines also give information concerning the structure of the molecule.

* There must be *at least* one intermediate level. Very often more than one intermediate level is involved in the production of a given Raman line.

19.23. **Quantum Theory of Polarized Light.**

In earlier discussions of Huygens' principle and the theory of diffraction, we showed that one and the same disturbance may be analysed in terms of a series of plane waves (by the usual Fourier method), or in a series of spherical waves (method of Huygens and Kirchhoff). In applying the quantum principles to the electromagnetic field, we analysed it into a series of plane waves polarized in two mutually perpendicular directions. This method gives the quantization of energy and of momentum. It shows that, *in regard to exchanges of energy and momentum*, an electromagnetic wave behaves like a set of particles, each with energy $h\nu$ and momentum $h\nu/c$. Application of the same method to an electromagnetic field analysed in terms of spherical waves leads to the quantization of the angular momentum of the field. The spherical waves are spherically symmetrical about a certain point. The field strengths are given by expressions of the form

$$P_l^m(\cos\theta)e^{im\phi}, \qquad \text{19(14)}$$

i.e. by spherical harmonics. The first harmonic gives the distribution corresponding to a dipole*; two higher harmonics enter into the distribution for quadripole radiation, and so on. We have seen that the problem of the application of quantum mechanics to the plane wave is mathematically similar to the problem of the simple harmonic oscillator. In a similar way, the application to the spherical wave is similar to quantization of angular momentum in the hydrogen atom. It is found that the z component of momentum for a single photon is given by

$$M_z = m\frac{h}{2\pi} \qquad \text{19(15)}$$

and the total momentum M (l and m being integers) by

$$M^2 = \frac{h^2}{4\pi^2}l(l+1). \qquad \text{19(16)}$$

19.24.—Experimentally we can detect angular momentum of a parallel beam of circularly polarized light in a " macroscopic " experiment.† In a less direct way, it can be shown that the angular momentum of the spherical wave as given above is just sufficient to allow conservation of angular momentum when radiation is emitted

* Cf. Appendix XIII B (p. 413). † § 17.24.

or absorbed by matter. These results suggest a very simple " picture ", in which photons are of two kinds—one without spin representing plane-polarized light, the other with either right- or left-hand spin representing circularly polarized light. This picture is adequate for the discussion of a number of the problems we have considered above, but it cannot be accepted as more than a convenient way of remembering certain groups of experimental results. An examination of its defects is worth while, because it forms a suitable starting point for the discussion of some important features of the quantum-mechanical theory of light.

19.25. **Superposition of States.**

We saw in Chapter XII that a circularly polarized wave can be regarded as the resultant of two waves plane-polarized in mutually perpendicular planes and with a phase difference of $\pi/2$. In a similar way, a beam of plane-polarized light can be regarded as the resultant of two kinds of circularly polarized light, or of two beams of light plane-polarized, but in other planes. The classical wave theory, which includes this " equivalence ", describes in an adequate way all the experiments on the analysis of plane-polarized, circularly polarized, and elliptically polarized light by means of combinations of quarter-wave plates and Nicol prisms. Therefore it must be included in any satisfactory theory of light. It is included in the theory of the radiation field given in Chapter XVIII. According to this theory, the electromagnetic radiation field has all the properties of classical waves in relation to its analysis into components by polarizers, etc., as well as in relation to interference and diffraction—within the limitations of the uncertainty relation. The quantization of the energy and momentum does not affect the wave properties, except that it recognizes the unavoidable errors in the experimental data. The quantum-mechanical rules limit the number of possible ways in which the energy may be distributed between the different simple harmonic waves into which, on purely classical principles, the field can be analysed.

19.26.—While this " resolution " of waves occasions no difficulty in the mathematical formulation of quantum mechanics, it shows the inadequacy of the simple concept suggested in § 19.24. If circularly polarized light " consists " of particles with spin, it cannot be regarded as a mixture of two sets of particles neither of which has any spin. In a similar way, if a photon corresponding to plane-polarized light is essentially a particle with some axis which defines its plane of polarization, it cannot be regarded as the " resultant " of two particles which have their axes in different planes. These and many similar considerations show that the word picture suggested in § 19.24 does not completely correspond with the situation described in the mathematical equations. In § 18.19 we saw that solutions of the wave equation can be superposed to form new solutions according to the equation

$$\psi_m = c_1\psi_1 + c_2\psi_2 + \ldots + c_n\psi_n, \quad . \quad . \quad 19(17)$$

where the coefficients c_1, c_2, etc., must be adjusted so that the function as a whole is normalized. We may describe this superposition in the following way.

Suppose that, as a result of an observation, we know that a system is in the state m. This means that we have the right to predict the probable result of certain types of experiments. Equation 19(17) says that our prediction for the state m is the same as that which we should make if there were a chance $c_1{}^2$ of the system being in state 1, a chance $c_2{}^2$ of it being in state 2, etc. In fact the state m implies the sum of all the possibilities represented by the states 1, 2, . . . n, each being correctly weighted. In this sense, the state m is a *superposition* of the states 1, 2, . . . n. For example, suppose that a beam of light has been passed through a Nicol prism A_1 (fig. 19.5) and is plane-polarized in

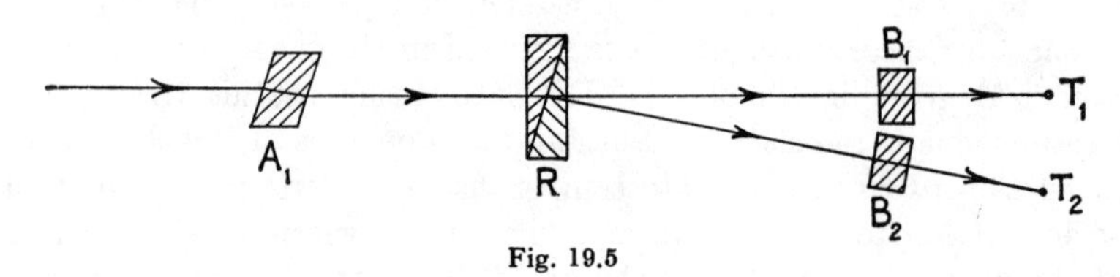

Fig. 19.5

a plane defined by $\theta = 0$. Then we can say that the photons in the beam are waves plane-polarized in the plane $\theta = 0$. If now the beam is passed through a Rochon prism (R), it may be divided into two beams. By inserting two Nicols B_1 and B_2, we may find the planes of polarization of the beams. For one angular position of the Rochon prism R these may be α and $\alpha + \pi/2$. It will then be found from the readings of thermocouples T_1 and T_2 that the numbers of photons in the beams are in the ratio $\cos^2 \alpha$ to $\sin^2 \alpha$. We cannot, however, take a simple particle view by saying that some of the photons in the beam which emerged from A were polarized parallel to the direction $\theta = \alpha$ and others parallel to $\theta = \alpha + \pi/2$, and that the Rochon prism separated these groups. This would immediately be in conflict with the observation that *all* the photons would pass a Nicol A_2 (fig. 19.6) with its principal plane set parallel to $\theta = 0$, provided that it was inserted before the Rochon prism. This difficulty does not appear if we say that the result of passing the light through the first Nicol prism A_1 is to prepare a state. The state so prepared is defined by the property that any one of the photons has a unit chance (i.e. a certainty) of passing a second

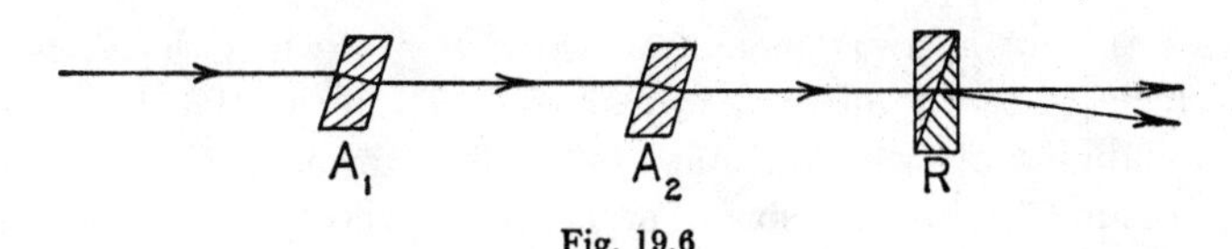

Fig. 19.6

Nicol A_2 placed immediately after the first and set with its plane at $\theta = 0$. This state, which we may call state 0, is represented by a certain wave function ψ_0. If we call ψ_α the wave function for a photon which is certain to pass a Nicol which is set with its plane at $\theta = \alpha$, we have

$$\psi_0 = \cos\alpha \,.\, \psi_\alpha + \sin\alpha \,.\, \psi_{(\alpha+\pi/2)}. \qquad 19(18)$$

Thus when we say that the photon is in the state 0 we mean that it is certain to pass a Nicol at an angle $\theta = 0$, and has chance $\cos^2\alpha$ of passing a Nicol oriented at α. This is the same thing as saying that it has chance $\cos^2\alpha$ of being in state α, which is by definition a state in which it is certain to pass the second Nicol oriented at $\theta = \alpha$.

19.27.—That this method of representation is consistent is shown by the following discussion. By the same argument as that which led to 19(18) we may show that

$$\psi_\alpha = \cos\alpha \,.\, \psi_0 + \sin\alpha \,.\, \psi_{\pi/2} \qquad 19(19)$$

and

$$\psi_{(\alpha+\pi/2)} = \sin\alpha \,.\, \psi_0 + \cos\alpha \,.\, \psi_{\pi/2}. \qquad 19(20)$$

Hence, substituting in 19(18),

$$\psi_0 = (\cos^2\alpha + \sin^2\alpha)\psi_0 + (\sin\alpha\cos\alpha - \sin\alpha\cos\alpha)\psi_{\pi/2} = \psi_0 + 0 \,.\, \psi_{\pi/2}. \qquad 19(21)$$

Thus, when a particle has a chance $\cos^2\alpha$ of being in state α, and $\sin^2\alpha$ of being in state $(\alpha + \pi/2)$, it has a total chance 1 of being in state 0, and a total chance 0 of being in state $\pi/2$, i.e. the state 0 is formed by a suitably weighted superposition of states (α) and $(\alpha + \pi/2)$.

19.28.—The superposition of *states*, and the associated superposition of *wave functions*, must be clearly distinguished from the superposition of *waves* first discussed in Chapter II. When two equal waves are superposed, the resultant is a wave of twice the initial amplitude and four times the initial energy at points where they are in phase, and zero energy when they are out of phase. When a quantum-mechanical state is superposed upon itself, we are only saying the same thing twice. We may write

$$\psi_0 = \frac{1}{2}\psi_0 + \frac{1}{2}\psi_0, \quad . \quad . \quad . \quad . \quad . \quad 19(22)$$

but this is only a trivial equation which says " when the system is certainly in state 0 it has a $\frac{1}{2}$ probability of being in the state 0 plus a $\frac{1}{2}$ probability of being in state 0 ". The normalization condition always secures that in this sort of situation the theory gives the correct —though unimportant—answer. It follows that the absolute amplitude of the ψ wave does not give the number of photons present. The quantum-mechanical equation determines the *relative* probabilities of certain results of experiments. The total energy determines the number of quanta.

19.29. **Empirical and Non-Empirical Questions.**

There is a certain type of question to which the quantum theory gives no direct answer, because the question is not related to experiment and possible observation. Let us consider two examples. First let us ask what happens to a photon of right-handed circularly polarized light as it passes through a half-wave plate. We know that it emerges as a photon of left-handed circularly polarized light, and the wave theory gives an account of the progress of the wave through the birefringent medium. If, however, we try to visualize the photon as a particle and ask for a mental picture of the way in which this particle alters as it passes through the plate, we come up against a blank wall. The theory has nothing to say unless we alter the question and propose an experiment designed to tell us what happens to the photon as it passes through the plate. Then the theory will predict the result of the experiment. Suppose we split the half-wave plate in two and pass the light through half of it, and then pass the light through a Nicol prism. The theory predicts that, if the Nicol has a certain orientation, the photon will certainly pass through. If it has another orientation making an angle α with the first one, there is a chance $\cos^2 \alpha$ that the photon will pass through, i.e. if N trials each with a single photon are made successively, then the photon will pass through on very approximately $N \cos^2 \alpha$ occasions (provided N is very large).* Also if a large

* The theory also gives the statistical fluctuation when N is not large.

number of photons is used in one trial, the fraction passing is $N \cos^2 \alpha / N = \cos^2 \alpha$. Thus the theory gives a prediction of the result of the experiment of cutting the plate—either that a particular result is certain, or that a specified distribution of results will be obtained if trials are made with a large number of photons. *It answers the original question in so far as it can be reduced to a series of experimental tests.*

19.30.—As another question let us consider a photon incident on the dividing mirror (M_1 in fig. 4.1) of a Michelson interferometer. Let us ask by which path through the apparatus a particular photon reaches the point in the system of interference fringes at which it is detected. If we know there is just one photon in the incident beam, then the uncertainty relation in the form given in equation 18(9) states that we can know nothing about the phase of the incident beam.

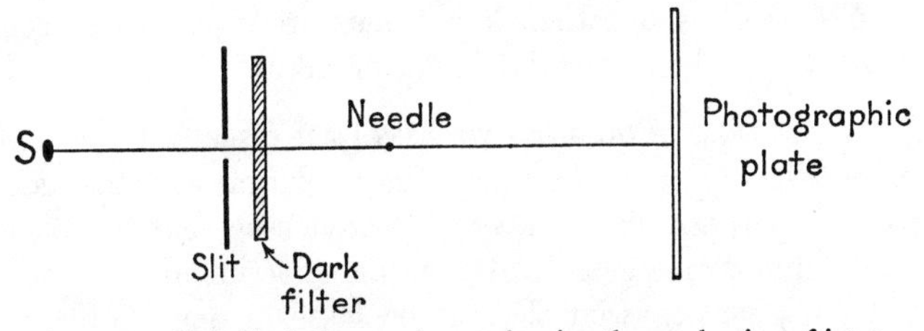

Fig. 19.7.—G. I. Taylor's experiment showing the production of interference patterns by successive light quanta

After the photons have passed the dividing mirror we can know the relative phases of the two beams, and calculation shows that they interfere to produce a certain pattern of interference fringes. A single photon can, of course, be detected at only one point; to obtain the pattern as a whole we must use a large number of photons, either all at once or one after another. The fact that they can be sent one after another and still produce interference is shown by an experiment due to G. I. Taylor. He set up an arrangement for photographing the diffraction pattern and reduced the illumination by interposing dark filters until very long exposures were needed to give the pattern (fig. 19.7). The chance of two or more energy quanta passing through the apparatus simultaneously was made very small, but the diffraction pattern was just the same as that obtained with a strong source of light.*

* The corresponding experiment with an interferometer has been made in the course of researches primarily designed for other objectives; for example, the night sky is an extremely feeble source of the sodium D lines, but perfectly clear interferometric patterns have been obtained.

19.31.—So long as we do not attempt to find out by which path each individual photon reaches the point of detection, we obtain the interference pattern. If we try to find the path of a given photon (e.g. by measuring the momentum transferred to one or other of the mirrors M_3 or M_2), the frequencies of the two beams are no longer quite the same (because the mirror must be free to move) and the interference pattern disappears. We can examine the particle property of the light only by altering the conditions of the experiment so that the wave property is no longer in evidence. If we attempt to evade this difficulty by successive experiments on particle and wave properties, we never reach a situation in which the question stated at the beginning of § 19.30 can be answered. It should, however, be pointed out that in the usual experiments on interference the number of quanta is very large. A one-per-cent error in the energy corresponds to an enormous value of ΔN, and we can then know phase relations very accurately without infringing the uncertainty relation 18(9).

19.32.—The type of question we have just discussed does not arise if we have full regard to the uncertainty relation and the associated principle of superposition of states. We have seen that the theory can answer any question which relates to an experiment, but not other questions. We can say that the photon is not a particle but a quantized wave. When asked what is a quantized wave, the theory replies by giving an account of its behaviour in all possible experiments, and this account is a correct and complete theory of optics. At first the reader will be tempted to say, " I know that light behaves sometimes like waves and sometimes like particles. Does the quantum mechanics say more than this? If not, what is the use of the elaborate mathematics?" The answer is that the simple statement that light behaves sometimes like particles and sometimes like waves does not enable us to predict the results of experiments. We need a way of saying *when and how far* light behaves like particles and *when and how far* it behaves like waves. If the theory is a formalism, it is a formalism so logical and so exactly suited to its purpose that it predicts the results of all optical experiments. The uncertainty relation is not a device invented by quantum physicists for the purpose of evading those difficult questions which concern the wave-particle conflict. It is directly deduced from observation. It must be included in any theory. The quantum mechanics includes it by the conjunction of the concept of a wave group with the hypothesis of a finite exchange of energy in measurement.

19.33. Historical Development of the Quantum Theory.

As stated at the beginning of Chapter XVII, our discussion of the quantum theory of radiation has followed an order chosen for ease of exposition and is not the historical order. We shall now state the historical order, partly in order to complete the outline of the history of the theory of light given in Chapter I, and partly because this may form a convenient recapitulation of the essential hypotheses.

At the beginning of the nineteenth century the wave theory was well established. During that century many beautiful experiments on interference, diffraction and polarized light were carried out. A corresponding advance in elegance of exposition reached its climax in the work of Rayleigh. It was perhaps inevitable that he, having the most complete understanding of the wave theory, should expose its limitations in the most fundamental way. In 1900 Rayleigh showed that the classical theory could not yield a correct radiation law.* Soon afterwards Planck introduced the quantum hypothesis that a fixed amount of energy ($E = h\nu$) must be exchanged whenever an atom interacts with the radiation field. Planck wished to leave the wave theory of the radiation field undisturbed and to confine his hypothesis to the interaction process. Einstein (about 1905) showed that many experimental results could be most easily described by a theory of photons. The Einstein photons were essentially light particles; they were localized concentrations of mass, momentum and energy. The years 1905–25 saw little advance in the theory of light except for the theory of the experiments described in Chapter XI (relativistic optics). In this period the quantum theory was applied in a tentative way to the atom. The most important advance was Bohr's concept of stationary states. About 1925 the ideas of wave mechanics began to appear, and by 1930 it had been shown that this theory gave a satisfactory account of most problems of atomic structure.

The theory of light was still in confusion. There appeared to be no possible reconciliation between the wave theory and the Einstein photons. Indeed there was no reconciliation until the understanding of the uncertainty principle (first stated by Heisenberg in 1927) had forced a modification of both wave and particle concepts. The next logical advance in the theory of radiation was, to a considerable extent, due to Dirac. In its developed form this theory:

* The complete proof was given by Jeans in 1909.

1. Expresses the energy of an electromagnetic radiation field in a form which is similar to the Hamiltonian energy function for a set of harmonic oscillators.

2. Applies the quantum theory of harmonic oscillators to this energy function and hence to the field. This gives a theory of quantized waves. These quantized waves are often called photons, but they do not have all the properties of the Einstein photons any more than they have all the properties of Maxwell's electromagnetic waves.

3. The photons are shown to have the properties of indistinguishability, etc., discussed in § 18.34. These properties are very closely related to the fundamental concepts of the superposition of quantum states and to the limitations implied in the recognition of the uncertainty principle.

4. When the theory has been applied to the radiation field it is possible to extend it to a system of radiation plus atoms. It is found that the radiation induces transitions corresponding to emission, absorption and the various types of scattering. This part of the theory can be joined to the classical theory by a concept of " equivalent oscillators " with "*f*-values ". In this way it is able to include that part of the wave theory which is in agreement with experiments on emission, absorption, scattering, and dispersion.

19.34.—We have said that the quantum mechanics is a complete theory of light and, apart from the difficulty of certain calculations, gives correct answers to all questions. We must be careful, however, not to claim that nothing lies beyond the present theory—that would be to repeat the mistake of some physicists of fifty years ago who were sure that the wave theory was perfect. Even now it is clear that the quantum mechanics, in its present form, has difficulty with the properties of material particles of extremely high energy and probably with very high-frequency radiation. It appears to the writer that, in the future, defects in the theory will be revealed and further progress engendered, not by laboratory experiments on optics, but by astronomical observations on effects like the nebular red shift or by experiments on high-energy photons. Yet even this prediction is only an opinion. Future experiments may reveal some purely optical phenomenon entirely different from anything which has so far been discovered.

REFERENCES

19.1. Rubinowicz, A.: *Reports on Progress in Physics*, 1949, Vol. XII, p. 233.

19.2. Heitler: *Quantum Theory of Radiation* (Oxford University Press).

19.3. Mott and Sneddon: *Wave Mechanics and its Applications* (Oxford University Press).

19.4. Ladenburg: *Reviews of Modern Physics*, 1933, Vol. 5, p. 243.

APPENDIX XIX A

Quantum Theory of Dispersion

1. Interaction of Radiation and Matter.

This appendix deals with the Dirac theory of the interaction between radiation and matter in its application to the absorption, emission and scattering of light, and the related problem of dispersion. Let us consider first the absorption and emission of light. The theory is discussed in two stages. In the first stage, the atoms are regarded as a quantum-mechanical system which is subject to a small disturbance due to the energy of interaction between the atoms and the electromagnetic field which represents the radiation. The disturbance is called a " perturbation " and this part of the theory is an extension of the perturbation theory which had previously been used to calculate the stationary states of atoms in steady magnetic and electric fields. The perturbation due to the radiation field has a high-frequency variation with time, and it is shown that this kind of perturbation leads to transitions between the stationary states of the atoms, provided that the Bohr frequency condition [equation 17(7)] is satisfied. The theory yields formulæ from which the numbers of transitions per unit time can be calculated, so that we obtain B_{nm} and B_{mn} directly, and A_{nm} by applying 17(53). The formulæ agree with those obtained in § 18.21 and this part of the calculation may be regarded as justifying the assumption stated there.

In the second stage of the Dirac theory, the atom and the radiation are regarded as one quantum-mechanical system. Transitions then involve transfer of energy from one part of the system to another, but there is no change in the energy of the system as a whole. Each of the coefficients † A_{nm}, B_{nm}, B_{mn} is calculated independently. This second stage is suitable for extension to many other important problems including Rayleigh scattering, dispersion theory and Raman effect. In both parts of the theory it is assumed that the perturbation is small and the whole theory is an approximation based on this assumption.

2. Radiation as a Perturbation.

Let us suppose that the wave equation for an unperturbed atom is

$$\mathscr{H}_0(\Psi) = -\frac{ih}{2\pi}\frac{\partial\Psi}{\partial t}, \qquad \qquad 19(23)$$

† If m is the state of higher energy, A_{mn} replaces A_{nm}.

where $\mathscr{H}_0(\Psi)$ stands for all the terms on the left-hand side of 18(49). We assume that solutions Ψ_1, Ψ_2, etc. corresponding to stationary states of the unperturbed atom have been found. A suitable physical measurement on the atom will either show that it is in one stationary state or will give a series of probabilities: P_1 that it is in state 1, P_2 that it is in state 2, and so on. In general, we may express the result of our measurement by saying that the state of the atom is represented by a solution Ψ of the wave equation and that †

$$\Psi = c_1\Psi_1 + c_2\Psi_2 + \ldots = \sum_l c_l\Psi_l, \qquad 19(24)$$

where

$$P_l = c_l c_l^*. \qquad 19(25)$$

The expansion 19(24) is always mathematically valid because Ψ_1, Ψ_2, etc. form an orthogonal set of functions.‡ The physical interpretation of this expansion we have given above is in accordance with the general principles of quantum mechanics (§§ 18.19, 18.33, and 18.35). Now, suppose that the atom is subject to an external field, so that the wave equation becomes

$$\mathscr{H}(\Psi) = (\mathscr{H}_0 + U)\Psi = -\frac{ih}{2\pi}\frac{\partial\Psi}{\partial t}. \qquad 19(26)$$

The symbol U stands for new terms in the wave equation. These may involve q, $\partial/\partial q$, and t. It is assumed that all new terms are small compared with corresponding terms in the wave equation for the unperturbed atom [19(23)].

3.—The result of a measurement on the atom in the radiation field is represented by a solution Ψ' of 19(26). We may express Ψ'

either (i) as a sum of functions Ψ_1', Ψ_2', etc. corresponding to the stationary states of the perturbed atom with constant coefficients c_1', c_2', etc.;

or (ii) as a sum of terms based on Ψ_1, Ψ_2, etc. (i.e. on the stationary states of the unperturbed atom) but with a new set of coefficients a_1, a_2, etc. instead of c_1, c_2, etc. We shall see that these new coefficients have to be functions of t in order to satisfy the wave equation.

These two methods of expanding Ψ' are equally valid from the mathematical point of view, since the solutions of 19(23) and 19(26) each form an orthogonal set. The choice between them is a matter of convenience in relation to the problem under discussion. The first expansion is convenient when the atom is perturbed by a *steady* electric or magnetic field. The second is suitable for our present discussion in which time variation is the essence of the problem. We therefore put

$$\Psi' = \sum_l a_l\Psi_l. \qquad 19(27)$$

Substituting from 19(27) into 19(26) we obtain

$$\sum_l a_l \mathscr{H}_0\Psi_l + \sum_l a_l U\Psi_l = -\frac{ih}{2\pi}\sum_l\left(\dot{a}_l\Psi_l + a_l\frac{\partial\Psi_l}{\partial t}\right). \qquad 19(28)$$

Since Ψ_1, Ψ_2, etc. are solutions of 19(23), this reduces to

$$\sum_l \dot{a}_l\Psi_l = \frac{2\pi i}{h}\sum_l a_l U\Psi_l. \qquad 19(29)$$

† Compare equation 18(52). ‡ See p. 375 of Reference 19.3

Multiplying both sides by Ψ_m^* and integrating over all space (using the orthogonality condition), we have

$$\dot{a}_m = \frac{2\pi i}{h} \sum_l a_l \int \Psi_m^* U \Psi_l \, d\tau. \qquad 19(30)$$

4.—Suppose that an observation made at time $t = 0$ has shown that the atom is in the nth state at time $t = 0$. Then $a_n = 1$ and all the other a's are zero when $t = 0$. Suppose that the radiation field operates from $t = 0$ to $t = t'$, and that a second observation is made immediately after the perturbation has ceased. Then the probability of finding (from the second observation) that the atom is in state m at time t' is

$$P_m(t') = a_m(t') a_m^*(t'). \qquad 19(31)$$

If t' is small and the radiation field is fairly weak, so that the probability of a transition is fairly small, we may put $a_n = 1$ and $a_l = 0$ (for all values of l except $l = n$) in 19(30). The sum then reduces to a single term and we have

$$\dot{a}_m = \frac{2\pi i}{h} \int \Psi_m^* U \Psi_n \, d\tau \qquad 19(32)$$

and, integrating with respect to t,

$$a_m(t') = \frac{2\pi i}{h} \int \int_0^{t'} \psi_m^* U \psi_n \exp \frac{2\pi i}{h} (W_n - W_m) t \, dt \, d\tau. \qquad 19(33)$$

The constant of integration is zero, since $a_m = 0$ when $t = 0$. The probability of a transition in unit time is $(1/t')P_m(t')$. This number may be calculated by obtaining a_m from 19(33) and substituting in 19(31), but before doing this it is convenient to derive an expression for U.

5. Calculation of the Interaction Energy.

Consider a parallel beam of plane-polarized light propagated along the z axis and with the electric vector along the x axis. Then at $z = 0$ we may put

$$E = E_x = 2E_{0x} \cos 2\pi\nu t = E_{0x}(e^{2\pi i\nu t} + e^{-2\pi i\nu t}). \qquad 19(34)$$

The mean value of E_x^2, taken over a cycle, is $2E_{0x}^2$ since the mean value of $\cos^2 2\pi\nu t$ is $\frac{1}{2}$. In free space the mean value of H^2 is equal to the mean value of E^2. Therefore

$$\rho_p = \frac{1}{8\pi}(\overline{E^2} + \overline{H^2}) = \frac{1}{4\pi}\overline{E^2} = \frac{1}{2\pi}E_{0x}^2, \qquad 19(35)$$

where ρ_p is the energy per unit volume.

We may now obtain an expression for U, using the assumption that the atom is small compared with the wavelength of the radiation. We regard the atom as an assembly of charges which are displaced by the field. An electric moment is created and its x component is

$$m_x = \sum e_j x_j, \qquad 19(36)$$

where x_j is the displacement of the jth charge from the position which it has when the field is zero. The interaction energy U is $E_x m_x$. In this simple case U may be regarded as an addition to the potential energy term (V) of the wave equation [18(17)]. In accordance with the notation of § 18.21, we put

$$ex_{mn} = \int \psi_m{}^* m_x \psi_n \, d\tau, \qquad 19(37)$$

so that

$$E_x e x_{mn} = \int \psi_m{}^* U \psi_n \, d\tau. \qquad 19(38)$$

6.—Substituting from 19(38) into 19(33) we have

$$a_m(t') = \frac{2\pi i}{h} \, ex_{mn} \int_0^{t'} E_x \exp \frac{2\pi i}{h} (W_n - W_m) t \, dt. \qquad 19(39)$$

We define a frequency ν_{mn} by the relation $h\nu_{mn} = W_m - W_n$. Inserting this symbol in 19(39), and also substituting for E_x from 19(34), we obtain

$$a_m(t') = \frac{2\pi i}{h} \, ex_{mn} E_{0x} \int_0^{t'} \left\{ \exp 2\pi i(\nu - \nu_{mn})t + \exp -2\pi i(\nu + \nu_{mn})t \right\} dt, \qquad 19(40)$$

and, on integration,

$$a_m(t') = \frac{ex_{mn} E_{0x}}{h} \left[\frac{1 - \exp -2\pi i(\nu + \nu_{mn})t'}{\nu + \nu_{mn}} - \frac{1 - \exp 2\pi i(\nu - \nu_{mn})t'}{\nu - \nu_{mn}} \right]. \qquad 19(41)$$

From this expression we see that $a_m(t')$ oscillates about a value near zero unless $(\nu - \nu_{mn})$ or $(\nu + \nu_{mn})$ is near zero. When we are considering absorption, W_m is greater than W_n and ν_{mn} is positive. The *second* term in 19(41) becomes large for a small group of frequencies near to the value $\nu = \nu_{mn}$, i.e. $h\nu = W_m - W_n$. Thus the probability of a transition corresponding to an absorption is appreciable if, and only if, the Bohr frequency condition is obeyed approximately. In a similar way, when W_n is greater than W_m, the *first* term in 19(41) becomes large for frequencies near to the value $\nu = -\nu_{mn}$ and we have stimulated emission. Note that, in each case, one term of 19(41) is important and the other is negligible.

7. The Absorption of Radiation from a Continuous Spectrum.

The energy in a range $d\nu$ of a continuous spectrum is $\rho_p(\nu)\, d\nu$, and we assume that $\rho_p(\nu)$ is nearly constant in a range of frequencies which includes the absorption line. Since ν_{mn} is positive, we consider only the *second* term in 19(41) and we have, for one frequency,

$$P_m(t') = a_m a_m{}^* = \frac{4}{h^2} e^2 x_{mn} x_{mn}{}^* E^2{}_{0x} \frac{\sin^2 \pi(\nu - \nu_{mn})t'}{(\nu - \nu_{mn})^2}. \qquad 19(42a)$$

Using 19(35) we have, for the range of frequencies,

$$P_m(t') = \frac{8\pi}{h^2} e^2 x_{mn} x_{mn}{}^* \int_0^\infty \rho_p(\nu) \frac{\sin^2 \pi(\nu - \nu_{mn})t'}{(\nu - \nu_{mn})^2} \, d\nu. \qquad 19(42b)$$

We know that the contribution to this integral from frequencies which differ from ν_{mn} by more than a few times the half-width of the line is small, and that

$\rho_p(\nu)$ is nearly independent of ν within this range. We may therefore take $\rho_p(\nu)$ outside the integral sign,† and we then have

$$P_m(t') = \frac{8\pi^3}{h^2} e^2 x_{mn} x_{mn}^* t' \rho_p(\nu). \quad . \quad . \quad . \quad . \quad 19(43)$$

But $B_{nm}\rho_p(\nu)t'$ is equal to the probability that a transition has taken place in time t' (§ 17.37) and is therefore equal to the probability $P_m(t')$ of finding a given atom in the mth state at time t'. Hence

$$B_{nm} = \frac{8\pi^3}{h^2} e^2 x_{mn} x_{mn}^*. \quad . \quad . \quad . \quad . \quad . \quad 19(44a)$$

In this equation $x_{mn}x_{mn}^*$ is proportional to the square of the electric moment *in the direction of the electric vector* of the light. We may regard this moment as a component of a moment proportional to $r_{mn}r_{mn}^*$ directed in some direction $\mathbf{r}$. This direction is fixed with respect to the atom. If, for a given atom, it makes an angle θ with the x axis, then $x_{mn}x_{mn}^* = r_{mn}r_{mn}^* \cos^2\theta$. In the absence of a steady magnetic field, the atoms are oriented at random and, since the mean value over a sphere of $\cos^2\theta = \frac{1}{3}$, we have

$$B_{nm} = \frac{8\pi^3}{3h^2} e^2 \mathbf{r}_{mn} \mathbf{r}_{mn}^*. \quad . \quad . \quad . \quad . \quad . \quad 19(44b)$$

The coefficient B_{mn} may be obtained by considering atoms initially in the state m (i.e. the state of higher energy). The *first* term in 19(41) then becomes important and the value of B_{mn} obtained is the same as that given by 19(44) for B_{nm}. The coefficient A_{mn} may be obtained by applying 17(53). The formulæ are in agreement with 18(62) and 18(63).

8.—It may appear, at first sight, that it should not be necessary to consider a range of frequencies and that we need only discuss the absorption for the frequency which satisfies the Bohr condition *exactly*. In any real experiment we could not use one frequency, because this would imply an infinite time of irradiation. If we used radiation which extended only over a range comparable with the width of the absorption line, the time of irradiation would be comparable with the life of the excited state and we should have to take account of re-emission and of other secondary processes which have been deliberately excluded from the above discussion by making t' very short. We have seen previously (§ 18.7) that it is not *necessary* for the Bohr frequency condition to be obeyed exactly in order to give conservation of energy. Note that, in the last paragraph, we have assumed that the transition probability for a group of frequencies acting simultaneously is the sum of the transition probabilities for the single frequencies. This is correct because the waves of different frequency are necessarily non-coherent. If we assume that a_m for all frequencies is the sum of the a_m's for each frequency, we should obtain (for P_m) the term given in equation 19(42b), together with a set of terms whose mean value is zero (see § 5.2 for discussion of a similar problem).

† Since $\int_{-\infty}^{\infty} \frac{\sin^2 x}{x^2}\, dx = \pi$. See note following eqn. 15(73a).

9. Theory of Emission and Absorption (Second Stage).

The preceding theory, although a considerable improvement on the rather crude assumption of § 18.21, is not completely satisfactory, because the quantum theory is applied to the atoms but not to the radiation. The latter appears as a purely classical field which " perturbs " the atom, and the quantization of the radiation field discussed in Chapter XVIII is disregarded. We shall now treat the atoms and the radiation as a single mechanical system and apply the quantum theory.

We consider an assembly of atoms and an *isotropic* field of radiation in an enclosure. There is no observable change in the energy of the system as a whole and no interaction with any external field. The system is not necessarily in thermodynamic equilibrium. We assume that, in the absence of the radiation, the atoms may be described by the wave equation 19(1). We take n as a typical quantum number for the atoms before transition and m as a typical* quantum number after transition. Corresponding numbers for the radiation are a and b. Let us suppose, for the moment, that the atoms and radiation do not interact. Then the wave equation for the combined system is

$$\mathscr{H}_0(\Psi) = (\mathscr{H}_p + \mathscr{H}_r)\Psi = -\frac{ih}{2\pi}\frac{\partial \Psi}{\partial t}, \quad . \quad . \quad . \quad 19(45)$$

where $\mathscr{H}_p$ stands for terms due to the atoms and $\mathscr{H}_r$ for terms due to the radiation. In accordance with the discussion of § 18.34, a wave function Ψ_{na} for two systems which do not interact is the product of the functions Ψ_n and Φ_a for the separate systems, i.e.

$$\Psi_{na} = \Psi_n \Phi_a. \quad . \quad . \quad . \quad . \quad . \quad . \quad . \quad . \quad 19(46)$$

The wave function for the combined system *with interaction* is

$$(\mathscr{H}_0 + \mathscr{H}_i)\Psi = \mathscr{H}(\Psi) = -\frac{ih}{2\pi}\frac{\partial \Psi}{\partial t}, \quad . \quad . \quad . \quad 19(47)$$

where $\mathscr{H}_i$ is the interaction energy.

10.—As a preliminary to our calculation, we need to derive an expression for $\mathscr{H}_i$ in terms of the vector potential $\mathbf{A}$ which is now used to describe the radiation.† Non-relativistic electromagnetic theory gives the following expression for the energy of a single electron in a field which is described by a vector potential $\mathbf{A}'$ and a scalar potential χ':

$$\mathscr{H}' = \frac{1}{2m}\left(\mathbf{P} - \frac{e}{c}\mathbf{A}'\right)^2 + e\chi' \quad . \quad . \quad . \quad . \quad . \quad 19(48)$$

[the bracket represents the *scalar* product of $(\mathbf{P} - e\mathbf{A}'/c)$ with itself]. This expression may be checked‡ by using it to calculate the forces acting on the electron, and showing that they agree with 13(11). The field acting on the electron is partly due to the atomic field, and partly due to the radiation. The latter contributes

* Obviously very large numbers of quantum numbers are needed, but we need consider only those which change.

† See Appendix XIII A (p. 410). ‡ See p. 39 of Reference 19.3.

only to the term $\mathbf{A}'$, and not to χ' [see equations 13(59) and 13(60) of Appendix XIII A]. We therefore put

$$\mathscr{H}_i = \frac{e}{mc}\mathbf{P}.\mathbf{A} + \frac{e^2}{2mc^2}\mathbf{A}^2. \quad . \quad . \quad . \quad . \quad 19(49a)$$

Since e/c is small, we take as an approximation

$$\mathscr{H}_i = \frac{e}{mc}\mathbf{P}.\mathbf{A} = \frac{e}{mc}\sum PA_s \cos\Theta_s; \quad . \quad . \quad . \quad 19(49b)$$

Θ_s is the angle between $\mathbf{P}$ and the unit vector $\mathbf{e}_s$ (which defines the direction of polarization of the radiation). Substituting from 18(85) and 18(88), we have

$$\mathscr{H}_i = \frac{2e\pi^{1/2}}{m}\sum_s\left\{q_s \exp(i\varkappa_s.\mathbf{r}) + q_s^* \exp(-i\varkappa_s.\mathbf{r})\right\}P\cos\Theta_s. \quad 19(50)$$

Since the atoms contain many electrons, the interaction energy contains many terms similar to the right-hand side of 19(50). We need consider only one.

11.—Now apply the procedure of §§ 2–8 to the joint system. Regard 19(45) as analogous to 19(23), and 19(47) as analogous to 19(26). We expand a solution of 19(47) in terms of a series of functions which are solutions of 19(45), and proceed as in § 4 of this appendix. If the system is initially in a state (n, a), the probability that at time t it will be in the state (m, b) is $P_{mb} = a_{mb}a_{mb}^*$, where

$$\dot{a}_{mb} = \frac{2\pi i}{h}\int \Psi_{mb}^*\mathscr{H}_i\Psi_{na}\,dx\,dq \quad . \quad . \quad . \quad . \quad 19(51)$$

[compare equation 19(32)].

In 19(51) we have replaced the volume element $d\tau$ by the element $dx\,dq$ in order to indicate that the integration must be carried out both for co-ordinates of the physical space (x, y, z) and for the q-co-ordinates of the "oscillators" which represent the radiation field.

Separating the time factors in the wave functions, we have

$$\dot{a}_{mb} = \frac{2\pi i}{h}U_{mn}\exp\frac{2\pi i}{h}(W_{na} - W_{mb})t, \quad . \quad . \quad . \quad 19(52)$$

where

$$U_{mn} = \int \psi_{mb}^*\mathscr{H}_i'\,\psi_{na}\,dx\,dq \quad . \quad . \quad . \quad . \quad . \quad . \quad . \quad . \quad 19(53)$$

and $\mathscr{H}_i'$ stands for $\mathscr{H}_i$ without time factors.

Hence

$$a_{mb} = U_{mn}\frac{\exp\left[\frac{2\pi i}{h}(W_{na} - W_{mb})t\right] - 1}{(W_{na} - W_{mb})}, \quad . \quad . \quad 19(54)$$

and

$$P_{mb} = 4U_{mn}U_{mn}^*\frac{\sin^2\frac{\pi}{h}(W_{na} - W_{mb})t}{(W_{na} - W_{mb})^2}. \quad . \quad . \quad 19(55)$$

This equation shows that the probability has a sharp maximum corresponding to transitions for which energy is conserved. It also shows that transitions for which energy is conserved, not exactly, but within the accuracy of measurement permitted by the uncertainty principle, must be included. It is necessary to take a sum corresponding to the integration of 19(42). Since the energy of the atom in the final state is always the same, the sum has to be taken over all oscillators whose frequencies are near the value which makes $W_{mb} = W_{na}$, i.e. near the frequency ν_{mn}, and whose directions are within a solid angle $d\Omega$. If $n(W)\,dW\,d\Omega$ is the number of oscillators *with a given direction of polarization* and with energies between W and $(W + dW)$, and if α is the number of transitions per second,

$$\alpha = \frac{d\Omega}{t}\int_0^\infty P_{mb}n(W)\,dW. \qquad \text{19(56)}$$

The variation of $n(W)$ over the small range of energies for which P_{mb} differs appreciably from zero may be ignored and $n(W)$ taken outside the integral sign. By substituting from 19(55) into 19(56) and integrating,† we obtain

$$\alpha = \frac{4\pi^2}{h}\,U_{mn}U_{mn}{}^*\,d\Omega n(W). \qquad \text{19(57)}$$

We substitute from 19(50) into 19(53), taking out the time factor $\exp(2\pi i\nu t)$ included in q and $\exp(-2\pi i\nu t)$ in q^*. Now we have already shown that only those transitions are permitted which approximately obey the conservation law. Therefore the term containing q has a significant value only when $W_m < W_n$ (i.e. when the atom emits energy) and the term containing q^* is negligible. Also only one oscillator changes its energy, so we take only one term in the summation. We then have, on substituting the first term of 19(50) into 19(53), and omitting the time factor,

$$U_{mn} = \frac{2e\pi^{\frac{1}{2}}}{m}\cos\Theta\int\phi_a q\phi_b{}^*\,dq\int\psi_m{}^*P\exp(i\boldsymbol{\kappa}_s\,.\,\mathbf{r})\,\psi_n\,d\tau. \qquad \text{19(58)}$$

The value of the first integral is given in § 18.24. It is zero unless $b = a \pm 1$. We are now considering emission so that $b = a + 1$, and the integral is given by putting $2\pi\nu_{nm}$ for $m_0\omega_0$ in 18(70). Making this substitution we have

$$U_{mn} = \frac{e}{m}\cos\Theta\left(\frac{a+1}{2\pi\nu_{nm}}\right)^{\frac{1}{2}}h^{\frac{1}{2}}\int\psi_m{}^*P\exp(i\boldsymbol{\kappa}_s\,.\,\mathbf{r})\,\psi_n\,d\tau. \qquad \text{19(59)}$$

The magnitude of $\boldsymbol{\kappa}_s$ is $2\pi/\lambda$ and the wave functions $\psi_m{}^*$ and ψ_n have appreciable values only when r is very small compared with λ (i.e. within the radius of the atom). Hence, to a first approximation, we put the exponential ‡ equal to 1. It is shown in textbooks on quantum mechanics that

$$\left|\int\psi_m{}^*P\psi_n\,d\tau\right|^2 = 4\pi^2m^2\nu_{nm}{}^2\,\mathbf{r}_{mn}\,\mathbf{r}_{mn}{}^*, \qquad \text{19(60)}$$

where $\mathbf{r}_{mn}$ is defined by 18(61). Hence, for emission,

$$(U_{mn}U_{mn}{}^*)_e = 2\pi e^2h\nu_{nm}(a+1)\mathbf{r}_{mn}\,\mathbf{r}_{mn}{}^*\cos^2\Theta. \qquad \text{19(61)}$$

† See footnote on p. 651.

‡ This is equivalent to expanding the exponential as a power series and taking only the first term. The second term corresponds to quadripole radiation, and higher terms are involved in the calculation of other types of multipole radiation.

Substituting in 19(57), we obtain α_e (the number of quanta emitted per second) and

$$\alpha_e = 8\pi^3 e^2 \nu_{nm}(a+1)n(W)\cos^2\Theta\,\mathbf{r}_{mn}\mathbf{r}_{mn}^*\,d\Omega. \quad . \quad . \quad 19(62)$$

12.—From equation 18(108) we have †

$$n(W)\,d\Omega = \frac{\nu^2}{hc^3}\,d\Omega. \quad . \quad . \quad . \quad . \quad . \quad . \quad 19(63)$$

Now suppose that the radiation is isotropic and has energy density $\rho(\nu)\,d\nu$ in the range ν to $(\nu + d\nu)$. Then

$$\rho(\nu) = 2\int n(\nu)ah\nu\,d\Omega = 8\pi h^2 an(W)\nu, \quad . \quad . \quad . \quad 19(64)$$

the factor of 2 being required to include both types of polarization. (If the quantum numbers for the oscillators in the range are not all the same, the mean value of a must be used.) Now integrate 19(62) over the whole solid angle, taking both directions of polarization into account, and separate the two parts of the bracket $(a + 1)$. In the first substitute from 19(64), and in the second from 19(63). Then we obtain for the total number of quanta emitted per second in all directions

$$(\alpha_e)_T = \frac{8\pi^3}{3h^2}e^2\mathbf{r}_{mn}\mathbf{r}_{mn}^*\rho(\nu) + \frac{64\pi^4}{3hc^3}e^2\nu_{nm}{}^3\,\mathbf{r}_{mn}\mathbf{r}_{mn}^*$$

$$\equiv B_{nm}\rho(\nu) + A_{nm}. \quad . \quad . \quad . \quad . \quad . \quad . \quad 19(65)$$

The calculation of α_a (the number of transitions per second when radiation is being absorbed) differs ‡ in that we use the term in q^* in 19(40) and that we have $b = a - 1$ for one of the oscillators; consequently, we must use 18(71) instead of 18(70), and obtain the factor $a^{\frac{1}{2}}$ instead of $(a + 1)^{\frac{1}{2}}$ in 19(59), so that

$$\alpha_a = \frac{a}{a+1}\alpha_e, \quad . \quad . \quad . \quad . \quad . \quad . \quad 19(66a)$$

and, in view of 19(62), the total number of transitions per second when radiation is absorbed is

$$(\alpha_a)_T = \frac{8\pi^3}{3h^2}e^2\mathbf{r}_{mn}\mathbf{r}_{mn}^*\rho(\nu)$$

$$\equiv B_{mn}\rho(\nu). \quad . \quad . \quad . \quad . \quad . \quad . \quad . \quad 19(66b)$$

The theory thus gives the values of the Einstein coefficients independently, and the values obtained comply with equation 17(53).

13.—It sometimes happens that certain groups of states have the same energy. Suppose that there is a group of g_n states each of energy W_n and another group of

† The equation quoted gives g_s, which is the number of states (for both directions of polarization) with energies between ν and $(\nu + d\nu)$ and with directions varying over the whole sphere. If $n(\nu)\,d\nu\,d\Omega$ is the number corresponding to one direction of polarization and solid angle $d\Omega$, we have $g_s = 8\pi n(\nu)$ and $hn(W) = n(\nu)$.

‡ This leads to the factor $\exp(-i\boldsymbol{\varkappa}_s\,.\,\mathbf{r})$ instead of $\exp(+i\boldsymbol{\varkappa}_s\,.\,\mathbf{r})$, but this makes no difference in the approximation to which we are working.

g_m states each of energy W_m and that $W_n > W_m$. We have no means of knowing which members of the groups are involved. We assume that g_n and g_m are known from other experiments (usually from observations of the Zeeman effect). Then let B_{NM}, B_{MN}, A_{NM} stand for the probabilities of transition which are obtained from observations, e.g. by measuring the integrated absorption and applying 17(57), or by measuring τ directly and applying 17(56*b*). Let B_{nm} stand for the probability calculated for one particular pair of states from equation 19(42*a*). Then B_{nm} gives the transition probability from one of the upper states to one of the lower states and

$$B_{NM} = g_m B_{nm}. \qquad \text{19(67}a\text{)}$$

B_{NM} is the total probability for transition from *all* the upper states to *all* the lower states. The fact that there are g_n upper states does not affect the number of transitions, because there are only $1/g_n$ of the atoms in each of the upper states.

Similarly

$$B_{MN} = g_n B_{mn} \qquad \text{19(67}b\text{)}$$

and

$$\frac{1}{\tau_{NM}} = A_{NM} = g_m A_{nm}, \qquad \text{19(67}c\text{)}$$

so that

$$g_n B_{NM} = g_m B_{MN}. \qquad \text{19(68)}$$

Note that 19(68) agrees with 17(53), which is deduced from thermodynamic considerations.

The f-value is related to observations by 17(59) or 17(60) and we have

$$f = \frac{mc^3}{8\pi^2\nu^2e^2} A_{NM} = \frac{g_m mc^3}{8\pi^2\nu^2e^2} A_{nm}. \qquad \text{19(69)}$$

Thus measured f-values may be compared with values calculated from the wave equation. The agreement obtained is satisfactory (§ 18.22).

14. The Scattering of Light.

Let us now consider the interaction between an atom and a photon when the Bohr condition ($h\nu = W_m \sim W_n$) is not even approximately obeyed. From the experimental side, we know that the photon can be:

(*a*) scattered without change of wavelength (Rayleigh scattering);

(*b*) scattered with a small change of wavelength (Compton effect);

(*c*) scattered with a comparatively large change of wavelength (Raman effect).

In processes (*a*) and (*b*) the atom does not change its state, though it may gain or lose small amounts of kinetic energy and momentum. In Raman scattering,* the atom changes its state, and if ν_i is the frequency of the incident and ν_e of the emitted quantum, we have $W_m - W_n = h(\nu_i - \nu_e)$. We may describe the scattering of light by an atom as a process in which one of the oscillators which represent the radiation field loses a quantum and another oscillator gains a quantum. In Rayleigh scattering these two oscillators have the same frequency, but correspond to waves travelling in different directions, i.e. the vectors $\boldsymbol{\varkappa}_s$ and $\mathbf{e}_s$ of § 18.25 are different. In the Compton and Raman scattering, the two oscillators differ in frequency as well as in the direction vectors.

* The experiments on Raman scattering are described in § 19.21.

15. Rayleigh Scattering of Light.

Let us now consider the scattering of light (without change of wavelength) from a parallel beam of light propagated in the OZ direction and with electric vector parallel to OX. The incident beam is represented by an oscillator s (quantum number initially a_s). For this oscillator $\boldsymbol{\varkappa}_s$ is parallel to OZ and $\mathbf{e}_s$ to OX. The scattered beam corresponds to an oscillator r (quantum number a_r, initially zero). It is convenient to assume that initially there is no radiation in the direction $\boldsymbol{\varkappa}_r$ (i.e. we do not take account of secondary scattering). The scattering process is made up of two stages:

(1) The oscillator s loses energy $h\nu$ and the atom changes from state n to state m (but ν_{nm} is not even approximately equal to ν).

(2) The oscillator r gains energy $h\nu$ and the atom changes from state m to state n.

Energy is not even approximately conserved in either stage taken by itself, but it is conserved in the process as a whole. This is all that we need require, since the intermediate stage is not observable. Suppose that the initial state of the system (atom plus radiation) is denoted by f, the intermediate state by g, and the final state by h. Then, applying the discussion of § 4 of this appendix to the two-stage process,

$$\dot{a}_g = \frac{2\pi i}{h} a_f U_{fg} \exp \frac{2\pi i}{h}(W_f - W_g)t \quad . \quad . \quad . \quad 19(70a)$$

and

$$\dot{a}_h = \frac{2\pi i}{h} a_g U_{gh} \exp \frac{2\pi i}{h}(W_g - W_h)t. \quad . \quad . \quad . \quad 19(70b)$$

If t is small, we put $a_f = 1$, integrate 19(70a) (using the initial condition $a_g = 0$ when $t = 0$) and substitute in 19(70b) to obtain

$$a_h = \frac{2\pi i}{h}\frac{U_{fg}U_{gh}}{W_f - W_g}\left[\exp \frac{2\pi i}{h}(W_f - W_g)t - \exp \frac{2\pi i}{h}(W_f - W_h)t\right]. \quad 19(71)$$

Integrating the *second* term † in 19(71) we obtain

$$P_h = 4UU^* \frac{\sin^2 \frac{\pi}{h}(W_f - W_h)t}{(W_f - W_h)^2}, \quad . \quad . \quad . \quad . \quad 19(72)$$

where

$$U = \frac{U_{fg}U_{gh}}{W_f - W_g}. \quad . \quad . \quad . \quad . \quad . \quad . \quad . \quad . \quad 19(73)$$

We are interested in the possibility that a quantum of energy may be taken from one particular oscillator s and that an oscillator r (which corresponds to a direction included in the solid angle $d\Omega$ and to a small range of frequencies round the frequency for which $W_f = W_h$) may gain a quantum of energy. Since 19(72) is like 19(55), we may use 19(57) and, substituting for $n(W)$ from 19(63), we obtain

$$\alpha' = \frac{4\pi^2}{h^2}\frac{\nu^2}{c^3} UU^* \, d\Omega. \quad . \quad . \quad . \quad . \quad . \quad . \quad 19(74)$$

† Since W_f is not even approximately equal to W_g, the term omitted contributes only a negligible oscillatory term to P_h.

16.—The calculation of U_{fg} and U_{gh} is similar to the calculation of U_{nm} [see equation 19(61) and equations leading thereto], but we have to allow for the fact that the incident beam is now a plane-polarized beam whereas 19(61), etc., apply to isotropic radiation. Since $\mathbf{e}_s$ is parallel to OX we may substitute P_x for $P \cos \Theta_s$ in 19(49*b*). Carrying this change forward, we obtain for the first interaction product

$$U_{fg}U_{fg}^* = 2\pi e^2 h\nu_{nm} a_s x_{mn} x_{mn}^*. \qquad 19(75)$$

This expression should be compared with 19(61). It contains a_s instead of $(a + 1)$, because we are considering an absorption process, and x_{mn} instead of r_{mn} for the reasons stated above. When we consider the emission process, we have to remember that the " oscillator " excited by the incident beam is parallel to OX. If the scattered radiation is emitted in some direction which makes an angle χ with OX (i.e. with the electric vector of the incident beam), then the interaction energy corresponding to 19(49*b*) is proportional to $P_x \sin \chi$ (the sine appearing instead of the cosine because the electric vector is perpendicular to the direction of the scattered beam). We then obtain

$$U_{gh}U_{gh}^* = 2\pi e^2 h\nu_{nm}(a_r + 1) x_{mn} x_{mn}^* \sin^2 \chi. \qquad 19(76)$$

Putting $a_r = 0$, and substituting from 19(75) and 19(76) into 19(73), we have

$$UU^* = \frac{4\pi^2 e^4 \nu_{nm}^2}{(\nu - \nu_{nm})^2} a_s (x_{mn} x_{mn}^*)^2 \sin^2 \chi. \qquad 19(77)$$

If the atoms or molecules are oriented at random, we have

$$x_{mn} x_{mn}^* = \tfrac{1}{3} r_{mn} r_{mn}^*. \qquad 19(78)$$

Inserting this in 19(77) and substituting the result in 19(74), we have

$$\alpha = \frac{16}{9} \frac{\pi^4 e^4 \nu_{nm}^2 \nu^2 a_s}{h^2 c^3 (\nu - \nu_{nm})^2} (r_{mn} r_{mn}^*)^2 \sin^2 \chi \, d\Omega. \qquad 19(79)$$

This gives the number of quanta scattered per second per atom when $a_s c$ quanta are incident per second. As explained in § 15.44, the scattered energy is non-coherent. Therefore the ratio of scattered to incident energy for N atoms is

$$k' = \frac{16}{9} \frac{\pi^4 e^4 \nu_{nm}^2 \nu^2 N}{h^2 c^4 (\nu - \nu_{nm})^2} (r_{mn} r_{mn}^*)^2 \sin^2 \chi \, d\Omega. \qquad 19(80)$$

17.—In addition to the two stages considered above, the following two-stage process produces the same final result:

(1) The atom moves from state n to state m, and the oscillator r gains a quantum.

(2) The atom reverts to state n and the oscillator s loses a quantum.

This two-stage process, in which the scattered quantum appears " before " the incident one has been absorbed, appears very " artificial "—if we attempt to separate the two stages; but, considering the equations for the two-stage process as a whole, we see that it must be taken into account. The calculation proceeds as before, except that $W_g - W_f$ is now equal to $h(\nu + \nu_{nm})$. We have to add the term for this process to a_h (not to P_h) because the transition probabilities belong to the same wave (see § 19.8). Adding it in this way, we obtain

$$k'' = \frac{16\pi^4 e^4 \nu^2 \nu_{nm}{}^2 N}{9h^2c^4} (r_{mn}r_{mn}{}^*)^2 \sin^2\chi \left[\frac{1}{\nu - \nu_{nm}} + \frac{1}{\nu + \nu_{nm}}\right]^2 d\Omega, \quad 19(81)$$

and, integrating over the whole solid angle,

$$k = \frac{512}{27} \frac{\pi^5 e^4}{h^2 c^4} \frac{N \nu_{nm}{}^2 \nu^4}{(\nu_{nm}{}^2 - \nu^2)^2} (r_{mn}r_{mn}{}^*)^2. \quad . \quad . \quad . \quad 19(82)$$

The relation between dispersion and scattering in the quantum theory is, for present purposes, the same as in classical theory and, using 15(97), we have

$$n - 1 = \frac{4\pi e^2 \nu_{nm} N}{3h(\nu_{nm}{}^2 - \nu^2)} r_{mn}r_{mn}{}^*, \quad . \quad . \quad . \quad . \quad 19(83)$$

or, if f is given by 18(64),

$$n - 1 = \frac{Ne^2}{2\pi m} \frac{f}{(\nu_{nm}{}^2 - \nu^2)}. \quad . \quad . \quad . \quad . \quad . \quad 19(84)$$

This agrees with 15(63). Note that in obtaining 15(63) it is assumed that $(n - 1)$ is small compared with 1.

If several transitions are possible, the right-hand side of 19(84) is replaced by a summation. Equation 19(4a) of § 19.11 follows if we put $\omega_s = 2\pi\nu_{nm}$. Equation 19(4b) follows if some of the transitions involved are to lower states. Equation 19(84) is related to the classical equation 15(63) in the way explained in § 19.11.

18.—The above discussion shows that the quantum theory of scattering leads to a dispersion formula identical in form with that given by the classical theory. There are two important differences in application: (a) in classical theory f is a constant to be determined by experiment, whereas in quantum theory it can be calculated and (b) the quantum theory gives the " negative terms " discussed in § 19.11. Both quantum theory and classical theory give an extra term which is discussed in the following paragraph. The magnitude of this term is small in the optical region but it is very important at shorter wavelengths. It may be seen that 19(84) does not contain the damping term γ^2 in the denominator. This is because, in order to reduce the mathematical complexity, we have neglected the effect of damping in this appendix. We did this when we put $a_m = 1$ in § 3 and at corresponding points in other paragraphs. To improve on this approximation, we should put $a_m = \exp(-\gamma t/2)$. With this substitution we obtain a dispersion equation which agrees exactly with the classical dispersion equation with damping. The classical theory gives the value shown in equation 13(86). This value is not in agreement with experiment. The quantum theory makes γ equal to the Einstein A, so that it is calculable in terms of the functions $r_{mn}r_{mn}{}^*$. The values so calculated agree with the results of measurements on line widths.

19. Relative Orders of Magnitude of Transition Probabilities.

The function U which appears in 19(72) is the product of two interaction energies. Each contains e^2, so that 19(82) contains e^4 as a factor. The transition probability derived from 19(72) is therefore of a much lower order of magnitude than the usual values of transition probabilities for absorption and emission obtained from 19(55), which depends on one interaction energy and contains e^2 as a factor. This is in accord with experimental observation. The absorption of resonance radiation by a column of mercury vapour one centimetre long at 10^{-6} atmosphere pressure is easily measurable. The Rayleigh scattering (i.e. scattering in a region remote from an absorption line) is very small even at a pressure of one atmosphere. In a corresponding way the value of $(n - 1)$ for a gas at 10^{-6} atmosphere is of the order 10^{-10} except near an absorption line. Thus for a medium in which the number of molecules per cubic centimetre is of order 10^{13}, single-quantum transition processes are readily observable, but processes involving two quanta are not.

20. **Scattering by Free Electrons.**

Since the probabilities for two quantum processes are of low order, it is necessary to reconsider the approximations of the preceding theory. The term $e^2A^2/2mc^2$ which was omitted from 19(49) may need to be included, since we are now discussing terms containing e^4. This term leads to a probability for direct transition from state n to state l without intermediate states, and hence to an addition to the rate of scattering. Using 18(85) we have

$$A^2 = \sum_{rs} [q_s q_r A_s A_r + q_s^* q_r^* A_s^* A_r^* + q_s^* q_r A_s^* A_r + q_s q_r^* A_s A_r^*]. \quad 19(85)$$

Note that this term does not contain **P**. Let us call the additional term in the interaction energy $\mathscr{H}_i$ and the corresponding function U_{nl}'. The case in which the rth oscillator loses a quantum and the sth oscillator gains a quantum corresponds to the $q_s q_r^*$ term in 19(85). Substituting from 18(88), we have

$$A^2 = \mathbf{e}_s \mathbf{e}_r 4\pi c^2 q_s^* q_r \exp i(\boldsymbol{\varkappa}_s - \boldsymbol{\varkappa}_r) \cdot \mathbf{r}. \quad . \quad . \quad . \quad 19(86)$$

If Θ is the angle of scattering and if the wavelength is great compared with the radius of the scattering particle,

$$A^2 = 4\pi c^2 q_r^* q_s \cos \Theta. \quad . \quad . \quad . \quad . \quad . \quad 19(87)$$

The contribution to the interaction energy is

$$\mathscr{H}_i = \frac{2\pi e^2}{m} q_r^* q_s \cos \Theta. \quad . \quad . \quad . \quad . \quad . \quad 19(88)$$

The function U_{nl}' (corresponding to a *direct* transition from n to l) is obtained by substituting from 18(70) for a transition in which the quantum number a_s for the sth oscillator increases by 1, and the number a_r decreases by 1. We have

$$U_{nl}' = \frac{e^2 h}{2\pi m \nu} (a_s + 1)^{\frac{1}{2}} a_r^{\frac{1}{2}} \cos \Theta \int \psi_n^* \psi_l \, d\tau, \quad . \quad . \quad 19(89)$$

if $\nu_s = \nu_r = \nu$. A factor of 2 enters in 19(89) for reasons discussed in Reference 19.2. n and l are states of the atom and are the same for the process considered. The value of the integral is therefore unity. We are interested in the number of transitions per second from a given state r (light going in one direction) to a set of states s, which includes a range of frequencies and also a solid angle $d\Omega$. This rate is given by 19(57). Using 19(57) and 19(89), we obtain the number of transitions per second. Calling this number α', we have

$$\alpha' = \frac{e^4 h}{m^2 \nu^2} n(W)(a_s + 1)a_r \cos^2 \Theta \, d\Omega. \quad . \quad . \quad . \quad 19(90)$$

Putting $a_s = 0$ as before, and substituting from 19(63) for $n(W)$, we have

$$\alpha' = \frac{e^4}{m^2 c^3} a_r \cos^2 \Theta \, d\Omega. \quad . \quad . \quad . \quad . \quad . \quad . \quad 19(91)$$

The number of quanta scattered per second divided by the number of incident quanta is then

$$\frac{e^4}{m^2 c^4} \cos^2 \Theta \, d\Omega. \quad . \quad . \quad . \quad . \quad . \quad . \quad . \quad . \quad . \quad 19(92)$$

Integrating over all values of Θ, we have †

$$k = \frac{8\pi e^4}{3m^2 c^4}. \quad . \quad . \quad . \quad . \quad . \quad . \quad . \quad . \quad . \quad 19(93)$$

21.—According to quantum mechanics, the scattering process discussed in the preceding paragraph applies to all electrons—not merely to free electrons. It is therefore necessary to add a term inside the square bracket in equation 19(81). In practice this term is of no importance in the scattering and dispersion of radiation of visible wavelengths. This type of scattering is very important at X-ray wavelengths and 19(93) is confirmed by the experimental data on X-ray scattering, in agreement with 13(97), which was derived from classical electromagnetic theory.

† A similar integration is given on p. 106 of Reference 19.2.

LIST OF SYMBOLS

The *paragraph numbers* indicate the places where symbols are introduced or defined. Use of various symbols as auxiliary constants (e.g. α in § 18.16) is not included in the list.

A, a Real amplitude or Fourier coefficient, 2.4, 2.6, 4.17.
A_{mn}, B_{mn}, etc. Einstein coefficients, 17.37.
B Fourier coefficient, 2.6.
b Phase velocity in a material medium, 2.9, 2.11, 4.30.
c Velocity of light in vacuum, ch. X.
E Relative energy, 2.18, 5.2.
e (i) Charge of the electron.
e (ii) Thickness of a film, 5.13.
f f-value, 15.22.
g Ray velocity, 16.9.
H, $\mathscr{H}$ Energy (Hamiltonian function), 2.7, 18.25.
h Planck's constant, 17.1.
k Boltzmann's constant, 17.30.
$\mathscr{N}$ Number of dipoles per unit volume, 15.21.
N Number of molecules per unit volume, 15.22.
m (i) Order of spectrum, 6.31.
m (ii) Mass; relativistic mass, 11.32.
m_0 Rest mass, 11.32.
n Index of refraction to vacuum, 3.11.
$\mathbf{n}$ Complex index, 15.5.
P (i) Complex amplitude, 2.26.
P (ii) Relativistic momentum, 11.34.
p (i) Momentum.
p (ii) Order of interference, 4.9.
q Co-ordinate, 2.3, 18.4.
T (i) Period, 2.4.
T (ii) Absolute temperature, 17.27.
U Group velocity, 4.29.
U_{mn} Interaction energy, 19.3.
V Potential energy, 2.4.
v Velocity of a material body, 2.22.
W Total energy (especially of e.m. field), 13.13.
U, W Special functions, 6.27–8.

GREEK LETTERS

α (i) Relativity constant $= (1 - v^2/c^2)^{1/2}$, 11.12.
α (ii) Angle between **E** and **D** in crystal, 16.7.
α (iii) Constant of Gaussian curve, 4.21.
α (iv) $2\alpha =$ Absorption coefficient, 15.5.
γ Damping constant, 4.25, 15.20.
δ Phase difference, 2.4.
ϵ Dielectric constant, 13.7, 16.2.
Θ Principal angle of incidence, 15.10.
θ (1) Angle of reflection in a thin film, 5.13.
θ_1 Angle of incidence in medium (1), 3.13, 14.5.
θ_1' Angle of reflection in medium (1), 3.13, 14.5.
θ_2 Angle of refraction in medium (2), 3.13, 14.5.
κ Wavelength constant, 2.11.
$\varkappa$ Extinction coefficient, 15.5.
$\varkappa_s$ Wave vector, 18.25.
λ Wavelength in the medium,* 2.11.
μ Refractive index to air, 3.11.
ν Frequency, 2.11.
ξ Disturbance or fluctuation, 2.9.
ρ (i) Reflection coefficient, 5.26.
ρ (ii) Energy density in e.m. field, 17.19.
$\boldsymbol{\rho}$ (i) Current density, 13.5.
$\boldsymbol{\rho}$ (ii) Unit vector in ray direction, 16.5.
σ (i) Electrical conductivity, 13.5.
σ (ii) Stefan's constant, 17.27.
τ Life of an excited state, 17.38.
ϕ (i) Phase, 2.4, 4.30.
Φ, ϕ (ii) Scalar potential, Appendix XIII A.
Φ, ϕ (iii) Wave functions for e.m. field, 18.27, Appendix XIX A, 9.
χ Scalar potential, Appendix XIII A.
ψ (i) Azimuth, 12.10, 15.11.
Ψ (i) Principal azimuth, 15.11.
ψ (ii) Wave function without time, 18.11.
Ψ (ii) Wave function with time, 18.18.
ω Circular (or angular) frequency, 2.4.
Ω Angular momentum, 17.24.

* In Chap. VIII, λ is used for wavelength in air and λ' for wavelength in medium.

VECTORS

Small letters stand for unit vectors (see §§ 13.3, 16.5), except for **v**, **j**, and **ρ** (ii) in Chap. XIII.

A	Vector potential, Appendix XIII A.
D	Electric displacement, 13.6.
E	Electric field, 13.4.
F	Mechanical force, 13.4.
G	Poynting vector, 13.14.
H	Magnetic field, 13.5.
j	Current, 13.5.
i, **j**, **k**	Unit vectors, 13.3.
$\boldsymbol{\varkappa}_s$	Wave vector, 18.25.
M	Dipole moment, Appendix XIII B.
P	(i) Electrical polarization, 13.6.
P	(ii) Momentum, 17.21, 18.29.
Π	Hertzian vector, Appendix XIII A.
ρ	(i) Ray direction, 16.5.
ρ	(ii) Current density, 13.5.
s	Wave normal, 16.5.
U	Group velocity, 4.29.
V, **v**	Velocity, 13.14, 13.16.

TABLE OF CONSTANTS

c velocity of light *in vacuo* =

(optical measurements to 1945) 299,773 km. sec.$^{-1}$

(radio measurements and Bergstrand's method) 299,792·5 km. sec.$^{-1}$

h Planck's constant = $6{\cdot}62 \times 10^{-27}$ erg. sec.

e electronic charge = $4{\cdot}802 \times 10^{-10}$ e.s.u.

m electronic mass = $9{\cdot}106 \times 10^{-28}$ g.

e/m electronic charge-mass ratio = $5{\cdot}274 \times 10^{17}$ e.s.u.g.$^{-1}$

m_H mass of H atom = $1{\cdot}673 \times 10^{-24}$ g.

R_H Rydberg constant (for hydrogen) = 109,677·58 cm.$^{-1}$

N Avogadro's number = $6{\cdot}023 \times 10^{23}$ mole^{-1}

c_2 second radiation constant (see § 17.32) = 1·438 cm. deg.

k Boltzmann's constant = $1{\cdot}380 \times 10^{-16}$ erg deg.$^{-1}$

σ Stefan's constant = $5{\cdot}672 \times 10^{-5}$ erg. cm.$^{-2}$ deg.$^{-4}$ sec.$^{-1}$

Wavelength of the Cd red line in dry air at 760 mm. pressure and 15° C. = 6438·4696 Å.

Energy of 1 electron-volt = $1{\cdot}602 \times 10^{-12}$ ergs.

The energy associated with a photon of wavelength 12,394 Å. is 1 electron-volt. Hence we obtain the following mnemonic for deriving the quantum energy associated with a given wavelength (correct to 1 per cent):

Divide the number 12,345 by the wavelength (in Å.) to find the energy (in electron-volts).

INDEX

References are to paragraphs.—References in heavy type either give the definition of a quantity or are more important than other references under the same heading.

A CATALOG OF SELECTED

DOVER BOOKS

IN SCIENCE AND MATHEMATICS

A CATALOG OF SELECTED

DOVER BOOKS

IN SCIENCE AND MATHEMATICS

QUALITATIVE THEORY OF DIFFERENTIAL EQUATIONS, V.V. Nemytskii and V.V. Stepanov. Classic graduate-level text by two prominent Soviet mathematicians covers classical differential equations as well as topological dynamics and ergodic theory. Bibliographies. 523pp. 5⅜ × 8½. 65954-2 Pa. $10.95

MATRICES AND LINEAR ALGEBRA, Hans Schneider and George Phillip Barker. Basic textbook covers theory of matrices and its applications to systems of linear equations and related topics such as determinants, eigenvalues and differential equations. Numerous exercises. 432pp. 5⅜ × 8½. 66014-1 Pa. $10.95

QUANTUM THEORY, David Bohm. This advanced undergraduate-level text presents the quantum theory in terms of qualitative and imaginative concepts, followed by specific applications worked out in mathematical detail. Preface. Index. 655pp. 5⅜ × 8½. 65969-0 Pa. $13.95

ATOMIC PHYSICS (8th edition), Max Born. Nobel laureate's lucid treatment of kinetic theory of gases, elementary particles, nuclear atom, wave-corpuscles, atomic structure and spectral lines, much more. Over 40 appendices, bibliography. 495pp. 5⅜ × 8½. 65984-4 Pa. $12.95

ELECTRONIC STRUCTURE AND THE PROPERTIES OF SOLIDS: The Physics of the Chemical Bond, Walter A. Harrison. Innovative text offers basic understanding of the electronic structure of covalent and ionic solids, simple metals, transition metals and their compounds. Problems. 1980 edition. 582pp. 6⅛ × 9¼. 66021-4 Pa. $15.95

BOUNDARY VALUE PROBLEMS OF HEAT CONDUCTION, M. Necati Özisik. Systematic, comprehensive treatment of modern mathematical methods of solving problems in heat conduction and diffusion. Numerous examples and problems. Selected references. Appendices. 505pp. 5⅜ × 8½. 65990-9 Pa. $12.95

A SHORT HISTORY OF CHEMISTRY (3rd edition), J.R. Partington. Classic exposition explores origins of chemistry, alchemy, early medical chemistry, nature of atmosphere, theory of valency, laws and structure of atomic theory, much more. 428pp. 5⅜ × 8½. (Available in U.S. only) 65977-1 Pa. $10.95

A HISTORY OF ASTRONOMY, A. Pannekoek. Well-balanced, carefully reasoned study covers such topics as Ptolemaic theory, work of Copernicus, Kepler, Newton, Eddington's work on stars, much more. Illustrated. References. 521pp. 5⅜ × 8½. 65994-1 Pa. $12.95

PRINCIPLES OF METEOROLOGICAL ANALYSIS, Walter J. Saucier. Highly respected, abundantly illustrated classic reviews atmospheric variables, hydrostatics, static stability, various analyses (scalar, cross-section, isobaric, isentropic, more). For intermediate meteorology students. 454pp. 6⅛ × 9¼. 65979-8 Pa. $14.95

DE RE METALLICA, Georgius Agricola. The famous Hoover translation of greatest treatise on technological chemistry, engineering, geology, mining of early modern times (1556). All 289 original woodcuts. 638pp. 6¾ × 11.
60006-8 Pa. $18.95

SOME THEORY OF SAMPLING, William Edwards Deming. Analysis of the problems, theory and design of sampling techniques for social scientists, industrial managers and others who find statistics increasingly important in their work. 61 tables. 90 figures. xvii + 602pp. 5⅜ × 8½. 64684-X Pa. $15.95

THE VARIOUS AND INGENIOUS MACHINES OF AGOSTINO RAMELLI: A Classic Sixteenth-Century Illustrated Treatise on Technology, Agostino Ramelli. One of the most widely known and copied works on machinery in the 16th century. 194 detailed plates of water pumps, grain mills, cranes, more. 608pp. 9 × 12.
28180-9 Pa. $24.95

LINEAR PROGRAMMING AND ECONOMIC ANALYSIS, Robert Dorfman, Paul A. Samuelson and Robert M. Solow. First comprehensive treatment of linear programming in standard economic analysis. Game theory, modern welfare economics, Leontief input-output, more. 525pp. 5⅜ × 8½. 65491-5 Pa. $14.95

ELEMENTARY DECISION THEORY, Herman Chernoff and Lincoln E. Moses. Clear introduction to statistics and statistical theory covers data processing, probability and random variables, testing hypotheses, much more. Exercises. 364pp. 5⅜ × 8½. 65218-1 Pa. $9.95

THE COMPLEAT STRATEGYST: Being a Primer on the Theory of Games of Strategy, J.D. Williams. Highly entertaining classic describes, with many illustrated examples, how to select best strategies in conflict situations. Prefaces. Appendices. 268pp. 5⅜ × 8½. 25101-2 Pa. $7.95

MATHEMATICAL METHODS OF OPERATIONS RESEARCH, Thomas L. Saaty. Classic graduate-level text covers historical background, classical methods of forming models, optimization, game theory, probability, queueing theory, much more. Exercises. Bibliography. 448pp. 5⅜ × 8¼. 65703-5 Pa. $12.95

CONSTRUCTIONS AND COMBINATORIAL PROBLEMS IN DESIGN OF EXPERIMENTS, Damaraju Raghavarao. In-depth reference work examines orthogonal Latin squares, incomplete block designs, tactical configuration, partial geometry, much more. Abundant explanations, examples. 416pp. 5⅜ × 8¼.
65685-3 Pa. $10.95

THE ABSOLUTE DIFFERENTIAL CALCULUS (CALCULUS OF TENSORS), Tullio Levi-Civita. Great 20th-century mathematician's classic work on material necessary for mathematical grasp of theory of relativity. 452pp. 5⅜ × 8½.
63401-9 Pa. $9.95

VECTOR AND TENSOR ANALYSIS WITH APPLICATIONS, A.I. Borisenko and I.E. Tarapov. Concise introduction. Worked-out problems, solutions, exercises. 257pp. 5⅜ × 8¼. 63833-2 Pa. $7.95